What Every Supervisor Must Know about OSHA — General Industry

Joe Teeples

CCH INCORPORATED
Chicago

A WoltersKluwer Company

Vice President, Business Compliance Unit: Paul Gibson, J.D.
Executive Editor: Jeff Reinholtz, J.D.
Managing Editor: Joy Waltemath, J.D.
Coordinating Editor: Debra Levin, J.D.
Contributing Editor: Kari Smith
Production Coordinators: Theresa J. Jensen, Jennifer Wintczak
Cover Design: Craig Arritola
Interior Design: Jason Wommack
Layout: Craig Arritola

This publication is designed to provide accurate and authoritative information in regard to the subject matter covered. It is sold with the understanding that the publisher is not engaged in rendering legal, accounting, or other professional service. If legal advice or other expert assistance is required, the services of a competent professional person should be sought.

ISBN 0-8080-1444-7
©2005 **CCH** INCORPORATED
4025 W. Peterson Ave.
Chicago, IL 60646-6085
1 800 248 3248
www.safety.cch.com

No claim is made to original government works; however, the gathering, compilation, magnetic translation, digital conversion, and arrangement of such materials, the historical, statutory, and other notes and references, as well as commentary and materials in this Product or Publication, are subject to CCH's copyright.

All Rights Reserved
Printed in the United States of America

AUTHOR BIOGRAPHY

Author Joe Teeples has over 25 years in the field of occupational safety. He has served as the safety manager of a federal installation where he was instrumental in reducing lost time injuries by 66 percent. Teeples also served the government in the role of director of one of OSHA's Training Institute education centers. In his role as a "train the trainer" in OSHA's Outreach program for various educational centers, he has prepared written training materials, slides, movies, online courses and books for other trainers to use. His company has taught safety around the world in locations such as Saudi Arabia, Trinidad, Guam, Saipan and the Virgin Islands. In the United States, he has trained students on federal requirements from the East Coast to the West Coast, from Alaska to Hawaii.

Teeples holds an M.B.A. from the University of Wisconsin and is a certified member of the World Safety Organization that serves a consultative role to the United Nations. He is also a member of the American Society of Safety Engineers. As a measure of his dedication to safety, Teeples maintains a web site at www.wisafety.com with many useful items for safety professionals to download and use at no charge.

Table of Contents

Chapter 1: Introduction to OSHA 1
General Health and Safety Section 5(a) 1
Section 5 (b) 1
The Act's Coverage 2
Code of Federal Regulations 3
Federal Employees 7
How New Standards are Developed 7
Recordkeeping and Reporting 10
Keeping Employees Informed 10
Workplace Inspections 11
Citations and Penalties 15
Appeals Process 16
OSHA-Approved State Programs 17
Services Available 18
Employer Responsibilities and Rights 19
Employee Responsibilities and Rights 21
Construction vs. General Industry 23
Safety Committees 23
Injured Workers May Sue Government 24
Your Safety Committee 25
Most Frequently Cited Standards 26
Hazard Violation Search Workshop #1 27
Hazard Violation Workshop #2 29

Chapter 2: Recording and Reporting Occupational Injuries and Illnesses 33
Major Changes To OSHA's Recordkeeping Rule 33

Chapter 3: Walking and Working Surfaces 91
General Requirements— 1910.22 91
Guarding Floor and Wall Openings and Holes— 1910.23 92
Fixed Industrial Stairs— 1910.24 93
Portable Ladders— 1910.25, 1910.26 94
Fixed Ladders— 1910.27 94
Safety Requirements for Scaffolding— 1910.28 95
Manually Propelled Mobile Ladder Stands and Scaffolds (Towers)— 1910.29 96
Other Working Surfaces— 1910.30 96

Chapter 4: Occupational Health and Industrial Hygiene 103
Industrial Hygiene 104
OSHA and Industrial Hygiene 105
Routes of Entry 108
Types of Air Contaminants 109
Threshold Limit Values 111
Federal Occupational Safety and Health Standards 112
Bloodborne Pathogens 113

Asbestos and Lead Awareness 119
Significant Changes in the Asbestos Standard for General Industry,
29 CFR 1910.1001 (through June 29, 1995) 121
Occupational Noise Exposure 122
Hearing Conservation Glossary 125

Chapter 5: Hazard Communication 129
Hazard Evaluation 130
Written Program 130
Labels and Other Forms of Warnings 130
Material Safety Data Sheets 131
List of Hazardous Chemicals 132
Employee Training and Information 132
Trade Secrets 132
Medical Emergency 132
Nonemergency Situation 133
Guidelines for Compliance 133
Preparing and Implementing a Hazard Communication Program 135
Eye And Body Flushing 141
Hazard Communication Glossary 141
Guide for Reviewing MSDS Completeness 147
Sample MSDS 148
Sample Letter Requesting an MSDS 151
Material Safety Data Sheet Checklist 152

Chapter 6: Personal Protective Equipment 155
General Requirements—1910.132 155
Eye and Face Protection—1910.133 156
Respiratory Protection—1910.134 158
Occupational Head Protection—1910.135 162
Occupational Foot Protection—1910.136 164
Electrical Protective Devices—1910.137 164
Hand Protection—1910.138 165
Hearing Protection—1910.95 165
29 CFR Part 1910, Subpart I Personal Protective Equipment 165
Questions and Answers Concerning the PPE Standards 166
Fact Sheet—Personal Protective Equipment (PPE) Standards 168
Hazard Assessment Form 169
Employee Safety And Health Record 171

Chapter 7: Hazardous Waste Operations and Process Safety Management ... 175
Hazardous Waste Operations and Emergency Response 175
Process Safety Management of Highly Hazardous Chemicals 179

Chapter 8: Permit-Required Confined Spaces 187
Alternative Protection Procedures 189
Permit-Required Confined Spaces—1910.146 189

Chapter 9: Electrical 195
General Requirements—1910.303 196
Wiring Design and Protection—1910.304 199
Wiring Methods—Components and Equipment for General Use— 1910.305 207
Hazardous (Classified) Locations—1910.307 214
Covered Work by Both Qualified and Unqualified Persons—1910.331 215
Training—1910.332 216
Selection And Use Of Work Practices—1910.333 217
Use of Equipment—1910.334 220
Safeguards For Personnel Protection—1910.335 221
Hazards of Electricity 222
Effects of Electricity on the Human Body 223
Electrical Protective Devices 224
The National Electric Code and the NFPA 70E 226

Chapter 10: Hazardous Materials 231
Flammable and Combustible Liquids—1910.106 231
Spray Finishing Using Flammable and Combustible Materials—1910.107 243
Hazardous (Classified) Locations 254

Chapter 11: Fire Protection 257
Scope, Application and Definitions—1910.155 257
Standpipe systems 258
Fire Brigades—1910.156 258
Portable Fire Extinguishers—1910.157 258
Standpipe and Hose Systems—1910.158 260
Automatic Sprinkler Systems—1910.159 260
Fixed Extinguishing Systems, General—1910.160 261
Total Flooding Systems with Potential Health and Safety Hazards to Employees 261
Fixed Extinguishing Systems, Dry Chemical—1910.161 261
Fixed Extinguishing Systems, Gaseous Agent—1910.162 261
Fixed Extinguishing Systems, Water Spray and Foam—1910.163 262
Fire Detection Systems—1910.164 262
Employee Alarm Systems—1910.165 262
Common Fire Extinguishing Agents 264
Portable Fire Extinguishers 266

Chapter 12: Exit Routes, Emergency Action Plans, and Fire Prevention Plans 269
Coverage and Definitions—1910.34 270
Compliance with NFPA 101-2000, Life Safety Code—1910.35 271
Design and Construction Requirements for Exit Routes—1910.36 271
Maintenance, Safeguards and Operational Features for Exit Routes—1910.37 273
Emergency Action Plans—1910.38 274
Fire Prevention Plan—1910.39 275
Emergency Escape Route and Procedures 277
Fact Sheet No. OSHA 92-19 277
Emergency Preparedness and Response 278
Final Rule 29 CFR Part 1910 Subpart E 284

Chapter 13: Machine Guarding 313
Where Mechanical Hazards Occur 313
Hazardous Mechanical Motions and Actions 313
Requirements for Safeguards 315
Methods of Machine Safeguarding 316
Machinery and Machine Guarding 325
General Requirements for All Machines—1910.212 325
Woodworking Machinery Requirements—1910.213 325
Abrasive Wheel Machinery—1910.215 326
Mills and Calendars—1910.216 328
Mechanical Power Presses—1910.217 328
Forging Machines—1910.218 328
Mechanical Power-Transmission Apparatus—1910.219 328
Control of Hazardous Energy (Lockout/Tagout) 329
Employee Training 333
Lockout/Tagout Glossary 335

Chapter 14: Hand and Portable Powered Tools 351
Hand Tools 351
Power Tools 352
Guards 352
Safety Switches 352
Electric Tools 353
Powered Abrasive Wheel Tools 353
Pneumatic Tools 354
Powder-Actuated Tools 354
Fasteners 355
Hydraulic Power Tools 355
Jacks 355

Chapter 15: Materials Handling 359
Handling Material—General—1910.176 359
Servicing Multi-Piece and Single-Piece Rim Wheels—1910.177 359
Safe Operating Procedures: Multi-Piece Rim Wheels 361
Safe Operating Procedures: Single-Piece Rim Wheels 362
Powered Industrial Trucks—1910.178 362
Overhead and Gantry Cranes—1910.179 365
Crawler Locomotive and Truck Cranes—1910.180 369
Derricks—1910.181 371
Helicopters—1910.183 373
Slings—1910.184 373
Safe Lifting Practices 380
Maintenance of Slings 382
Summary 382

Chapter 16: Welding, Cutting, and Brazing 385
Compressed Gases General Requirements—1910.101 385
Welding, Cutting and Brazing General Requirements—1910.252 386
Oxygen-Fuel Gas Welding and Cutting—1910.253 388
Arc Welding and Cutting—1910.254 393

Resistance Welding—1910.255 395
Welding Health Hazards 396
Care of Gas Cylinders 398

Chapter 17: Safety Programs 403
Major Elements of a Safety Program 404
Management Commitment and Employee Involvement 404
Work Site Analysis 405
Hazard Prevention and Control 405
Safety and Health Training 406
Elements of a Written Safety Program 406
State Programs 407
Sample Plan 409
Sample Safety and Health Policy 409
How this Safety Program Meets Federal and State Requirements 414
Checklist to Assess Your Safety Program 414
Appendices and Local Requirements as Needed for the Job 415

Chapter 18: Risk Management 417
Introduction 417
Appendix A: Risk Management Steps 422
Appendix B: The Risk Management Evaluation Profile (RMEP) 430

Chapter 19: Safety and Health Hazards in the Office 445
Leading Types of Disabling Accidents 445
Common Safety and Health Hazards in the Office 445
Some Do's and Dont's of Office Safety 448
Checklist for Office Safety Concerns 450
Personal Office Safety and Health Checklist 451

Chapter 20: OSHA Programs 453
Voluntary Protection Programs 453
OSHA Outreach Program 455
OSHA Outreach Training Program Report 459
Shortcut Procedures for Outreach Trainers with ID Numbers 460
Further Information 465
OSHA Training Institute Education Centers 466
OSHA Outreach Training Program Fact Sheet—West Coast 469

Chapter 21: Training Techniques 471
Adult Learners 471
Classroom Training in General 472
About Your Session 473
Handling Questions 473
Presenting Material 474

Chapter 22: Inspection Procedures 475
Purpose of the Inspection 475
Preparation for the Inspection 475
Conduct of the Inspection 475
Examples of What to Checkin a Safety Inspection 476
Common Occupational Safety Inconsistencies Found During Inspections 477
Sample Inspection Form 479

Chapter 23: Accident Investigation 481
Accident Prevention 481
Investigative Procedures 482
Fact-Finding 483
Interviews 483
Problem Solving Techniques 484
Report of Investigation 485
Summary 486

Chapter 24: Field Inspection Reference Manual 487
FIRM Table of Contents 493
CHAPTER I Pre-inspection Procedures 498
CHAPTER II Inspection Procedures 502
CHAPTER III Inspection Documentation 510
CHAPTER IV Post-Inspection Procedures 520

Appendix: Answers to Test and Workshops 535

OCCUPATIONAL SAFETY AND HEALTH PRE-TEST
(See appendix for answers)

1. **What are the two basic purposes of the OSH Act of 1970?**
   a. f and h
   b. g and i
   c. g and h
   d. f and I
   f. To eliminate or minimize occupational hazards in the workplace.
   g. To provide a safe and healthful workplace.
   h. To recognize, avoid and control hazards in the workplace.
   i. To conserve human resources.

2. **What is one important effect that Executive Order 12196 has on Federal Agencies?**
   a. Mandates compliance with all standards applicable to private industry, except where alternative standards are approved by the Secretary of Labor.
   b. Mandates that Federal Agencies' programs be consistent with private industry.
   c. Provides for OSHA response to reports of hazardous conditions from employees of any Federal Agency.
   d. Mandates the establishment of Occupational Safety and Health Committees at the national and other appropriate levels.

3. **A confined space supervisor must ensure the following before allowing entry into a confined space.**
   a. Employee's are trained and qualified
   b. The atmosphere has been tested and is safe for entry
   c. A standby attendant is present
   d. All of the above

4. **OSHA regulates the use of compressed air for cleaning purposes. What is the main reason for this and how is it regulated?**
   a. Compressed air can cause increased noise exposure, and this is controlled by the use of sound-dampening nozzles.
   b. Air hoses can rupture causing the hose to "whip" and strike workers. Airlines must be equipped with pressure loss closure valves.
   c. Compressed air blows particles and dust causing eye and pressure with guarding.
   d. Gases in compressed air can be toxic. Filters and gas absorbents must be used on air compressors.

5. **Air purifying respirators are:**
   a. Permitted in oxygen deficient atmospheres
   b. Not permitted in oxygen deficient atmospheres
   c. Worn with full beards if the employee will not shave
   d. Designed to cool the respirator user

6. **Arc welding refers to the use of an electrical current to produce the weld.**
   a. True
   b. False

7. **What is the purpose of ring-testing abrasive grinding wheels?**
   a. A method of detecting cracks in the wheels.
   b. Checks the wheel diameter to assure proper size and mounting.
   c. Tests the rated capacity (in rpm) for the compatibility of the wheel and grinder.
   d. Checks the abrasive action to assure that older wheels are not overworked.

8. **The 29 CFR 19 10, Subpart N addresses storage and handling of materials. Which of the following does Subpart N not address?**
   a. Flammable storage
   b. Fork lift trucks
   c. Hoisting equipment
   d. Railroad cars

9. **Doors, passageways, and stairways, which are not a means of egress, shall be:**
   a. Clearly marked "NOT AN EXIT"
   b. Locked to prevent entry
   c. Removed to prevent confusion
   d. No action required

10. **The most important factor in the use of hearing protectors is:**
    a. The Noise Reduction Factor published by the vendor
    b. The case hardened shell
    c. The polymeric expanded foam adhesive
    d. The fitting to an employee

11. **The permissible exposure limit set by OSHA for lead is:**
    a. 200 ugh/m3
    b. 20 ugh/m3
    c. 50 ugh/m3
    d. 100 ugh/m3

12. **What is a safety can?**
    a. A tightly sealed metal container used to store corrosive material.
    b. A spring-loaded, self-closing container of 10 gallons or less which will not allow the escape of vapor when heated.
    c. A glass container used to store materials that are corrosive to metal containers.
    d. A spring-loaded, self-closing metal container, 5 gallons or less, that will allow the escape of vapor when heated.

13. **Rolling scaffolds must have guardrails at what height?**
    a. 4 feet
    b. 6 feet
    c. 10 feet
    d. 12 feet

14. **Flammable materials have a flash point of less than _____ degrees F.**
    a. 200
    b. 110
    c. 160
    d. 100

15. Guardrails must be 42" high with a midrail between the top rail and floor.
   a. True
   b. False

16. A spreader must be used whenever cross cutting stock.
   a. True
   b. False

17. High pressure compressed gas cylinders can be handled by their protective caps:
   a. Anytime, that's what the caps are made for
   b. Only when lifting vertically
   c. During transportation and storage
   d. Never, the cap is only designed to protect the valve.

18. Which of the following is not one of the storage requirement flammable and combustible liquids?
   a. The quantity of material stored
   b. The types of containers
   c. The stacking and arrangement of containers
   d. Continuous monitoring for vapor concentration

19. OSHA makes only three basic requirements of employers with regard to Medical and First Aid. What are they?
   a. f, g, and h
   b. f, g, and j
   c. g, h, and j
   d. g, h, and I
   e. h, 1, and j
   f. Availability of ambulance service
   g. First Aid
   h. Availability of medical personnel
   i. Availability of drenching & flushing
   j. Personal Protection Equipment

20. A chemical with a skin designation indicates:
   a. The chemical may be absorbed into the bloodstream through the skin, the mucous membranes, and/or the eye. This absorption contributes to overall exposure.
   b. Skin absorption is the only significant route of exposure; inhalation potential is minimal.
   c. All chemicals are readily absorbed through the skin; those chemicals with a skin designation are corrosive or irritating to the skin.
   d. All of the above

21. Reverse polarity would be a serious OSHA citation.
   a. True
   b. False

22. A class "A" GFCI trips at 5 mA plus or minus 1 mA.
   a. True
   b. False

23. "Medical personnel" as defined by OSHA, may be physicians, nurses, or trained (Red Cross or equivalent) First Aid personnel.
   a. True
   b. False

24. Which of the following is not a function of foot protection?
   a. Protect the foot from crushing
   b. Protect the sole of the foot from puncture by nails
   c. Electrical protection
   d. High visibility with reflective footwear

25. A good disinfectant mixture is 1 part chlorine bleach to _ part(s) water:
   a. 1
   b. 7
   c. 5
   d. 10

26. A "Point of Operation" can be described as:
   a. The area where a belt and pulley create a nip point
   b. The location where the operator stands.
   c. The point at which cutting, shaping, boring, or forming is accomplished upon the stock
   d. The area of the press where material is actually positioned and work is being performed during any process such as shearing, punching, forming, or assembling.

27. Which of the following devices may be used for overcurrent protection?
   a. Circuit Breaker
   b. Main Disconnect
   c. Fuse
   d. GFCI
   e. Both A and C

28. Which of the following best describes the term egress as applied by OSHA?
   a. The way to, through, and away from an exit.
   b. A specially designated exit or escape.
   c. A function of the fire protection requirements of a building.
   d. Non-portable fire protection, such as a sprinkling system.

29. Water is a good extinguishing agent but should never be used on a Class___ fire.
   a. A
   b. B
   c. C
   d. It can be used on all fires

30. 29 CFR 19 10, Subpart D, covers the safe use of walking and working surfaces. Which of the following is not considered under Subpart D?
   a. Stairways
   b. Ladders
   c. Handrails
   d. Slippery floors
   e. Workbenches

31. **Egress (emergency exiting) fundamentals include:**
    a. Sufficient exits for occupant load
    b. Illuminated exit markings
    C. Exits not locked or obstructed
    d. All of the above

32. **The Hazard Communication standard requires which of the following:**
    a. Employees must be provided with a comprehensive medical examination.
    b. Employees must receive training on the hazardous chemicals in their work area.
    c. Each employee must be provided with his/her own set of MSDSs.
    d. All of the above

33. **Tongue guards on bench grinders must be adjusted within:**
    a. 1/8"
    b. 1/4"
    c. 3/8"
    d. 1/2"

34. **Which of the following are not points of operation guards or devices**
    a. Photoelectric light curtain
    b. Pullbacks
    c. Emergency pull cord
    d. Two hand controls

35. **A worker is using a ladder improperly. Which of the following situations constitutes improper use of a ladder?**
    a. An extension ladder is being used in place of a single ladder.
    b. An employee is using a ten-foot stepladder and standing on the next to-the-top step,
    c. A single ladder is leaning against a roof, extending one foot above the roof surface.
    d. An extension ladder is extended to a length of forty feet and the footing is about ten feet from the supporting wall.

COURSE OBJECTIVES

- Become familiar with the Occupational Safety and Health Act, 29 CFR 1910 and relevant interpretations and applications
- Identify common causes of accidents and fatalities in hazardous areas of industry
- Identify abatement strategies for hazards found in industry
- Understand basic instructional methods and techniques
- Be able to present 10 and 30 hour training programs using the OSHA regulations and guidelines

Note: This 2005 version of the course material has been revised based upon comments from instructors and students in the field. It was felt that the original material provided by the OSHA Training Institute in Illinois was of such high quality that the number of changes should be kept to a minimum. Wherever possible the original intent, text and artwork of the 1995 manual was retained in order to allow current instructors to remain familiar with the manual by maintaining the "look and feel" that they have been used to as well as to keep the material in the public domain.

It was felt that the material needed to be updated (the last complete update was in 1995) to:
- Create a flow of educational topics that tended to enhance the educational process as opposed to skipping from one section to another in the book. For example, following a discussion of Industrial Hygiene where the student learns about PEL's and TWA's, the discussion leads to asbestos, lead, hazard communication and hearing conservation programs to reinforce the use of those IH terms. Rewriting the class matrix and ensuring that the book followed the matrix accomplished this. The four day flow was designed to lead the student from a discovery of OSHA and its jurisdictions to specific regulations, and then cover how to train an adult population and finally how to use the OSHA Outreach program to its fullest.
- Update the manual to reflect changes made in Federal regulations since 1995.
- Reduce the number of pages and bulk of the overall book in order to make it more useable and less expensive to print, publish, distribute and ship.
- Include page numbers for easy reference.
- Include student activities (the pretest is printed in the book, along with a safety committee work activity and the answers.
- Sample PPE Assessment forms and Training forms have been included.
- OSHAs FIRM and Outreach Training Catalog have been included.
- Sections on Safety Committees, Office Safety, Training and Accident Investigation have been added.

What Every Supervisor Must Know About OSHA General Industry

Suggested Matrix

TIME	MONDAY	TUESDAY	WEDNESDAY	THURSDAY
8:00 8:50	Opening Remarks Bio Sheets Objectives	Bloodborne Pathogens Asbestos Update	Hazardous Materials	Welding
9:00 9:50	Pre test/review	Hazard Communication	Fire Protection	Vpp OSHA Outreach
10:00 10:50	Intro to OSHA, Inspections, Penalties Intro to Standards	Hazard Communication	Routes of Exit	Training Techniques
11:00 11:45	Standards Workshop	Personal Protective Equipment	Machine Guarding	Q & A OSHARep.
11:45 12:45	Lunch	Lunch	Lunch	Lunch
12:45 1:35	Record Keeping	Hazwoper And Process Safety Management	Lockout Tagout	Review for Exam
1:45 2:35	Record Keeping	Permit Required Confined Spaces	Hand and Power Tools	Exam
2:45 3:35	Walking and Working Surfaces	Power Generation and Electrical	Materials Handling	Exam and Course Evaluations
3:45 4:30	Introduction to Occupational Health	Electrical Concluded	Forklift	Certificate Awards Course Closes

No claim is made to original government works; however, the gathering, compilation, translation, digital conversion and arrangement of such materials, the additional material, note and reference as well as commentary and materials in this product or publication are subject to We're Into Safety copyright.

All Rights Reserved
Printed in the United States of America

Introduction to OSHA

More than 90 million Americans spend their days on the job. They are our most valuable national resource. Yet, until 1970, no uniform, comprehensive provisions existed for their protection against workplace safety and health hazards.

In 1970, the U.S. Congress considered annual figures such as these:
- Job-related accidents accounted for more than 14,000 worker deaths;
- Ten times as many person-days were lost from job-related disabilities as from strikes;
- Estimated new cases of occupational diseases totaled 300,000; and
- Nearly 2½ million workers were disabled.

In terms of lost production and wages, medical expenses and disability compensation, the burden on the nation's commerce was staggering. Human cost was beyond calculation. Therefore, the Occupational Safety and Health Act of 1970 was passed by a bipartisan Congress *"to assure so far as possible every working man and woman in the Nation safe and healthful working conditions and to preserve our human resources."*

This Act requires employers to provide a safe and healthful work place.

General Health and Safety Section 5 (a)

Each employer:

1) shall furnish to each of their employees employment and a place of employment which are free from recognized hazards that are causing or are likely to cause serious physical harm to their employees.

2) shall comply with occupational health standards promulgated under this act.

The Act also requires that employees follow the rules:

Section 5 (b)

Each employee
1) shall comply with occupational safety and health standards and all rules, regulations, and orders issued pursuant to this act which are applicable to their own actions and conduct.

Under the Act, the Occupational Safety and Health Administration (OSHA) was created to:
- Encourage employers and employees to reduce workplace hazards and to implement new or improve existing safety and health programs;
- Provide for research in occupational safety and health to develop innovative ways of dealing with occupational safety and health problems;
- Establish "separate but dependent responsibilities and rights" for employers and employees for the achievement of better safety and health conditions;
- Maintain a reporting and recordkeeping system to monitor job-related injuries and illnesses;
- Establish training programs to increase the number and competence of occupational safety and health personnel;
- Develop mandatory job safety and health standards and enforce them effectively; and
- Provide for the development, analysis, evaluation and approval of state occupational safety and health programs.

While OSHA continually reviews and redefines specific standards and practices, its basic purposes remain constant. OSHA strives to implement its mandate fully and firmly with fairness to all concerned. In all of its procedures, from standards development through implementation and enforcement, OSHA guarantees employers and employees the right to be fully informed, to participate actively and to appeal actions.

The Act's Coverage

In general, coverage of the Act extends to all employers and their employees in the 50 states, the District of Columbia, Puerto Rico and all other territories under federal government jurisdiction. Coverage is provided either directly by federal OSHA or through an OSHA-approved state program (see section on OSHA-Approved State Programs).

As defined by the Act, an employer is any "person engaged in a business affecting commerce who has employees, but does not include the United States or any State or political subdivision of a State." Therefore, the Act applies to employers and employees in such varied fields as manufacturing, construction, long shoring, agriculture, law, medicine, charity, disaster relief, organized labor and private education. Such coverage includes religious groups to the extent that they employ workers for secular purposes.

The following are not covered under the Act:
- Self-employed persons;
- Farms at which only immediate members of the farm employer's family are employed; and
- Working conditions regulated by other federal agencies under other federal statutes.

However, when another federal agency is authorized to regulate safety and health working conditions in a particular industry, if it does not do so in specific areas, OSHA standards then apply.

As OSHA develops effective safety and health standards of its own, standards issued under the following laws administered by the Department of Labor are the Safety Act, the Arts and Humanities Act and the Longshoremen's and Harbor Workers' Compensation Act.

Origin of OSHA Standards

The OSHA standards were taken from three sources:
- Consensus standards;
- Proprietary standards; and
- Federal laws in effect when the Occupational Safety and Health Act became law.

Consensus standards are developed by industry-wide, standard-developing organizations and are discussed and substantially agreed upon through consensus by industry. OSHA has incorporated the standards of the two primary standards groups, the American National Standards Institute (ANSI) and the National Fire Protection Association (NFPA), into its set of standards.

For example, ANSI Standard B56.1-1969, Standard for Powered Industrial Trucks, covers the safety requirements relating to the elements of design, operation and maintenance of powered industrial trucks.

Another consensus standard source is the NFPA standards. NFPA No. 301969, Flammable and Combustible Liquids Code was the source standard for Part 1910, Section 106. It covers the storage and use of flammable and combustible liquids with flash points below 200°F.

Professional experts within specific industries, professional societies, and associations prepare proprietary standards. The proprietary standards are determined by a straight membership vote, not by consensus. An example of these would be the "Compressed Gas Association, Pamphlet P-1, Safe Handling of Compressed Gases." This proprietary standard covers requirements for safe handling, storage and use of compressed gas cylinders.

Some federal laws in effect before the OSH Act are enforced by OSHA, including the Federal Supply Contracts Act (Walsh-Healey); Federal Service Contracts Act (McNamara-O'Hara); Contract Work Hours and Safety Standards Act (Construction Safety Act); and the National Foundation on the Arts and Humanities Act.

Horizontal and Vertical Standards

Standards are sometimes referred to as being either "horizontal" or "vertical" in their application. Most standards are horizontal or "general," which means they apply to any employer in any industry. Standards relating to fire protection, working surfaces and first aid are examples of horizontal standards.

Some standards, however, are relevant only to a particular industry, and are called vertical, or "particular" standards. Examples are standards applying to the long shoring industry or the construction industry, and to the special industries covered in Subpart R.

OSHA is responsible for promulgating legally enforceable standards. OSHA standards may require conditions or the adoption or use of one or more practices, means, methods or processes reasonably necessary and appropriate to protect workers on the job. It is the responsibility of employers to become familiar with standards applicable to their establishments and to ensure that employees have and use personal protective equipment when required for safety.

Employees must comply with all rules and regulations that are applicable to their own actions and conduct. Where OSHA has not promulgated specific standards, employers are responsible for following the Act's general duty clause.

The general duty clause of the Act states that each employer "shall furnish ... a place of employment which is free from recognized hazards that are causing or are likely to cause death or serious physical harm to his employees."

States with OSHA-approved occupational safety and health programs must set standards that are at least as effective as the federal standards. Many state plan states adopt standards identical to their federal counterparts. OSHA standards fall into four categories: General industry, maritime, construction and agriculture.

Code of Federal Regulations

One of the most common complaints from people who must comply with the Part 1910 standards is "How do you wade through hundreds of pages of standards and make sense out of them?" From time to time you may have experienced this frustration and been tempted to toss the standards in the "round file."

The Code of Federal Regulations is a codification of the general and permanent rules published in the *Federal Register* by the executive departments and agencies of the federal government. The Code is divided into 50 titles that represent broad areas subject to federal regulation. Each title is divided into chapters that usually bear the name of the issuing agency. Each chapter is further subdivided into parts covering specific regulatory areas. Based on this breakdown, the Occupational Safety and Health Administration is designated Title 29 – Labor, Chapter XVII.

OSHA's regulations (Title 29) are issued as of July 1. The approximate revision date is printed on the cover of each volume. Each volume of the Code is revised at least once each calendar year and issued on a quarterly basis approximately as follows:
- Title 1 – Title 16 as of January 1
- Title 17 – Title 27 as of April 1
- Title 28 – Title 41 as of July 1
- Title 42 – Title 50 as of October 1

The Code of Federal Regulations is kept up to date by individual issues of the *Federal Register*. These two publications (the CFR and the *Federal Register*) must be used together to determine the latest version of any given rule.

To determine whether there have been any amendments since the revision date of the Code volume in which the user is interested, the following two lists must be consulted: the "Cumulative List of CFR Sections Affected" issued monthly and the "Cumulative List of Parts Affected" which appears daily in the *Federal Register*. These two lists will refer you to the *Federal Register* page where you may find the latest amendment of any given rule. The pages in the *Federal Register* are numbered sequentially from January 1 to January 1 of the next year. The *Federal Register* is one of the best sources of information on standards, since all OSHA standards are published there when adopted, as are all amendments, corrections, insertions or deletions. Each year the Office of Federal Register publishes all current regulations and standards in the *Code of Federal Regulations* 29 (CFR), available at many libraries and from the Government Printing Office. OSHA's regulations are in Title 29, Part 1900 - 9999. The *Federal Register* is available in many public libraries. Annual subscriptions are available from the Superintendent of Documents, U.S. Government Printing Office, Washington, D.C. 20402.

Under Chapter XVII, the regulations are broken down into Parts. Part 1910, for example, is the standard you are all familiar with, "Occupational Safety and Health Standards," commonly known as the "General Industry Standards." Under each part, such as Part 1910, major blocks of information are broken down into Subparts.

Where to Obtain Copies of Standards

To assist the public in keeping current with OSHA standards, "OSHA Regulations, Documents and Technical Information on CD-ROM" was developed. This CD contains the full text of all OSHA regulations (standards), selected documents and technical information from the OSHA Computerized Information System (OCIS). The OSHA CD is available from the Superintendent of Documents. However, only the *Federal Register* can be used as the official source for OSHA regulations.

Major Subparts in the 1910 standards include:
- Subpart D, Walking-Working Surfaces;
- Subpart E, Exit Routes, Emergency Action Plans, and Fire Prevention Plans;
- Subpart F, Powered Platforms, Manlifts, and Vehicle-Mounted Platforms;
- Subpart G, Occupational Health and Environmental Control;
- Subpart H, Hazardous Materials;
- Subpart I, Personal Protective Equipment;
- Subpart J, General Environmental Controls;
- Subpart K, Medical and First Aid;
- Subpart L, Fire Protection;
- Subpart M, Compressed Gas and Compressed Air Equipment;
- Subpart N, Materials Handling and Storage;
- Subpart O, Machinery and Machine Guarding;
- Subpart P, Hand/ Portable Powered Tools – Other Hand-Held Equipment;
- Subpart Q, Welding, Cutting and Brazing;
- Subpart R, Special Industries;
- Subpart S, Electrical; and
- Subpart Z, Toxic and Hazardous Substances.

Each Subpart is further broken down into sections. The index of Subpart D, Walking-Working Surfaces is shown below.
- 1910.21   Definitions.
- 1910.22   General requirements.
- 1910.23   Guarding floor and wall openings and holes.
- 1910.24   Fixed industrial stairs.
- 1910.25   Portable wood ladders.
- 1910.26   Portable metal ladders.
- 1910.27   Fixed ladders.
- 1910.28   Safety requirements for scaffolding.
- 1910.29   Manually propelled mobile ladder stands and scaffolds (towers).
- 1910.30   Other working surfaces.
- 1910.31   Sources of standards.
- 1910.32   Standards organizations.

Subpart D is broken down into sections. Section 21 (1910.21), called "Definitions," provides the user with definitions of terms used in each of the Subpart D sections. Section 22 covers general requirements, which include housekeeping, aisle marking, open pit protection and floor loading designation. Sections 23 through 30 give specific requirements for such things as floor openings, ladders, scaffolding, mobile ladder stands and dock boards. Section 31 and 32 provide the user with basic information on the source standard used to prepare each section and the standard organizations referenced in the subpart. Each subsequent subpart repeats these last two items (i.e., source of standard and standards organization).

Subparts E and L go together. All of the requirements for an exit, access to an exit, and exit discharge are contained in Subpart E. Employee emergency plans and fire prevention plans are also included in this subpart.

Subpart L, Fire Protection, contains requirements for fire brigades, whenever an employer establishes them, and provides standards relating to different types of fire extinguishing systems. These include portable and fixed fire-suppression equipment, as well as other fire protection systems, including fire detection and employee alarm systems.

Subpart F, Powered Platforms, Manlifts, and Vehicle Mounted Work Platforms, provides standards for devices used to lift personnel to elevated work positions.

Subparts G and Z go together as they are health related. Subpart G contains ventilation requirements and limitations for noise and radiation exposures. Subpart Z contains limitations for air contaminants as well as specific standards for certain toxic health hazards (e.g., asbestos, lead, inorganic arsenic and acrylonitrile). OSHA's "right-to-know" rule, titled "Hazard Communication," is also contained in Subpart Z.

Subpart H is titled "Hazardous Materials." This subpart has many sections common to all industry.

One of these sections is 101 and contains general requirements relating to handling, storage, and use of compressed gases. Section 106 is widely used because it gives requirements for storage of flammable liquids in portable tanks or other portable containers and safety standards for industrial operations and service station operations. Section 106 also addresses the design of flammable liquid storage cabinets and the interior design of flammable liquid storage rooms. Another commonly used section is 107, which discusses the basic requirements for the design, construction, electrical wiring, ventilation, illumination, operations and maintenance of paint spray booths.

Section 110 is applicable to any operation where liquefied petroleum gas is used to fuel forklifts or for other industrial purposes. The requirements for companies using anhydrous ammonia for any of its operations are detailed in Section 111. Standards for Hazardous Waste Operations and Emergency Response are contained in Section 120.

Subpart I contains requirements for the use of personal protective equipment. Requirements are set forth for protection of all parts of the body, including the eye, face, head and extremities. Section 132 also mentions use of protective clothing, shields and barriers. Section 134 contains requirements for the use of respiratory protection.

Subpart J, General Environmental Controls, contains requirements for sanitation, temporary labor camps and "non-water carriage disposal systems." Section 144 contains safety color-coding requirements for marking physical hazards, such as portable containers of flammable liquids. For example, in Section 106 where the standard discusses the use of approved cans, the color of that approved can is actually described in Section 144. Colors and specifications for signs, aisle marking, tags, machine controls and other items are in Section 144 and 145.

Section 147 covers the control of hazardous energy (lockout/tagout) during the servicing and maintenance of machines and equipment.

Subpart K contains requirements for first aid and consists of only three paragraphs. The first paragraph requires the ready availability of medical consultation on worker health. The second paragraph requires first aid-trained personnel when the company is not geographically close to a hospital. The third paragraph requires eyewashes and deluge showers to be in the work area where employees might be exposed to injurious corrosive materials.

Subpart M, "Compressed Gas and Compressed Air Equipment," sets forth the basic requirements for air receivers in Section 169. These include such things as an indicating gauge, a safety valve, and a drainpipe and valve.

Subpart N is titled "Materials Handling and Storage." Section 176 discusses general storage requirements. Section 178 covers powered industrial trucks used to transport, load, unload and stack these materials. Other materials handling devices, such as overhead cranes, derricks and helicopters, are also covered in Sections 179 through 183. Sling standards are in Section 184.

Subparts O and P go together. Basically, Subpart O contains the guarding requirements for larger pieces of machinery. Subpart P then contains guarding requirements for hand-held portable power tools. If you wanted to know the guarding requirements for a mechanical power press, you would use Subpart O, Section 217. If you wanted to know the requirements for a hand-held circular saw, you would use Subpart P, Section 242.

Subpart Q is a specific standard on welding, cutting and brazing. Any welding operation should meet these standards. Specific types of operations covered by this standard include oxygen-fuel gas, arc and resistance welding.

Subpart R contains specific industry standards or "vertical" standards as we discussed earlier. For example, if a company's main business were in pulpwood logging, any hazards noted in those operations should be referenced to the vertical standard.

Subpart S is the electrical standard. These are among the most frequently cited general industry standards.

Paragraph Numbering System

The OSHA standards are easier to deal with once you become familiar with the paragraph numbering system in the *Federal Register*. For example, Part 1910, Section 110 will appear as:

29 CFR 1910.110

The first number, 29, stands for the Title, which in this case relates to Labor. Next, CFR, stands for Code of Federal Regulations. Next, 1910, is the Part number. After 1910 is a period, followed by an Arabic number. This is the Section number; in this case, the number is 110. If you followed this system to its place in the standards, you would find information on the storage and handling of liquefied petroleum gases.

After the Section level, there will be more numbers and letters within parentheses. These represent Subsections. The following example refers to a specific Subsection within the standard:

29 CFR 1910.110(b)(13)(ii)

Like Parts and Sections, Subsections appear in a specific order as well. As noted in the example above, the first level consists of lowercase letters. If the section has three major paragraphs of information in it, they will be numbered 110(a), 110(b) and 110(c). The letter b in the example above refers to the second paragraph or subsection in Section 110.

The next level of numbering involves Arabic numbers. If there are three paragraphs of information between subsections (a) and (b), they will be numbered (a)(1), (a)(2) and (a)(3). The number 13 appears in the example, representing the 13th paragraph or subsection in Subsection (b).

The next level uses the lowercase roman numeral. If there are three paragraphs of information between subsections (a)(1) and (a)(2), they will be numbered (a)(1)(i), (a)(1)(ii) and (a)(1)(iii). The lowercase Roman numeral ii appears in the example, representing the second paragraph or subsection in subsection (b)(13).

Fortunately, the system complexity ends here. If there are more than three levels of subsections, the three-paragraph numbering sequence will be repeated. The one difference is that they will appear in italics:

29 CFR 1910.110(b)(13)(ii)(*b*)(7)(*iii*)

It should be noted, however, that standards promulgated after 1979 may be identified differently when there are more than three levels of subsections. Rather than repeat and italicize the lowercase letter, a capital letter will be used in the fourth set of parentheses. The next subsections also will be Arabic and lowercase Roman numerals, but not italicized:

29 CFR 1910.304(f)(5)(iv)(F)(1)

Color Coding

To further simplify the reading of standards, it is suggested that you color code your standards book. For example, go through the book and find all the section numbers and highlight them in pink. Also highlight every section head full column width in pink. Then go through the book and highlight the subsections (i.e., lowercase letter, Arabic number, lowercase Roman numeral in parentheses) in yellow:

		PINK		**YELLOW**
Code of				
Title	Fed. Reg.	Part	Section	Subsections
29	CFR	1910	.110	(b)(13)(ii)(*b*)(7)(*iii*)

Also note there is a subject index in the back of the standards book; this index can be very helpful when trying to locate specific standards based on a key word. If you try to locate information within the standards by using the Table of Contents, remember that the particular section number contained on each page is printed in the upper corner of that page.

Federal Employees

Under the Act, federal agency heads are responsible for providing safe and healthful working conditions for their employees. The Act requires agencies to comply standards consistent with those OSHA issues for private sector employers. OSHA conducts federal workplace inspections in response to employees' reports of hazards and as part of a special program that identifies federal workplaces with higher than average rates of injuries and illnesses.

Federal agency heads are required to operate comprehensive occupational safety and health programs that include recording and analyzing injury/illness data, providing training to all personnel and conducting self-inspections to ensure compliance with OSHA standards. OSHA conducts comprehensive evaluations of these programs to assess their effectiveness.

OSHA's federal sector authority is different from that in the private sector in several ways. The most significant difference is that OSHA cannot propose monetary penalties against another federal agency for failure to comply with OSHA standards. Instead, compliance issues unresolved at the local level are raised to higher organizational levels until resolved. Another significant difference is that OSHA does not have authority to protect federal employee "whistleblowers." However, the Whistleblower Protection Act of 1989 affords present and former federal employees (other than those in the U.S. Postal Service and certain intelligence agencies) an opportunity to file their reports using their internal Inspector General lines of communication.

Provisions for State and Local Governments

OSHA provisions do not apply to state and local governments in their role as employers. The Act does provide that any state desiring to gain OSH approval for its private sector occupational safety and health program (see section on State Plans) must provide a program that covers its state and local government workers and that is at least as effective as its program for private employees. State plans may also cover only public sector employees.

How New Standards are Developed

OSHA can begin standards-setting procedures on its own initiative or in response to petitions from other parties, including the Secretary of Health and Human Services (HHS), the National Institute for Occupational Safety and Health (NIOSH), state and local governments, any nationally recognized standards-producing organization, employer or labor representatives or any other interested person.

Advisory Committees

If OSHA determines that a specific standard is needed, any of several advisory committees may be called upon to develop specific recommendations. There are two standing committees, and ad hoc committees may be appointed to examine special areas of concern. All advisory committees, standing or ad hoc, must have members representing management, labor and state agencies, as well as one or more designees of the Secretary of HHS. The occupational safety and health professions and the general public also may be represented. The two standing advisory committees are the:

- National Advisory Committee on Occupational Safety and Health (NACOSH), which advises, consults with, and makes recommendations to the Secretary of HHS and to the Secretary of Labor on matters regarding administration of the Act; and
- Advisory Committee on Construction Safety and Health, which advises the Secretary of Labor on formulation of construction safety and health standards and other regulations.

NIOSH Recommendations

Recommendations for standards also may come from the National Institute for Occupational Safety and Health (NIOSH), established by the Act as an agency of the Department of HHS. NIOSH conducts research on various safety and health problems, provides technical assistance to OSHA and recommends standards for OSHA's adoption. While conducting its research, NIOSH may conduct workplace investigations, gather testimony from employers and employees and require that employers provide information to them. NIOSH also may require employers to provide medical examinations and tests to determine the incidence of occupational illness among employees. When such examinations and tests are required by NIOSH for research purposes, they may be paid for by NIOSH rather than the employer.

Standards Adoption
Once OSHA has developed plans to propose, amend or revoke a standard, it publishes these intentions in the *Federal Register* as a "Notice of Proposed Rulemaking" or often as an earlier "Advance Notice of Proposed Rulemaking."

An "Advance Notice" or a "Request for Information" is used, when necessary, to solicit information that can be used in drafting a proposal. The Notice of Proposed Rulemaking will include the terms of the new rule and provide a specific time (at least 30 days from the date of publication, usually 60 days or more) for the public to respond. Interested parties who submit written arguments and pertinent evidence may request a public hearing on the proposal when none has been announced in the notice. When such a hearing is requested, OSHA will schedule one and will publish, in advance, the time and place for it in the *Federal Register*.

After the close of the comment period and public hearing, if one is held, OSHA must publish in the *Federal Register* the full final text of any standard amended or adopted and the date it becomes effective, along with an explanation of the standard and the reasons for implementing it. OSHA may also publish a determination that no standard or amendment needs to be issued.

Emergency Temporary Standards
Under certain limited conditions, OSHA is authorized to set emergency temporary standards that take effect immediately. First, OSHA must determine that workers are in grave danger due to exposure to toxic substances or agents determined to be toxic or physically harmful or to new hazards. Then, if OSHA determines an emergency standard is needed, it will publish the emergency temporary standard in the *Federal Register*, where it also serves as a proposed permanent standard. It is then subject to the usual procedure for adopting a permanent standard, except that a final ruling must be made within six months. The validity of an emergency temporary standard may be challenged in an appropriate U.S. Court of Appeals.

Appealing a Standard
No decision on a permanent standard is ever reached without due consideration of the arguments and data received from the public in written submissions and at hearings. Any person who may be adversely affected by a final or emergency standard, however, may file a petition (within 60 days of the rule's promulgation) for judicial review of the standard with the U.S. Court of Appeals for the circuit in which the objector lives or has his or her principal place of business. Filing an appeals petition, however, will not delay the enforcement of a standard, unless the Court of Appeals specifically orders it.

Variances
Employers may ask OSHA for a variance from a standard or regulation if they cannot fully comply by the effective date, due to shortages of materials, equipment or professional or technical personnel or can prove their facilities or methods of operation provide employee protection "at least as effective." Employers located in states with their own occupational safety and health programs should apply to the state for a variance. If however, an employer operates facilities in states under federal OSHA jurisdiction and also in state-plan states, the employer may apply directly to federal OSHA for a single variance applicable to all the establishments in question. OSHA will then work with the state-plan states involved to determine if a variance can be granted that will satisfy state as well as federal OSHA requirements.

Temporary Variance
A temporary variance may be granted to an employer who cannot comply with a standard or regulation by its effective date due

to unavailability of professional or technical personnel, materials or equipment, or because the necessary construction or alteration of facilities cannot be completed in time.

Employers must demonstrate to OSHA that they are taking all available steps to safeguard employees in the meantime and that they have put in force an effective program for coming into compliance with the standard or regulation as quickly as possible.

A temporary variance may be granted for the period needed to achieve compliance or for one year, whichever is shorter. It is renewable twice, each time for six months. An application for a temporary variance must identify the standard or portion of a standard from which the variance is requested and the reasons why the employer cannot comply with the standard. The employer must document those measures already taken and to be taken (including dates) to comply with the standard.

The employer must certify that workers have been informed of the variance application, that a copy has been given to the employees' authorized representative and that a summary of the application has been posted wherever notices are normally posted. Employees also must be informed that they have the right to request a hearing on the application. The temporary variance will not be granted to an employer who simply cannot afford to pay for the necessary alterations equipment or personnel.

Permanent Variance

A permanent variance (alternative to a particular requirement or standard) may be granted to employers who prove their conditions, practices, means, methods operations or processes provide a safe and healthful workplace as effectively as would compliance with the standard.

In making a determination, OSHA weighs the employer's evidence and arranges a variance inspection and hearing where appropriate. If OSHA finds the request valid, it prescribes a permanent variance detailing the employer's specific exceptions and responsibilities under the ruling. When applying for a permanent variance, the employer must inform employees of the application and of their right to request a hearing. Anytime after six months from the issuance of a permanent variance, the employer or employees may petition OSHA to modify or revoke it. OSHA also may do this of its own accord.

Interim Order

So that employers may continue to operate under existing conditions until a variance decision is made, they may apply to OSHA for an interim order. Application for an interim order may be made either at the same time as, or after, application for a variance. Reasons why the order should be granted may be included in the interim order application. If OSHA denies the request, the employer will be notified of the reason for denial.

If the interim order is granted, the employer and other concerned parties will be informed of the order, and the terms of the order are published in the *Federal Register*. The employer must inform employees of the order by giving a copy to the authorized employee representative and by posting a copy wherever notices are normally posted.

Experimental Variance

If an employer is participating in an experiment to demonstrate or validate new job safety and health techniques, and that experiment has been approved by either the Secretary of Labor or the Secretary of HHS, a variance may be granted to permit the experiment. In addition to temporary, permanent and experimental variances, the Secretary of Labor also may find certain variances justified when the national defense is impaired. Variances are not retroactive. An employer who has been cited for a standards violation may not seek relief from that citation by applying for a variance. The fact that a citation is outstanding, however, does not prevent an employer from filing a variance application.

Public Petitions

OSHA continually reviews its standards to keep pace with developing and changing industrial technology. Therefore, employers and employees should be aware that just as they may petition OSHA for the development of standards, they may also petition OSHA for the modification or revocation of standards.

Recordkeeping and Reporting

Before the Act became effective, no centralized, systematic method existed for monitoring occupational safety and health problems. Statistics on job injuries and illnesses were collected by some states and by some private organizations; national figures were based on not-altogether-reliable projections. With OSHA came the first basis for consistent, nationwide procedures — a vital requirement for gauging problems and solving them.

Employers of 11 or more employees must maintain records of occupational injuries and illnesses as they occur. The purpose of keeping records is to permit survey material to be compiled, to help define high-hazard industries, and to inform employees of the status of their employer's record. Employers in state-plan states are required to keep the same records as employers in other states.

OSHA recordkeeping is not required for certain retail trades and some service industries. Exempt employers, like nonexempt employers, must comply with OSHA standards, display the OSHA poster and report to OSHA within eight hours of the occurrence of any accident that results in one or more fatalities or the hospitalization of three or more employees. If an on-the-job accident occurs that results in the death of an employee or in the hospitalization of three or more employees, all employers, regardless of the number of employees, must report the accident, in detail, to the nearest OSHA office within eight hours. In states with approved plans, employers report such accidents to the state agency responsible for safety and health programs.

Injury and Illness Records

Recordkeeping forms are maintained on a calendar-year basis. They are not sent to OSHA or any other agency. They must be maintained for five years at the place of business and must be available for inspection by representatives of OSHA, HHS or the designated state agency. Many specific OSHA standards have additional recordkeeping and reporting requirements. (See Chapter 20, Recordkeeping.)

Keeping Employees Informed

Employers are responsible for keeping employees informed about OSHA and the various safety and health matters with which they are involved. Federal OSHA and states with their own occupational safety and health programs require that each employer post certain materials at a prominent location in the workplace. These include:

- The Job Safety and Health Protection workplace poster (OSHA 2203 or state equivalent) informing employees of their rights and responsibilities under the Act. Any official edition of the poster is acceptable. In addition displaying the workplace poster, the employer must make available to employees, upon request, copies of the Act and copies of relevant OSHA rules and regulations.
- Summaries of petitions for variances from standards or recordkeeping procedures.
- Copies of all OSHA citations for violations of standards. These copies must remain posted at or near the location of alleged violations for three days or until the violations are corrected, whichever is longer.
- The Log and Summary of Occupational Injuries and Illnesses (OSHA No. 300). The log must be maintained and a summary (OSHA No. 300A) must be signed by an officer of the company and posted no later than February 1. It must remain in place until May 1.

All employees have the right to examine any records kept by their employers regarding their exposure to hazardous materials or the results of medical surveillance. Occasionally, OSHA standards or NIOSH research activities will require an employer to measure and record employee exposure to potentially harmful substances. Employees have the right (in person or through their authorized representative) to be present during the measuring as well as to examine records of the results.

Under these substance-specific requirements, each employee or former employee has the right to see his or her examination records, and must be told by the employer if his or her exposure has exceeded the levels set by OSHA. The employee must also be told what corrective measures are being taken.

In addition to having access to records, employees in manufacturing facilities must be provided with information about all of the hazardous chemicals in their work areas.

Employers are to provide this information via labels on containers, material safety data sheets and training programs.

Workplace Inspections

Authority to Inspect

To enforce its standards, OSHA representatives are authorized under the Act to conduct workplace inspections. Every establishment covered by the Act is subject to inspection by OSHA compliance safety and health officers, who are chosen for their knowledge and experience in the occupational safety and health field. Compliance officers are vigorously trained in OSHA standards and in recognition of safety and health hazards. Similarly, states with their own occupational safety and health programs conduct inspections using qualified compliance safety and health officers.

Under the Act, "upon presenting appropriate credentials to the owner, operator or agent in charge," an OSHA compliance officer is authorized to:

Enter without delay and at reasonable times any factory, plant, establishment, construction site or other areas, workplace, or environment where work is performed by an employee of an employer....

Inspect and investigate during regular working hours, and at other reasonable times, and within reasonable limits and in a reasonable manner, any such place of employment and all pertinent conditions, structures, machines, apparatus, devices, equipment and materials therein, and to question privately any such employer, owner, operator, agent or employee.

Inspections are conducted without advance notice. There are, however, special circumstances under which OSHA may indeed give notice to the employer, but even then, such a notice will be less than 24 hours. These special circumstances include:
- Imminent-danger situations that require correction as soon as possible;
- Inspections that must take place after regular business hours or that require special preparation;
- Cases where notice is required to ensure that the employer and employee representative or other personnel will be present; and
- Situations in which the OSHA area director determines that advance notice would produce a more thorough or effective inspection.

Employers receiving advance notice of an inspection must inform their employees' representative or arrange for OSHA to do so. If an employer refuses to admit an OSHA compliance officer or if an employer attempts to interfere with the inspection, the Act permits appropriate legal action to be taken.

Based on a 1978 Supreme Court ruling (Marshall v. Barlows, Inc.), OSHA may not conduct warrantless inspections without an employer's consent. It may however inspect after acquiring a judicially authorized search warrant based upon administrative probable cause or evidence of a violation.

Inspection Priorities

Obviously, not all 6 million workplaces covered by the Act can be inspected immediately. The worst situations need attention first. Therefore, OSHA has established a system of inspection priorities.

Imminent Danger

Imminent-danger situations are given top priority. An imminent danger is any condition where there is reasonable certainty that a danger exists that can be expected to cause death or serious physical harm immediately or before the danger can be eliminated through normal enforcement procedures.

Serious physical harm is any type of harm that could cause permanent or prolonged damage to the body or while not damaging the body on a prolonged basis, could cause such temporary disability as to require inpatient hospital treatment. OSHA considers that "permanent or prolonged damage" has occurred when, for example, a part of the body is crushed or severed; an arm, leg or finger is amputated; or sight in one or both eyes is lost. This kind of damage also includes that which renders a part of the body either functionally useless or substantially reduced in efficiency on or off the job. For example, bones in a limb shattered so severely that mobility or dexterity will be permanently reduced.

Temporary disability requiring in-patient hospital treatment includes injuries, such as simple fractures, concussions, burns, or wounds involving substantial loss of blood and requiring extensive suturing or other healing aids. Injuries or illnesses that are difficult to observe are classified as serious if they inhibit a person in performing normal functions, cause reduction in physical or mental efficiency or shorten life.

Health hazards may constitute imminent-danger situations when they present a serious and immediate threat to life or health. For a health hazard to be considered an imminent danger, there must be a reasonable expectation that toxic substances, such as dangerous fumes, dusts or gases are present and that exposure to them will cause immediate and irreversible harm to such a degree as to shorten-life or cause a reduction in physical or mental efficiency, even though the resulting harm is not immediately apparent.

Employees should inform the supervisor or employer immediately if they detect or even suspect an imminent-danger situation in the workplace. If the employer takes no action to eliminate the danger, an employee or an authorized employee representative may notify the nearest OSHA office and request an inspection. The request should identify the workplace location, detail the hazard or condition and include the employee's name, address and telephone number. Although the employer has the right to see a copy of the complaint if an inspection results, the name of the employee will be withheld if the employee so requests.

If a request for inspection is made, the OSHA area director will review the information and immediately determine whether there is a reasonable basis for the allegation. If it is decided the case has merit, the area director will assign a compliance officer to conduct an immediate inspection of the workplace.

Upon inspection, if an imminent danger situation is found, the compliance officer will ask the employer to voluntarily abate the hazard and to remove endangered employees from exposure. Should the employer fail to do so OSHA, through the regional solicitor, may apply to the nearest federal district court for appropriate legal action to correct the situation. Before the OSHA inspector leaves the workplace, he or she will advise all affected employees of the hazard and post an imminent-danger notice; judicial action can produce a temporary restraining order (immediate shutdown) of the operation or section of the workplace where the imminent danger exists. Should OSHA "arbitrarily or capriciously" decline to bring court action, the affected employees may sue the Secretary of Labor to compel the Secretary to do so.

Walking off the job because of potentially unsafe workplace conditions is not ordinarily an employee right. To do so may result in disciplinary action by the employer. However, an employee does have the right to refuse (in good faith) to be exposed to an imminent danger. OSHA rules protect employees from discrimination if:

- Where possible, he or she asked the employer to eliminate the danger and the employer failed to do so;
- The danger is so imminent that there is not sufficient time to have the danger eliminated through normal enforcement procedures;
- The danger facing the employee is so grave that "a reasonable person" in the same situation would conclude there is a real danger of death or serious physical harm; or
- The employee has no reasonable alternative to refusing to work under such conditions (e.g., asking for reassignment to another area).

Catastrophes and Fatal Accidents

Second priority is given to investigation of fatalities and catastrophes resulting in hospitalization of three or more employees. The employer must report such situations to OSHA within eight hours of their occurrence. Investigations are made to determine if OSHA standards were violated and to avoid recurrence of similar accidents

Employee Complaints

Third priority is given to employee complaints of an alleged violation of standard or of unsafe or unhealthful working conditions. Also included in this category are serious referrals of unsafe or unhealthful working conditions from other sources, such as local or state agencies or departments.

The Act gives each employee the right to request an OSHA inspection when the employee feels he or she is in imminent danger from a hazard or when he or she feels that there is a violation of an OSHA standard that threatens physical harm. OSHA will maintain confidentiality if requested, will inform the employee of any action it takes regarding the complaint and, if requested, will hold an informal review of any decision not to inspect. Just as in situations of imminent danger, the employee's name will be withheld from the employer, if the employee so requests.

Programmed High-Hazard Inspections

Next in priority are programmed, or planned, inspections aimed at specific high-hazard industries, occupations or health substances. Industries are selected for inspection on the basis of factors such as death, injury and illness incidence rates and employee exposure to toxic substances. Special emphasis may be regional or national in scope, depending on the distribution of the workplaces involved. States with their own occupational safety and health programs may use somewhat different systems to identify high-hazard industries for inspection.

Follow-up Inspections

A follow-up inspection determines whether previously cited violations have been corrected. If an employer has failed to abate a violation, the compliance officer informs the employer that he or she is subject to alleged "Notification of Failure to Abate" violations and may face additional proposed daily penalties while such failure or violations continues.

Inspection Process

Prior to inspection, the compliance officer will become familiar with as many relevant facts as possible about the workplace, taking into account such things as the history of the establishment, the nature of the business and the particular standards likely to apply. Preparing for the inspection also involves selecting appropriate equipment for detecting and measuring such things as fumes, gases, toxic substances and noise.

Inspector's Credentials

An inspection begins when the OSHA compliance officer arrives at the establishment. He or she will display official credentials and ask to meet an appropriate employer representative. Employers should always insist upon seeing the compliance officer's credentials. An OSHA compliance officer carries U.S. Department of Labor credentials bearing his or her photograph and a serial number that can be verified by phoning the nearest OSHA office. Anyone who tries to collect a penalty at the time of inspection or promotes the sale of a product or service at any time is not an OSHA compliance officer. Posing as a compliance officer is a violation of law; suspected impostors should be promptly reported to local law enforcement agencies.

Opening Conference

In the opening conference, the compliance officer (CSHO) will explain why the establishment was selected. The CSHO also will ascertain whether an OSHA-funded consultation program is in progress or whether the facility is pursuing or has received an inspection exemption; if programmed, the inspection is usually terminated. The employer will then be asked to select an employer representative to accompany the CSHO during the inspection.

The CSHO will explain the purpose of the visit, the scope of the inspection and the standards that apply. The employer will be given a copy of any employee complaint that may be involved.

An authorized employee representative also will be given the opportunity to attend the opening conference and to accompany the compliance officer during inspection. If a recognized bargaining representative represents the employees, the union ordinarily will designate the employee representative to accompany the compliance officer. Similarly, if there is a plant safety committee, the employee members of the committee will designate the employee representative (in the absence of a recognized bargaining representative). Where neither employee group exists, the employees themselves may select an employee representative or the compliance officer will determine if any employee suitably represents the interest of other employees. Under no circumstances may the employer select the employee representative for the walk-around.

The Act does not require that there be an employee representative for each inspection. Where there is no authorized employee representative, however, the compliance officer must consult with a reasonable number of employees concerning safety and health matters in the workplace; such consultations may be held privately.

Inspection Tour
After the opening conference, the compliance officer and accompanying representatives will proceed through the establishment, inspecting work areas for compliance with OSHA standards. The compliance officer will determine the route and duration of the inspection. When speaking with employees, the compliance officer will make every effort to minimize any work interruptions. The compliance officer will observe conditions, consult with employees, take instrument readings, examine records and possibly tale photos for record purposes.

Trade secrets observed by the compliance officer will be kept confidential. An inspector who releases confidential information without authorization is subject to a $1,000 fine and/or one year in jail. The employer may require that the employee representative have a security clearance for any area in question.

Employees will be consulted during the inspection tour. The compliance officer may stop and question workers in private about safety and health conditions and practices in their workplaces. Each employee is protected under the Act from discrimination for exercising their safety and health rights.

Posting and recordkeeping will be checked. The compliance officer will inspect records of deaths, injuries and illnesses that the employer is required to keep. He or she will check to see that a copy of the totals from the last page of OSHA No. 300 has been posted and that the OSHA workplace poster (OSHA 2203) is prominently displayed. Where records of employee exposure to toxic substances and harmful physical agents are required, they will be examined.

The CSHO also will explain the requirements of the Hazard Communication Standard, which requires employers to establish a written, comprehensive hazard communication program that includes provisions for container labeling, material safety data sheets and an employee-training program. The program must provide a list of the hazardous chemicals in each work area and the means the employer will use to inform employees of the hazards of non-routine tasks.

During the course of the inspection, the CSHO will point out to the employer any unsafe or unhealthful working conditions he or she observes. At the same time, the CSHO will discuss possible corrective action if the employer so desires. Some apparent violations detected by the compliance officer can be corrected immediately. However, even if immediately corrected the apparent violations may still serve as the basis for a citation and/or notice of proposed penalty. An inspection tour may cover part or all of an establishment, even if the inspection was in response to a specific complaint, accident or fatality.

Closing Conference
After the inspection tour, a closing conference will be held between the compliance officer and the employer or the employer representative. It is a time for free discussion of problems and needs — a time for frank questions and answers.

The compliance officer discusses with the employer all unsafe or unhealthful conditions observed on the inspection and indicates all apparent violations for which a citation may be issued or recommended. The employer is told of appeal rights. The compliance officer does not indicate any proposed penalties. Only the OSHA area director has that authority, and only after having received a full report.

During the closing conference, the employer may wish to produce records to show compliance efforts and to provide information that can help OSHA determine how much time may be needed to abate an alleged violation. When appropriate, more than one closing conference may be held. This is usually necessary when health hazards are being evaluated or when laboratory reports are required.

A closing discussion will be held with the employees, or their representative if requested, to discuss matters of direct interest to employees. The employees' representative may be present at the closing conference.

The CSHO will explain that OSHA area offices are full-service resource centers that provide a number of services, including guest speakers, handout materials that can be distributed to interested persons, and training and technical materials on safety and health matters.

Citations and Penalties

The OSHA area director issues citations. After the compliance officer reports his or her findings, the area director will determine what citations, if any, will be issued and what penalties, if any, will be proposed.

Citations inform the employer and employees of the regulations and standard alleged to have been violated and of the proposed length of time set for the abatement. The employer will receive citations and notices of propose penalties by certified mail. The employer must post a copy of each citation near the place the violation occurred for three days or until the violation is abated, whichever is longer.

Penalties

There are several types of violations that may be cited and penalties that may be proposed. Citation and penalty procedures may differ somewhat in states that administer their own occupational safety and health programs.

Other Than Serious Violation. A violation that has a dire relationship to job safety and health, but probably would not cause death or serious physical harm. A proposed penalty of up to $7,000 for each violation is discretionary.

A penalty for an other-than-serious violation may be adjusted downward by as much as 95 percent, depending on the employer's good faith (demonstrated efforts to comply with the Act), history of previous violations and business size. When the adjusted penalty amounts to less than $100, no penalty is proposed.

Serious Violation. A violation where there is substantial probability that death or serious physical harm could result and that the employer knew, or should have known, of the hazard.

A mandatory penalty of up to $7,000 for each violation is proposed. A penalty for a serious violation may be adjusted downward, based on the employer's good faith, history of previous violations, the gravity of the alleged violation, and size of business.

Willful Violation. A violation that the employer knowingly commits or commits with plain indifference to the law. The employer either knows that what it is doing constitutes a violation, or is aware that a hazardous condition existed and made no reasonable effort to eliminate it.

Penalties of up to $70,000 may be proposed for each willful violation, with a minimum penalty of $5,000 for each violation. A proposed penalty for a willful violation may be adjusted downward, depending on the size of the business and its history of previous violations. Usually, no credit is given for good faith.

If an employer is convicted of a willful violation of a standard that has resulted in the death of an employee, the offense is punishable by a court-imposed fine, imprisonment for up to six months, or both. A fine of up to $250,000 for an individual or $500,000 for a corporation may be imposed for a criminal conviction.

Repeated Violation. A violation of any standard, regulation, rule or order where, upon reinspection, is found to be substantially similar to a previous violation. A repeated violation brings a fine of up to $70,000. To be the basis of a repeated citation, the original citation must be final; a citation under contest may not serve as the basis for a subsequent repeated citation.

Failure to Abate Prior Violation. Failure to abate a prior violation may bring a civil penalty of up to $7,000 for each day the violation continues beyond the prescribed abatement date.

De Minimis Violation. De minimis violations are violations of standards that have no direct or immediate relationship to safety or health. When de minimis conditions are found during an inspection, they are documented in the same way as any other violation, but are not included on the citation.

Additional violations for which citations and proposed penalties may be issued upon conviction:
- Falsifying records, reports or applications can bring a fine of $10,000, up to six months in jail, or both.
- Violations of posting requirements can bring a civil penalty of up to $7,000.

- Assaulting a compliance officer or otherwise resisting, opposing, intimidating or interfering with a compliance officer while he or she is engaged in the performance of his or her duties is a criminal offense, subject to a fine of not more than $5,000 and imprisonment for not more than three years.

**Table 1-1.
OSHA PENALTIES**

Willful	
– Maximum	$70,000
– Minimum	$5,000
Repeated – Maximum	$70,000
Serious, Other-than-Serious, Other Specific Violations	
– Maximum	$7,000
Failure to Abate for each calendar day beyond abatement date	
– Maximum	$7,000
OSHA Notice	$1,000
Posting of OSHA Record Keeping Summary	$1,000
Posting of Citation	$3,000
Maintaining OSHA 300, OSHA Record Keeping	$1,000
Reporting Fatality/Catastrophe	$5,000
Access to Records under 1904	$1,000
Notification Requirements under 1903.6 (Advance Notice)	$2,000

Appeals Process

Appeals by Employees

If an inspection was initiated due to an employee complaint, the employee or authorized employee representative may request an informal review of any decision not to issue a citation.

Employees may not contest citations, amendments to citations, penalties or lack of penalties. They may contest the time in the citation for abatement of a hazardous condition. They also may contest an employer's Petition for Modification of Abatement (PMA), which requests an extension of the abatement period. Employees must contest the PMA within 10 working days of its posting or within 10 working days after an authorized employee representative has received a copy. Within 15 working days of the employer's receipt of the citation, the employee may submit a written objection to OSHA. The OSHA area director will forward the objection to the Occupational Safety and Health Review Commission, which operates independently of OSHA.

Employees may request an informal conference with OSHA to discuss any issues raised by an inspection, citation, and notice of proposed penalty or employer's notice of intention to contest.

Appeals by Employers

When issued a citation or notice of a proposed penalty, an employer may request an informal meeting with OSHA's area director to discuss the case. Employee representatives may be invited to attend the meeting. The area director is authorized to enter into settlement agreements that revise citations and penalties to avoid prolonged legal disputes.

Petition for Modification of Abatement
Upon receiving a citation, the employer must correct the cited hazard by the prescribed date unless it contests the citation or abatement date. Factors beyond the employer's reasonable control may prevent the completion of corrections by that date.

The written petition should specify all steps taken to achieve compliance, the additional time needed to achieve complete compliance, the reasons such additional time is needed, all temporary steps being taken to safeguard employees against the cited hazard during the intervening period, that a copy of the PMA was posted in a conspicuous place at or near each place where a violation occurred and that the employee representative (if there is one) received a copy of the petition.

Notice of Contest
If the employer decides to contest either the citation, the time set for abatement or the proposed penalty, it has 15 working days from the time the citation and proposed penalty are received in which to notify the OSHA area director in writing. An orally expressed disagreement will not suffice.

There is no specific format for the Notice of Contest; however, it must clearly identify the employer's basis for filing the citation, notice of proposed penalty, abatement period or notification of failure to correct violations.

A copy of the Notice of Contest must be given to the employees' authorized representative. If a recognized bargaining agent does not represent any affected employees, a copy of the notice must be posted in a prominent location in the workplace or personally served upon each unrepresented employee.

Review Procedure

If the written Notice of Contest has been filed within the required 15 working days, the OSHA Area Director will forward the case to the Occupational Safety and Health Review Commission. The commission is an independent agency not associated with OSHA or the Department of Labor. The commission will assign the case to an administrative law judge.

The judge may schedule a hearing in a public place near the employer's workplace or, if it is found to be legally invalid, disallow the contest. The employer and the employees have the right to participate in the hearing; the commission does not require that attorneys represent them. Once the administrative law judge has ruled, any party to the case may request a further review by OSHRC. Any of the three OSHRC commissioners also may, at his or her own motion, bring a case before the commission for review. Commission rulings may be appealed to the appropriate U.S. Court of Appeals.

Appeals in State-Plan States

States with their own occupational safety and health programs have a state system for review and appeal of citations, penalties and abatement periods. The procedures are generally similar to federal OSHA's, but a state review board or equivalent authority hears cases.

OSHA-Approved State Programs

The Act encourages states to develop and operate, under OSHA guidance, state job safety and health plans.

Once a state plan is approved, OSHA funds up to 50 percent of the program's operating costs. State plans are required to provide standards and enforcement programs, as well as voluntary compliance activities that are at least as effective as the federal program. State plans developed for the private sector also must, to the extent permitted by state law, provide coverage for state and local government employees. OSHA rules also permit states to develop plans limited in coverage for state and local government, or public schools, employees only; in such cases, private sector employment remains under federal jurisdiction.

To gain OSHA approval as a developmental plan, a state must have adequate legislative authority and must demonstrate that within three years it will provide standards-setting, enforcement and appeals procedures; public employee protection; a sufficient number of competent enforcement personnel; and training, education and technical assistance programs.

If, at any time during this period or later, it appears that the state is capable of enforcing standards in accordance with the above requirements, OSHA may enter into an operational status agreement with the state. OSHA generally limits its enforcement activity to areas not covered by the state in the agreement and suspends all concurrent federal enforcement.

Scheduled accident and complaint inspections generally become the primary responsibility of the state.

When all developmental steps concerning resources, procedures, and other requirements have been completed and approved, OSHA then certifies that a state has the legal, administrative, and enforcement means necessary to operate effectively. This action renders no judgment on how well or poorly a state is actually operating its program, but merely attests to the structural completeness of its program. After this certification, there is a period of at least one year to determine if a state is effectively providing safety and health protection.

Employers and employees should find out if their state operates an OSHA-approved state program and, if so, become familiar with it (See Exhibit 1-2.). State safety and health standards under approved plans must keep pace with federal standards, and state plans must guarantee employer and employee rights as does OSHA. If it is found that the state is operating at least as effectively as federal OSHA and other requirements including compliance staffing levels are met, final approval of the plan may be granted and federal authority will cease in those areas over which the state has jurisdiction.

Exhibit 1-2. State-plan states.

The following states and territories have OSHA-approved state plans:

Alaska	New Mexico
Arizona	New York*
California	North Carolina
Connecticut*	Oregon
Hawaii	Puerto Rico
Indiana	South Carolina
Iowa	Tennessee
Kentucky	Utah
Maryland	Vermont
Michigan	Virgin Islands*
Minnesota	Virginia
Nevada	Washington
New Jersey*	Wyoming

*Plans only cover public-sector employees.

Services Available

Consultation Assistance

Consultation assistance is available to employers who want help in establishing and maintaining a safe and healthful workplace. Largely funded by OSHA, this service is provided at no cost to the employer. No penalties are proposed or citations issued for hazards identified by the consultant. Consultation is provided to the employer with the assurance that his or her name and firm and any information about the workplace will not be routinely reported to the OSHA inspection staff.

Primarily developed for smaller employers with more hazardous operations, the consultation service is delivered by state government agencies or universities employing professional safety consultants and health consultants. When delivered at the worksite, consultation assistance includes an opening conference with the employer to explain the ground rules for consultation, a walk through the workplace to identify any specific hazards and to examine those aspects of the employer's safety and health program that relate to the scope of the visit, and a closing conference followed by a written report to the employer of the consultant's finding and recommendations.

This process begins with the employer's request for consultation and the commitment to correct any serious job safety and health hazards identified by the consultant. Possible violations of OSHA standards will not be reported to OSHA enforcement staff unless the employer fails or refuses to eliminate or control worker exposure to any identified serious hazard or imminent danger situation. In such unusual circumstances, OSHA may investigate and begin enforcement action.

Employers who receive a comprehensive consultation visit, correct all identified hazards, and demonstrate that an effective safety and health program is in operation may be exempted from OSHA programmed enforcement inspections (not complaint or accident investigations) for a period of one year. Comprehensive consultation assistance includes an appraisal of all mechanical, physical, work practice, and environmental hazards of the workplace and all aspects of the employer's present job safety and health program.

Voluntary Protection Programs

The voluntary protection programs (VPPs) represent one part of OSHA's effort to extend worker protection beyond the minimum required by OSHA standards. These programs as well as expanded onsite consultation services and full-service area offices are cooperative approaches that when coupled with an effective enforcement program, expand worker protection to help meet the goals of the Occupational Safety and Health Act of 1970.

The three VPPs-Star, Merit, and Demonstration-are designed to:

- Recognize outstanding achievement of those who have successfully incorporated comprehensive safety and health programs into their total management systems;
- Motivate others to achieve excellent safety and health results in the same outstanding way; and
- Establish a relationship between employers, employees, and OSHA that is based on cooperation rather than coercion.

Star Program. This program is the most demanding and the most prestigious. It is open to employers in any industry who have successfully managed a comprehensive safety and health program to reduce injury rates below the national average for that industry. Specific requirements for the program include:

- Management commitment and employee participation;
- A high quality worksite analysis program;

- Hazard prevention and control programs; and
- Comprehensive safety and health training for all employees.

These requirements must all be in place and operating effectively.

Merit Program. This program is primarily a stepping stone to Star Program. An employer with a basic safety and health program built around the Star requirement who is committed to improving the company's program and has the resources to do so within a specified period of time may work with OSHA to meet Star qualifications.

Demonstration Program. This program is for companies that provide Star-quality worker protection in industries where certain Star requirements may not be appropriate or effective. It allows OSHA both the opportunity to recognize outstanding safety and health programs that would otherwise be unreached by the VPP and to determine if general Star requirements can be changed to include these companies as Star participants.

OSHA reviews an employer's VPP application and conducts an on-site review to verify that the safety and health program described is in operation at the site. Evaluations are conducted on a regular basis: annually for Merit and Demonstration programs and triennially for Star. All participants must send their injury information annually to their OSHA regional office. Sites participating in the VPP are not scheduled for programmed inspection; however, any employee complaints, serious accidents or significant chemical releases that may occur are handled according to routine enforcement procedures.

An employer may make application for any VPP at the nearest OSHA regional office. Once OSHA is satisfied that, on paper, the employer qualifies for the program, an on-site review will be scheduled. The review team will present its findings in a written report for the company's review prior to submission to the Assistant Secretary of Labor, who heads OSHA. If approved, the employer will receive a letter from the Assistant Secretary informing the site of its participation in the VPP. A certificate of approval and flag are presented at a ceremony held at or near the approved worksite. Star sites receiving reapproval after each triennial evaluation receive plaques at similar ceremonies. The VPPs described are available in states under federal jurisdiction.

Training and Education

OSHA's area offices are full-service centers offering a variety of informational services, such as availability for speaking engagements, publication, audio-visual aids on workplace hazards and technical advice.

The OSHA Training Institute in Arlington Heights, Ill., provides basic and advanced training and education in safety and health for federal and state compliance officers; state consultants; other federal agency personnel; and private sector employers, employees and their representatives. Institute courses cover areas such as electrical hazards, machine guarding, ventilation and ergonomics. The Institute facility includes classrooms, laboratories, a library, and an audiovisual unit. The laboratories contain various demonstrations and equipment, such as power presses, woodworking and welding shops, a complete industrial ventilation unit and a sound demonstration laboratory.

OSHA also provides funds to nonprofit organizations to conduct workplace training and education in subjects where OSHA identifies areas of unmet needs for safety and health.

Current grant subjects include agricultural health and safety, and hazard communication programs for small businesses that do not have safety and health staff to assist them. Organizations awarded grants use funds to develop training and educational programs, reach out to workers and employers for whom their program is appropriate and provide these programs to workers and employers.

Grants are awarded annually, with a one-year renewal possible. Grant recipients are expected to contribute 20 percent of the total grant cost. Information on grants can be found at the OSHA web site at www.osha.gov. Look for the Susan Harwood Grants. These grants run annually from September to September with the submissions being due to the OSHA Training Institute in Arlington Heights, Ill. in June or July.

Employer Responsibilities and Rights

Employers have certain responsibilities and rights under the Occupational Safety and Health

Act of 1970. Employer responsibilities and rights in states with their own occupational safety and health programs are generally the same as in federal OSHA states. Employers must cooperate with the OSHA compliance officer by furnishing names of authorized employee representatives who may be asked to accompany the compliance officer during an inspection. If none, the compliance officer will consult with a reasonable number of employees concerning safety and health in the workplace.

All employers must:
- Meet the general duty responsibility to provide a workplace free from recognized hazards that are causing or are likely to cause death or serious physical harm to employees and comply with standards, rules and regulations issued under the Act.
- Be familiar with mandatory OSHA standards and make copies available to employees for review upon request.
- Inform all employees about OSHA.
- Examine workplace conditions to make sure they conform to applicable standards.
- Minimize or reduce hazards.
- Make sure employees have and use safe tools and equipment (including appropriate personal protective equipment) and that such equipment is properly maintained.
- Use color codes, posters, labels or signs when needed to warn employees of potential hazards.
- Establish or update operating procedures and communicate them so that employees follow safety and health requirements.
- Provide training required by OSHA standards (e.g., hazard communication, lead).
- Report to the nearest OSHA office within eight hours any fatal accident or one that results in the hospitalization of three or more employees.
- Keep OSHA-required records of work-related injuries and illnesses and post a copy of the OSHA 300A from February through April each year. (This applies to employers with 11 or more employees.)
- Post, at a prominent location within the workplace, the OSHA poster (OSHA 2203) informing employees of their rights and responsibilities.
- Provide employees, former employees and their representatives' access to the Log and Summary of Occupational Injuries and Illnesses (OSHA 300) at a reasonable time and in a reasonable manner.
- Provide access to employee medical records and exposure records to employees or their authorized representatives.
- Not discriminate against employees who properly exercise their rights under the Act.
- Post OSHA citations at or near the worksites involved. Each citation, or copy thereof, must remain posted until the violation has been abated or for three working days, whichever is longer.
- Abate cited violations within the prescribed period.

Employers have the right to:
- Seek advice and off-site consultation as needed by writing, calling or visiting the nearest OSHA office. (OSHA will not inspect merely because an employer requests assistance.)
- Be active in an industry association's job safety and health programs.
- Request and receive proper government-issued identification from OSHA compliance officers prior to inspections.
- Be advised of the reason for the inspection.
- Have an opening conference with the compliance officer.
- Accompany the compliance officer during the inspection.
- Have a closing conference at the end of the inspection.
- File a written Notice of Contest with the OSHA area director within 15 working days of the receipt of a notice of citation or a proposed penalty.
- Apply to OSHA for a temporary variance from a standard if unable to comply because of the unavailability of materials, equipment or personnel to make necessary changes within the required time.
- Apply to OSHA for a permanent variance from a standard if able to furnish proof its facilities or method of operation provide employee protection at least as effective as that required by the standard.
- Take an active role in developing safety and health standards through participation in

OSHA Standard Advisory Committees, nationally recognized standards-setting organizations, and evidence and views presented in writing or at hearings.
- Be assured of the confidentiality of any trade secrets observed by an OSHA compliance officer during an inspection.
- Submit a written request to NIOSH for information on whether any substance in its workplace has potentially toxic effects in the concentrations being used.

Employee Responsibilities and Rights

Although OSHA does not cite employees for violations of their responsibilities, each employee "shall comply with all occupational safety and health standards and all rules, regulations, and orders Issued under the Act" that are applicable. Employee responsibilities and rights in states with their own occupational safety and health programs are generally the same as for workers in federal OSHA states.

Employees should:
- Read the OSHA poster at the job site.
- Comply with all applicable OSHA standards.
- Follow all employer safety and health rules and regulations and wear or use prescribed protective equipment while engaged in work.
- Report hazardous conditions to the supervisor.
- Cooperate with the OSHA compliance officer conducting an inspection if he or she inquires about safety and health conditions in the workplace.

11(c) Rights: Protection for Using Rights

Employees have a right to seek safety and health on the job without fear of punishment as described in Section 11(c) of the Act. The law says employers shall not punish or discriminate against workers for exercising rights such as:
- Complaining to an employer, union, OSHA or any other government agency about job safety and health hazards;
- Filing safety or health grievances;
- Participating on a workplace safety and health committee or in union activities concerning job safety and health; and
- Participating in OSHA inspections, conferences, hearings, or other OSHA-related activities.

If an employee is exercising these or other OSHA rights, the employer is not allowed to discriminate against that worker in any way, such as through firing, demotion, taking away seniority or other earned benefits, transferring the worker to an undesirable job or shift, or threatening or harassing the worker.

If the employer has knowingly allowed the employee to do something in the past (e.g., leaving work early), he or she may be violating the law by punishing the worker for doing the same thing following a protest of hazardous conditions. If the employer knows that a number of workers are doing the same thing wrong, it cannot legally single out for punishment the worker who has taken part in safety and health activities.

Workers believing they have been punished for exercising safety and health rights must contact the nearest OSHA office within 30 days of the time they learn of the alleged discrimination. A union representative can file a 11(c) complaint for a worker. The worker does not have to complete any forms. An OSHA staff member will complete the forms, asking what happened and who was involved.

Following an 11(c) complaint, OSHA will investigate. If an employee has been illegally punished for exercising safety and health rights, OSHA will ask the employer to restore that worker's job earning and benefits. If necessary, and if it can prove discrimination, OSHA will take the employer to court. In such cases, the worker does not pay any legal fees. If a state agency has an OSHA-approved state program, employees may file their complaint with either federal OSHA or a state agency under its laws.

Other Rights

Employees have the right to:
- Review copies of appropriate OSHA standards, rules, regulations and requirements that the employer should have available at the workplace.
- Request information from the employer on safety and health hazards in the area, precautions that may be taken and procedures to be followed if an employee is involved in an accident or is exposed to toxic substances.
- Receive adequate training and information on workplace safety and health hazards.

- Request the OSHA area director to investigate if you believe hazardous conditions or violations of standards exist in your workplace.
- Have your name withheld from your employer, upon request to OSHA, if you file a written and signed complaint.
- Be advised of OSHA actions regarding your complaint and have an informal review, if requested, of any decision not to inspect or to issue a citation.
- Have your authorized employee representative accompany the OSHA compliance officer during the inspection tour.
- Respond to questions from the OSHA compliance officer, particularly if there is no authorized employee representative accompanying the compliance officer.
- Observe any monitoring or measuring of hazardous materials and have the right to see these records and your medical records, as specified under the Act.
- Have your authorized representative, or yourself, review the Log and Summary of Occupational Injuries (OSHA 300) at a reasonable time and in a reasonable manner.
- Request a closing discussion with the compliance officer following an inspection.
- Submit a written request to NIOSH for information on whether any substance in your workplace has potentially toxic effects in the concentration being used and have your name withheld from your employer if you so request.
- Object to the abatement period set in the citation issued to your employer by writing to the OSHA area director within 15 working days of the issuance of the citation.
- Participate in hearings conducted by the Occupational Safety and Health Review Commission.
- Be notified by your employer if it applies for a variance from an OSHA standard and testify at a variance hearing and appeal the final decision.
- Submit information or comment to OSHA on the issuance, modification or revocation of OSHA standards and request a public hearing.

While OSHA normally does not issue fines and penalties to federal agencies, those agencies must recognize the financial cost of this type of enforcement action. Congress has been exploring the possibility of allowing OSHA to fine agencies of the federal government and has enacted the Postal Employee Safety Act. The U.S. Postal Service recently came under the auspices of OSHA and is now subject to penalties and fines that must come out of their operating budget.

Construction vs. General Industry

Because OSHA is responsible for ensuring the safety and health of workers in a variety of industries, it has developed regulations for each industry, such as maritime, construction and general industry (manufacturing). Often these regulations may seem to overlap and the worker must understand what regulation applies to his or her work. OSHA considers the work being performed as the key element to this "entry argument" into the regulations. So even though a worker has a job title of construction operator, he or she may fall under the general industry rules depending on the job that is being performed.

For example, OSHA defines "construction work" as "work for construction, alteration, and/or repair, including painting and decorating."

The General Industry Standard § 1910.23(c)(1) for fall protection starts at 4 ft. and requires a guardrail:

> Every open-sided floor or platform 4 feet or more above adjacent floor or ground level shall be guarded by a standard railing (or the equivalent as specified in paragraph (e)(3) of this section) on all open sides except where there is entrance to a ramp, stairway, or fixed ladder. The railing shall be provided with a toeboard wherever, beneath the open sides, workers can pass.

The Construction Standard § 1926.501(b)(15) for fall protection starts at 6 ft. and allows for such protective devices as guardrails and personal fall arrest systems:

> Each employee on a walking/working surface (horizontal and vertical surface) with an unprotected side or edge which is 6 feet (1.8 m) or more above a lower level shall be protected from falling by the use of guardrail systems, safety net systems, or personal fall arrest systems.

For example, a company has one maintenance worker who fixes heating and ventilation systems and a building with a roof that is 5 ft. tall. On Monday the worker goes to the rooftop to replace a bad motor. That work would be considered an alteration of the system, so it would fall under the construction standard. The rule that applies is the construction rules that takes effect at 6 ft. The worker needs no fall protection. Two weeks later, the same worker returns to the same heating unit to change filters. He is not altering, constructing, painting or decorating, he is performing routine maintenance. At this point, he must be protected by the general industry standard, which takes effect at 4 ft. Now he must be protected by a guardrail system. Same person, same roof, same heating system. Different rule.

Safety Committees

The principal function of committees shall be to consult and provide policy advice on and monitor the performance of the safety and health program.

Committees shall be established at establishments or groupings of establishments consistent with the mission, size and organization of the company and its collective bargaining configuration. The employer shall form committees at the lowest practicable local level. The principal function of the establishment (or local) committees is to monitor and assist in the execution of the employer's safety and health policies and program at the workplaces.

- Committee members should serve overlapping terms. Such terms should be of at least two years' duration, except when the committee is initially organized.
- The committee chairperson shall be nominated from among the committee's members and shall be elected by the committee members. Management and nonmanagement members should alternate in this position. Maximum service time as chairperson should be two consecutive years.
- Committees shall establish a regular schedule of meetings and special meetings shall be held as necessary.
- Adequate advance notice of committee meetings shall be furnished to employees and each meeting shall be conducted pursuant to a prepared agenda.
- Written minutes of each committee meeting shall be maintained and distributed to each committee member, and upon request, shall be made available to employees and to the Secretary.

- Employers shall provide all committee members appropriate training as required.
- The safety and health committee is an integral part of the safety and health program and helps ensure effective implementation of the program at the establishment level.
- Monitor and assist the safety and health program at establishments under its jurisdiction and make recommendations to the official in charge on the operation of the program;
- Monitor findings and reports of workplace inspections to confirm that appropriate corrective measures are implemented

Injured Workers May Sue Government

A construction worker injured on an Army installation after a contracting officer failed to enforce contractual safety standards may sue the government under the federal Tort Claims Act, a federal district court ruled June 4, 1999. In *Pelham v. U.S.* (No. 84-1395), the U.S. District Court for the District of New Jersey affirmed its earlier ruling that while a contracting officer's implementation of a safety inspection program is discretionary, once the officer suspects safety deficiency, he or she must act upon it.

The accident occurred when a forklift was used to lift a garbage dumpster in which Jerry Pelham was standing in order to attach cables to a 20-ft. roof-support suspension beam. When the forklift lurched, Pelham lost his balance and caught his hand between the forklift mast chain and pulley, amputating parts of three fingers and severely cutting a fourth.

Shortly before Pelham was injured, the contracting officer observed a worker in a metal container being hoisted by a forklift. The contracting officer testified that he "raised the issue" with the job superintendent, who told him that it was standard practice.

Under the terms of the contract, the contractor was required to take safety measures prescribed by the government's contracting officer. The contracting officer was required to notify the contractor of any non-compliance with safety provisions and had authority to issue a stop-work order if the contractor failed to comply promptly.

Discretionary Function Exception

Lawsuits against the United States for injuries caused by government employees in situations where a private person, would be liable are authorized by the FTCA. Under an exception to the act, the government is not liable for claims based on a discretionary function, whether or not the discretion involved is abused.

The court distinguished the Pelham case from two other Third Circuit decisions that applied the discretionary function exemption to the government's alleged failure to inspect a radioactive extraction facility and remove asbestos. Negligence in the development of safety provisions or a spot check program is not the kind of conduct at issue here the court stated, where the issue is "ministerial implementation of safety regulations" rather than "the discretionary authority to develop such provisions."

The contracting officer accepted the contractor's response rather than assume his ministerial responsibility under the contract to enforce compliance. The court found that under the contract "there is nothing discretionary about this responsibility." Once the contracting officer suspected a safety violation, his duty to check safety requirements was ministerial and not discretionary under the FTCA, the court ruled. Decisions made at the operational level, as distinguished from the planning level, are not protected by the exception, the court found.

Your Safety Committee

Name of the Committee (be creative)

_____

Committee Motto:

_____

Measurable Goals For The Committee:

_____

_____

_____

Elected Leader:

_____

Recorder:

_____

Name of members: _____

Name of members: _____

Name of members: _____

Name of members: _____

Name of members: _____

Name of members: _____

Name of members: _____

Name of members: _____

INTRODUCTION TO OSHA

Most Frequently Cited Standards

Manufacturing

Listed below are the standards which were cited by OSHA during the period October 2004 through September 2005.

Standard	Cited	Insp	Penalty ($)	Description
Total	33121	5711	$20,319,065	
19100147	2954	1590	2,375,887	The Control of Hazardous Energy, Lockout/Tagout
19101200	2768	1458	550,653	Hazard Communication
19100212	2232	1803	2,753,640	Machines, General Requirements
19100134	1894	890	493,366	Respiratory Protection
19100305	1672	1079	724,657	Electrical, Wiring Methods, Components and Equipment
19100219	1528	810	948,639	Mechanical Power-Transmission Apparatus
19100178	1525	1027	687,755	Powered Industrial Trucks
19100303	1222	930	691,474	Electrical Systems Design, General Requirements
19100213	960	471	500,575	Woodworking Machinery Requirements
19100215	950	583	323,454	Abrasive Wheel Machinery
19100132	837	613	463,934	Personal Protective Equipment, General Requirements
19100095	768	382	452,790	Occupational Noise Exposure
19100266	764	125	318,152	Pulpwood Logging
19100023	752	588	765,749	Guarding Floor and Wall Openings and Holes
19100217	747	271	725,076	Mechanical Power Presses
19100157	630	487	168,058	Portable Fire Extinguishers
19100146	620	239	396,889	Permit-Required Confined Spaces
19100107	546	214	254,911	Spray Finishing with Flammable/Combustible Materials
19100022	536	462	297,061	Walking-Working Surfaces, General Requirements
19100037	503	392	160,483	Means of Egress, General
19100106	459	284	214,733	Flammable and Combustible Liquids
19100179	437	224	282,525	Overhead and Gantry Cranes
19100304	407	364	199,438	Electrical, Wiring Design and Protection
19040029	398	362	108,868	Record Keeping
19100151	390	381	300,824	Medical Services and First Aid
19100242	384	376	181,488	Hand and Portable Powered Tools and Equipment, General
19100253	358	294	146,357	Oxygen-Fuel Gas Welding and Cutting
19100133	298	278	144,523	Eye and Face Protection
5A0001	298	264	740,219	General Duty Clause
19101025	247	89	227,042	Lead
19101030	237	127	95,037	Bloodborne Pathogens
19100184	223	138	107,416	Slings
19101000	216	110	156,666	Air Contaminants
19100119	182	37	406,655	Process Safety Management, Highly Hazardous Chem's
19100252	175	139	160,226	Welding, Cutting and Brazing, General Requirements
19100120	162	55	283,196	Hazardous Waste Operations and Emergency Response
19100036	158	147	113,767	Means of Egress, General Requirements
19100141	155	111	49,663	Sanitation
19100024	151	124	85,570	Fixed Industrial Stairs
19040032	146	117	17,375	
19100038	145	130	49,675	Employee Emergency Plans and Fire Prevention Plans
19040002	135	134	23,475	Log and Summary of Occupational Injuries and Illnesses
19100334	134	119	50,806	Electrical, Use of Equipment
19100176	133	127	161,067	Materials Handling, General
19100138	117	114	59,846	Hand Protection
19100265	116	52	59,293	Sawmills
19100110	111	90	50,627	Storage and Handling of Liquified Petroleum Gases
19100333	108	97	170,400	Electrical, Selection and Use of Work Practices
19101027	96	23	55,450	Cadmium
19100243	80	78	42,873	Guarding of Portable Powered Tools
19100101	75	73	29,651	Compressed Gases, General Requirements
19100332	75	67	54,894	Electrical, Training
19100254	67	52	44,668	Arc Welding and Cutting
19100027	64	43	48,231	Fixed Ladders
19100335	64	44	70,820	Electrical, Safeguards for Personnel Protection
19100307	62	59	40,487	Electrical, Hazardous (Classified) Locations

2004 Most Frequently Cited General Industry Standards

Standard	Description	Citations
212(a)(1)	General Machine Guarding	1675
1200(e)(1)	Hazard Communication Written Program	1084
212(a)(3)(ii)	Guarding Point of Operations	882
23(c)(1)	Guarding open sided floors	707
147(c)(1)	Lockout/Tagout - Program	689
215(b)(9)	Tongue Guards on Grinders	669
151(c)	Eye and body flushing facilities	651
147(c)(4)(i)	Lockout/Tagout procedures	596
219(d)(1)	Guarding pulleys	596
1200(h)(1)	Hazard Communication – Information and Training	589

Plus 1,273 General Duty Clause Citations

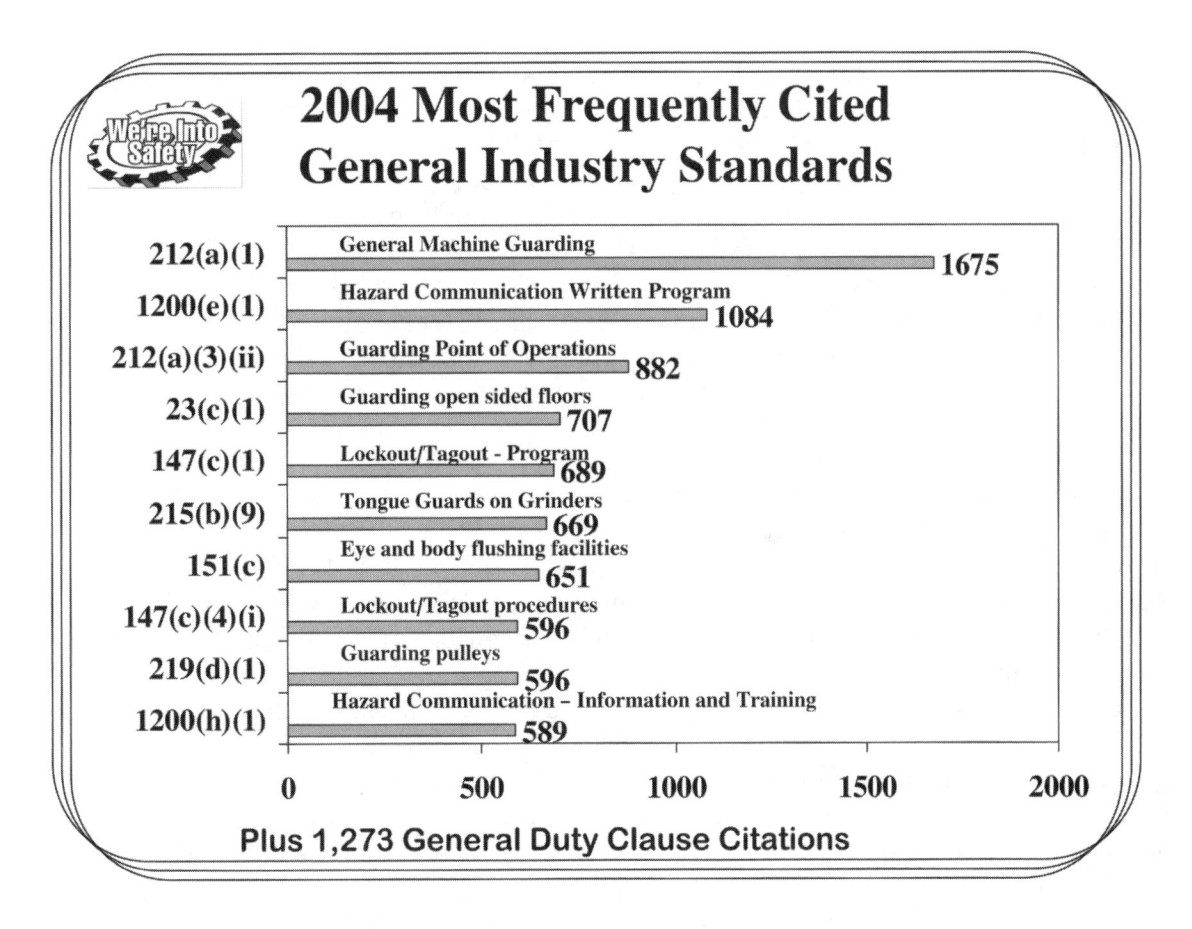

Hazard Violation Search Workshop #1
(See appendix for answers.)

Description of Hazard
1. Exit access blocked by pallet. Less than 20 in. width.
2. Fire extinguisher (charged) found lying on floor.
3. Type BC fire extinguisher used near bales of shredded paper – no type A available.
4. At some locations in large facility, directions of travel to exit are not apparent.
5. No portable fire extinguisher inspection/maintenance program.
6. Fixed carbon dioxide extinguishing system not being inspected and tested annually.
7. No exit signs anywhere in plant employing 50 employees.
8. Exit door blocked from outside.
9. Sprinkler heads in storage area blocked by stacked bags within 5 in. of heads.
10. Fire extinguishers obstructed by containers.
11. Exit access door swings against exit travel; considered high hazard occupancy room.
12. No eyewash facility in battery charging area — battery acid being used.
13. Employee observed smoking in a flammable liquid storage area.
14. Eye protection required — items being used do not meet ANSI requirements.
15. Respirators required — no training program for users.
16. Baseball caps being worn by maintenance employees replacing take-up roller bearing on conveyor while standing on a maintenance catwalk.
17. Equipment ground prong on portable electric drill broken off.
18. Forklift observed being used as a man lift using only standard pallet.
19. Permanent aisles in container storage area not marked.
20. Containers being used for storing flammable liquids — not of the approved type.
21. Paint spray booth with electric motor and fan placed inside exhaust duct.
22. End attachments on a wire rope sling are cracked and deformed.

23. General industrial plant: 150 gallons of toluene stored outside of an inside storage room (Flashpoint = 45°F, B.P. = 232°F).
24. Welding operation being done in paint spray booth.
25. Forklift purchased in 1966 performing high lifts without overhead guard.
26. LP gas service station dispensing area without "No Smoking" signs.
27. 150 gallons of Class I liquid stored in a flammable liquid storage cabinet.
28. Forklift operator giving a ride to two personnel on a standard pallet on forks.
29. The dry filter spray booth had an exhaust of less than 50 LFM.
30. Containers used to transfer flammables were not bonded to the dispensing drum.
31. Safety cans containing flammable liquids not in red can nor contents identified.
32. Stairway — 6 risers — no handrail.
33. Hospital 20 miles away, no employee trained in first aid.
34. Open-sided floor — 6-ft. drop; no guardrails.
35. Fixed ladder — 30 ft. high with no cage guard.
36. Wooden extension ladder – ladder safety shoes broken.
37. Stairway more than 44 in. wide does not have stair rails on open sides.
38. Opened-sided work platform (less than 4 ft. high) adjacent to dangerous equipment not guarded with standard railing and toeboard.
39. Elevated area used for storage has no approved floor-loading signs posted in a conspicuous place.
40. Portable dockboard without any securing device to prevent slipping.
41. Stairway with steam pipe 6 ft. above stair tread causing overhead obstruction.
42. Defective portable wooden ladder with split side rail being used in building maintenance operation.
43. Compressed gas cylinders in storage with no valve protection caps.
44. Oxygen and fuel-gas cylinders stored together.
45. Respirators not stored in a clean, sanitary location.
46. Arc welding area — adjacent work area personnel exposed to welding rays.
47. Welding electrode cable lead is damaged and frayed, exposing bare wire.
48. Air receiver not equipped with indicating pressure gauge.
49. There is a loose conduit feeding into a fuse box.
50. Powered industrial truck idling while its operator drinks coffee 200 ft. away.
51. Steel frame of sorter machine not grounded. Machine located 3 ft. from water pipe.
52. Hand-held portable electric grinder with no guard.
53. Hand-held electric circular saw equipped with a lock-on control switch.
54. 157 psi air being used for cleaning purposes in machine shop.
55. Hand-held circular saw being used with no lower blade guard.
56. Portable electric drill owned by employee had defective switch.
57. Radial arm saw has lower portion of blade unguarded.
58. Explosive-actuated fastening tool without a protective shield.
59. Work rest on bench grinder has ½ in. space between wheel and rest.
60. Employee working in a mechanical equipment room daily for eight hours is exposed to 92 dBA.
61. Employer had no energy control (lockout/tagout) procedures.
62. No guarding to protect employees from flying chips of a metal turning lathe.

Hazard Violation Workshop #2

1. Identify what is wrong in each of the following scenarios.
2. Identify what Subpart of the OSHA regulations has been violated.
3. Propose corrective action.

Scenario 1.

Scenario 2.

INTRODUCTION TO OSHA

Scenario 3.

Scenario 4.

WHAT EVERY SUPERVISOR MUST KNOW ABOUT OSHA-GENERAL

Scenario 5.

Scenario 6.

Scenario 7.

Scenario 8.

CHAPTER 2

RECORDING AND REPORTING OCCUPATIONAL INJURIES AND ILLNESSES

The following is an abstract of Section VII, Summary and Explanation of the Final Rule, of 29 CFR 1904, Occupational Injury and Illness Recording and Reporting Requirements (66 Fed. Reg. 5916). Subparts covered include:
  A. Purpose
  B. Scope
  C. Recordkeeping Forms and Recording Criteria
  D. Other OSHA Injury and Illness Recordkeeping Requirements
  E. Reporting Fatality, Injury and Illness Information to the Government.
  F. Transition From the Former Rule
  G. Definitions

No attempt has been made to discuss every detail of the regulation. Readers are encouraged to consult the *Federal Register* for the complete text. The corresponding page numbers of major paragraphs as they appear in the *Federal Register* are provided throughout this document to facilitate further reading.

A list of major changes to the recordkeeping rule, released by the Occupational Safety and Health Administration, also is included.

Major Changes To OSHA's Recordkeeping Rule

This document provides a list of the major changes from OSHA's old 1904 recordkeeping rule to the new rule employers began using in 2002. This list summarizes the major differences between the old and new recordkeeping rules to help people who are familiar with the old rule to learn the new rule quickly.

Scope
- The list of service and retail industries that are partially exempt from the rule has been updated. Some establishments that were covered under the old rule are not required to keep OSHA records under the new rule and some formerly exempted establishments will now have to keep records. (§1904.2)
- The new rule continues to provide a partial exemption for employers who had 10 or fewer workers at all times in the previous calendar year. (§1904.1)

Forms

- The new OSHA Form 300 (Log of Work-Related Injuries and Illnesses) has been simplified and can be printed on smaller legal-sized paper.
- The new OSHA Form 301 (Injury and Illness Incident Report) includes more data about how the injury or illness occurred.
- The new OSHA Form 300A (Summary of Work-Related Injuries and Illnesses) provides additional data to make it easier for employers to calculate incidence rates.
- Maximum flexibility has been provided so employers can keep all the information on computers, at a central location, or on alternative forms, as long as the information is compatible and the data can be produced when needed. (§1904.29 and §1904.30)

Work-related

- A "significant" degree of aggravation is required before a preexisting injury or illness becomes work-related. (§1904.5(a))
- Additional exceptions have been added to the geographic presumption of work relationship; cases arising from eating and drinking of food and beverages, blood donations, exercise programs, etc., no longer need to be recorded. Common cold and flu cases also no longer need to be recorded. (§1904.5(b)(2))
- Criteria for deciding when mental illnesses are considered work-related have been added. (§1904.5(b)(2))
- Sections have been added clarifying work relationship when employees travel or work out of their home. (§1904.5(b)(6) and §1904.5(b)(7))

Recording criteria

- Different criteria for recording work-related injuries and work-related illnesses are eliminated; one set of criteria is used for both. (The former rule required employers to record all illnesses, regardless of severity). (§1904.4)
- Employers are required to record work-related injuries or illnesses if they result in one of the following: death; days away from work; restricted work or transfer to another job; medical treatment beyond first aid; loss of consciousness; or diagnosis of a significant injury/illness by a physician or other licensed health care professional. (§1904.7(a))
- New definitions are included for medical treatment and first aid. First aid is defined by treatments on a finite list. All treatment not on this list is medical treatment. (§1904.7(b)(5))
- The recording of "light duty" or restricted work cases is clarified. Employers are required to record cases as restricted work cases when the injured or ill employee only works partial days or is restricted from performing their "routine job functions" (defined as work activities the employee regularly performs at least once weekly). (§1904.7(b)(4))
- Employers are required to record all needlestick and sharps injuries involving contamination by another person's blood or other potentially infectious material. (§1904.8)
- Musculoskeletal disorders (MSDs) are treated like all other injuries or illnesses: they must be recorded if they result in days away, restricted work, transfer to another job, or medical treatment beyond first aid.
- Special recording criteria are included for cases involving the work-related transmission of tuberculosis or medical removal under OSHA standards. (§1904.9 and §1904.11)

Day counts

- The term "lost workdays" is eliminated and the rule requires recording of days away, days of restricted work, or transfer to another job. Also, new rules for counting that rely on calendar days instead of workdays are included. (§1904.7(b)(3))
- Employers are no longer required to count days away or days of restriction beyond 180 days. (§1904.7(b)(3))
- The day on which the injury or illness occurs is not counted as a day away from work or a day of restricted work. (§1904.7(b)(3) and §1904.7(b)(4))

Annual Summary
- Employers must review the 300 Log information before it is summarized on the 300A form. (§1904.32(a))
- The new rule includes hours worked data to make it easier for employers to calculate incidence rates. (§1904.32(b)(2))
- A company executive is required to certify the accuracy of the summary. (§1904.32(b)(3))
- The annual summary must be posted for three months instead of one. (§1904.32(b)(6))

Employee involvement
- Employers are required to establish a procedure for employees to report injuries and illnesses and to tell their employees how to report. (§1904.35(a))
- The new rule informs employers that the OSH Act prohibits employers from discriminating against employees who do report. (§1904.36)
- Employees are allowed to access the 301 forms to review records of their own injuries and illnesses. (§1904.35(b)(2))
- Employee representatives are allowed to access those parts of the OSHA 301 form relevant to workplace safety and health. (§1904.35(b)(2))

Protecting privacy
- Employers are required to protect employee's privacy by withholding an individual's name on Form 300 for certain types of sensitive injuries/illnesses (e.g., sexual assaults, HIV infections, mental illnesses, etc.). (§1904.29(b)(6) to §1904.29(b)(8))
- Employers are allowed to withhold descriptive information about sensitive injuries in cases where not doing so would disclose the employee's identity. (§1904.29(b)(9))
- Employee representatives are given access only to the portion of Form 301 that contains information about the injury or illness, while personal information about the employee and his or her health care provider is withheld. (§1904.35(b)(2))
- Employers are required to remove employees' names before providing injury and illness data to persons who do not have access rights under the rule. (§1904.29(b)(10))

Reporting information to the government
- Employers must call in all fatal heart attacks occurring in the work environment. (§1904.39(b)(5))
- Employers do not need to call in public street motor vehicle accidents except those in a construction work zone. (§1904.39(b)(3))
- Employers do not need to call in commercial airplane, train, subway or bus accidents. (§1904.39(b)(4))
- Employers must provide records to an OSHA compliance officer who requests them within 4 hours. (§1904.40(a))

VII. Summary and Explanation [p. 5932]
The following sections discuss the contents of the final 29 CFR Part 1904 and section 1952.4 regulations. OSHA has written these regulations using the plain language guidance set out in a Presidential Memo to the heads of executive departments and agencies on June 1, 1998. The Agency also used guidance from the Plain Language Action Network (PLAN), which is a government-wide group working to improve communications from the Federal government to the public, with the goals of increasing trust in government, reducing government costs, and reducing the burden on the public. For more information on PLAN, see their Internet site at http://www.plainlanguage.gov/. The plain language concepts encourage government agencies to adopt a first person question and answer format, which OSHA used for the Part

1904 rule. The rule contains several types of provisions. Requirements are described using the "you must * * *" construction, prohibitions are described using "you may not * * *", and optional actions that are not requirements or prohibitions are preceded by "you may * * *." OSHA has also included provisions to provide information to the public in the rule.

Subpart A. Purpose [p. 5933]
The Purpose section of the final rule explains why OSHA is promulgating this rule. The Purpose section contains no regulatory requirements and is intended merely to provide information. A Note to this section informs employers and employees that recording a case on the OSHA recordkeeping forms does

not indicate either that the employer or the employee was at fault in the incident or that an OSHA rule has been violated. Recording an injury or illness on the Log also does not, in and of itself, indicate that the case qualifies for workers' compensation or other benefits. Although any specific work-related injury or illness may involve some or all of these factors, the record made of that injury or illness on the OSHA recordkeeping forms only shows three things: (1) that an injury or illness has occurred; (2) that the employer has determined that the case is work-related (using OSHA's definition of that term); and (3) that the case is non-minor, i.e., that it meets one or more of the OSHA injury and illness recording criteria. OSHA has added the Note to this first subpart of the rule because employers and employees have frequently requested clarification on these points.

The following paragraphs describe the changes OSHA has made to the Purpose provisions in Subpart A of the final rule, and discusses the Agency's reasons for these changes.

[The final rule's Purpose paragraph states "The purpose of this rule (Part 1904) is to require employers to record and report work-related fatalities, injuries and illnesses." It clearly and succinctly states OSHA's reasons for issuing the final rule.]

Employers have frequently asked OSHA to explain the relationship between workers' compensation reporting systems and the OSHA injury and illness recording and reporting requirements. As NYNEX (Ex. 15: 199) noted,

[t]he issue of confusion between OSHA recordkeeping and workers' compensation/insurance requirements cannot be totally eliminated as the workers' compensation criteria vary somewhat from state to state. There will always be some differences between OSHA recordability and compensable injuries and illnesses. The potential consequences of these differences can be minimized, however, if all stakeholders in the recordkeeping process (i.e., employers, employees, labor unions, OSHA compliance officials) are well informed that OSHA recordability does not equate to compensation eligibility. This can be facilitated by printed reminders on all of the OSHA recordkeeping documents (e.g., forms, instructions, pamphlets, compliance directives, etc.).

As NYNEX observed, employers must document work-related injuries and illnesses for both OSHA recordkeeping and workers' compensation purposes. Many cases that are recorded in the OSHA system are also compensable under the State workers' compensation system, but many others are not. However, the two systems have different purposes and scopes. The OSHA recordkeeping system is intended to collect, compile and analyze uniform and consistent nationwide data on occupational injuries and illnesses. The workers' compensation system, in contrast, is not designed primarily to generate and collect data but is intended primarily to provide medical coverage and compensation for workers who are killed, injured or made ill at work, and varies in coverage from one State to another.

Although the cases captured by the OSHA system and workers' compensation sometimes overlap, they often do not. For example, many injuries and illnesses covered by workers' compensation are not required to be recorded in the OSHA records. Such a situation would arise, for example, if an employee were injured on the job, sent to a hospital emergency room, and was examined and x-rayed by a physician, but was then told that the injury was minor and required no treatment. In this case, the employee's medical bills would be covered by workers' compensation insurance, but the case would not be recordable under Part 1904.

Conversely, an injury may be recordable for OSHA's purposes but not be covered by workers' compensation. For example, in some states, workers' compensation does not cover certain types of injuries (e.g., certain musculoskeletal disorders) and certain classes of workers (e.g., farm workers, contingent workers). However, if the injury meets OSHA recordability criteria it must be recorded even if the particular injury would not be compensable or the worker not be covered. Similarly, some injuries, although technically compensable under the state compensation system, do not result in the payment of workers' compensation benefits. For example, a worker who is injured on the job, receives treatment from the company physician, and returns to work without loss of wages would generally not receive workers' compensation because the company would usually absorb the costs. However, if the case meets the OSHA recording criteria, the employer would nevertheless be required to record the injury on the OSHA forms.

As a result of these differences between the two systems, recording a case does not mean that the case is compensable, or vice versa. When an injury or illness occurs to an employee, the employer must independently analyze the case in light of both the OSHA recording criteria and the requirements of the State workers' compensation system to determine whether the case is recordable or compensable, or both.

The American Federation of Labor and Congress of Industrial Organizations (AFL-CIO) urged OSHA to emphasize the no-fault philosophy of the Agency's recordkeeping system.

...

OSHA believes that the note to the Purpose paragraph of the final rule will allay any fears employers and employees may have about recording injuries and illnesses, and thus will encourage more accurate reporting. Both the Note to Subpart A of the final rule and the new OSHA Form 300 expressly state that recording a case does not indicate fault, negligence, or compensability.

...

OSHA has rejected the suggestion made by these commenters to limit the admissibility of the forms as evidence in a court proceeding. Such action is beyond the statutory authority of the agency, because OSHA has no authority over the courts, either Federal or State.

...

OSHA notes that many circumstances that lead to a recordable work-related injury or illness are "beyond the employer's control," at least as that phrase is commonly interpreted. Nevertheless, because such an injury or illness was caused, contributed to, or significantly aggravated by an event or exposure at work, it must be recorded on the OSHA form (assuming that it meets one or more of the recording criteria and does not qualify for an exemption to the geographic presumption). This approach is consistent with the no-fault recordkeeping system OSHA has adopted, which includes work-related injuries and illnesses, regardless of the level of employer control or non-control involved. The issue of whether different types of cases are deemed work-related under the OSHA recordkeeping rule is discussed in the Legal Authority section, above, and in the work-relationship section (section 1904.5) of this preamble.

Subpart B. Scope [p. 5935]

The coverage and partial exemption provisions in Subpart B of the final rule establish which employers must keep OSHA injury and illness records at all times, and which employers are generally exempt but must keep records under specific circumstances. This subpart contains sections 1904.1 through 1904.3 of the final rule.

OSHA's recordkeeping rule covers many employers in OSHA's jurisdiction but continues to exempt many employers from the need to keep occupational injury and illness records routinely. This approach to the scope of the rule is consistent with that taken in the former recordkeeping rule. Whether a particular employer must keep these records routinely depends on the number of employees in the firm and on the Standard Industrial Classification, or SIC code, of each of the employer's establishments. Employers with 10 or fewer employees are not required to keep OSHA records routinely. In addition, employers whose establishments are classified in certain industries are not required to keep OSHA records under most circumstances. OSHA refers to establishments exempted by reason of size or industry classification as "partially exempt," for reasons explained below.

The final rule's size exemption and the industry exemptions listed in non-mandatory Appendix A to Subpart B of the final rule do not relieve employers with 10 or fewer employees or employers in these industries from all of their recordkeeping obligations under 29 CFR Part 1904. Employers qualifying for either the industry exemption or the employment size exemption are not routinely required to record work-related injuries and illnesses occurring to their employees, that is, they are not normally required to keep the OSHA

Log or OSHA Form 301. However, as sections 1904.1(a)(1) and 1904.2 of this final recordkeeping rule make clear, these employers must still comply with three discrete provisions of Part 1904. First, all employers covered by the Act must report work-related fatalities or multiple hospitalizations to OSHA under Sec. 1904.39. Second, under Sec. 1904.41, any employer may be required to provide occupational injury and illness reports to OSHA or OSHA's designee upon written request. Finally, under Sec. 1904.42, any employer may be required to respond to the Survey of Occupational Injuries and Illnesses conducted by the Bureau of Labor Statistics (BLS) if asked to do so. Each of these requirements is discussed in greater detail in the relevant portion of this summary and explanation.

Section 1904.1 Partial Exemption for Employers With 10 or Fewer Employees [p. 5935]

In Sec. 1904.1 of the final rule, OSHA has retained the former rule's size-based exemption, which exempts employers with 10 or fewer employees in all industries covered by OSHA from most recordkeeping requirements. Section 1904.1, "Partial exemption for employers with 10 or fewer employees," states that:

(a) Basic requirement.
  (1) If your company had ten (10) or fewer employees at all times during the last calendar year, you do not need to keep OSHA injury and illness records unless OSHA or the BLS informs you in writing that you must keep records under Sec. 1904.41 or Sec. 1904.42. However, as required by Sec. 1904.39, all employers covered by the OSH Act must report to OSHA any workplace incident that results in a fatality or the hospitalization of three or more employees.
  (2) If your company had more than ten (10) employees at any time during the last calendar year, you must keep OSHA injury and illness records unless your establishment is classified as a partially exempt industry under Sec. 1904.2.

(b) Implementation.
  (1) Is the partial exemption for size based on the size of my entire company or on the size of an individual business establishment?
  The partial exemption for size is based on the number of employees in the entire company.
  (2) How do I determine the size of my company to find out if I qualify for the partial exemption for size?

To determine if you are exempt because of size, you need to determine your company's peak employment during the last calendar year. If you had no more than 10 employees at any time in the last calendar year, your company qualifies for the partial exemption for size.

The Size-Based Exemption in the Former Rule

The final rule published today maintains the former rule's partial exemption for employers in all covered industries who have 10 or fewer employees. Under the final rule (and the former rule), an employer in any industry who employed no more than 10 employees at any time during the preceding calendar year is not required to maintain OSHA records of occupational illnesses and injuries during the current year unless requested to do so in writing by OSHA (under Sec. 1904.41) or the BLS (under Sec. 1904.42). If an employer employed 11 or more people at a given time during the year, however, that employer is not eligible for the size-based partial exemption.

[Size Exemption Threshold for Construction Companies]

In the final rule, OSHA has decided to continue the Agency's longstanding practice of partially exempting employers with 10 or fewer employees from most recordkeeping requirements, but not to extend the exemption to non-construction businesses with 19 or fewer employees, as was proposed.

Section 1904.2 Partial Exemption for Establishments in Certain Industries [p. 5939]

Section 1904.2 of the final rule partially exempts employers with establishments classified in certain lower-hazard industries. The final rule updates the former rule's listing of partially exempted lower-hazard industries. Lower-hazard industries are those Standard Industrial Classification (SIC) code industries within SICs 52-89 that have an average Days Away, Restricted, or Transferred (DART) rate at or below 75% of the national average DART rate. The former rule also contained such a list based on data from 1978-1980. The final rule's list differs from that of the former rule in two respects: (1) the hazard information supporting the final rule's lower-hazard industry exemptions is based on the most recent three years of BLS statistics (1996, 1997, 1998), and (2) the exception is calculated at the 3-digit rather than 2-digit level.

The changes in the final rule's industry exemptions are designed to require more employers in higher-hazard industries to keep records all of the time and to exempt employers in certain lower-hazard industries from keeping OSHA injury and illness records routinely. For example, compared with the former rule, the final rule requires many employers in the 3-digit industries within retail and service sector industries that have higher rates of occupational injuries and illnesses to keep these records but exempts employers in 3-digit industries within those industries that report a lower rate of occupational injury and illness. Section 1904.2 of the final rule, "Partial exemption for establishments in certain industries," states:

(a) Basic requirement.
  (1) If your business establishment is classified in a specific low hazard retail, service, finance, insurance or real estate industry listed in Appendix A to this Subpart B, you do not need to keep OSHA injury and illness records unless the government asks you to keep the records under Sec. 1904.41 or Sec. 1904.42. However, all employers must report to OSHA any workplace incident that results in a fatality or the hospitalization of three or more employees (see Sec. 1904.39).
  (2) If one or more of your company's establishments are classified in a non-exempt industry, you must keep OSHA injury and illness records for all of such establishments unless your company is partially exempted because of size under Sec. 1904.1.

(b) Implementation.
  (1) Does the partial industry classification exemption apply only to business establishments in the retail, services, finance, insurance or real estate industries (SICs 52-89)?
Yes. Business establishments classified in agriculture; mining; construction; manufacturing; transportation; communication, electric, gas and sanitary services; or wholesale trade are not eligible for the partial industry classification exemption.

  (2) Is the partial industry classification exemption based on the industry classification of my entire company or on the classification of individual business establishments operated by my company?
The partial industry classification exemption applies to individual business establishments. If a company has several business establishments engaged in different classes of business activities, some of the company's establishments may be required to keep records, while others may be exempt.

  (3) How do I determine the Standard Industrial Classification code for my company or for individual establishments?
You determine your Standard Industrial Classification (SIC) code by using the Standard Industrial Classification Manual, Executive Office of the President, Office of Management and Budget. You may contact your nearest OSHA office or State agency for help in determining your SIC.

Employers with establishments in those industry sectors shown in Appendix A are not required routinely to keep OSHA records for their establishments. They must, however, keep records if requested to do so by the Bureau of Labor Statistics in connection with its Annual Survey (section 1904.42) or by OSHA in connection with its Data Initiative (section 1904.41). In addition, all employers covered by the OSH Act must report a work-related fatality, or an accident that results in the hospitalization of three or more employees, to OSHA within 8 hours (section 1904.39).

...

Evaluating industries at the 3-digit level allows OSHA to identify 3-digit industries with high LWDI rates (DART rates in the terminology of the final rule) that are located within 2-digit industries with relatively low rates. Conversely, use of this approach allows OSHA to identify lower-hazard 3-digit industries within a 2-digit industry that have relatively high LWDI (DART) rates. Use of LWDI (DART) rates at the more detailed level of SIC coding increases the specificity of the targeting of the exemptions and makes the rule more equitable by exempting workplaces in lower-hazard industries and requiring employers in more hazardous industries to keep records.

...

For multi-establishment firms, the industry exemption is based on the SIC code of each establishment, rather than the industrial classification of a firm as a whole. For example, some larger corporations have establishments that engage in different business activities. Where this is the case, each establishment could fall into a different SIC code, based on its business activity. The Standard Industrial Classification manual states that the establishment, rather than the firm, is the appropriate unit for determining the SIC code. Thus, depending on the SIC code of the establishment, one establishment of a firm may be exempt from routine recordkeeping under Part 1904, while another establishment in the same company may not be exempt.

...

For those States with OSHA-approved State plans, the state is generally required to adopt Federal OSHA rules, or a State rule that is at least as effective as the Federal OSHA rule. States with approved plans do not need to exempt employers from recordkeeping, either by employer size or by industry classification, as the final Federal OSHA rule does, although they may choose to do so. For example, States with approved plans may require records from a wider universe of employers than Federal OSHA does. These States cannot exempt more industries or employers than Federal OSHA does, however, because doing so would result in a State rule that is not as effective as the Federal rule. A larger discussion of the effect on the State plans can be found in Section VIII of this preamble, State Plans.

Recordkeeping Under the Requirements of Other Federal Agencies

Section 1904.3 of the final rule provides guidance for employers who are subject to the occupational injury and illness recording and reporting requirements of other Federal agencies. Several other Federal agencies have similar requirements, such as the Mine Safety and Health Administration (MSHA), the Department of Energy (DOE), and the Federal Railroad Administration (FRA). The final rule at section 1904.3 tells the employer that OSHA will accept these records in place of the employer's Part 1904 records under two circumstances: (1) if OSHA has entered into a memorandum of understanding (MOU) with that agency that specifically accepts the other agency's records, the employer may use them in place of the OSHA records, or (2) if the other agency's records include the same information required by Part 1904, OSHA would consider them an acceptable substitute.

...

Subpart C. Recordkeeping Forms and Recording Criteria [p. 5945]

Subpart C of the final rule sets out the requirements of the rule for recording cases in the recordkeeping system. It contains provisions directing employers to keep records of the recordable occupational injuries and illnesses experienced by their employees, describes the forms the employer must use, and establishes the criteria that employers must follow to determine which work-related injury and illness cases must be entered onto the forms. Subpart C contains sections 1904.4 through 1904.29.

Section 1904.4 provides an overview of the requirements in Subpart C and contains a flowchart describing the recording process. How employers are to determine whether a given injury or illness is work-related is set out in section 1904.5. Section 1904.6 provides the requirements employers must follow to determine whether or not a work-related injury or illness is a new case or the continuation of a previously recorded injury or illness. Sections 1904.7 through 1904.12 contain the recording criteria for determining which new work-related injuries and illnesses must be recorded on the

OSHA forms. Section 1904.29 explains which forms must be used and indicates the circumstances under which the employer may use substitute forms.

Section 1904.4 Recording Criteria [p. 5945]

Section 1904.4 of the final rule contains provisions mandating the recording of work-related injuries and illnesses that must be entered on the OSHA 300 (Log) and 301 (Incident Report) forms. It sets out the recording requirements that employers are required to follow in recording cases.

Paragraph 1904.4(a) of the final rule mandates that each employer who is required by OSHA to keep records must record each fatality, injury or illness that is work-related, is a new case and not a continuation of an old case, and meets one or more of the general recording criteria in section 1904.7 or the additional criteria for specific cases found in sections 1904.8 through 1904.12. Paragraph (b) contains provisions implementing this basic requirement.

Paragraph 1904.4(b)(1) contains a table that points employers and their recordkeepers to the various sections of the rule that determine which work-related injuries and illnesses are to be recorded. These sections lay out the requirements for determining whether an injury or illness is work-related, if it is a new case, and if it meets one or more of the general recording criteria. In addition, the table contains a row addressing the application of these and additional criteria to specific kinds of cases (needlestick and sharps injury cases, tuberculosis cases, hearing loss cases, medical removal cases, and musculoskeletal disorder cases). The table in paragraph 1904.4(b)(1) is intended to guide employers through the recording process and to act as a table of contents to the sections of Subpart C.

Paragraph (b)(2) is a decision tree, or flowchart, that shows the steps involved in determining whether or not a particular injury or illness case must be recorded on the OSHA forms. It essentially reflects the same information as is in the table in paragraph 1904.4(b)(1), except that it presents this information graphically.

...

Section 1904.5 Determination of Work-Relatedness [p. 5946]

This section of the final rule sets out the requirements employers must follow in determining whether a given injury or illness is work-related. Paragraph 1904.5(a) states that an injury or illness must be considered work-related if an event or exposure in the work environment caused or contributed to the injury or illness or significantly aggravated a pre-existing injury or illness. It stipulates that, for OSHA recordkeeping purposes, work relationship is presumed for such injuries and illnesses unless an exception listed in paragraph 1904.5(b)(2) specifically applies.

Implementation requirements are set forth in paragraph (b) of the final rule. Paragraph (b)(1) defines "work environment" for recordkeeping purposes and makes clear that the work environment includes the physical locations where employees are working as well as the equipment and materials used by the employee to perform work.

Paragraph (b)(2) lists the exceptions to the presumption of work-relatedness permitted by the final rule; cases meeting the conditions of any of the listed exceptions are not considered work-related and are therefore not recordable in the OSHA recordkeeping system.

This section of the preamble first explains OSHA's reasoning on the issue of work relationship, then discusses the exceptions to the general presumption and the comments received on the exceptions proposed, and then presents OSHA's rationale for including paragraphs (b)(3) through (b)(7) of the final rule, and the record evidence pertaining to each.

...

Final Rule's Exceptions to the Geographic Presumption

Paragraph 1904.5(b)(2) of the final rule contains eight exceptions to the work environment presumption that are intended to exclude from the recordkeeping system those injuries and illnesses that occur or manifest in the work environment, but have been identified by OSHA, based on its years of experience with recordkeeping, as cases that do not provide information useful to the identification of occupational injuries and illnesses and would thus tend to skew national injury and illness statistics. These eight exceptions are the only exceptions to the presumption permitted by the final rule.

(i) Injuries or illnesses will not be considered work-related if, at the time of the injury or illness, the employee was present in the work environment as a member of the general public rather than as an employee. This exception, which is codified at paragraph 1904.5(b)(2)(i), is based on the fact that no employment relationship is in place at the time an injury or illness of this type occurs. A case exemplifying this exception would occur if an employee of a retail store patronized that store as a customer on a non-work day and was injured in a fall. This exception allows the employer not to record cases that occur outside of the employment relationship when his or her establishment is also a public place and a worker happens to be using the facility as a member of the general public. In these situations, the injury or illness has nothing to do with the employee's work or the employee's status as an employee, and it would therefore be inappropriate for the recordkeeping system to capture the case.

...

(ii) Injuries or illnesses will not be considered work-related if they involve symptoms that surface at work but result solely from a non-work-related event or exposure that occurs outside the work environment. OSHA's recordkeeping system is intended only to capture cases that are caused by conditions or exposures arising in the work environment. It is not designed to capture cases that have no relationship with the work environment. For this exception to apply, the work environment cannot have caused, contributed to, or significantly aggravated the injury or illness. This exception is consistent with the position followed by OSHA for many years and reiterated in the final rule: that any job-related contribution to the injury or illness makes the incident work-related, and its corollary—that any injury or illness to which work makes no actual contribution is not work-related. An example of this type of injury would be a diabetic incident that occurs while an employee is working. Because no event or exposure at work contributed in any way to the diabetic incident, the case is not recordable. This exception allows the employer to exclude cases where an employee's non-work activities are the sole cause of the injury or illness.

...

(iii) Injuries and illnesses will not be considered work-related if they result solely from voluntary participation in a wellness program or in a medical, fitness, or recreational activity such as blood donation, physical, flu shot, exercise classes, racquetball, or baseball. This exception allows the employer to exclude certain injury or illness cases that are related to personal medical care, physical fitness activities and voluntary blood donations. The key words here are "solely" and "voluntary." The work environment cannot have contributed to the injury or illness in any way for this exception to apply, and participation in the wellness, fitness or recreational activities must be voluntary and not a condition of employment. This exception allows the employer to exclude cases that are related to personal matters of exercise, recreation, medical examinations or participation in blood donation programs when they are voluntary and are not being undertaken as a condition of work. For example, if a clerical worker was injured while performing aerobics in the company gymnasium during his or her lunch hour, the case would not be work-related. On the other hand, if an employee who was assigned to manage the gymnasium was injured while teaching an aerobics class, the injury would be work-related because the employee was working at the time of the injury and the activity was not voluntary. Similarly, if an employee suffered a severe reaction to a flu shot that was administered as part of a voluntary inoculation program, the case would not be considered work-related; however, if an employee suffered a reaction to medications administered to enable the employee to travel overseas on business, or the employee had an illness reaction to a medication administered to treat a work-related injury, the case would be considered work-related.

...

(iv) Injuries and illnesses will not be considered work-related if they are solely the result of an employee eating, drinking, or preparing food or drink for personal consumption (whether bought on the premises or brought in). This exception responds to a situation that has given rise to many letters of interpretation and caused employer concern over the years. An example of the application of this exception would be a case where the employee injured himself or herself by choking on a sandwich brought from home but eaten in the employer's establishment; such a case would not be considered work-related under this exception. On the other hand, if the employee was injured by a trip or fall hazard present in the employer's lunchroom, the case would be considered work-related. In addition, a note to the exception makes clear that if an employee becomes ill as a result of ingesting food contaminated by workplace contaminants such as lead, or contracts food poisoning from food items provided by the employer, the case would be considered work-related. As a result, if an employee contracts food poisoning from a sandwich brought from home or purchased in the company cafeteria and must take time off to recover, the case is not considered work related. On the other hand, if an employee contracts food poisoning from a meal provided by the employer at a business meeting or company function and takes time off to recover, the case would be considered work related. Food provided or supplied by the employer does not include food purchased by the employee from the company cafeteria, but does include food purchased by the employer from the company cafeteria for business meetings or other company functions. OSHA believes that the number of cases to which this exception applies will be few.

...

(v) Injuries and illnesses will not be considered work-related if they are solely the result of employees doing personal tasks (unrelated to their employment) at the establishment outside of their assigned working hours. This exception, which responds to inquiries received over the years, allows employers limited flexibility to exclude from the recordkeeping system situations where the employee is using the employer's establishment for purely personal reasons during his or her off-shift time. For example, if an employee were using a meeting room at the employer's establishment outside of his or her assigned working hours to hold a meeting for a civic group to which he or she belonged, and slipped and fell in the hallway, the injury would not be considered work-related. On the other hand, if the employee were at the employer's establishment outside his or her assigned working hours to attend a company business meeting or a company training session, such a slip or fall would be work-related. OSHA also expects the number of cases affected by this exception to be small.

...

(vi) Injuries and illnesses will not be considered work-related if they are solely the result of personal grooming, self-medication for a non-work-related condition, or are intentionally self-inflicted. This exception allows the employer to exclude from the Log cases related to personal hygiene, self-administered medications and intentional self-inflicted injuries, such as attempted suicide. For example, a burn injury from a hair dryer used at work to dry the employee's hair would not be work-related. Similarly, a negative reaction to a medication brought from home to treat a non-work condition would not be considered a work-related illness, even though it first manifested at work. OSHA also expects that few cases will be affected by this exception.

...

(vii) Injuries will not be considered work-related if they are caused by motor vehicle accidents occurring in company parking lots or on company access roads while employees are commuting to or from work. This exception allows the employer to exclude cases where an employee is injured in a motor vehicle accident while commuting from work to home or from home to work or while on a personal errand. For example, if an employee was injured in a car accident while arriving at work or while leaving the company's property at the end of the day, or while driving on his or her lunch hour to run an errand, the case would not be considered work-related. On the other hand, if an employee was injured in a car accident while leaving the property to purchase supplies for the employer, the case would be work-related. This exception represents a change from the position taken under the former rule, which was that no injury or illness occurring in a company parking lot was considered work-related. As explained further below, OSHA has concluded, based on the evidence in the record, that some injuries and illnesses that occur in company parking lots are clearly caused by work conditions or activities—e.g., being struck by a car while painting parking space indicators on the pavement of the lot, slipping on ice permitted to accumulate in the lot by the employer—and by their nature point to conditions that could be corrected to improve workplace safety and health.

(viii) Common colds and flu will not be considered work-related. Paragraph 1904.5(b)(2)(viii) allows the employer to exclude cases of common cold or flu, even if contracted while the employee was at work. However, in the case of other infectious diseases such as tuberculosis, brucellosis, and hepatitis C, employers must evaluate reports of such illnesses for work relationship, just as they would any other type of injury or illness.

(ix) Mental illness will not be considered work-related unless the employee voluntarily provides the employer with an opinion from a physician or other licensed health care professional with appropriate training and experience (psychiatrist, psychologist, psychiatric nurse practitioner, etc.) stating that the employee has a mental illness that is work-related.

...

OSHA agrees that recording work-related mental illnesses involves several unique issues, including the difficulty of detecting, diagnosing and verifying mental illnesses; and the sensitivity and privacy concerns raised by mental illnesses. Therefore, the final rule requires employers to record only those mental illnesses verified by a health care professional with appropriate training and experience in the treatment of mental illness, such as a psychiatrist, psychologist, or psychiatric nurse practitioner. The employer is under no obligation to seek out information on mental illnesses from its employees, and employers are required to consider mental illness cases only when an employee voluntarily presents the employer with an opinion from the health care professional that the employee has a mental illness and that it is work related. In the event that the employer does not believe the reported mental illness is work-related, the employer may refer the case to a physician or other licensed health care professional for a second opinion.

OSHA also emphasizes that work-related mental illnesses, like other illnesses, must be recorded only when they meet the severity criteria outlined in Sec. 1904.7. In addition, for mental illnesses, the employee's identity must be protected by omitting the employee's name from the OSHA 300 Log and instead entering "privacy concern case" as required by Sec. 1904.29.

...

Determining Whether the Precipitating Event or Exposure Occurred in the Work Environment or Elsewhere
Paragraph 1904.5(b)(3) of the final rule provides guidance on applying the geographic presumption when it is not clear whether the event or exposure that precipitated the injury or illness occurred in the work environment or elsewhere. If an employee reports pain and swelling in a joint but cannot say whether the symptoms first arose during work or during recreational activities at home, it may be difficult for the employer to decide whether the case is work-related. The same problem arises when an employee reports symptoms of a contagious disease that affects the public

at large, such as a staphylococcus infection ("staph" infection) or Lyme disease, and the workplace is only one possible source of the infection. In these situations, the employer must examine the employee's work duties and environment to determine whether it is more likely than not that one or more events or exposures at work caused or contributed to the condition. If the employer determines that it is unlikely that the precipitating event or exposure occurred in the work environment, the employer would not record the case. In the staph infection example given above, the employer would consider the case work-related, for example, if another employee with whom the newly infected employee had contact at work had been out with a staph infection. In the Lyme disease example, the employer would determine the case to be work-related if, for example, the employee was a groundskeeper with regular exposure to outdoor conditions likely to result in contact with deer ticks.

In applying paragraph 1904.5(b)(3), the question employers must answer is whether the precipitating event or exposure occurred in the work environment. If an event, such as a fall, an awkward motion or lift, an assault, or an instance of horseplay, occurs at work, the geographic presumption applies and the case is work-related unless it otherwise falls within an exception. Thus, if an employee trips while walking across a level factory floor, the resulting injury is considered work-related under the geographic presumption because the precipitating event—the tripping accident—occurred in the workplace.

The case is work-related even if the employer cannot determine why the employee tripped, or whether any particular workplace hazard caused the accident to occur. However, if the employee reports an injury at work but cannot say whether it resulted from an event that occurred at work or at home, as in the example of the swollen joint, the employer might determine that the case is not work-related because the employee's work duties were unlikely to have caused, contributed to, or significantly aggravated such an injury.

Significant Workplace Aggravation of a Pre-existing Condition

In paragraph 1904.5(b)(4), the final rule makes an important change to the former rule's position on the extent of the workplace aggravation of a preexisting injury or illness that must occur before the case is considered work-related. In the past, any amount of aggravation of such an injury or illness was considered sufficient for this purpose. The final rule, however, requires that the amount of aggravation of the injury or illness that work contributes must be "significant," i.e., non-minor, before work-relatedness is established. The preexisting injury or illness must be one caused entirely by non-occupational factors.

...

Paragraph 1904.5(b)(4) of the final rule defines aggravation as significant if the contribution of the aggravation at work is such that it results in tangible consequences that go beyond those that the worker would have experienced as a result of the preexisting injury or illness alone, absent the aggravating effects of the workplace. Under the final rule, a preexisting injury or illness will be considered to have been significantly aggravated, for the purposes of OSHA injury and illness recordkeeping, when an event or exposure in the work environment results in: (i) Death, providing that the preexisting injury or illness would likely not have resulted in death but for the occupational event or exposure; (ii) Loss of consciousness, providing that the preexisting injury or illness would likely not have resulted in loss of consciousness but for the occupational event or exposure; (iii) A day or days away from work or of restricted work, or a job transfer that otherwise would not have occurred but for the occupational event or exposure; or (iv) Medical treatment where no medical treatment was needed for the injury or illness before the workplace event or exposure, or a change in the course of medical treatment that was being provided before the workplace event or exposure. OSHA's decision not to require the recording of cases involving only minor aggravation of preexisting conditions is consistent with the Agency's efforts in this rulemaking to require the recording only of non-minor injuries and illnesses; for example, the final rule also no longer requires employers to record minor illnesses on the Log.

Preexisting Conditions

Paragraph 1904.5(b)(5) stipulates that pre-existing conditions, for recordkeeping purposes, are conditions that resulted solely from a non-work-related event or exposure that occurs outside the employer's work environment. Pre-existing conditions also include any injury or illness that the employee experienced while working for another employer.

Off Premises Determinations

Employees may be injured or become ill as a result of events or exposures away from the employer's establishment. In these cases, OSHA proposed to consider the case work-related only if the employee was engaged in a work activity or was present as a condition of employment (61 FR 4063). In the final rule, (paragraph 1904.5(b)(1)) the same concept is carried forward in the definition of the work environment, which defines the environment as including the establishment and any other location where one or more employees are working or are present as a condition of their employment.

Thus, when employees are working or conducting other tasks in the interest of their employer but at a location away from the employer's establishment, the work-relatedness of an injury or illness that arises is subject to the same decision making process that would occur if the case had occurred at the establishment itself. The case is work-related if one or more events or exposures in the work environment either caused or contributed to the resulting condition or significantly aggravated a pre-existing condition, as stated in paragraph 1904.5(a). In addition, the exceptions for determining work relationship at paragraph 1904.5(b)(2) and the requirements at paragraph 1904.5(b)(3) apply equally to cases that occur at or away from the establishment.

As an example, the work-environment presumption clearly applies to the case of a delivery driver who experiences an injury to his or her back while loading boxes and transporting them into a building. The worker is engaged in a work activity and the injury resulted from an event—loading/unloading—occurring in the work environment. Similarly, if an employee is injured in an automobile accident while running errands for the company or traveling to make a speech on behalf of the company, the employee is present at the scene as a condition of employment, and any resulting injury would be work-related.

Employees on Travel Status

The final rule continues (at Sec. 1904.5(b)(6)) OSHA's longstanding practice of treating injuries and illnesses that occur to an employee on travel status as work-related if, at the time of the injury or illness, the employee was engaged in work activities "in the interest of the employer." Examples of such activities include travel to and from customer contacts, conducting job tasks, and entertaining or being entertained if the activity is conducted at the direction of the employer.

The final rule contains three exceptions for travel-status situations. The rule describes situations in which injuries or illnesses sustained by traveling employees are not considered work-related for OSHA recordkeeping purposes and therefore do not have to be recorded on the OSHA 300 Log. First, when a traveling employee checks into a hotel, motel, or other temporary residence, he or she is considered to have established a "home away from home." At this time, the status of the employee is the same as that of an employee working at an establishment who leaves work and is essentially "at home". Injuries and illnesses that occur at home are generally not considered work related. However, just as an employer may sometimes be required to record an injury or illness occurring to an employee working in his or her home, the employer is required to record an injury or illness occurring to an employee who is working in his or her hotel room (see the discussion of working at home, below).

Second, if an employee has established a "home away from home" and is reporting to a fixed worksite each day, the employer does not consider injuries or illnesses work-related if they occur while the employee is commuting between the temporary residence and the job location. These cases are parallel to those involving employees commuting to and from work when they are at their home location, and do not have to be recorded, just as injuries and illnesses that occur during normal commuting are not required to be recorded.

Third, the employer is not required to consider an injury or illness to be work-related if it occurs while the employee is on a personal detour from the route of business travel. This exception allows the employer to exclude injuries and illnesses that occur when the worker has taken a side trip for personal reasons while on a business trip, such as a vacation or sightseeing excursion, to visit relatives, or for some other personal purpose.

...

Working at Home

The final rule also includes provisions at Sec. 1904.5(b)(7) for determining the work-relatedness of injuries and illnesses that may arise when employees are working at home. When an employee is working on company business in his or her home and reports an injury or illness to his or her employer, and the employee's work activities caused or contributed to the injury or illness, or significantly aggravated a pre-existing injury, the case is considered work-related and must be further evaluated to determine whether it meets the recording criteria. If the injury or illness is related to non-work activities or to the general home environment, the case is not considered work-related.

...

OSHA has recently issued a compliance directive (CPL 2-0.125) containing the Agency's response to many of the questions raised by this commenter. That document clarifies that OSHA will not conduct inspections of home offices and does not hold employers liable for employees' home offices. The compliance directive also notes that employers required by the recordkeeping rule to keep records "will continue to be responsible for keeping such records, regardless of whether the injuries occur in the factory, in a home office, or elsewhere, as long as they are work-related, and meet the recordability criteria of 29 CFR Part 1904."

With more employees working at home under various telecommuting and flexible workplace arrangements, OSHA believes that it is important to record injuries and illnesses attributable to work tasks performed at home. If these cases are not recorded, the Nation's injury and illness statistics could be skewed. For example, placing such an exclusion in the final rule would make it difficult to determine if a decline in the overall number or rate of occupational injuries and illnesses is attributable to a trend toward working at home or to a change in the Nation's actual injury and illness experience. Further, excluding these work-related injuries and illnesses from the recordkeeping system could potentially obscure previously unidentified causal connections between events or exposures in the work environment and these incidents. OSHA is unwilling to adopt an exception that would have these potential effects.

...

Section 1904.6 Determination of New Cases [p. 5962]

Employers may occasionally have difficulty in determining whether new signs or symptoms are due to a new event or exposure in the workplace or whether they are the continuation of an existing work-related injury or illness. Most occupational injury and illness cases are fairly discrete events, i.e., events in which an injury or acute illness occurs, is treated, and then resolves completely. For example, a worker may suffer a cut, bruise, or rash from a clearly recognized event in the workplace, receive treatment, and recover fully within a few weeks. At some future time, the worker may suffer another cut, bruise or rash from another workplace event. In such cases, it is clear that the two injuries or illnesses are unrelated events, and that each represents an injury or illness that must be separately evaluated for its recordability.

However, it is sometimes difficult to determine whether signs or symptoms are due to a new event or exposure, or are a continuance of an injury or illness that has already been recorded. This is an important distinction, because a new injury or illness requires the employer to make a new entry on the OSHA 300 Log, while a continuation of an old recorded case requires, at most, an updating of the original entry.

Section 1904.6 of the final rule being published today explains what employers must do to determine whether or not an injury or illness is a new case for recordkeeping purposes.

The basic requirement at Sec. 1904.6(a) states that the employer must consider an injury or illness a new case to be evaluated for recordability if (1) the employee has not previously experienced a recorded injury or illness of the same type that affects the same part of the body, or (2) the employee previously experienced a recorded injury or illness of the same type that affected the same part of the body but had recovered completely (all signs and symptoms of the previous injury or illness had disappeared) and an event or exposure in the work environment caused the injury or illness, or its signs or symptoms, to reappear.

The implementation question at Sec. 1904.6(b)(1) addresses chronic work-related cases that have already been recorded once and distinguishes between those conditions that will progress even in the absence of workplace exposure and those that are triggered by events in the workplace. There are some conditions that will progress even in the absence of further exposure, such as some occupational cancers, advanced asbestosis, tuberculosis disease, advanced byssinosis, advanced silicosis, etc. These conditions are chronic; once the disease is contracted it may never be cured or completely resolved, and therefore the case is never "closed" under the OSHA recordkeeping system, even though the signs and symptoms of the condition may alternate between remission and active disease.

However, there are other chronic work-related illness conditions, such as occupational asthma, reactive airways dysfunction syndrome (RADs), and sensitization (contact) dermatitis, that recur if the ill individual is exposed to the agent (or agents, in the case of cross-reactivities or RADs) that triggers the illness again. It is typical, but not always the case, for individuals with these conditions to be symptom-free if exposure to the sensitizing or precipitating agent does not occur.

The final rule provides, at paragraph (b)(1), that the employer is not required to record as a new case a previously recorded case of chronic work-related illness where the signs or symptoms have recurred or continued in the absence of exposure in the workplace. This paragraph recognizes that there are occupational illnesses that may be diagnosed at some stage of the disease and may then progress without regard to workplace events or exposures. Such diseases, in other words, will progress without further workplace exposure to the toxic substance(s) that caused the disease. Examples of such chronic work-related diseases are silicosis, tuberculosis, and asbestosis. With these conditions, the ill worker will show signs (such as a positive TB skin test, a positive chest roentgenogram, etc.) at every medical examination, and may experience symptomatic bouts as the disease progresses.

Paragraph 1904.6(b)(2) recognizes that many chronic occupational illnesses, however, such as occupational asthma, RADs, and contact dermatitis, are triggered by exposures in the workplace. The difference between these conditions and those addressed in paragraph 1904.6(b)(1) is that in these cases exposure triggers the recurrence of symptoms and signs, while in the chronic cases covered in the previous paragraph, the symptoms and signs recur even in the absence of exposure in the workplace.

Paragraph 1904.6(b)(3) addresses how to record a case for which the employer requests a physician or other licensed health care professional (HCP) to make a new case/continuation of an old case determination. Paragraph (b)(3) makes clear that employers are to follow the guidance provided by the HCP for OSHA recordkeeping purposes. In cases where two or more HCPs make conflicting or differing recommendations, the employer is required to base his or her decision about recordation based on the most authoritative (best documented, best reasoned, or most persuasive) evidence or recommendation.

...

Section 1904.7 General Recording Criteria [p. 5968]

Section 1904.7 contains the general recording criteria for recording work-related injuries and illnesses. This section describes the recording of cases that meet one or more of the following six criteria: death, days away from work, restricted work or transfer to another job, medical treatment beyond first aid, loss of consciousness, or diagnosis as a significant injury or illness by a physician or other licensed health care professional.

Paragraph 1904.7(a)

Paragraph 1904.7(a) describes the basic requirement for recording an injury or illness in the OSHA recordkeeping system. It states that employers must record any work-related injury or illness that meets one or more of the final rule's general recording criteria. There are six such criteria: death, days away from work, days on restricted work or on job transfer, medical treatment beyond first aid, loss of consciousness, or diagnosis by a physician or other licensed heath care professional as a significant injury or illness. Although most cases are recorded because they meet one of these criteria, some cases may meet more than one criterion as the case continues. For example, an injured worker may initially be sent home to recuperate (making the case recordable as a "days away" case) and then subsequently return to work on a restricted ("light duty") basis (meeting a second criterion, that for restricted work). (see the discussion in Section 1904.29 for information on how to record such cases.)

Paragraph 1904.7(b)

Paragraph 1904.7(b) tells employers how to record cases meeting each of the six general recording criteria and states how each case is to be entered on the OSHA 300 Log. Paragraph 1904.7(b)(1) provides a simple decision table listing the six general recording criteria and the paragraph number of each in the final rule. It is included to aid employers and recordkeepers in recording these cases.

1904.7(b)(2) Death

Paragraph 1904.7(b)(2) requires the employer to record an injury or illness that results in death by entering a check mark on the OSHA 300 Log in the space for fatal cases. This paragraph also directs employers to report work-related fatalities to OSHA within 8 hours and cross references the fatality and catastrophe reporting requirements in Sec. 1904.39 of the final rule, Reporting fatalities and multiple hospitalizations to OSHA. Paragraph 1904.7(b)(2) imple- ments the OSH Act's requirements to record all cases resulting in work-related deaths.
...

Paragraph 1904.7(b)(3) Days Away From Work

Paragraph 1904.7(b)(3) contains the requirements for recording work-related injuries and illnesses that result in days away from work and for counting the total number of days away associated with a given case. Paragraph 1904.7(b)(3) requires the employer to record an injury or illness that involves one or more days away from work by placing a check mark on the OSHA 300 Log in the space reserved for day(s) away cases and entering the number of calendar days away from work in the column reserved for that purpose. This paragraph also states that, if the employee is away from work for an extended time, the employer must update the day count when the actual number of days away becomes known. This requirement continues the day counting requirements of the former rule and revises the days away requirements in response to comments in the record.

Paragraphs 1904.7(b)(3)(i) through (vi) implement the basic requirements. Paragraph 1904.7(b)(3)(i) states that the employer is not to count the day of the injury or illness as a day away, but is to begin counting days away on the following day. Thus, even though an injury or illness may result in some loss of time on the day of the injurious event or exposure because, for example, the employee seeks treatment or is sent home, the case is not considered a days-away-from-work case unless the employee does not work on at least one subsequent day because of the injury or illness. The employer is to begin counting days away on the day following the injury or onset of illness. This policy is a continuation of OSHA's practice under the former rule, which also excluded the day of injury or onset of illness from the day counts.

Paragraphs 1904.7(b)(3)(ii) and (iii) direct employers how to record days-away cases when a physician or other licensed health care professional (HCP) recommends that the injured or ill worker stay at home or that he or she return to work but the employee chooses not to do so. As these paragraphs make clear, OSHA requires employers to follow the physician's or HCP's recommendation when recording the case.

Further, whether the employee works or not is in the control of the employer, not the employee. That is, if an HCP recommends that the employee remain away from work for one or more days, the employer is required to record the injury or illness as a case involving days away from work and to keep track of the days; the employee's wishes in this case are not relevant, since it is the employer who controls the conditions of work. Similarly, if the HCP tells the employee that he or she can return to work, the employer is required by the rule to stop counting the days away from work, even if the employee chooses not to return to work. These policies are a continuation of OSHA's previous policy of requiring employees to follow the recommendations of health care professionals when recording cases in the OSHA system. OSHA is aware that there may be situations where the employer obtains an opinion from a physician or other health care professional and a subsequent HCP's opinion differs from the first. (The subsequent opinion could be that of an HCP retained by the employer or the employee.) In this case, the employer is the ultimate recordkeeping decision-maker and must resolve the differences in opinion; he or she may turn to a third HCP for this purpose, or may make the recordability decision himself or herself.

Paragraph 1904.7(b)(3)(iv) specifies how the employer is to account for weekends, holidays, and other days during which the employee was unable to work because of a work-related injury or illness during a period in which the employee was not scheduled to work. The rule requires the employer to count the number of calendar days the employee was unable to work because of the work-related injury or illness, regardless of whether or not the employee would have been scheduled to work on those calendar days. This provision will ensure that a measure of the length of disability is available, regardless of the employee's work schedule. This requirement is a change from the former policy, which focused on scheduled workdays missed due to injury or illness and excluded from the days away count any normal days off, holidays, and other days the employee would not have worked.

Paragraph 1904.7(b)(3)(v) tells the employer how to count days away for a case where the employee is injured or becomes ill on the last day of work before some scheduled time off, such as on the Friday before the weekend or the day before a scheduled vacation, and returns to work on the next day that he or she was scheduled to work. In this situation, the employer must decide if the worker would have been able to work on the days when he or she was not at work. In other words, the employer is not required to count as days away any of the days on which the employee would have been able to work but did not because the facility was closed, the employee was not scheduled to work, or for other reasons unrelated to the injury or illness. However, if the employer determines that the employee's injury or illness would have kept the employee from being able to work for part or all of time the employee was away, those days must be counted toward the days away total.

Paragraph 1904.7(b)(3)(vi) allows the employer to stop counting the days away from work when the injury or illness has resulted in 180 calendar days away from work. When the injury or illness results in an absence of more than 180 days, the employer may enter 180 (or 180+) on the Log. This is a new provision of the final rule; it is included because OSHA believes that the "180" notation indicates a case of exceptional severity and that counting days away beyond that point would provide little if any additional information.

Paragraph 1904.7(b)(3)(vii) specifies that employers whose employees are away from work because of a work-related injury or illness and who then decide to leave the company's employ or to retire must determine whether the employee is leaving or retiring because of the injury or illness and record the case accordingly. If the employee's decision to leave or retire is a result of the injury or illness, this paragraph requires the employer to estimate and record the number of calendar days away or on restricted work/job transfer the worker would have experienced if he or she had remained on the employer's payroll. This provision also states that, if the employee's decision was unrelated to the injury or illness, the employer is not required to continue to count and record days away or on restricted work/job transfer.

Paragraph 1904.(b)(3)(viii) directs employers how to handle a case that carries over from one year to the next. Some cases occur in one calendar year and then result in days away from work in the next year.

For example, a worker may be injured on December 20th and be away from work until January 10th. The final rule directs the employer only to record this type of case once, in the year that it occurred. If the employee is still away from work when the annual summary is prepared (before February 1), the employer must either count the number of days the employee was away or estimate the total days away that are expected to occur, use this estimate to calculate the total days away during the year for the annual summary, and then update the Log entry later when the actual number of days is known or the case reaches the 180-day cap allowed in Sec. 1904.7(b)(3)(v).

...

Final Rule's Restricted Work and Job Transfer Provisions, and OSHA's Reasons for Adopting Them

Paragraph 1904.7(b)(4) contains the restricted work and job transfer provisions of the final rule. These provisions clarify the definition of restricted work in light of the comments received and continue, with a few exceptions, most of the former rule's requirements with regard to these kinds of cases. OSHA finds, based on a review of the record, that these provisions of the final rule will increase awareness among employers of the importance of recording restricted work activity and job transfer cases and make the recordkeeping system more accurate and the process more efficient.

OSHA believes that it is even more important today than formerly that the definition of restricted work included in the final rule be clear and widely understood, because employers have recently been relying on restricted work (or "light duty") with increasing frequency, largely in an effort to encourage injured or ill employees to return to work as soon as possible. According to BLS data, this category of cases has grown by nearly 70% in the last six years. In 1992, for example, 9% of all injuries and illnesses (or a total of 622,300 cases) recorded as lost workday cases were classified in this way solely because of restricted work days, while in 1998, nearly 18% of all injury and illness cases (or a total of 1,050,200 cases) were recorded as lost workday cases only because they involved restricted work [BLS Press Release 99-358, 12-16-99]. The return-to-work programs increasingly being relied on by employers (often at the recommendation of their workers' compensation insurers) are designed to prevent exacerbation of, or to allow recuperation from, the injury or illness, rehabilitate employees more effectively, reintegrate injured or ill workers into the workplace more rapidly, limit workers' compensation costs, and retain productive workers. In addition, many employees are eager to accept restricted work when it is available and prefer returning to work to recuperating at home.

The final rule's requirements in paragraph 1904.10(b)(4) of the final rule state:

(4) How do I record a work-related injury or illness that involves restricted work or job transfer? When an injury or illness involves restricted work or job transfer but does not involve death or days away from work, you must record the injury or illness on the OSHA 300 Log by placing a check mark in the space for job transfer or restricted work and entering the number of restricted or transferred days in the restricted work column.

(i) How do I decide if the injury or illness resulted in restricted work? Restricted work occurs when, as the result of a work-related injury or illness:

(A) You keep the employee from performing one or more of the routine functions of his or her job, or from working the full workday that he or she would otherwise have been scheduled to work; or

(B) A physician or other licensed health care professional recommends that the employee not perform one or more of the routine functions of his or her job, or not work the full workday that he or she would otherwise have been scheduled to work.

(ii) What is meant by "routine functions"? For recordkeeping purposes, an employee's routine functions are those work activities the employee regularly performs at least once per week.

(iii) Do I have to record restricted work or job transfer if it applies only to the day on which the injury occurred or the illness began?

No. You do not have to record restricted work or job transfers if you, or the physician or other licensed health care professional, impose the restriction or transfer only for the day on which the injury occurred or the illness began.

(iv) If you or a physician or other licensed health care professional recommends a work restriction, is the injury or illness automatically recordable as a "restricted work" case?

No. A recommended work restriction is recordable only if it affects one or more of the employee's routine job functions. To determine whether this is the case, you must evaluate the restriction in light of the routine functions of the injured or ill employee's job. If the restriction from you or the physician or other licensed health care professional keeps the employee from performing one or more of his or her routine job functions, or from working the full workday the injured or ill employee would otherwise have worked, the employee's work has been restricted and you must record the case.

(v) How do I record a case where the worker works only for a partial work shift because of a work-related injury or illness?

A partial day of work is recorded as a day of job transfer or restriction for recordkeeping purposes, except for the day on which the injury occurred or the illness began.

(vi) If the injured or ill worker produces fewer goods or services than he or she would have produced prior to the injury or illness but otherwise performs all of the activities of his or her work, is the case considered a restricted work case?

No. The case is considered restricted work only if the worker does not perform all of the routine functions of his or her job or does not work the full shift that he or she would otherwise have worked.

(vii) How do I handle vague restrictions from a physician or other licensed health care professional, such as that the employee engage only in "light duty" or "take it easy for a week"?

If you are not clear about a physician or other licensed health care professional's recommendation, you may ask that person whether the employee can perform all of his or her routine job functions and work all of his or her normally assigned work shift. If the answer to both of these questions is "Yes," then the case does not involve a work restriction and does not have to be recorded as such. If the answer to one or both of these questions is "No," the case involves restricted work and must be recorded as a restricted work case. If you are unable to obtain this additional information from the physician or other licensed health care professional who recommended the restriction, record the injury or illness as a case involving job transfer or restricted work.

(viii) What do I do if a physician or other licensed health care professional recommends a job restriction meeting OSHA's definition but the employee does all of his or her routine job functions anyway?

You must record the injury or illness on the OSHA 300 Log as a restricted work case. If a physician or other licensed health care professional recommends a job restriction, you should ensure that the employee complies with that restriction. If you receive recommendations from two or more physicians or other licensed health care providers, you may make a decision as to which recommendation is the most authoritative, and record the case based upon that recommendation.

...

As the regulatory text for paragraph (b)(4) makes clear, the final rule's requirements for the recording of restricted work cases are similar in many ways to those pertaining to restricted work under the former rule. First, like the former rule, the final rule only requires employers to record as restricted work cases those cases in which restrictions are imposed or recommended as a result of a work-related injury or illness. A work restriction that is made for another reason, such as to meet reduced production demands, is not a recordable restricted work case. For example, an employer might "restrict" employees from entering the area in which a toxic chemical spill has occurred or make an accommodation for an employee who is disabled as a result of a non-work-related injury or illness. These cases would not be recordable as restricted work cases because they are not associated with a work-related injury or illness. However, if an employee has a work-related injury or illness, and that employee's work is restricted by the employer to prevent exacerbation of, or to allow recuperation from, that injury or illness, the case is recordable as a restricted work case because the restriction was necessitated by the work-related injury or illness. In some cases, there may be more than one reason for imposing or recommending a work restriction, e.g., to prevent an injury or illness from becoming worse or to prevent entry into a contaminated area. In such cases, if the employee's work-related illness or injury played any role in the restriction, OSHA considers the case to be a restricted work case.

Second, for the definition of restricted work to apply, the work restriction must be decided on by the employer, based on his or her best judgment or on the recommendation of a physician or other licensed health care professional. If a work restriction is not followed or implemented by the employee, the injury or illness must nevertheless be recorded on the Log as a restricted case. This was also the case under the former rule.

Third, like the former rule, the final rule's definition of restricted work relies on two components: whether the employee is able to perform the duties of his or her pre-injury job, and whether the employee is able to perform those duties for the same period of time as before.

The principal differences between the final and former rules' concept of restricted work cases are these: (1) the final rule permits employers to cap the total number of restricted work days for a particular case at 180 days, while the former rule required all restricted days for a given case to be recorded; (2) the final rule does not require employers to count the restriction of an employee's duties on the day the injury occurred or the illness began as restricted work, providing that the day the incident occurred is the only day on which work is restricted; and (3) the final rule defines work as restricted if the injured or ill employee is restricted from performing any job activity the employee would have regularly performed at least once per week before the injury or illness, while the former rule counted work as restricted if the employee was restricted in performing any activity he or she would have performed at least once per year.

In all other respects, the final rule continues to treat restricted work and job transfer cases in the same manner as they were treated under the former rule, including the counting of restricted days. Paragraph 1904.7(b)(4)(xi) requires the employer to count restricted days using the same rules as those for counting days away from work, using Sec. 1904.7(b)(3)(i) to (viii), with one exception. Like the former rule, the final rule allows the employer to stop counting restricted days if the employee's job has been permanently modified in a manner that eliminates the routine functions the employee has been restricted from performing. Examples of permanent modifications would include reassigning an employee with a respiratory allergy to a job where such allergens are not present, or adding a mechanical assist to a job that formerly required manual lifting. To make it clear that employers may stop counting restricted days when a job has been permanently changed, but not to eliminate the count of restricted work altogether, the rule makes it clear that at least one restricted workday must be counted, even if the restriction is imposed immediately. A discussion of the desirability of counting days of restricted work and job transfer at all is included in the explanation for the OSHA 300 form and the Sec. 1904.29 requirements. The revisions to this category of cases that have been made in the final rule reflect the views of commenters, suggestions made by the Keystone report (Ex. 5), and OSHA's experience in enforcing the former recordkeeping rule.

Paragraph 1904.7(b)(5) Medical Treatment Beyond First Aid

The definitions of first aid and medical treatment have been central to the OSHA recordkeeping scheme since 1971, when the Agency's first recordkeeping rule was issued. Sections 8(c)(2) and 24(a) of the OSH Act specifically require employers to record all injuries and illnesses other than those "requiring only first aid treatment and which do not involve medical treatment, loss of consciousness, restriction of work or motion, or transfer to another job." Many injuries and illnesses sustained at work do not result in death, loss of consciousness, days away from work or restricted work or job transfer. Accordingly, the first aid and medical treatment criteria may be the criteria most frequently evaluated by employers when deciding whether a given work-related injury must be recorded.

In the past, OSHA has not interpreted the distinction made by the Act between minor (i.e., first aid only) injuries and non-minor injuries as applying to occupational illnesses, and employers have therefore been required to record all occupational illnesses, regardless of severity. As a result of this final rule, OSHA will now apply the same recordability criteria to both injuries and illnesses (see the discussion of this issue in the Legal Authority section of this preamble). The Agency believes that doing so will simplify the decision-making process that employers carry out when determining which work-related injuries and illnesses to record and will also result in more complete data on occupational illness, because employers will know that they must record these cases when they result in medical treatment beyond first aid, regardless of whether or not a physician or other licensed health care professional has made a diagnosis.

The former recordkeeping rule defined first aid as "any one-time treatment and any follow-up visit for the purpose of observation, of minor scratches, cuts, burns, splinters, and so forth, which do not ordinarily require medical care." Medical treatment was formerly defined as "treatment administered by a physician or by registered professional personnel under the standing orders of a physician."

To help employers determine the recordability of a given injury, the Recordkeeping Guidelines, issued by the Bureau of Labor Statistics (BLS) in 1986, provided numerous examples of medical treatments and of first aid treatments (Ex. 2). These examples were published as mutually exclusive lists, i.e., a treatment listed as a medical treatment did not also appear on the first-aid list. Thus, for example, a positive x-ray diagnosis (fractures, broken bones, etc.) was included among the treatments generally considered medical treatment, while a negative x-ray diagnosis (showing no fractures) was generally considered first aid. Despite the guidance provided by the Guidelines, OSHA continued to receive requests from employers for interpretations of the recordability of specific cases, and a large number of letters of interpretation addressing the distinction between first aid and medical treatment have been issued.

...

[The final rule] includes a list of first-aid treatments that is inclusive, and defines as medical treatment any treatment not on that list. OSHA recognizes, as several commenters pointed out, that no one can predict how medical care will change in the future. However, using a finite list of first aid treatments—knowing that it may have to be amended later based on new information—helps to limit the need for individual judgment about what constitutes first aid treatment. If OSHA adopted a more open-ended definition or one that relied on the judgment of a health care professional, employers and health care professionals would inevitably interpret different cases differently, which would compromise the consistency of the data. Under the system adopted in the final rule, once the employer has decided that a particular response to a work-related illness or injury is in fact treatment, he or she can simply turn to the first aid list to determine, without elaborate analysis, whether the treatment is first aid and thus not recordable.

OSHA finds that this simple approach, by providing clear, unambiguous guidance, will reduce confusion for employers and improve the accuracy and consistency of the data.

...

Final Rule
The final rule, at Sec. 1904.7(b)(5)(i), defines medical treatment as the management and care of a patient for the purpose of combating disease or disorder. For the purposes of Part 1904, medical treatment does not include:
(A) Visits to a physician or other licensed health care professional solely for observation or counseling;
(B) The conduct of diagnostic procedures, such as x-rays and blood tests, including the administration of prescription medications used solely for diagnostic purposes (e.g., eye drops to dilate pupils); or
(C) "first aid" as defined in paragraph (b)(5)(ii) of this section.

The final rule, at paragraph (b)(5)(ii), defines first aid as follows:
(A) Using a nonprescription medication at nonprescription strength (for medications available in both prescription and non-prescription form, a recommendation by a physician or other licensed health care professional to use a non-prescription medication at prescription strength is considered medical treatment for recordkeeping purposes).
(B) administering tetanus immunizations (other immunizations, such as hepatitis B vaccine or rabies vaccine, are considered medical treatment).
(C) Cleaning, flushing or soaking wounds on the surface of the skin;
(D) Using wound coverings, such as bandages, Band-Aids, gauze pads, etc.; or using butterfly bandages or Steri-Strips (other wound closing devices, such as sutures, staples, etc. are considered medical treatment);
(E) Using hot or cold therapy;
(F) Using any non-rigid means of support, such as elastic bandages, wraps, non-rigid back belts, etc. (devices with rigid stays or other systems designed to immobilize parts of the body are considered medical treatment for recordkeeping purposes);
(G) Using temporary immobilization devices while transporting an accident victim (e.g. splints, slings, neck collars, back boards, etc.)
(H) Drilling of a fingernail or toenail to relieve pressure, or draining fluid from a blister;
(I) Using eye patches;
(J) Removing foreign bodies from the eye using only irrigation or a cotton swab;
(K) Removing splinters or foreign material from areas other than the eye by irrigation, tweezers, cotton swabs, or other simple means;
(L) Using finger guards;
(M) Using massages (physical therapy or chiropractic treatment are considered medical treatment for recordkeeping purposes);
(N) Drinking fluids for relief of heat stress.

This list of first aid treatments is comprehensive, i.e., any treatment not included on this list is not considered first aid for OSHA recordkeeping purposes. OSHA considers the listed treatments to be first aid regardless of the professional qualifications of the person providing the treatment; even when these treatments are provided by a physician, nurse, or other health care professional, they are considered first aid for recordkeeping purposes.

...

The three listed exclusions from the definition—visits to a health care professional solely for observation or counseling; diagnostic procedures, including prescribing or administering of prescription medications used solely for diagnostic purposes; and procedures defined in the final rule as first aid—clarify the applicability of the definition and are designed to help employers in their determinations of recordability.

...

In making its decisions about the items to be included on the list of first aid treatments, OSHA relied on its experience with the former rule, the advice of the Agency's occupational medicine and occupational nursing staff, and a thorough review of the record comments. In general, first aid treatment can be distinguished from medical treatment as follows:
- First aid is usually administered after the injury or illness occurs and at the location (e.g., workplace) where the injury or illness occurred.
- First aid generally consists of one-time or short-term treatment.
- First aid treatments are usually simple and require little or no technology.
- First aid can be administered by people with little training (beyond first aid training) and even by the injured or ill person.

- First aid is usually administered to keep the condition from worsening, while the injured or ill person is awaiting medical treatment.

The final rule's list of treatments considered first aid is based on the record of the rulemaking, OSHA's experience in implementing the recordkeeping rule since 1986, a review of the BLS Recordkeeping Guidelines, letters of interpretation, and the professional judgment of the Agency's occupational physicians and nurses.

...

In the final rule, OSHA has not included prescription medications, whether given once or over a longer period of time, in the list of first aid treatments. The Agency believes that the use of prescription medications is not first aid because prescription medications are powerful substances that can only be prescribed by a licensed health care professional, and for the majority of medications in the majority of states, by a licensed physician. The availability of these substances is carefully controlled and limited because they must be prescribed and administered by a highly trained and knowledgeable professional, can have detrimental side effects, and should not be self-administered.

...

OSHA has decided to retain its long-standing policy of requiring the recording of cases in which a health care professional issues a prescription, whether that prescription is filled or taken or not. The patient's acceptance or refusal of the treatment does not alter the fact that, in the health care professional's judgment, the case warrants medical treatment. In addition, a rule that relied on whether a prescription is filled or taken, rather than on whether the medicine was prescribed, would create administrative difficulties for employers, because such a rule would mean that the employer would have to investigate whether a given prescription had been filled or the medicine had actually been taken. Finally, many employers and employees might well consider an employer's inquiry about the filling of a prescription an invasion of the employee's privacy. For these reasons, the final rule continues OSHA's longstanding policy of considering the giving of a prescription medical treatment. It departs from former practice with regard to the administration of a single dose of a prescription medicine, however, because there is no medical reason for differentiating medical treatment from first aid on the basis of the number of doses involved. This is particularly well illustrated by the recent trend toward giving a single large dose of antibiotics instead of the more traditional pattern involving several smaller doses given over several days.

Yet another issue raised by commenters about medications involved the use of non-prescription medications at prescription strength. In recent years, many drugs have been made available both as prescription and "over-the-counter" medications, depending on the strength or dosage of the product. Some examples include various non-steroidal anti-inflammatory drugs (NSAIDs), such as ibuprofen, and cortisone creams.

...

The final rule does not consider the prescribing of non-prescription medications, such as aspirin or over-the-counter skin creams, as medical treatment. However, if the drug is one that is available both in prescription and nonprescription strengths, such as ibuprofen, and is used or recommended for use by a physician or other licensed health care professional at prescription strength, the medical treatment criterion is met and the case must be recorded. There is no reason for one case to be recorded and another not to be recorded simply because one physician issued a prescription and another told the employee to use the same medication at prescription strength but to obtain it over the counter. Both cases received equal treatment and should be recorded equally. This relatively small change in the recordkeeping rule will improve the consistency and accuracy of the data on occupational injuries and illnesses and simplify the system as well.

...

OSHA believes that cleaning, flushing or soaking of wounds on the skin surface is the initial emergency treatment for almost all surface wounds and that these procedures do not rise to the level of medical treatment. This relatively simple type of treatment does not require technology, training, or even a visit to a health care professional. More serious wounds will be captured as recordable cases because they will meet other recording criteria, such as prescription medications, sutures, restricted work, or days away from work. Therefore, OSHA has included cleaning, flushing or soaking of wounds on the skin surface as an item on the first aid list. As stated previously, OSHA does not believe that multiple applications of first aid should constitute medical treatment; it is the nature of the treatment, not how many times it is applied, that determines whether it is first aid or medical treatment.

...

In the final rule, OSHA has included hot and cold treatment as first aid treatment, regardless of the number of times it is applied, where it is applied, or the injury or illness to which it is applied. The Agency has decided that hot or cold therapy must be defined as either first aid or medical treatment regardless of the condition being treated, a decision that departs from the proposal. It is OSHA's judgment that hot and cold treatment is simple to apply, does not require special training, and is rarely used as the only treatment for any significant injury or illness. If the worker has sustained a significant injury or illness, the case almost always involves some other form of medical treatment (such as prescription drugs, physical therapy, or chiropractic treatment); restricted work; or days away from work. Therefore, there is no need to consider hot and cold therapy to be medical treatment, in and of itself. Considering hot and cold therapy to be first aid also clarifies and simplifies the rule, because it means that employers will not need to consider whether to record when an employee uses hot or cold therapy without the direction or guidance of a physician or other licensed health care professional.

...

In the final rule, OSHA has included hot and cold treatment as first aid treatment, regardless of the number of times it is applied, where it is applied, or the injury or illness to which it is applied.

The Agency has decided that hot or cold therapy must be defined as either first aid or medical treatment regardless of the condition being treated, a decision that departs from the proposal. It is OSHA's judgment that hot and cold treatment is simple to apply, does not require special training, and is rarely used as the only treatment for any significant injury or illness. If the worker has sustained a significant injury or illness, the case almost always involves some other form of medical treatment (such as prescription drugs, physical therapy, or chiropractic treatment); restricted work; or days away from work. Therefore, there is no need to consider hot and cold therapy to be medical treatment, in and of itself. Considering hot and cold therapy to be first aid also clarifies and simplifies the rule, because it means that employers will not need to consider whether to record when an employee uses hot or cold therapy without the direction or guidance of a physician or other licensed health care professional.

...

[OSHA believes] that the use of these devices during an emergency to stabilize an accident victim during transport to a medical facility is not medical treatment. In this specific situation, a splint or other device is used as temporary first aid treatment, may be applied by non-licensed personnel using common materials at hand, and often does not reflect the severity of the injury. OSHA has included this item as G on the first aid list: "[u]sing temporary immobilization devices while transporting an accident victim (e.g. splints, slings, neck collars, etc.)"

...

[Item H on the first aid list is the "drilling of a fingernail or toenail to relieve pressure, or draining fluid from a blister."] These are both one time treatments provided to relieve minor soreness caused by the pressure beneath the nail or in the blister. These are relatively minor procedures that are often performed by licensed personnel but may also be performed by the injured worker. More serious injuries of this type will continue to be captured if they meet one or more of the other recording criteria. OSHA has specifically mentioned finger nails and toenails to provide clarity. These treatments are now included as item H on the first aid list.

...

In the final rule, OSHA has included the use of eye patches as first aid in item I of the first aid list. Eye patches can be purchased without a prescription, and are used for both serious and non-serious injuries and illnesses. OSHA believes that the more serious injuries to the eyes will that NIOSH refers to require medical treatment, such as prescription drugs or removal of foreign material by means other than irrigation or a cotton swab, and will thus be recordable.

...

In the final rule, OSHA has included as item J "Removing foreign bodies from the eye using only irrigation or a cotton swab." OSHA believes that it is often difficult for the health care professional to determine if the object is embedded or adhered to the eye, and has not included this suggested language in the final rule. In all probability, if the object is embedded or adhered, it will not be removed simply with irrigation or a cotton swab, and the case will be recorded because it will require additional treatment.

...

OSHA believes that it is appropriate to exclude those cases from the Log that involve a foreign body in the eye of a worker that can be removed from the eye merely by rinsing it with water (irrigation) or touching it with a cotton swab. These cases represent minor injuries that do not rise to the level requiring recording. More significant eye injuries will be captured by the records because they involve medical treatment, result in work restrictions, or cause days away from work.

...

[OSHA has included as item K "Removal of splinters or foreign material from areas other than the eyes by irrigation, tweezers, cotton swabs or other simple means."] The inclusion of the phrase "other simple means" will provide some flexibility and permit simple means other than those listed to be considered first aid. Cases involving more complicated removal procedures will be captured on the Log because they will require medical treatment such as prescription drugs or stitches or will involve restricted work or days away from work.

...

Paragraph 1904.7(b)(6) Loss of Consciousness

The final rule, like the former rule, requires the employer to record any work-related injury or illness resulting in a loss of consciousness. The recording of occupational injuries and illnesses resulting in loss of consciousness is clearly required by Sections 8(c) and 24 of the OSH Act. The new rule differs from the former rule only in clearly applying the loss of consciousness criterion to illnesses as well as injuries. Since the former rule required the recording of all illnesses, illnesses involving loss of consciousness were recordable, and thus OSHA expects that this clarification will not change recording practices. Thus, any time a worker becomes unconscious as a result of a workplace exposure to chemicals, heat, an oxygen deficient environment, a blow to the head, or some other workplace hazard that causes loss of consciousness, the employer must record the case.

...

Paragraph 1904.7(b)(7) Recording Significant Work-Related Injuries and Illnesses Diagnosed by a Physician or Other Licensed Health Care Professional

Paragraph 1904.7(b)(7) of this final rule requires the recording of any significant work-related injury or illness diagnosed by a physician or other licensed health care professional. Paragraph 1904.7(b)(7) clarifies which significant, diagnosed work-related injuries and illnesses OSHA requires the employer to record in those rare cases where a significant work-related injury or illness has not triggered recording under one or more of the general recording criteria, i.e., has not resulted in death, loss of consciousness, medical treatment beyond first aid, restricted work or job transfer, or days away from work. Based on the Agency's prior recordkeeping experience, OSHA believes that the great majority of significant occupational injuries and illnesses will be captured by one or more of the other general recording criteria in Section 1904.7. However, OSHA has found that there is a limited class of significant work-related injuries and illnesses that may not be captured under the other Sec. 1904.7 criteria. Therefore, the final rule stipulates at paragraph 1904.7(b)(7) that any significant work-related occupational injury or illness that is not captured by any of the general recording criteria but is diagnosed by a physician or other licensed health care professional be recorded in the employer's records.

Under the final rule, an injury or illness case is considered significant if it is a work-related case involving occupational cancer (e.g., mesothelioma), chronic irreversible disease (e.g., chronic beryllium disease), a fractured or cracked bone (e.g., broken arm, cracked rib), or a punctured eardrum. The employer must record such cases within 7 days of receiving a diagnosis from a physician or other licensed health care professional that an injury or illness of this kind has occurred. As explained in the note to paragraph 1904.7(b)(7), OSHA believes that the great majority of significant work-related injuries and illnesses will be recorded because they meet one or more of the other recording criteria listed in Sec. 1904.7(a): death, days away from work, restricted work or job transfer, medical treatment beyond first aid, or loss of consciousness. However, there are some significant injuries, such as a punctured eardrum or a fractured toe or rib, for which neither medical treatment nor work restrictions may be administered or recommended.

There are also a number of significant occupational diseases that progress once the disease process begins or reaches a certain point, such as byssinosis, silicosis, and some types of cancer, for which medical treatment or work restrictions may not be recommended at the time of diagnosis, although medical treatment and loss of work certainly will occur at later stages. This provision of the final rule is designed to capture this small group of significant work-related cases. Although the employer is required to record these illnesses even if they manifest themselves after the employee leaves employment (assuming the illness meets the standards for work-relatedness that apply to all recordable incidents), these cases are less likely to be recorded once the employee has left employment. OSHA believes that work-related cancer, chronic irreversible diseases, fractures of bones or teeth and punctured eardrums are generally recognized as constituting significant diagnoses and, if the condition is work-related, are appropriately recorded at the time of initial diagnosis even if, at that time, medical treatment or work restrictions are not recommended.

As discussed in the Legal Authority section, above, OSHA has modified the Agency's prior position so that, under the final rule, minor occupational illnesses no longer are required to be recorded on the Log. The requirement pertaining to the recording of all significant diagnosed injuries and illnesses in this paragraph of the final rule, on the other hand, will ensure that all significant (non-minor) injuries and illnesses are in fact captured on the Log, as required by the OSH Act. Requiring significant cases involving diagnosis to be recorded will help to achieve several of the goals of this rulemaking. First, adherence to this requirement will produce better data on occupational injury and illness by providing for more complete recording of significant occupational conditions. Second, this requirement will produce more timely records because it provides for the immediate recording of significant disorders on first diagnosis. Many occupational illnesses manifest themselves through gradual onset and worsening of the condition. In some cases, a worker could be diagnosed with a significant illness, such as an irreversible respiratory disorder, not be given medical treatment because no effective treatment was available, not lose time from work because the illness was not debilitating at the time, and not have his or her case recorded on the Log because none of the recording criteria had been met. If such a worker left employment or changed employers before one of the other recording criteria had been met, this serious occupational illness case would never be recorded. The requirements in paragraph 1904.7(b)(7) remedy this deficiency and will thus ensure the capture of more complete and timely data on these injuries and illnesses.

...

Section 1904.8 Additional Recording Criteria for Needlestick and Sharps Injuries [p. 5998]

Section 1904.8 of the final rule being published today deals with the recording of a specific class of occupational injuries involving punctures, cuts and lacerations caused by needles or other sharp objects contaminated or reasonably anticipated to be contaminated with blood or other potentially infectious materials that may lead to bloodborne diseases, such as Acquired Immunodeficiency Syndrome (AIDS), hepatitis B or hepatitis C. The final rule uses the terms "contaminated," "other potentially infectious material," and "occupational exposure" as these terms are defined in OSHA's Bloodborne Pathogens standard (29 CFR 1910.1030). These injuries are of special concern to healthcare workers because they use needles and other sharp devices in the performance of their work duties and are therefore at risk of bloodborne infections caused by exposures involving contaminated needles and other sharps. Although healthcare workers are at particular risk of bloodborne infection from these injuries, other workers may also be at risk of contracting potentially fatal bloodborne disease. For example, a worker in a hospital laundry could be stuck by a contaminated needle left in a patient's bedding, or a worker in a hazardous waste treatment facility could be occupationally exposed to bloodborne pathogens if contaminated waste from a medical facility was not treated before being sent to waste treatment.

Section 1904.8(a) requires employers to record on the OSHA Log all work-related needlestick and sharps injuries involving objects contaminated (or reasonably anticipated to be contaminated) with another person's blood or other potentially infectious material (OPIM). The rule prohibits the employer from entering the name of the affected employee on the Log to protect the individual's privacy; employees are understandably sensitive about others knowing that they may have contracted a bloodborne disease. For these cases, and other types of privacy concern cases, the employer simply enters "privacy concern case" in the space reserved for the employee's name. The employer then keeps a separate, confidential list of privacy concern cases with the case number from the Log and the employee's name; this list is used by the employer to keep track of the injury or illness so that the Log can later be updated, if necessary, and to ensure that the information will be available if a government representative needs information about injured or ill employees during a workplace inspection (see Sec. 1904.40). The regulatory text of Sec. 1904.8 refers recordkeepers and others to Sec. 1904.29(b)(6) through Sec. 1904.29(b)(10) of the rule for more information about how to record privacy concern cases of all types, including those involving needlesticks and sharps injuries. The implementation section of Sec. 1904.8(b)(1) defines "other potentially infectious material" as it is defined in OSHA's Bloodborne Pathogens Standard (29 CFR Sec. 1910.1030, paragraph (b)). Other potentially infectious materials include (i) human bodily fluids, human tissues and organs, and (ii) other materials infected with the HIV or hepatitis B (HBV) virus such as laboratory cultures or tissues from experimental animals. (For a complete list of OPIM, see paragraph (b) of 29 CFR 1910.1030.)

Although the final rule requires the recording of all workplace cut and puncture injuries resulting from an event involving contaminated sharps, it does not require the recording of all cuts and punctures. For example, a cut made by a knife or other sharp instrument that was not contaminated by blood or OPIM would not generally be recordable, and a laceration made by a dirty tin can or greasy tool would also generally not be recordable, providing that the injury did not result from a contaminated sharp and did not meet one of the general recording criteria of medical treatment, restricted work, etc. Paragraph (b)(2) of Sec. 1904.8 contains provisions indicating which cuts and punctures must be recorded because they involve contaminated sharps and which must be recorded only if they meet the general recording criteria.

Paragraph (b)(3) of Sec. 1904.8 contains requirements for updating the OSHA 300 Log when a worker experiences a wound caused by a contaminated needle or sharp and is later diagnosed as having a bloodborne illness, such as AIDS, hepatitis B or hepatitis C. The final rule requires the employer to update the classification of such a privacy concern case on the OSHA 300 Log if the outcome of the case changes, i.e., if it subsequently results in death, days away from work, restricted work, or job transfer. The employer must also update the case description on the Log to indicate the name of the bloodborne illness and to change the classification of the case from an injury (i.e., the needlestick) to an illness (i.e., the illness that resulted from the needlestick). In no case may the employer enter the employee's name on the Log itself, whether when initially recording the needlestick or sharp injury or when subsequently updating the record.

The privacy concern provisions of the final rule make it possible, for the first time, for the identity of the bloodborne illness caused by the needlestick or sharps injury to be included on the Log. By excluding the name of the injured or ill employee throughout the recordkeeping process, employee privacy is assured. This approach will allow OSHA to gather valuable data about the kinds of bloodborne illnesses healthcare and other workers are contracting as a result of these occupational injuries, and will provide the most accurate and informative data possible, including the seroconversion status of the affected worker, the name of the illness he or she contracted, and, on the OSHA 301 Form for the original case, more detailed information about how the injury occurred, the equipment and materials involved, and so forth. Use of the privacy case concept thus meets the primary objective of this rulemaking, providing the best data possible, while simultaneously ensuring that an important public policy goal—the protection of privacy about medical matters—is met. OSHA recognizes that requiring employers to treat privacy cases differently from other cases adds some complexity to the recordkeeping system and imposes a burden on those employers whose employees experience such injuries and illnesses, but believes that the gain in data quality and employee privacy outweigh these disadvantages considerably.

The last paragraph (paragraph (c)) of Sec. 1904.8 deals with the recording of cases involving workplace contact with blood or other potentially infectious materials that do not involve needlesticks or sharps, such as splashes to the eye, mucous membranes, or non-intact skin. The final recordkeeping rule does not require employers to record these incidents unless they meet the final rule's general recording criteria (i.e., death, medical treatment, loss of consciousness, restricted work or motion, days away from work, diagnosis by an HCP) or the employee subsequently develops an illness caused by bloodborne pathogens. The final rule thus provides employers, for the first time, with regulatory language delineating how they are to record injuries caused by contaminated needles and other sharps, and how they are to treat other exposure incidents (as defined in the Bloodborne Pathogens standard) involving blood or OPIM. "Contaminated" is defined just as it is in the Bloodborne Pathogens standard: "Contaminated means the presence or the reasonably anticipated presence of blood or other potentially infectious materials on an item or surface."

...

Section 1904.9 Additional Recording Criteria for Cases Involving Medical Removal Under OSHA Standards [p. 6003]

The final rule, in paragraph 1904.9(a), requires an employer to record an injury or illness case on the OSHA 300 Log when the employee is medically removed under the medical surveillance requirements of any OSHA standard. Paragraph 1904.9(b)(1) requires each such case to be recorded as a case involving days away from work (if the employee does not work during the medical removal) or as a case involving restricted work activity (if the employee continues to work but in an area where exposures are not present.) This paragraph also requires any medical removal related to chemical exposure to be recorded as a poisoning illness.

Paragraph 1904.9(b)(2) informs employers that some OSHA standards have medical removal provisions and others do not. For example, the Bloodborne Pathogen Standard (29 CFR 1910.1030) and the Occupational Noise Standard (29 CFR 1910.95) do not require medical removal. Many of the OSHA standards that contain medical removal provisions are related to specific chemical substances, such as lead (29 CFR 1901.1025), cadmium (29 CFR 1910.1027), methylene chloride (29 CFR 1910.1052), formaldehyde (29 CFR 1910.1048), and benzene (29 CFR 1910.1028). Paragraph 1904.9(b)(3) addresses the issue of medical removals that are not required by an OSHA standard. In some cases employers voluntarily rotate employees from one job to another to reduce exposure to hazardous substances; job rotation is an administrative method of reducing exposure that is permitted in some OSHA standards. Removal (job transfer) of an asymptomatic employee for administrative exposure control reasons does not require the case to be recorded on the OSHA 300 Log because no injury or illness—the first step in the recordkeeping process—exists. Paragraph 1904.9(b)(3) only applies to those substances with OSHA mandated medical removal criteria. For injuries or illnesses caused by exposure to other substances or hazards, the employer must look to the general requirements of paragraphs 1910.7(b)(3) and (4) to determine how to record the days away or days of restricted work.

The provisions of Sec. 1904.9 are not the only recording criteria for recording injuries and illnesses from these occupational exposures. These provisions merely clarify the need to record specific cases, which are often established with medical test results, that result in days away from work, restricted work, or job transfer. The Sec. 1904.9 provisions are included to produce more consistent data and provide needed interpretation of the requirements for employers. However, if an injury or illness results in the other criteria of Sec. 1904.7 (death, medical treatment, loss of consciousness, days away from work, restricted work, transfer to another job, or diagnosis as a significant illness or injury by a physician or other licensed health care professional) the case must be recorded whether or not the medical removal provisions of an OSHA standard have been met.

...

OSHA has therefore included section 1904.9 in the final rule to provide a uniform, simple method for recording a variety of serious disorders that have been addressed by OSHA standards. The Sec. 1904.9 provisions of the final rule cover all of the OSHA standards with medical removal provisions, regardless of whether or not those provisions are based on medical tests, physicians' opinions, or a combination of the two. Finally, by relying on the medical removal provisions in any OSHA standard, section 1904.9 of the final rule establishes recording criteria for future standards, and avoids the need to amend the recordkeeping rule whenever OSHA issues a standard containing a medical removal level.

...

Section 1904.10 Recording Criteria for Cases Involving Occupational Hearing Loss [p. 6004]

[The Department of Labor has proposed that the criteria for recording work-related hearing loss not be implemented for one year pending further investigation into the level of hearing loss that should be recorded as a "significant" health condition. Paragraph 1904.10(b)(1) of the final rule had required that a standard threshold shift of an average of 10 decibels (dB) or more at 2000, 3000, or 4000 hertz in one or both ears would be recordable.

Until more information is obtained, a standard threshold shift of 25 decibels, as required by the former regulation, is to be recorded.]

Section 1904.11 Additional Recording Criteria for Work-Related Tuberculosis Cases [p. 6013]

Section 1904.11 of the final rule being published today addresses the recording of tuberculosis (TB) infections that may occur to workers occupationally exposed to TB. TB is a major health concern, and nearly one-third of the world's population may be infected with the TB bacterium at the present time. There are two general stages of TB, tuberculosis infection and active tuberculosis disease. Individuals with tuberculosis infection and no active disease are not infectious; tuberculosis infections are asymptomatic and are only detected by a positive response to a tuberculin skin test. Workers in many settings are at risk of contracting TB infection from their clients or patients, and some workers are at greatly increased risk, such as workers exposed to TB patients in health care settings. Outbreaks have also occurred in a variety of workplaces, including hospitals, prisons, homeless shelters, nursing homes, and manufacturing facilities (62 FR 54159).

The text of Sec. 1904.11 of the final rule states:

(a) **Basic requirement.** If any of your employees has been occupationally exposed to anyone with a known case of active tuberculosis (TB), and that employee subsequently develops a tuberculosis infection, as evidenced by a positive skin test or diagnosis by a physician or other licensed health care professional, you must record the case on the OSHA 300 Log by checking the "respiratory condition" column.

(b) Implementation.
(1) Do I have to record, on the Log, a positive TB skin test result obtained at a pre-employment physical?

No, because the employee was not occupationally exposed to a known case of active tuberculosis in your workplace.

(2) May I line-out or erase a recorded TB case if I obtain evidence that the case was not caused by occupational exposure?

Yes. you may line-out or erase the case from the Log under the following circumstances:
   (i) The worker is living in a household with a person who has been diagnosed with active TB;
   (ii) The Public Health Department has identified the worker as a contact of an individual with a case of active TB unrelated to the workplace; or
   (iii) A medical investigation shows that the employee's infection was caused by exposure to TB away from work, or proves that the case was not related to the workplace TB exposure.

Section 1904.12 Recording Criteria for Cases Involving Work-Related Musculoskeletal Disorders [p. 6017]

[This section provides requirements for recording work-related musculoskeletal disorders (MSDs). OSHA has proposed to delay for one year the definition of MSD and the requirement to check the MSD column. However, the same recording criteria applies to musculoskeletal disorders (MSDs) as to all other injuries and illnesses.]

Section 1904.29 Forms [p. 6022]

Section 1904.29, titled "Forms," establishes the requirements for the forms (OSHA 300 Log, OSHA 300A Annual Summary, and OSHA 301 Incident Report) an employer must use to keep OSHA Part 1904 injury and illness records, the time limit for recording an injury or illness case, the use of substitute forms, the use of computer equipment to keep the records, and privacy protections for certain information recorded on the OSHA 300 Log.

Paragraph 1904.29(a) sets out the basic requirements of this section. It directs the employer to use the OSHA 300 (Log), 300A (Summary), and 301 (Incident Report) forms, or equivalent forms, to record all recordable occupational injuries and illnesses. Paragraph 1904.29(b) contains requirements in the form of questions and answers to explain how employers are to implement this basic requirement. Paragraph 1904.29(b)(1) states the requirements for: (1) Completing the establishment information at the top of the OSHA 300 Log, (2) making a one- or two-line entry for each recordable injury and illness case, and (3) summarizing the data at the end of the year. Paragraph 1904.29(b)(2) sets out the requirements for employers to complete the OSHA 301 Incident Report form (or equivalent) for each recordable case entered on the OSHA 300 Log. The requirements for completing the annual summary on the Form 300A are found at Section 1904.32 of the final rule.

Required Forms

In addition to establishing the basic requirements for employers to keep records on the OSHA 300 Log and OSHA 301 Incident Report and providing basic instructions on how to complete these forms, this section of the rule states that employers may use two lines of the OSHA 300 Log to describe an injury or illness, if necessary.

...

OSHA believes that most injury and illness cases can be recorded using only one line of the Log. However, for those cases requiring more space, this addition to the Log makes it clear that two lines may be used to describe the case. The OSHA 300 Log is designed to be a scannable document that employers, employees and government representatives can use to review a fairly large number of cases in a brief time, and OSHA believes that employers will not need more than two lines to describe a given case. Employers should enter more detailed information about each case on the OSHA 301 form, which is designed to accommodate lengthier information.

Deadline for Entering a Case

Paragraph 1904.29(b)(3) establishes the requirement for how quickly each recordable injury or illness must be recorded into the records. It states that the employer must enter each case on the OSHA 300 Log and OSHA 301 Form within 7 calendar days of receiving information that a recordable injury or illness has occurred. In the vast majority of cases, employers know immediately or within a short time that a recordable case has occurred. In a few cases, however, it may be several days before the employer is informed that an employee's injury or illness meets one or more of the recording criteria.

...

Accordingly, paragraphs 1904.29(b)(4) and (b)(5) of the final rule make clear that employers are permitted to record the required information on electronic media or on paper forms that are different from the OSHA 300 Log, provided that the electronic record or paper forms are equivalent to the OSHA 300 Log. A form is deemed to be "equivalent" to the OSHA 300 Log if it can be read and understood as easily as the OSHA form and contains at least as much information as the OSHA 300 Log. In addition, the equivalent form must be completed in accordance with the instructions used to complete the OSHA 300 Log. These provisions are intended to balance OSHA's obligation, as set forth in Section 8(d) of the OSH Act, to reduce information collection burdens on employers as much as possible, on the one hand, with the need, on the other hand, to maintain uniformity of the data recorded and provide employers flexibility in meeting OSHA's recordkeeping requirements. These provisions also help to achieve one of OSHA's goals for this rulemaking: to allow employers to take full advantage of modern technology and computers to meet their OSHA recordkeeping obligations.

...

The final rule does not include a requirement that certain questions on an equivalent form be asked in the same order and be phrased in language identical to that used on the OSHA 301 form. Instead, OSHA has decided, based on a review of the record evidence, that employers may use any substitute form that contains the same information and follows the same recording directions as the OSHA 301 form, and the final rule clearly allows this. Although the consistency of the data on the OSHA 301 form might be improved somewhat if the questions asking for further details were phrased and positioned in an identical way on all employers' forms, OSHA has concluded that the additional burden such a requirement would impose on employers and workers' compensation agencies outweighs this consideration.

...

Handling of Privacy Concern Cases

Paragraphs 1904.29(b)(6) through (b)(10) of the final rule are new and are designed to address privacy concerns raised by many commenters to the record. Paragraph 1904.29(b)(6) requires the employer to withhold the injured or ill employee's name from the OSHA 300 Log for injuries and illnesses defined by the rule as "privacy concern cases" and instead to enter "privacy concern case" in the space where the employee's name would normally be entered if an injury or illness meeting the definition of a privacy concern case occurs. This approach will allow the employer to provide OSHA 300 Log data to employees, former employees and employee representatives, as required by Sec. 1904.35, while at the same time protecting the privacy of workers who have experienced occupational injuries and illnesses that raise privacy concerns. The employer must also keep a separate, confidential list of these privacy concern cases, and the list must include the employee's name and the case number from the OSHA 300 Log. This separate listing is needed to allow a government representative to obtain the employee's name during a workplace inspection in case further investigation is warranted and to assist employers to keep track of such cases in the event that future revisions to the entry become necessary.

Paragraph 1904.29(b)(7) defines "privacy concern cases" as those involving: (i) An injury or illness to an intimate body part or the reproductive system; (ii) an injury or illness resulting from a sexual assault; (iii) a mental illness; (iv) a work-related HIV infection, hepatitis case, or tuberculosis case; (v) needlestick injuries and cuts from sharp objects that are contaminated with another person's blood or other potentially infectious material, or (vi) any other illness, if the employee independently and voluntarily requests that his or her name not be entered on the log. Paragraph 1904.29(b)(8) establishes that these are the only types of occupational injuries and illnesses that the employer may consider privacy concern cases for recordkeeping purposes.

Paragraph 1904.29(b)(9) permits employers discretion in recording case information if the employer believes that doing so could compromise the privacy of the employee's identity, even though the employee's name has not been entered. This clause has been added because OSHA recognizes that, for specific situations, coworkers who are allowed to access the log may be able to deduce the identity of the injured or ill worker and obtain inappropriate knowledge of a privacy-sensitive injury or illness. OSHA believes that these situations are relatively infrequent, but still exist. For example, if knowing the department in which the employee works would inadvertently divulge the person's identity, or recording the gender of the injured employee would identifying that person (because, for example, only one woman works at the plant), the employer has discretion to mask or withhold this information both on the Log and Incident Report.

The rule requires the employer to enter enough information to identify the cause of the incident and the general severity of the injury or illness, but allows the employer to exclude details of an intimate or private nature. The rule includes two examples; a sexual assault case could be described simply as "injury from assault," or an injury to a reproductive organ could be described as "lower abdominal injury." Likewise, a work-related diagnosis of post traumatic stress disorder could be described as "emotional difficulty." Reproductive disorders, certain cancers, contagious diseases and other disorders that are intimate and private in nature may also be described in a general way to avoid privacy concerns. This allows the employer to avoid overly graphic descriptions that may be offensive, without sacrificing the descriptive value of the recorded information.

Paragraph 1904.29(b)(10) protects employee privacy if the employer decides voluntarily to disclose the OSHA 300 and 301 forms to persons other than those who have a mandatory right of access under the final rule. The paragraph requires the employer to remove or hide employees' names or other personally identifying information before disclosing the forms to persons other than government representatives, employees, former employees or authorized representatives, as required by paragraphs 1904.40 and 1904.35, except in three cases. The employer may disclose the forms, complete with personally identifying information, (2) only: (i) to an auditor or consultant hired by the employer to evaluate the safety and health program; (ii) to the extent necessary for processing a claim for workers' compensation or other insurance benefits; or (iii) to a public health authority or law enforcement agency for uses and disclosures for which consent, an authorization, or opportunity to agree or object is not required under section 164.512 of the final rule on Standards for Privacy of Individually Identifiable Health Information, 45 CFR 164.512.

These requirements have been included in Sec. 1904.29 rather than in Sec. 1904.35, which establishes requirements for records access, because waiting until access is requested to remove identifying information from the OSHA 300 Log could unwittingly compromise the injured or ill worker's privacy and result in unnecessary delays. The final rule's overall approach to handling privacy issues is discussed more fully in the preamble discussion of the employee access provisions in Sec. 1904.35.

The Treatment of Occupational Illness and Injury Data on the Forms

The treatment of occupational injury and illness data on the OSHA forms is a key issue in this rulemaking. Although the forms themselves are not printed in the Code of Federal Regulations (CFR), they are the method OSHA's recordkeeping regulation uses to meet the Agency's goal of tracking and reporting occupational injury and illness data. As such, the forms are a central component of the recordkeeping system and mirror the requirements of the Part 1904 regulation. The final Part 1904 rule requires employers to use three forms to track occupational injuries and illnesses: the OSHA 300, 300A, and 301 forms, which replace the OSHA 200 and 101 forms called for under the former recordkeeping rule, as follows:

1. The OSHA Form 300, Log of Work-Related Injuries and Illnesses, replaces the Log portion of the former OSHA Form 200 Log and Summary of Occupational Injuries and Illnesses. The OSHA 300 Log contains space for a description of the establishment name, city and state, followed by a one-line space for the entry for each recordable injury and illness.
2. The OSHA Form 300A, Summary of Work-Related Injuries and Illnesses, replaces the Summary portion of the former OSHA Form 200 Log and Summary of Occupational Injuries and Illnesses. The Form 300A is used to summarize the entries from the Form 300 Log at the end of the year and is then posted from February 1 through April 30 of the following year so that employees can be aware of the occupational injury and illness experience of the establishment in which they work. The form contains space for entries for each of the columns from the Form 300, along with information about the establishment, and the average number of employees who worked there the previous year, and the recordkeeper's and corporate officer's certification of the accuracy of the data recorded on the summary. (These requirements are addressed further in Section 1904.32 of the final rule and its associated preamble.)
3. The OSHA Form 301, Injury and Illness Report, replaces the former OSHA 101 Form. Covered employers are required to fill out a one-page form for each injury and illness recorded on the Form 300. The form contains space for more detailed information about the injured or ill employee, the physician or other health care professional who cared for the employee (if medical treatment was necessary), the treatment (if any) of the employee at an emergency room or hospital, and descriptive information telling what the employee was doing when injured or ill, how the incident occurred, the specific details of the injury or illness, and the object or substance that harmed the employee. (Most employers use a workers' compensation form as a replacement for the OSHA 301 Incident Report.)

...

The OSHA recordkeeping forms required by the final Part 1904 recordkeeping rule are printed on legal size paper (8½" x 14"). The former rule's Log was an 11 by 17-inch form, the equivalent of two standard 8½ by 11-inch pages. The former 200 Log was criticized because it was unwieldy to copy and file and contained 12 columns for recording occupational injury and occupational illness cases.

Accordingly, OSHA has redesigned the OSHA 300 Log to fit on a legal size (8½ x 14 inches) piece of paper and to clarify that employers may use two lines to enter a case if the information does not fit easily on one line. The OSHA forms 300A and 301, and the remainder of the recordkeeping package, have also been designed to fit on the same-size paper as the OSHA 300 Log. For those employers who use computerized systems (where handwriting space is not as important) equivalent computer-generated forms can be printed out on 8½ x 11 sheets of paper if the printed copies are legible and are as readable as the OSHA forms.

...

Defining Lost Workdays
In the final rule, OSHA has eliminated the term "lost workdays" on the forms and in the regulatory text. The use of the term has been confusing for many years because many people equated the terms ``lost workday" with "days away from work" and failed to recognize that the former OSHA term included restricted days. OSHA finds that deleting this term from the final rule and the forms will improve clarity and the consistency of the data.

The 300 Log has four check boxes to be used to classify the case: death, day(s) away from work, days of restricted work or job transfer; and case meeting other recording criteria. The employer must check the single box that reflects the most severe outcome associated with a given injury or illness. Thus, for an injury or illness where the injured worker first stayed home to recuperate and then was assigned to restricted work for several days, the employer is required only to check the box for days away from work (column I). For a case with only job transfer or restriction, the employer must check the box for days of restricted work or job transfer (Column H). However, the final

Log still allows employers to calculate the incidence rate formerly referred to as a "lost workday injury and illness rate" despite the fact that it separates the data formerly captured under this heading into two separate categories. Because the OSHA Form 300 has separate check boxes for days away from work cases and cases where the employee remained at work but was temporarily transferred to another job or assigned to restricted duty, it is easy to add the totals from these two columns together to obtain a single total to use in calculating an injury and illness incidence rate for total days away from work and restricted work cases.

Counting Days of Restricted Work or Job Transfer
Although the final rule does not use the term "lost workday" (which formerly applied both to days away from work and days of restricted or transferred work), the rule continues OSHA's longstanding practice of requiring employers to keep track of the number of days on which an employee is placed on restricted work or is on job transfer because of an injury or illness.

...

In the final rule, OSHA has decided to require employers to record the number of days of restriction or transfer on the OSHA 300 Log. From the comments received, and based on OSHA's own experience, the Agency finds that counts of restricted days are a useful and needed measure of injury and illness severity. OSHA's decision to require the recording of restricted and transferred work cases on the Log was also influenced by the trend toward restricted work and away from days away from work. In a recent article, the BLS noted that occupational injuries and illnesses are more likely to result in days of restricted work than was the case in the past. From 1978 to 1986, the annual rate in private industry for cases involving only restricted work remained constant, at 0.3 cases per 100 full-time workers. Since 1986, the rate has risen steadily to 1.2 cases per 100 workers in 1997, a fourfold increase. At the same time, cases with days away from work declined from 3.3 in 1986 to 2.1 in 1997 (Monthly Labor Review, June 1999, Vol. 122. No. 6, pp. 11-17). It is clear that employers have caused this shift by modifying their return-to-work policies and offering more restricted work opportunities to injured or ill employees. Therefore, in order to get an accurate picture of the extent of occupational injuries and illnesses, it is necessary for the OSHA Log to capture counts of days away from work and days of job transfer or restriction.

The final rule thus carries forward OSHA's longstanding requirement for employers to count and record the number of restricted days on the OSHA Log. On the Log, restricted work counts are separated from days away from work counts, and the term "lost workday" is no longer used. OSHA believes that the burden on employers of counting these days will be reduced somewhat by the simplified definition of restricted work, the counting of calendar days rather than work days, capping of the counts at 180 days, and allowing the employer to stop counting restricted days when the employees job has been permanently modified to eliminate the routine job functions being restricted (see the preamble discussion for 1904.7 General Recording Criteria).

Separate 300 Log Data on Occupational Injury and Occupational Illness
OSHA proposed (61 FR 4036-4037) to eliminate any differences in the way occupational injuries, as opposed to occupational illnesses, were recorded on the forms.

After a thorough review of the comments in the record, however, OSHA has concluded that the proposed approach, which would have eliminated, for recording purposes, the distinction between work-related injuries and illnesses, is not workable in the final rule. The Agency finds that there is a continuing need for separately identifiable information on occupational illnesses and injuries, as well as on certain specific categories of occupational illnesses. The published BLS statistics have included separate estimates of the rate and number of occupational injuries and illnesses for many years, as well as the rate and number of different types of occupational illnesses, and employers, employees, the government, and the public have found this information useful and worthwhile. Separate illness and injury data are particularly useful at the establishment level, where employers and employees can use them to evaluate the establishment's health experience and compare it to the national experience or to the experience of other employers in their industry or their own prior experience. The data are also useful to OSHA personnel performing worksite inspections, who can use this information to identify potential health hazards at the establishment.

Under the final rule, the OSHA 300 form has therefore been modified specifically to collect information on five types of occupational health conditions: musculoskeletal disorders, skin diseases or disorders, respiratory conditions, poisoning, and hearing loss. There is also an "all other illness" column on the Log. To record cases falling into one of these categories, the employer simply enters a check mark in the appropriate column, which will allow these cases to be separately counted to generate establishment-level summary information at the end of the year.

...

In the final rule, two of the illness case columns on the OSHA 300 Log are identical to those on the former OSHA Log: a column to capture cases of skin diseases or disorders and one to capture cases of systemic poisoning. The single column for respiratory conditions on the new OSHA Form 300 will capture data on respiratory conditions that were formerly captured in two separate columns, i.e., the columns for respiratory conditions due to toxic agents (formerly column 7c) and for dust diseases of the lungs (formerly column 7b). Column 7g of the former OSHA Log provided space for data on all other occupational illnesses, and that column has also been continued on the new OSHA 300 Log. On the other hand, column 7e from the former OSHA Log, which captured cases of disorders due to physical agents, is not included on the new OSHA Log form. The cases recorded in former column 7e primarily addressed heat and cold disorders, such as heat stroke and hypothermia; hyperbaric effects, such as caisson disease; and the effects of radiation, including occupational illnesses caused by x-ray exposure, sun exposure and welder's flash. Because space on the form is at a premium, and because column 7e was not used extensively in the past (recorded column 7e cases accounted only for approximately five percent of all occupational illness cases),

OSHA has not continued this column on the new OSHA 300 Log.

...

Column on the Log for Musculoskeletal Disorders

[OSHA has proposed to delay for one year the definition of MSD and the requirement to check the MSD column. However, the same recording criteria applies to musculoskeletal disorders (MSDs) as to all other injuries and illnesses.]

The OSHA 301 Form

Although the final OSHA 300 Log presents information on injuries and illnesses in a condensed format, the final OSHA 301 Incident Record allows space for employers to provide more detailed information about the affected worker, the injury or illness, the workplace factors associated with the accident, and a brief description of how the injury or illness occurred. Many employers use an equivalent workers' compensation form or internal reporting form for the purpose of recording more detailed information on each case, and this practice is allowed under paragraph 1904.29(b)(4) of the final rule.

...

OSHA has also added several items to the OSHA Form 301 that were not on the former OSHA No. 101:

- The date the employee was hired;
- The time the employee began work;
- The time the event occurred;
- Whether the employee was treated at an emergency room; and
- Whether the employee was hospitalized overnight as an in-patient (the form now requires a check box entry rather than the name and address of the hospital).

OSHA concludes that these data fields will provide safety and health professionals and researchers with important information regarding the occurrence of occupational injuries and illnesses. The questions pertaining to what the employee was doing, how the injury or illness occurred, what the injury or illness was, and what object or substance was involved have been reworded somewhat from those contained on the former OSHA No. 101, but do not require employers or employees to provide additional information.

...

The final 301 form contains four questions eliciting case detail information (i.e., what was the employee doing just before the incident occurred?, what happened?, what was the injury or illness?, and what object or substance directly harmed the employee?). The language of these questions on the final 301 form has been modified slightly from that used in the proposed questions to be consistent with the language used on the BLS Survey of Occupational Injuries and Illnesses collection form. The BLS performed extensive testing of the language used in these questions while developing its survey form and has subsequently used these questions to collect data for many years. The BLS has found that the order in which these questions are presented and the wording of the questions on the survey form elicit the most complete answers to the relevant questions. OSHA believes that using the time-tested language and ordering of these four questions will have the same benefits for employers using the OSHA Form 301 as they have had for employers responding to the BLS Annual Survey. Matching the BLS wording and order will also result in benefits for those employers selected to participate in the BLS Annual Survey. To complete the BLS survey forms, employers will only need to copy information from the OSHA Injury and Illness Incident Report to the BLS survey form. This should be easier and less confusing than researching and rewording responses to the questions on two separate forms.

...

Subpart D. Other OSHA injury and illness recordkeeping requirements [p. 6035]

Subpart D of the final rule contains all of the 29 CFR Part 1904 requirements for keeping OSHA injury and illness records that do not actually pertain to entering the injury and illness data on the forms. The nine sections of Subpart D are:
- Section 1904.30, which contains the requirements for dealing with multiple business establishments;
- Section 1904.31, which contains the requirements for determining which employees' occupational injuries and illnesses must be recorded by the employer;
- Section 1904.32, which requires the employer to prepare and post the annual summary;
- Section 1904.33, which requires the employer to retain and update the injury and illness records;
- Section 1904.34, which requires the employer to transfer the records if the business changes owners;
- Section 1904.35, which includes requirements for employee involvement, including employees' rights to access the OSHA injury and illness information;
- Section 1904.36, which prohibits an employer from discriminating against employees for exercising their rights under the Act;
- Section 1904.37, which sets out the state recordkeeping regulations in OSHA approved State-Plan states; and
- Section 1904.38, which explains how an employer may seek a variance from the recordkeeping rule.

Section 1904.30 Multiple Establishments [p. 6035]

Section 1904.30 covers the procedures for recording injuries and illnesses occurring in separate establishments operated by the same business. For many businesses, these provisions are irrelevant because the business has only one establishment. However, many businesses have two or more establishments, and thus need to know how to apply the recordkeeping rule to multiple establishments. In particular, this section applies to businesses where separate work sites create confusion as to where injury and illness records should be kept and when separate records must be kept for separate work locations, or establishments. OSHA recognizes that the recordkeeping system must accommodate operations of this type, and has adopted language in the final rule to provide some flexibility for employers in the construction, transportation, communications, electric and gas utility, and sanitary services industries, as well as other employers with geographically dispersed operations. The final rule provides, in part, that operations are not considered separate establishments unless they continue to be in operation for a year or more. This length-of-site-operation provision increases the chances of discovering patterns of occupational injury and illness, eliminates the burden of creating OSHA 300 Logs for transient work sites, and ensures that useful records are generated for more permanent facilities.

...

The basic requirement of Sec. 1904.30(a) of this final rule states that employers are required to keep separate OSHA 300 Logs for each establishment that is expected to be in business for one year or longer. Paragraph 1904.30(b)(1) states that for short-term establishments, i.e., those that will exist for less than a year, employers are required to keep injury and illness records, but are not required to keep separate OSHA 300 Logs. They may keep one OSHA 300 Log covering all short-term establishments, or may include the short-term establishment records in logs that cover individual company divisions or geographic regions. For example, a construction company with multi-state operations might have separate OSHA 300 Logs for each state to show the injuries and illnesses of its employees engaged in short-term projects, as well as a separate OSHA 300 Log for each construction project expected to last for more than one year. If the same company had only one office location and none of its projects lasted for more than one year, the company would only be required to have one OSHA 300 Log.

Paragraph 1904.30(b)(2) allows the employer to keep records for separate establishments at the business' headquarters or another central location, provided that information can be transmitted from the establishment to headquarters or the central location within 7 days of the occurrence of the injury or illness, and provided that the employer is able to produce and send the OSHA records to each establishment when Sec. 1904.35 or Sec. 1904.40 requires such transmission. The sections of the final rule are consistent with the corresponding provisions of the proposed rule.

Paragraph 1904.30(b)(3) states that each employee must be linked, for recordkeeping purposes, with one of the employer's establishments. Any injuries or illnesses sustained by the employee must be recorded on his or her home establishment's OSHA 300 Log, or on a general OSHA 300 Log for short-term establishments. This provision ensures that all employees are included in a company's records. If the establishment is in an industry classification partially exempted under Sec. 1904.2 of the final rule, records are not required. Under paragraph 1904.30(b)(4), if an employee is injured or made ill while visiting or working at another of the employer's establishments, then the injury or illness must be recorded on the 300 Log of the establishment at which the injury or illness occurred.

...

Recording Injuries and Illnesses Where They Occur

For the vast majority of cases, the place where the injury or illness occurred is the most useful recording location. The events or exposures that caused the case are most likely to be present at that location, so the data are most useful for analysis of that location's records. If the case is recorded at the employee's home base, the injury or illness data have been disconnected from the place where the case occurred, and where analysis of the data may help reveal a workplace hazard. Therefore, OSHA finds that it is most useful to record the injury or illness at the location where the case occurred. Of course, if the injury or illness occurs at another employer's workplace, or while the employee is in transit, the case would be recorded on the OSHA 300 Log of the employee's home establishment.

For cases of illness, two types of cases must be considered. The first is the case of an illness condition caused by an acute, or short term workplace exposure, such as skin rashes, respiratory ailments, and heat disorders. These illnesses generally manifest themselves quickly and can be linked to the workplace where they occur, which is no different than most injury cases. For illnesses that are caused by long-term exposures or which have long latency periods, the illness will most likely be detected during a visit to a physician or other health care professional, and the employee is most likely to report it to his or her supervisor at the home work location.

Recording these injuries and illnesses could potentially present a problem with incidence rate calculations. In many situations, visiting employees are a minority of the workforce, their hours worked are relatively inconsequential, and rates are thus unaffected to any meaningful extent. However, if an employer relies on visiting labor to perform a larger amount of the work, rates could be affected. In these situations, the hours of these personnel should be added to the establishment's hours of work for rate calculation purposes.

Section 1904.31 Covered employees [p. 6037]

Final Rule Requirements and Legal Background Section 1904.31 requires employers to record the injuries and illnesses of all their employees, whether classified as labor, executive, hourly, salaried, part-time, seasonal, or migrant workers. The section also requires the employer to record the injuries and illnesses of employees they supervise on a day-to-day basis, even if these workers are not carried on the employer's payroll. Implementing these requirements requires an understanding of the Act's definitions of "employer" and "employee." The statute defines "employer," in relevant part, to mean "a person engaged in a business affecting interstate commerce who has employees." 29 U.S.C. 652 (5). The term "person" includes "one or more individuals, partnerships, associations, corporations, business trusts, legal representatives, or any organized group of persons." 29 U.S.C. 652 (4). The term "employee" means "an employee of an employer who is employed in a business of his employer which affects interstate commerce." 29 U.S.C. 652(6). Thus, any individual or entity having an employment relationship with even one worker is an employer for purposes of this final rule, and must fulfill the recording requirements for each employee.

The application of the coverage principles in this section presents few issues for employees who are carried on the employer's payroll, because the employment relationship is usually well established in these cases. However, issues sometimes arise when an individual or entity enters into a temporary relationship with a worker. The first question is whether the worker is an employee of the hiring party. If an employment relationship exists, even if temporary in duration, the employee's injuries and illnesses must be recorded on the OSHA 300 Log and 301 form. The second question, arising in connection with employees provided by a temporary help service or leasing agency, is which employer—the host firm or the temporary help service—is responsible for recordkeeping.

Whether an employment relationship exists under the Act is determined in accordance with established common law principles of agency. At common law, a self-employed "independent contractor" is not an employee; therefore, injuries and illnesses sustained by independent contractors are not recordable under the final Recordkeeping rule. To determine whether a hired party is an employee or an independent contractor under the common law test, the hiring party must consider a number of factors, including the degree of control the hiring party asserts over the manner in which the work is done, and the degree of skill and independent judgment the hired party is expected to apply. Loomis Cabinet Co. v. OSHRC, 20 F.3d 938, 942 (9th Cir. 1994).

Other individuals, besides independent contractors, who are not considered to be employees under the OSH Act are unpaid volunteers, sole proprietors, partners, family members of farm employers, and domestic workers in a residential setting. See 29 CFR Sec. 1975.4(b)(2) and Sec. 1975.6 for a discussion of the latter two categories of workers. As is the case with independent contractors, no employment relationship exists between these individuals and the hiring party, and consequently, no recording obligation arises.

A related coverage question sometimes arises when an employer obtains labor from a temporary help service, employee leasing firm or other personnel supply service. Frequently the temporary workers are on the payroll of the temporary help service or leasing firm, but are under the day-to-day supervision of the host party. In these cases, Section 1904.31 places the recordkeeping obligation upon the host, or utilizing, employer. The final rule's allocation of recordkeeping responsibility to the host employer in these circumstances is consistent with the Act for several reasons.

First, the host employer's exercise of day-to-day supervision of the temporary workers and its control over the work environment demonstrates a high degree of control over the temporary workers consistent with the presence of an employment relationship at common law. See Loomis Cabinet Co., 20 F.3d at 942. Thus, the temporary workers will ordinarily be the employees of the party exercising day-to-day control over them, and the supervising party will be their employer.

Even if daily supervision is not sufficient alone to establish that the host party is the employer of the temporary workers, there are other reasons for the final rule's allocation of recordkeeping responsibility. Under the OSH Act, an employer's duties and responsibilities are not limited only to his own employees. Cf. Universal Constr. Co. v. OSHRC, 182 F.3d 726, 728-731 (10th Cir. 1999). Assuming that the host is an employer under the Act (because it has an employment relationship with someone) it reasonably should record the injuries of all employees, whether or not its own, that it supervises on a daily basis. This follows because the supervising employer is in the best position to obtain the necessary injury and illness information due to its control over the worksite and its familiarity with the work tasks and the work environment.

Section 1904.32 Annual Summary [p. 6042]

At the end of each calendar year, section 1904.32 of the final rule requires each covered employer to review his or her OSHA 300 Log for completeness and accuracy and to prepare an Annual Summary of the OSHA 300 Log using the form OSHA 300-A, Summary of Work-Related Injuries and Illnesses, or an equivalent form. The summary must be certified for accuracy and completeness and be posted in the workplace by February 1 of the year following the year covered by the summary. The summary must remain posted until April 30 of the year in which it was posted. Preparing the Annual Summary requires four steps: reviewing the OSHA 300 Log, computing and entering the summary information on the Form 300-A, certification, and posting. First, the employer must review the Log as extensively as necessary to make sure it is accurate and complete.

Second, the employer must total the columns on the Log; transfer them to the summary form; and enter the calendar year covered, the name of the employer, the name and address of the establishment, the average number of employees on the establishment's payroll for the calendar year, and the total hours worked by the covered employees. If there were no recordable cases at the establishment for the year covered, the summary must nevertheless be completed by entering zeros in the total for each column of the OSHA 300 Log. If a form other than the OSHA 300-A is used, as permitted by paragraph 1904.29(b)(4), the alternate form must contain the same information as the OSHA 300-A form and include identical statements concerning employee access to the Log and Summary and employer penalties for falsifying the document as are found on the OSHA 300-A form.

Third, the employer must certify to the accuracy and completeness of the Log and Summary, using a two-step process. The person or persons who supervise the preparation and maintenance of the Log and Summary (usually the person who keeps the OSHA records) must sign the certification statement on the form, based on their direct knowledge of the data on which it was based. Then, to ensure greater awareness and accountability of the recordkeeping process, a company executive, who may be an owner, a corporate officer, the highest ranking official working at the establishment, or that person's immediate supervisor, must also sign the form to certify to its accuracy and completeness. Certification of the summary attests that the individual making the certification has a reasonable belief, derived from his or her knowledge of the process by which the information in the Log was reported and recorded, that the Log and summary are "true" and "complete."

Fourth, the Summary must be posted no later than February 1 of the year following the year covered in the Summary and remain posted until April 30 of that year in a conspicuous place where notices are customarily posted. The employer must ensure that the Summary is not defaced or altered during the 3 month posting period.

Changes from the former rule. Although the final rule's requirements for preparing the Annual Summary are generally similar to those of the former rule, the final rule incorporates four important changes that OSHA believes will strengthen the recordkeeping process by ensuring greater completeness and accuracy of the Log and Summary, providing employers and employees with better information to understand and evaluate the injury and illness data on the Annual Summary, and facilitating greater employer and employee awareness of the recordkeeping process.

1. **Company Executive Certification of the Annual Summary.** The final rule carries forward the proposed rule's requirement for certification by a higher ranking company official, with minor revision. OSHA concludes that the company executive certification process will ensure greater completeness and accuracy of the Summary by raising accountability for OSHA recordkeeping to a higher managerial level than existed under the former rule. OSHA believes that senior management accountability is essential if the Log and Annual Summary are to be accurate and complete. The integrity of the OSHA recordkeeping system, which is relied on by the BLS for national injury and illness statistics, by OSHA and employers to understand hazards in the workplaces, by employees to assist in the identification and control of the hazards identified, and by safety and health professionals everywhere to analyze trends, identify emerging hazards, and develop solutions, is essential to these objectives. Because OSHA cannot oversee the preparation of the Log and Summary at each establishment and cannot audit more than a small sample of all covered employers' records, this goal is accomplished by requiring employers or company executives to certify the accuracy and completeness of the Log and Summary.

The company executive certification requirement imposes different obligations depending on the structure of the company. If the company is a sole proprietorship or partnership, the certification may be made by the owner. If the company is a corporation, the certification may be made by a corporate officer. For any management structure, the certification may be made by the highest ranking company official working at the establishment covered by the Log (for example, the plant manager or site supervisor), or the latter official's supervisor (for example, a corporate or regional director who works at a different establishment, such as company headquarters).

The company executive certification is intended to ensure that a high ranking company official with responsibility for the recordkeeping activity and the authority to ensure that the recordkeeping function is performed appropriately has examined the records and has a reasonable belief, based on his or her knowledge of that process, that the records are accurate and complete.

The final rule does not specify how employers are to evaluate their recordkeeping systems to ensure their accuracy and completeness or what steps an employer must follow to certify the accuracy and completeness of the Log and Summary with confidence. However, to be able to certify that one has a reasonable belief that the records are complete and accurate would suggest, at a minimum, that the certifier is familiar with OSHA's recordkeeping requirements, and the company's recordkeeping practices and policies, has read the Log and Summary, and has obtained assurance from the staff responsible for maintaining the records (if the certifier does not personally keep the records) that all of OSHA's requirements have been met and all practices and policies followed. In most if not all cases, the certifier will be familiar with the details of some of the injuries and illnesses that have occurred at the establishment and will therefore be able to spot check the OSHA 300 Log to see if those cases have been entered correctly. In many cases, especially in small to medium establishments, the certifier will be aware of all of the injuries and illnesses that have been reported at the establishment and will thus be able to inspect the forms to make sure all of the cases that should have been entered have in fact been recorded.

The certification required by the final rule may be made by signing and dating the certification section of the OSHA 300-A form, which replaces the summary portion of the former OSHA 200 form, or by signing and dating a separate certification statement and appending it to the OSHA Form 300-A. A separate certification statement must contain the identical penalty warnings and employee access information as found on the OSHA Form 300-A. A separate statement may be needed when the certifier works at another location and the certification is mailed or faxed to the location where the Summary is posted.

The certification requirement modifies the certification provision of the former rule (former paragraph 1904.5(c)), which required a certification of the Annual Summary by the employer or an officer or employee who supervised the preparation of the Log and Summary. The former rule required that individual to sign and date the year-end summary on the OSHA Form 200 and to certify that the summary was true and complete. Alternatively, the recordkeeper could, under the former rule, sign a separate certification statement rather than signing the OSHA form.

...

2. **Number of employees and hours worked.** Injury and illness records provide a valuable tool for OSHA, employers, and employees to determine where and why injuries and illnesses occur, and they are crucial in the development of prevention strategies. The final rule requires employers to include in the Annual Summary (the OSHA Form 300-A) the annual average number of employees covered by the Log and the total hours worked by all covered employees.

...

OSHA's view is that the value of the total hours worked and average number of employees information requires its inclusion in the Summary, and the final rule reflects this determination. Having this information will enable employers and employees to calculate injury and illness incidence rates, which are widely regarded as the best statistical measure for the purpose of comparing an establishment's injury and illness experience with national statistics, the records of other establishment, or trends over several years. Having the data available on the Form 300-A will also make it easier for the employer to respond to government requests for the data, which occurs when the BLS and OSHA collect the data by mail, and when an OSHA or State inspector visits the facility. In particular, it will be easier for the employer to provide the OSHA inspector with the hours worked and employment data for past years.

OSHA does not believe that this requirement creates the time and cost burden some commenters to the record suggested, because the information is readily available in payroll or other records required to be kept for other purposes, such as income tax, unemployment, and workers' compensation insurance records. For the approximately 10% of covered employers who participate in the BLS's Annual Survey of Occupational Injuries and Illnesses, there will be no additional burden because this information must already be provided to the BLS. Moreover, the rule does not require employers to use any particular method of calculating the totals, thus providing employers who do not maintain certain records—for example the total hours worked by salaried employees—or employers without sophisticated computer systems, the flexibility to obtain the information in any reasonable manner that meets the objectives of the rule. Employers who do not have the ability to generate precise numbers can use various estimation methods. For example, employers typically must estimate hours worked for workers who are paid on a commission or salary basis. Additionally, the instructions for the OSHA 300-A Summary form include a worksheet to help the employer calculate the total numbers of hours worked and the average number of employees.

3. **Extended posting period.** The final rule's requirement increasing the summary Form 300-A posting period from one month to three months is intended to raise employee awareness of the recordkeeping process (especially that of new employees hired during the posting period) by providing greater access to the previous year's summary without having to request it from management. The additional two months

of posting will triple the time employees have to observe the data without imposing additional burdens on the employer. The importance of employee awareness of and participation in the recordkeeping process is discussed in the preamble to sections 1904.35 and 1904.36. The requirement to post the Summary on February 1 is unchanged from the posting date required by the former rule.

...

After a review of all the comments received and its own extensive experience with the recordkeeping system and its implementation in a variety of workplaces, OSHA has decided to adopt a 3-month posting period. The additional posting period will provide employees with additional opportunity to review the summary information, raise employee awareness of the records and their right to access them, and generally improve employee participation in the recordkeeping system without creating a "wallpaper" posting of untimely data. In addition, OSHA has concluded that any additional burden on employers will be minimal at best and, in most cases, insignificant. All the final rule requires the employer to do is to leave the posting on the bulletin board instead of removing it at the end of the one-month period. In fact, many employers preferred to leave the posting on the bulletin board for longer than the required one-month period in the past, simply to provide workers with the opportunity to view the Annual Summary and increase their awareness of the recordkeeping system in general and the previous year's injury and illness data in particular. OSHA agrees that the 3-month posting period required by the final rule will have these benefits which, in the Agency's view, greatly outweigh any minimal burden that may be associated with such posting. The final rule thus requires that the Summary be posted from February 1 until April 30, a period of three months; OSHA believes that the 30 days in January will be ample, as it has been in the past, for preparing the current year's Summary preparatory to posting.

4. **Review of the records.** The provisions of the final rule requiring the employer to review the Log entries before totaling them for the Annual Summary are intended as an additional quality control measure that will improve the accuracy of the information in the Annual Summary, which is posted to provide information to employees and is also used as a data source by OSHA and the BLS. Depending on the size of the establishment and the number of injuries and illnesses on the OSHA 300 Log, the employer may wish to cross-check with any other relevant records to make sure that all the recordable injuries and illnesses have been included on the Summary. These records may include workers' compensation injury reports, medical records, company accident reports, and/or time and attendance records.

...

Section 1904.33 Retention and Updating [p. 6048]

Section 1904.33 of the final rule deals with the retention and updating of the OSHA Part 1904 records after they have been created and summarized. The final rule requires the employer to save the OSHA 300 Log, the Annual Summary, and the OSHA 301 Incident Report forms for five years following the end of the calendar year covered by the records. The final rule also requires the employer to update the entries on the OSHA 300 Log to include newly discovered cases and show changes that have occurred to previously recorded cases. The provisions in section 1904.33 state that the employer is not required to update the 300A Annual Summary or the 301 Incident Reports, although the employer is permitted to update these forms if he or she wishes to do so.

As this section makes clear, the final rule requires employers to retain their OSHA 300 and 301 records for five years following the end of the year to which the records apply. Additionally, employers must update their OSHA 300 Logs under two circumstances. First, if the employer discovers a recordable injury or illness that has not previously been recorded, the case must be entered on the forms. Second, if a previously recorded injury or illness turns out, based on later information, not to have been recorded properly, the employer must modify the previous entry. For example, if the description or outcome of a case changes (a case requiring medical treatment becomes worse and the employee must take days off work to recuperate), the employer must remove or line out the original entry and enter the new information. The employer also has a duty to enter the date of an employee's return to work or the date of an injured worker's death on the Form 301; OSHA considers the entering of this information an integral part of the recordkeeping for such cases. The Annual Summary and the Form 301 need not be updated, unless the employer wishes to do so. The requirements in this section 1904.33 do not affect or supersede any longer retention periods specified in other OSHA standards and regulations, e.g., in OSHA health standards such as Cadmium, Benzene, or Lead (29 CFR 1910.1027, 1910.1028, and 1910.1025, respectively).

...

Section 1904.34 Change in Business Ownership [p. 6050]

Section 1904.34 of the final rule addresses the situation that arises when a particular employer ceases operations at an establishment during a calendar year, and the establishment is then operated by a new employer for the remainder of the year. The phrase "change of ownership," for the purposes of this section, is relevant only to the transfer of the responsibility to make and retain OSHA-required injury and illness records. In other words, if one employer, as defined by the OSH Act, transfers ownership of an establishment to a different employer, the new entity becomes responsible for retaining the previous employer's past OSHA-required records and for creating all new records required by this rule.

The final rule requires the previous owner to transfer these records to the new owner, and it limits the recording and recordkeeping responsibilities of the previous employer only to the period of the prior owner. Specifically, section 1904.34 provides that if the business changes ownership, each employer is responsible for recording and reporting work-related injuries and illnesses only for that period of the year during which each employer owned the establishment. The selling employer is required to transfer his or her Part 1904 records to the new owner, and the new owner must save all records of the establishment kept by the prior owner. However, the new owner is not required to update or correct the records of the prior owner, even if new information about old cases becomes available.

...

Sections 1904.35 Employee Involvement, and 1904.36, Prohibition Against Discrimination [p. 6050]

One of the goals of the final rule is to enhance employee involvement in the recordkeeping process. OSHA believes that employee involvement is essential to the success of all aspects of an employer's safety and health program. This is especially true in the area of recordkeeping, because free and frank reporting by employees is the cornerstone of the system. If employees fail to report their injuries and illnesses, the "picture" of the workplace that the employer's OSHA forms 300 and 301 reveal will be inaccurate and misleading. This means, in turn, that employers and employees will not have the information they need to improve safety and health in the workplace. Section 1904.35 of the final rule therefore establishes an affirmative requirement for employers to involve their employees and employee representatives in the recordkeeping process. The employer must inform each employee of how to report an injury or illness, and must provide limited access to the injury and illness records for employees and their representatives. Section 1904.36 of the final rule makes clear that Sec. 11(c) of the Act prohibits employers from discriminating against employees for reporting work-related injuries and illnesses. Section 1904.36 does not create a new obligation on employers. Instead, it clarifies that the OSH Act's anti-discrimination protection applies to employees who seek to participate in the recordkeeping process.

Under the employee involvement provisions of the final rule, employers are required to let employees know how and when to report work-related injuries and illnesses. This means that the employer must establish a procedure for the reporting of work-related injuries and illnesses and train its employees to use that procedure. The rule does not specify how the employer must accomplish these objectives. The size of the workforce, employees' language proficiency and literacy levels, the workplace culture, and other factors will determine what will be effective for any particular workplace.[footnote omitted]

Employee involvement also requires that employees and their representatives have access to the establishment's injury and illness records. Employee involvement is further enhanced by other parts of the final rule, such as the extended posting period provided in section 1904.32 and the access statements on the new 300 and 301 forms.

...

Employee access to OSHA injury and illness records The Part 1904 final rule continues OSHA's long-standing policy of allowing employees and their representatives access to the occupational injury and illness information kept by their employers, with some limitations. However, the final rule includes several changes to improve employees' access to the information, while at the same time implementing several measures to protect the privacy interests of injured and ill employees. Section 1904.35 requires an employer covered by the Part 1904 regulation to provide limited access to the OSHA recordkeeping forms to current and former employees, as well as to two types of employee representatives. The first is a personal representative of an employee or former employee, who is a person that the employee or former employee designates, in writing, as his or her personal representative, or is the legal representative of a deceased or legally incapacitated employee or former employee. The second is an authorized employee representative, which is defined as an authorized collective bargaining agent of one or more employees working at the employer's establishment.

Section 1904.35 accords employees and their representatives three separate access rights. First, it gives any employee, former employee, personal representative, or authorized employee representative the right to a copy of the current OSHA 300 Log, and to any stored OSHA 300 Log(s), for any establishment in which the employee or former employee has worked. The employer must provide one free copy of the OSHA 300 Log(s) by the end of the next business day. The employee, former employee, personal representative or authorized employee representative is not entitled to see, or to obtain a copy of, the confidential list of names and case numbers for privacy cases. Second, any employee, former employee, or personal representative is entitled to one free copy of the OSHA 301 Incident Report describing an injury or illness to that employee by the end of the next business day. Finally, an authorized employee representative is entitled to copies of the right-hand portion of all OSHA 301 forms for the establishment(s) where the agent represents one or more employees under a collective bargaining agreement. The right-hand portion of the 301 form contains the heading "Tell us about the case," and elicits information about how the injury occurred, including the employee's actions just prior to the incident, the materials and tools involved, and how the incident occurred, but does not contain the employee's name. No information other than that on the right-hand portion of the form may be disclosed to an authorized employee representative. The employer must provide the authorized employee representative with one free copy of all the 301 forms for the establishment within 7 calendar days.

Employee privacy is protected in the final rule in paragraphs 1904.29(b)(7) to (10). Paragraph 1904.29(b)(7) requires the employer to enter the words "privacy case" on the OSHA 300 Log, in lieu of the employee's name, for recordable privacy concern cases involving the following types of injuries and illnesses: (i) an injury from a needle or sharp object contaminated by another person's blood or other potentially infectious material; (ii) an injury or illness to an intimate body part or to the reproductive system; (iii) an injury or illness resulting from a sexual assault; (iv) a mental illness; (v) an illness involving HIV, hepatitis; or tuberculosis, or (vi) any other illness, if the employee independently and voluntarily requests that his or her name not be entered on the log. Musculoskeletal disorders (MSDs) are not considered privacy concern cases, and thus employers are required to enter the names of employees experiencing these disorders on the log. The employer must keep a separate, confidential list of the case numbers and employee names for privacy cases. The employer may take additional action in privacy concern cases if warranted. Paragraph 1904.29(b)(9) allows the employer to use discretion in describing the nature of the injury or illness in a privacy concern case, if the employer has a reasonable basis to believe that the injured or ill employee may be identified from the records even though the employee's name has been removed. Only the six types of injuries and illnesses listed in Paragraph 1904.29(b)(7) may be considered privacy concern cases, and thus the additional protection offered by paragraph 1904.29(b)(9) applies only to such cases.

Paragraph 1904.29(b)(10) protects employee privacy if the employer decides voluntarily to disclose the OSHA 300 and 301 forms to persons other than those who have a mandatory right of access under the final rule. The paragraph requires the employer to remove or hide employees' names or other personally identifying information before disclosing the forms to persons other than government representatives, employees, former employees or authorized representatives, as required by paragraphs 1904.40 and 1904.35, except in three cases. The employer may disclose the forms, complete with personally identifying information, (2) only: (i) to an auditor or consultant hired by the employer to evaluate the safety and health program; (ii) to the extent necessary for processing a claim for workers' compensation or other insurance benefits; or (iii) to a public health authority or law enforcement agency for uses and disclosures for which consent, an authorization, or opportunity to agree or object is not required under section 164.512 of the final rule on Standards for Privacy of Individually Identifiable Health Information, 45 CFR 164.512.

Section 1904.37 State Recordkeeping Regulations [p. 6060]

Section 1904.37 addresses the consistency of the recordkeeping and reporting requirements between Federal OSHA and those States where occupational safety and health enforcement is provided by an OSHA-approved State Plan. Currently, in 21 States and 2 territories, the State government has been granted authority to operate a State OSHA Plan covering both the private and public (State and local government) sectors under section 18 of the OSH Act (see the State Plan section of this preamble for a listing of these States). Two additional States currently operate programs limited in scope to State and local government employees only. State Plans, once approved, operate under authority of State law and provide programs of standards, regulations and enforcement which must be "at least as effective" as the Federal program. (State Plans must extend their coverage to State and local government employees, workers not otherwise covered by Federal OSHA regulations.) Section 1904.37 of the final rule describes what State Plan recordkeeping requirements must be identical to the Federal requirements, which State regulations may be different, and provides cross references to the State Plan regulations codified in Section 1902.3(k), 1952.4, and 1956.10(i). The provisions of Subpart A of 29 CFR part 1952 specify the regulatory discretion of the State Plans in general, and section 1952.4 spells out the regulatory discretion of the State Plans specifically for the recordkeeping regulation.

...

Under Section 18 of the OSH Act, a State Plan must require employers in the State to make reports to the Secretary in the same manner and to the same extent as if the Plan were not in effect. Final section 1904.37 makes clear that States with approved State Plans must promulgate new regulations that are substantially identical to the final Federal rule. State Plans must have recording and reporting regulations that impose identical requirements for the recordability of occupational injuries and illnesses and the manner in which they are entered. These requirements must be the same for employers in all the States, whether under Federal or State Plan jurisdiction, and for State and local government employers covered only through State Plans, to ensure that the occupational injury and illness data for the entire nation are uniform and consistent so that statistics that allow comparisons between the States and between employers located in different States are created.

For all of the other requirements of the Part 1904 regulations, the regulations adopted by the State Plans may be more stringent than or supplemental to the Federal regulations, pursuant to paragraph 1952.4(b). This means that the States' recording and reporting regulations could differ in several ways from their Federal Part 1904 counterparts. For example, a State Plan could require employers to keep records for the State, even though those employers are within an industry exempted by the Federal rule. A State Plan could also require employers to keep additional supplementary injury and illness information, require employers to report fatality and multiple hospitalization incidents within a shorter timeframe than Federal OSHA does, require other types of incidents to be reported as they occur, or impose other requirements. While a State Plan must assure that all employee participation and access rights are assured, the State may provide broader access to records by employees and their representatives. However, because of the unique nature of the national recordkeeping program, States must secure Federal OSHA approval for these enhancements.

...

OSHA understands the advantages to multi-State businesses of following identical OSHA rules in both Federal and State Plan jurisdictions, but also recognizes the value of allowing the States to have different rules to meet the needs of each State, as well as the States' right to impose different rules as long as the State rule is at least as effective as the Federal rule. Accordingly, the Part 1904 rules impose identical requirements where they are needed to create consistent injury and illness statistics for the nation and allows the States to impose supplemental or more stringent requirements where doing so will not interfere with the maintenance of comprehensive and uniform national statistics on workplace fatalities, injuries and illnesses.

Section 1904.38 Variances From the Recordkeeping Rule [p. 6061]

Section 1904.38 of the final rule explains the procedures employers must follow in those rare instances where they request that OSHA grant them a variance or exception to the recordkeeping rules in Part 1904.

The rule contains these procedures to allow an employer who wishes to maintain records in a manner that is different from the approach required by the rules in Part 1904 to petition the Assistant Secretary.

Section 1904.8 allows the employer to apply to the Assistant Secretary for OSHA and request a Part 1904 variance if he or she can show that the alternative recordkeeping system: (1) Collects the same information as this Part requires; (2) Meets the purposes of the Act; and (3) Does not interfere with the administration of the Act.

The variance petition must include several items, namely the employer's name and address; a list of the State(s) where the variance would be used; the addresses of the business establishments involved; a description of why the employer is seeking a variance; a description of the different recordkeeping procedures the employer is proposing to use; a description of how the employer's proposed procedures will collect the same information as would be collected by the Part 1904 requirements and achieve the purpose of the Act; and a statement that the employer has informed its employees of the petition by giving them or their authorized representative a copy of the petition and by posting a statement summarizing the petition in the same way notices are posted under paragraph 1903.2(a).

The final rule describes how the Assistant Secretary will handle the variance petition by taking the following steps:

- The Assistant Secretary will offer employees and their authorized representatives an opportunity to comment on the variance petition. The employees and their authorized representatives will be allowed to submit written data, views, and arguments about the petition.
- The Assistant Secretary may allow the public to comment on the variance petition by publishing the petition in the Federal Register. If the petition is published, the notice will establish a public comment period and may include a schedule for a public meeting on the petition.

- After reviewing the variance petition and any comments from employees and the public, the Assistant Secretary will decide whether or not the proposed recordkeeping procedures will meet the purposes of the Act, will not otherwise interfere with the Act, and will provide the same information as the Part 1904 regulations provide. If the procedures meet these criteria, the Assistant Secretary may grant the variance subject to such conditions as he or she finds appropriate.
- If the Assistant Secretary grants the variance petition, OSHA will publish a notice in the Federal Register to announce the variance. The notice will include the practices the variance allows, any conditions that apply, and the reasons for allowing the variance.

The final rule makes clear that the employer may not use the proposed recordkeeping procedures while the Assistant Secretary is processing the variance petition and must wait until the variance is approved. The rule also provides that, if the Assistant Secretary denies the petition, the employer will receive notice of the denial within a reasonable time and establishes that a variance petition has no effect on the citation and penalty for a citation that has been previously issued by OSHA and that the Assistant Secretary may elect not to review a variance petition if it includes an element which has been cited and the citation is still under review by a court, an Administrative Law Judge (ALJ), or the OSH Review Commission.

The final rule also states that the Assistant Secretary may revoke a variance at a later date if the Assistant Secretary has good cause to do so, and that the procedures for revoking a variance will follow the same process as OSHA uses for reviewing variance petitions. Except in cases of willfulness or where necessary for public safety, the Assistant Secretary will: Notify the employer in writing of the facts or conduct that may warrant revocation of a variance and provide the employer, employees, and authorized employee representatives with an opportunity to participate in the revocation procedures.

...

Like the former variance section of the rule, the final rule does not specifically note that the states operating OSHA-approved state plans are not permitted to grant recordkeeping variances. Paragraph (b) of former section 1952.4, OSHA's rule governing the operation of the State plans, prohibited the states from granting variances, and paragraph (c) of that rule required the State plans to recognize any Federal recordkeeping variances. The same procedures continue to apply to variances under section 1904.37 and section 1952.4 of this final rule. OSHA has not included the provisions from these two sections in the variance sections of this recordkeeping rule, because doing so would be repetitive.

The final rule adds several provisions to those of the former rule. They include (1) the identification of petitioning employers' pending citations in State plan states, (2) the discretion given to OSHA not to consider a petition if a citation on the same subject matter is pending, (3) the clarification that OSHA may provide additional notice via the Federal Register and opportunity for comment, (4) the clarification that variances have only prospective effect, (5) the opportunity of employees and their representatives to participate in revocation procedures, and (6) the voiding of all previous variances and exceptions.

Variance procedures were not discussed in the Recordkeeping Guidelines (Ex. 2), nor have there been any letters of interpretations or OSHRC or court decisions on recordkeeping variances. As noted in the proposal, at 61 FR 4039, only one recordkeeping variance has ever been granted by OSHA. This variance was granted to AT&T and subsequently expanded to its Bell subsidiaries to enable them to centralize records maintenance for workers in the field.

...

Subpart E. Reporting Fatality, Injury and Illness Information to the Government [p. 6062]

Subpart E of this final rule consolidates those sections of the rule that require employers to give recordkeeping information to the government. In the proposed rule, these sections were not grouped together. OSHA believes that grouping these sections into one Subpart improves the overall organization of the rule and will make it easier for employers to find the information when needed. The four sections of this subpart of the final rule are:

(a) Section 1904.39, which requires employers to report fatality and multiple hospitalization incidents to OSHA.

(b) Section 1904.40, which requires an employer to provide his or her occupational illness and injury records to a government inspector during the course of a safety and health inspection.

(c) Section 1904.41, which requires employers to send their occupational illness and injury records to OSHA when the Agency sends a written request asking for specific types of information.

(d) Section 1904.42, which requires employers to send their occupational illness and injury records to the Bureau of Labor Statistics (BLS) when the BLS sends a survey form asking for information from these records.

Each of these sections, and the record evidence pertaining to them, is discussed below.

Section 1904.39 Reporting Fatality or Multiple Hospitalization Incidents to OSHA [p. 6062]

...

Making oral reports of fatalities or multiple hospitalization incidents and the OSHA 800 number.

...

It is essential for OSHA to speak promptly to any employer whose employee(s) have experienced a fatality or multiple hospitalization incident to determine whether the Agency needs to begin an investigation. Therefore, the final rule does not permit employers merely to leave a message on an answering machine, send a fax, or transmit an e-mail message. None of these options allows an Agency representative to interact with the employer to clarify the particulars of the catastrophic incident. Additionally, if the Area Office were closed for the weekend, a holiday, or for some other reason, OSHA might not learn of the incident for several days if electronic or facsimile transmission were permitted. Paragraph 1904.39(b)(1) of the final rule makes this clear.

As noted, OSHA allows the employer to report a fatality or multiple hospitalization incident by speaking to an OSHA representative at the local Area Office either on the phone or in person, or by using the 800 number. This policy gives the employer flexibility to report using whatever mechanism is most convenient. The employer may use whatever method he or she chooses, at any time, as long as he or she is able to speak in person to an OSHA representative or the 800 number operator. Therefore, there is no need to define business hours or otherwise add additional information about when to use the 800 number; it is always an acceptable option for complying with this reporting requirement.

This final rule also includes the 800 number in the text of the regulation. OSHA has decided to include the number in the regulatory text at this time to provide an easy reference for employers. OSHA will also continue to include the 800 number in any interpretive materials, guidelines or outreach materials that it publishes to help employers comply with the reporting requirement.

...

Motor vehicle and public transportation accidents.

...

[OSHA believes] that there is no need for an employer to report a fatality or multiple hospitalization incident when OSHA is clearly not going to make an investigation. When a worker is killed or injured in a motor vehicle accident on a public highway or street, OSHA is only likely to investigate the incident if it occurred in a highway construction zone. Likewise, when a worker is killed or injured in an airplane crash, a train wreck, or a subway accident, OSHA does not investigate, and there is thus no need for the employer to report the incident to OSHA. The text of paragraphs 1904.39(b)(3) and (4) of the final rule clarifies that an employer is not required to report these incidents to OSHA. These incidents are normally investigated by other agencies, including local transit authorities, local or State police, State transportation officials, and the U.S. Department of Transportation.

However, although there is no need to report these incidents to OSHA under the 8-hour reporting requirement, any fatalities and hospitalizations caused by motor vehicle accidents, as well as commercial or public transportation accidents, are recordable if they meet OSHA's recordability criteria. These cases should be captured by the Nation's occupational fatality and injury statistics and be included on the employer's injury and illness forms. The statistics need to be complete, so that OSHA, BLS, and the public can see where and how employees are being made ill, injured and killed. Accordingly, the final rule includes a sentence clarifying that employers are still required to record work-related fatalities and injuries that occur as a result of public transportation accidents and injuries.

Although commenters are correct that OSHA only rarely investigates motor vehicle accidents, the Agency does investigate motor vehicle accidents that occur at street or highway construction sites. Such accidents are of concern to the Agency, and OSHA seeks to learn new ways to prevent these accidents and protect employees who are exposed to them. For example, OSHA is currently participating in a Local Emphasis Program in the State of New Jersey that is designed to protect highway construction workers who are exposed to traffic hazards while performing construction work. Therefore, the final rule provides provisions that require an employer to report a fatality or multiple hospitalization incident that occurs in a construction zone on a public highway or street.

...

Section 1904.40 Providing Records to Government Representatives [p. 6065]

Under the final rule, employers must provide a complete copy of any records required by Part 1904 to an authorized government representative, including the Form 300 (Log), the Form 300A(Summary), the confidential listing of privacy concern cases along with the names of the injured or ill privacy case workers, and the Form 301 (Incident Report), when the representative asks for the records during a workplace safety and health inspection. This requirement is unchanged from the corresponding requirement in OSHA's former recordkeeping rule.

However, the former rule combined the requirements governing both government inspectors' and employers' rights of access to the records into a single section, section 1904.7 "Access to Records." The final rule separates the two. It places the requirements governing access to the records by government inspectors in Subpart E, along with other provisions requiring employers to submit their occupational injury and illness records to the government or to provide government personnel access to them. Provisions for employee access to records are now in section 1904.35, Employee Involvement, in Subpart D of this final rule.

The final regulatory text of paragraph (a) of section 1904.40 requires an employer to provide an authorized government representative with records kept under Part 1904 within four business hours. As stated in paragraph 1904.40(b)(1), the authorized government representatives who have a right to obtain the Part 1904 records are a representative of the Secretary of Labor conducting an inspection or investigation under the Act, a representative of the Secretary of Health and Human Services (including the National Institute for Occupational Safety and Health (NIOSH) conducting an investigation under Section 20(b) of the Act, or a representative of a State agency responsible for administering a State plan approved under section 18 of the Act. The government's right to ask for such records is limited by the jurisdiction of that Agency. For example, a representative of an OSHA approved State plan could only ask for the records when visiting an establishment within that state.

The final rule allows the employer to take into account difficulties that may be encountered if the records are kept at a location in a different time zone from the establishment where the government representative has asked for the records. If the employer maintains the records at a location in a different time zone, OSHA will use the business hours of the establishment at which the records are located when calculating the deadline, as permitted by paragraph 1904.40(b)(2).

...

Government Representatives. Each employer shall provide, upon a request made in person or in writing, copies of the OSHA Forms 300 and 301 or equivalents, and year-end summaries for their own employees, and injury and illness records for "subcontractor employees" as required under this Part to any authorized representative of the Secretary of Labor or Secretary of Health and Human Services or to any authorized representative of a State accorded jurisdiction for occupational safety and health for the purposes of carrying out the Act.

(1) When the request is made in person, the information must be provided in hard copy (paper printout) within 4 hours. If the information is being transmitted to the establishment from some other location, using telefax or other electronic transmission, the employer may provide a copy to the government representative present at the establishment or to the government representative's office.

(2) When the request is made in writing, the information must be provided within 21 days of receipt of the written request, unless the Secretary requests otherwise.

...

Privacy of medical records.

...

This section of the final rule does not give unfettered access to the records by the public, but simply allows a government inspector to use the records during the course of a safety and health inspection. As discussed above in the section covering access to the records for employees, former employees, and employee representatives (Section 1904.35), OSHA does not consider the Forms 300 and 301 to be medical records, for the following reasons. First, they do not have to be completed by a physician or other licensed health care professional. Second, they do not contain the detailed diagnostic and treatment information usually found in medical records. Finally, the injuries and illnesses found in the records are usually widely known among other employees at the workplace where the injured or ill worker works; in fact, these co-workers may even have witnessed the accident that gave rise to the injury or illness.

OSHA does not agree that its inspectors should be required to obtain permission from all injured or ill employees before accessing the full records. Gaining this permission would make it essentially impossible to obtain full access to the records, which is needed to perform a meaningful workplace investigation. For example, an inspector would not be able to obtain the names of employees who were no longer working for the company to perform follow-up interviews about the specifics of their injuries and illnesses. The names of the injured or ill workers are needed to allow the government inspector to interview the injured and ill workers and determine the hazardous circumstances that led to their injury or illness. The government inspector may also need the employee's names to access personnel and medical records if needed (medical records can only be accessed after the inspector obtains a medical access order). Additionally, refusing the inspector access to the names of the injured and ill workers would effectively prohibit any audit of the Part 1904 records by the government, a practice necessary to verify the accuracy of employer recordkeeping in general and to identify problems that employers may be having in keeping records under OSHA's recordkeeping rules. Adopting the inefficient access method suggested by these commenters would also place a substantial administrative burden on the employer, the employees, and the government. Further, since OSHA inspectors do not allow others to see the medical records they have accessed, the privacy of employees is not compromised by CSHO access to the records.

Time for response to requests for records. Paragraphs 1904.40(a) and (b) of the final rule require records to be made available to a government inspector within 4 business hours of an oral request for the records, using the business hours of the establishment at which the records are located.

...

OSHA has concluded that 4 hours is a reasonable and workable length of time for employers to respond to governmental requests for records. The 4-hour time period for providing records from a centralized source strikes a balance between the practical limitations inherent in record maintenance and the government official's need to obtain these records and use the information to conduct a workplace inspection.

...

OSHA believes that it is essential for employers to have systems and procedures that can produce the records within the 4-hour time. However, the Agency realizes that there may be unusual or unique circumstances where the employer cannot comply. For example, if the records are kept by a health care professional and that person is providing emergency care to an injured worker, the employer may need to delay production of the records. In such a situation, the OSHA inspector may allow the employer additional time.

If a government representative requests records of an establishment, but those records are kept at another location, the 4-hour period can be measured in accordance with the normal business hours at the location where the records are being kept.

...

OSHA has designed the final rule to give each employer considerable flexibility in maintaining records. It permits an employer to centralize its records, to use computer and facsimile technologies, and to hire a third party to keep its records. However, an employer who chooses these options must also ensure that they are sufficiently reliable to comply with this rule. In other words, the flexibility provided to employers for recordkeeping must not impede the Agency's ability to obtain and use the records.

Provide copies.

...

OSHA's experience has been that the vast majority of employers willingly provide copies to government representatives during safety and health inspections. Making copies is a routine office function in almost every modern workplace. With the widespread availability of copying technology, most workplaces have copy machines on-site or readily available. The cost of providing copies is minimal, usually less than five cents per copy. In addition, the government representative needs to obtain copes of records promptly, so that he or she can analyze the data and identify workplace hazards. Therefore, in this final rule, OSHA requires the employer to provide copies of the records requested to authorized government representatives.

Section 1904.41 Annual OSHA Injury and Illness Survey of Ten or More Employers [p. 6069]

Section 1904.41 of this final rule replaces section 1904.17, "Annual OSHA Injury and Illness Survey of Ten or More Employers," of the former rule issued on February 11, 1997. The final rule does not change the contents or policies of the corresponding section of the former rule in any way. Instead, the final rule simply rephrases the language of the former rule in the plain language question-and-answer format used in the rest of this rule.

...

Section 1904.42 Requests From the Bureau of Labor Statistics for Data [p. 6069]

...

Section 1904.42 of the final rule derives from the subpart of the former rule titled "Statistical Reporting of Occupational Injuries and Illnesses." The former rule described the Bureau of Labor Statistics annual survey of occupational injuries and illnesses, discussed the duty of employers to answer the survey, and explained the effect of the BLS survey on the States operating their own State plans.

Both OSHA and the BLS collect occupational injury and illness information, each for separate purposes. The BLS collects data from a statistical sample of employers in all industries and across all size classes, using the data to compile the occupational injury and illness statistics for the Nation. The Bureau gives each respondent a pledge of confidentiality (as it does on all BLS surveys), and the establishment-specific injury and illness data are not shared with the public, other government agencies, or OSHA. The BLS's sole purpose is to create statistical data.

OSHA collects data from employers from specific size and industry classes, but collects from each and every employer within those parameters. The establishment-specific data collected by OSHA are used to administer OSHA's various programs and to measure the performance of those programs at individual workplaces.

...

Paragraph 1904.42(a) states the general obligation of employers to report data to the BLS or a BLS designee. Paragraph 1904.42(b)(1) states that some employers will receive a BLS survey form and others will not, and that the employer should not send data unless asked to do so. Paragraph 1904.42(b)(2) directs the employer to follow the instructions on the survey form when completing the information and return it promptly.

Paragraph 1904.42(b)(3) of this final rule notes that the BLS is authorized to collect data from all employers, even those who would otherwise be exempt, under section 1904.1 to section 1904.3, from keeping OSHA injury and illness records. This enables the BLS to produce comprehensive injury and illness statistics for the entire private sector. Paragraph 1904.42(b)(3) combines the requirements of former rule paragraphs 1904.15(b) and 1904.16(b) into this paragraph of the final rule.

In response to the question "Am I required to respond to a BLS survey form if I am normally exempt from keeping OSHA injury and illness records?," the final rule states

"Yes. Even if you are exempt from keeping injury and illness records under Sec. 1904.1 to Sec. 1904.3, the BLS may inform you in writing that it will be collecting injury and illness information from you in the coming year. If you receive such a survey form, you must keep the injury and illness records required by Sec. 1904.4 to Sec. 1904.12 and make survey reports for the year covered by the survey."

Paragraph 1904.42(b)(4) of this final rule replaces section 1904.22 of the former rule. It provides that employers in the State-plan States are also required to fill out and submit survey forms if the BLS requests that they do so. The final rule thus specifies that the BLS has the authority to collect information on occupational fatalities, injuries and illnesses from: (1) employers who are required to keep records at all times; (2) employers who are normally exempt from keeping records; and (3) employers under both Federal and State plan jurisdiction. The information collected in the annual survey enables BLS to generate consistent statistics on occupational death, injury and illness for the entire Nation.

Subpart F. Transition From the Former Rule to the New Rule [p. 6070]

The transition interval from the former rule to the new rule involves several issues, including training and outreach to familiarize employers and employees about the new forms and requirements, and informing employers in newly covered industries that they are now required to keep OSHA Part 1904 records. OSHA intends to make a major outreach effort, including the development of an expert software system, a forms package, and a compliance assistance guide, to assist employers and recordkeepers with the transition to the new rule. An additional transition issue for employers who kept records under the former system and will also keep records under the new system is how to handle the data collected under the former system during the transition year. Subpart F of the final rule addresses some of these transition issues.

Subpart F of the new rule (sections 1904.43 and 1904.44), addresses what employers must do to keep the required OSHA records during the first five years the new system required by this final rule is in effect. This five-year period is called the transition period in this subpart. The majority of the transition requirements apply only to the first year, when the data from the previous year (collected under the former rule) must be summarized and posted during the month of February. For the remainder of the transition period, the employer is simply required to retain the records created under the former rule for five years and provide access to those records for the government, the employer's employees, and employee representatives, as required by the final rule at sections 1904.43 and 44.

...

The transition also raises questions about what should be done in the year 2002 with respect to posting, updating, and retaining the records employers compiled in 2001 and previous years. In the transition from the former rule to the present rule, OSHA intends employers to make a clean break with the former system. The new rule will replace the old rule on the effective date of the new rule, and OSHA will discontinue the use of all previous forms, interpretations and guidance on that date (see, e.g., Exs. 21, 22, 15: 184, 423). Employers will be required to prepare a summary of the OSHA Form 200 for the year 2001 and to certify and post it in the same manner and for the same time (one month) as they have in the past. The following time table shows the sequence of events and postings that will occur:

Date	Activity
2001	Employers keep injury and illness information on the OSHA 200 form
January 1, 2002	Employers begin keeping data on the OSHA 300 form
February 1, 2002	Employers post the 2001 data on the OSHA 200 Form
March 1, 2002	Employers may remove the 2001 posting
February 1, 2003	Employers post the 2002 data on the OSHA 300A form
May 1, 2003	Employers may remove the 2002 posting

The final rule's new requirements for dual certification and a 3-month posting period will not apply to the Year 2000 Log and summary. Employers still must retain the OSHA records from 2001 and previous years for five years from the end of the year to which they refer. The employer must provide copies of the retained records to authorized government representatives, and to his or her employees and employee representatives, as required by the new rule.

However, OSHA will no longer require employers to update the OSHA Log and summary forms for years before the year 2002. The former rule required employers to correct errors to the data on the OSHA 200 Logs during the five-year retention period and to add new information about recorded cases. The former rule also required the employer to adjust the totals on the Logs if changes were made to cases on them (Ex. 2, p. 23). OSHA believes it would be confusing and burdensome for employers to update and adjust previous years' Logs and Summaries under the former system at the same time as they are learning to use the new OSHA occupational injury and illness recordkeeping system.

Subpart G. Definitions [p. 6071]

The Definitions section of the final rule contains definitions for five terms: "the Act," "establishment," "health care professional," "injury and illness," and "you." To reduce the need for readers to move back and forth from the regulatory text to the Definitions section of this preamble, all other definitions used in the final rule are defined in the regulatory text as the term is used. OSHA defines the five terms in this section here because they are used in several places in the regulatory text.

The Act

The Occupational Safety and Health Act of 1970 (the "OSH Act") is defined because the term is used in many places in the regulatory text. The final rule's definition is essentially identical to the definition in the proposal. OSHA received no comments on this definition. The definition of "the Act" follows:

The Act means the Occupational Safety and Health Act of 1970 (84 Stat. 1590 et seq., 29 U.S. 651 et seq.), as amended. The definitions contained in section (3) of the Act and related interpretations shall be applicable to such terms when used in this Part 1904.

Establishment

The final rule defines an establishment as a single physical location where business is conducted or where services or industrial operations are performed. For activities where employees do not work at a single physical location, such as construction; transportation; communications, electric, gas and sanitary services; and similar operations, the establishment is represented by main or branch offices, terminals, stations, etc. that either supervise such activities or are the base from which personnel carry out these activities.

The final rule also addresses whether one business location can include two or more establishments. Normally, one business location has only one establishment. However, under limited conditions, the employer may consider two or more separate businesses that share a single location to be separate establishments for recordkeeping purposes. An employer may divide one location into two or more establishments only when: each of the proposed establishments represents a distinctly separate business; each business is engaged in a different economic activity; no one industry description in the Standard Industrial Classification Manual (1987) applies to the joint activities of the proposed establishments; and separate reports are routinely prepared for each establishment on the number of employees, their wages and salaries, sales or receipts, and other business information. For example, if an employer operates a construction company at the same location as a lumber yard, the employer may consider each business to be a separate establishment.

The final rule also deals with the opposite situation, and explains when an establishment includes more than one physical location. An employer may combine two or more physical locations into a single establishment only when the employer operates the locations as a single business operation under common management; the locations are all located in close proximity to each other; and the employer keeps one set of business records for the locations, such as records on the number of employees, their wages and salaries, sales or receipts, and other kinds of business information. For example, one manufacturing establishment might include the main plant, a warehouse serving the plant a block away, and an administrative services building across the street. The final rule also makes it clear that when an employee telecommutes from home, the employee's home is not a business establishment for recordkeeping purposes, and a separate OSHA 300 Log is not required.

The definition of "establishment" is important in OSHA's recordkeeping system for many reasons. First, the establishment is the basic unit for which records are maintained and summarized. The employer must keep a separate injury and illness Log (the OSHA Form 300), and prepare a single summary (Form 300A), for each establishment. Establishment-specific records are a key component of the recordkeeping system because each separate record represents the injury and illness experience of a given location, and therefore reflects the particular circumstances and hazards that led to the injuries and illnesses at that location. The establishment-specific summary, which totals the establishment's injury and illness experience for the preceding year, is posted for employees at that establishment and may also be collected by the government for statistical or administrative purposes. Second, the definition of establishment is important because injuries and illnesses are presumed to be work-related if they result from events or exposures occurring in the work environment, which includes the employer's establishment. The presumption that injuries and illnesses occurring in the work environment are by definition work-related may be rebutted under certain circumstances, which are listed in the final rule and discussed in the section of this preamble devoted to section 1904.5, Determination of work-relatedness. Third, the establishment is the unit that determines whether the partial exemption from recordkeeping requirements permitted by the final rule for establishments of certain sizes or in certain industry sectors applies (see Subpart B of the final rule). Under the final rule's partial exemption, establishments classified in certain Standard Industrial Classification codes (SIC codes) are not required to keep injury and illness records except when asked by the government to do so. Because a given employer may operate establishments that are classified in different SIC codes, some employers may be required to keep OSHA injury and illness records for some establishments but not for others, e.g. if one or more of the employer's establishments falls under the final rule's partial exemption but others do not.

Fourth, the definition of establishment is used to determine which records an employee, former employee, or authorized employee representative may access. According to the final rule, employees may ask for, and must be given, injury and illness records for the establishment they currently work in, or one they have worked in, during their employment.

...

Subpart G of the final rule defines "establishment" as "a single physical location where business is conducted or where services or industrial operations are performed. For activities such as construction; transportation; communications, electric and gas utility, and sanitary services; and similar operations, the establishment is represented for recordkeeping purposes by main or branch offices, terminals, stations, etc. that either supervise such activities or are the base from which personnel carry out these activities." This part of the definition of "establishment" provides flexibility for employers whose employees (such as repairmen, meter readers, and construction superintendents) do not work at the same workplace but instead move between many different workplaces, often in the course of a single day.

How the definition of "establishment" must be used by employers for recordkeeping purposes is set forth in the answers to the questions posed in this paragraph of Subpart G:

(1) Can one business location include two or more establishments?
(2) Can an establishment include more than one physical location?
(3) If an employee telecommutes from home, is his or her home considered a separate establishment?

The employer may consider two or more economic activities at a single location to be separate establishments (and thus keep separate OSHA Form 300s and Form 301s for each activity) only when: (1) Each such economic activity represents a separate business, (2) no one industry description in the Standard Industrial Classification Manual (1987) applies to the activities carried out at the separate locations; and (3) separate reports are routinely prepared on the number of employees, their wages and salaries, sales or receipts, and other business information. This part of the definition of "establishment" allows for separate establishments when an employer uses a common facility to house two or more separate businesses, but does not allow different departments or divisions of a single business to be considered separate establishments. However, even if the establishment meets the three criteria above, the employer may, if it chooses, consider the physical location to be one establishment.

The definition also permits an employer to combine two or more physical locations into a single establishment for recordkeeping purposes (and thus to keep only one Form 300 and Form 301 for all of the locations) only when (1) the locations are all geographically close to each other, (2) the employer operates the locations as a single business operation under common management, and (3) the employer keeps one set of business records for the locations, such as records on the number of employees, their wages and salaries, sales or receipts, and other business information. However, even for locations meeting these three criteria, the employer may, if it chooses, consider the separate physical locations to be separate establishments. This part of the definition allows an employer to consider a single business operation to be a single establishment even when some of his or her business operations are carried out on separate properties, but does not allow for separate businesses to be joined together. For example, an employer operating a manufacturing business would not be allowed to consider a nearby storage facility to be a separate establishment, while an employer who operates two separate retail outlets would be required to consider each to be a separate establishment.

...

Health Care Professional

The final rule defines health care professional (HCP) as "a physician or other state licensed health care professional whose legally permitted scope of practice (i.e. license, registration or certification) allows the professional independently to provide or be delegated the responsibility to provide some or all of the health care services described by this regulation."

...

OSHA recognizes that injured employees may be treated by a broad range of health care practitioners, especially if the establishment is located in a rural area or if the worker is employed by a small company that does not have the means to provide on-site access to an occupational nurse or a physician. Although the rule does not specify what medical specialty or training is necessary to provide care for injured or ill employees, the rule's use of the term health care professional is intended to ensure that those professionals providing treatment and making determinations about the recordability of certain complex cases are operating within the scope of their license, as defined by the appropriate state licensing agency.

Injury or Illness

The final rule's definition of injury or illness is based on the definitions of injury and illness used under the former recordkeeping regulation, except that it combines both definitions into a single term "injury or illness." Under the final rule, an injury or illness is an abnormal condition or disorder. Injuries include cases such as, but not limited to, a cut, fracture, sprain, or amputation. Illnesses include both acute and chronic illnesses, such as, but not limited to, a skin disease, respiratory disorder, or systemic poisoning. The definition also includes a note to inform employers that some injuries and illnesses are recordable and others are not, and that injuries and illnesses are recordable only if they are new, work-related cases that meet one or more of the final rule's recording criteria.

...

"You"

The last definition in the final rule, of the pronoun "you," has been added because the final rule uses the "you" form of the question-and-answer plain-language format recommended in Federal plain-language guidance. "You," as used in this rule, mean the employer, as that term is defined in the Act. This definition makes it clear that employers are responsible for implementing the requirements of this final rule, as mandated by the Occupational Safety and Health Act of 1970 (29 U.S.C. 651 et seq.)

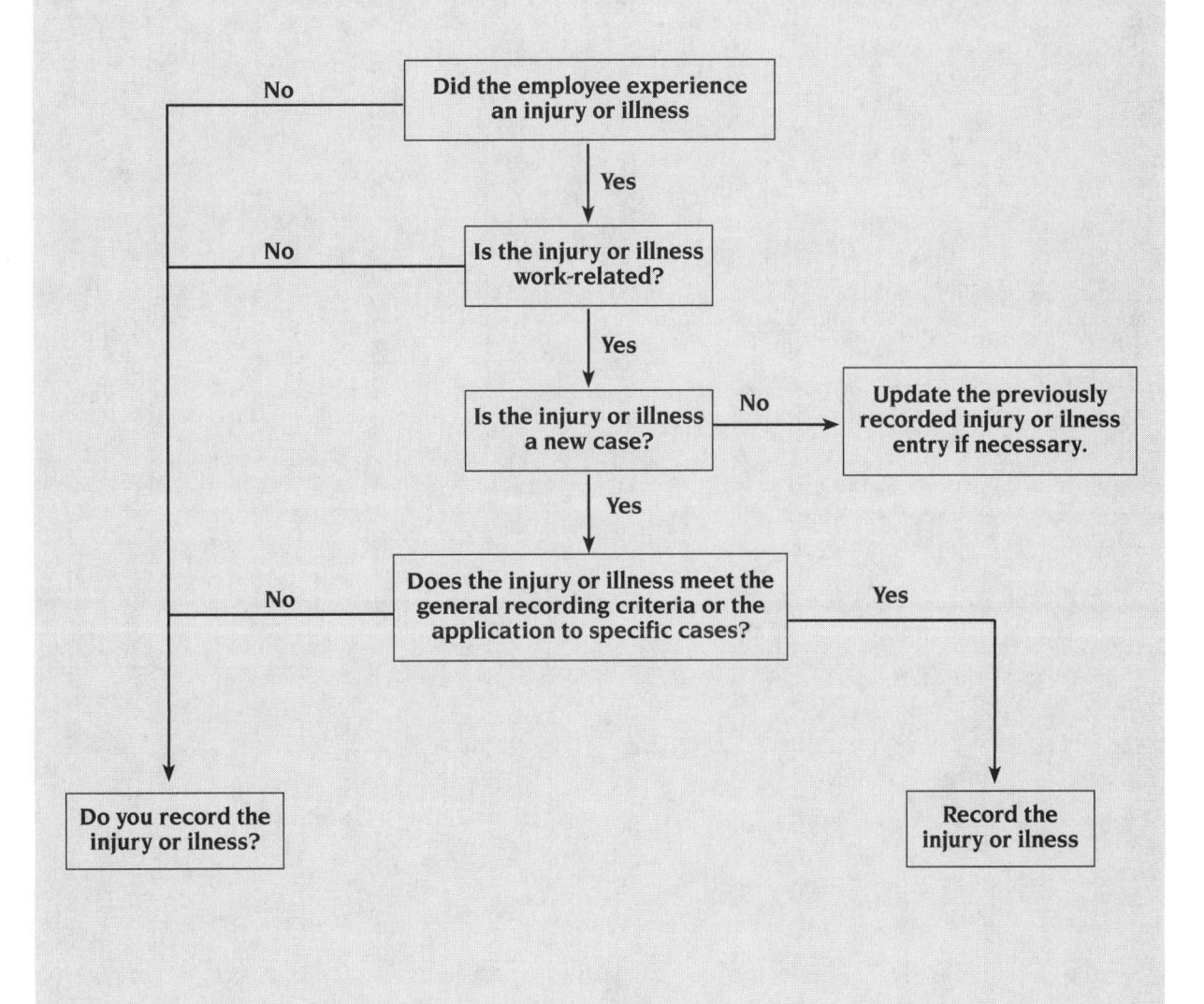

CHAPTER 3

WALKING AND WORKING SURFACES

Slips, trips and falls constitute the majority of general industry accidents. They cause 15 percent of all accidental deaths and are second only to motor vehicles as a cause of fatalities. The OSHA standards for walking and working surfaces apply to all permanent places of employment, except where only domestic, mining or agricultural work is performed.

General Requirements—1910.22

Housekeeping
All places of employment, passageways, storerooms and service rooms shall be kept clean, orderly and in a sanitary condition. The floor of every workroom shall be in a clean and, so far as possible, a dry condition. Where wet processes are used, drainage shall be maintained and gratings, mats or raised platforms shall be provided. Every floor, work area and passageway shall be kept free from protruding nails, splinters, holes or loose boards.

Aisles and Passageways
Aisles and passageways shall be kept clear and in good repair with no obstruction across or in aisles that could create a hazard. Permanent aisles and passageways shall be appropriately marked. Where mechanical handling equipment is used, aisles shall be sufficiently wide. Improper aisle widths coupled with poor housekeeping and vehicle traffic can cause injury to employees, damage the equipment and material, and can limit egress in emergencies.

Covers and Guardrails
Covers and/or guardrails shall be provided to protect personnel from such hazards as open pits, tanks, vats and ditches.

Floor Loading Protection
Load rating limits shall be marked on plates and conspicuously posted. It shall be unlawful to place, cause or permit to be placed on any floor or roof of a building or other structure, a load greater than that for which such floor or roof is approved.

Exhibit 3-1.
A clean, appropriately marked aisle that is wide enough for mechanical handling equipment to travel safely.

Guarding Floor and Wall Openings and Holes—1910.23

Floor openings and holes, wall openings and holes, and the open sides of platforms may create hazards. People may fall through the openings or over the sides to the level below. Objects, such as tools or parts, may fall through the holes and strike people or damage machinery on lower levels.

OSHA standards for guarding openings and holes use the following definitions:

Floor hole. An opening measuring less than 12 in., but more than 1 in., in its least dimension, in any floor, platform, pavement or yard through which materials, but not persons, may fall.

Floor opening. An opening measuring 12 in. or more in its least dimension, in any floor, platform, pavement or yard through which persons may fall.

Platform. A working space for persons, elevated above the surrounding floor or ground.

Wall hole. An opening less than 30 in., but more than 1 in. high, of unrestricted width, in any wall or partition.

Wall opening. An opening at least 30 in. high and 18 in. wide in any wall or partition through which persons may fall.

Protection for Floor Openings

Standard railings shall be provided on all exposed sides of a stairway opening, except at the stairway entrance. For infrequently used stairways, where traffic across the opening prevents the use of a fixed standard railing, the guard shall consist of a hinged floor opening cover of standard strength and construction, along with removable standard railings on all exposed sides, except at the stairway entrance.

A "standard railing" consists of top rail, mid-rail, and posts, and shall have a vertical height of 42 in. nominal from the upper surface of top rail to floor, platform, runway or ramp level. Nominal height of mid-rail is 21 in. (See Exhibit 3-2.).

A "standard toeboard" is 4 in. nominal in vertical height, with not more than 1/4-inch clearance above floor level.

Floor openings may be covered rather than guarded with rails. When the floor opening cover is removed, a temporary guardrail shall be in place or an attendant shall be stationed at the opening to warn personnel.

Either a standard railing with toeboard or a floor hole cover of standard strength or construction shall guard every floor hole into which persons can accidentally walk. While the cover is not in place, the floor hole shall be constantly attended by someone or shall be protected by a removable standard railing.

Exhibit 3-2.
A standard railing consists of a top rail, mid-rail and posts.

Protection of Open-Sided Floors, Platforms and Runways

One of the most frequently cited violations in Subpart D is the requirement that every open-sided floor or platform 4 ft. or more above adjacent floor or ground level shall be guarded by a standard railing (or the equivalent as specified in paragraph (e)(3) of this section) on all open sides, except where there is an entrance to a ramp, stairway, or fixed ladder. The railing shall be provided with a toeboard wherever, beneath the open sides:

- Persons can pass;
- There is moving machinery; or
- There is equipment with which falling materials could create a hazard.

A standard railing (or the equivalent as specified in paragraph (e)(3) of this section) shall guard every runway on all sides 4 ft. or more above floor or ground level. Wherever tools, machine parts or materials are likely to be used on the runway, a toeboard shall also be provided on each exposed side.

Regardless of height, open-sided floors, walkways, platforms or runways above or adjacent to dangerous equipment, pickling or galvanizing tanks, degreasing units and similar hazards shall be guarded with a standard railing and toeboard.

Stairway Railings and Guards

Every flight of stairs with four or more risers shall have standard stair railings or standard handrails as specified below. Stair width is measured clear of all obstructions except handrails.

- On stairways less than 44 in. wide having both sides enclosed, at least one handrail shall be affixed, preferably on the right side descending.
- On stairways less than 44 in. wide with one open side, at least one stair rail shall be affixed on the open side.
- On stairways less than 44 in. wide having both sides open, two stair rails shall be provided, one for each side.
- On stairways more than 44 in. wide, but less than 88 in., one handrail shall be provided on each enclosed side and one stair rail on each open side.
- On stairways 88 in. or more in width, one handrail shall be provided on each enclosed side, one stair rail on each open side, and one intermediate stair rail placed approximately in the middle of the stairs.

A "standard stair railing" (stair rail) shall be of construction similar to a standard railing, but the vertical height shall be not more than 34 in. nor less than 30 in. from the upper surface of the top rail to the surface of the tread in line with the face of the riser at the forward edge of the tread.

A "standard handrail" consists of a lengthwise member mounted directly on a wall or partition by means of brackets attached to the lower side of the handrail in order to keep a smooth, unobstructed surface along the top and both sides of the handrail. The brackets shall hold the rail 3 in. from the wall and be no more than 8 ft. apart

The height of handrails shall be no more than 34 in. nor less than 30 in. from the upper surface of the handrail to the surface of the tread in line with the face of the riser or to the surface of the ramp.

Winding stairs shall have a handrail that is offset to prevent people from walking on any portion of the treads where the width is less than 6 in.

Fixed Industrial Stairs—1910.24

This section contains specifications for the safe design and construction of fixed general industrial stairs. This includes interior and exterior stairs around machinery, tanks and other equipment as well as stairs leading to or from floors, platforms or pits. This section does not apply to stairs used for fire exit purposes, to construction operations, to private residences, or to articulated stairs, such as may be installed on floating roof tanks, the angle of which changes with the rise and fall of the base support. Intermediate landings and platforms on stairways shall be no less than the stair width and a minimum of 30 in. in length measured in the direction of travel.

Fixed industrial stairs shall be provided for access to and from places of work where operations necessitate regular travel between levels. OSHA requirements include:

- Fixed industrial stairs shall be strong enough to carry five times the normal anticipated live load.
- At the very minimum, any fixed stairway shall be able to safely carry a moving concentrated load of 1,000 lbs.
- All fixed stairways shall have a minimum width of 22 in.
- Fixed stairs shall be installed at angles to the horizontal of between 30° and 50°.
- Vertical clearance above any stair tread to an overhead obstruction shall be at least 7 ft., measured from the leading edge of the tread.

When inspecting the condition of stairways, items to look for include:
- Handrails and stair rails:
    - Lack of handrails;
    - Placement of handrails;
    - Smoothness of surface;
    - Strength; and
    - Clearance between rail and wall or other object.
- Treads:
    - Strength;
    - Slip resistance;
    - Dimensions;
    - Evenness of surface; and
    - Visibility of leading edge.

WALKING AND WORKING SURFACES

- General inspection points:
  - Improper or inadequate design, construction or location of staircases;
  - Wet, slippery or damaged walking or grasping surfaces;
  - Improper illumination (Note: There is no horizontal OSHA standard for illumination levels; the Illuminating Engineering Society publications should be consulted for recommendations.);
  - Poor housekeeping; and
  - Length of a staircase—long flights of steps without landings should be avoided whenever possible.

Portable Ladders—1910.25, 1910.26

The chief hazard when using a ladder is falling. A poorly designed, maintained or improperly used ladder may collapse under the load placed upon it and cause the employee to fall. A ladder is an appliance consisting of two side rails joined at regular intervals by crosspieces on which a person may step to ascend or descend. Types of portable ladders include:

Stepladder. A self-supporting portable ladder, non-adjustable in length, having flat steps and hinged back.

Single ladder. A non-self-supporting portable ladder, nonadjustable in length, consisting of one section. Its size is designed by the overall length of the side rail.

Extension ladder. A non-self-supporting portable ladder adjustable in length.

OSHA's requirements for portable ladders include:

- Portable stepladders longer than 20 ft. shall not be used.
- Stepladders shall be equipped with a metal spreader or locking device of sufficient size and strength to securely hold the front and back sections in an open position.
- Single ladders longer than 30 ft. shall not be used.
- Extension ladders longer than 60 ft. shall not be used.
- Ladders shall be maintained in good condition at all times.

Ladders shall be inspected frequently and those that have developed defects shall be withdrawn from service for repair or destruction and tagged or marked as "Dangerous, Do Not Use." Portable ladders should extend 3 ft. past the level to which the employee is climbing. The 1-in-4 rule should also be used: For every 4 ft. in height, the base of the ladder should come out 1 ft. from the building (See Exhibit 3-3.). Ladders shall be placed with a secure footing or they shall be lashed, or held in position.

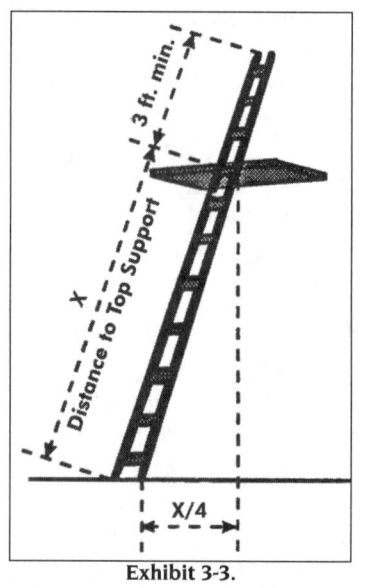

Exhibit 3-3.
The base and top the ladder must extend a certain distance from the climbing surface.

The foot of a ladder shall, where possible, be used at such a pitch that the horizontal distance from the top support to the foot of the ladder is one-quarter of the working length of the ladder (the length along the ladder between the foot and the support). OSHA standards also require that:

- The worker shall always face the ladder when climbing up or down;
- Short ladders shall not be spliced together to make long ladders;
- Ladders shall never be used in the horizontal position as scaffolds or work platforms;
- The top of a regular stepladder shall not be used as a step;
- Use both hands when climbing and descending ladders; and
- Metal ladders shall never be used near electrical equipment.

Fixed Ladders—1910.27

A fixed ladder is a ladder permanently attached to a structure, building or equipment. A point to remember is that fixed ladders, with a length of more than 20 ft. to a maximum unbroken length of 30 ft. shall be equipped with cages or a ladder safety device.

A "cage" is a guard that is fastened to the side rails of the fixed ladder or to the structure to encircle the climbing space of the ladder for the safety of the person who must climb the ladder. Cages shall extend a minimum of 42 in. above the top of a landing, unless other acceptable protection is provided. Cages shall extend down the ladder to a point not less than 7 ft. nor more than 8 ft. above the base of the ladder.

A ladder safety device is any device, other than a cage or well, designed to eliminate or reduce the possibility of accidental falls and may incorporate such features as life belts, friction brakes and sliding attachments. Another feature of fixed ladders is the landing platform that provides a means of interrupting a free fall and serves as a resting-place during long climbs.

When fixed ladders are used to ascend to heights exceeding 20 ft., except on landing, platforms shall be provided. Where no cage, well or ladder safety device is provided, a landing platform shall be provided for each 20 ft. of height or fraction thereof.

Ladder safety devices may be used on tower, water tank and chimney ladders more than 20 ft. in unbroken length in lieu of cage protection. No landing platform is required in these cases.

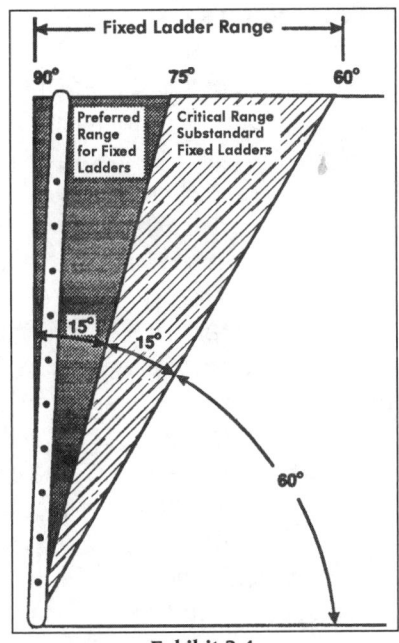

Exhibit 3-4.
Fixed ladders must be installed within a certain pitch range. The substandard range shown above may sometimes be permitted.

The preferred pitch of fixed ladders shall be considered to come in the range of 75° and 90° with the horizontal. Fixed ladders shall be considered to be substandard if they are installed within the pitch range of 60° and 75° with the horizontal. Substandard fixed ladders are permitted only when it is found necessary to meet conditions of installation. This substandard pitch range shall be considered a critical range to be avoided, if possible. Ladders having a pitch in excess of 90° with the horizontal are prohibited.

As with all ladders, fixed ladders shall be maintained in a safe condition and inspected regularly.

Safety Requirements for Scaffolding—1910.28

This section establishes safety requirements for the construction, operation, maintenance and use of scaffolds used in the maintenance of buildings and structures. It is important to note some of the general requirements of 1910.28(a), which apply to all scaffolds:

> The footing or anchorage for scaffolds shall be sound, rigid and capable of carrying the maximum intended load without settling or displacement. Unstable objects, such as barrels, boxes, loose brick, or concrete blocks shall not be used to support scaffolds or planks.

Scaffolds and their components shall be capable of supporting at least four times the maximum intended load. Scaffolds shall be maintained in a safe condition and shall not be altered or moved horizontally while they are in use or occupied. Damaged or weakened scaffolds shall be immediately repaired and shall not be used until repairs have been completed. A safe means must be provided to gain access to the working platform level through the use of a ladder, ramp or other such item. Overhead protection must be provided for personnel on a scaffold exposed to overhead hazards.

Guardrails, mid-rails and toeboards must be installed on any open sides and ends of platforms more than 10 ft. above the ground or floor. Wire mesh must be installed between the toeboard and the guardrail along the entire opening, except where persons are required to work or pass under the scaffolds.

WALKING AND WORKING SURFACES

Employees shall not work on scaffolds during storms or high winds or if the scaffolds are covered with ice or snow.

Manually Propelled Mobile Ladder Stands and Scaffolds (Towers)—1910.29

This section contains requirements for the design, construction and use of mobile work platforms (including ladder stands excluding aerial ladders) and rolling (mobile) scaffolds (towers). As in the previous section, there are a variety of materials and design possibilities involved, and no attempt will be made to discuss detailed design criteria.

General requirements include:
- All exposed surfaces of mobile ladder stands and scaffolds shall be free from sharp edges, burrs or other safety hazards.
- The maximum work height shall not exceed four times the minimum base dimensions unless outriggers, guys or braces are added to provide stability.
- Mobile scaffolds are mounted on casters with fixed working levels. The working height is limited to four times the minimum base dimensions to achieve a stable work platform.
- Access is required to be provided by ladder and minimum platform width is 18 in.

This standard requires guardrails and toeboards for work levels 10 ft. or more above the ground or floor.

Other Working Surfaces—1910.30

Portable dock boards (bridge plates) shall be secured in position, either by being anchored or equipped with devices that will prevent their slipping. Movement of the dock boards during material handling operations has resulted in forklifts overturning or falling off the dock, often with serious injury or death to the driver and damage to equipment and material.

Many accidents occur during material handling. Handholds shall be provided on portable dock boards to permit safe handling when the dock board must be repositioned or relocated.

OSHA Instruction OCT 30, 1978 STD 1-1.4
OSHA PROGRAM DIRECTIVE # 100-60
TO: REGIONAL ADMINISTRATORS/OSHA
SUBJECT: **29 CFR 1910.22(b)(2), Markings For Aisles and Passageways**

1. **Purpose.** To provide an interpretation of "appropriately marked" as applies to permanent aisles and passageways where there are dirt floors or floors having continuous concentrations of sand or fine dusts.
2. **Documentation Affected.** None.
3. **Background.**
   a. In some instances, 29 CFR. 1910.22(b)(2) has been narrowly interpreted to mean that aisles and passageways must be marked by painted floor lines.
   b. The intent of "appropriately marked" is not to restrict the markings to one method only. It would be impractical to paint lines on dirt floors or floors that have continuous concentrations of sand or other dusts. These conditions may exist in such industries as foundries, scrap salvage operations or motor winding facilities.
4. **Action.** Painted lines remain the most feasible method of marking, where practical, since they may last several years without maintenance or repainting. Other appropriate methods such as marking pillars, powder stripping, flags, traffic cones or barrels are acceptable, when the training programs for vehicle operators and employees include the recognition of such markings.
5. **Effective Date.** This directive is effective immediately upon receipt and will remain in effect until canceled or superseded.

Richard P. Wilson Deputy Director, Federal Compliance and State Programs

OSHA Instruction STD 1-1.12 JUN 20, 1983 Office of Compliance Programming
SUBJECT: **Application of 29 CFR 1910.27, Fixed Ladders, to Fixed Ladders Used in Emergency Situations**

A. **Purpose.** This instruction clarifies the meaning of 29 CFR 1910.27 as it applies to the protection of employees exposed to falling from fixed ladders used only as a means of escape from fire and other emergency situations.
B. **Scope.** This instruction applies OSHA-wide.
C. **Reference.** OSHA Instruction STD 1-1.3, October 30, 1978, dated January 18, 1977.
D. **Action.** Regional Administrators and Area Directors shall ensure that the interpretative guidelines given in this instruction are addressed when inspecting sites with fixed ladders used only during fire and other emergency situations.
E. **Federal Program Change.** This instruction describes a Federal program change which affects State programs. Each Regional Administrator shall:
   1. Ensure that this change is forwarded to each State designee.
   2. Explain the technical content of the change to the State designee as requested.
   3. Ensure that State designees are asked to acknowledge receipt of this Federal program change in writing, within 30 days of notification, to the Regional Administrator. This acknowledgment should include a description either of the State's plan to implement the change or of the reasons why the change should not apply to that State.
   4. Review policies, instructions and guidelines issued by the State to determine that this change has been communicated to State program personnel. Routine monitoring activities (accompanied inspections and case file reviews) shall also be used to determine if this change has been implemented in actual performance.
F. **Background.** OSHA has historically established that the requirements of 29 CFR 1910.27 for cages, platforms, or similar fall prevention protection devices are not appropriate for fixed ladders on structures where the fixed ladders are used only as a means of access by fire fighters, other emergency personnel, or escape for employees in fire and other emergency situations. Sometimes these ladders are not provided with employee protection as presently required in 29 CFR 1910.27, when they are intended to be used only in an emergency. In these circumstances, it is sometimes more hazardous to install a cage, well, landing platform or ladder safety device pursuant to the standard than it is not to comply. A cage or well, etc., may interfere with fire fighting or other rescue equipment, or employee escape from fire or other emergency situations.
G. **Guidelines.** This instruction provides performance criteria for fixed ladders used only as a means of access for fire fighters and other emergency personnel, or escape for employees in fire and other emergency situations.
   1. Employers must establish and implement adequate administrative controls such as barricades and signs to prevent nonemergency use of fixed ladders which are meant for fire fighter use and emergency escape only.
   2. In the event the employer does not provide adequate administrative controls such as barricades or signs and employees use an emergency ladder for other than its intended purpose, the employer may be appropriately cited under 29 CFR 1910.27.
   3. Fixed ladders not equipped with cages, landing platforms, ladder safety devices, or other forms of employee protection, in some situations may be allowed as a means of access for fire fighters and other emergency personnel, or escape for employees in fire and other emergency situations. These guidelines are provided because it may be more hazardous to comply with 29 CFR 1910.27 than not to comply.

Thomas G Auchter, Assistant Secretary

DEC 27, 1984
Memorandum FOR: R.A. Clark, REGIONAL Administration
From JOHN B. MILES, Jr., Director of Field Operations
SUBJECT: **Interpretation of 1910.22(b)(l), Aisles and passageways**

This is in response to your memorandum of November 2, 1984. The requested interpretations follow:

1. 1910.22(b)(1) applies to mobile material handling equipment such as carts, wagons, loaders and industrial trucks. Indeed the Review Commission has repeatedly held that the standard is designed to protect only the operators of such equipment rather than pedestrians. Floor operated overhead hoists and cranes are regulated by 1910.179 or by Section 5(a)(1) of the Act, with reference to an applicable ANSI standard.

2. 1910.22(b)(1) does not apply when the aisles and passageways are not used by material handling equipment. Therefore where hazards are created by objects other than material handling equipment citations of 1910.22(a)(1) and 1910.37 should be considered.

May 21, 1984
MEMORANDUM FOR MR. BYRON R. CHADWICK, REGIONAL ADMINISTRATOR
FROM JOHN B. Miles Director DIRECTORATE OF FIELD OPERATIONS
SUBJECT: **Interpretation: 1910.178(m)(12)**

This is in response to your memorandum dated May 7, 1984, requesting an interpretation of 1910.178(m)(12).

29 CFR 1910.178(m)(12) is exclusively applicable to lift trucks equipped with elevatable controls. The standard does not require or mandate that such controls must be installed on trucks not so equipped. The current industry consensus standard ANSI B56.1-1975 helps to clear up the confusion because it addresses the use of Personnel safety platforms on trucks without elevatable controls as well as those with such controls. Sections 427 and 513A of the 1975 ANSI standard (copy attached) specify the safety requirements for fabrication and use of elevatable personnel platforms for use with powered industrial trucks. Since the OSHA standard does not require elevatable controls on such platforms employers who comply with these sections of the 1975 ANSI standard would be compliance with the OSHA Act and are not subject to citation unless an unusually' hazardous situation exists.

In this instance, it appears that the employer's position is correct. Under the circumstances described in the Area Director's memo to you, the employer is not required to have an elevatable power shut-off when there are no elevatable controls on the platform. However, he is required to have a vehicle operator remain at the controls whenever employees are elevated.

OSHA Standards Interpretation and Compliance Letters
02/09/1983 - Perimeter protection at setback roof levels.
Subject: **Perimeter protection at setback roof levels.**

February 9, 1983
Lawrence R. Stafford, P.E.
Consulting Engineer, 8 Gracemore St., Albany, New York 12203

Dear Mr. Stafford:

This is in response to your letter of January 28, 1983, concerning perimeter protection at setback roof levels. A parapet height of 29 inches, where employees are exposed to falls from a roof, does not comply with the height requirement in 29 CFR 1910.23(e)(1) and cannot be considered acceptable by OSHA. The employer may install a temporary portable section of guardrail which will comply with 29 CFR 1910.23(e)(3)(v) and provide a minimum height of 36 inches at the exposure locations. Employers may also use a safety belt and rope tie-off system in this type of exposure. If I may be of further assistance, please feel free to contact me.

Sincerely,
John K. Barto
Chief, Division of Occupational
Safety Programming

OSHA Standards Interpretation and Compliance Letters
04/05/1978 - Clarification of 1910.23(c)(1).
Subject: **Clarification of 1910.23(c)(1).**

April 5, 1978
Mr. John R. Reilly
Corporate Director   Fisher Scientific Company
711 Forbes Avenue   Pittsburgh, Pennsylvania 15219

Dear Mr. Reilly:

This is in response to your letter dated February 20, 1978, requesting a clarification of 29 CFR 1910.23(c)(1) and confirms your telephone conversation with Mr. William Simms of my staff.

The aforementioned General Industry Safety Standard requires that every open-sided floor or platform more than four feet above the adjacent level shall be provided with a standard railing. This requirement is applied to areas where employees are required to work or walk as part of fulfilling conditions of their employment. If the roof platform on top of a walk-in refrigerator is visited even a few days each year to store materials, employees required to work adjacent to or near the edge shall be provided with some means of protection.

Such protection could be a life belt and lanyard tied off of a structural member so as to prevent the employee from walking off or falling from the exposed edges of the refrigerator roof top used as a storage platform.

If I may be of any further assistance, please feel free to contact me.

Sincerely,
Janet H. Sprickman,
Acting Chief
Division of Occupational Safety Programming

WALKING AND WORKING SURFACES

02/16/1984 - Guarding requirements for skylights.
Record Type: Interpretation Standard Number: 1910.23(a)(4); 1910.23(e)(8)
Subject: **Guarding requirements for skylights.**

February 16, 1984
Mr. Joseph P. Stanton
Iovine & Woods, P.C., Suite 200, 7908 Frankford Ave., Philadelphia, PA 19136

Dear Mr. Stanton:

This is in response to your letter of December 21, 1983, regarding skylights as regulated by the Occupational Safety and Health Administration (OSHA). This response provides an interpretation and clarification of the General Industry Standard 29 CFR 1910.23(a)(4) and (e)(8).

These regulations are included in 29 CFR 1910, Subpart D—Waling-Working Surfaces. 29 CFR 1910.21(a)(1) of the same Subpart defines floor opening as: An opening measuring 12 inches or more in its least dimension, in any floor, platform, pavement, or yard through which persons may fall, such as a hatchway, stair or ladder opening, pit, or large manhole.

Moreover, a definition given in Webster's New Collegiate Dictionary 1977 edition) for "hatch" is "an opening in the...floor or roof of a building;" the same entry gives "hatchway" as a synonym.

Using these definitions, therefore, OSHA concludes that a skylight should be regarded as a hatchway, i.e., an opening in the roof of a building through which persons may fall. 29 CFR 1910.23(a)(4) therefore requires that skylights in the roof of buildings through which persons may fall while walking or working shall be guarded by a standard skylight screen or a fixed standard railing on all exposed sides.

When a skylight screen is selected for safeguarding the opening, and in the event the skylight is constructed of plastic material subject to fracture (as glass would be), then the skylight must at a minimum be provided with a skylight screen capable of withstanding a load of at least 200 pounds applied perpendicularly at any one area on the screen. On the other hand, a plastic skylight which can provide the necessary structural integrity to support the 200-pound load would not be required to be further safeguarded, since it would meet the intended function of a screen as well.

As expressed in 29 CFR 1910.23(e)(8), the primary function of the screen is to support at least a 200-pound load such as a person may place upon it. This provision further relates that the screen shall provide a minimum deflection so as not to break the glass; but that portion of the requirement may be inapplicable when no glass is present. (The concern for breaking the glass results from the possible fragment exposure to persons beneath the skylight.)

We hope this information is helpful to you. If we may be of further assistance, please contact us.

Sincerely,
John B. Miles, Jr., Director
Directorate of Field Operations

[OSHA Instruction STD 1-1.13 April 16, 1984 Office of General Industry Compliance Assistance
Subject: **Protection in General Industry: 29 CFR 1910.23(c)(1), (c)(3), and 29 CFR 1910.132(a)**

A. **Purpose.** This instruction clarifies the applicability of 29 CFR 1910.23(c)(1), (c)(3) and 1910.132(a) where employees are exposed to falling hazards while performing various tasks including maintenance from elevated surfaces.

B. **Scope.** This instruction applies OSHA-wide.

C. **Action.** Regional Administrators and Area Directors shall ensure that the interpretations in F. and the guidelines in G. of this instruction are adhered to when inspecting general industry facilities where employees are exposed to the hazard of falling from elevated surfaces.

D. **Federal Program Change.** This instruction describes a Federal program change which affects State programs. Each Regional Administrator shall:
   1. Ensure that this change is forwarded to each State designee.
   2. Explain the technical content of the change to the State designee as requested.
   3. Ensure that State designees are asked to acknowledge receipt of this Federal program change in writing, within 30 days of notification, to the Regional Administrator. This acknowledgment should include a description either of the State's plan to implement the change or of the reasons why the change should not apply to that State.
   4. Review policies, instructions and guidelines issued by the State to determine that this change has been communicated to State program personnel. Routine monitoring shall also be used to determine if this change has been implemented in actual performance.

E. **Background.** Adjudicated decisions concerning employee exposures to falls from elevated surfaces have been inconsistent. As a result, OSHA has cited employers for violations of 29 CFR 1910.23(c)(1) or of Section 5(a)(1) of the OSH Act when employees have been engaged in various tasks which include inspections, service, repairs and maintenance on elevated surfaces such as, but not limited to, conveyers, tops of machinery and other structures not normally considered "walking and working" surfaces.
   1. Although 29 CFR 1910.23(c)(1) requires the safeguarding of "platforms" used by employees, there has been disagreement as to when an "elevated surface" constitutes a platform within the meaning of the standard.
   2. In at least one instance (General Electric Company v. OSHRC, 583 F. 2d 61 (2d Cir. 1978) the court noted the need for increased clarity of definition by OSHA regarding its intended meaning of the term "platform". Therefore, this instruction clarifies and defines the conditions and circumstances under which a "platform" is deemed to exist, and where the requirements of 29 CFR 1910.23(c) apply.

F. **Interpretation.** The following interpretations are established for uniform enforcement and application of G. of this instruction.
   1. Platforms are interpreted to be any elevated surface designed or used primarily as a walking or working surface, and any other elevated surfaces upon which employees are required or allowed to walk or work while performing assigned tasks on a predictable and regular basis (See 29 CFR 1910.21(a)(4) for definition of "platform".)
   2. Predictable and regular basis means employee functions such as, but not limited to, inspections, service, repair and maintenance which are performed:
      a. At least once every 2 weeks, or
      b. For a total of 4 man-hours or more during any sequential 4-week period (e.g., 2 employees once every 4 weeks for 2 hours = 4 man-hours per 4-week period).

G. **Guidelines.** The following guidelines are established for the uniform enforcement of 29 CFR 1910.23(c)(1), 1910.23(c)(3) and 1910.132(a) regarding employee exposures to falls from elevated surfaces.
   1. Employee exposures to falls from platforms (interpreted in F.1.) are regulated by the following OSHA standards:

a. 29 CFR 1910.23(c)(1), or
   b. 29 CFR 1910.23(c)(3).
2. In situations where the safeguarding requirements of G.1. are not applicable because employees are exposed to falls from an elevated surface other than a predictable and regular basis, personal protective equipment as required by 29 CFR 1910.132(a) or other effective fall protection shall be provided.

Thorne G. Auchter Assistant Secretary

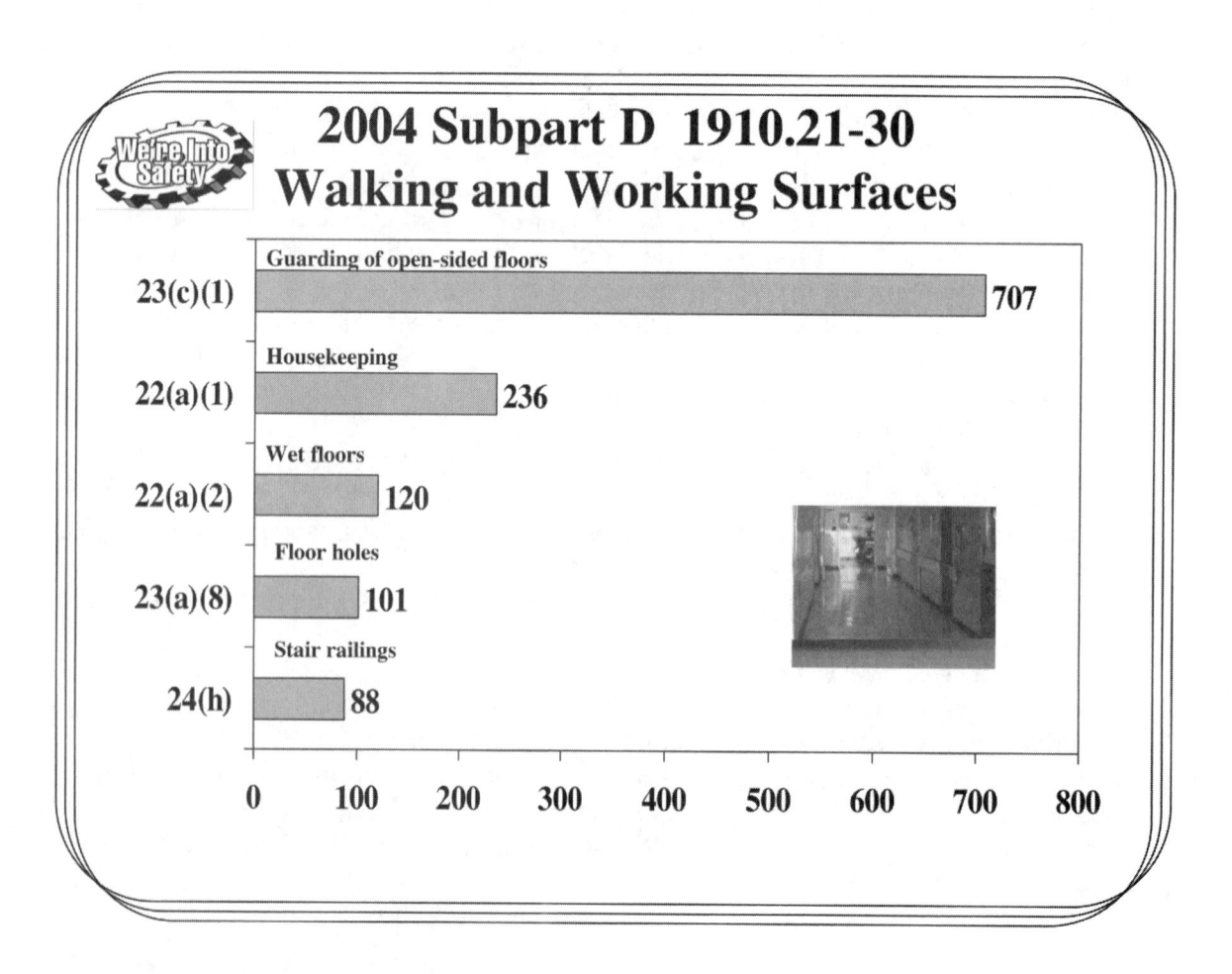

2004 Subpart D 1910.21-30 Walking and Working Surfaces

Citation	Category	Count
23(c)(1)	Guarding of open-sided floors	707
22(a)(1)	Housekeeping	236
22(a)(2)	Wet floors	120
23(a)(8)	Floor holes	101
24(h)	Stair railings	88

OCCUPATIONAL HEALTH AND INDUSTRIAL HYGIENE

Some significant events in occupational safety in the United States include:

In 1812, the Embargo of the War of 1812 spurred the development of the New England textile industry and the founding of factory mutual companies. These early insurance companies inspected properties for hazards and suggested loss control and prevention methods in order to secure low rates for their policyholders.

In 1864 the Pennsylvania Mine Safety Act (PMSA) was passed into law.

In 1864 North America's first accident insurance policy was issued.

In 1867 the state of Massachusetts instituted the first government-sponsored factory inspection program.

In 1877 the state of Massachusetts passed a law requiring guarding for dangerous machinery, and took authority for enforcement of factory inspection programs.

In 1878 the first recorded call by a labor organization for a federal occupational safety and health law is heard.

In 1896 an association to prevent fires and write codes and standards, the National Fire Protection Association (NFPA), was founded.

In 1902 the state of Maryland passed the first workers' compensation law.

In 1904 the first attempt by a state government to force employers to compensate their employees for on-the-job injuries was overturned when the U.S. Supreme Court declared Maryland's workers' compensation law to be unconstitutional.

In 1911 a professional, technical organization responsible for developing safety codes for boilers and elevators, the American Society of Mechanical Engineers (ASME), was founded.

From 1911 to 1915 30 states passed workers' compensation laws.

In 1911 the American Society of Safety Engineers (ASSE) was founded. The ASSE was dedicated to the development of accident prevention techniques and the advancement of safety engineering as a profession.

In 1912 a group of engineers representing insurance companies, industry and government met in Milwaukee to exchange data on accident prevention. The organization formed at this meeting was to become the National Safety Council (NSC). (Today, the NSC carries on major safety campaigns for the general public, as well as assists industry in the development of safety promotion programs.)

In 1916 the U.S. Supreme Court upheld the constitutionality of state workers' compensation laws.

In 1918 the American Standards Association was founded. Responsible for the development of many voluntary safety standards, some of which are referenced into laws it is now called the American National Standards Institute (ANSI).

In 1936 Frances Perkins, Secretary of Labor, called for a federal occupational safety and health law. This action came a full 58 years after organized labor's first recorded request for a law of this nature.

In 1936 the Walsh-Healey (Public Contracts) Act passed. This law required that all federal contracts be fulfilled in a healthful and safe working environment.

In 1948 all 48 states had workers' compensation laws.

In 1952 Coal Mine Safety Act (CMSA) was passed into law.

In 1960 specific safety standards were promulgated for the Walsh-Healey Act.

In 1966 the Metal and Nonmetallic Mines Safety Act (MNMSA) was passed.

In 1966 the U.S. Department of Transportation (DOT) and its sections, the National Highway Traffic Safety Administration (NHTSA) and the National Transportation Safety Board (NTSB), were established.

In 1968 President Lyndon Johnson called for a federal occupational safety and health law.

In 1969 the Construction Safety Act (CSA) was passed.

In 1969 the Board of Certified Safety Professionals (BCSP) was established. This organization certifies practitioners in the safety profession.

In 1970 President Richard Nixon signed into law the Occupational Safety and Health Act thus creating the Occupational Safety and Health Administration (OSHA) and the National Institute for Occupational Safety and Health (NIOSH).

Industrial Hygiene

Industrial hygiene has been defined as

> that science and art devoted to the anticipation, recognition, evaluation, and control of those environmental factors or stresses arising in or from the workplace, which may cause sickness, impaired health and well-being, or significant discomfort among workers or among the citizens of the community.

Industrial hygienists use environmental monitoring and analytical methods to detect the extent of worker exposure and employ engineering, work practice controls and other methods to control potential health hazards.

There has been an awareness of industrial hygiene since antiquity. The environment and its relation to worker health was recognized as early as the fourth century B.C. when Hippocrates noted lead toxicity in the mining industry. In the first century A.D., Pliny the Elder, a Roman scholar, perceived health risks to those working with zinc and sulfur. He devised a face mask made from an animal bladder to protect workers from exposure to dust and lead fumes. In the second century A.D., the Greek physician Galen accurately described the pathology of lead poisoning and also recognized the hazardous exposures of copper miners to acid mists.

In the Middle Ages, guilds worked at assisting sick workers and their families.

In 1556, the German scholar Agricola advanced the science of industrial hygiene even further when in his book *De Re Metallica*, he described the diseases of miners and prescribed preventive measures. The book included suggestions for mine ventilation and worker protection, discussed mining accidents and described diseases associated with mining occupations such as silicosis.

Industrial hygiene gained further respectability in 1700 when in Italy, Bernardo Ramazzini, known as the "father of industrial medicine," published the first comprehensive book on industrial medicine, *De Morbis Artificum Diatriba* (*The Diseases of Workmen*). The book contained accurate descriptions of the occupational diseases of most of the workers of his time. Ramazzini greatly affected the future of industrial hygiene because he asserted that occupational diseases should be studied in the work environment rather than in hospital wards.

Industrial hygiene received another major boost in 1743 when Ulrich Ellenborg published a pamphlet on occupational diseases and injuries among gold miners. Ellenborg also wrote about the toxicity of carbon monoxide, mercury, lead, and nitric acid.

In England in the 18th century, Percival Pott, as a result of his findings on the insidious effects of soot on chimney sweepers, was a major force in getting the British

Parliament to pass the *Chimney-Sweepers Act of 1788*. The passage of the English Factory Acts beginning in 1833 marked the first effective legislative acts in the field of industrial safety. The acts, however, were intended to provide compensation for accidents rather than to control their causes. Later, several other European nations developed workers' compensation acts, which stimulated the adoption of increased factory safety precautions and the establishment of medical services within industrial plants.

In the early 20th century in the United States, Dr. Alice Hamilton led efforts to improve industrial hygiene. She observed industrial conditions first hand and startled mine owners, factory managers and state officials with evidence that there was a correlation between worker illness and exposure to toxins. She also presented definitive proposals for eliminating unhealthful working conditions.

At about the same time, U.S. federal and state agencies began investigating health conditions in industry. In 1908 public awareness of occupationally related diseases stimulated the passage of compensation acts for certain civil employees. States passed the first workers' compensation laws in 1911. In 1913 the New York Department of Labor and the Ohio Department of Health established the first state industrial hygiene programs. All states enacted such legislation by 1948. In most states, there is some compensation coverage for workers who contract occupational diseases.

The U.S. Congress has passed three landmark pieces of legislation related to safeguarding workers' health:
- The Metal and Nonmetallic Mines Safety Act of 1966;
- The Federal Coal Mine Safety and Health Act of 1969; and
- The Occupational Safety and Health Act of 1970.

Today nearly every employer is required to implement the elements of an industrial hygiene and safety, occupational health, or hazard communication program and to be responsive to the Occupational Safety and Health Administration (OSHA) and its regulations.

OSHA and Industrial Hygiene

Under the OSH Act, OSHA develops and sets mandatory occupational safety and health requirements applicable to the more than 6 million workplaces in the United States. OSHA relies on, among many others, industrial hygienists to evaluate jobs for potential health hazards. Developing and setting mandatory occupational safety and health standards involves determining the extent of employee exposure to hazards and deciding what is needed to control these hazards to protect workers. Industrial hygienists are trained to anticipate, recognize, evaluate and recommend controls for environmental and physical hazards that can affect the health and well-being of workers.

More than 40 percent of the OSHA compliance officers who inspect U.S. workplaces are industrial hygienists. Industrial hygienists also play a major role in developing and issuing OSHA standards to protect workers from health hazards associated with toxic chemicals, biological hazards and harmful physical agents. They also provide technical assistance and support to the agency's national and regional offices. OSHA also employs industrial hygienists who assist in setting up field enforcement procedures and issue technical interpretations of OSHA regulations and standards.

Industrial hygienists analyze, identify and measure workplace hazards or stresses that can cause sickness, impaired health or significant discomfort in workers through chemical, physical, ergonomic or biological exposures. The two main roles of the OSHA industrial hygienist are to spot those conditions and help eliminate or control them through appropriate measures.

Worksite Analysis

A worksite analysis is an essential first step that helps an industrial hygienist determine the jobs and workstations where problems could arise. During the worksite analysis, the industrial hygienist measures and identifies exposures, problem tasks and risks. The most-effective worksite analyses include all jobs, operations and work activities. The industrial hygienist inspects, researches or analyzes how the particular chemicals or physical hazards at that worksite affect worker health. If a situation hazardous to health is discovered, the industrial hygienist recommends the appropriate corrective actions.

Recognizing and Controlling Hazards

Industrial hygienists recognize that engineering, work practice and administrative controls are the primary means of reducing employee exposure to occupational hazards. Engineering controls minimize employee exposure by either reducing or removing the hazard at the source or isolating the worker from the hazard. Engineering controls include eliminating toxic chemicals and substituting nontoxic chemicals, enclosing work processes or confining work operations, and installing general and local ventilation systems.

Work practice controls alter the manner in which a task is performed. Some fundamental and easily implemented work practice controls include:

- changing existing work practices to follow proper procedures that minimize exposures when operating production and control equipment;
- inspecting and maintaining process and control equipment on a regular basis;
- implementing good housekeeping procedures;
- providing good supervision; and
- mandating that eating, drinking, smoking, chewing tobacco or gum and applying cosmetics in regulated areas is prohibited.

Administrative controls include controlling employees' exposure by scheduling production and tasks, or both, in ways that minimize exposure levels. For example, the employer might schedule operations with the highest exposure potential during periods when the fewest employees are present.

When effective work practices or engineering controls are not feasible or while such controls are being instituted, appropriate personal protective equipment must be used. Examples of personal protective equipment are gloves, safety goggles, helmets, safety shoes, protective clothing and respirators. To be effective, personal protective equipment must be:

- individually selected;
- properly fitted and periodically refitted;
- conscientiously and properly worn;
- regularly maintained; and
- replaced, as necessary.

Examples of Job Hazards

To be effective in recognizing and evaluating on-the-job hazards and recommending controls, industrial hygienists must be familiar with the hazards' characteristics. Potential hazards can include air contaminants and chemical, biological, physical and ergonomic hazards.

Air Contaminants

Air contaminants are commonly classified as either particulate or gas and vapor contaminants. The most common particulate contaminants include dusts, fumes, mists, aerosols and fibers. Dusts are solid particles generated by handling, crushing, grinding, colliding, exploding and heating organic or inorganic materials such as rock, ore, metal, coal, wood and grain. Any process that produces dust fine enough to remain in the air long enough to be inhaled or ingested should be regarded as hazardous until proven otherwise.

Fumes are formed when material from a volatilized solid condenses in cool air. In most cases, the solid particles resulting from the condensation react with air to form an oxide.

The term mist is applied to liquid suspended in the atmosphere. Mists are generated by liquids condensing from a vapor back to a liquid or by a liquid being dispersed by splashing or atomizing. Aerosols are also a form of a mist characterized by highly respirable, minute liquid particles.

Fibers are solid particles whose length is several times greater than their diameter, such as asbestos.

Gases are formless fluids that expand to occupy the space or enclosure in which they are confined. They are atomic, diatomic or molecular in nature as opposed to droplets or particles that are made up of millions of atoms or molecules.

Through evaporation, liquids change into vapors and mix with the surrounding atmosphere. Vapors are the volatile form of substances that are normally in a solid or liquid state at room temperature and pressure. Vapors are gases in that true vapors are atomic or molecular in nature.

Chemical Hazards

Harmful chemical compounds in the form of solids, liquids, gases, mists, dusts, fumes and vapors exert toxic effects by inhalation (breathing), absorption (through direct contact with the skin) or ingestion (eating or drinking). Airborne chemical hazards exist as concentrations of mists, vapors, gases, fumes or solids. Some are toxic through inhalation and some of them irritate the skin on contact; some can be toxic by absorption through the skin or through ingestion; and some are corrosive to living tissue.

The degree of worker risk from exposure to any given substance depends on the nature and potency of the toxic effects and the magnitude and duration of exposure.

Information on the risk to workers from chemical hazards can be obtained from the material safety data sheet (MSDS) that OSHA's Hazard Communication Standard requires be supplied by the manufacturer or importer to the purchaser of all hazardous materials (29 CFR 1910.1200). The MSDS is a summary of the important health, safety, and toxicological information on the chemical or the mixture's ingredients. Other provisions of the Hazard Communication Standard require that all containers of hazardous substances in the workplace have appropriate warning and identification labels.

Biological Hazards

These include bacteria, viruses, fungi and other living organisms that can cause acute and chronic infections by entering the body either directly or through breaks in the skin. Occupations that deal with plants, animals or their products, or with food and food processing may expose workers to biological hazards. Laboratory and medical personnel also can be exposed to biological hazards. Any occupations that result in contact with bodily fluids pose a risk to workers from biological hazards.

In occupations where animals are involved, biological hazards are dealt with by preventing and controlling diseases in the animal population as well as properly caring for and handling infected animals. Also, effective personal hygiene, particularly proper attention to minor cuts and scratches, especially on the hands and forearms, helps keep worker risks to a minimum.

In occupations where there is potential exposure to biological hazards, workers should practice proper personal hygiene, particularly hand washing. Hospitals should provide proper ventilation, proper personal protective equipment such as gloves and respirators, adequate infectious waste disposal systems, and appropriate controls, including isolation in instances of particularly contagious diseases such as tuberculosis.

Physical Hazards

These include excessive levels of ionizing and nonionizing electromagnetic radiation, noise, vibration, illumination and temperature.

In occupations where there is exposure to ionizing radiation, time, distance and shielding are important tools in ensuring worker safety. Danger from radiation increases with the amount of time one is exposed to it; hence, the shorter the time of exposure, the smaller the radiation danger.

Distance also is a valuable tool in controlling exposure to both ionizing and nonionizing radiation. Radiation levels from some sources can be estimated by comparing the squares of the distances between the worker and the source. For example, at a reference point of 10 ft. from a source, the radiation is 1/100 of the intensity at 1 ft. from the source.

Shielding also is a way to protect against radiation. The greater the protective mass between a radioactive source and the worker, the lower the radiation exposure.

Similarly, shielding workers from nonionizing radiation can also be an effective control method. For example, workers exposed to radiant heat can be protected by installing reflective shields and providing them with protective clothing. In some instances, however, limiting exposure to or increasing distance from certain forms of nonionizing radiation, such as lasers, is not effective. For example, an exposure to laser radiation that is faster than the blinking of an eye can be hazardous and would require workers to be miles from the laser source before being adequately protected.

Noise, another significant physical hazard, can be controlled by various measures. Noise can be reduced by:

- Installing equipment and systems that have been engineered, designed, and built to operate quietly;

- Enclosing or shielding noisy equipment;
- Making certain that equipment is in good repair and properly maintained with all worn or unbalanced parts replaced;
- Mounting noisy equipment on special mounts to reduce vibration; and
- Installing silencers, mufflers or baffles.

Substituting quiet work methods for noisy ones is another significant way to reduce noise (e.g., welding parts rather than riveting them). Also, treating floors, ceilings and walls with acoustical material can reduce reflected or reverberant noise. In addition, erecting sound barriers at adjacent workstations around noisy operations will reduce those workers' exposure to noise.

It is also possible to reduce noise exposure by:
- Increasing the distance between the source and the receiver;
- Isolating workers in acoustical booths;
- Limiting workers' exposure time to noise; and
- Providing hearing protection.

OSHA also requires that workers in noisy surroundings be periodically tested as a precaution against hearing loss.

Ergonomic Hazards

The science of ergonomics is the study and evaluation of a range of tasks including, but not limited to, lifting, holding, pushing, walking and reaching. Many ergonomic problems result from technological changes, such as increased assembly line speeds, added specialized tasks and increased repetition. However, some problems arise from poorly designed job tasks. Any of those conditions can cause ergonomic hazards such as excessive vibration and noise, eye strain, repetitive motion and heavy lifting problems. Improperly designed tools or work areas also can be ergonomic hazards. Repetitive motions or repeated shocks over prolonged periods of time as in jobs involving sorting, assembling and data entry can often cause irritation and inflammation of the tendon sheath of the hands and arms, a condition known as carpal tunnel syndrome.

Ergonomic hazards are avoided primarily by the effective design of a job or job site and by better designed tools or equipment that meet workers' needs in terms of physical environment and job tasks. Through thorough worksite analyses, employers can set up procedures to correct or control ergonomic hazards by:
- Using the appropriate engineering controls (e.g., designing or redesigning workstations, lighting, tools, equipment);
- Teaching correct work practices (e.g., proper lifting methods);
- Employing proper administrative controls (e.g., shifting workers among several different tasks, reducing production demand and increasing rest breaks); and
- Providing and mandating personal protective equipment, if necessary.

Evaluating working conditions from an ergonomics standpoint involves looking at the total physiological and psychological demands of the job on the worker. Overall, the benefits of a well-designed, ergonomic work environment can include increased efficiency, fewer accidents, lower operating costs and more effective use of personnel.

Routes of Entry

For a harmful agent to exert its toxic effect, it must come into contact with a body cell and must enter the body through inhalation, skin absorption or ingestion. Chemical compounds in the form of liquids, gases, mists, dusts, fumes and vapors can cause problems by inhalation, absorption or ingestion.

Inhalation

Inhalation involves those airborne contaminants that can be inhaled directly into the lungs and can be physically classified as gases, vapors and particulate matter that includes dusts, fumes, smokes and mists.

Inhalation, as a route of entry, is particularly important because of the rapidity with which a toxic material can be absorbed in the lungs, pass into the bloodstream and reach the brain. Inhalation is the major route of entry for hazardous chemicals in the work environment.

Absorption

Penetration through the skin can occur quite rapidly if the skin is cut or abraded. Intact skin, however, offers a reasonably good barrier to chemicals. Unfortunately, there are many compounds that can be absorbed through intact skin.

Some substances are absorbed via hair follicle openings and others dissolve in the fats and oils of the skin, such as organic lead compounds, many nitro compounds and organic phosphate pesticides. Compounds that are good solvents for fats (e.g., toluene, xylene) also can cause problems by being absorbed through the skin.

Many organic compounds, such as cyanides and most aromatic aniines, amides and phenols, can produce systemic poisoning by direct contact with the skin. Absorption of toxic chemicals through the skin and eyes is the second most important route of entry.

Ingestion

In the workplace, people can unknowingly eat or drink harmful chemicals if they store drinking containers in the workplace or do not wash before eating. Toxic compounds are capable of being absorbed from the gastrointestinal tract into the blood stream. Lead oxide can cause serious problems if people working with this material are allowed to eat or smoke in work areas. In this situation, careful and thorough washing is required both before eating and at the end of every shift.

Inhaled toxic dusts can also be ingested in amounts that may cause trouble. If the toxic dust swallowed with food or saliva is not soluble in digestive fluids, it is eliminated directly through the intestinal tract. Toxic materials that are readily soluble in digestive fluids can be absorbed into the blood from the digestive system.

In addition to studying all routes of entry when evaluating the work environment (e.g., snack foods or lunches in the work area, solvents being used to clean work clothing and hands), specific types of air contaminants must be identified.

Types of Air Contaminants

There are precise meanings of certain words commonly used in industrial hygiene. These must be used correctly in order to:
- Understand the requirements of OSHA's regulations;
- Effectively communicate with other workers in the field of industrial hygiene; and
- Intelligently prepare purchase orders to procure health services and personal protective equipment.

For example, a fume respirator is worthless as protection against gases or vapors. Too frequently, terms (e.g., gases, vapors, fumes, mists) are used interchangeably. Each term has a definite meaning and describes a certain state of matter. Air contaminants are commonly classified as either particulate contaminants or gas and vapor contaminants.

Particulate Contaminants

The most common particulate contaminants include dusts, fumes, mists and fibers.

Dusts

Dusts are solid particles generated by handling crushing grinding, rapid impact, detonation and decrepitation (breaking apart by heating) of organic or inorganic materials, such as rock, ore, metal, coal, wood and grain.

Dust is a term used in industry to describe airborne solid particles that range in size from 0.1 to 25 micrometers. One micrometer is a unit of length equal to one millionth of a meter. A micrometer is also referred to as a "micron" and is equal to 1/25,400 of an inch. Dust can enter the air from various sources, such as the handling of dusty materials, or during processes such as grinding, crushing, blasting and shaking.

Most industrial dusts consist of particles that vary widely in size, with the small particles greatly outnumbering the large ones. Consequently (with few exceptions), when dust is noticeable in the air near a dusty operation, more invisible dust particles than visible ones are probably present. A process that produces dust fine enough to remain in the air long enough to be inhaled should be regarded as hazardous until proven otherwise.

An airborne dust of a potentially toxic material will not cause pulmonary illness if its particle size is too large to gain access to the lungs. Particles 10 ppm in diameter and larger are known as nonrespirable These particles will be deposited in the respiratory system long before they reach the alveolar sacs — the most important area in the lungs.

Particles less than 10 ppm in diameter are known as respirable. Because the particles are likely to reach the alveoli in great quantities, they are potentially more harmful than larger particles.

By using a size-selective device (e.g., a cyclone) ahead of a filter at a specific airflow-sampling rate, it is possible to collect respirable-sized particles on the filter. This allows one to determine the dust concentration of respirable particles.

Fumes

Fumes are formed when the material from a volatilized solid condenses in cool air. The solid particles that are formed make up a fume that is extremely fine — usually less than 1.0 micron in diameter. In most cases, the hot vapor reacts with the air to form an oxide. Gases and vapors are not fumes, although the terms are often mistakenly used interchangeably.

Welding, metalizing and other operations involving vapors from molten metals may produce fumes; these may be harmful under certain conditions. Arc welding volatilizes metal vapor that condenses as the metal or its oxide in the air around the arc. In addition, the rod coating is partially volatilized. Because these fumes are extremely small, they are readily inhaled.

Other toxic fumes, such as those formed when welding structures that have been painted with lead-based paints or when welding galvanized metal, can produce severe symptoms of toxicity rather rapidly in the absence of good ventilation or proper respiratory protection.

Mists

Mists are suspended liquid droplets generated by condensation of liquids from the vapor back to the liquid state or by breaking up a liquid into a dispersed state, such as by splashing or atomizing. The term mist is applied to a finely divided liquid suspended in the atmosphere. Examples include oil mist produced during cutting and grinding operations, acid mists from electroplating acid or alkali mists from pickling operations, and spray mist from spray-finishing operations.

Fibers

Fibers are solid particles that have a slender, elongated structure with length several times as great as their diameter. Examples include asbestos, fibrous talc, and fiberglass. Airborne fibers may be found in construction activities, mining friction product manufacturing and fabrication, and demolition operations.

Gas and Vapor Contaminants

Gases

Gases are formless fluids that expand to occupy the space or enclosure in which they are confined. Gases are a state of matter in which the molecules are unrestricted by cohesive forces. Examples are arc-welding gases, internal combustion engine exhaust gases and air.

Vapors

Vapors are the volatile form of substances that are normally in the solid or liquid state at room temperature and pressure. Evaporation is the process by which a liquid is changed into the vapor state and mixed with the surrounding atmosphere. Some of the most common exposures to vapors in industry occur from organic solvents. Solvents with low boiling points readily form vapors at room temperature. Solvent vapors enter the body mainly by inhalation, although some skin absorption can occur.

Units of Concentration

In addition to the definitions concerning states of matter that find daily usage in the vocabulary of the industrial hygienist, other terms used to describe degree of exposure include the following:

ppm. Parts per million. Parts of contaminated air on a volumetric basis. It is used for expressing the concentration of a gas or vapor.

mg/m3. Milligrams of a substance per cubic meter of air. The term is most commonly used for expressing concentrations of dusts, metal fumes or other particles in the air.

mppcf. Millions of particles of a particulate per cubic foot of air.

f/cc. The number of fibers per cubic centimeter of air. This term is used for expressing the concentration of airborne asbestos fibers.

The concentration of a gas or vapor in air is usually expressed in parts per million (ppm), but may be converted to mg/m3 at a temperature of 25°C and a pressure of 760 mm Hg:

mg/m3 = ppm × (Molecular Weight/24.45)

Example: A 50-ppm concentration of carbon monoxide (molecular weight = 28) is equivalent to a concentration of 57.26 mg/m3 at 25°C and 760 mm Hg. Note also that:

Concentration (ppm) = Concentration (%) × 10,000

Example: A concentration of a gas or vapor equal to 0.01 percent is equivalent to a concentration of 100 ppm. The health and safety professional recognizes that air contaminants exist as a gas, dust, fumes, mist or vapor in the workroom air. In evaluating the degree of exposure, the measured concentration of the air is compared to limits or exposure guidelines.

Threshold Limit Values

Threshold limit values (TLVs) have been established for airborne concentrations of many chemical compounds. It is important to understand something about TLVs and the terminology in which their concentrations are expressed.

The American Conference of Governmental Industrial Hygienists (ACGIH) publishes annually the list of "Threshold Limit Values and Biological Exposure Indices." The ACGIH is not an official government agency. Membership is limited to professional personnel in government agencies or educational institutions engaged in occupational safety and health programs.

The data for establishing TLVs comes from animal studies, human studies and industrial experience. The limit may be selected for several reasons; it may be based on the fact that a substance is very irritating to the majority of people exposed, or, other substances may be asphyxiants. Other reasons for establishing a TLV include the fact that certain chemical compounds are anesthetic, or fibrogenic, or can cause allergic reactions or malignancies. Some additional TLVs have been established because exposure above a certain airborne concentration is a nuisance.

The basic idea of TLVs is fairly simple. They refer to airborne concentrations of substances and represent conditions under which it is believed that nearly all workers may be repeatedly exposed, day after day, without adverse effect.

Because individual susceptibility varies widely, an occasional exposure of an individual at (or even below) the threshold limit may not prevent discomfort, aggravation of a preexisting condition or occupational illness. In addition to the TLVs set for chemical compounds, there are limits for physical agents, such as noise, microwaves, and heat stress.

Several important points should be noted concerning TLVs. First, "TLV" is a copyrighted trademark of the ACGIH. It should not be used to refer to the values published in OSHA or other standards. OSHA's limits are known as permissible exposure limits and will be discussed later. The ACGIH TLVs are not mandatory federal or state employee exposure standards. These limits are not fine lines between safe and dangerous concentrations nor are they a relative index of toxicity. Three categories of TLVs are specified.

Time-weighted average (TLV-TWA) is the time-weighted average concentration for a normal eight-hour workday or 40-hour workweek to which nearly all workers may be repeatedly exposed, day after day, without adverse effect. Time-weighted averages permit excursions above the limit, provided they are compensated by equivalent excursions below the limit during the workday.

Short-term exposure limit (TLV-STEL) is the maximal concentration to which workers can be exposed continuously for a short period of time without suffering from any of the following:
- Irritation;
- Chronic or irreversible tissue change; or
- Narcosis of sufficient degree to increase accident proneness, impair self-rescue or materially reduce work efficiency.

The STEL is a 15-minute TWA exposure that should not be exceeded at any time during a workday, even if the eight-hour time weighted average, is within the TLV-TWA. Exposures above the TLV-TWA up to the STEL should not be longer than 15 minutes and should not occur more than four times per day. There should be at least 60 minutes between successive exposures in this range.

The STEL is not a separate independent exposure limit; rather it supplements the time-weighted average limit where there are recognized acute effects from a substance whose toxic effects are primarily of a chronic nature. STELs are recommended only where toxic effects have been reported from high short-term exposures in humans or animals.

Ceiling (TLV-C) is the concentration that should not be exceeded even instantaneously. Although the time-weighted average concentration provides the most satisfactory, practical way of monitoring airborne agents for compliance with the limits, there are certain substances for which it is inappropriate. In the latter group are substances that are predominantly fast acting and whose threshold limit is more appropriately based on this particular response. A ceiling limit that should not be exceeded best controls substances with this type of response.

For some substances (e.g., irritant gases), only one category, the TLV-C, may be relevant. For other substances, either two or three categories may be relevant, depending upon their physiologic action. It is important to observe that if any one of these three TLVs is exceeded, a potential hazard from that substance is presumed to exist.

Skin Notation

Nearly one-fourth of the substances in the TLV list are followed by the designation "Skin." This refers to the potential significant contribution to the overall exposure by the cutaneous route, including mucous membranes and the eyes, usually by direct contact with the substance. This designation is intended to suggest appropriate measures for the prevention of cutaneous absorption.

Federal Occupational Safety And Health Standards

The first compilation of health and safety standards promulgated by the Department of Labor's OSHA in 1970 was derived from the then-existing federal standards and national consensus standards. Thus, many of the 1968 TLVs established by the ACGIH became federal standards or permissible exposure limits (PEL).

Also, certain workplace quality standards known as maximal acceptable concentrations of the American National Standards Institute (ANSI) were incorporated as federal health standards in 29 CFR 1910.1000 as national consensus standards. These PEL values for general industry were subsequently updated in 1989. Unlike the TLVs, OSHA's PELs are enforceable by law. Employers must keep employee exposure levels below the PELs of regulated substances. As with TLVs, there are three types of PELs.

Time-Weighted Average

In adopting the TLVs of the ACGIH, OSHA also adopted the concept of the time-weighted average concentration for a workday. The eight-hour time-weighted average (TWA) is the average concentration of a chemical in air over an eight-hour exposure period. In general:

$$TWA = \frac{C_aT_a + C_bT_b + C_cT_c \ldots C_rT_r}{8}$$

Where:
- T_a is the time of the first exposure period.
- C_a is the concentration of contaminant in period "a."
- T_b is another time period during the shift.
- C_b is the concentration during period "b."
- C_r is the concentration during the "no" time period.
- T_r is the "nth" time period.

To illustrate the formula prescribed above, assume that a substance has an eight-hour TWA PEL of 100 ppm. Assume that an employee is subject to the following exposure:
- Two hours exposure at 150 ppm;
- Two hours exposure at 75 ppm; and
- Four hours exposure at 50 ppm.

Substituting this information in the formula, we have:

$$TWA = \frac{(150)(2) + (75)(2) + (50)(4)}{8} = 81.25 \text{ ppm}$$

Since 81.25 ppm is less than 100 ppm, the eight-hour TWA limit, the exposure is acceptable.

Amendments to OSHA's Air Contaminant Standard

On January 19, 1989, OSHA amended its Air Contaminant standard, 1910.1000 (54 Fed. Reg. 2332). New limits were established for many substances and many new PELs were set for substances previously not regulated by OSHA.

The Eleventh Circuit Court of Appeals issued a decision vacating the "Final Rules" of the Air Contaminants Standard (29 CFR 1910.1000) on July 7, 1992. The court's decision struck down the entire standard. However, PELs specified in 29 CFR 1910.1001 through the end of Subpart Z of Part 1910 are unaffected by the court's decision.

Effective March 22, 1993, OSHA is enforcing only the following Permissible Exposure Limits (PELs) in 29 CFR 1910.1000:
- Those limits specified in the "Transitional Limits" column of Table Z-1-A;
- All limits in Table Z-2; and
- All limits in Table Z-3.

Expanded Health—Work Practice Standards

OSHA also has promulgated expanded health standards at 1910.1001 through 1910.1101 for substances including asbestos, vinyl chloride, arsenic, lead, benzene, coke oven emissions, cotton dust, dibromocloropropane, acrylonitrile, ethylene oxide and formaldehyde. Additionally, 13 identified carcinogens are regulated.

These standards contain work practice requirements such as exposure monitoring protective equipment, housekeeping hygiene facilities, medical surveillance, and employee training in addition to permissible limits. Additional information regarding industrial hygiene, including sampling instrumentation and methods, can be found in OSHA Technical Manual, CPL 2-2.20B.

Early in its history, OSHA recognized industrial hygiene as an integral part of a healthful work setting. OSHA places a high priority on using industrial hygiene concepts in its health standards and as a tool for effective enforcement of job safety and health regulations. By recognizing and applying the principles of industrial hygiene to the work environment, U.S. workplaces will become more healthful and safer.

The National Institute for Occupational Safety and Health publishes a Pocket Guide to Chemical Hazards. It may be obtained at no charge by calling 1-800-35NIOSH (1-800-356-4674) or by logging on to www.cdc.gov/niosh/homepage.html.

Bloodborne Pathogens

Bloodborne pathogens are pathogenic microorganisms that are present in human blood and can infect and cause disease in humans. These pathogens include, but are not limited to, Hepatitis B Virus (HBV), which causes Hepatitis B, a serious liver disease and Human Immune Deficiency Virus (HIV), which causes Acquired Immunodeficiency Syndrome (AIDS).

OSHA has determined that certain employees (particularly healthcare employees) face a significant health risk as a result of occupational exposure to blood and other potentially infectious materials (OPIM) because they may contain bloodborne pathogens.

To minimize or eliminate the risk of occupational exposure to bloodborne pathogens, OSHA issued the Occupational Exposure to Bloodborne Pathogens Standard (29 CFR 1910.1030). This standard prescribes actions that employers must take to reduce the risk of exposure to bloodborne pathogens in the workplace. These actions include the use of:
- Engineering and work practice controls;
- Personal protective equipment;
- Training;
- Medical surveillance;
- Hepatitis B vaccinations; and
- Signs and labels.

Scope and Application

The standard applies to all employees with occupational exposure to blood and OPIM. Occupational exposure means "a reasonably anticipated skin, eye, mucous membrane or parenteral contact with blood or OPIM that may result from the performance of the employees duties." Blood is defined as "human blood, human blood components, and products made from human blood."

OPIM includes the following human body fluids:
- Semen;
- Vaginal secretions;
- Cerebrospinal fluid;
- Synovial fluid;
- Pleural fluid;

- Perinicardial fluid;
- Peritoneal fluid;
- Amniotic fluid;
- Saliva in dental procedures;
- Any body fluid visibly contaminated with blood; and
- All body fluids in situations where it is difficult or impossible to differentiate between body fluids.

OPIM also includes:
- Unfixed tissue or organs from a human;
- FHV-containing cells or tissue cultures;
- Organ cultures and HIV- or HBV-containing culture medium or other solutions; and
- Blood, organs or other tissue from experimental animals infected with HIV or HBV.

The Bloodborne Pathogens Standard covers many types of employees including those in healthcare, non-healthcare, and permanent and temporary worksites. Examples of employees in healthcare facilities include physicians and surgeons, nurses, dentists and dental workers, and laboratory personnel. Non-healthcare facilities employees include those who service and repair medical and dental equipment, infectious waste disposal employees and employees in law enforcement and correctional institutions.

Exposure Control Plan

The Exposure Control Plan (ECP) is the key provision of the standard. It requires the employer to identify employees who will receive training, protective equipment, vaccination and other provisions of the standard.

The ECP requires employers to identify, in writing, tasks and procedures as well as job classifications where occupational exposure to blood occurs without regard to personal protective equipment. The plan must also set forth the schedule for implementing other provisions of the standard and specify the procedure for evaluating circumstances surrounding exposure incidents. It must be accessible to employees and available to OSHA and National Institute for Occupational Safety and Health (NIOSH) representatives. Employers must review and update the plan annually or more often if changes in exposure occur.

Methods of Compliance

The standard describes various methods of compliance that the employer must take to protect their employees from exposure to bloodborne pathogens. These methods include universal precautions, engineering and work practice controls, personal protective equipment and housekeeping.

Universal precautions is an approach to infection control in which all human blood and certain body fluids are treated as if known to be infectious for HIV, HBV, and other bloodborne pathogens. Universal precautions are OSHA's required method of control to protect employees from exposure to all human blood and OPIM.

Engineering controls are controls that isolate or remove the bloodborne pathogens hazard from the workplace. Examples of engineering controls include puncture-resistant sharps containers; mechanical needle recapping devices; and biosafety cabinets. To ensure effectiveness, engineering controls must be examined and maintained or replaced on a regularly scheduled basis.

Work practice controls reduce the likelihood of exposure by altering the manner in which a task is performed. Some examples of work practice requirements in the standard include not bending or breaking contaminated sharps and hand washing when gloves are removed and as soon as possible after contact with body fluids.

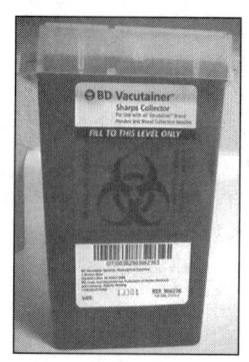

Exhibit 4-1. Puncture-resistant sharps containers are an effective engineering control.

Personal protective equipment must be used if occupational exposure remains after instituting engineering and work practice controls or if these controls are not feasible. Personal protective equipment is specialized clothing or equipment that is worn by an employee for protection against a hazard. Employers must provide, at no cost, and require employees to use appropriate personal protective equipment such as gloves, gowns, masks, mouthpieces, and resuscitation devices. The employer must clean, repair and replace these safety items when necessary.

In addition to other compliance methods, the standard requires that the employer maintain the work site in a clean and sanitary condition. Employees must follow certain procedures for cleaning and decontaminating the environment, equipment and work surfaces and for handling contaminated laundry and regulated waste.

Contaminated work surfaces must be decontaminated with a disinfectant upon completion of procedures or when contaminated by splashes, spills or contact with blood or OPIM. The employer must develop a written schedule for cleaning and decontaminating the work site based on the location within the facility, type of surface to be cleaned, amount of soil and the task being performed. Reusable trash containers must also be cleaned on a regular basis and after contamination.

Contaminated laundry is any laundry that may contain blood or OPIM or may contain sharps. The standard requires that contaminated laundry be handled as little as possible with a minimum of agitation. It must be bagged or containerized at the location where it was used and not be sorted or rinsed where it was used. Contaminated laundry must also be placed and transported in bags or containers and properly labeled. When a facility uses universal precautions in handling soiled laundry, alternative labeling or color-coding is sufficient if it permits all employees to recognize the containers as requiring compliance with universal precautions.

Regulated waste is defined in the standard as:
- Liquid or semi-liquid blood or OPIM;
- Contaminated items that would release blood or OPIM in a liquid or semi-liquid state if compressed;
- Items that are caked with dried blood or OPIM and are capable of releasing these materials during handling;
- Contaminated sharps; and
- Pathological and microbiological waste containing blood or OPIM.

Regulated waste must be placed in closeable, leak-proof containers designed to contain all contents during handling, storing, transporting, or shipping and labeled appropriately.

HIV and HBV Research Laboratories and Production Facilities

The standard describe various requirements for HIV and HBV research laboratories and production facilities. A research laboratory produces or uses research laboratory-scale amounts of HIV or HBV. Research laboratories may produce high concentrations of HIV or HBV, but not in the volume found in production facilities that produce high volumes and high concentrations of HIV and HBV.

Some requirements in the standard specific to these facilities include that regulated waste must be incinerated, autoclaved or decontaminated before disposal and that laboratory doors must be closed when work involves HIV and HBV hazards.

In addition, warning signs must be placed on access doors and biological safety cabinets must be used when working with potentially infectious material. Employees in these facilities also require additional training.

Hepatitis B Vaccination and Post-Exposure Evaluation and Follow-up

The employer is required to make available the Hepatitis B vaccine and vaccination series to all employees who may be exposed in the workplace and post-exposure evaluation and follow-up to all employees who have had an exposure incident. All medical evaluations and procedures must be provided:
- At no cost to the employee;
- At a reasonable time and place;
- Under the supervision of a licensed physician or health care professional; and
- According to current recommendations of the U.S. Public Health Service.

The Hepatitis B vaccine and vaccination series must be offered within 10 working days of initial assignment to employees who have occupational exposure to blood or OPIM. Exceptions to this requirement are:
- When employees have previously completed the Hepatitis B vaccination series;
- Immunity is confirmed through anti-body testing; or
- The vaccine is contraindicated for medical reasons.

Employees must sign a declination form if they choose not to be the vaccinated; however, they may request and obtain the vaccination at a later date at no cost.

A confidential medical evaluation and follow-up must be made available to an employee involved in an exposure incident. An exposure incident is a specific eye, mouth, other mucous membrane, nonintact skin or parenteral contact with blood or OPIM that results from the performance of an employee's duties.

The evaluation and follow-up procedures must include

- Documenting the circumstances of exposure;
- Identifying and testing the source individual, if feasible;
- Testing the exposed employee's blood if he or she consents; and
- Conducting post-exposure prophylaxis, counseling and evaluation of reported illnesses.

Following the post-exposure evaluation, the healthcare professional must provide a written opinion to the employer. This opinion is limited to a statement that the employee has been informed of the need, if any, for further evaluation or treatment. All other findings are confidential. The employer must provide a copy of the written opinion to the employee within 15 days of the evaluation.

Communication of Hazards to Employees

The hazards of bloodborne pathogens must be communicated to employees through signs, labels and training. The standard requires that warning labels be attached to containers of regulated waste, refrigerators or freezers containing blood or OPIM, and other containers used to store, transport, or ship blood or OPIM.

The warning label must include the universal biohazard symbol followed by the term "BIOHAZARD" in a fluorescent orange or orange-red color with lettering or symbols in a contrasting color.

The labels are not required when:

- Red bags or red containers are used for regulated waste;
- Regulated waste has been decontaminated;
- Individual containers of blood or OPIM are placed in a labeled container during storage, transport, shipment or disposal; and
- Containers of blood or blood products are labeled as to their content and have been released for transfusion or other clinical use.

Contaminated equipment must also be marked with a label that states which portion of the equipment remains contaminated.

The biohazard label must also be posted at the entrance to HIV and HBV research laboratories and production facilities work areas. The employer must ensure that all employees with occupational exposure participate in an effective training program. Training must be provided within 90 days after the effective date of the standard and annually thereafter.

An individual who is knowledgeable in the subject matter must provide training at no cost to employees during a time that is regularly accessible to employees. The training program must include an:

- Accessible copy of the regulatory text and explanation of its contents;
- Explanation of the modes of transmitting and epidemiology of HBV and HIV;
- Explanation of the written exposure control plan and how to obtain a copy;
- Explanation of use and limitations of engineering controls, work practices and personal protective equipment.

The training materials must be appropriate in content, language and vocabulary to the educational, literacy, and language background of the employees.

Recordkeeping

Employers must establish and maintain accurate records for each employee with occupational exposure. The standard requires employers to maintain two types of employee-related records: medical and training. The medical records must include:

- Employee's name and social security number,
- Employee's Hepatitis B vaccination status;
- Post-exposure evaluation and follow-up procedures results, as necessary;
- The healthcare professional's written opinion, as necessary; and
- Other specific information that has been provided to the healthcare professional.

The medical record must be kept confidential and retained for the duration of employment plus 30 years in accordance with OSHA's Access to Employee Exposure and Medical Records Standard. These records must be made available upon request to employees, anyone with the written consent of the employee and OSHA and NIOSH representatives.

Training records must include:
- Training dates;
- Contents or summary of the training session;
- Names and qualifications of the trainers; and
- Names of job titles of trainees.

All training records must be kept for three years from the date of the training. If the employer ceases to do business, medical and training records must be transferred to the successor employer. If there is no successor employer, the employer must notify the Director of NIOSH for instructions regarding the disposal of records. This must be done at least three months prior to the disposing of the records.

Work Activities Involving Potential Exposure to Bloodborne Pathogens

Below are listed the tasks and procedures in our facility where human blood and other potentially infectious materials are handled that may result in exposure to bloodborne pathogens:

Procedure/Task	Job Classification	Location/Department

Vaccination Declination Form

DATE:

Employee Name:

Employee ID #:

I understand that due to my occupational exposure to blood or other potential infectious materials I may be at risk of acquiring Hepatitis B virus (HBV) infection. I have been given the opportunity to be vaccinated with Hepatitis B vaccine at no charge to myself.

However, I decline the Hepatitis B vaccination at this time. I understand that by declining this vaccine, I continue to be at risk of acquiring Hepatitis B, a serious disease. If, in the future, I continue to have occupational exposure to blood or other potentially infectious materials and I want to be vaccinated with Hepatitis B vaccine, I can receive the vaccination series at no charge to me.

Employee Signature   Date

Facility Representative Signature   Date

Asbestos and Lead Awareness

OSHA regulations apply for industrial applications and construction uses of asbestos and lead. OSHA enforces safety rules through their regional offices . OSHA can issue citations to federal agencies, but cannot issue fines.

Asbestos Exposure

Asbestos is a naturally occurring mineral that is used in vehicle brakes and building materials. It is highly resistant to heat and corrosion and has been used extensively in the past. Auto workers may be exposed to asbestos during brake and clutch repairs.

Inhaling asbestos fibers can cause serious lung disease that may not appear for years. Asbestos can cause a buildup of scar-like tissue in the lungs, resulting in a loss of lung functions and often disability and death.

Simply being exposed to asbestos does not mean a person will become ill or die. In an eight-hour work shift, employees can be exposed to 0.1 fiber per cubit centimeter of air. This small amount of asbestos is normally filtered by the body's system of defense, such as mucous membranes and nose hairs.

Lead Exposure

Like asbestos, lead is very resistant to corrosion and has a low melting temperature. For these reasons it is used extensively in manufacturing, construction work, welding, cutting, brazing and some painting. Lead also can be found in radiator repair shops, motor gaskets and firing ranges.

Once lead gets into the blood, it settles in the blood and soft tissues, such as the kidney, bone marrow, liver and brain. It also settles in the bones and teeth of workers. Overexposure to lead is a leading cause of workplace illnesses. Lead has a permissible exposure level based on an eight-hour day of less than 50 micrograms per cubic meter of air.

Worker Protection

Both asbestos and lead must enter the body to cause damage. The primary route of entry is inhalation. In some small cases, they may enter via ingestion when a worker eats or smokes in an area where airborne lead or asbestos is found.

The best way to protect workers is to keep the material from getting into the air, typically by using a wet method or process. If the

OCCUPATIONAL HEALTH AND INDUSTRIAL HYGIENE

asbestos or lead is wet, it cannot float in the air and thus cannot enter workers' lungs. For example, if materials contain lead or asbestos, the worker may wet the material using a spray bottle or mister to reduce or eliminate the possibility of inhalation.

Employees should never use compressed air to clean asbestos or lead particles because the resulting blast of air throws the contaminant into the air where it will float for a long time, often until someone inhales it.

Personal protective equipment, gloves and respirators with a HEPA filter provide protection as well. If there is a possibility of eye irritation, employees must wear the goggles and face shields (See Exhibit 4-2).

**Exhibit 4-2.
Respirators and goggles
can provide protection
from asbestos.**

After working with asbestos or lead, the employee should immediately wash his or her hands when finished with the procedure. Naturally, smoking should not be allowed in any area where these contaminants are used.

Medical Considerations

Before any work with asbestos began, the facility should have been monitored to determine if the level of asbestos exceeded the permissible exposure limit of 0.1 fibers per cubic centimeter. If a process has changed and more asbestos is possibly being released into the air through a particular process, contact the safety office to arrange for another evaluation.

If the company's work requires that workers be exposed beyond the PEL, they should alert their personal physicians so that the physicians can monitor them for possible asbestos-related problems.

If exposed to lead or asbestos on a routine basis beyond the PEL, workers also should contact the local safety office and enroll in a medical surveillance program.

Significant Changes in the Asbestos Standard for General Industry, 29 CFR 1910.1001 (through June 29, 1995)

Topic	Change
PEL	The PEL has been reduced to 0.1 fibers/cc from 0.2 fibers/cc as a TWA. The Excursion Limit remains 1.0 fibers/cc averaged over 30 minutes.
PACM	Installed thermal system insulation and sprayed-on and troweled-on surfacing materials found in buildings constructed no later than 1980 are presumed to be asbestos-containing materials (greater than 1% asbestos).
	Asphalt and vinyl flooring material installed no later than 1980 also must be treated as asbestos-containing.
	These presumptions may be rebutted by specified inspection and testing.
Asbestos-containing flooring material	Sanding of asbestos-containing flooring material is prohibited.
	Specific procedures for floor care are mandated.
Brake/clutch repair	Employers must use specific controls and methods specified in Appendix F, unless another method is demonstrated to achieve equivalent results.
Duties of building owners (identification, recordkeeping, notification, signs and labels)	Building and facility owners must determine the presence, location, and quantity of ACM/PACM and keep records of asbestos-containing material and presumed asbestos-containing material.
	They must inform other employers, and their own employees who will perform housekeeping activities, of the presence and location of such materials.
	They must post signs at entrances to mechanical rooms/areas that contain ACM/PACM and that employees may enter.
	Previously installed ACM/PACM must be identified by labels or signs.
Asbestos awareness training	Employers must provide an asbestos awareness training course to employees who will perform housekeeping activities in an area containing ACM or PACM.
Medical surveillance	A pre-employment exam may not be used as a "recent exam" to fulfill any of the standard's medical surveillance requirements, unless the employer paid for the pre-employment exam.

Occupational Noise Exposure

It is estimated by OSHA that there are 2.9 million workers in U.S. production industries who experience eight-hour noise exposures in excess of 90 dBA (46 Fed. Reg. 4078). An additional 2.3 million experience exposure levels in excess of 85 dBA. The Hearing Conservation Amendment (HCA) applies to all 5.2 million employees except for those in oil and gas well drilling and service industries, which are specifically exempted. Additionally, the amendment does not apply to those engaged in construction and agriculture, although a construction industry noise standard exists (29 CFR 1926.52, 1926.101). This standard is identical to paragraphs (a) and (b) of the general industry noise standard (29 CFR 1910.95).

Prior to promulgation of the HCA, the existing noise standard (29 CFR 1910.95(a), (b)) established a permissible noise exposure level of 90 dBA for eight hours and required the employer to reduce exposure to that level by use of feasible engineering and administrative controls. In all cases in which sound levels exceeded the permissible exposure, regardless of the use of hearing-protective devices, "a continuing, effective hearing conservation program" was required. However, the details of such a program were never mandated. Paragraphs (c) through (p) of the HCA replaced paragraph (b)(3) of 29 CFR 1910.95 and supplemented OSHA's definition of an "effective hearing conservation program."

All employees whose noise exposures equal or exceed an eight-hour TWA of 85 dBA must be included in a hearing conservation program composed of five basic components:

- Exposure monitoring;
- Audiometric testing;
- Hearing protection;
- Employee training; and
- Recordkeeping.

Note that although the eight-hour TWA permissible exposure remains 90 dBA, a hearing conservation program becomes mandatory at an eight-hour TWA exposure of 85 dBA.

Monitoring

The HCA requires employers to monitor noise exposure levels in a manner that will accurately identify employees who are subjected to an eight-hour TWA exposure of 85 dBA or more. The exposure measurement must include all noise within an 80 to 130 dBA range. The requirement is performance oriented and allows employers to choose the monitoring method that best suits each situation.

Employees are entitled to observe monitoring procedures and they must be notified of the results of exposure monitoring. However, the method used to notify employees is left to the discretion of the employer.

Employers must re-monitor workers' exposures whenever changes in exposures are sufficient to require new hearing protectors or whenever employees not previously included in the program because they were not exposed to an eight-hour TWA of 85 dBA are included in the program.

Instruments used for monitoring employee exposures must be calibrated to ensure the measurements are accurate. Since calibration procedures are unique to specific instruments, employers should follow the manufacturer's instructions to determine when and how extensively to calibrate.

Audiometric Testing Program

Audiometric testing not only monitors employee hearing acuity over time but also provides an opportunity for employers to educate employees about their hearing and the need to protect it. The audiometric testing program includes obtaining baseline audiograms and annual audiograms and initiating training and follow-up procedures. The audiometric testing program should indicate whether hearing loss is being prevented by the employer's hearing conservation program. Audiometric testing must be made available to all employees who have an average exposure level of 85 dBA. The program shall be provided at no cost to employees

A professional (e.g., audiologist, otolaryngologist, physician) must be responsible for the program, but he or she does not have to be present when a qualified technician is actually conducting the testing. Professional responsibilities include overseeing the program and the work of the technicians, reviewing problem audiograms and determining whether referral is necessary. Either a professional or a trained technician may conduct audiometric testing.

In addition to administering audiometric tests, the tester (or the supervising professional) is also responsible for:
- Ensuring the tests are conducted in an appropriate test environment;
- Seeing that the audiometer works properly;
- Reviewing audiograms for standard threshold shifts (as defined in the HCA); and
- Identifying audiograms that require further evaluation by a professional.

Baseline and Annual Audiograms

There are two types of audiograms required in the hearing conservation program: baseline and annual audiograms. The baseline audiogram is the reference audiogram against which future audiograms are compared. Baseline audiograms must be provided within six months of an employee's first exposure at or above a TWA of 85 dBA.

Testing to establish a baseline audiogram shall be preceded by at least 14 hours without exposure to workplace noise. Hearing protectors may be used as a substitute for this requirement.

Employees shall be notified regarding the need to avoid high levels of nonoccupational noise exposure during the 14-hour period immediately preceding the audiometric examination.

Table 4-1. How Loud Is It?
Typical A-Weighted Sound Levels
(dB, re: 20 micropascals)

150	Loud rock concert
140	Fire arms and air raid sirens
130	Jack hammer
120	Jet plane take-off, car stereo, bands
110	Discotheque
106	Timpani and bass drum rolls
100	Snowmobiles, chain saws, pneumatics
90	Lawnmower, subway noise
70	Vacuum cleaner at 3 meters
60	Conversation at 1 meter
50	Urban Residence
30	North Rim of Grand Canyon
0	Threshold of Hearing (1000 Hz)

When employers obtain audiograms in mobile test vans, baseline audiograms must be completed within one year after an employee's first exposure to workplace noise at or above an average of 85 dBA. Additionally, when mobile vans are used and employers are allowed to delay baseline testing for up to one year, those employees exposed to levels of 85 dBA or more must be issued and fitted with hearing protectors six months after initial exposure. The hearing protectors are to be worn until the baseline audiogram is obtained. Baseline audiograms taken before the effective date of the amendment are acceptable as baselines in the program if the professional supervisor determines that the audiogram is valid. The annual audiogram must be conducted within one year of the baseline. It is important to test hearing on an annual basis to identify changes in hearing acuity so that protective follow-up measures can be initiated before hearing loss progresses.

Evaluation of Audiograms

Annual audiograms must be routinely compared to baseline audiograms to determine whether the audiogram is accurate and whether the employee has lost hearing ability (i.e., to determine whether a standard threshold shift (STS) has occurred). An effective program depends on a uniform definition of an STS. An STS is defined in the amendment as an average shift (or loss) in either ear of 10 dB or more at the 2,000, 3,000, and 4,000 Hertz (Hz) frequencies. A method of determining an STS by computing an average was chosen because it diminishes the number of persons identified as having an STS who are later shown not to have had a significant change in hearing ability.

If an STS is identified, the employee must be fitted or refitted with adequate hearing protectors, instructed in how to use them and required to wear them. In addition, employees must be notified in writing within 21 days from the time the determination is made that their audiometric test results indicate an STS. Some employees with an STS may need to be referred for further testing if the professional determines that their test results are questionable or if they have an ear problem of a medical nature caused or aggravated by wearing hearing protectors. If the suspected medical problem is not thought to be related to wearing protectors, employees must merely be informed that they should see a physician. If subsequent audiometric tests show that the STS identified on a previous audiogram is not persistent, employees exposed to an average level of 90 dBA may discontinue wearing hearing protectors.

OCCUPATIONAL HEALTH AND INDUSTRIAL HYGIENE

A subsequent audiogram may be substituted for the original baseline audiogram if the professional supervising the program determines that the employee has experienced a persistent STS. The substituted audiogram becomes known as the revised baseline audiogram. This substitution will ensure that the same shift is not repeatedly identified. The professional may also need the baseline audiogram after an improvement in hearing has occurred, which will ensure that the baseline reflects actual thresholds as much as possible. When a baseline audiogram is revised, the employer must also retain the original audiogram.

Audiometric Test Requirements
Audiometric tests shall be pure tones, air conduction and hearing threshold examinations, with test frequencies including, at a minimum, 500, 1000, 2000, 3000, 4000, and 6000 Hz. Tests at each frequency shall be taken separately for each ear.

To obtain valid audiograms, audiometers must be used, maintained and calibrated according to the specifications detailed in appendices C and E of the standard.

Hearing Protectors

Hearing protectors must be made available to all workers exposed to an eight-hour TWA of 85 dBA or more at no cost to the employees. This requirement will ensure that employees have access to protectors before they experience a loss in hearing. When baseline audiograms are delayed because it is inconvenient for mobile test vans to visit the workplace more than once a year, protectors must be worn by employees for any period exceeding six months from the time they are first exposed to eight-hour average noise levels of 85 dBA or more until their baseline audiograms are obtained. The use of hearing protectors is also mandatory for employees who have experienced STSs, since these workers are particularly susceptible to noise.

With the help of a person who is trained in fitting hearing protectors, employees shall be given the opportunity to select their hearing protectors from a suitable variety provided by the employer. The protector selected should be comfortable to wear and offer sufficient attenuation to prevent hearing loss. Employees must be shown how to use and care for their protectors; they also must be supervised on the job to ensure that they continue to wear them correctly. Hearing protectors shall be replaced as necessary.

Hearing protectors must provide adequate attenuation in each employee's work environment. The employer must reevaluate the suitability of the employee's present protector whenever there is a change in working conditions that may render the hearing protector inadequate. If workplace noise levels increase, employees must be given more effective protectors. The protector must reduce the level of exposure to at least 90 dBA or 85 dBA when an STS has occurred. A variety of hearing protection is available, ranging from earplugs to ear bands to earmuffs. Some even have radios attached so that workers can communicate in loud environments (See Exhibit 4-4.).

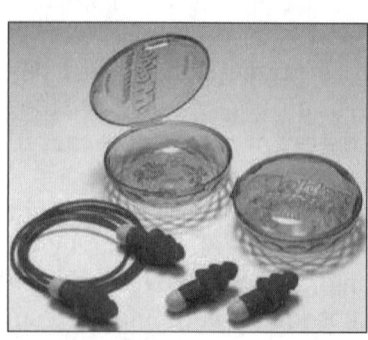

Exhibit 4-3.
Employees must be able to select from a variety of hearing protectors. Here, a popular choice is shown.

Exhibit 4-4.
Headphones with a radio attached to allow worker communication as well as hearing protection.

Training

Employee training is important because when workers understand the hearing conservation programs requirements and why it is necessary to protect their hearing, they will be better motivated to actively participate in the program. Employees will be more willing to cooperate by wearing their protectors and by undergoing audiometric tests. Employees exposed to

TWAs of 85 dBA or more must undergo at least annual training in the following:
- Effects of noise;
- Purpose, advantages, disadvantages and attenuation characteristics of various types of hearing protectors;
- Selection, fitting and care of protectors; and
- Purposes and procedures of audiometric testing.

Training does not have to be accomplished in one session. The program may be structured in any format and different parts may be conducted by different individuals as long as the required topics are covered. For example, audiometric procedures could be discussed immediately prior to audiometric testing. The training requirements are such that employees must be reminded on an annual basis that noise is hazardous to hearing and that they can prevent damage by wearing hearing protectors, when appropriate, and by participating in audiometric testing.

Recordkeeping

Noise exposure measurement records must be kept for two years. Records of audiometric test results must be maintained for the duration of the affected employee's employment. Audiometric test records must include the name and job classification of the employee, the date the test was performed, the examiner's name, the date of acoustic or exhaustive calibration, measurements of the background sound pressure levels in audiometric test rooms and the employee's most recent noise exposure measurement. All records required by this section shall be provided upon request to employees, former employees, representatives designated by the individual employee and OSHA.

The effectiveness of a hearing conservation program depends on the cooperation of employers, supervisors, employees and others concerned. Management's responsibility in this type of program includes:
- Taking noise measurements;
- Initiating noise-control measures;
- Undertaking the audiometric testing of employees;
- Providing hearing-protective equipment where it is required;
- Enforcing the use of such protective equipment with sound policies and by example;
- Informing employees of the benefits to be derived from a hearing conservation program; and
- Providing annual training.

OSHA also requires employers to make available to affected employees or their representatives copies of the OSHA noise standard and post a copy in the workplace.

It is the employee's responsibility to make proper use of the protective equipment provided by management. It is also the employee's responsibility to observe any rules or regulations in the use of equipment designed to minimize noise exposure.

Detailed references to noise, its management, effects, and control can be found in a great many books and periodicals. For those employers needing assistance in establishing hearing conservation programs, consultation services are available in a number of professional areas through private consultation, insurance, and governmental groups.

Hearing Conservation Glossary

Acoustic trauma. A hearing injury produced by exposure to sudden intense acoustic energy, such as from blasts and explosions, or from direct trauma to the head or ear. It should be thought of as one single incident relating to the onset of hearing loss.

Airborne gap. The difference in decibels between the hearing levels for sound at a particular frequency as determined by air conduction and bone conduction threshold measurements.

Airborne sound. Sound transmitted through air as a medium.

Air conduction. The process by which sound is conducted to the inner ear through the air in the outer ear canal using the tympanic membrane (eardrum) and the ossicles as part of the pathway.

Audible. Capable of being heard.

Audible range. The frequency range over which normal ears hear (approximately 20 Hz through 20,000 Hz).

Audiogram. A chart or table relating hearing level (for pure tones) to frequency. Referred to in OSHA standards as a "pure tone, air conduction, hearing threshold examination."

OCCUPATIONAL HEALTH AND INDUSTRIAL HYGIENE

Audiologist. A person trained in the specialized problems of hearing and deafness.

Audiometer, pure tone. An electro-acoustical generator that provides pure tones of selected frequencies and of calibrated output for the purpose of determining an individual's threshold of audibility. OSHA's standard requires testing at frequencies including as a minimum 500, 1000, 2000, 3000, 4000 and 6000 Hz.

Audiometric reference level. That sound pressure level (ASA, ISO, or ANSI) to which the audiometer is calibrated. A declared value, at a particular frequency, of the threshold of hearing for normal persons within a given age range, normally 18 to 25 years.

Bone conduction. The process by which sound is conducted to the inner ear through the cranial bones.

Cycle per second (cps). A unit of frequency. The preferred terminology is hertz, abbreviated Hz.

Decibel (dB). A nondimensional unit used to express sound levels. It is a logarithmic expression of the ratio of a measured quantity to a reference quantity. In audiometry, a level of zero decibels roughly represents the weakest sound that can be heard by a person with good hearing.

Where a weighted network filter is employed in making sound pressure measurements, this is indicated by a suffix added to the unit symbol. For example, a sound-level reading in decibels made on the A-weighted network of a sound level meter is designated as "dBA."

Dosimeter. A device worn on the person for determining the accumulated sound exposure with regard to level and time.

Eustachian tube. A tube approximately 2½ in. long leading from the back of the throat to the middle ear. It equalizes the pressure of air in the middle ear with that outside the eardrum.

Hearing conservation. The prevention or minimization of noise-induced hearing loss through the use of hearing protection devices and the control of noise through engineering methods or administrative procedures.

Hearing loss. An increase in the threshold of audibility, at specific frequencies, as the result of normal aging, disease or injury to the hearing organs.

Hearing threshold level (HTL). Amount (in decibels) by which the threshold of audibility for that ear exceeds a standard audiometric threshold.

Hertz (Hz). Synonymous term for cycles per second. It is the preferred unit of frequency.

Kilohertz (kHz). 1000 Hz

Meniere's Disease. The combination of deafness, tinnitus, nausea and vertigo.

Noise-induced hearing loss. Usually restricted to mean the slowly progressive inner ear hearing loss that results from exposure to noise over a long period of time as contrasted to acoustic trauma or physical injury to the ear.

Ossicles. Any one of the small bones (i.e., malleus, incus, stapes) that forms a chain for the transmission of sound from the tympanic membrane to the oval window.

Otolaryngologist. A physician or surgeon specializing in the practice of otology (ear disease), rhinology (nose disease) and laryngology (throat and larynx diseases).

Otologist. A physician or surgeon who specializes in the diagnosis and treatment of the disorders and diseases of the ear.

Presbycusis. Decline in hearing acuity that normally occurs as part of the aging process.

Standard threshold shift (STS). Defined in OSHA's hearing conservation regulation as an average shift (or loss) in either ear of 10 dB or more at the 2,000, 3,000, and 4,000 Hertz (Hz) frequencies.

Temporary threshold shift (TTS). The component of threshold shift which shows progressive reduction with the passage of time when the apparent cause has been removed.

Threshold of hearing of a continuous sound. The value of the sound pressure that excites the sensation of hearing.

Threshold of pain for a specified signal. The minimum effective sound pressure level of that signal, which in a specified fraction of the trials, will stimulate the ear to a point at which there is a sensation of pain that is distinct from a feeling of discomfort (usually above 120 dBA).

Tinnitus. A subjective sense of noises in the head or ringing in the ears for which there is no observable external cause.

2004 Subpart G 1910.94-98
Occupational Health

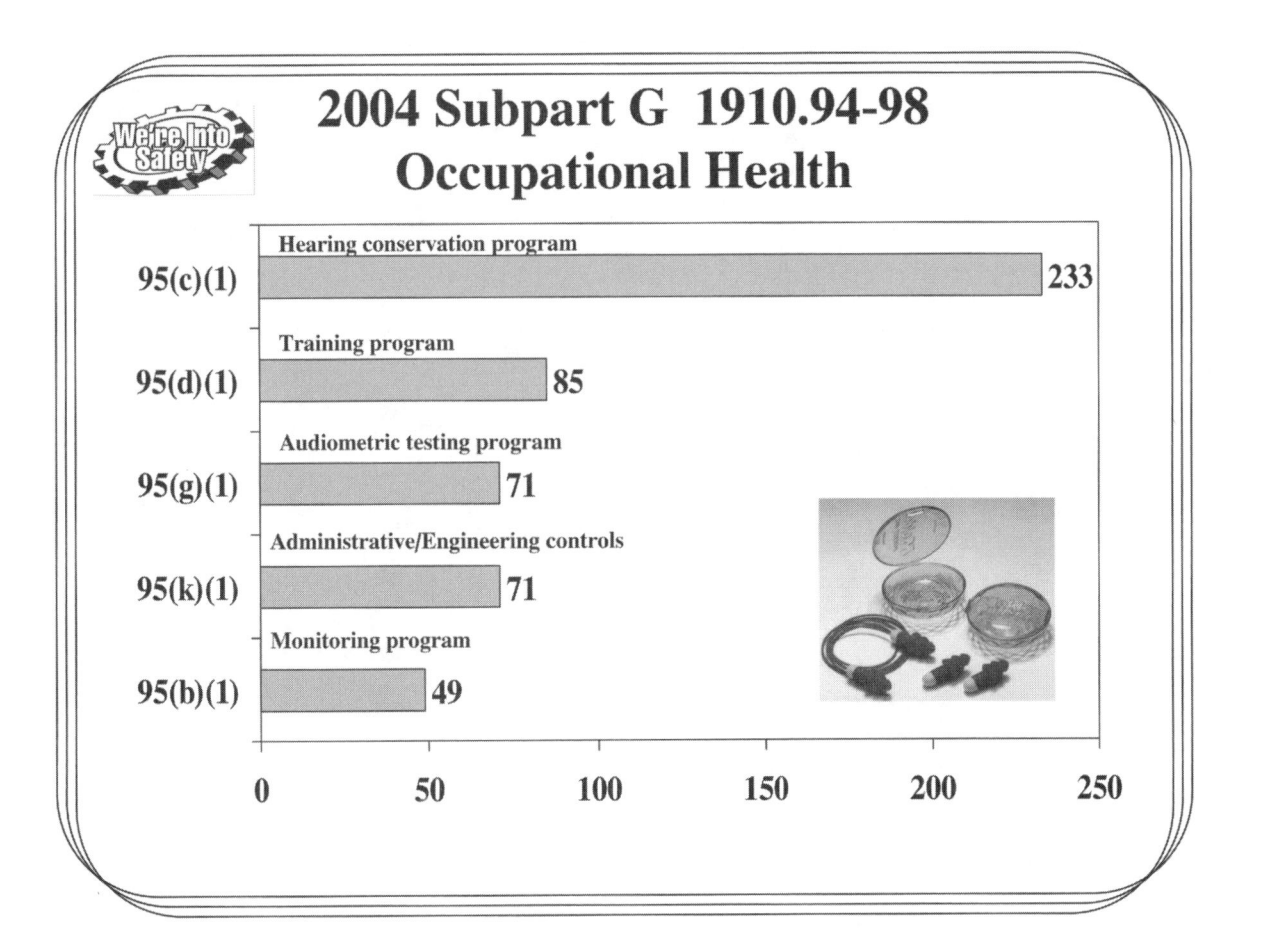

Citation	Description	Count
95(c)(1)	Hearing conservation program	233
95(d)(1)	Training program	85
95(g)(1)	Audiometric testing program	71
95(k)(1)	Administrative/Engineering controls	71
95(b)(1)	Monitoring program	49

CHAPTER 5

HAZARD COMMUNICATION

Nearly 32 million workers are potentially exposed to one or more chemical hazards. There are an estimated 575,000 existing chemical products, and hundreds of new ones being introduced annually. This poses a serious problem for exposed workers and their employers.

Chemical exposure may cause or contribute to many serious health effects such as heart ailments, kidney and lung damage, sterility, cancer, burns and rashes. Some chemicals may also be safety hazards and have the potential to cause fires and explosions and other serious accidents.

Because of the seriousness of these safety and health problems, and because many employers and employees know little or nothing about them, the Occupational Safety and Health Administration (OSHA) has issued a rule called "Hazard Communication." The basic goal of the standard is to be sure employers and employees know about work hazards and how to protect themselves — which should help to reduce the incidence of chemical source illness and injuries.

The Hazard Communication Standard establishes uniform requirements to make sure that the hazards of all chemicals imported into, produced or used in U.S. workplaces are evaluated, and the subsequent hazard information is transmitted to affected employers and exposed employees.

Chemical manufacturers and importers must convey the hazard information they learn from their evaluations to downstream employers by means of labels on containers and material safety data sheets (MSDSs). In addition, all covered employers must have a hazard communication program to get this information to their employees through labels on containers, MSDSs and training.

This program ensures that all employers receive the information they need to inform and train their employees properly and to design and put in place employee protection programs. It also provides necessary hazard information to employees, so they can participate in, and support, the protective measures in place at their workplaces.

The Hazard Communication Standard is different from other OSHA health rules as it covers all hazardous chemicals. The rule also incorporates a "downstream flow of information," which means that producers of chemicals have the primary responsibility of generating and disseminating information, while users of chemicals must obtain the information and transmit it to their own employees.

In general, it works like this:
- Chemical Manufacturers/Importers determine the hazards of each product.
- Chemical Manufacturers/ Importers/Distributors communicate the hazard information and associated measures downstream to customers through labels and MSDSs.

- Employers
  - Identify and list hazardous chemicals in their workplaces.
  - Obtain MSDSs and labels for each hazardous chemical.
  - Develop and implement a written hazard communication program, including labels, MSDSs, and employee training based on the list of chemicals, and the MSDS and label information.
  - Communicate hazard information to their employees through labels, MSDSs and formal training programs.

Hazard Evaluation

The quality of the hazard communication program depends on the adequacy and accuracy of the hazard assessment Chemical manufacturers and importers are required to review available scientific evidence concerning the hazards of chemicals they produce or import, and to report the information they find their employees and to employers who distribute or use their products Downstream employers can rely on the evaluations performed by the chemical manufactures or importers to establish the hazards of the chemicals they use.

Each chemical must be evaluated for its potential to cause adverse health effects and to pose physical hazards such as flammability. (Definitions of hazards covered are included in the standard.) Chemicals that are listed in one of the following sources are to be considered hazardous in all cases:

- 29 CFR 1910, Subpart Z, and Hazardous Substances, Occupational Safety and Health Administration, (OSHA); and
- Threshold Limit Values for Chemical Substances and Physical Agents in the Work Environment, American Conference of Governmental Industrial Hygienists (ACGIH).

OSHA considers chemical to be hazardous if it is:
- A carcinogen;
- Corrosive;
- Highly toxic;
- Toxic;
- An irritant;
- A sensitizer; or
- Targets organ effects.

Written Program

Employers must develop, implement and maintain at the workplace a written, comprehensive hazard communication program that includes provisions for container labeling, collection and availability of material safety data sheets and an employee-training program. It also must contain a list of the hazardous chemicals in each work area, the means the employer will use to inform employees of the hazards of non-routine tasks (for example, the cleaning of reactor vessels) and the hazards associated with chemicals in unlabeled pipes. If the workplace has multiple employers onsite (e.g., a construction site), the rule requires these employers to ensure that information regarding hazards and protective measures be made available to the other employers' onsite, where appropriate.

The written program does not have to be lengthy or complicated, and some employers may be able to rely on existing hazard communication programs to comply with the above requirements. The written program must be available to employees, their designated representatives, the Assistant Secretary of Labor for Occupational Safety and Health and the Director of the National Institute for Occupational Safety and Health (NIOSH).

Labels and Other Forms of Warnings

Chemical manufacturers, importers and distributors must be sure that containers of hazardous chemicals leaving the workplace are labeled, tagged or marked with the identity of the chemicals, appropriate hazard warnings and the name and address of the manufacturer or other responsible party.

Each container must be labeled, tagged or marked with the identity of hazardous chemicals contained therein, and must show hazard warnings appropriate for employee protection. The hazard warning can be any type of message, words, pictures or symbols that convey the hazards of the chemicals in the container (See Exhibit 5-1.). Labels must be legible, in English (plus other languages, if desired) and prominently displayed.

Table 5-1. Information found on MSDSs.

Section	Contents
I	Product Identity
II	Hazardous Ingredients
III	Physical/Chemical Characteristics/Fire Hazards
IV	Health Hazards
V	Routes of Entry
VI	Exposure Limits/IH information
VII	Precautions for Safe Handling and Use
VIII	Control Measures/Protection Information
IX	First Aid
X	Date of preparation and manufacturer

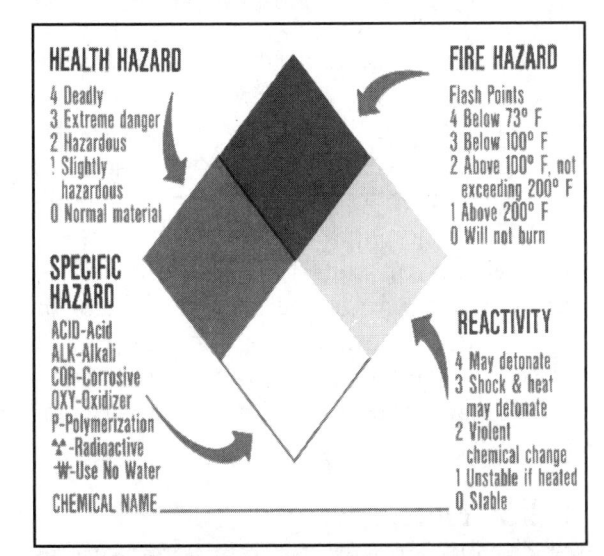

Exhibit 5-1.
Labels on containers must show the hazards associated with the chemicals therein.

Exemptions to the requirement for in-plant individual container labels are as follows:

- Employers can post signs or placards that convey the hazard information if there are a number of stationary containers within a work area that have similar contents and hazards.
- Employers can substitute various types of standard operating procedures, process sheets, batch tickets, blend tickets and similar written materials for container labels on stationary process equipment if they contain the same information and are readily available to employees in the work area.
- Employers are not required to label portable containers when the hazardous chemicals are being transferred from labeled containers and the portable container is only for the immediate use of the employee who is making the transfer. Employers are not required to label pipes or piping systems.

Material Safety Data Sheets

Chemical manufacturers and importers must develop an MSDS for each hazardous chemical they produce or import, and must automatically provide the MSDS with the initial shipment of a hazardous chemical to a downstream distributor or user. Distributors must also ensure that downstream employers are provided an MSDS. Each MSDS must be in English and include information regarding the specific chemical identity of the hazardous chemicals involved as well as their common names. In addition, information must be provided on:

- the physical and chemical characteristics of the hazardous chemical;
- known acute and chronic health effects and related health information;
- exposure limits;
- whether the chemical is considered to be a carcinogen by the NTP, IARC, or OSHA;
- precautionary measures;
- emergency and first-aid procedures; and
- the identification of the organization responsible for preparing the sheet.

Copies of the MSDS for hazardous chemicals in a given work site are to be readily accessible to employees in that area.

HAZARD COMMUNICATION

List of Hazardous Chemicals

Employers must prepare a list of all hazardous chemicals in the workplace. When the list is complete, it should be checked against the collected MSDSs that the employer has been sent. If there are hazardous chemicals used for which no MSDS has been received, the employer must write to the supplier, manufacturer or importer to obtain the missing MSDS. If employers do not receive the MSDS within a reasonable period of time, they should contact the nearest OSHA office.

Employee Training and Information

Employers must establish a training and information program for employees who are exposed to hazardous chemicals in their work area at the time of initial assignment and whenever a new hazard is introduced into their work area.

At a minimum, the discussion topics must include:

- The existence and requirements of the hazard communication standard;
- The components of the hazard communication program in the employees' workplaces;
- Operations in work areas where hazardous chemicals are present; and
- Where the employer will keep the written hazard evaluation procedures, communications program, lists of hazardous chemicals and required MSDS forms.

The employee-training plan must include:

- How the hazard communication program is implemented in that workplace, how to read and interpret information on labels and MSDSs and how employees can obtain and use the available hazard information;
- The hazards of the chemicals in the work area. Some hazards may be discussed by individual chemical or by hazard categories such as flammability;
- Measures employees can take to protect themselves from the hazards;
- Specific procedures put into effect by the employer to provide protection such as engineering controls, work practices and the use of personal protective equipment; and
- Methods and observations, such as visual appearance or smell, workers can use to detect the presence of a hazardous chemical to which they may be exposed.

Trade Secrets

A "trade secret" is something that gives an employer an opportunity to obtain an advantage over competitors who do not know about the trade secret or who do not use it. For example, a trade secret may be a confidential device, pattern, information or chemical make-up. Chemical industry trade secrets are generally formulas, process data, or a "specific chemical identity." The latter is the type of trade secret information referred to in the hazard communication standard. The term includes the chemical name, the Chemical Abstracts Services (CAS) registry number, or any other specific information that reveals the precise designation. It does not include common names.

The standard strikes a balance between the need to protect exposed employees and the employer's need to maintain the confidentiality of a bona fide trade secret. This is achieved by providing for limited disclosure to health professionals who are furnishing medical or other occupational health services to exposed employees, employees and their designated representatives, under specified conditions of need and confidentiality.

Medical Emergency

The chemical manufacturer, importer or employer must immediately disclose the specific chemical identity of a hazardous chemical to a treating physician or nurse when the information is needed for proper emergency or first aid treatment. As soon as circumstances permit, the chemical manufacturer, importer or employer may obtain a written statement of need and a confidentiality agreement.

Under the contingency described here, the treating physician or nurse has the ultimate responsibility for determining that a medical emergency exists. At the time of the emergency, the professional judgment of the physician or nurse regarding the situation must form the basis for triggering the immediate disclosure requirement. Because the chemical manufacturer, importer or employer can demand a written statement of need and a confidentiality agreement to be completed after the emergency is abated, further disclosure of the trade secret can be effectively controlled.

Nonemergency Situation

In nonemergency situations, chemical manufacturers, importers or employers must disclose the withheld specific chemical identity to health professionals providing medical or other occupational health services to exposed employees and to employees and their designated representatives, if certain conditions are met. In this context, "health professionals" include physicians, occupational health nurses, industrial hygienists, toxicologists and epidemiologists.

The request for information must be in writing and must describe with reasonable detail the medical or occupational health need for the information. The request will be considered if the information will be used to:
- Assess the hazards of the chemicals to which employees will be exposed;
- Conduct or assess sampling of the workplace atmosphere to determine employee exposure levels;
- Conduct pre-assignment or periodic medical surveillance of exposed employees;
- Provide medical treatment to exposed employees;
- Select or assess appropriate personal protective equipment for exposed employees;
- Design or assess engineering controls or other protective measures for exposed employees; or
- Conduct studies to determine the health effects of exposure.

The health professional, employee or designated representative must also specify why alternative information is insufficient. The request for information must explain in detail why disclosure of the specific chemical identity is essential, and include the procedures that will be used to protect the confidentiality of the information. The request must include an agreement not to use the information for any purpose other than the health need stated or to release the information under any circumstances, except to OSHA.

The standard details the steps to be followed in the event that an employer decides not to disclose the specific chemical identity requested by the health professional, employee or designated representative.

Guidelines for Compliance

The Hazard Communication Standard (HCS) is based on a simple concept — that employees have both a need and a right to know the hazards and identities of the chemicals they are exposed to when working. They also need to know what protective measures are available to prevent adverse effects from occurring. Knowledge acquired under the HCS will help employers provide safer workplaces for their employees. When employers have information about the chemicals being used, they can take steps to reduce exposures, substitute less hazardous materials and establish proper work practices. These efforts will help prevent the occurrence of work-related illnesses and injuries caused by chemicals.

The HCS addresses the issues of evaluating and communicating hazards to workers. The evaluation of chemical hazards involves a number of technical concepts, and is a process that requires the professional judgment of experienced experts. Therefore, the HCS is designed so that employers who simply use chemicals, rather than produce or import them, are not required to evaluate the hazards of those chemicals. Hazard determination is the responsibility of the producers and importers of the materials, who are then required to provide the hazard information to employers who purchase their products.

Employers who do not produce or import chemicals need only focus on those parts of the rule that deal with establishing a workplace program and communicating information to their workers. This lesson provides a general guide for such employers to help determine what is required under the rule. It does not supplant or substitute for the regulatory provisions, but rather provides a simplified outline of the steps an average employer would follow to meet those requirements.

Become Familiar with the Rule

The standard is long, and some parts of it are technical, but the basic concepts are simple. In fact, the requirements reflect what many employers have been doing for years. An employer may find that it is already largely in compliance with many of the provisions and will simply have to modify existing programs somewhat. The rule requires information to be prepared and transmitted regarding all hazardous chemicals. The regulation covers both physical hazards (e.g., flammability) and health hazards (e.g., irritation, lung damage, cancer). Most chemicals used in the workplace have some hazard potential, and thus will be covered by the rule.

One difference between this rule and many others adopted by OSHA is that this one is performance oriented. That means that the employer has the flexibility to adapt the rule to the needs of the workplace, rather than having to follow specific, rigid requirements. It also means that the employer has to exercise more judgment to implement an appropriate and effective program.

The standard's design is simple. Chemical manufacturers and importers must evaluate the hazards of the chemicals they produce or import. Using that information, they must then prepare labels for containers and MSDSs.

Chemical manufacturers, importers and distributors of hazardous chemicals are required to provide the appropriate labels and MSDSs to the employers to whom they ship the chemicals. Every container of hazardous chemicals the company receives must be labeled, tagged or marked with the required information. Suppliers must also send the company a properly completed MSDS at the time of the first shipment of the chemical, and with the next shipment after the MSDS is updated with new and significant information about the hazards. Employers can rely on the information received from suppliers. They have no independent duty to analyze the chemical or evaluate the hazards of it.

Employers that use hazardous chemicals must have a program to ensure the information is provided to exposed employees. "Use" means to package, handle, react or transfer. This is an intentionally broad scope, and includes any situation where a chemical is present in such a way that employees may be exposed under normal conditions of use or in a foreseeable emergency.

The requirements of the rule that deal specifically with the hazard communication program are found in the standard in paragraphs (e) Written Hazard Communication Programs; (f) Labels and Other Forms of Warning; (g) Material Safety Data Sheets; and (h) Employee Information and Training. The requirements of these paragraphs should be the focus of your attention. Concentrate on becoming familiar with them using paragraphs (b), Scope and Application, and (c), Definitions, as references when needed to help explain the provisions.

There are two types of work operations where the coverage of the rule is limited. These are laboratories and operations where chemicals are only handled in sealed containers (e.g., a warehouse). The limited provisions for these workplaces can be found in paragraph (b), Scope and Application. Basically, employers having these types of work operations need only keep labels on containers as they are received; maintain material safety data sheets that are received and give employees access to them; and provide information and training for employees. Employers do not have to have written hazard communication programs and lists of chemicals for these types of operations.

The limited coverage of laboratories and sealed container operations addresses the obligation of an employer to the workers in the operations involved, and does not affect the employer's duties as a distributor of chemicals. For example, a distributor may have warehouse operations where employees would be protected under the limited sealed container provisions. In this situation, requirements for obtaining and maintaining MSDSs are limited to providing access to those received with containers while the substance is in the workplace, and requesting MSDSs when employees request access for those not received with the containers. However, as a distributor of hazardous chemicals, that employer will still have responsibilities for providing MSDSs to downstream customers at the time of the first shipment and when the MSDS is updated. Therefore, although they may not be required for the employees in the work operation, the distributor may, nevertheless, have to have MSDSs to satisfy other requirements of the rule.

Identify Responsible Staff

Hazard communication is going to be a continuing program in your facility. Compliance with the HCS is not a "one shot deal." In order to have a successful program, it will be necessary to assign responsibility for both the initial and ongoing activities that have to be undertaken to comply with the rule. In some cases, these activities may already be part of current job assignments. For example, site supervisors are frequently responsible for on-the-job training sessions. Early identification of the responsible employees, and involvement of them in the development of the plan of action, will result in a more effective program design. Evaluation of the effectiveness of the program will also be affected by the involvement of affected employees.

Identify Hazardous Chemicals in the Workplace

The standard requires a list of hazardous chemicals in the workplace as part of the written hazard communication program. The list will eventually serve as an inventory of everything for which an MSDS must be maintained. At this point, however, preparing the list will help complete the rest of the program since it will give some idea of the scope of the program required for compliance in a facility.

The best way to prepare a comprehensive list is to survey the workplace. Purchasing records may also help, and certainly employers should establish procedures to ensure that in the future purchasing procedures result in MSDSs being received before a material is used in the workplace.

The broadest possible perspective should be taken when doing the survey. Sometimes people think of "chemicals" as being only liquids in containers. The HCS covers chemicals in all physical forms — liquids, solids, gases, vapors, fumes and mists — whether they are "contained" or not. The hazardous nature of the chemical and the potential for exposure are the factors that determine whether a chemical is covered. If it is not hazardous, it is not covered. If there is no potential for exposure (e.g., the chemical is inextricably bound and cannot be released), the rule does not cover the chemical.

Look around. Identify chemicals in containers, including pipes, but also think about chemicals generated in the work operations. For example, welding fumes, dusts and exhaust fumes are all sources of chemical exposures. Read labels provided by the suppliers for hazard information. Make a list of all chemicals in the workplace that are potentially hazardous. For your own information and planning, the employer may also want to note on the list the locations of the products within the workplace and an indication of the hazards as found on the label. This will help as the employer prepares the rest of the program.

Paragraph (b), Scope and Application, includes exemptions for various chemicals and workplace situations. After compiling the complete list of chemicals, you should review paragraph (b) to determine if any of the items can be eliminated from the list because they are exempt materials. For example, food, drugs and cosmetics brought into the workplace for employee consumption are exempt; rubbing alcohol in the first-aid kit also is not covered.

Once an employer has compiled as complete a list as possible of the potentially hazardous chemicals in the workplace, the next step is to determine if you have received MSDSs for all of them. Check files against the inventory just compiled. If any are missing, contact your supplier and request one. It is a good idea to document these requests, either by copy of a letter or a note regarding telephone conversations. If you have MSDSs for chemicals that are not in your list, figure out why. Maybe you don't use the chemical anymore. Or maybe you missed it in your survey. Some suppliers do provide MSDSs for products that are not hazardous. These do not have to be maintained by you.

Do not allow employees to use any chemicals for which you have not received an MSDS. The MSDS provides information you need to ensure proper protective measures are implemented prior to exposure.

Preparing and Implementing a Hazard Communication Program

All workplaces where employees are exposed to hazardous chemicals must have a written plan that describes how the standard will be implemented in that facility. Preparation of a plan is not just a paper exercise — all of the elements must be implemented in the workplace in order to be in compliance with the rule. See paragraph (e) of the standard for the specific requirements regarding written hazard communication programs. The only work operations that do not have to comply with the written plan requirements are laboratories and work operations where employees only handle chemicals in sealed containers. See paragraph (b), Scope and Application, for the specific requirements for these two types of workplaces.

The plan does not have to be lengthy or complicated. It is intended to be a blueprint for implementation of your program — an assurance that all aspects of the requirements have been addressed. Many trade associations and other professional groups have provided sample programs and other assistance materials to affected employers. These have been very helpful to many employers since they tend to be tailored to the particular industry involved. Although such general guidance may be helpful, keep in mind that the written program must be specific to your workplace.

Therefore, if you use a generic program it must be adapted to address the facility it covers. For example, the written plan must list the chemicals present at the site, indicate who is to be responsible for the various aspects of the program in your facility and indicate where written material will be made available to employees.

If OSHA inspects your workplace for compliance with the HCS, the OSHA compliance officer will ask to see the written plan at the outset of the inspection. In general, the following items will be considered in evaluating your program.

The written program must describe how the requirements for labels and other forms of warnings material safety data sheets, and employee information and training, are going to be met in your facility. The following discussion provides the type of information compliance officers will be looking for to decide whether these elements of the hazard communication program have been properly addressed.

Labels and Other Forms of Warning

In-plant containers of hazardous chemicals must be labeled, tagged or marked with the identity of the material and appropriate hazard warnings. Chemical manufacturers, importers and distributors are required to ensure that every container of hazardous chemicals they ship is appropriately labeled with such information and with the name and address of the producer or other responsible party. Employers purchasing chemicals can rely on the labels provided by their suppliers. If the employer subsequently transfers the material from a labeled container to another container, the employer will have to label that container unless it is subject to the portable container exemption.

The information needed on a label is the identity for the material and appropriate hazard warnings. The identity is any term that appears on the label, MSDS and list of chemicals, thus linking these three sources of information. The identity used by the supplier may be a common or trade name (e.g., Black Magic Formula) or a chemical name (e.g., trichloroethane). The hazard warning is a brief statement of the hazardous effects of the chemical (e.g., flammable, causes lung damage). Labels frequently contain other information, such as precautionary measures (e.g., do not use near open flame); however, this information is provided voluntarily and is not required by the rule. Labels must be legible and prominently displayed. There are no specific requirements for size, color or text.

With these requirements in mind, the OSHA compliance officer will be looking for the following types of information to ensure that labeling will be properly implemented in your facility:
- Designation of person(s) responsible for ensuring labeling of in-plant containers;
- Designation of person(s) responsible for ensuring labeling of any shipped containers;
- Description of labeling system(s) used;
- Description of written alternatives to labeling of in-plant containers (if used); and
- Procedures to review and update label information when necessary.

Employers that are purchasing and using hazardous chemicals — rather than producing or distributing them — will primarily be concerned with ensuring that every purchased container is labeled. If materials are transferred into other containers, the employer must ensure that these are labeled as well, unless they fall under the portable container exemption (paragraph (f)(7)). In terms of labeling systems, an employer can simply choose to use the labels provided by suppliers on the containers. These will generally be verbal text labels and do not usually include numerical rating systems or symbols that require special training. The most important thing to remember is that this is a continuing duty — all in-plant containers of hazardous chemicals must always be labeled. Therefore, it is important to designate someone to be responsible for ensuring that the labels are maintained as required on the containers in your facility and that newly purchased materials are checked for labels prior to use.

Material Safety Data Sheets

Chemical manufacturers and importers are required to obtain or develop a material safety data sheet for each hazardous chemical they produce or import. Distributors are responsible for ensuring that their customers are provided a copy of these MSDSs. Employers must have an MSDS for each hazardous chemical that they use. Employers may rely on the information received from their suppliers.

There is no specified format for the MSDS under the rule, although there are specific information requirements. OSHA has developed a non-mandatory format, OSHA Form 174, that may be used by chemical manufacturers and importers to comply with the rule. The MSDS must be in English. An employer is entitled to receive from its supplier a data sheet that includes all of the information required under the rule. If you do not receive one automatically, you should request one. If you receive one that is obviously inadequate, with, for example, blank spaces that are not completed, you should request an appropriately completed one.

The role of MSDSs under the rule is to provide detailed information on each hazardous chemical, including its potential hazardous effects; its physical and chemical characteristics; and recommendations for appropriate protective measures. This information should be useful to you as the employer responsible for designing protective programs, as well as to the workers. If you are not familiar with material safety data sheets or chemical terminology, you may need to learn to use them yourself. A glossary of MSDS terms may be helpful in this regard. Generally speaking most employers using hazardous chemicals will primarily be concerned with MSDS information regarding hazardous effects and recommended protective measures. Focus on the sections of the MSDS that are applicable to your situation.

MSDSs must be readily accessible to employees when they are in their work areas during their work shifts. This may be accomplished in several ways. You must decide what is appropriate for your particular workplace. Some employers keep the MSDSs in a binder in a central location (e.g., in the pick-up truck on a construction site). Others, particularly in workplaces with large numbers of chemicals, computerize the information and provide access through terminals. As long as employees can get the information when they need it, any approach may be used. The employees must have access to the MSDSs themselves — simply having a system where the information can be read to employees over the phone is only permitted under the mobile worksite provision, paragraph (g)(9), when employees must travel between workplaces during the shift. In this situation, employees have access to the MSDSs at the primary worksite so the telephone system is simply an emergency arrangement.

To ensure that you have a current MSDS for each chemical in the facility and employee access is provided, the OSHA compliance officer will be looking for the following types of information in your written program:

- Designation of person(s) responsible for obtaining and maintaining the MSDSs;
- How such sheets are to be maintained in the workplace (e.g., in notebooks in the work area(s) or in a computer with terminal access) and how employees can obtain access to them when they are in their work areas during the work shifts;
- Procedures to follow when the MSDS is not received at the time of the first shipment;
- For producers, procedures to update the MSDS when new and significant health information is found; and
- Description of alternatives to actual MSDSs in the workplace, if necessary.

For employers using hazardous chemicals, the most important aspect of the written program in terms of MSDSs is to ensure that someone is responsible for obtaining and maintaining the MSDSs for every hazardous chemical in the workplace. The list of hazardous chemicals required to be maintained as part of the written program will serve as an inventory. As new chemicals are purchased, the list should be updated. Many companies have found it convenient to include on their purchase orders the name and address of the person designated in their company to receive MSDSs.

Employee Information and Training

Each employee who may be exposed to hazardous chemicals when working must be provided information and be trained prior to initial assignment to work with a hazardous chemical and whenever the hazard changes. "Exposure" or "exposed" under the rule means "an employee is subjected to a hazardous chemical in the course of employment through any route of entry (inhalation, ingestion, skin contact or absorption, etc.) and includes potential (e.g., accidental or possible) exposure." Information and training may be done either by individual chemical or by categories of hazards (e.g., flammability, carcinogenicity). If there are only a few chemicals in the workplace, then you may want to discuss each one individually. Where there are large numbers of chemicals or the chemicals change frequently, you will probably want to train generally based on the hazard categories (e.g., flammable liquids, corrosive materials, and carcinogens). Employees must have access to the substance-specific information on the labels and MSDSs.

Information regarding hazards and protective measures is provided to workers through written labels and MSDSs. However, through effective information and training, workers will learn to read and understand such information, determine how it can be obtained and used in their own work areas, and understand the risks of exposure to the chemicals in their work areas as well as the ways to protect themselves. Properly conducted training programs will ensure comprehension and understanding. It is not sufficient to either read material to the workers or hand them material to read. Create a climate where workers feel free to ask questions. This will help to ensure that the information is understood. Remember that the underlying purpose of the HCS is to reduce the incidence of chemical source illnesses and injuries. If your employees' comprehension is increased, proper work practices will be implemented and used, resulting in fewer injuries. The procedures you establish regarding, for example, purchasing, storage and handling of these chemicals will improve and thereby reduce the risks posed to employees exposed to the chemical hazards involved.

In reviewing your written program with regard to information and training consider the following:

- Designation of person(s) responsible for conducting training;
- Format of the program to be used (e.g., audiovisuals, classroom instruction);
- Elements of the training program (should be consistent with the elements in paragraph (h) of the HCS); and
- Procedures to train new employees at the time of their initial assignment to work with a hazardous chemical and to train employees when a new hazard is introduced into the workplace.

The written program should provide enough details about the employer's plans in this area to assess whether or not a good-faith effort is being made to train employees. OSHA does not expect that every worker will be able to recite all of the information about each chemical in the workplace. In general, the most important aspects of training under the HCS are to ensure that employees are aware that they are being exposed to hazardous chemicals, that they know how to read and use labels and MSDSs and that as a consequence of learning this information, they are following the appropriate protective measures established by the employer. OSHA compliance officers will speak to employees to determine whether they have received training and understand those aspects discussed previously.

The rule does not require employers to maintain records of employee training, but many employers choose to do so. This may help you monitor your own program to ensure that all employees are appropriately trained. If you already have a training program, you may simply have to supplement it with whatever additional information is required under the HCS. For example, construction employers that are already in compliance with the construction standard (29 CFR 1926.21) will have little extra training to do.

An employer can provide employees with information and training through whatever means it deems appropriate. Although there should always be some on-site training (e.g., informing employees of the location and availability of the written program and MSDSs), employee training may be satisfied in part by general training regarding the requirements of the HCS and chemical hazards on the job. This general training could be provided by, for example, trade associations, unions, colleges and professional schools. Similarly, the previous training, education and experience of a worker may relieve the employer of some of the burdens of training that worker. Regardless of the method relied upon; however, the employer is ultimately responsible for ensuring that employees are adequately trained. If the OSHA compliance officer finds the training is deficient, the employer will be cited for the deficiency regardless of who actually provided the training.

Other Requirements

In addition to these specific items, OSHA compliance officers will also ask the following questions when assessing the adequacy of the program:

- Does a list of the hazardous chemicals exist in each work area or at a central location?
- Are methods the employer will use to inform employees of the hazards of non-routine tasks outlined?
- Are employees informed of the hazards associated with chemicals contained in unlabeled pipes in their work areas?
- On multi-employer worksites, has the employer provided other employers with information about labeling systems and precautionary measures where the other employers have employees exposed to the initial employer's chemicals?
- Is the written program made available to employees and their designated representatives?

If your program adequately addresses the means of communicating information to employees in your workplace and provides answers to the basic questions outlined above, it will be found to be in compliance with the rule.

Table 5-2.
Chemical Hazard Communication Checklist
- ❑ Obtain a copy of the rule.
- ❑ Read and understand the requirements.
- ❑ Assign responsibility for tasks by name or job title.
- ❑ Prepare an annual inventory of chemicals.
- ❑ Ensure all containers of hazardous chemicals are labeled.
- ❑ Obtain MSDS for each chemical.
- ❑ Prepare written program
- ❑ Make MSDSs available to affected workers.
- ❑ Conduct initial training of workers.
- ❑ Retrain workers when new chemicals are introduced.
- ❑ Establish procedures to maintain current program.

Establish procedures to evaluate effectiveness. The hazard evaluation procedures required by the standard are performance-oriented. Basically, OSHA's concern is that the information on labels and data sheets, and in the training program, is adequate and accurate. Although specific procedures to follow and number of sources to be consulted cannot be established, general guidance can be provided. The hazard evaluation process can be characterized as a "tiered" approach--the extent to which a chemical must be evaluated depends to a large degree upon the common knowledge regarding the chemical, whether its health effects are under review, and how prevalent it is in the workplace.

The first step for a compliance officer's evaluating chemicals is to determine whether the chemical is part of the "floor" of chemicals to be considered hazardous in all situations.

The floor of chemicals consists of three sources. They are:

- Any substance for which OSHA has a permissible exposure limit (PEL) in 1910.1000, or a comprehensive substance-specific standard in Subpart Z. This includes any compound of such substances where OSHA would sample to determine compliance with the PEL.
- Any substance for which the American Conference of Governmental Industrial Hygienists (ACGIH) has a Threshold Limit Value (TLV) in the latest edition of their annual list is to be included in the Hazard Communication Program. Any mixture or combination of these substances would also be included.

- Any substance, which the National Toxicology Program (NTP) or the International Agency for Research on Cancer (IARC) has found to be a suspect, or confirmed carcinogen, or that OSHA regulates as a carcinogen is to be included in the Hazard Communication Program.

Sources to generally establish hazards of the chemicals that are part of the floor of hazardous chemicals covered by the standard include:
- The OSHA Chemical Information Manual, OSHA Instruction CPL 2-2.43, October 1987.
- NIOSH/OSHA Occupational Health Guidelines.
- Documentation for the Threshold Limit Values.
- NTP Summary of the Annual Report on carcinogens.
- IARC Monographs

Check the NIOSH *Registry of Toxic Effects of Chemical Substances* (RTECS) to see if any hazards are indicated which do not appear in these sources. If there are, further study should be done to evaluate the hazards. RTECS should never be considered a definitive source for establishing a hazard since it consists of data that has not been evaluated. It is, however, a useful screening resource.

The second step is to consult other generally available sources to see what has been published regarding the chemical. Patty's *Industrial Hygiene and Toxicology* would be one such source. OCIS contains a number of other chemical information sources. Material Safety Data Sheets available through information services would also be useful.

The third step, for those chemicals where information is not readily available or where such available information is not complete, is to perform searches of bibliographic, databases. In general, the National Library of Medicine (NLM) services should be used. These include the Toxicology Data Bank (TDB), TOXLINE, and MEDLARS. The information generated by these data bases should be evaluated using the criteria in Appendix B of the HCS; i.e., to qualify as an acceptable study, it must be conducted according to scientific principles (e.g., in animal studies, number of subjects is adequate to do statistical analyses of the results; control group is used, and the study must show statistically significant results indicating an adverse health effect). This evaluation obviously requires a subjective, professional assessment. Any questions should be referred to the Directorate of Compliance Programs, office of Health Compliance Assistance (through the Regional Office) for assistance. In general, uncorroborated cases reports and in vitro studies, such as Ames tests, are useful pieces of information, but not definitive findings of hazards. Animal studies involving species other than those indicated in the acute hazard definitions must be evaluated as well.

The acute hazard definitions are not included in the standard to "categorize" chemicals but rather to establish that chemicals meeting those definitions fall under the coverage of the standard.

In some cases, the only information available on a substance may be employer-generated data. If the employer indicates that such information is the basis for the hazard evaluation, the Compliance Safety and Health Officer (CSHO) shall ask to see it to complete the OSHA evaluation.

In cases where the employer denies the CSHO access to its own hazard data and no published data on the chemical can be found to review the sufficiency of the hazard determination, the Regional Office shall be contacted for assistance in obtaining an administrative subpoena. The Directorate of Compliance Programs shall be contacted if assistance is required in order to obtain unpublished chemical hazard information available from other Federal agencies such as Environmental Protection Agency.

If an employer has found any chemical to be non-hazardous, and the CSHO has reason to believe it is hazardous, further investigation is required. The definitions of hazard in the standard are very broad, and it is not expected that many chemicals can be considered non-hazardous under this approach. Those most likely to be exempted would be chemicals that pose no physical hazards, and which have lethal dose findings above the limits found in the acute hazard definitions.

In some cases, the employer may not have addressed in the Hazard Communication Program a specific chemical that the CSHO knows to be present through knowledge of the process or through sampling or other investigation of the workplace. This situation should also be further investigated. If the CSHO has information to indicate that there is a hazard, the employer must be able to defend the finding of no hazard.

Eye And Body Flushing

OSHA Regulation 29 CFR 1910.151(c) states "Where the eyes or body of any person may be exposed to injurious corrosive materials, suitable facilities for quick drenching or flushing of the eyes and body shall be provided within the work area for immediate emergency use."

The ANSI standard Z358.1-1998 specifies requirements for eyewash and shower units:
- Sole purpose must be an emergency eyewash or shower unit
- If shower is needed, eyewash must be available (combination unit is ok)
- Valves must be quick opening- equal or less than one second
- Unit must operate with both hands free
- Water temperature- tepid or lukewarm (60-95 degrees F)
- Travel time to unit less than eleven seconds
- Unit on same level as the hazardous substance
- Immediate access means unobstructed route to unit
- Compliance with ANSI Z358 must be verified annually

Requirements for eyewashes only:
- Nozzles are 33-45 inches from floor
- Minimum of 1.5 liters per minute (.4 gallons) potable water or commercial flush for 15 minutes
- Weekly checks to flush lines for three minutes
- Nozzles protected from contamination by covers
- Squeeze bottles do not replace a fixed or portable eyewash

Requirements for showers only:
- Overhead mounting of head height of 82-96 inches from floor or platform
- Minimum flow of 75.7 liters (20 gallons) per minute
- Weekly checks to flush lines
- Valve actuator 69 inches above floor or platform

Requirements for training:
- Employees must be trained on the location and proper use
- Training shall address holding eyelids open and rolling eyeballs to flush entire eye
- If squeeze bottles are provided, employees must be trained on proper use in conjunction with eyewash stations

Exhibit 5-3.
Example of an eyewash station.

Hazard Communication Glossary

ACGIH. American Conference of Governmental Industrial Hygienists, which develops and publishes recommended occupational exposure limits for hundreds of chemical substances and physical agents. See also Threshold limit value (TLV).

Acid. Any chemical with a low pH that in water solution can burn the skin or eyes. Acids turn litmus paper red and have pH values of 0 to 6.

Action level. Term used by OSHA and NIOSH to express the level of toxicant that requires medical surveillance, usually one half of the permissible exposure limit.

Activated charcoal. Charcoal is an amorphous form of carbon formed by burning wood, nutshells, animal bones and other carbonaceous materials. Heating it with steam to between 800°C and 900°C activates charcoal. During this treatment, a porous, submicroscopic internal structure is formed that gives it an extensive internal surface area. Activated charcoal is commonly used as a gas or vapor adsorbent in air-purifying respirators and as a solid sorbent in air sampling.

Acute effect. Adverse effect on a human or animal that has severe symptoms developing rapidly and coming quickly to a crisis. See also Chronic effect.

Adsorption. The condensation of gases, liquids or dissolved substances on the surfaces of solids.

AIHA. American Industrial Hygiene Association.

Air. The mixture of gases that surround the earth; its major components are as follows: 78.08% nitrogen, 20.95 percent oxygen, 0.03 percent carbon dioxide and 0.93 percent argon. Water vapor (humidity) varies.

Airline respirator. A respirator that is connected to a compressed breathing air source by a hose of small inside diameter. The air is delivered continuously or intermittently in a sufficient volume to meet the wearer's breathing requirements.

Air-purifying respirator. A respirator that uses chemicals to remove specific gases and vapors from the air or uses a mechanical filter to remove particulate matter. An air-purifying respirator must only be used when there is sufficient oxygen to sustain life and the air contaminant level is below the concentration limits of the device.

Alkali. Any chemical with a high pH that in water solution is irritating or caustic to the skin. Strong alkalis in solution are corrosive to the skin and mucous membranes. Example: Sodium hydroxide, referred to as caustic soda or lye. Alkalis turn litmus paper blue and have pH values from 8 to 14. Another term for alkali is base.

Allergy. An abnormal response of a hypersensitive person to chemical and physical stimuli. Allergic manifestations of major importance occur in about 10 percent of the population.

ANSI. The American National Standards Institute is a voluntary membership organization (run with private funding) that develops consensus standards nationally for a variety of devices and procedures.

Asphyxiant. A vapor or gas that can cause unconsciousness or death by suffocation (lack of oxygen). Asphyxiation is one of the principal potential hazards of working in confined spaces.

ASTM. American Society for Testing and Materials.

Atmosphere-supplying respirator. A respirator that provides breathing air from a source independent of the surrounding atmosphere. There are two types: airline and self-contained breathing apparatus.

Atmospheric pressure. The pressure exerted in all directions by the atmosphere. At sea level, mean atmospheric pressure is 29.92 in. Hg, 14.7 psi, or 407 in. Hg.

Base. A compound that reacts with an acid to form a salt. It is another term for alkali.

Benign. Not malignant. A benign tumor is one that does not metastasize or invade tissue. Benign tumors may still be lethal, due to pressure on vital organs.

Biohazard. A combination of the words *biological* and *hazard*. Organisms or products of organisms that present a risk to humans.

Boiling point. The temperature at which the vapor pressure of a liquid equals atmospheric pressure.

Carbon monoxide. A colorless, odorless toxic gas produced by any process that involves the incomplete combustion of carbon-containing substances. It is emitted through the exhaust of gasoline-powered vehicles.

Carcinogen. A substance or agent capable of causing or producing cancer in mammals, including humans. A chemical is considered to be a carcinogen if: (a) it has been evaluated by the International Agency for Research on Cancer (IARC) and found to be a carcinogen or potential carcinogen; (b) it is listed as a carcinogen or potential carcinogen in the Annual Report on Carcinogens published by the National Toxicology Program (NTP) (latest edition) or (c) it is regulated by OSHA as a carcinogen.

CAS. Chemical Abstracts Service is an organization under the American Chemical Society. CAS abstracts and indexes chemical literature from all over the world in Chemical Abstracts. CAS Numbers are used to identify specific chemicals or mixtures.

Ceiling limit (C). An airborne concentration of a toxic substance in the work environment, which should never be exceeded.

CERCLA. Comprehensive Environmental Response, Compensation and Liability Act of 1980. Commonly known as Superfund (U.S. EPA).

Chemical cartridge respirator. A respirator that uses various chemical substances to purify inhaled air of certain gases and vapors. This type of respirator is effective for concentrations no more than 10 times the TLV of the contaminant if the contaminant has warning properties (odor or irritation below the TLV).

CHEMTREC. Chemical Transportation Emergency Center. Public service of the Chemical Manufacturers Association that provides immediate advice for those at the scene of hazardous materials emergencies. CHEMTREC has a 24-hour toll-free telephone number (1-800-424-9300) to help respond to chemical transportation emergencies.

Chronic effect. An adverse effect on a human or animal body, with symptoms that develop slowly over a long period of time or recur frequently. See also Acute effect.

Combustible liquid. Those liquids having a flash point at or above 37.8°C.

Concentration. The amount of a given substance in a stated unit of measure. Common methods of stating concentration include percent by weight or by volume, weight per unit volume and normality.

Corrosive. A substance that causes visible destruction or permanent changes in human skin tissue at the site of contact.

CFR. Code of Federal Regulations. A collection of the regulations that have been promulgated under United States Law.

Cutaneous. Pertaining to or affecting the skin.

Degrees Celsius (Centigrade). The temperature on a scale in which the freezing point of water is 0°C and the boiling point is 100°C. To convert to Degrees Fahrenheit, use the following formula: °F = (C x 1.8) + 32.

Degrees Fahrenheit. The temperature on a scale in which the boiling point of water is 212°F and the freezing point is 32°F.

Density. The mass per unit volume of a substance. For example, lead is much denser than aluminum.

Dermatitis. Inflammation of the skin from any cause.

Dermatosis. A broader term than dermatitis; it includes any cutaneous abnormality, thus encompassing folliculitis, pigmentary changes, and nodules and tumors.

Dose-response relationship Correlation between the amount of exposure to an agent or toxic chemical and the resulting effect on the body.

DOL. U.S. Department of Labor. OSHA and MSHA are part of the DOL.

DOT. U.S. Department of Transportation.

Dusts. Solid particles generated by handling, crushing, grinding, rapid impact, detonation, and decrepitation of organic or inorganic materials, such as rock, ore, metal, coal, wood and grain. Dusts do not tend to flocculate, except under electrostatic forces; they do not diffuse in air, but settle under the influence of gravity.

Dyspnea. Shortness of breath, difficult or labored breathing.

Evaporation. The process by which a liquid is changed into the vapor state.

Evaporation rate. The ratio of the time required to evaporate a measured volume of a liquid to the time required to evaporate the same volume of a reference liquid (e.g., butyl acetate, ethyl ether) under ideal test conditions. The higher the ratio, the slower the evaporation rates. The evaporation rate can be useful in evaluating the health and fire hazards of a material.

Federal Register. Publication of U.S. government documents officially promulgated under the law documents whose validity depends upon such publication. It is published on each day following a government working day. It is, in effect, the daffy supplement to the Code of Federal Regulations.

Fire point. The lowest temperature at which a material can evolve vapors fast enough to support continuous combustion.

First aid. Emergency measures to be taken when a person is suffering from overexposure to a hazardous material, before regular medical help can be obtained.

Flammable limits. Flammables have a minimum concentration below which propagation of flame does not occur on contact with a source of ignition. This is known as the lower flammable explosive limit (LEL). There is also a maximum concentration of vapor or gas in air above which propagation of flame does not occur. This is known as the upper flammable explosive limit (UEL). These units are expressed in percent of gas or vapor in air by volume.

HAZARD COMMUNICATION

Flammable liquid. Any liquid having a flash point below 37.8°C (100°F), except any mixture having components with flashpoints of 100°F or higher, the total of which make up 99 percent or more of the total volume of the mixture.

Flammable range. The difference between the lower and upper flammable limits, expressed in terms of percentage of vapor or gas in air by volume. Also referred to as explosive range.

Flash point. The minimum temperature at which a liquid gives off vapor within a test vessel in sufficient concentration to form an ignitable mixture with air near the surface of the liquid. Two tests are used: open cup and closed cup.

Fume. Airborne particulate formed by the evaporation of solid materials (e.g. metal fume emitted during welding). Usually less than 1 micron in diameter.

Gage pressure. Pressure measured with respect to atmospheric pressure.

Gas. A state of matter in which the material has very low density and viscosity; can expand and contract greatly in response to changes in temperature and pressure; easily diffuses into other gases; and readily and uniformly distributes itself throughout any container. A gas can be changed to the liquid or solid state only by the combined effect of increased pressure and decreased temperature. Examples include sulfur dioxide, ozone and carbon monoxide.

Gram (g). A metric unit of weight. One ounce equals 28.4 grams.

HEPA filter. High-efficiency particulate air filter. A disposable, extended medium, dry type filter with a particle removal efficiency of no less than 99.97 percent for 0.3pm particles.

IARC. International Agency for Research on Cancer.

IDLH. Immediately dangerous to life and health. An atmospheric concentration of any toxic, corrosive or asphyxiant substance that poses an immediate threat to life, would cause irreversible or delayed adverse health effects or would interfere with an individual's ability to escape from a dangerous atmosphere. Also known as "I Don't Leave Here."

Ignition source. Anything that provides heat, sparks or flame sufficient to cause combustion or explosion.

Ignition temperature. The minimum temperature to initiate or cause self-sustained combustion in the absence of any source of ignition.

Impervious. A material that does not allow another substance to pass through or penetrate it. Frequently used to describe gloves.

Inches of mercury column. A unit used in measuring pressures. One inch of mercury column equals a pressure of 1.66 kPa (0.491 psi).

Inches of water column. A unit used in measuring pressures. One inch of water column equals a pressure of 0.25 kPa (0.036 psi).

Incompatible. Materials that could cause dangerous reactions from direct contact with one another.

Ingestion. Taking in by the mouth.

Inhalation. Breathing of a substance in the form of a gas, vapor, fume, mist or dust.

Insoluble. Incapable of being dissolved in a liquid.

Irritant. A chemical, which is not corrosive, but which causes a reversible inflammatory effect on living tissue by chemical action at the site of contact.

Latent period. The time that elapses between exposure and the first manifestation of damage.

LC. Lethal concentration that will kill 50 percent of the test animals within a specified time. See LD50.

LD50. The dose required to produce the death in 50 percent of the exposed species within a specified time.

Liter (L). A measure of capacity. One quart equals 0.9L.

Lower explosive limit (LEL). The lower limits of flammability of a gas or vapor at ordinary ambient temperatures expressed in percent of the gas or vapor in air by volume. This limit is assumed constant for temperatures up to 120°C (250°F). Above this, it should be decreased by a factor of 0.7 because explosability increases with higher temperatures.

Malignant. As applied to a tumor. Cancerous and capable of undergoing metastasis or invasion of surrounding tissue.

Metastasis. Transfer of the causal agent (cell or microorganism) of a disease from a primary focus to a distant one through the blood or lymphatic vessels. Also, spread of malignancy from site of primary cancer to secondary sites.

Meter. A metric unit of length, equal to about 39 inches.

Micron Micrometer (pW). A unit of length equal to one millionth of a meter, approximately 1/25 of an inch.

Milligram (mg). A unit of weight in the metric system. One thousand equals 1 gram.

Milligrams per cubic meter (mg/m). Unit used to measure air concentrations of dusts, gases, mists and fumes.

Milliliter (mL). A metric unit used to measure volume. One milliliter equals one cubic centimeter.

Millimeter of mercury (mmHg). The unit of pressure equal to the pressure exerted by a column of liquid mercury 1mm high at a standard temperature.

Mists. Suspended liquid droplets generated by condensation from the gaseous to the liquid state or by breaking up a liquid into a dispersed state, such as by splashing, foaming or atomizing. Mist is formed when a finely divided liquid is suspended in air.

MSDS. Material safety data sheet.

MSHA. Mine Safety and Health Administration, U.S. Department of Labor.

Mucous membranes. Lining of the hollow organs of the body, notably the nose, mouth, stomach, intestines, bronchial tubes and urinary tract.

NFPA. The National Fire Protection Association is a voluntary membership organization whose aim is to promote and improve fire protection and prevention. The NFPA publishes 16 volumes of codes known as the National Fire Codes.

NIOSH. The National Institute for Occupational Safety and Health is a federal agency. It conducts research on health and safety concerns, tests and certifies respirators, and trains occupational health and safety professionals.

NTP. National Toxicology Program. The NTP publishes an annual report on carcinogens.

Nuisance dust. Dust having a long history of little adverse effect on the lungs and does not produce significant organic disease or toxic effect when exposures are kept under reasonable control.

OSHA. U.S. Occupational Safety and Health Administration, U.S. Department of Labor.

Oxidizer. A substance that gives up oxygen readily. Presence of an oxidizer increases the risk of fire.

Oxygen deficiency. That concentration of oxygen by volume below which atmosphere supplying respiratory protection must be provided. It exists in atmospheres where the percentage of oxygen by volume is less than 19.5 percent oxygen.

Oxygen-enriched atmosphere. An atmosphere containing more than 23.5 percent oxygen by volume.

Particulate matter. A suspension of fine solid or liquid particles in air, such as dust, fog, fume, dust, smoke or sprays. Particulate matter suspended in air is commonly known as an aerosol.

Permissible exposure limit (PEL). An exposure limit that is published and enforced by OSHA as a legal standard.

Personal protective equipment (PPE). Devices worn by the worker to protect against hazards in the environment. Respirators, gloves and hearing protectors are examples.

pH. Means used to express the degree of acidity or alkalinity of a solution with neutrality indicated as seven.

Polymerization. A chemical reaction in which two or more small molecules (monomers) combine to form larger molecules (polymers) that contain repeating structural units of the original molecules. A hazardous polymerization is the above reaction, with an uncontrolled release of energy.

ppm. Parts per million parts of air by volume of vapor or gas or other contaminant. Used to measure air concentrations of vapors and gases.

psi. Pounds per square inch (for MSDS purposes) is the pressure a material exerts on the walls of a confining vessel or enclosure. For technical accuracy, pressure must be expressed as psig (pounds per square inch gauge) or psia (pounds per square absolute; that is, gauge pressure plus sea level atmospheric pressure or psig plus approximately 14.7 lbs. per square inch).

RCRA. Resource Conservation and Recovery Act of 1976 (U.S. EPA).

Reactivity (chemical). A substance's susceptibility to undergo a chemical reaction or change that may result in dangerous side effects, such as an explosion, burns and corrosive or toxic emission.

Respirable size particulates. Particulates in the size range that permits them to penetrate deep into the lungs upon inhalation.

Respirator (approved). A device designed to protect the wearer from harmful atmospheres, meets the requirements of 30 CFR Part 11 and has been approved by the National Institute for Occupational Safety and Health and the Mine Safety and Health Administration.

Respiratory system. Consists of (in descending order) the nose, mouth, nasal passages, nasal pharynx, pharynx, larynx, trachea, bronchi, bronchioles, air sacs (alveoli) of the lungs and muscles of respiration.

Route of entry. The path by which chemicals can enter the body. There are three main routes of Entry: inhalation, ingestion and skin absorption.

SARA. Superfund Amendments and Reauthorization Act of 1986 (U.S. EPA).

SCBA. Self-contained breathing apparatus.

Sensitizer. A substance that on first exposure causes little or no reaction but which on repeated exposure may cause a marked response not necessarily limited to the contact site. Skin sensitization is the most common form of sensitization in the industrial setting.

Short-term exposure limit (STEL). ACGIH-recommended exposure limit. Maximum concentration to which workers can be exposed for a short period of time (15 minutes) for only four times throughout the day with at least one hour between exposures.

"Skin." A notation (sometimes used with PEL or TLV exposure data) that indicates the stated substance may be absorbed by the skin, mucous membranes, and eyes — either airborne or by direct contact — and that this additional exposure must be considered part of the total exposure to avoid exceeding the PEL or TLV for that substance.

Solubility in water. A term expressing the percentage of a material (by weight) that will dissolve in water at ambient temperature. Solubility information can be useful in determining spill-cleanup methods and re-extinguishing agents and methods for a material

Solvent. A substance, usually a liquid, in which other substances are dissolved. The most common solvent is water.

Sorbent. (1) A material that removes toxic gases and vapors from air inhaled through a canister or cartridge. (2) Material used to collect gases and vapors during air sampling.

Specific gravity. The ratio of the mass of a unit volume of a substance to the mass of the same volume of a standard substance at a standard temperature. Water at 4°C (39.2°F) is the standard usually referred to for liquids; for gases, dry air (at the same temperature and pressure as the gas) is often taken as the standard substance. See also Density.

Stability. An expression of the ability of a material to remain unchanged. A material is stable if it remains in the same form under expected and reasonable conditions of storage or use. Conditions that may cause instability (dangerous change) are stated. Examples are temperatures above 150°F, shock from dropping.

Synergistic. Cooperative action of substances whose total effect is greater than the sum of their separate effects.

Systemic. Spread throughout the body. Affecting all body systems and organs, not localized in one spot or area.

Threshold. The lowest dose or exposure to a chemical at which a specific effect is observed.

Time-weighted average concentration (TWA). Refers to concentrations of airborne toxic materials that have been weighted for certain time duration, usually eight hours.

Threshold limit value (TLV). A time-weighted average concentration under which most people can work consistently for eight hours a day, day after day, with no harmful effects. A table of these values and accompanying precautions is published annually by the American Conference of Governmental Industrial Hygienists.

Toxicity. A relative property of a chemical agent. Refers to a harmful effect on some biologic mechanism and the conditions under which this effect occurs.

Upper explosive limit (UEL). The highest concentration (expressed in percent vapor or gas in the air by volume) of a substance that will burn or explode when an ignition source is present.

Vapor pressure. Pressure (measured in pounds per square inch absolute (psia) exerted by a vapor. If a vapor is kept in confinement over its liquid so that the vapor can accumulate above the liquid (the temperature being held constant), the vapor pressure approaches a fixed limit called the maximum (or saturated) vapor pressure, dependent only on the temperature and the liquid.

Vapors. The gaseous form of substances that are normally in the solid or liquid state (at room temperature and pressure). The vapor can be changed back to the solid or liquid state either by increasing the pressure or decreasing the temperature alone. Vapors also diffuse. Evaporation is the process by which a liquid is changed into the vapor state and mixed with the surrounding air. Solvents with low boiling points will volatilize readily. Examples include benzene, methyl alcohol, mercury and toluene.

Viscosity. The property of a fluid that resists internal flow by releasing counteracting forces.

Volatility. The tendency or ability of a liquid to vaporize. Such liquids as alcohol and gasoline, because of their well-known tendency to evaporate rapidly, are called volatile liquids.

Water column. A unit used in measuring pressure. See also Inches of water column.

Guide for Reviewing MSDS Completeness

While reviewing the MSDS, the compliance officer may ask the following questions:
1. Do chemical manufacturers and importers have an MSDS for each hazardous chemical produced or imported into the United States?
2. Do employers have an MSDS for each hazardous chemical used?
3. Is each MSDS in at least English?
4. Does each MSDS contain at least the:
   (a) Identity used on the label?
   (b) Chemical and common name(s) for single substance hazardous chemicals?
   (c) For mixtures tested as a whole:
      (1) Chemical and common name(s) of the ingredients that contribute to the known hazards?
      (2) Common name(s) of the mixture itself?
   (d) For mixtures not tested as a whole:
      (1) Chemical and common name(s) of all ingredients that are health hazards (1 percent concentration or greater), including carcinogens (0.1 percent concentration or greater)?
      (2) Chemical and common name(s) of all ingredients that are health hazards and present a risk to employees, even though they are present in the mixture in concentrations of less than 1 percent or 0.1 percent for carcinogens?
   (e) Chemical and common name(s) of all ingredients that have been determined to present a physical hazard when present in the mixture?
   (f) Physical and chemical characteristics of the hazardous chemical (e.g., vapor pressure, flash point)?
   (g) Physical hazards of the hazardous chemical, including the potential for fire, explosion and reactivity?
   (h) Health hazards of the hazardous chemical, including signs, symptoms and medical conditions aggravated?
   (i) Primary routes of entry? OSHA permissible exposure limit? The American Conference of Governmental Industrial Hygienists threshold limit value? Other exposure limit(s), including ceiling and other short-term limits?
   (k) Information on carcinogen listings (reference OSHA-regulated carcinogens, those indicated in the National Toxicology Program Annual Report on Carcinogens and/or those listed by the International Agency for Research on Carcinogens)?
   (l) Generally applicable procedures and precautions for safe handling and use of the chemical (e.g., hygienic practices, maintenance and spill procedures)?
   (m) Generally applicable control measures (e.g., engineering controls, work practices, personal protective equipment)?
   (n) Pertinent emergency and first-aid procedures?
   (o) Date that the MSDS was prepared or the date of the last change?
   (p) Name, address and telephone number of the responsible party?
5. Are all sections of the MSDS completed?

Sample MSDS

OATEY ALL PURPOSE CEMENT 4 & 8 oz   Latest Revision Date   06/21/93

SECTION 1	IDENTITY OF MATERIAL
TRADE NAME	CEMENT 001 – OATEY ALL PURPOSE CEMENT
PRODUCT NUMBERS	30818 4 oz., 30821 8 oz.
FORMULA	PVC Resin in Solvent Solution
CHEMICAL FAMILY	PVC Organisol
SYNONYMS	PVC Plastic Pipe Cement

SECTION 2	HAZARDOUS INGREDIENTS		
INGREDIENTS	%	CAS NUMBER	C 313
CPVC Resin (non-hazardous)	12-17%	686-48-B2-8	No
Tetrahydrofuran	40-55%	109-99-9	No
Acetone	<0.25%	67-W-1	Yes
Methyl Ethyl Ketone	25-35%	7S-93-3	Yes
Cyclohexanone	11-16%	108-94-1	No
Fumed Silica (non-hazardous)	1-2%	7631-86-9	No

SECTION 3	KNOWN HAZARDS UNDER 29 CFR 1910.1200				
HAZARDS	YES	NO	HAZARDS	YES	NO
Combustible liquid	X		Skin hazard	X	
Flammable liquid	X		Eye hazard	X	
Pyrophoric material		X	Toxic agent	X	
Explosive material		X	Highly toxic agent		X
Unstable material		X	Sensitizer		X
Water reactive material		X	Kidney toxin	X	
Oxidizer		X	Reproductive toxin		X
Organic peroxide		X	Blood toxin	X	
Corrosive material		X	Nervous system toxin	X	
Compressed gas		X	Lung toxin	X	
Irritant	X		Liver toxin	X	

SECTION 4	REGULATIONS		
	TLV (TEA)	PEL (Transitional Limit) STEL	Hazard Action Level
Tetrahydrofuran	200ppm, 590mg/cm	200ppm, 590mg/cm	250ppm, 35mg/cm
Acetone	750ppm, 1800mg/cm	1000ppm, 2400mg/cm	1000ppm, 2400mg/cm
Cyclohexanone	25ppm, 100mg/cm (skin)	50ppm, 200mg/cm	N/A
Methyl ethyl ketone	200ppm, 5909mg/cm	200ppm, 5990mg/cm	300ppm, 885mg/cm

SECTION 5	REGULATED IDENTIFICATION
DOT PROPER SHIPPING NAME	CEMENT
DOT HAZARD CLASS	Flammable Liquid
SHIPPING ID NUMBER	NA 1133 (Gallons Only)
EPA HAZARDOUS WASTE ID NUMBER	D-001
EPA HAZARD WASTE CLASS	Ignitable Waste/Toxic Waste

SECTION 6 — EFFECTS OF EXPOSURE

ENTRY ROUTE	INHALE-YES	INGES-YES	SKIN-YES	EYE-YES
INHALATION	May cause irritation of mucous membranes, nose and throat, headache, dizziness, nausea, numbness of the extremities and narcosis in high concentrations. Has caused CNS depression and liver damage in animals, high concentrations have caused retardation of fetal development in rats.			
SKIN	Chronic contact may lead to irritation and dermatitis. Chronic exposure to vapors of high concentration may cause dermatitis. May be absorbed through the skin.			
EYE	Vapors or direct contact may cause irritation.			
INGESTION	May be aspirated into the lungs/cause effects described under inhalation.			
TARGET ORGANS	Eye, Skin, Kidney, Lung, Liver, Central Nervous System			

SECTION 7 — EMERGENCY AND FIRST-AID PROCEDURES—303/623-5716

SKIN	If irritation arises, wash thoroughly with soap and water. Seek medical attention if irritation persists.
EYES	If fumes cause irritation, move to fresh air and irrigate eyes with water for 15 minutes. If irritation persists, seek medical attention. If eye is struck with wire, seek medical attention.
INHALATION	Move to fresh air. It breathing is difficult, give oxygen. If not breathing, give artificial respiration. Keep victim quiet and warm. Call a poison control center or physician immediately.
INGESTION	Drink water and call a poison control center or physician immediately. Avoid alcoholic beverages. Never give anything by mouth to an unconscious person.

SECTION 8 — PHYSICAL AND CHEMICAL PROPERTIES

NFPA HAZARD SIGNAL	HEALTH STABILITY FLAMMABILITY
BOILING POINT	15° F/66° C
MELTING POINT	N/A
VAPOR PRESSURE	145 nmHg 2 20° psi
VAPOR DENSITY (AIR = 1)	2.5
VOLATILE COMPONENTS	84-88X
SOLUBILITY IN WATER	Negligible
PH	N/A
SPECIFIC GRAVITY	0.90 +/- 0.02
EVAPORATION RATE	(8UAC = 1) = 5.5 - 8.0
APPEARANCE	Milky liquid
COLOR	Ether-like
WILL DISSOLVE IN	Tetrahydrofuran
MATERIAL IS	Liquid

Hazard Communication

SECTION 9	FIRE AND EXPLOSION HAZARD DATA
FLAMMABILITY	LEL =1.8 X Volume LIEL= 11.8 X Volt
FLASHPOINT AND METHOD USED	0-5°F/ PMCC
STABILITY	Stable
CONDITIONS TO AVOID:	Heat, sparks and open flame
HAZARDOUS DECOMPOSITION POTENTIAL:	Carbon monoxide/ carbon dioxide/hydrogen chloride/smoke.
HAZARDOU S POLYMERIZATION	Will Not Occur.
CONDITIONS TO AVOID:	None
INCOMPATIBILITY/MATERIAL TO AVOID	Acids, oxidizing materials, alkalis, chlorinated inorganics (potassium, calcium and sodium hypochlorite), copper or copper alloys.
SPECIAL FIRE FIGHTING PROCEDURE	FOR SMALL FIRES: Use dry chemical, CO_2, water or foam extinguisher. FOR LARGE FIRES: Evacuate area and call fire department immediately.

SECTION 10	SPILL AND DISPOSAL INFORMATION
SPILL OR LEAK PROCEDURES	Ventilate area. Stop leak if it can be done without risk. Take up with sand, earth, or other non-combustible absorbing material.
WASTE DISPOSAL	Dispose of according to local, state, and federal regulations.

SECTION 11	SAFE USAGE DATA

PROTECTIVE EQUIPMENT TYPES

EYES:	Safety glasses with side shields.
RESPIRATORY:	NIOSH-approved canister respirator in absence of adequate ventilation.
GLOVES:	Rubber gloves
OTHER:	Eyewash and safety shower should be available.

VENTILATION...

GENERAL MECHANICAL:	Exhaust ventilation capable of maintaining emissions at the point of use below PEL.
LOCAL EXHAUST:	Open doors and windows. If used in enclosed area, use exhaust fans.

PRECAUTIONS

HANDLING and STORAGE:	Keep away from heat, sparks, and flames. Store in cool, dry place.
OTHER:	Containers, even empties, still retain residue and vapors.

SECTION 12	MANUFACTURER, OR SUPPLIER DATA
FIRM NAME AND MAILING ADDRESS	OATEY COMPANY, P.O. BOX 35906, 4700 West 160th Street, Cleveland, Ohio 44135
OATEY PHONE NUMBER	(216) 267-7100
EMERGENCY PHONE NUMBER	For emergency first aid (303) 623-5716 (COLLECT)

SECTION 13	DISCLAIMER

The information herein has been compiled from sources believed to be reliable, up-to-date, and is accurate to the best of our knowledge. However, Oatey cannot give any guarantees regarding information from other sources and expressly does not make warranties nor assumes any liability for its use.

Sample Letter Requesting an MSDS

Blitz Manufacturing Company
1923 Oak Grove Lane
Springfield, Massachusetts 02110

Dear Sir:

The Occupational Safety and Health Administration (OSHA) Hazard Communication Standard (29 CFR 1910.1200) requires employers be provided Material Safety Data Sheets (MSDSs) for hazardous substances used in their facility, and to make these MSDSs available to employees potentially exposed to these hazardous substances.

We, therefore, request a copy of the MSDS for your product listed as Stock Number _____. We did not receive an MSDS with the initial shipment of the Blitz Solvent 90 we received from you on October 1st. We also request any additional information, supplemental MSDSs, or any other relevant data that your company or supplier has concerning the safety and health aspects of this product.

Please consider this letter as a standing request to your company for any information concerning the safety and health aspects of using this product that may become known in the future.

The MSDS and any other relevant information should be sent to us within 10, 20, 30 days (select appropriate time). Delays in receiving the MSDS information may prevent use of your product. Please send the requested information to Mr. Robert Smith, Safety and Health Manager, XYZ Company, Boston, Massachusetts 02109.

Please be advised that if we do not receive the MSDS on the above chemical by (date), we may have to notify OSHA of our inability to obtain this information. It is our intent to comply with all provisions of the Hazard Communication Standard (1910.1200) and the MSDSs are integral to this effort.

Your cooperation is greatly appreciated. Thank you for your timely response to this request. If you have any questions concerning this matter, please contact Mr. Smith at (555) 123-4567.

Sincerely,
George Rogers, President
XYZ Company

Material Safety Data Sheet Checklist

Ensure that each MSDS contains the following information:
1. Product or chemical identity used on the label.
2. Manufacturer's name and address.
3. Chemical and common names of each hazardous ingredient.
4. Name, address, and phone number for hazard and emergency information.
5. Preparation or revision date.
6. The hazardous chemical's physical and chemical characteristics, such as vapor pressure and flashpoint, physical hazards the chemical presents, including the potential for fire, explosion, and reactivity.
7. Known health hazards.
8. OSHA permissible exposure limit (PEL), ACGIH threshold limit value (TLV) or other exposure limits.
9. Emergency and first-aid procedures.
10. Whether OSHA, NTP or IARC lists the ingredient as a carcinogen.
11. Precautions for safe handling and use.
12. Control measures such as engineering controls, work practices, hygienic practices, or personal protective equipment required.
13. Primary routes of entry.
14. Procedures for spills, leaks, and clean-up.

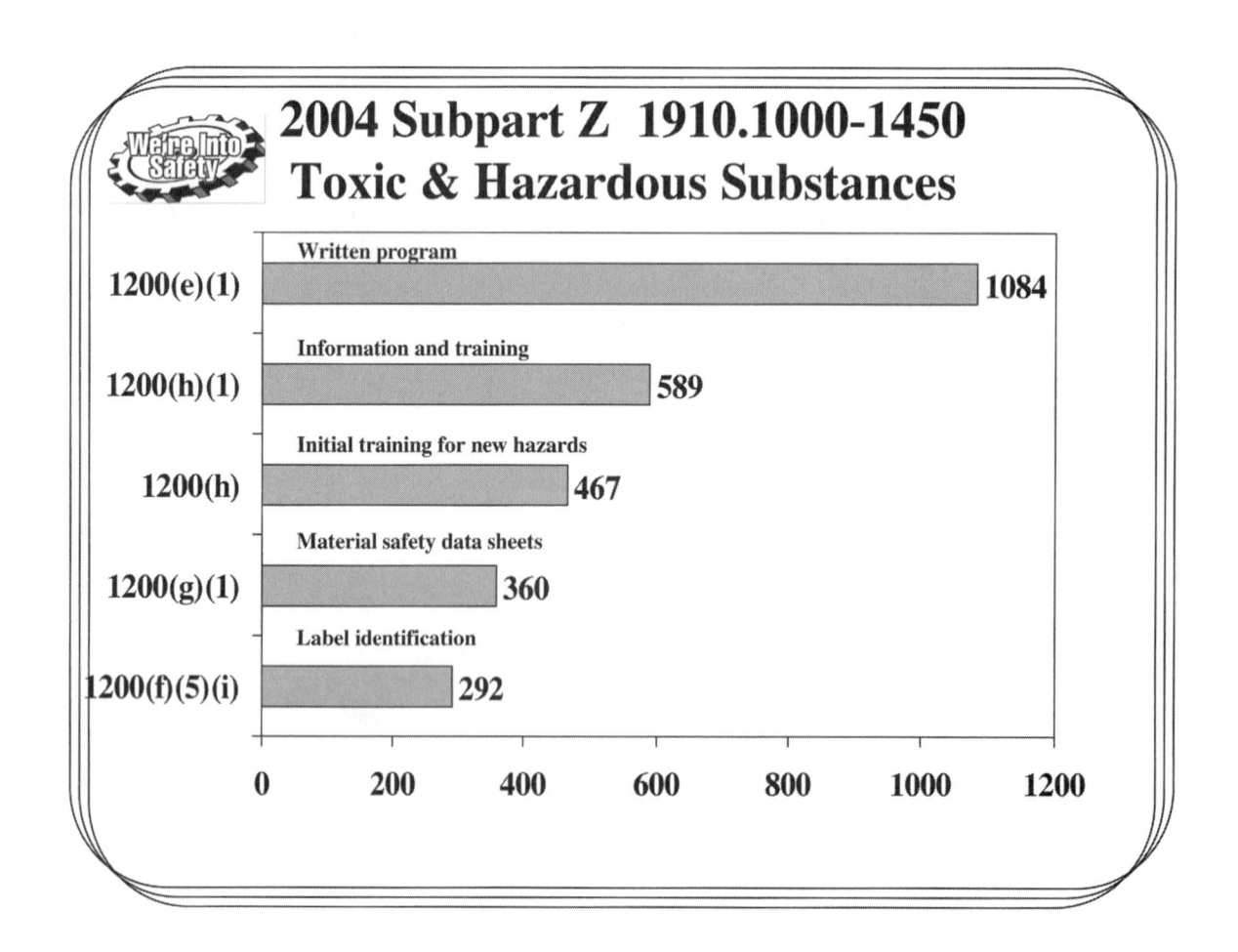

2004 Subpart Z 1910.1000-1450 Toxic & Hazardous Substances

Citation	Category	Count
1200(e)(1)	Written program	1084
1200(h)(1)	Information and training	589
1200(h)	Initial training for new hazards	467
1200(g)(1)	Material safety data sheets	360
1200(f)(5)(i)	Label identification	292

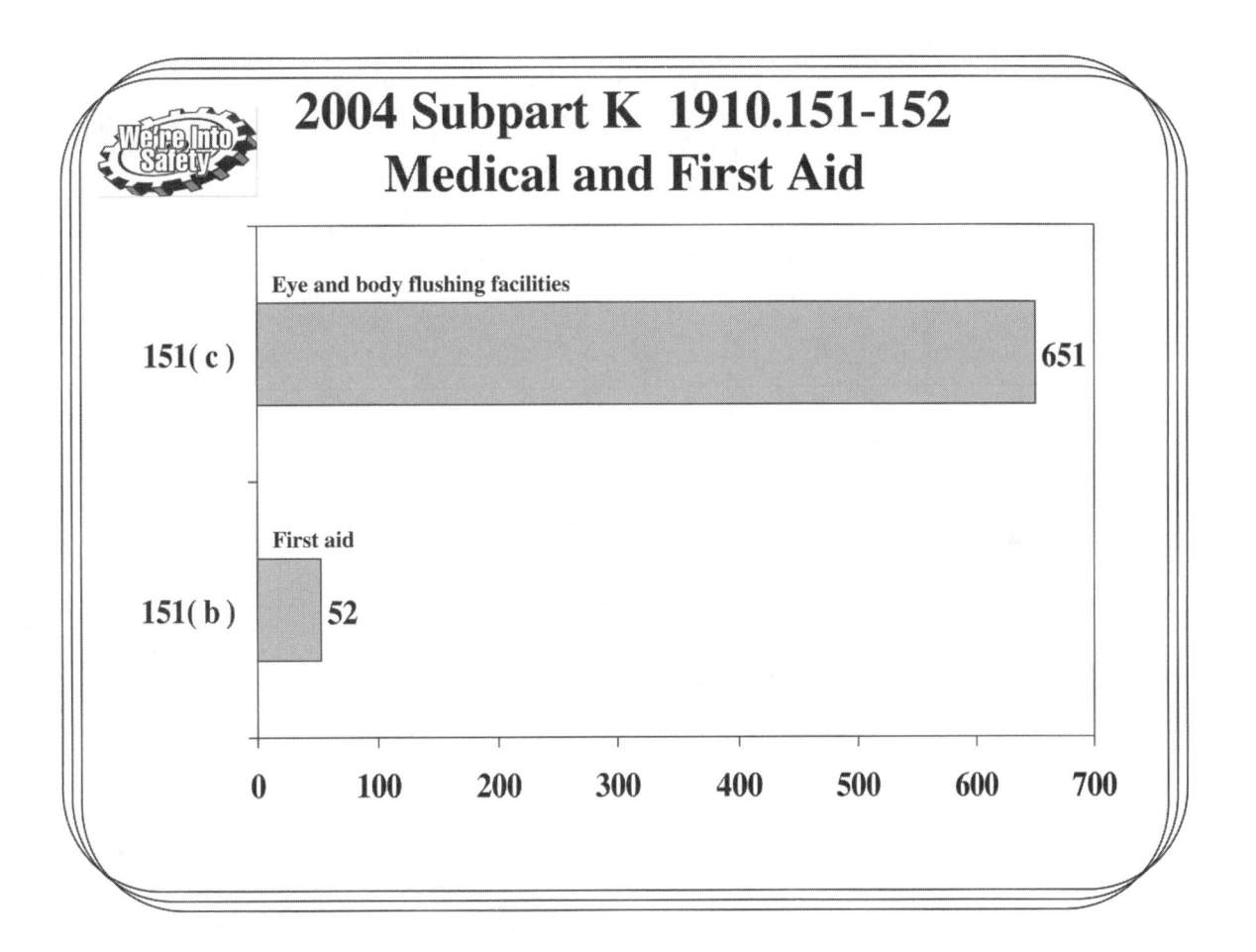

2004 Subpart K 1910.151-152
Medical and First Aid

Section	Topic	Count
151(c)	Eye and body flushing facilities	651
151(b)	First aid	52

HAZARD COMMUNICATION

CHAPTER 6

Personal Protective Equipment

Hard hats, goggles, face shields, steel-toed shoes, respirators, aprons, gloves and full body suits. These are all various forms of personal protective equipment (PPE).

Personal protective equipment should not be used as a substitute for engineering, work practice or administrative controls. To provide for employee safety and health in the workplace, personal protective equipment should be used in conjunction with these controls. Personal protective equipment includes all clothing and other work accessories designed to create a barrier against workplace hazards. The basic element of any management program for PPE should be an in-depth evaluation of the equipment needed to protect against the hazards at the workplace. Management dedicated to the safety and health of employees should use that evaluation to set a standard operating procedure for employees, and then train employees on the protective limitations of PPE, as well as the proper use and maintenance of PPE.

Using personal protective equipment requires hazard awareness and training on the part of the user. Employees must be aware that the equipment does not eliminate the hazard. If the equipment fails, exposure will occur. To reduce the possibility of failure, equipment must be properly fitted and maintained in a clean and serviceable condition.

Selection of the proper personal protective equipment for a job is important. Employers and employees must understand the equipment's purpose and its limitations. The equipment must not be altered or removed even though an employee may find it uncomfortable. In fact, sometimes equipment may be uncomfortable simply because it does not fit properly. Where possible, employees should test or try on equipment before using it during a shift to ensure proper fit.

General Requirements—1910.132

This regulation requires employers to ensure that personal protective equipment be "provided, used, and maintained in a sanitary and reliable condition wherever it is necessary" to prevent injury. This includes protection of any part of the body from hazards through absorption, inhalation or physical contact.

For example, many hazards can threaten the torso: heat, splashes from hot metals and liquids, impacts, cuts, acids and radiation. A variety of protective clothing is available: vests, jackets, aprons, coveralls, and full body suits.

Wool and specially treated cotton are two natural fibers that are fire resistant, comfortable and adaptable to a variety of workplace temperatures.

Duck, a closely woven cotton fabric, is good for light-duty protective clothing. It can protect against cuts and bruises on jobs where employees handle heavy, sharp or rough material.

Heat-resistant material, such as leather, is often used in protective clothing to guard against dry heat and flame. Rubber and rubberized fabrics, neoprene and plastics give protection against some acids and chemicals. It is important to refer to manufacturer's selection guides for the effectiveness of specific materials against specific chemicals.

Disposable suits of plastic-like or other similar synthetic material are particularly important for protection from dusty materials or materials that can splash. If the substance is extremely toxic, a completely enclosed chemical suit may be necessary. The clothing should be inspected to ensure proper fit and function for continued protection.

Employee-Owned Equipment
When employees provide their own equipment, the employer shall ensure the adequacy, including the proper maintenance and sanitation, of such equipment.

Design
All personal protective equipment must be of safe design and construction for the work to be performed.

Hazard Assessment and Equipment Selection
Employers are required to assess the workplace to determine if hazards that require the use of personal protective equipment are present or are likely to be present. If hazards or the likelihood of hazards are found, employers must select and have affected employees use properly fitted PPE suitable for protection from existing hazards.

Employers must certify in writing that a workplace hazard assessment has been performed. Defective and damaged equipment and defective or damaged personal protective equipment shall not be used.

Training
Before doing work requiring the use of personal protective equipment, employees must be trained to know when personal protective equipment is necessary, what type is necessary, how it is to be worn, what its limitations are and its proper care, maintenance, useful life, and disposal.

Employers are required to certify in writing that training has been carried out and that employees understand it. Each written certification shall contain the name of each employee trained, the date(s) of training and identify the subject in which the employee is certified.

Eye and Face Protection—1910.133
OSHA requires eye and face protective equipment where there is a reasonable probability of preventing injury when such equipment is used. Employers must provide a type of protector suitable for work to be performed and employees must use the protectors. These stipulations also apply to supervisors and management personnel and should apply to visitors while they are in hazardous areas.

Suitable eye protectors must be provided where there is a potential for injury to the eyes or face from flying particles, molten metal, liquid chemicals, acids or caustic liquids, chemical gases or vapors, potentially injurious light radiation or a combination of these. Protectors must meet the following minimum requirements:

- Provide adequate protection against the particular hazard;
- Be reasonably comfortable when worn under the designated conditions;
- Fit snugly without interfering with the movements or vision of the wearer;
- Be durable;
- Be capable of being disinfected;
- Be easily cleanable and kept clean and in good repair; and
- Every protector shall be distinctly marked to facilitate identification of the manufacturer.

Each affected employee shall use equipment with filter lenses that have a shade number appropriate for the work being performed for protection from injurious light radiation. Start with a shade that is too dark to see the weld zone. Then go to a lighter shade that gives sufficient view of the weld zone without going below the minimum. In oxyfuel gas welding or cutting where the torch produces a high yellow light, it is desirable to use a filter lens that absorbs the yellow or sodium line in the visible light of the (spectrum) operation.

Selection

Each eye, face or face-and-eye protector is designed for a particular hazard. In selecting a protector, consideration should be given to the kind and degree of hazard. Where a choice of protectors is given and the degree of protection required is not an important issue, worker comfort may be a deciding factor. When the manufacturer indicates limitations or precautions, they should be transmitted to the user and strictly observed.

Persons using corrective spectacles and those who are required by OSHA to wear eye protection must wear face shields, goggles or spectacles of one of the following types:
- Spectacles with protective lenses providing optical correction;
- Goggles worn over corrective spectacles that do not disturb the adjustment of the spectacles; or
- Goggles that incorporate corrective lenses mounted behind the protective lenses.

Goggles come in a number of different styles: eyecups, flexible or cushioned goggles, plastic eye shield goggles and foundrymen's goggles. Goggles are manufactured in several styles for specific uses such as protecting against dusts and splashes, in chippers, and for welders and cutters.

Safety spectacles require special frames. Combinations of normal streetwear frames with safety lenses are not in compliance. Many hard hats and nonrigid helmets are designed with face and eye protective equipment.

Design, construction, tests, and use of eye and face protection purchased prior to July 3, 1994, must be in accordance with ANSI Z87.1-1968 LISA Standard Practice for Occupational and Educational Eye and Face Protection. Protective eye and face devices purchased after July 5, 1994, must comply with ANSI Z87.1 1989, American National Standard Practice for Occupational and Educational Eye and Face Protection.

Fit

Someone skilled in the procedure should perform the fitting of goggles and safety spectacles. Only qualified optical personnel should fit prescription safety spectacles.

Inspection and Maintenance

It is essential that the lenses of eye protectors be kept clean. Continuous vision through dirty lenses can cause eyestrain—often an excuse for not wearing the eye protectors. Daily inspection and cleaning of the eye protector with soap and hot water or with a cleaning solution and tissue is recommended. Pitted lenses, like dirty lenses, can be a source of reduced vision. They should be replaced. Deeply scratched or excessively pitted lenses are apt to break more readily.

Slack, worn out, sweat-soaked, or twisted headbands do not hold the eye protector in proper position. Visual inspection can determine when the headband elasticity is reduced to a point beyond proper function.

Goggles should be kept in a case when not in use. Spectacles, in particular, should be given the same care as one's own glasses because the frame, nose pads and temples can be damaged by rough use.

Personal protective equipment that has been previously used should be disinfected before being issued to another employee. Also, when each employee is assigned protective equipment for extended periods, it is recommended that such equipment be cleaned and disinfected regularly.

Several methods for disinfecting eye-protective equipment are acceptable. The most effective method is to disassemble the goggles or spectacles and thoroughly clean all parts with soap and warm water. Equipment should be carefully rinsed to eliminate all traces of soap and defective parts should be replaced with new ones. All parts should be swabbed thoroughly and immersed for 10 minutes in a solution of germicidal deodorant fungicide. Once removed from the solution, equipment should be suspended in a clean place for air-drying at room temperature or with heated air (See Exhibit 6-1.). Equipment should not be rinsed after being removed from the solution because this will remove the germicidal residue that retains its effectiveness after drying.

PERSONAL PROTECTIVE EQUIPMENT

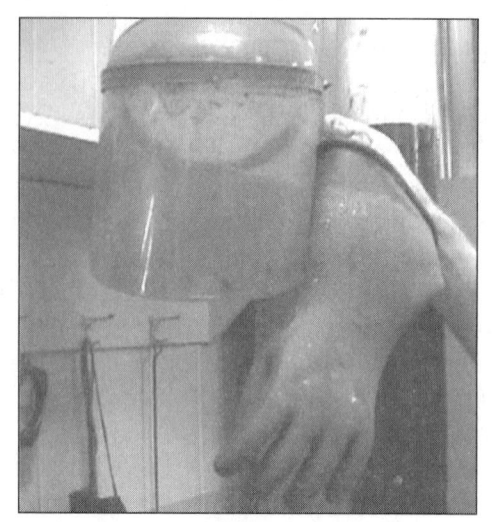

Exhibit 6.1.
Personal protective equipment should be cleaned and disinfected regularly.

The dry parts or items should be placed in a clean, dust-proof container, such as a box, bag, or plastic envelope to protect them until reissue.

Respiratory Protection—1910.134

OSHA standards require employers to establish and maintain a respiratory protective program whenever respirators are necessary to protect the health of employees. Respiratory protective devices fall into three classes:
- Air purifying;
- Atmosphere or air supplying; and
- Combination air-purifying and air-supplying devices.

Class 1: Air-Purifying Devices

The air-purifying device cleanses the contaminated atmosphere. Chemicals can be used to remove specific gases and vapors and mechanical filters can remove particulate matter. This type of respirator is limited in its use to those environments where the air contaminant level is within the specified concentration limitation of the device. These devices do not protect against oxygen deficiency.

"Oxygen deficiency" occurs in an atmosphere with an oxygen content below 19.5 percent by volume.

Types of air-purifying devices include:
- Mechanical-filter cartridge;
- Chemical-cartridge;
- Combination mechanical-filter/chemical-cartridge;
- Gas masks; and
- Powered air-purifying respirators.

Mechanical-filter respirators offer respiratory protection against airborne particulate matter, including dusts, mists, metal fumes and smoke, but do not provide protection against gases or vapors.

Chemical-cartridge respirators afford protection against low concentrations of certain gases and vapors by using various chemical filters to purify the inhaled air. They differ from mechanical-filter respirators in that they use cartridges containing chemicals to remove harmful gases and vapors.

Combination mechanical-filter/chemical-cartridge respirators use dust, mist or fume filters with a chemical cartridge for dual or multiple exposures.

Gas masks provide respiratory protection against certain gases, vapors and particulate matter. Gas masks are designed solely to remove specific contaminants from the air; therefore, it is essential that their use be restricted to atmospheres that contain sufficient oxygen to support life. Gas masks may be used for *escape only* from atmospheres that are immediately dangerous to life or health, and never for entry into such environments.

Immediately dangerous to life or health (IDLH) means an atmospheric concentration of any toxic, corrosive or asphyxiant substance that poses an immediate threat to life or would cause irreversible or delayed adverse health effects or would interfere with an individual's ability to escape from a dangerous atmosphere.

Canisters for gas masks are color-coded according to the contaminant against which they provide protection. This information is included in the standard.

Powered air-purifying respirators protect against particulates, gases and vapors, or both. The air-purifying element may be a filter, chemical cartridge, combination filter and chemical cartridge, or canister. The powered air-purifying respirator uses a power source (usually a battery pack) to operate a blower that passes air across the air-cleaning element to supply purified air to the respirator. The great advantage of the powered air-purifying respirator is that it usually supplies air at positive pressure (relative to atmospheric) so that any leakage is outward from the face piece. However, it is possible at high work rates to create a negative pressure in the face piece, thereby increasing face-piece leakage.

Class 2: Atmosphere- or Air-Supplying Devices

Atmosphere- or air-supplying devices are the class of respirators that provide a respirable atmosphere to the wearer, independent of the ambient air. Atmosphere-supplying respirators fall into three groups: supplied-air respirators, self-contained breathing apparatus (SCBA), and combination-SCBA and supplied-air respirators.

Supplied-air respirators deliver breathing air through a supply hose connected to the wearer's face piece or enclosure. The air delivered must be free of contaminants and must be from a source located in clean air. The OSHA requirements for compressed air used for breathing, including monitoring for carbon monoxide, are listed in 1910.134(d). Supplied-air respirators should only be used in non-IDLH atmospheres.

There are three types of supplied-air respirators, which are classified by type. Type A supplied-air respirators are also known as hose masks with blower. A motor-driven or hand-operated blower through a strong, large diameter hose supplies air. Type B supplied-air respirators are hose masks as in Type A, but without a blower. The wearer draws air through the hose by breathing. Type C supplied-air respirators are commonly referred to as airline respirators. An airline respirator must be supplied with respirable air conforming to Grade D Compressed Gas Association's Standard CGA G-7.1-73, Commodity Specification for Air, 1973. This standard requires air to have the oxygen content normally present in the atmosphere, no more than 5 mg/NV of condensed hydrocarbon contamination, no more than 20-ppm carbon monoxide, no pronounced odor and a maximum of 1,000 ppm of carbon dioxide.

There are three basic classes of airline respirators.

- **Continuous flow.** A continuous-flow unit has a regulated amount of air fed to the face piece and is normally used where there is an ample air supply such as that provided by an air compressor.
- **Demand flow.** These airline respirators deliver airflow only during inhalation. Such respirators are normally used when the air supply is restricted to high-pressure compressed air cylinders. A suitable pressure regulator is required to make sure that the air is reduced to the proper pressure for breathing.
- **Pressure-demand flow.** For those conditions where the possible inward leakage (caused by the negative pressure during inhalation that is always present in demand systems) is unacceptable and there cannot be the relatively high air consumption of the continuous-flow units, a pressure-demand airline respirator may be the best choice. It provides a positive pressure during both inhalation and exhalation.

Types A, B, and C that are approved for abrasive blasting are designated AE, BE and CE, respectively. These respirators are equipped with additional devices designed to protect the wearer's head and neck against impact and abrasion from rebounding abrasive material and with shielding to protect the windows of face pieces, hoods and helmets.

Self-contained breathing apparatus (SCBA) provide complete respiratory protection against toxic gases and oxygen deficiency. The wearer is independent of the surrounding atmosphere because he or she is breathing with a system that is portable and admits no outside air. The oxygen or air supply of the apparatus itself takes care of respiratory requirements.

There are two basic types of self-contained breathing apparatus: closed circuit and open-circuit. In a closed-circuit apparatus, the exhalation is rebreathed by the wearer after the carbon dioxide has been effectively removed and a suitable oxygen concentration restored from sources composed of compressed oxygen, chemical oxygen or liquid oxygen. In an open-circuit apparatus, exhalation is vented to the atmosphere and is not rebreathed. There are two types of open-circuit SCBAs: demand and pressure-demand.

Combination SCBA and supplied-air respirators are airline respirators with auxiliary self-contained air supplies. An auxiliary SCBA is an independent air supply that allows a person to evacuate an area or enter such an area for a very short period of time where a connection to an outside air supply can be made.

These devices are approved for use in IDLH atmospheres. The auxiliary air supply can be switched to in the event the primary air supply fails to operate. This allows the wearer to escape from the IDLH atmosphere. Combination airline respirators with auxiliary SCBA are designed to operate in three modes: continuous-flow, demand flow, and pressure-demand flow.

Class 3: Combination Air-Purifying and Atmosphere-Supplying Devices

These respirators are available in either continuous-flow or pressure-demand flow and are most often used with a high-efficiency filter as the air purifying element. Use in the filtering mode is allowed for escape only. Because of the positive pressure and escape provisions, these respirators have been recommended for asbestos work.

Exhibit 6-2. Classifications of Respiratory Protective Devices.

I. Air-Purifying Devices
   A. Mechanical-filter cartridge
   B. Chemical-cartridge
   C. Combination mechanical-filter/chemical cartridge
   D. Gas masks
   E. Powered air purifying
II. Atmosphere- or Air-Supplying Devices
   A. Supplied-air
      1. Types A and AE
      2. Types B and BE
      3. Types C and CE (airline)
         a. Continuous-flow
         b. Demand-flow
         c. Pressure-demand flow
   B. Self-contained breathing apparatus (SCBA)
      1. Closed-circuit
      2. Open-circuit
         a. Demand
         b. Pressure-demand
   C. Combination SCBA and supplied-air
      1. Continuous-flow
      2. Demand-flow
      3. Pressure-demand flow
III. Combination Air-Purifying and Atmosphere Supplying Devices
   A. Continuous-flow
   B. Pressure-demand flow

Requirements for a minimal acceptable respirator program are specified in 1910.134(b)(1) through 1910.134(b)(11). Other sections of the standard also refer to these requirements as listed in Exhibit 6-3.

Exhibit 6-3. Minimal Acceptable Respirator Program Requirements.

Requirement	Standard
Written Operating Procedures	1910.134(b)(1), (e)(1) and (e)(3)
Proper Selection	1910.134(b)(2), (c) and (e)(2)
Training and Fitting	1910.134(b)(3), (e)(5) and (e)(5)(i-iii)
Cleaning and Disinfecting	1910.134(b)(5) and (f)(3)
Storage	1910.134(b)(6), and (f)(5)(i-iii)
Inspection and Maintenance	1910.134(b)(7), (e)(4), (f)(2)(i-iv) and (f)(4)
Work Area Surveillance	1910.134(b)(8) only
Inspection/Evaluation of Program	1910.134(b)(9) only
Medical Examinations	1910.134(b)(10) only
Approved Respirators	1910.134(b)(11) only

Written Operating Procedures
OSHA standards state that the employer is responsible not only for providing appropriate respirators, but also for developing written standard operating procedures for their selection, use and care. The procedures must include a discussion or explanation of all items specified in 29 CFR 1910.134(b).

Program Selection
Respirators shall be selected on the basis of hazards to which the worker is exposed. A qualified individual supervising the respiratory protective program usually specifies the respirator type in the work procedures. The individual issuing them shall be adequately instructed to ensure that the correct respirator is issued.

In selecting the correct respirator for a given circumstance, many factors must be taken into consideration (e.g., the nature of the hazard, location of the hazardous area, employee's health, work activity, and respirator characteristics, capabilities and limitations).

In order to make subsequent decisions, the nature of the hazard must be identified to ensure that an overexposure does not occur. One important factor to consider is oxygen deficiency. NIOSH/MSHA approval for supplied-air and air-purifying respirators is valid only for atmospheres containing greater than 19.5 percent oxygen. If oxygen deficiency is not an issue, then the contaminant(s) and their concentrations must be determined. Exhibit 6-4 presents an outline for the selection process based on these criteria.

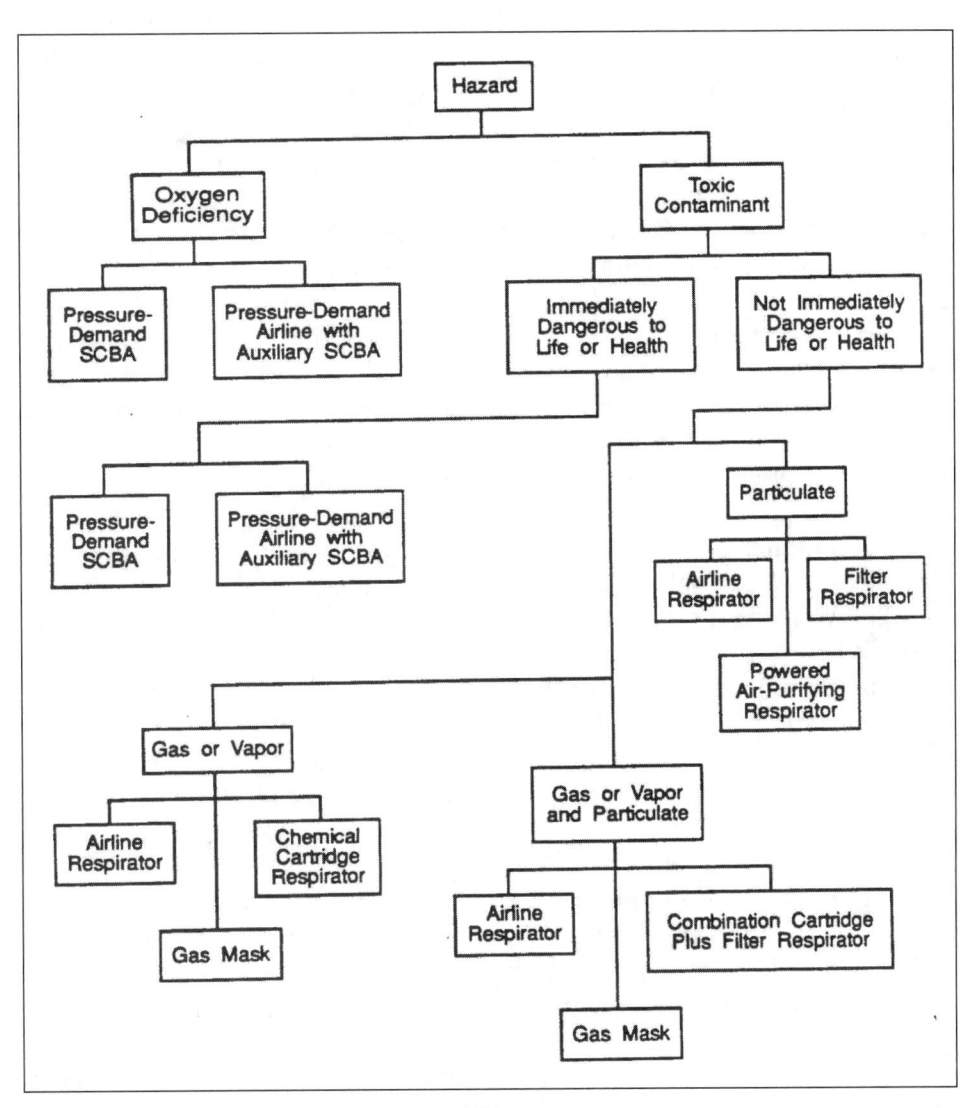

Exhibit 6-4.
Respirator Selection Criteria. The type of respirator used should be based on the contaminants to which the worker will be exposed as well as their concentrations.

PERSONAL PROTECTIVE EQUIPMENT

Training and Fitting
The user must be instructed and trained in the selection, use and maintenance of respirators. Every respirator user shall receive fitting instructions, including demonstrations and practice in how the respirator should be worn, how to adjust it, and how to determine if it fits properly.

Cleaning, Disinfecting, and Storage
Respirators must be regularly cleaned and disinfected. Those issued for the exclusive use of one worker should be cleaned after each day's use or more often if necessary. OSHA standards require that respirators be stored in a "convenient, clean, and sanitary location." The purpose of good respirator storage is to ensure that the respirator will function properly when used. Care must be taken to ensure that respirators are stored in such a manner as to protect against dust, harmful chemicals, sunlight, excessive heat or cold and moisture.

Inspection and Maintenance
Respirators used routinely shall be inspected during cleaning. Worn or deteriorated parts shall be replaced. Respirators for emergency use, such as self-contained devices, shall be thoroughly inspected at least once a month and after each use.

Work Area Surveillance
The OSHA standard requires that "appropriate surveillance of work area conditions and degree of employee exposure or stress be maintained." This should include identification of the contaminant, nature of the hazard and concentration at the breathing zone.

Inspection and Evaluation of the Program
The standard requires regular inspection and evaluation to determine the continued effectiveness of the respirator program. Many factors affect the employee's acceptance of respirators, including comfort, ability to breathe without objectionable effort, adequate visibility under all conditions, provisions for wearing prescription glasses (if necessary), ability to communicate, ability to perform all tasks without undue interference and confidence in the face piece fit. Failure to consider these factors is likely to reduce cooperation of the users in promoting a satisfactory program.

Medical Examinations
Persons should not be assigned to tasks requiring the use of respirators unless it has been determined that they are physically able to perform the work and use the equipment. A physician shall determine the health and physical conditions that are pertinent for an employee's ability to work while wearing a respirator. The user's medical status should be reviewed periodically.

Approved Respirators
The standard states that "approved or accepted respirators shall be used when they are available." A respirator is approved as the whole unit with specific components. OSHA recognizes a respirator as approved if NIOSH and the Mine Safety and Health Administration (MSHA) have jointly approved it.

Occupational Head Protection—1910.135
Prevention of head injuries is an important factor in every safety program. A survey by the Bureau of Labor Statistics of accidents and injuries noted that most workers who suffered impact injuries to the head were not wearing head protection. The majority of workers were injured while performing their normal jobs at their regular worksites.

The survey showed that in most instances where head injuries occurred, employers had not required their employees to wear head protection. Of those workers wearing hard hats, all but 5 percent indicated that they were required by their employers to wear them. It was found that the vast majority of those who wore hard hats all or most of the time at work believed that hard hats were practical for their jobs. According to the report, in almost half of the accidents involving head injuries, employees knew of no action taken by employers to prevent such injuries from recurring.

Elimination or control of a hazard leading to an accident should, of course, be given first consideration, but many accidents causing head injuries are of a type difficult to anticipate and control. Where these conditions exist, head protection must be provided to eliminate injury.

Head injuries are caused by falling or flying objects, or by bumping the head against a fixed object. Head protection, in the form of protective hats, must do two things: (1) resist penetration and (2) absorb the shock of a blow. This is accomplished by making the shell of the hat of a material hard enough to resist the blow, and by utilizing a shock-absorbing lining composed of headband and crown straps to keep the shell away from the wearers skull. Protective hats also are used to protect against electrical shock.

Criteria for Head Protection

The standards recognized by OSHA for head protection purchased prior to July 5, 1994, are contained in ANSI Requirements for Industrial Head Protection, Z89.1-1969, and ANSI Requirements for Industrial Protective Helmets for Electrical Workers, Z89.2-1971. These standards should be consulted for further details. The standards for protective helmets purchased after July 5, 1994, are contained in ANSI Personnel Protection-Protective Headwear for Industrial Workers-Requirements, Z89.1-1986. Later editions of these standards are available and acceptable for use.

Selection

Each type and class of head protectors is intended to provide protection against specific hazardous conditions. An understanding of these conditions will help in selecting the right protection for the particular situation. Head protection is made in the following types and classes:
- Type 1: Helmets with full brim, not less than 1¼ in. wide; and
- Type 2: Brimless helmets with a peak extending forward from the crown.

For industrial purposes, three classes are recognized:
- Class A: General service, limited voltage protection;
- Class B: Utility service, high-voltage helmets; and
- Class C: Special service, no voltage protection.

For firefighters, head protection must consist of a protective head device with ear flaps and a chin strap that meet the performance, construction and testing requirements stated in 29 CFR 1910.156(e)(5).

Helmets under Class A are intended for protection against impact hazards. They are used in mining, construction, shipbuilding, tunneling, lumbering and manufacturing.

Class B, utility service helmets protect the wearer's head from impact and penetration by falling or flying objects and from high-voltage shock and burn. Electrical workers use them extensively.

The safety helmets in Class C are designed specifically for lightweight comfort and impact protection. This class is usually manufactured from aluminum and offers no dielectric protection. Class C helmets are used in certain construction and manufacturing occupations, oil fields, refineries and chemical plants where there is not danger from electrical hazards or corrosion. They also are used on occasions where there is a possibility of bumping the head against a fixed object.

Materials used in helmets should be water resistant and slow burning. Each helmet consists essentially of a shell and suspension. Ventilation is provided by a space between the headband and the shell. Each helmet should be accompanied by instructions explaining the proper method of adjusting and replacing the suspension and headband.

The wearer should be able to identify the type of helmet by looking inside the shell for the manufacturer, ANSI designation and class. For example:
- Manufacturer's name;
- ANSI Z89.1-1969 (or later year);
- Class (A, B or C).

Fit

Headbands are adjustable in 1/8-size increments. When the headband is adjusted to the right size, it provides sufficient clearance between the shell and the headband. The removable or replaceable-type sweatband should cover at least the forehead portion of the headband. The shell should be of one-piece seamless construction and designed to resist the impact of a blow from falling material. The internal cradle of the headband and sweatband forms the suspension. Any part that comes into contact with the wearer's head must not be irritating to normal skin. Hard hats must be worn correctly for the headband to do its job. When workers wear baseball caps beneath their hard hats it defeats the purpose of the headband and does not provide adequate protection.

Inspection and Maintenance

Manufacturers should be consulted with regard to paint or cleaning materials for their helmets because some paints and thinners may damage the shell and reduce protection by physically weakening it or negating electrical resistance.

A common method of cleaning shells is to dip them in hot water containing a good detergent for at least one minute. Shells should then be scrubbed and rinsed in clear hot water. After rinsing, the shell should be carefully inspected for any signs of damage.

All components, shells, suspensions, headbands, sweatbands and any accessories should be visually inspected daily for signs of dents, cracks, penetration or any other damage that might reduce the degree of safety originally provided.

Users are cautioned that if unusual conditions occur (e.g., higher or lower extreme temperatures than described in the standards) or if there are signs of abuse or mutilation of the helmet or any component, the margin of safety may be reduced. If damage is suspected, helmets should be replaced or representative samples tested in accordance with procedures contained in ANSI Z89.1-1986. This discussion references national consensus standards, for example, ANSI standards that were adopted into OSHA regulations. Employers are encouraged to use up-to-date national consensus standards that provide employee protection equal to or greater than that provided by OSHA standards.

Helmets should not be stored or carried on the rear-window shelf of an automobile because sunlight and extreme heat may adversely affect the degree of protection.

Occupational Foot Protection— 1910.136

According to a survey by the Bureau of Labor Statistics, most of the workers in selected occupations who suffered foot injuries were not wearing protective footwear. Furthermore, most of their employers did not require them to wear safety shoes. These injuries occurred while workers were performing normal job activities at their worksites.

For protection of feet and legs from falling or rolling objects, sharp objects, molten metal, hot surfaces and wet slippery surfaces, workers should use appropriate foot guards, safety shoes, or boots and leggings. Leggings protect the lower leg and feet from molten metal or welding sparks. Safety snaps permit their rapid removal.

Aluminum alloy, fiberglass or galvanized steel foot guards can be worn over usual work shoes, although they may present the possibility of catching on something and causing workers to trip. Heat-resistant soled shoes protect against hot surfaces like those found in the roofing, paving and hot metal industries.

Safety shoes should be sturdy and have an impact-resistant toe. In some shoes, metal insoles protect against puncture wounds. Additional protection, such as metatarsal guards, may be found in some types of footwear. Safety shoes come in a variety of styles and materials, such as leather and rubber boots and oxfords.

Safety footwear is classified according to its ability to meet minimum requirements for both compression and impact tests. These requirements and testing procedures may be found in American National Standards Institute standards. Protective footwear purchased prior to July 5, 1994, must comply with ANSI Z41.1-1967, USA Standard for Men's Safety-Toe Footwear. Protective footwear purchased after July 5, 1994, must comply with ANSI Z41-1991, American National Standard for Personal Protection-Protective Footwear.

Electrical Protective Devices—1910.137

Insulating blankets, matting, covers, line hose, gloves and sleeves made of rubber shall meet specified requirements for manufacture, marking, electrical properties, workmanship and finish.

In-Service Care and Use

Electrical protective equipment shall be maintained in a safe, reliable condition. Specific requirements for in-service care and use are given for insulating blankets, covers, line hose, gloves and sleeves made of rubber.

Hand Protection—1910.138

Employers shall select and require employees to use appropriate hand protection when employees' hands are exposed to hazards such as those from skin absorption of harmful substances; severe cuts or lacerations; severe abrasions; punctures; chemical burns; thermal burns; and harmful temperature extremes.

Selection

There is a wide assortment of gloves, hand pads, sleeves and wristlets for protection against various hazardous situations. Employers need to determine what hand protection their employees need. The work activities of the employees should be studied to determine the degree of dexterity required, the duration, frequency, and degree of exposure to hazards and the physical stresses that will be applied.

Also, it is important to know the performance characteristics of gloves relative to the specific hazard anticipated (e.g., exposure to chemicals, heat or flames). Standard test procedures should assess gloves' performance characteristics.

Before purchasing gloves, the employer should request documentation from the manufacturer that the gloves meet the appropriate test standard(s) for the hazard(s) anticipated. For example, for protection against chemical hazards, the toxic properties of the chemicals must be determined, particularly the ability of the chemical(s) to pass through the skin and cause systemic effects.

The protective device should be selected to fit the job. For example, some gloves are designed to protect against specific chemical hazards. Employees may need to use gloves, such as wire mesh, leather and canvas, that have been tested and provide insulation from burns and cuts. The employee should become acquainted with the limitations of the clothing used.

Hearing Protection—1910.95

Exposure to high noise levels can cause hearing loss or impairment. It can create physical and psychological stress. There is no cure for noise-induced hearing loss, so the prevention of excessive noise exposure is the only way to avoid hearing damage. Specially designed protection is required, depending on the type of noise encountered.

A professional should individually fit preformed or molded earplugs. Waxed cotton, foam and fiberglass wool earplugs are self-forming. When properly inserted, they work as well as most molded earplugs.

Some earplugs are disposable—to be used one time and then thrown away. The non-disposable type should be cleaned after each use for proper protection. Plain cotton is ineffective as protection against hazardous noise.

Earmuffs need to make a perfect seal around the ear to be effective. Glasses, long sideburns, long hair and facial movements, such as chewing, can reduce protection. Special equipment is available for employees with glasses or beards.

For extremely noisy situations, earplugs should be worn in addition to earmuffs. When used together, earplugs and earmuffs change the nature of sounds; all sounds are reduced including one's own voice, but other voices or warning devices are easier to hear.

---

29 CFR Part 1910, Subpart I
Personal Protective Equipment

GENERAL REQUIREMENTS: The employer must assess the workplace to determine if hazards are present, or are likely to be present, which necessitate the use of personal protective equipment (PPE). If such hazards are present, or are likely to be present, the employer shall select, and have each affected employee use, the types of PPE that will protect against the identified hazards. PPE must properly fit each affected employee and the employer shall verify the hazard assessment in writing.

Damaged or defective equipment shall not be used.

The employer must provide training to each employee required to use PPE.

PERSONAL PROTECTIVE EQUIPMENT

Training will include when PPE is necessary, what PPE is necessary, how to wear PPE, and the proper care, maintenance, useful life, and disposal of the PPE. The employer has to certify in writing that the employee has received and understands the training.

EYE AND FACE PROTECTION: Employees must use appropriate eye or face protection when exposed to eye or face hazards from flying particles, molten metal, liquid chemicals, acids or caustic liquids, chemical gases or vapors, or potentially injurious light radiation. Requirements for side protection, prescription lenses, filter lenses, and identification of the manufacturer are spelled out. Protective eye and face devices purchased after July 5, 1994 must comply with ANSI Z87.1-1989 or be demonstrated to be equally effective. Devices purchased before that date must comply with ANSI Z87.1-1968 or are equally effective. The ANSI standard was updated in 2003.

HEAD PROTECTION: Employees must wear protective helmets when working in areas where there is a potential for injury to the head from falling objects. Each such affected employee when near exposed electrical conductors which could contact the head shall wear protective helmets designed to reduce electrical shock hazards. Protective helmets purchased after July 5, 1994 shall comply with ANSI Z89.1-1994 or be equally effective. Helmets purchased before that date shall comply with ANSI Z89-1-1969 or are equally effective.

FOOT PROTECTION: Employees must wear protective footwear when working in areas where there is a danger of foot injuries due to falling or rolling objects, or objects piercing the sole, and where employees' feet are exposed to electrical hazards. Protective footwear purchased after July 5, 1994 must comply with ANSI Z41-1991 or is equally effective. Protective footwear purchased before that date must comply with ANSI Z41.1-1967 or is equally effective.

HAND PROTECTION: Employers must select and require employees to use appropriate hand protection when employees' hands are exposed to hazards such as those from skin absorption of harmful substances; severe cuts or lacerations; severe abrasions; punctures; chemical burns; thermal burns and harmful temperature extremes. Employers shall base the selection of the appropriate hand protection on evaluation of the performance characteristics of the hand protection relative to the tasks to be performed, conditions present, duration of use and the hazards and potential hazards identified.

Questions and Answers Concerning the PPE Standards

Question: Does all PPE have to be replaced with PPE that meets the new revised standards?

Answer: No. The final standards "grandfather" existing PPE. That is, PPE that was purchased before the effective date of the standard can continue in use if it meets the pertinent ANSI standard in effect at the time of purchase, or if it can be demonstrated to be equally effective.

Q: Must all eye protective devices be equipped with side protectors?

A: No, not in all situations. However, side protectors, such as side shields, are required when it is determined (through the hazard assessment) that there is a hazard from flying objects.

Q: May side protectors be detachable?

A: Yes. Detachable side protectors (i.e., clip-on or slide-on side shields) are permitted if they meet the pertinent requirements of 1910.133.

Q: Is it permissible to wear contact lenses with eye protection?

A: Yes. OSHA believes that contact lenses do not pose additional hazards to the wearer. However, it is important to note that contact lenses are not eye protective devices. If eye hazards are present, appropriate eye protection must be worn instead of, or in conjunction with, contact lenses.

Q: Do PPE manufacturers have to obtain third-party certification that their equipment has been tested in accordance with, and meets the requirements of, the OSHA PPE standards?

A: No. However, many PPE manufacturers already voluntarily obtain third-party certification of their equipment.

Q: Since PPE manufacturers are not required to obtain third-party certification, what should employers do when purchasing PPE?

A: Employers can request the manufacturer to document that the PPE has been tested to meet specified criteria and employers can determine if the PPE is marked as meeting the pertinent ANSI standard. For instance, head protective devices should be marked as meeting ANSI Z41.1. In addition, as stated previously, many PPE manufacturers already obtain third-party certification. Therefore, when purchasing PPE, employers should determine whether or not the PPE has such certification.

Q: Is an employer required to have a written hazard assessment?

A: No, although a written hazard assessment is certainly recommended. Employers are, however, required to verify that the hazard assessment has been performed through a written certification that identifies the workplace evaluated, the person certifying that the evaluation has been performed, the date(s) of the hazard assessment and identifies the document as a certification of hazard assessment.

Q: Is PPE required for certain SIC codes or certain job titles? For example, is foot protection required for employees who work in a warehouse?

A: The PPE standard contains performance-oriented requirements and does not specify SIC codes or job titles to mandate the use of certain types of PPE. Instead, the employer is required to perform a hazard assessment of the workplace to determine what hazards, if any, are present. The employer must then make a selection of the appropriate PPE that will protect employees from the hazards identified by the hazard assessment.

Q: Many female employees have complained that they have to wear PPE that does not fit properly because the PPE has been sized to fit only male employees. Is this situation going to continue?

A: No. One of the new requirements of the PPE standard is that employers are now required to select PPE that properly fits employees (29 CFR 1910.132(d)(1)(iii)). PPE is now available in many different sizes or can be adjusted to many sizes (such as head protective devices). Even protective footwear is now available in sizes that will properly fit female employees.

Q: How often must employees be trained?

A: Employees must be trained before being allowed to perform work requiring the use of PPE. Employees must also be retrained: when there are changes in the workplace that impact the use of PPE; there are changes in the types of PPE to be used or when inadequacies in an employee's knowledge or use of assigned PPE indicate that the employee has not retained understanding or skill.

Q: The final standards reference specific editions of the ANSI standards that different types of PPE must meet. For example, protective eye and face devices must meet the 1989 edition of ANSI Z87.1. In the future, what happens if an employer wants to purchase PPE that meets a later edition of the same ANSI standard?

A: OSHA will accept PPE as complying with the standard if the PPE is demonstrated to be as effective as the PPE meeting the specific ANSI standard referenced by the final standard. For example, eye and face protective devices meeting a subsequent edition of the same ANSI standard would be acceptable to OSHA if it could be demonstrated that they were as effective as those meeting the 1989 edition. Employers would need to establish either that there was no substantive difference between a subsequent edition of Z87.1 and the 1989 edition, or that PPE that satisfied subsequently modified test criteria provided protection equivalent to that provided by PPE that satisfied the 1989 edition.

Q: What constitutes an adequate hazard assessment?

A: The final standard is performance oriented to allow the flexibility needed for employers to perform a hazard assessment that best reflects their particular workplace. An adequate hazard assessment will result as long as the employer meets the requirements of § 1910.132(d)(1). Additionally, Appendix B contains an example of procedures that satisfy the hazard assessment requirement.

Fact Sheet—Personal Protective Equipment (PPE) Standards

- The previous standards for head, face, eye and foot protection were adopted from the 1971 ANSI standards. These standards are outdated and do not reflect current technology and improvements in PPE.
- The final rule is a revision that updates the standards to be more consistent with the latest editions of the ANSI standards.
- The final rule provides guidance for the selection and use of PPE, as well as performance-oriented requirements, where appropriate.
- The standards also contain some new requirements. Employers will now have to perform a hazard assessment of the workplace to determine if any hazards are present that would necessitate the need for PPE. Based on this assessment, the employer would then select the appropriate PPE for the hazards that were found.
- Employers will also have to provide PPE training to employees.
- The new rule contains a section addressing hand protection. This fills a gap in employee protection since the prior standards had no requirement for hand protection.
- The proposal was published in August 1989. Hearings were held in Washington, D.C. on April 3 and 4, 1990. The rulemaking record contains 173 comments, 577 pages of testimony and 53 exhibits.
- The final rule was published in the *Federal Register* on April 6, 1994. It became effective on July 5, 1994.
- The new standards will cover 1.1 million work establishments and 11.7 million employees.
- It is estimated that compliance with the new provisions of the final rule will cost $52.4 million annually and will prevent four fatalities and 712,000 lost workdays annually.

HAZARD ASSESSMENT FORM

Instructions: This is a two-part form printed back to back. The first part is on the front page; the second part is on the back page. Fill in the heading to identify the area being assessed, and then conduct a walk-through survey of the work area. Fill in the front page to identify the hazards during a normal operation in the area. This form is an assessment of workplace hazards as required by OSHA.

AREA: _____

OPERATION: _____

YOUR NAME: _____

DATE OF ASSESSMENT: _____

Head Hazards – Working beneath material or other workers that could fall and strike the head; working on electrical equipment, chemicals or other hazards that could affect the head of the worker.
- ☐ Impact
- ☐ Electrical shock
- ☐ Chemical splash

Describe hazard here: _____
_____
_____

Eye and Face Hazards – Working with chemicals, acids, grinding, sanding, welding, woodworking operations or anytime there is the possibility of something flying through the air and striking the worker in the face or eyes.
- ☐ Impact
- ☐ Dust
- ☐ Chemical splash
- ☐ Light/radiation

Describe hazard here: _____
_____
_____

Hand Hazards – Cutting stock material, working with hot or cold items or chemicals.
- ☐ Cuts
- ☐ Puncture
- ☐ Chemical splash
- ☐ Hot/cold temperature

Describe hazard here: _____
_____
_____

Foot Hazards – Look for the potential for tools or stock to be dropped on the foot of workers or chemical spills or spike-like items that may puncture the bottom of the foot.
- ☐ Impact
- ☐ Puncture
- ☐ Compression
- ☐ Chemical splash

Describe hazard here: _____
_____
_____

PERSONAL PROTECTIVE EQUIPMENT

Based on the hazard assessment conducted on the previous page, the following personal protective equipment is required for the job classification of_____.

Head Hazards

Job	Identified Hazard	PPE to be Worn

Eye and Face Hazards

Job	Identified Hazard	PPE to be Worn

Hand Hazards

Job	Identified Hazard	PPE to be Worn

Foot Hazards

Job	Identified Hazard	PPE to be Worn

Ensure that workers are issued the correct PPE and are fitted and trained on how to use and maintain the equipment. Workers also must be trained on the limitations of the equipment as well as the proper disposal and replacement of the equipment.

EMPLOYEE SAFETY AND HEALTH RECORD

Name: _____ Permanent ___
Job Title: _____ Temporary ___
Department: _____ Date Hired: _____

List Hazards Associated with Present Duties:

_____

Occupational Health Medical Examination Required:
Asbestos: _____ Hearing: _____
Lead: _____ Other (specify): _____

MANDATORY TRAINING TOPICS (To be briefed to all personnel; initial when conducted)

_____ Location of required and appropriate safety bulletin boards, OSHA Poster
_____ OSHA and Company Safety Rules that apply to the job and workplace
_____ Hazards of the work area to include physical, physiological and chemical.
_____ Reason for specific medical evaluation
_____ Hazards of the assigned job or tasks and safety procedures to be followed
_____ Personal protective equipment (PPE), required, how, when and where to use it
_____ Emergency procedures for evacuation, fire reporting, emergency equipment
_____ Location of alarms and extinguishers
_____ Emergency telephone numbers
_____ Location of medical facilities or first aid kits
_____ Need to report of unsafe/unhealthful equipment, conditions or procedures
_____ How to report unsafe/unhealthful equipment, conditions or procedures
_____ Requirements for documentation and notification of on-the-job injury or illness
_____ Hazard communications program
_____ Company safety program

INITIAL INDIVIDUAL TRAINING TOPICS (As needed; initial when conducted)

_____ Respirator protection program   _____ Lockout/tagout
_____ Permit-required confined space program   _____ Back injury prevention
_____ Hearing conservation   _____ Bloodborne pathogens
_____ Jewelry safety   _____ PPE use and care
_____ Fall protection   _____ Others (list and initial)

PERSONAL PROTECTION ISSUED (Initial when issued)

_____ Head protection   _____ Eye protection   _____ Face protection
_____ Arm/hand protection   _____ Foot protection   _____ Hearing protection
_____ Respiratory protection _____ Other (specify)   _____ Other (specify)

PERSONAL PROTECTIVE EQUIPMENT

PERSONAL PROTECTION PROVIDED IN THE WORK AREA

_____ Head protection   _____ Eye protection   _____ Face protection
_____ Arm/hand protection   _____ Foot protection   _____ Hearing protection
_____ Respiratory protection   _____ Other (specify)   _____ Other (specify)

RECORD OF CONTINUOUS JOB SAFETY TRAINING

DATE	INITIAL	ANNUAL	SUPERVISOR SIGNATURE	EMPLOYEE SIGNATURE

RECORD OF JOB SAFETY NON-COMPLIANCE

Verbal Warning for _____ Date: _____ Supervisor: _____
Written Warning for _____ Date: _____ Supervisor: _____
Written Warning for _____ Date: _____ Supervisor: _____
Written Warning for _____ Date: _____ Supervisor: _____
Written Warning for _____ Date: _____ Supervisor: _____
Notice of intent to Terminate: _____ Date: _____ Human Resources

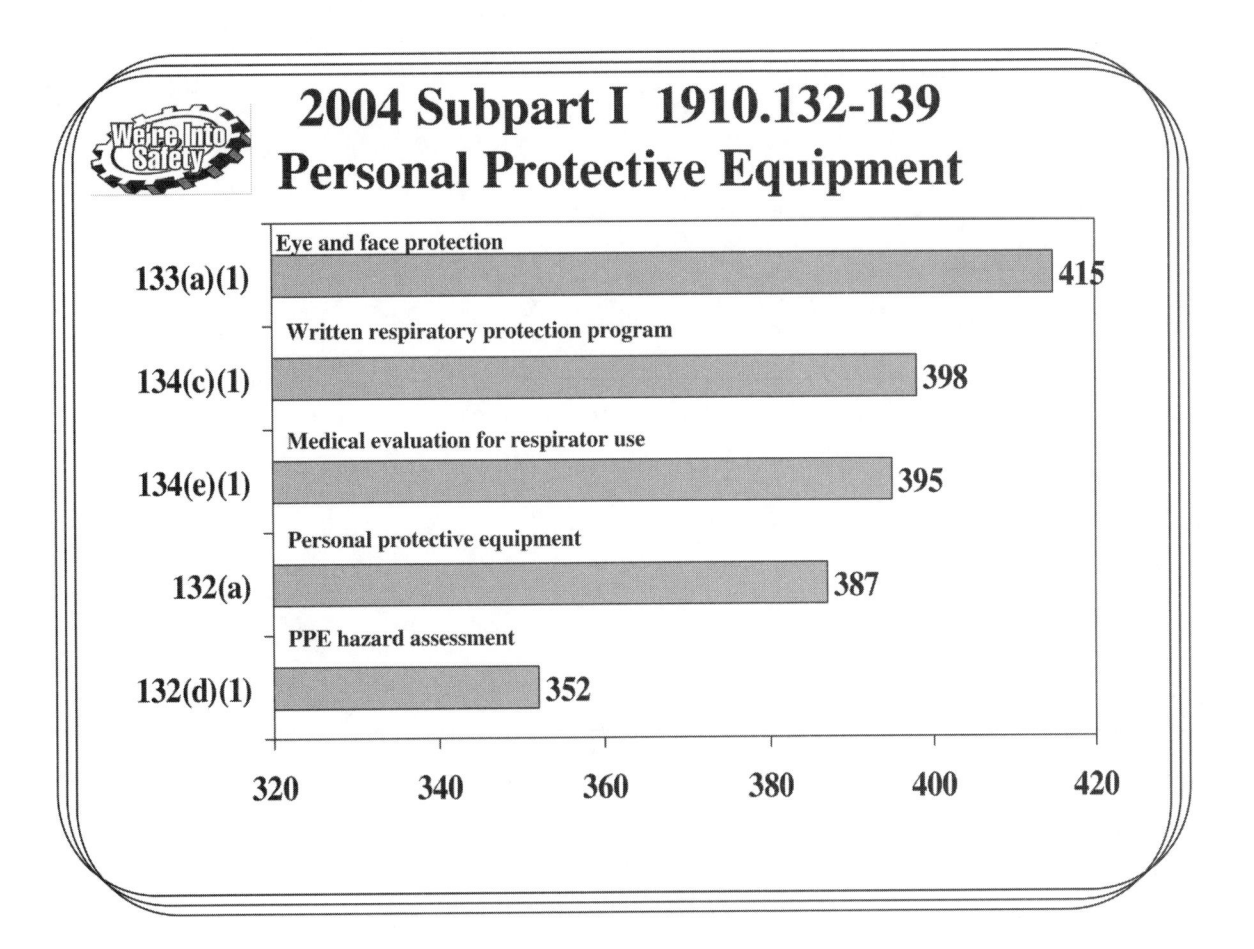

2004 Subpart I 1910.132-139
Personal Protective Equipment

Citation	Description	Count
133(a)(1)	Eye and face protection	415
134(c)(1)	Written respiratory protection program	398
134(e)(1)	Medical evaluation for respirator use	395
132(a)	Personal protective equipment	387
132(d)(1)	PPE hazard assessment	352

CHAPTER 7

Hazardous Waste Operations and Process Safety Management

Hazardous Waste Operations and Emergency Response

This standard, set forth at 29 CFR 1910.120, regulates hazardous waste clean up, treatment, and emergency response for general industry. This standard applies to employees involved in:

- **Clean-up operations**
  1. Clean-up operations required by a governmental body, whether federal, state, local, or other, involving hazardous substances that are conducted at uncontrolled hazardous waste sites.
  2. Corrective actions involving clean-up operations at sites covered by the Resource Conservation and Recovery Act (RCRA).
  3. Voluntary clean-up operations at sites recognized by federal, state, local, or other governmental bodies as uncontrolled hazardous waste sites.
- **Treatment, storage, and disposal of hazardous wastes**
  1. Operations involving hazardous wastes that are conducted at treatment, storage, and disposal facilities licensed under the RCRA.
- **Emergency response operations**
  1. Emergency response operations for the release of, or substantial threats of the release of, hazardous substances. This is the type of operation with which the majority of employers will be concerned. Exceptions are permitted if the employer can demonstrate that the operation does not involve employee exposure or a reasonable possibility of such exposure to hazards.

General Requirements

- Development by each hazardous waste site employer of a safety and health program designed to identify, evaluate, and control safety and health hazards and to provide for emergency response.
- A preliminary evaluation of the site's characteristics prior to entry by a trained person to identify potential site hazards and to aid in the selection of appropriate employee protection methods. Included would be all suspected conditions immediately dangerous to life or health or that may cause serious harm.
- Implementation of a site control program to protect employees from hazardous contamination. At a minimum, it must have a site map, site work zones, site communications, safe work practices, and identification of the nearest medical assistance. The "buddy system" also is required as a protective measure in particularly hazardous situations so that employees can keep watch on one another to provide quick aid if needed.

- Training of employees before they are allowed to engage in hazardous waste or emergency response operations that could expose them to safety and health hazards. However, experienced workers will be allowed to continue operations and then be given refresher courses when appropriate. Specific training requirements are listed for clean-up personnel, equipment operators, general laborers and supervisory employees and for various levels of emergency response personnel. Persons completing specified training for hazardous waste operations shall be certified. Those neither certified nor with proper experience shall be prohibited from engaging in those operations specified by the standard.

Training requirements vary with the type of operation involved. The various operations and their dependent training requirements are:
- Uncontrolled hazardous waste operations mandated by the government.
  - These workers must have 40 hours of initial training before they may enter the site and at least three days of actual field experience under a trained, experienced supervisor.
  - Employees visiting the site occasionally need only 24 hours of training and one day of supervised field experience before they may enter.
  - Managers and supervisors directly responsible for clean-up operations must have an additional eight hours of specialized training in waste management.
- Annual refresher training of eight hours is required for site workers and managers.
- Sites licensed under the RCRA. Employees must have 24 hours of training plus eight hours of annual refresher training.

Medical surveillance is required at least annually and at the end of employment for all employees exposed to any particular hazardous substance at or above established exposure levels and/or those who wear approved respirators for 30 days or more on site. Surveillance also will be conducted if a worker is exposed by unexpected or emergency releases as well.

Engineering controls, work practices, and personal protective equipment, or a combination of these methods, must be implemented to reduce exposure below established exposure levels for the hazardous substance involved. Air monitoring must be initiated periodically to identify and quantify levels of hazardous substances to ensure that proper protective equipment is being used.

An informational program that includes the names of key personnel and their alternates who are responsible for site safety and health, and the requirements of the standard must be provided to employees.

Before any employee or equipment may leave an area where there is a potential for hazardous exposure, a decontamination procedure must be implemented. Standard operating procedures to minimize exposure through contact with exposed equipment, employees, or used clothing must be established and showers and change rooms must be provided where needed. The employer must establish an emergency response plan to handle possible onsite emergencies before any hazardous waste operations may begin. The emergency response plan must address:
- Personnel roles;
- Lines of authority;
- Training and communications;
- Emergency recognition and prevention;
- Safe places of refuge;
- Site security;
- Evacuation routes and procedures;
- Emergency medical treatment; and
- Methods of sounding the emergency alert.

Employers also must have an off-site emergency plan to better coordinate emergency action by local emergency services and to implement appropriate control actions.

Emergency Response to Hazardous Substance Releases

In the event of a release of hazardous material or in response to an emergency or potential emergency, employees must comply with 1910.120(q). This section concerns more employers than any other part of the standard.

The regulation defines "emergency response" or "responding to emergencies" as:

[A] response effort by employees from outside the immediate release area or by other designated responders (i.e., mutual-aid groups, local fire departments, etc.) to an occurrence which results, or is likely to result, in an uncontrolled release of a hazardous substance."

Incidental releases of hazardous substances where the substance can be absorbed, neutralized, or otherwise controlled at the time of release by employees in the immediate release area or by maintenance personnel are not considered to be emergency releases within the scope of this standard. Responses to releases of hazardous substances where there is no potential safety or health hazard (i.e., fire, explosion, chemical exposure) are not considered to be emergency responses.

If there is a potential safety or health hazard, there may be an emergency and hence an emergency response that requires compliance with 1910.120(q). This applies whether the release is responded to by employees from outside the work area, outside groups such as the fire department, or employees from the immediate work areas who have been designated by the employer to respond to emergencies.

On the other hand, if there is no potential safety or health hazard, there is no emergency. This applies whether the release is cleaned up by personnel within the immediate work area or from outside the work area. Generally, employees will be trained under the hazard communication standard to deal with such incidental releases as 1910.120(q) does not apply.

Situations generally resulting in emergency responses include:
- The response comes from outside the immediate release area;
- The release requires evacuation of employees in the area;
- The release poses, or has the potential to pose, conditions that are immediately dangerous to life and health (IDLH);
- The release poses a serious threat of fire or explosion;
- The release requires immediate attention because of imminent danger;
- The release may cause high levels of exposure to toxic substances;
- There is uncertainty that the employees in the work area can handle the severity of the hazard with the personal protective equipment and other equipment that has been provided and the exposure limit could easily be exceeded; and
- The situation is unclear or data is lacking key factors.

The companies and employers whose employees will be engaged in emergency responses must develop and implement an emergency response plan. Employers may be exempt from developing and implementing a plan in the following circumstances:
- When an emergency occurs, they evacuate all of their employees from the danger area and do not permit any of their employees to assist in handling the emergency. This action should be outlined in the company emergency action plan.
- Those organizations that have developed programs for handling releases of hazardous substances in accordance with the Superfund Amendments and the Reauthorization Act of 1986 (Emergency Planning and Community Right-to-Know Act of 1986) may use those programs to meet the requirements of 1910.120 as long as they are equivalent. These firms may use the local emergency response or state emergency response plans as part of their plans to avoid duplication.

The emergency response plan must be in writing and contain the following elements:
- Pre-emergency planning and coordination with outside parties;
- Personnel roles, lines of authority, and communication;
- Emergency recognition and prevention;
- Safe distances and places of refuge;
- Site security and control;
- Evacuation routes and procedures;
- Decontamination procedures that are not covered by the site safety and health plan;
- Emergency medical treatment and first aid;
- Emergency alerting and response procedures;

- Critique of response and follow up (after-action review procedures); and
- Personnel protective equipment and emergency equipment needed.

Procedures for Handling Emergency Response
The senior emergency response official responding to an emergency shall become the individual in charge of a site-specific Incident Command System (ICS). All emergency responders and their communications must be coordinated and controlled through the individual in charge of the ICS who is assisted by the senior official present for each employer. The Incident Commander is normally the responding fire department chief who establishes a Command Post with which the employer should be familiar. All operations in the hazardous area must be performed using the buddy system where workers are organized to work in groups of two or more so they can observe each other for contamination. Back-up personnel must stand by with equipment ready to provide assistance or rescue. Advanced first-aid support personnel, as a minimum, also must stand by with medical equipment and transportation capability.

Training

Different levels of initial training are required depending on the duties and functions of the employee-responder. Demonstrated competence or annual refresher training is required to maintain that competence.

First Responders at the Awareness Level
These individuals are likely to witness or discover the hazardous substance release and initiate the emergency response. They must demonstrate competency in such areas as recognizing the presence of hazardous materials in an emergency, the risks involved, and the role they are expected to perform in such situations.

First Responders at the Operations Level
These employees respond to protect people, property and the nearby environment while not actually trying to stop the release. They must have eight hours of training as well as the training required of first responders at the awareness level, or a demonstrated competence in their roles.

Hazardous Materials Technicians
These workers respond to stop the release. They must have 24 hours of training equal to the operations level and demonstrate their competence in several specific areas.

Hazardous Materials Specialists
These workers support the technicians, but require a more specific knowledge of the substances being contained. They must have 24 hours of training equal to the technical level and demonstrate competence in certain areas.

On-Scene Incident Commander
This individual assumes control of the incident scene beyond the awareness level. This person must have 24 hours of training equal to the operations level and must demonstrate competence in specific areas.

Other Personnel
Skilled support personnel such as heavy earth-moving equipment operators who are needed on a temporary basis are not required to meet the training required for an employer's regular employees. They must be given an initial briefing at the site prior to their participation in an emergency response. This briefing must include instruction in the wearing of appropriate personal protective equipment, what chemical hazards are involved, and what duties are to be performed.

Specialist employees who work regularly with and are trained in hazards of specific hazardous substances and will provide technical advice or assistance at a hazardous substance release incident must receive training or demonstrate their competency on an annual basis.

Medical Surveillance and Consultation

Members of designated hazardous materials (HAZMAT) teams and HAZMAT specialists must have a baseline physical examination and be provided with medical surveillance as required for employees at uncontrolled hazardous waste sites. Any emergency response employees who exhibit signs or symptoms that may have resulted from exposure during the course of an emergency incident must be provided with medical consultation.

Post-Emergency Response Operations

After the emergency response is over, clean up of released hazardous substances, health hazards, or contaminated materials may be necessary. In this case, the employer may either comply with all of the requirements for uncontrolled hazardous waste sites or clean up the plant property using plant or workplace employees. If the employer elects to clean up its material using its own employees and equipment, the employer must ensure its employees have completed the training requirements of the following sections:
- Emergency action plan training, 1910.38;
- Respirator training, 1910.134;
- Hazard communication training, 1910.1200; and
- Any other appropriate safety training such as personal protective equipment or decontamination procedures.

Process Safety Management of Highly Hazardous Chemicals

Unexpected releases of toxic, reactive, or flammable liquids and gases in processes involving highly hazardous chemicals have been reported for many years. Incidents continue to occur in various industries that use highly hazardous chemicals that may be toxic, reactive, flammable, explosive, or may exhibit a combination of these properties. Regardless of the industry that uses these highly hazardous chemicals, there is a potential for an accidental release any time 'they are not properly controlled. This, in turn, creates the possibility of disaster.

Recent major disasters include the 1984 Bhopal, India, incident resulting in more than 2,000 deaths; the October 1989 Phillips Petroleum Company, Pasadena, Texas, incident resulting in 23 deaths and 132 injuries; the July 1990 BASF, Cincinnati, Ohio, incident resulting in two deaths, and the May 1991 IMC, Sterlington, Louisiana, incident resulting in 8 deaths and 128 injuries.

Although these major disasters involving highly hazardous chemicals drew national attention to the potential for major catastrophes, the public record is replete with information concerning many other less notable releases of highly hazardous chemicals. Hazardous chemical releases continue to pose a significant threat to employees and provide impetus, internationally and nationally, for authorities to develop or consider developing legislation and regulations to eliminate or minimize the potential for such events. OSHA's standard emphasized the management of hazards associated with highly hazardous chemicals and established a comprehensive management program that integrated technologies, procedures, and management practices.

Approximately four months after the publication of OSHA's proposed standard for process safety management of highly hazardous chemicals, the Clean Air Act Amendments (CAAA) were enacted into law (November 15, 1990). Section 304 of the CAAA requires that the Secretary of Labor, in coordination with the Administrator of the Environmental Protection Agency (EPA), promulgate, pursuant to the Occupational Safety and Heath Act of 1970, a chemical process safety standard to prevent the accidental release of chemicals that could pose a threat to employees.

The CAAA requires that the standard include a list of highly hazardous chemicals that includes toxic, flammable, highly reactive, and explosive substances. The CAAA also specified minimum elements that the OSHA standard must require of employers:
- Develop and maintain written safety information identifying workplace chemical and process hazards, equipment used in the processes, and technology used in the processes;
- Perform a workplace hazard assessment, including, as appropriate, identification of potential sources of accidental releases, identification of any previous release within the facility that had a potential for catastrophic consequences in the workplace, estimation of workplace effects of a range of releases, and estimation of the health and safety effects of such a range on employees;
- Consult with employees and their representatives on the development and conduct of hazard assessments and the development of chemical accident prevention plans and

HAZARDOUS WASTE OPERATIONS AND PROCESS SAFETY MANAGEMENT

provide access to these and other records required under the standard;
- Establish a system to respond to the workplace hazard assessment findings, which shall address prevention, mitigation, and emergency responses;
- Review periodically the workplace hazard assessment and response system;
- Develop and implement written operating procedures for the chemical processes, including procedures for each operating phase, operating limitations, and safety and health considerations;
- Provide written safety and operating information for employees and employee training in operating procedures by emphasizing hazards and safe practices that must be developed and made available;
- Ensure contractors and contract employees are provided with appropriate information and training;
- Train and educate employees and contractors in emergency response procedures in a manner as comprehensive and effective as that required by the regulation promulgated pursuant to section 126(d) of the Superfund Amendments and Reauthorization Act;
- Establish a quality assurance program to ensure that initial process-related equipment, maintenance materials, and spare parts are fabricated and installed consistent with design specifications;
- Establish maintenance systems for critical process-related equipment, including written procedures, employee training, appropriate inspections, and testing of such equipment to ensure ongoing mechanical integrity;
- Conduct pre-startup safety reviews of all newly installed or modified equipment;
- Establish and implement written procedures managing changes to process chemicals, technology, equipment and facilities; and
- Investigate incidents that result in or could have resulted in a major accident in the workplace, with any findings and modifications made to be reviewed by operating personnel.

How the Standard Works

The OSHA final process safety management (PSM) standard applies mainly to manufacturing industries, particularly those pertaining to chemicals, transportation equipment, and fabricated metal products. Other affected sectors include natural gas liquids; farm product warehousing; electric, gas, and sanitary services; and wholesale trade. It also applies to pyrotechnics and explosives manufacturers covered under other OSHA rules and has special provisions for contractors working in covered facilities.

In each industry, the PSM standard applies to those companies that deal with any of more than 130 specific toxic and reactive chemicals in listed quantities; it also includes flammable liquids and gases in quantities of 10,000 lbs. (4,535.9 kg) or more.

Subject to the rules and procedures set forth in OSHA's hazard communication standard (29 CFR 1910.1200), employees and their designated representatives must be given access to trade secret information contained within the process hazard analysis and other documents required to be developed by the PSM standard.

The key provision of the PSM standard is process hazard analysis—a careful review of what could go wrong and what safeguards must be implemented to prevent the release of hazardous chemicals. Covered employers must identify those processes that pose the greatest risks and begin evaluating those first. The PSM standard clarifies the responsibilities of employers and contractors involved in work that affects or takes place near covered processes to ensure the safety of both plant and contractor employees is considered. The standard also mandates written operating procedures; employee training; pre-startup safety reviews; the evaluation of mechanical integrity of critical equipment; and written procedures for managing change. The PSM standard specifies a permit system for hot work; the investigation of incidents involving releases or near misses of covered chemicals; emergency action plans; compliance audits at least every three years; and trade secret protection.

Process means any activity involving a highly hazardous chemical, including using, storing, manufacturing, handling, or moving such chemicals at the site, or any combination of these activities. For purposes of this definition, any group of vessels that are interconnected and separate vessels located in a way that could involve a highly hazardous chemical in a potential release are considered a single process.

Process Safety Information

Employers must complete a compilation of written process safety information before conducting any process hazard analysis required by the standard. The compilation of written process safety information will help the employer and the employees involved in operating the process to identify and understand the hazards that the processes pose.

Process safety information must include information on the hazards of the highly hazardous chemicals used or produced by the process, information on the technology of the process, and information on the equipment in the process.

Information on the hazards of the highly hazardous chemicals in the process shall consist of at least the following:
- Toxicity;
- Permissible exposure limits;
- Physical data;
- Reactivity data;
- Corrosivity data;
- Thermal and chemical stability data; and
- The hazardous effects of inadvertently mixing different materials.

Information on the technology of the process must include at least the following:
- A block flow diagram or simplified process flow diagram;
- Process chemistry;
- Maximum intended inventory; and
- Safe upper and lower limits for such items as temperatures, pressures, and flows.

NOTE: Material safety data sheets (MSDSs) meeting the requirements of the hazard communication standard (29 CFR 1910.1200) may be used to comply with this requirement to the extent they contain the required information.

Where the original technical information no longer exists, such information may be developed in conjunction with the process hazard analysis in sufficient detail to support the analysis.

Information on the equipment in the process must include the following:

- Materials of construction;
- Piping and instrument diagrams;
- Electrical classification;
- Relief system design and design basis;
- Ventilation system design;
- Design codes and standards employed;
- Material and energy balances for processes built after May 26, 1992; and
- Safety systems (e.g., interlocks, detection, suppression systems).

The employer shall document that equipment complies with recognized and generally accepted good engineering practices. For existing equipment designed and constructed in accordance with codes, standards, or practices that are no longer in general use, the employer shall determine and document that the equipment is designed, maintained, inspected, tested, and operated in a safe manner.

Process Hazard Analysis

The process hazard analysis is a thorough, orderly, systematic approach for identifying, evaluating, and controlling the hazards of processes involving highly hazardous chemicals. The employer must perform an initial process hazard analysis (hazard evaluation) on all processes covered by this standard. The process hazard analysis methodology selected must be appropriate to the complexity of the process and must identify, evaluate, and control the hazards involved in the process.

First, employers must determine and document the priority order for conducting process hazard analyses based on a rationale that includes such considerations as the extent of the process hazards, the number of potentially affected employees, the age of the process, and the operating history of the process. All initial process hazard analyses should be conducted as soon as possible.

Process hazard analyses completed after May 26, 1987, that meet the requirements of the PSM standard are acceptable as initial process hazard analyses. All process hazard analyses must be updated and revalidated, based on their completion date, at least every five years.

The employer must use one or more of the following methods, as appropriate, to determine and evaluate the hazards of the process being analyzed:

- What-if;
- Checklist;
- What-If/checklist;
- Hazard and operability study (HAZOP);
- Failure mode and effects analysis (FMEA);
- Fault tree analysis; or
- An appropriate equivalent methodology.

Whichever method(s) are used, the process hazard analysis must address the following:
- The hazards of the process;
- The identification of any previous incident that had a potential for catastrophic consequences in the workplace;
- Engineering and administrative controls applicable to the hazards and their interrelationships, such as the appropriate application of detection methodologies to provide early warning of releases. Acceptable detection methods might include process monitoring and control instrumentation with alarms and detection hardware such as hydrocarbon sensors;
- Consequences of failure of engineering and administrative controls;
- Facility site;
- Human factors; and
- Qualitative evaluation of a range of the possible safety and health effects on employees if there is a failure of controls.

For best results, the process hazard analysis should be conducted by a team that includes employees with expertise in engineering and process operations, at least one employee who has experience with and knowledge of the process being evaluated, and at least one employee who has knowledge of the specific analysis methods being used.

The employer must:
- Establish a system to address promptly the team's findings and recommendations;
- Ensure the recommendations are resolved in a timely manner and the resolutions are documented;
- Document what actions are to be taken;
- Develop a written schedule of when these actions are to be completed;
- Complete actions as soon as possible; and
- Communicate the actions to operating, maintenance, and other employees whose work assignments are in the process and who may be affected by the recommendations or actions.

At least every five years after the completion of the initial process hazard analysis, the process hazard analysis must be updated and revalidated by a team meeting the standard's requirements to ensure the hazard analysis is consistent with the current process.

Employers must keep on file and make available to OSHA, on request, process hazard analyses and updates or revalidation for each process covered by the PSM standard, as well as the documented resolution of recommendations, for the life of the process.

Operating Procedures

The employer must develop and implement written operating procedures, consistent with the process safety information, that provide clear instructions for safely conducting activities involved in each covered process. The tasks and procedures related to the covered process must be appropriate, clear, consistent, and most importantly, well communicated to employees. The procedures must address at least the following elements:
- Steps for each operating phase:
  - Initial startup;
  - Normal operations;
  - Temporary operations;
  - Emergency shutdown, including the conditions under which emergency shutdown is required and the assignment of shutdown responsibility to qualified operators to ensure emergency shutdown is executed in a safe and timely manner;
  - Emergency operations;
  - Normal shutdown; and
  - Startup following a turnaround or emergency shutdown; and
- Operating limits:
  - Consequences of deviation; and
  - Steps required to correct or avoid deviation; and
- Safety and health considerations:
  - Properties of and hazards presented by the chemicals used in the process;
  - Precautions necessary to prevent exposure, including engineering controls,

administrative controls, and personal protective equipment;
- Control measures to be taken if physical contact or airborne exposure occurs;
- Quality control for raw materials and control of hazardous chemical inventory levels;
- Any special or unique hazards; and
- Safety systems (e.g., interlocks, detection, suppression systems) and their functions.

To ensure that a ready and up-to-date reference is available and to form a foundation for needed employee training, operating procedures must be readily accessible to employees who work in or maintain a process. The operating procedures must be reviewed as often as necessary to ensure they reflect current operating practices, including changes in process chemicals, technology and equipment, and facilities. To guard against outdated or inaccurate operating procedures, the employer must certify annually that its operating procedures are current and accurate.

The employer must develop and implement safe work practices to provide for the control of hazards during work activities such as lockout/tagout; confined space entry; opening process equipment or piping; and control over entrance into a facility by maintenance, contractor, laboratory, or other support personnel. These safe work practices must apply to both employees and to contractor-employees.

Employee Participation

Employers must develop a written plan of action to implement the employee participation required by the PSM standard. Employers must consult with employees and their representatives on the conduct and development of process hazard analyses and on the development of the other elements of process management. They also must provide access to process hazard analyses and to all other information required to be developed by the standard to employees and their representatives.

Training

Initial Training

The implementation of an effective training program is one of the most important steps that an employer can take to enhance employee safety. Accordingly, the PSM standard requires that each employee presently involved in operating a process or a newly assigned process must be trained in an overview of the process and its operating procedures. The training must include emphasis on the specific safety and health hazards of the process, emergency operations, including shutdown, and other safe work practices that apply to the employee's job tasks.

Refresher Training

Refresher training must be provided at least every three years to each employee involved in operating a process to ensure he or she understands and adheres to the current operating procedures of the process. The employer, in consultation with the employees involved in operating the process, must determine the appropriate frequency of refresher training.

Training Documentation

The employer must determine whether each employee operating a process has received and understands the training required by the PSM standard. A record must be kept containing the identity of the employee, the date of training, and how the employer verified that the employee understood the training.

Contractors

Application

Many categories of contract labor may be present at a job site; such workers may actually operate the facility or perform only certain tasks because they have specialized knowledge or skill. Others work only for short periods when there is need for increased staff quickly, such as in turnaround operations. The PSM standard includes special provisions for contractors and their employees so they do nothing to endanger those working nearby.

The PSM standard therefore applies to contractors performing maintenance or repair, turnaround, major renovation, or specialty work on or adjacent to a covered process. It does not apply, however, to contractors providing incidental services that do not influence process safety, such as janitorial, food and drink, laundry, delivery, or other supply services.

Employer Responsibilities

When selecting a contractor, the employer must obtain and evaluate information regarding the contract employer's safety performance and programs. The employer must:
- Inform contract employers of the known potential fire, explosion, or toxic release hazards related to the contractor's work and the process;
- Explain to contract employers the applicable provisions of the emergency action plan;
- Develop and implement safe work practices to control the presence, entrance, and exit of contract employers and employees in covered process areas;
- Evaluate periodically the performance of contract employers in fulfilling their obligations; and
- Maintain an employee injury and illness log related to the contractor's work in the process areas.

Contract Employer Responsibilities

The contract employer must:
- Ensure contract employees are trained in the work practices necessary to perform their jobs safely;
- Ensure contract employees are instructed in the known potential for fire, explosion, or toxic release hazards related to their jobs and the processes, and in the applicable provisions of the emergency action plan;
- Document that each contract employee has received and understands the training required by the standard by preparing a record that contains the identity of the contract employee, the date of his or her training, and the means used to verify that he or she understood the training;
- Ensure each contract employee follows the safety rules of the facility, including the required safe work practices required in the operating procedures section of the standard; and
- Advise the employer of any unique hazards presented by the contract employer's work.

Pre-Startup Safety Review

It is important that the employer conduct a safety review before any highly hazardous chemical is introduced into a process. The PSM standard therefore requires the employer to perform a pre-startup safety review for new facilities and modified facilities when the modification is significant enough to require a change in the process safety information. Before a highly hazardous chemical is introduced to a process, the pre-startup safety review must confirm the following:
- Construction and equipment are in accordance with design specifications;
- Safety, operating, maintenance, and emergency procedures are in place and adequate;
- A process hazard analysis has been performed for new facilities and recommendations have been resolved or implemented before startup, and modified facilities meet the management of change requirements; and
- Each employee involved in operating a process has been trained.

Mechanical Integrity

It is important to maintain the mechanical integrity of critical process equipment to ensure it is designed and installed correctly and operates properly. The PSM mechanical integrity requirements apply to the following equipment:
- Pressure vessels and storage tanks;
- Piping systems, including piping components such as valves;
- Relief and vent systems and devices;
- Emergency shutdown systems;
- Controls, including monitoring devices and sensors, alarms, and interlocks; and
- Pumps.

The employer must establish and implement written procedures to maintain the ongoing integrity of process equipment. Employees involved in maintaining the ongoing integrity of process equipment must be trained in an overview of that process and its hazards as well as the procedures applicable to his or her specific job tasks. Inspection and testing must be performed on process equipment, using procedures that follow recognized

and generally accepted good engineering practices. The frequency of inspections and tests of process equipment must conform to the manufacturer's recommendations and good engineering practices, or more frequently if determined to be necessary by previous operating experience. Each inspection and test on process equipment must be documented, identifying:
- The date of the inspection or test;
- The name of the person who performed the inspection or test;
- The serial number or other identifier of the equipment on which the inspection or test was performed;
- A description of the inspection or test performed; and
- The results of the inspection or test.

Equipment deficiencies outside the acceptable limits defined by the process safety information must be corrected before further use. When other necessary steps are taken to ensure safe operation, it may not be necessary to correct deficiencies before further use, as long as the deficiencies are corrected in a safe and timely manner.

When constructing new plants and equipment, the employer must ensure that equipment as it is fabricated is suitable for the process application for which it will be used. Appropriate checks and inspections must be performed to ensure the equipment is installed properly and is consistent with design specifications and the manufacturer's instructions. The employer also must ensure that maintenance materials, spare parts, and equipment are suitable for the process application for which they will be used.

Hot Work Permit

A permit must be issued for hot work operations conducted on or near a covered process. The permit must document that the OSHA fire prevention and protection regulations (29 CFR 1910.252(a)) were implemented before the hot work operations commenced; indicate the date(s) authorized for hot work; and identify the object on which hot work is to be performed. The permit must be kept on file until the hot work is completed.

Management of Change

Before changes to a process are implemented, they must be thoroughly evaluated to fully assess their impact on employee safety and health and to determine whether other changes to the operating procedures are necessary. To this end, the PSM standard contains a section on procedures for managing changes to processes. Written procedures to manage changes (except for "replacements in kind") to process chemicals, technology, equipment, and procedures and changes to facilities that affect a covered process must be established and implemented. These written procedures must ensure the following considerations are addressed prior to any change:
- The technical basis for the proposed change;
- The impact of the change on employee safety and health;
- Modifications to operating procedures;
- Necessary time period for the change; and
- Authorization requirements for the proposed change.

Employees who operate a process and maintenance and contract employees whose job tasks will be affected by a change in the process must be informed of and trained in the change before startup of the process or startup of the affected part of the process. If a change covered by these procedures results in a change in the required process safety information, such information also must be updated accordingly. If a change covered by these procedures changes the required operating procedures or practices, they also must be updated.

Incident Investigation

A crucial part of the process safety management program is a thorough investigation of incidents to identify the chain of events and causes so that corrective measures can be developed and implemented. Accordingly, the PSM standard requires the investigation of each incident that resulted in, or could have reasonably resulted in, a catastrophic release of a highly hazardous chemical in the workplace. Such an investigation must be initiated as promptly as possible, but no later than 48 hours following the incident.

The investigation must be conducted by a team consisting of at least one person knowledgeable in the process involved, including a contract employee if the incident involved the work of a contractor, and other persons with appropriate knowledge and experience to investigate and analyze the incident thoroughly.

The team must submit a written investigation report to the employer that includes, at a minimum:
- The date of the incident;
- The date the investigation began;
- A description of the incident;
- Factors that contributed to the incident; and
- Recommendations resulting from the investigation.

A system must be established to promptly address and resolve the incident report's findings and recommendations. Resolutions and corrective actions must be documented and the report reviewed by all affected personnel whose job tasks are relevant to the incident findings (including contract employees when applicable).

The employer must keep an investigation report for five years after the incident's occurrence.

Emergency Planning and Response

If despite the best planning an incident occurs, it is essential that emergency pre-planning and training make employees aware of, and able to execute, proper actions. For this reason, an emergency action plan for the entire site must be developed and implemented in accordance with the provisions of 29 CFR 1910.38. In addition, the emergency action plan must include procedures for handling small releases of hazardous chemicals. Employers covered under the PSM standard also may be subject to the hazardous waste and emergency response standards set forth at 29 CFR 1910.120(a), (p), and (q).

Compliance Audits

To be certain process safety management is effective, employers must certify that they have evaluated their compliance with the PSM standard at least every three years. This audit will verify that the procedures and practices developed under the standard are adequate and are being followed. The compliance audit must be conducted by at least one person knowledgeable in the process. The employer must determine and document an appropriate response to each of the findings of the compliance audit and document whether (and how) deficiencies have been corrected. The two most recent compliance audit reports must be kept on file.

Trade Secrets

Employers must make available all information necessary to comply with the PSM standard to those persons responsible for compiling the process safety information; developing the process hazard analysis; developing the operating procedures; and performing incident investigations, emergency planning and response, and compliance audits, without regard to the possible trade secret status of such information. Nothing in the PSM standard however precludes the employer from requiring those persons to enter into confidentiality agreements not to disclose any proprietary information.

CHAPTER 8

PERMIT-REQUIRED CONFINED SPACES

The regulation covers general industry workers including 1.6 million who enter confined spaces annually and an additional 10.6 million employed at the 240,000 worksites covered by the standard (29 CFR 1910.146). This standard is expected to prevent nearly 85 percent of deaths and injuries (i.e., 54 deaths and 10,949 injuries) each year.

A confined space is defined as an area that:
- Has adequate size and configuration for employee entry;
- Has limited means of access or egress; and
- Is not designed for continuous employee occupancy.

A permit-required confined space is a confined space that presents or has the potential for hazards related to atmospheric conditions (e.g., toxic, flammable, asphyxiating), engulfment, configuration or any other recognized serious hazard.

A prohibited condition is defined as any condition not allowed by permit during entry operations.

The standard requires employers initially to evaluate their workplaces and determine if there are any permit-required confined spaces, inform employees through signs or other equally effective means and prevent unauthorized entry.

The permit-required confined space program:
- Mandates a written program to:
  - prevent unauthorized entry;
  - identify and evaluate hazards; and
  - establish procedures and practices for safe entry, including testing and monitoring conditions;
- Calls for:
  - an attendant be stationed outside permit spaces during entry;
  - procedures to summon rescuers and prevent unauthorized personnel from attempting rescue; and
  - a system for preparing, issuing, using and canceling entry permits; and
- Requires:
  - coordinated entry for more than one employer;
  - procedures for concluding entry operations and canceling entry permits; and
  - review of permit program at least annually and additionally as necessary.

A permit system must be established that requires an entry supervisor to authorize entry, prepare and sign written permits, order corrective measures if necessary and cancel permits when work is completed. Permits must be available to all employees and extend only for duration of the task. Permits must be retained for one year to facilitate review of the confined space program. Permits must include:

- Identification of space;
- Purpose of entry;
- Date and duration of permit;
- List of authorized entrants;
- Names of current attendants and entry supervisor;
- List of hazards in the permit space;
- List of measurements to isolate permit space and eliminate or control hazards;
- The acceptable entry conditions;
- Results of tests initialed by the person(s) performing tests;
- Rescue emergency services and means to summon;
- Communication procedures for attendants or entrants;
- Required equipment (e.g., respirators, communications, alarm);
- Any additional permits (e.g., for hot work).

Training mandates initial and refresher training (when duties change, hazards in spaces change or whenever evaluation determines inadequacies in employees' knowledge) to provide employees understanding, skills and knowledge to do jobs safely. Employer certification of training must include employee's name, signature or initials of trainer and date of training.

Authorized entrants must know the hazards they may face, be able to recognize signs or symptoms of exposure and understand the consequences of exposure to hazards. Entrants must know how to use any needed equipment, communicate with attendants as necessary, alert attendants when a warning symptom or other hazardous condition exists and exit as quickly as possible whenever ordered or alerted (by alarm, warning sign or prohibited condition) to do so.

Attendants must know the hazards of confined spaces, be aware of behavioral effects of potential exposures, maintain continuous count and identification of authorized attendants, remain outside space until relieved, communicate with entrants as necessary to monitor entrant status. Attendants also must monitor activities inside and outside the permit space and order exit if required, summon rescuers if necessary, prevent unauthorized entry into confined space, and perform non-entry rescues if required. They may not perform other duties that interfere with their primary duties to monitor and protect the safety of authorized entrants.

Entry supervisors must know hazards of confined spaces, verify that all tests have been conducted and all procedures and equipment are in place before endorsing permit, terminate entry and cancel permits and verify that rescue services are available and the means for summoning them are operable. Supervisors are to remove unauthorized individuals who enter confined space. They also must determine at least when shifts and entry supervisors change—that acceptable conditions as specified in the permit continue.

Rescue services may be on-site or off-site. Rescue teams are to use employee retrieval systems whenever possible. On-site rescue teams must be properly equipped and receive the same training as authorized entrants plus training to use personal protective and rescue equipment and first aid, including CPR. They must practice simulated rescues at least once every 12 months. Outside rescue services must be made aware of hazards, receive access to comparable permit spaces to develop rescue plans and practice rescues. Employers must provide hospitals or treatment facilities with any material safety data sheets (MSDSs) or other information on a permit-required space hazard exposure situation that may aid in the treatment of rescued employees.

Contractors call for host employers to provide information to contractors on permit-required confined spaces, the permit space program, and procedures and likely hazards that the contractor might encounter. Joint entries must be coordinated and the contractor debriefed at the conclusion of entry operations.

Alternative Protection Procedures

For permit spaces where the only hazard is atmospheric and ventilation alone can control the hazard, employers may use alternative procedures for entry.

To qualify for alternative procedures employers must:
- Ensure it is safe to remove the entrance cover;
- Determine that ventilation alone is sufficient to maintain the permit space safe for entry and work to be performed within the permit-required space must introduce no additional hazards;
- Gather monitoring and inspection data to support the above two items;
- If entry is necessary to conduct initial data gathering, perform such entry under the full permit program; and
- Document the determinations and supporting data and make them available to employees.

Entry can take place after:
- It has been determined safe to remove the entrance cover;
- Any openings have been guarded to protect against falling and falling objects;
- Internal atmospheric testing has been performed;
- Air remains without hazard whenever any employee is inside the space;
- Continuous forced air ventilation has eliminated any hazardous atmosphere; and
- The space is tested periodically.

Employees must exit immediately if a hazardous atmosphere is detected during entry Afterward. the space must be evaluated to determine how the hazardous atmosphere developed.

Permit-Required Confined Spaces— 1910.146

Many workplaces contain spaces that are considered to be "confined" because their configurations hinder the activities of any employees who must enter into, work in and exit from them. In many instances, employees who work in confined spaces also face increased risk of exposure to serious physical injury from hazards such as entrapment, engulfment and hazardous atmospheric conditions. Confinement itself may pose entrapment hazards, and work in confined spaces may keep employees closer to hazards, such as an asphyxiating atmosphere, than they would be otherwise. For example, confinement, limited access and restricted airflow can result in hazardous conditions that would not arise in an open workplace.

The term "permit-required confined space" refers to those spaces that meet the definition of a "confined space" and pose health or safety hazards, thereby requiring a permit for entry.

A confined space may include, but are not limited to, underground vaults, tanks, storage bins, pits and diked areas, vessels, and silos that have the following characteristics:
- Limited or restricted means of entry or exit;
- Is large enough for an employee to enter and perform assigned work; and
- Is not designed for continuous occupancy by the employee.

A permit-required confined space is one that meets the definition of a confined space and has one or more of the following characteristics:
- Contains or has the potential to contain a hazardous atmosphere;
- Contains a material that has the potential for engulfing an entrant;
- Has an internal configuration that might cause an entrant to be trapped or asphyxiated by inwardly converging walls or by a floor that slopes downward and tapers to a smaller cross-section; or
- Contains any other recognized serious safety or health hazards.

In general, employers must evaluate the workplace to determine if spaces are permit-required confined spaces. If there are permit spaces in the workplace, the employer must inform exposed employees of the existence, location and danger posed by the spaces. This can be accomplished by posting danger signs or by another equally effective means (See Exhibit 8-1.). The following language would satisfy the requirements for such a sign:

**DANGER
PERMIT-REQUIRED CONFINED SPACE**

AUTHORIZED ENTRANTS ONLY

If employees are not to enter and work in permit spaces, employers must take effective measures to prevent their employees from entering the permit spaces.

Exhibit 8-1.
Permit-required confined spaces should be clearly marked as dangerous.

If employees are to enter permit spaces, the employer must develop a written permit space program that shall be made available to employees or their representatives. Under certain conditions, the employer may use alternate procedures for worker entry into a permit space. For example, if employers can demonstrate with monitoring and inspection data that the only hazard is an actual or potential hazardous atmosphere that can be made safe for entry by the use of continuous forced-air ventilation, they may be exempted from some requirements, such as permits and attendants. Even in such circumstances, however, the internal atmosphere of the space must be tested first for oxygen content, second for flammable gases and vapors, and third for potential toxic air contaminants before any employee enters.

Written Program

The employer who allows employee entry must develop and implement a written program for permit-required confined spaces.

The OSHA standard requires the employer's program to:

- Identify and evaluate permit space hazards before allowing employee entry;
- Test conditions in the permit space before entry operations and monitor the space during entry;
- Perform appropriate testing for atmospheric hazards in the following order:
  - Oxygen;
  - Combustible gases or vapors; and
  - Toxic gases or vapors; and
- Implement necessary measures to prevent unauthorized entry;
- Establish and implement the means, procedures and practices, such as specifying acceptable entry conditions, isolating the permit space, providing barriers, verifying acceptable entry conditions, purging, making inert, flushing, or ventilation of the permit space to eliminate or control hazards necessary for safe permit-space entry operations;
- Identify employee job duties;
- Provide, maintain and require, at no cost to the employee, the use of personal protective equipment and any other equipment necessary for safe entry (e.g., testing, monitoring, ventilating, communications, and lighting equipment; barriers; shields and ladders);
- Ensure that at least one attendant is stationed outside the permit space for the duration of entry operations;
- Coordinate entry operations when employees of more than one employer are to be working in the permit space;
- Implement appropriate procedures for summoning rescue and emergency services;
- Establish, in writing, and implement a system for the preparation, issuance, use and cancellation of entry permits;
- Review established entry operations and annually revise the permit-space entry program; and
- When an attendant is required to monitor multiple spaces, implement the procedures to be followed during an emergency in one or more of the permit spaces being monitored.

If hazardous conditions are detected during entry, employees must immediately leave the space and the employer must evaluate the space to determine the cause of the hazardous atmospheres.

When entry to permit spaces is prohibited, the employer must take effective measures to prevent unauthorized entry. Non-permit confined spaces must be reevaluated when there are changes in their use or configuration and, where appropriate, must be reclassified.

If testing and inspection data prove that a permit-required confined space no longer poses hazards, that space may be reclassified as a non-permit confined space. If entry is required to eliminate hazards and to obtain the data, the employer must follow procedures as set forth under sections (d) through (k) of the standard. A certificate documenting the data must be made available to employees entering the space. The certificate must include the date, location of the space and the signature of the person making the certification.

Contractors must be informed of permit spaces and permit-space entry requirements, any identified hazards, the employer's experience with the space (i.e., the knowledge of hazardous conditions), and precautions or procedures to be followed when in or near permit spaces.

When employees of more than one employer are conducting entry operations, the affected employers must coordinate entry operations to ensure that affected employees are appropriately protected from permit space hazards. Contractors also must be given any other pertinent information regarding hazards and operations in permit spaces and be debriefed at the conclusion of entry operations.

Permit System

A permit must be posted at entrances or otherwise made available to entrants before they enter a permit space. The permit must be signed by the entry supervisor, attest that pre-entry preparations have been completed and that the space is safe to enter.

The duration of entry permits must not exceed the time required to complete an assignment. Also, the entry supervisor must terminate entry and cancel permits when an assignment has been completed or when new conditions exist. New conditions must be noted on the canceled permit and used in revising the permit space program. The standard also requires the employer to keep all canceled entry permits for at least one year.

Entry Permits

Entry permits must include the following information:
- Test results;
- Tester's initials or signature;
- Name and signature of the supervisor who authorizes entry;
- Name of permit space to be entered, authorized entrant(s), eligible attendants and individual(s) authorized to be entry supervisor(s);
- Purpose of entry and known space hazards;
- Measures to be taken to isolate permit spaces and to eliminate or control space hazards (i.e., locking out or tagging of equipment and procedures for purging, making inert, ventilating and flushing permit spaces);
- Name and telephone numbers of rescue and emergency services;
- Date and authorized duration of entry;
- Acceptable entry conditions;
- Communication procedures and equipment to maintain contact during entry;
- Additional permit(s), such as for hot work, that have been issued to authorize work in the permit space;
- Special equipment and procedures, including personal protective equipment and alarm systems; and
- Any other information needed to ensure employee safety.

Training and Education

Before an initial work assignment begins, the employer must provide proper training for all workers who are required to work in permit spaces. Upon completing this training, employers must ensure that employees have acquired the understanding, knowledge and skills necessary for the safe performance of their duties. Additional training is required when:
- The job duties change;
- There is a change in the permit-space program or the permit-space operation presents a new hazard; or
- When an employee's job performance shows deficiencies.

Training also is required for rescue team members, including cardiopulmonary resuscitation (CPR) and first aid (see Emergencies).

Employers must certify this training has been accomplished.

Upon completion of training, employees must receive a certificate of training that includes the employee's name, signature or initials of trainer(s), and dates of training. The certification must be made available for inspection by employees and their authorized representatives. In addition, the employer also must ensure that employees are trained in their assigned duties.

Authorized Entrant's Duties

- Know space hazards, including information on the mode of exposure (e.g., inhalation or dermal absorption), signs or symptoms and consequences of the exposure;
- Use appropriate personal protective equipment properly (e.g., face and eye protection and other forms of barrier protection such as gloves, aprons and coveralls);
- As necessary, maintain communication (i.e., telephone, radio, visual observation) with attendants to enable the attendant to monitor the entrant's status as well as to alert the entrant to evacuate;
- Exit from permit space as soon as possible when ordered by an authorized person, the entrant recognizes the warning signs or symptoms of exposure exist, a prohibited condition exists, or an automatic alarm is activated; and
- Alert the attendant when a prohibited condition exists or warning signs or symptoms of exposure exist.

Attendant's Duties

- Remain outside permit space during entry operations unless relieved by another authorized attendant;
- Perform no-entry rescues when specified by employer's rescue procedure;
- Know existing and potential hazards, including information on the mode of exposure, signs or symptoms, consequences of the exposure and their physiological effects;
- Maintain communication with and keep an accurate account of those workers entering the permit-required space;
- Order evacuation of the permit space when a prohibited condition exists, a worker shows signs of physiological effects of hazardous exposure, an emergency outside the confined space exists, or the attendant cannot effectively and safely perform required duties;
- Summon rescue and other services during an emergency;
- Ensure that unauthorized persons stay away from permit spaces or exit immediately if they have entered the permit space;
- Inform authorized entrants and entry supervisor of entry by unauthorized persons; and;
- Perform no other duties that interfere with the attendant's primary duties.

Entry Supervisor's Duties

- Know space hazards, including information on the mode of exposure, signs or symptoms and consequences of exposure;
- Verify emergency plans and specified entry conditions such as permits, tests, procedures and equipment before allowing entry;
- Terminate entry and cancel permits when entry operations are completed or if a new condition exists;
- Take appropriate measures to remove unauthorized entrants; and
- Ensure that entry operations remain consistent with the entry permit and that acceptable entry conditions are maintained.

Emergencies

The standard requires the employer to ensure that rescue service personnel are provided with and trained in the proper use of personal protective and rescue equipment, including respirators; trained to perform assigned rescue duties; and have had authorized entrant's training. The standard also requires that rescuers be trained in first aid and CPR and, at a minimum, one rescue team member be currently certified in first aid and in CPR. The employer also must ensure that practice rescue exercises are performed annually, and that rescue services are provided access to permit spaces so they can practice rescue operations. Rescuers also must be informed of the hazards of the permit space.

Also, when appropriate, authorized entrants who enter a permit space must wear a chest or full body harness with a retrieval line attached to the center of their backs near shoulder level or above their heads. Wristlets may be used if the employer can demonstrate that the use of a chest or full body harness is infeasible or creates a greater hazard. Also, the employer must ensure that the other end of the retrieval line is attached to a mechanical device or to a fixed point outside the permit space. A mechanical device must be available to retrieve personnel from vertical type permit spaces more than 5 ft. deep.

In addition, if an injured entrant is exposed to a substance for which an MSDS or other similar written information is required to be kept at the worksite, that MSDS or other written information must be made available to the medical facility treating the exposed entrant.

CHAPTER 9

ELECTRICAL

Electricity has become an essential of modern life, both at home and on the job. Some employees work with electricity directly, as is the case with engineers, electricians or people who do wiring, such as overhead lines, cable harnesses or circuit assemblies. Others, such as office workers and salespeople, work with it indirectly. As a source of power, electricity is accepted without much thought to the hazards encountered. Perhaps because it has become such a familiar part of our surroundings, it often is not treated with the respect it deserves.

The Bureau of Labor Statistics reported that work-related death occurred in private-sector workplaces employing 11 workers or more have been reduced since 1992 when 334 electrocutions were reported. Still, more than 200 workers are electrocuted each year.

Exhibit 9-1.
Annual workplace electrocutions as reported by the Bureau of Labor Statistics.

Year	Annual
1992	334
1993	325
1994	348
1995	348
1996	281
1997	298
1998	334
1999	280
2000	256 (p)
(p): preliminary	

OSHA's electrical standards address the government's concern that electricity has long been recognized as a serious workplace hazard, exposing employees to such dangers as electric shock, electrocution, fires and explosions. The objective of the standards is to minimize such potential hazards by specifying design characteristics of safety in use of electrical equipment and systems.

OSHA's electrical standards were carefully developed to cover only those parts of any electrical system that an employee would normally use or contact. The exposed and/or operating elements of an electrical installation (e.g., lighting equipment, motors, machines, appliances, switches, controls, enclosures) must be so constructed and installed as to minimize electrical dangers to people in any workplace.

The OSHA electrical standards were based on the National Fire Protection Association's standard NFPA 70E, Electrical Safety Requirements for Employee Workplaces, and the NFPA 70 Committee derived Part I of their document from the 1978 edition of the National Electrical Code (NEC). The standards extracted from the NEC were those considered to most directly apply to employee safety and least likely to change with each new edition of the NEC. OSHA's electrical standards are performance oriented; therefore, they contain few direct references to the NEC. However, the NEC contains specific information as to how the required performance can be obtained.

General Requirements—1910.303

The conductors and equipment required or permitted by this subpart shall be acceptable only if approved.

Examination, Installation and Use of Equipment

Examination

Electrical equipment shall be free from recognized hazards that are likely to cause death or serious physical harm to employees. Safety of equipment shall be determined using the following considerations:

- Suitability for installation and use in conformity with the provisions of this subpart;
- Suitability of equipment for an identified purpose may be evidenced by listing or labeling for that identified purpose;
- Mechanical strength and durability, including, for parts designed to enclose and protect other equipment, the adequacy of the protection thus provided;
- Electrical insulation;
- Heating effects under conditions of use;
- Arcing effects;
- Classification by type, size, voltage, current capacity, and specific use; and
- Other factors that contribute to the practical safeguarding of employees using or likely to come in contact with the equipment.

Installation and Use

Listed or labeled equipment shall be used or installed in accordance with any instructions included in the listing or labeling.

Splices

Conductors shall be spliced or joined with splicing devices suitable for such use or by brazing, welding or soldering with a fusible metal or alloy. Soldered splices shall be so spliced or joined as to be mechanically and electrically secure without solder and then soldered. All splices and joints and the free ends of conductors shall be covered with insulation equivalent to that of the conductors or with an insulating device suitable for the purpose.

Arcing Parts

Parts of electric equipment that in ordinary operation produce arcs, sparks, flames or molten metal shall be enclosed or separated and isolated from all combustible material.

Marking

Electrical equipment may not be used unless the manufacturer's name, trademark or other descriptive marking that may identify the organization responsible for the product is placed on the equipment. Other markings shall be provided giving voltage, current, wattage or other ratings as necessary. The markings shall be of sufficient durability to withstand the environment involved.

Sometimes this section is confused with the section "Identification of Disconnecting Means." The difference is that *identification* (as used in OSHA standards) means marked to indicate the purpose of a disconnecting means, rather than its rating.

Marking is very important. If an item of equipment is connected to a voltage higher than its rating, the chances are that a violent failure will result. If it is connected to a voltage below its rating, it may attempt to perform its intended function, but may overheat badly and eventually fail. If AC equipment is energized with the wrong frequency, or with direct current, it will probably fail violently. These occurrences can cause burns to employees and may cause fires. If equipment is connected to a circuit not adequate for the load, conductors may overheat and deteriorate. Overcurrent protective devices should prevent serious damage in such a case, but sometimes these devices have been tampered with and do not protect properly.

If the manufacturer's name is not marked on the nameplate, there will be difficulty in tracing the reasons for faulty performance and in preventing future failures. Sometimes it is necessary to know the manufacturer in order to determine if equipment is approved for a particular purpose. The manufacturer's name is also necessary in many cases so that the user can obtain information or replacement parts.

The marking described above is almost always provided on the nameplate of the equipment of reputable manufacturers. Violations usually appear when the nameplate is covered by some part of the installation, is removed or is obliterated by painting or other abuse.

Identification of Disconnecting Means and Circuits

Each disconnecting means required by this subpart for motors and appliances shall be legibly marked to indicate its purpose, unless located and arranged so the purpose is evident. Each service, feeder and branch circuit, at its disconnecting means or overcurrent device, shall be legibly marked to indicate its purpose, unless located and arranged so the purpose is evident. These markings shall be of sufficient durability to withstand the environment involved.

A disconnecting means is a switch that is used to disconnect the conductors of a circuit from the source of electric current. Having the ability to stop electricity provides better protection for workers and equipment.

Each disconnect switch or overcurrent device required for a service, feeder, or branch circuit must be clearly labeled to indicate the circuit's function, and the label or marking should be located at the point where the circuit originates.

For example, on a panel that controls several motors or on a motor control center, each disconnect must be clearly marked to indicate the motor to which each circuit is connected. In Exhibit 9-2, the Number 2 circuit breaker in the panel box supplies current only to disconnect Number 2, which in turn controls the current to motor Number 2. The current to motor Number 2 can be shut off by the Number 2 circuit breaker or the Number 2 disconnect.

If the purpose of the circuit is obvious, no identification of the disconnect is required.

All labels and markings must be durable enough to withstand weather, chemicals, heat, corrosion, or any other environment to which they may be exposed.

600 Volts, Nominal, or Less

Working Space About Electric Equipment

Note that this particular section is concerned with the safety of a person qualified to work on the equipment (presumably an electrician). Obviously, the hazard must be treated in a different way if the person will remove guards and enclosures and actually work on the live parts. Sufficient access and working space shall be provided and maintained about all electric equipment to permit ready and safe operation and maintenance of such equipment.

Working Clearances

Except as required or permitted elsewhere in this subpart, the dimension of the working space in the direction of access to live parts operating at 600 volts or less and likely to require examination, adjustment, servicing or maintenance while alive may not be less

Panel Schedule
1. Motor No. 1
2. Motor No. 2
3. Motor No. 3
4. Motor No. 4

Motor No. 1 is Controlled by Disconnect No. 1 and Circuit Breaker No. 1

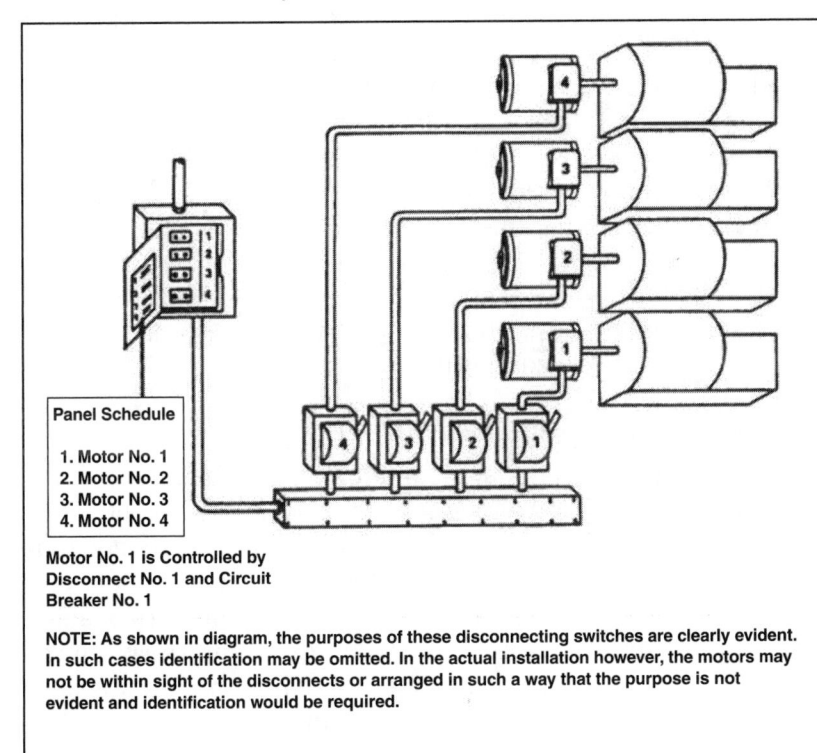

NOTE: As shown in diagram, the purposes of these disconnecting switches are clearly evident. In such cases identification may be omitted. In the actual installation however, the motors may not be within sight of the disconnects or arranged in such a way that the purpose is not evident and identification would be required.

Exhibit 9-2.
Each disconnect and circuit requires identification.

than indicated in Table S-1 of 1910.303(g)(1)(i). In addition, workspace may not be less than 30 inches wide in front of the electrical equipment. Distances shall be measured from the live parts if they are exposed or from the enclosure front or opening if the live parts are enclosed. Concrete, brick or tile walls are considered to be grounded. Working space is not required in back of assemblies such as dead-front switchboards or motor control centers where there are no renewable or adjustable parts, such as fuses or switches, on the back and where all connections are accessible from locations other than the back.

The remaining paragraphs of this section require that the working spaces shall not be used for storage, that access to the working space be maintained, and that reasonable illumination and headroom be maintained.

Guarding of Live Parts

It should be noted that the purpose of this requirement is to protect *any person* who may be in the vicinity of electrical equipment against accidental contact. These people are presumably not electricians working on the equipment and are not qualified or trained to be in close proximity to live parts.

Except as required or permitted elsewhere in this subpart, live parts of electric equipment operating at 50 volts or more shall be guarded against accidental contact by approved cabinets, other forms of approved enclosures or by any of the following means:

- By location in a room, vault or similar enclosure that is accessible only to qualified persons.
- By suitable permanent substantial partitions or screens so arranged that only qualified persons would have access to the space within reach of the live parts. Any openings in such partitions or screens shall be so sized and located that persons are not likely to come into accidental contact with the live parts or to bring conducting objects into contact with them.
- By location on a suitable balcony, gallery or platform.
- By elevation of 8 ft. or more above the floor or other working surface.

- In locations where electric equipment would be exposed to physical damage, enclosures or guards shall be so arranged and of such strength as to prevent such damage.

Over 600 Volts, Nominal

Enclosure for Electrical Installations

Electrical installations in a vault, room, closet or in an area surrounded by a wall, screen, or fence, access to which is controlled by lock and key or other approved means, are considered accessible to qualified persons only. A wall, screen or fence less than 8 ft. in height is not considered to prevent access unless it has other features that provide a degree of isolation equivalent to an 8-ft. fence. The entrances to all buildings, rooms or enclosures containing exposed live parts or exposed conductors operating at over 600 volts, nominal, shall be kept locked or under the observation of a qualified person at all times.

Workspace About Equipment

Electrical installations having exposed live parts shall be accessible to qualified persons only. Sufficient space shall be provided and maintained about electric equipment to permit ready and safe operation and maintenance of such equipment. Where energized parts are exposed, the minimum clear workspace may not be less than 6 ft. 6 in. high (measured vertically from the floor or platform) or less than 3 ft. wide (measured parallel to the equipment). The minimum depth of clear working space in front of electric equipment is given in Table S-2 of 1910.303(h)(3)(i).

Illumination

Adequate illumination shall be provided for all working spaces about electric equipment. The lighting outlets shall be so arranged that live parts or other equipment will not endanger persons changing lamps or making repairs on the lighting system. The points of control shall be so located that persons are not likely to come in contact with any live part or moving part of the equipment while turning on the lights.

Elevation of Unguarded Live Parts
Unguarded live parts above working space shall be maintained at elevations not less than specified in Table S-3 of 1910.303(h)(3)(iii).

Entrance and Access to Workspace
At least one entrance not less than 24 in. wide and 6 ft. 6 in. high shall be provided to give access to the workspace about electric equipment. On switchboard and control panels exceeding 48 in. in width, there shall be one entrance at each end of such board where practicable. Where bare energized parts at any voltage or insulated energized parts above 600 volts are located adjacent to such entrance, they shall be suitably guarded.

Permanent ladders or stairways shall be provided to give safe access to the working space around electric equipment installed on platforms, balconies, mezzanine floors, or in attic or roof rooms or spaces.

Wiring Design and Protection—1910.304

Use and Identification of Grounded and Grounding Conductors

Identification of Conductors
A conductor used as a grounded conductor shall be identifiable and distinguishable from all other conductors. A conductor used as an equipment-grounding conductor shall be identifiable and distinguishable from all other conductors.

The grounded conductor is an energized circuit conductor that is connected to earth through the system ground. It is commonly referred to as the neutral. The equipment-grounding conductor is not an energized conductor under normal conditions. The equipment-grounding conductor acts as a safeguard against insulation failure or faults in the other circuit conductors. The equipment-grounding conductor is energized only if there is a leak or fault in the normal current path and it directs this current back to the source. Directing the fault current back to the source enables protective devices, such as circuit breakers or fuses, to operate thus preventing fires and reducing the hazard of electrical shocks.

The grounded and equipment-grounding conductors of an electrical circuit must be marked or color-coded in a way that allows employees to identify them and tell them apart from each other and from the other conductors in the circuit (See Exhibit 9-3).

Acceptable color-coding includes the method required by the National Electrical Code, Section 210-5. The Code states:

> [t]he grounded conductor of a branch circuit shall be identified by a continuous white or natural gray color.... The equipment grounding conductor of a branch circuit shall be identified by a continuous green color or a continuous green color with one or more yellow stripes unless it is bare.

Polarity of Connections
No grounded conductor may be attached to any terminal or lead so as to reverse designated polarity.

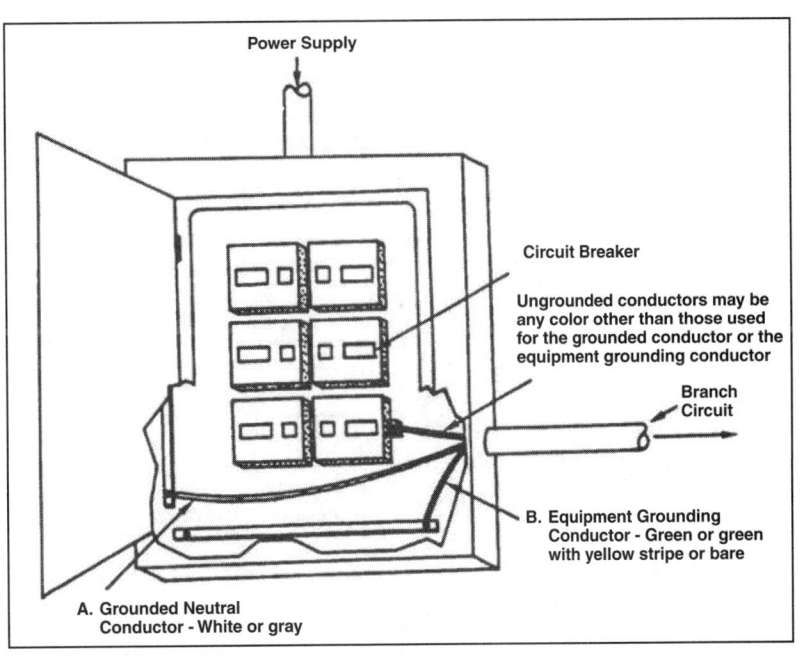

Exhibit 9-3.
Color-coding is an effective way to identify and distinguish conductor types.

Use of Grounding Terminals and Devices

A grounding terminal or grounding-type device on a receptacle, cord connector or attachment plug may not be used for purposes other than grounding.

Reversed Polarity

The above two subparagraphs dealing with polarity of connections and use of grounding terminals and devices address one potentially dangerous aspect of alternating current: Many pieces of equipment will operate properly even though the supply wires are not connected in the order designated by design or the manufacturer. Improper connection of these conductors is most prevalent on the smaller branch circuit typically associated with standard 120 volt receptacle outlets, lighting fixtures and cord- and plug-connected equipment. When plugs, receptacles and connectors are used in an electrical branch circuit, correct polarity between the ungrounded (hot) conductor, the grounded (neutral) conductor and the grounding conductor must be maintained.

Reversed polarity is a condition when the identified circuit conductor (the grounded conductor or neutral) is incorrectly connected to the ungrounded or "hot" terminal of a plug, receptacle or other type of connector. Exhibit 9-4 shows the correct wiring for the common 120-volt outlet with a portable hand tool attached.

Suppose now that the black (ungrounded) and white (grounded) conductors are reversed as shown in Exhibit 9-5. This is the traditional reversed

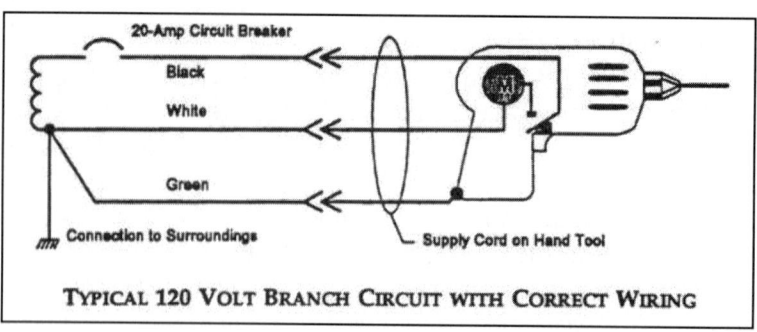

Exhibit 9-4.
Typical 120-volt branch circuit with correct wiring.

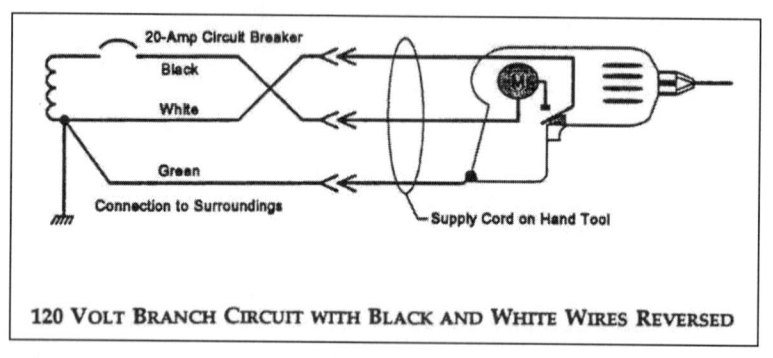

Exhibit 9-5.
A 120-volt branch circuit with black and white wires reversed.

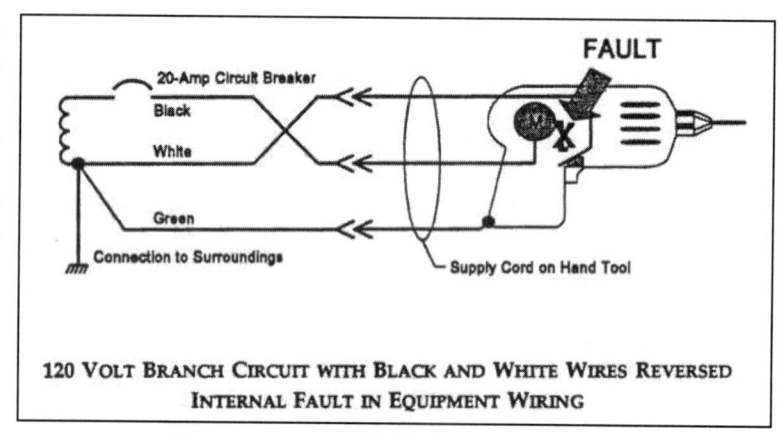

Exhibit 9-6.
A 120-volt branch circuit with an internal fault.

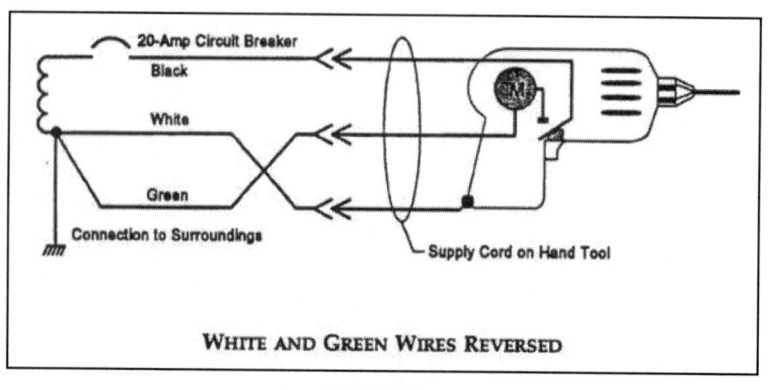

Exhibit 9-7.
A 120-volt branch circuit with white and green wires reversed.

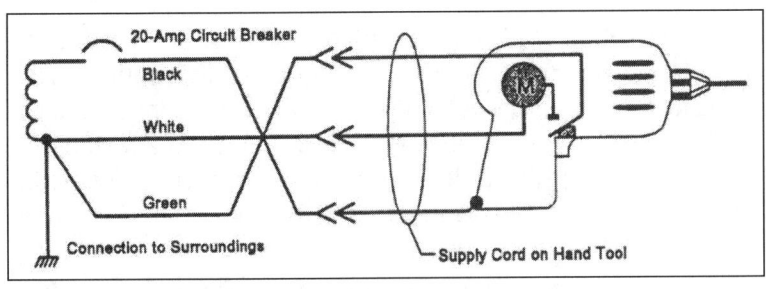

Exhibit 9-8.
A 120-volt branch circuit with black and green wires reversed.

polarity. Although a shock hazard may not exist, there are other mechanical hazards that can occur.

For example, if an internal fault should occur in the wiring as shown in Exhibit 9-6, the equipment would not stop when the switch is released or would start as soon as a person plugged the supply cord into the improperly wired outlet. This could result in serious injury.

Exhibit 9-7 shows the white (grounded) and green (grounding) conductors reversed. Although it is not accurate, considering OSHA or code terminology, to call this reversed polarity, a hazard can still exist. In this case, due to the wiring error, the white wire is being used to provide equipment grounding. Under certain conditions, this could be dangerous.

Exhibit 9-8 shows an extremely dangerous situation. In this example, the black (ungrounded) and green (grounding) conductors have been reversed. The metal case of the equipment is at 120 volts with reference to the surroundings. As soon as a person picks up the equipment and touches a conductive surface in his or her surroundings, he or she will receive a serious, perhaps deadly, shock.

Although the equipment will not work with this wiring error, it would not be unusual for a person to pick up the equipment before realizing this. The person may even attempt to troubleshoot the problem before unplugging the power cord.

Correct polarity is achieved when the grounded conductor is connected to the corresponding grounded terminal and the ungrounded conductor is connected to the corresponding ungrounded terminal. The reverse of the designated polarity is prohibited. Exhibit 9-9 illustrates a duplex receptacle correctly wired. Terminals are designated and identified to avoid confusion. An easy way to remember the correct polarity is:

- White to light: The white (grounded) wire should be connected to the light or nickel-colored terminal;
- Black to brass: The black or multi-colored (ungrounded) wire should be connected to the brass terminal; and
- Green to green: The green or bare (grounding) wire should be connected to the green hexagonal head terminal screw.

Services

Disconnecting Means

Means shall be provided to disconnect all conductors in a building or other structure from the service-entrance conductors. The disconnecting means shall plainly indicate whether it is in the open or closed position and shall be installed at a readily accessible location nearest the point of entrance of the service-entrance conductors.

A readily accessible means of disconnecting conductors is required to be located at a point near the service entrance. The service entrance is the location where the serving conductors enter a building. The disconnecting means can be a switch or circuit breaker and must be capable of interrupting the circuit from the source of supply.

Exhibit 9-9.
A correctly wired duplex receptacle.

ELECTRICAL

This will disconnect the electrical equipment within the building from its source of supply in the event of an emergency or during normal servicing operations.

Services Over 600 Volts, Nominal

The following additional requirements apply to services over 600 volts, nominal:

- Guarding. Service-entrance conductors installed as open wires shall be guarded to make them accessible only to qualified persons.
- Warning signs. Signs warning of high voltage shall be posted where other than qualified employees might come in contact with live parts.

Overcurrent Protection

600 Volts, Nominal, or Less

The following requirements apply to overcurrent protection of circuits rated 600 volts, nominal, or less.

Protection of Conductors and Equipment

Conductors and equipment shall be protected from overcurrent in accordance with their ability to safely conduct current.

Electric current is the flow of electrons through a conductor. The size of the wire is the main determining factor as to how much current can safely flow through a conductor. The larger the wire, the more current can flow safely. If too much current flows through a conductor excess heat is produced. If the circuit is not protected, the heat may continue to build and reach a temperature high enough to destroy the insulation and cause a fire.

Fuses and circuit breakers are protective devices designed to disconnect a circuit from its source of supply when a maximum allowable heat level is reached. The basic idea of a protective device is to make a weak link in the circuit. In the case of a fuse, the fuse is destroyed before another part of the system is destroyed. In the case of a circuit breaker, a set of contacts opens the circuit. Unlike a fuse, a circuit breaker can be reused by reclosing the contacts. Fuses and circuit breakers are designed to protect equipment and facilities, and in so doing, they provide considerable protection against shock in most situations. It is important to ensure that overcurrent devices have adequate interrupting ratings in order to protect employees from shock.

The National Electrical Code specifies the allowable current flow permitted in certain-sized conductors. Ampacity is the term used to describe the current-carrying capacity of a conductor. The size of the fuse or circuit breaker required to provide protection is determined by the ampacity of the conductor in the circuit to be protected and the type of load that is on the circuit.

Grounded Conductors

Except for motor-running overload protection, overcurrent devices may not interrupt the continuity of the grounded conductor unless all conductors of the circuit are opened simultaneously. Unless excepted, overcurrent devices are always placed in the "hot" side of a circuit (usually a black wire) and in series with the load so that all the current in the circuit must flow through them.

Disconnection of Fuses and Thermal Cutouts

Except for service fuses, all cartridge fuses that are accessible to other than qualified persons and all fuses and thermal cutouts on circuits over 150 volts to ground shall be provided with disconnecting means. This disconnecting means shall be installed so that the fuse or thermal cutout can be disconnected from its supply without disrupting service to equipment and circuits unrelated to those protected by the overcurrent device.

Location in or on Premises

Overcurrent devices shall be readily accessible to each employee or authorized building management personnel. These overcurrent devices may not be located where they will be exposed to physical damage or in the vicinity of easily ignitable material.

The functions that overcurrent devices perform require proper operation and maintenance because of the tremendous amounts of energy likely to be handled during a fault. Also, when trouble occurs, it may be necessary to reach these devices quickly.

Physical damage often results when overcurrent devices are located where they can be struck by lift trucks or other vehicles, by crane hooks, by materials being handled, etc.

Easily ignitable material should not be stored in the vicinity of overcurrent devices. Most fuses and circuit breakers operate at elevated temperatures, at least some of the time, and they may on occasion emit flashes or sparks.

Arcing or Suddenly Moving Parts

Fuses and circuit breakers shall be so located or shielded that employees will not be burned or otherwise injured by their operation.

Circuit Breakers

Circuit breakers shall clearly indicate whether they are in the open (off) or closed (on) position.

Where circuit breaker handles on switchboards are operated vertically rather than horizontally or rotationally, the up position of the handle shall be the closed (on) position. If used as switches in 120-volt, fluorescent lighting circuits, circuit breakers shall be approved for the purpose and marked "SWD."

Over 600 Volts, Nominal

Feeders and branch circuits over 600 volts, nominal, shall have short-circuit protection.

Grounding

This section contains grounding requirements for systems, circuits and equipment. Grounding electrical circuits and electrical equipment is required to protect employees from electrical shock and fire and electrical equipment from damage. There are two kinds of grounding: (1) electrical circuit or system grounding and (2) electrical equipment grounding (See Exhibit 9-10.).

Electrical system grounding is accomplished when one conductor of the circuit is intentionally connected to earth. This is done to protect the circuit should lightning strike or other high voltage contact occur. Grounding a system also stabilizes the voltage in the system so "expected voltage levels" are not exceeded under normal conditions.

Electrical equipment grounding is accomplished when all metal frames of equipment

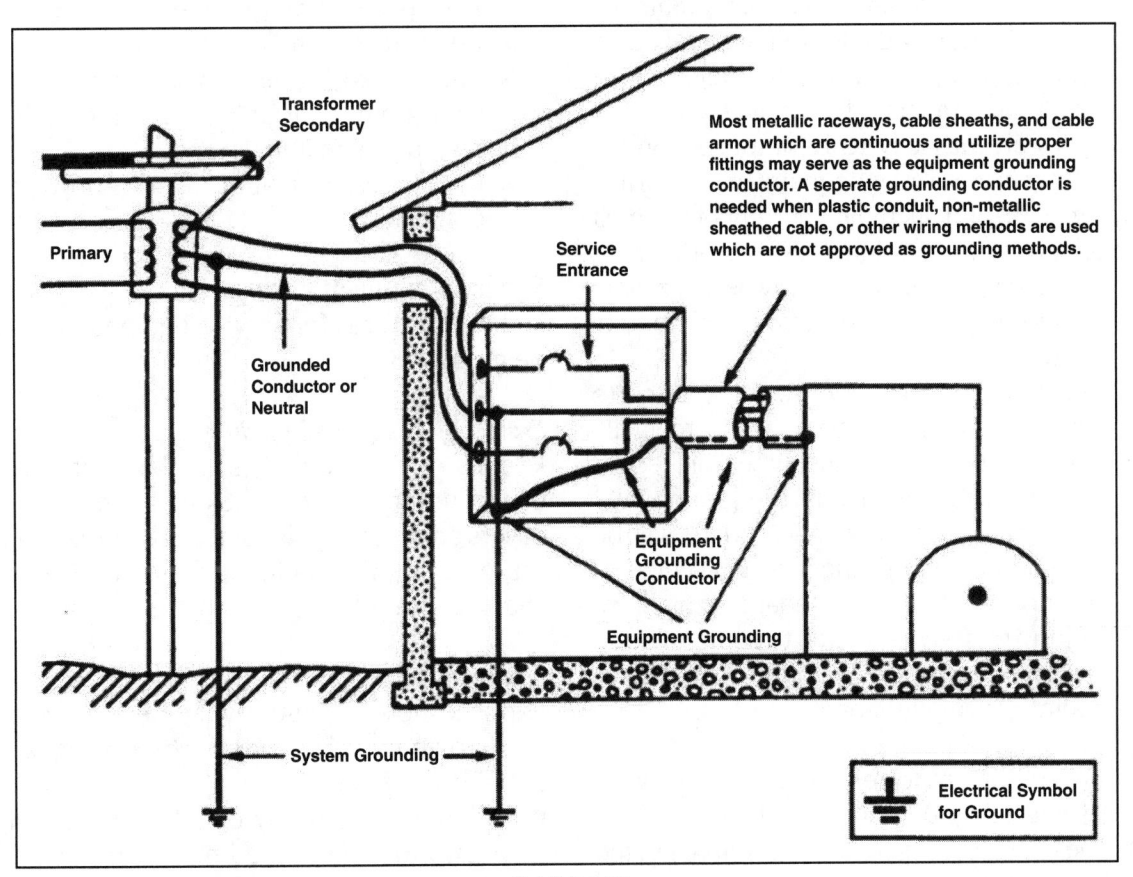

Exhibit 9-10.
There are two types of grounding: system grounding and equipment grounding.

and enclosures containing electrical equipment or conductors are grounded by means of a permanent and continuous connection or bond. The equipment grounding conductor provides a path for dangerous fault current to return to the system ground at the supply source of the circuit should an insulation failure take place. If installed properly, the equipment-grounding conductor is the current path that enables protective devices, such as circuit breakers and fuses, to operate when a fault occurs.

Grounding Path
The path to ground from circuits, equipment and enclosures shall be permanent and continuous.

This requirement was extracted from NEC 250-51, Effective Grounding Path, which is more complete and fundamental to the understanding of electrical safety. It states that the path to ground:
- "Shall be permanent and continuous." (If the path is installed in such a way that damage, corrosion, loosening, etc. may impair the continuity during the life of the installation, shock and burn hazards will develop.)
- "Shall have capacity to conduct safely any fault current likely to be imposed on it." (Fault currents may be many times normal currents, and such high currents may melt or burn metal at points of poor conductivity. These high temperatures may be a hazard in themselves and may destroy the continuity of the ground-fault path.)
- "Shall have sufficiently low impedance to limit the voltage to ground and to facilitate the operation of the circuit protective devices in the circuit." (If the ground-fault path has a high impedance, there will be hazardous voltages whenever fault currents attempt to flow. If the impedance is high, the fault current will be limited to some value so low that the fuse or circuit breaker will not operate promptly, if at all.)

It is important to remember the following regarding safe grounding paths:
- The fault current in AC circuits will be limited by the sum of resistance and reactance; the only low-reactance path is that which closely follows the circuit conductors.
- If a metallic raceway system is used, the metallic system must be continuous and permanent.
- In cases where a metallic raceway system is not used, a green or bare equipment-grounding conductor must be close to the supply conductors to ensure all enclosures are bonded together and to the source.

Supports, Enclosures and Equipment to be Grounded

Supports and Enclosures for Conductors
Metal cable trays, metal raceways and metal enclosures for conductors shall be grounded, except:
- Metal enclosures, such as sleeves, that are used to protect cable assemblies from physical damage need not be grounded; and
- Metal enclosures for conductors added to existing installations of open wire, knob-and-tube wiring and nonmetallic-sheathed cable need not be grounded if all of the following conditions are met:
  - Runs are less than 25 ft.;
  - Enclosures are free from probable contact with ground, grounded metal, or other conductive materials; and
  - Enclosures are guarded against employee contact.

Service Equipment Enclosures
Metal enclosures for service equipment shall be grounded.

Frames of Ranges and Clothes
Frames of electric ranges, wall-mounted ovens, counter-mounted cooking units, clothes dryers and metal outlet or junction boxes that are part of the circuit for these appliances shall be grounded.

Fixed Equipment
Exposed noncurrent-carrying metal parts of fixed equipment that may become energized shall be grounded if:
- Within 8 ft. vertically or 5 ft. horizontally of ground or grounded metal objects and subject to employee contact;
- Located in a wet or damp location and not isolated;

- In electrical contact with metal;
- In a hazardous (classified) location; or
- Supplied by a metal-clad, metal-sheathed or grounded metal raceway wiring method.
- Equipment operates with any terminal at over 150 volts to ground; however, the following need not be grounded:
  - Enclosures for switches or circuit breakers used for other than service equipment and accessible to qualified persons only;
  - Metal frames of electrically heated appliances that are permanently and effectively insulated from ground; and
  - The cases of distribution apparatus such as transformers and capacitors mounted on wooden poles at a height exceeding 8 ft. above ground or grade level.

Equipment Connected by Cord and Plug

Under any of the conditions described below, exposed noncurrent-carrying metal parts of cord- and plug-connected equipment that may become energized shall be grounded if:

- In a hazardous (classified) location;
- Operated at over 150 volts to ground, except for guarded motors and metal frames of electrically heated appliances if the appliance frames are permanently and effectively insulated from ground; or
- The equipment is of the following type:

- Refrigerators, freezers and air conditioners;
- Clothes-washing, clothes-drying and dishwashing machines, sump pumps and electrical aquarium equipment;
- Hand-held motor-operated tools;
- Motor-operated appliances of the following types: hedge clippers, lawn mowers, snow blowers and wet scrubbers;
- Cord- and plug-connected appliances used in damp or wet locations or by employees standing on the ground or metal floors or working inside of metal tanks or boilers;
- Portable and mobile X-ray and associated equipment;
- Tools likely to be used in wet and conductive locations; and
- Portable hand lamps.

Under the conditions described above, exposed noncurrent-carrying metal parts of cord- and plug-connected equipment must be grounded. Grounding metal parts is not required where the equipment is supplied through an isolating transformer with an ungrounded secondary of not over 50 volts or if portable tools are protected by an approved system of double insulation. To ground cord- and plug-connected equipment, a third wire is commonly provided in the cord set and a third prong in the plug. The

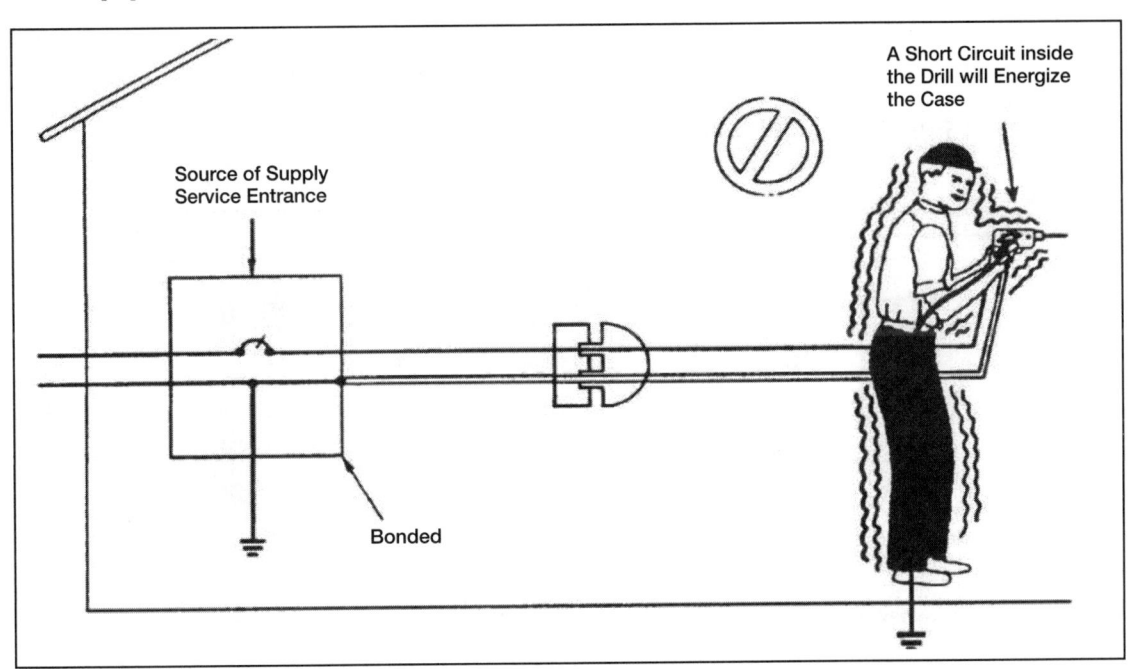

Exhibit 9-11.
If no grounding conductor is present, some portion of the fault current will return to ground through the worker.

ELECTRICAL

third wire serves as an equipment-grounding conductor that is connected to the metal housing of a portable tool and a metal grounding bus inside the service entrance equipment. The service entrance equipment is located at the entrance point of the electric supply for a building or plant and contains, or serves other panel boards that contain, branch circuit protective devices such as fuses and circuit breakers. The third wire provides a path for fault current should an insulation failure occur. In this manner, dangerous fault current will be directed back to the source, the service entrance, and will enable circuit breakers or fuses to operate thus opening the circuit and stopping the current flow.

Exhibit 9-11 illustrates the potential shock hazard that exists when no third wire, grounding conductor, is used. If a fault occurs, most of the current will follow the path of least resistance. If the worker provides a path to ground as shown, some portion of the current will flow away from the grounded white conductor (neutral) and return to ground through the worker. The severity of the shock received will depend on the amount of current that flows through the worker.

To keep this from occurring it is important to ensure that the electrical system is properly grounded. Exhibit 9-12 illustrates the advantage of a properly connected grounded conductor.

It should be noted that properly bonded conduit and associated metal enclosures can also serve as a grounding conductor.

Tools likely to be used in wet and conductive locations need not be grounded if supplied through an isolating transformer with an ungrounded secondary of not over 50 volts. Listed or labeled portable tools and appliances protected by an approved system of double insulation, or its equivalent, need not be grounded. If such a system is employed, the equipment shall be distinctively marked to indicate that the tool or appliance uses an approved system of double insulation.

Nonelectrical Equipment
The metal parts of the following nonelectrical equipment shall be grounded: frames and tracks of electrically operated cranes; frames of nonelectrically driven elevator cars to which electric conductors are attached; hand-operated, metal-shifting ropes or cables of electric elevators, and metal partitions, grill work, and similar metal enclosures around equipment of over 750 volts between conductors.

Methods of Grounding Fixed Equipment
Noncurrent-carrying metal parts of fixed equipment, if required to be grounded by this subpart (Electrical), shall be grounded by an

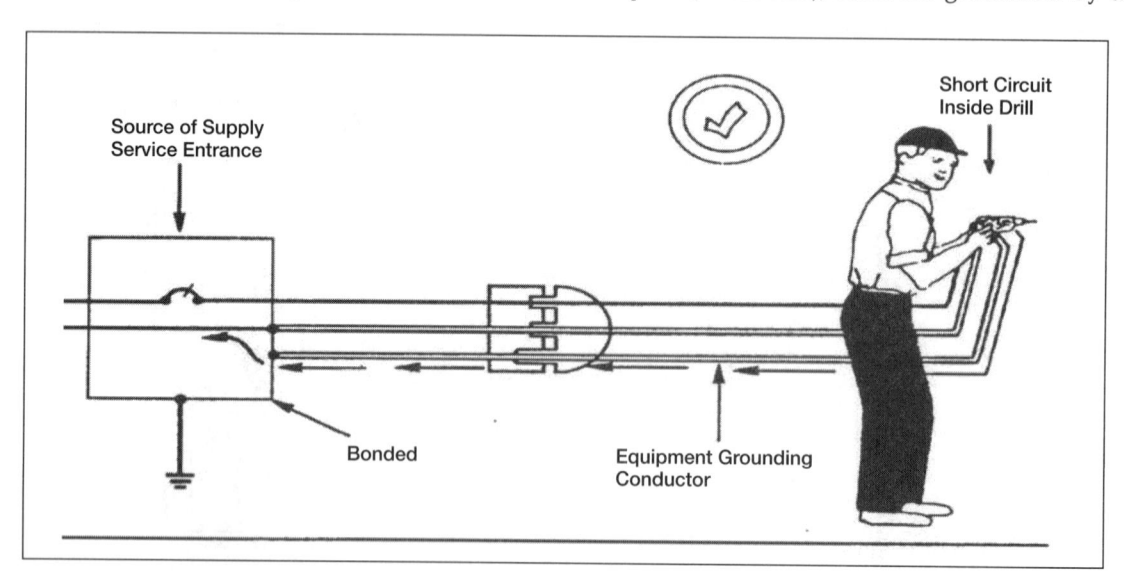

Exhibit 9-12.
Cord- and plug-connected equipment with a grounding connector will protect the worker.

Figure 1

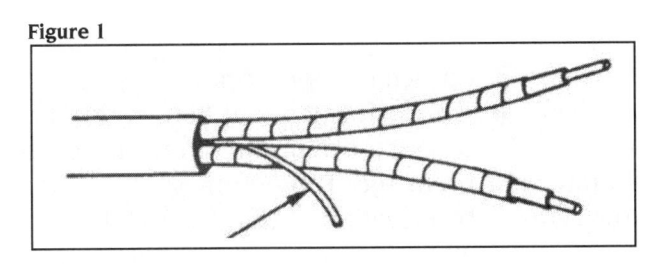

Figure 2

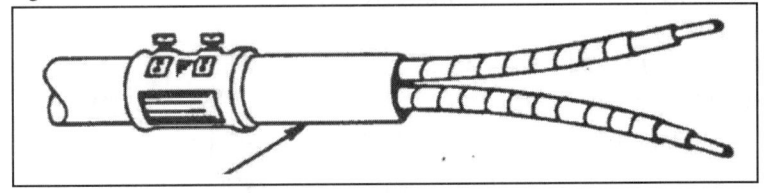

Exhibit 9-13.
The equipment-grounding conductor must run with the circuit conductors powering fixed equipment (Fig. 1) or the metal conduit that encloses the power conductors servicing fixed equipment may serve as the equipment-grounding conductor for that equipment (Fig. 2).

equipment-grounding conductor that is contained within the same raceway, cable, or cord, or runs with or encloses the circuit conductors. For DC circuits only, the equipment-grounding conductor may be run separately from the circuit conductors (See Exhibit 9-13.).

Grounding of Systems and Circuits of 1000 Volts and Over (High Voltage)

If high voltage systems are grounded, they shall comply with all applicable provisions previously discussed in this subpart (Electrical) as well as additional requirements and modifications contained in this section.

Wiring Methods—Components and Equipment for General Use— 1910.305
Wiring Methods
The provisions of this section do not apply to the conductors that are an integral part of factory-assembled equipment.

General Requirements for Electrical Continuity of Metal Raceways and Enclosures
Metal raceways, cable armor and other metal enclosures for conductors shall be metallically joined together into a continuous electric conductor and shall be so connected to all boxes, fittings and cabinets as to provide effective electrical continuity.

Wiring in Ducts
No wiring systems of any type shall be installed in ducts used to transport dust, loose stock or flammable vapors. No wiring system of any type may be installed in any duct used for vapor removal or for ventilation of commercial-type cooking equipment or in any shaft containing only such ducts.

Temporary Wiring
Temporary electrical power and lighting wiring methods may be of a class less than would be required for a permanent installation. Except as specifically modified in this paragraph, all other requirements of this subpart (Electrical) for permanent wiring shall apply to temporary wiring installations.

Uses Permitted, 600 Volts, Nominal, or Less
Temporary electrical power and lighting installations 600 volts, nominal, or less may be used only:
- During and for remodeling, maintenance, repair, or demolition of buildings, structures, or equipment, and similar activities;
- For experimental or development work; and
- For a period not to exceed 90 days for Christmas decorative lighting, carnivals and similar purposes.

Uses Permitted, Over 600 Volts, Nominal
Temporary wiring over 600 volts, nominal may be used only during periods of tests, experiments or emergencies.

General Requirements for Temporary Wiring
- Feeders shall originate in an approved distribution center. The conductors shall be run as multi-conductor cord or cable assemblies or, where not subject to physical damage, they may be run as open conductors or insulators not more than 10 ft. apart.
- Branch circuits shall originate in an approved power outlet or panel board. Conductors shall be multi-conductor cord or cable assemblies or open conductors. If run

ELECTRICAL

as open conductors, they shall be fastened at ceiling height every 10 ft. No branch-circuit conductor may be laid on the floor. Each branch circuit that supplies receptacles or fixed equipment shall contain a separate equipment-grounding conductor if run as open conductors.
- Receptacles used in temporary wiring circuits must provide a connection for an equipment-grounding conductor (See Exhibit 9-14.). Unless the receptacle is supplied by a metallic raceway that provides a continuous grounding path back to the source, a separate equipment-grounding conductor must be placed in the branch circuit. There must be good electrical connection between the receptacle grounding terminal and the equipment-grounding conductor.
- No bare conductors nor earth returns may be used for the wiring of any temporary circuit. Bare conductors are conductors that do not have any coverings whatsoever.
- Earth returns use the earth itself to provide a current path back to the supply source. Implanting a grounding electrode at the equipment being served and connecting the equipment to the ungrounded conductor and to the grounding electrode do this. Because one side of the supply source is also connected to ground through a grounding electrode, a return path exists. However, its effectiveness is dependent on varying soil conditions. Earth returns must not be used for wiring temporary circuits because they are not always effective and may present a serious hazard on temporary work sites. Exhibit 9-15 shows an earth return, which is not allowed, in contrast to a proper return system.
- In addition, a separate equipment-grounding conductor must be used to provide a low-impedance path to the source. This path will allow sufficient current to flow to operate the circuit breaker when a fault occurs.
- Suitable disconnecting switches or plug connectors shall be installed to permit the disconnection of all ungrounded conductors of each temporary circuit.
- Lamps for general illumination shall be protected from accidental contact or breakage. Protection shall be provided by elevation of at least 7 ft. from normal working surface or by a suitable fixture or lamp holder with a guard.

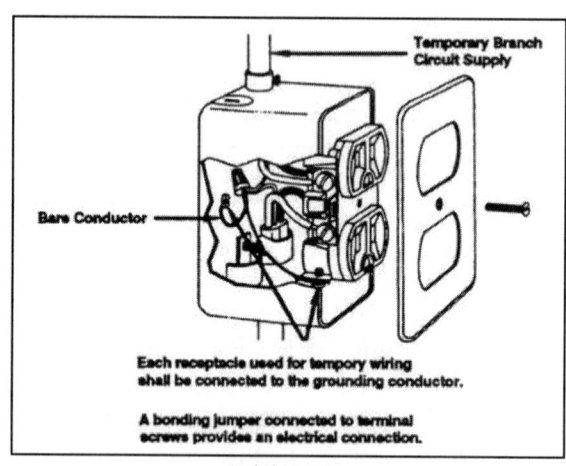

Exhibit 9-14.
Temporary branch circuit with connections for equipment-grounding conductors.

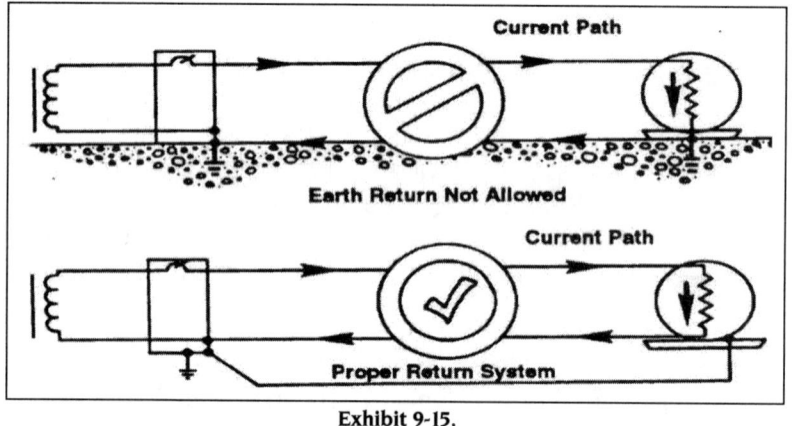

Exhibit 9-15.
Examples of an earth return and a proper return system.

- Flexible cords and cables shall be protected from accidental damage. Sharp corners and projections shall be avoided. Where passing through doorways or other pinch points, flexible cords and cables shall be provided with protection to avoid damage.

Cabinets, Boxes and Fittings

Conductors Entering Boxes, Cabinets or Fittings

Because conductors can be damaged if they rub against the sharp edges of cabinets, boxes or fittings, they must be protected from damage where they enter. To protect the conductors, some type of clamp or rubber grommet must be used. The device used must close the hole through which the conductor passes as well as provide protection from abrasion. If the conductor is in a conduit and the conduit fits tightly in the opening, additional sealing is not required.

The knockouts in cabinets, boxes and fittings should be removed only if conductors are to be run through them. However, if a knockout is missing or if there is another hole in the box, the hole or opening must be closed.

Covers and Canopies
All pull boxes, junction boxes and fittings shall be provided with covers approved for the purpose. If metal covers are used, they shall be grounded. In completed installations, each outlet box shall have a cover, faceplate or fixture canopy. Covers of outlet boxes having holes through which flexible cord pendants pass shall be provided with bushings designed for the purpose or shall have smooth, well-rounded surfaces on which the cords may bear.

Pull and Junction Boxes for Systems over 600 Volts, Nominal
Boxes shall provide a complete enclosure for the contained conductors or cables. Suitable covers securely fastened in place shall close boxes. Underground box covers that weigh more than 100 lbs. meet this requirement. Covers for boxes shall be permanently marked "HIGH VOLTAGE." The marking shall be on the outside of the box cover and shall be readily visible and legible.

Switches

Knife Switches
Single-throw knife switches have one energized (closed or "ON") position and one open (dead or "OFF") position. The switch must be designed so that when it is in the open position, the blades are not energized (i.e., the blades must be connected to the load side, not the supply side of the circuit). The switches must also be installed so that if the switch falls downward, it will not fall into its energized position. However, some single-throw knife switches

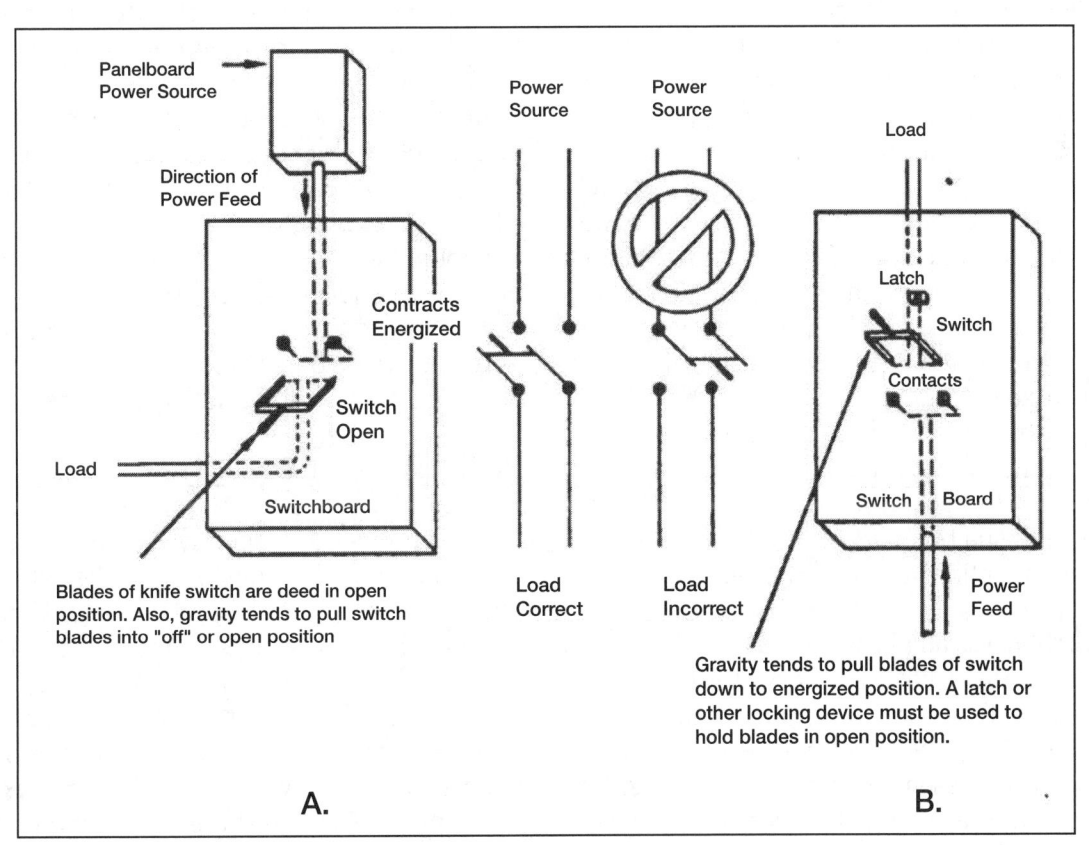

Exhibit 9-16.
Examples of single-throw knife switches.

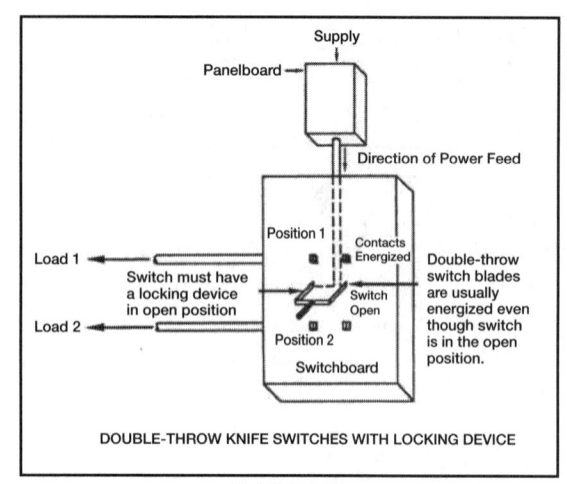

Exhibit 9-17.
A double-throw knife switch with locking device.

are designed to be installed so that they open upward. To be approved for this type of installation, they must have a latch or other locking device (e.g., a spring-loaded device) used to secure the switch in the open position. In Exhibit 9-16, Illustration A shows a single-throw knife switch connected so that the blades are dead when the switch is open. Illustration B shows a latch arrangement that holds the blade in the open position and will prevent gravity from pulling the switch closed.

Double-throw knife switches are knife switches that have two energized (closed or "ON") positions and one open (dead or "OFF") position. These switches can be mounted vertically so that they are moved up and down or horizontally so that they are moved back and forth. If switches are mounted vertically, they must have a locking device (e.g., a spring-loaded device) that will hold the switchblades in the open position (See Exhibit 9-17.).

Flush snap switches that are mounted in ungrounded metal boxes and located within reach of conducting floors or other conducting surfaces shall be provided with faceplates of nonconducting, noncombustible material.

Switchboards and Panel Boards

A switchboard that has exposed live parts must be located in an area that is not subject to wetness or dampness. One purpose of this regulation is to lessen the chance of severe shock if a worker accidentally comes into contact with the five parts. Additionally, only qualified persons may have access to switchboards with exposed live parts. To limit access, the switchboard should be located in a locked room or within a locked cage or fenced area. Keys to the locks should be controlled to ensure that only properly trained personnel are allowed to enter the area.

Enclosures for Damp or Wet Locations

Cabinets, cutout boxes, fittings, boxes and panel board enclosures in damp or wet locations shall be installed so as to prevent moisture or water from entering and accumulating within the enclosures. In wet locations, the enclosures shall be weatherproof. Switches, circuit breakers and switchboards installed in wet locations shall be enclosed in weatherproof enclosures.

Conductors for General Wiring

To provide adequate protection against shock and fire hazards, conductors must be insulated with approved materials. Insulating material should be the appropriate composition and thickness for the voltage and current the conductor will carry, for the temperature extremes and other environmental factors to which it will be subjected, and for the location in which it will be placed.

Insulated conductors must also be easily identifiable; color-coding is used most often. Neutral, or grounded, conductors should be white or natural gray.

Grounding conductors such as equipment-grounding conductors should be green or green with yellow stripes. Grounding conductors are permitted to be bare wires. Other types of circuit wires may be any colors except these.

Flexible Cords and Cables

This standard for safe use of flexible cords is one of the most frequently violated electrical standards, particularly in smaller plants. There is a definite need and place for cords, but there is also a temptation to misuse them because they seem to offer a quick and easy way to carry electricity to where it is needed. The basic problem is that flexible cords in general are more vulnerable than the fixed wiring of the building. Therefore, cords should not be used if one of the recognized wiring methods could be used instead.

Use of Flexible Cords and Cables

Flexible cords and cables shall be approved and suitable for conditions of use and location. The standard lists specific situations in which flexible cords may be used. Flexible cords and cables shall be used only for:

- Pendants (a lamp holder or cord-connector body suspended by a length of cord properly secured and terminated directly above the suspended device);
- Wiring of fixtures;
- Connection of portable lamps or appliances;
- Elevator cables;
- Wiring of cranes and hoists (where flexibility is necessary);
- Connection of stationary equipment to facilitate their frequent interchange (equipment that is not normally moved from place to place, but might be on occasion);
- Prevention of the transmission of noise or vibration (in some cases, vibration might fatigue fixed wiring and result in a situation more hazardous than flexible cord);
- Appliances where the fastening means and mechanical connections are designed to permit removal for maintenance and repair (e.g., water coolers, exhaust fans); and
- Data processing cables approved as a part of the data processing system.

Note that all of the above situations involve conditions where flexibility is necessary. Unless specifically permitted by one of these situations, flexible cords and cables may not be used:

- As a substitute for the fixed wiring of the structure;
- Where run through holes in walls, ceilings, or floors;
- Where run through doorways, windows or similar openings;
- Where attached to building surfaces; or
- Where concealed behind building walls, ceilings or floors.

There is usually not much question about the use of the short length of cord that is furnished as part of an approved appliance or tool; the use of an extension cord to permit the temporary use of an appliance or tool in its intended manner at some distance from a fixed outlet, but there are questions when the use obviously is not temporary and the cord is extended to some distant outlet to avoid providing a fixed outlet where needed.

Flexible cord used in violation of this standard is likely to be damaged by activities in the area, door or window edges, staples or fastenings, abrasion from adjacent materials, or simply by aging. If the conductors become partially exposed over a period of time, there will be an increased chance of shocks, burns or fire.

Identification, Splices and Termination

Flexible cords shall be used only in continuous lengths without splice or tap. Hard service flexible cords, No. 12 or larger, may be repaired if spliced so that the splice retains the insulation, outer sheath properties and usage characteristics of the cord being spliced. Flexible cords shall be connected to devices and fittings so that strain relief is provided that will prevent pull from being directly transmitted to joints or terminal screws.

Portable Cables Over 600 Volts, Nominal

Multi-conductor portable cable for use in supplying power to portable or mobile equipment at over 600 volts, nominal, shall consist of No. 8 or larger conductors employing flexible stranding. Cables operated at over 2000 volts shall be shielded for the purpose of confining the voltage stresses to the insulation. Grounding conductors shall be provided. Connectors for these cables shall be of a locking type, with provisions to prevent their opening or closing while energized. Strain relief shall be provided at connections and terminations. Portable cables may not be operated with splices unless the splices are of the permanent molded, vulcanized or other approved type. Termination enclosures shall be suitably marked with a high voltage hazard warning, and terminations shall be accessible only to authorized and qualified personnel.

Equipment for General Use

Lighting Fixtures, Lamp Holders, Lamps and Receptacles

Fixtures, lamp holders, lamps, rosettes, and receptacles may have no live parts normally exposed to employee contact. However, rosettes and cleat-type lamp holders and receptacles located at least 8 ft. above the floor may have exposed parts.

Portable-type hand lamps supplied through flexible cords shall be equipped with a handle of molded composition or other material approved for the purpose. A substantial guard also shall be attached to the lamp holder or the handle.

Screw-shell-type lamp holders shall be installed for use as lamp holders only and must not be used with screw-base socket adapters. These adapters screw into the existing lamp socket and convert lamp holders into receptacles.

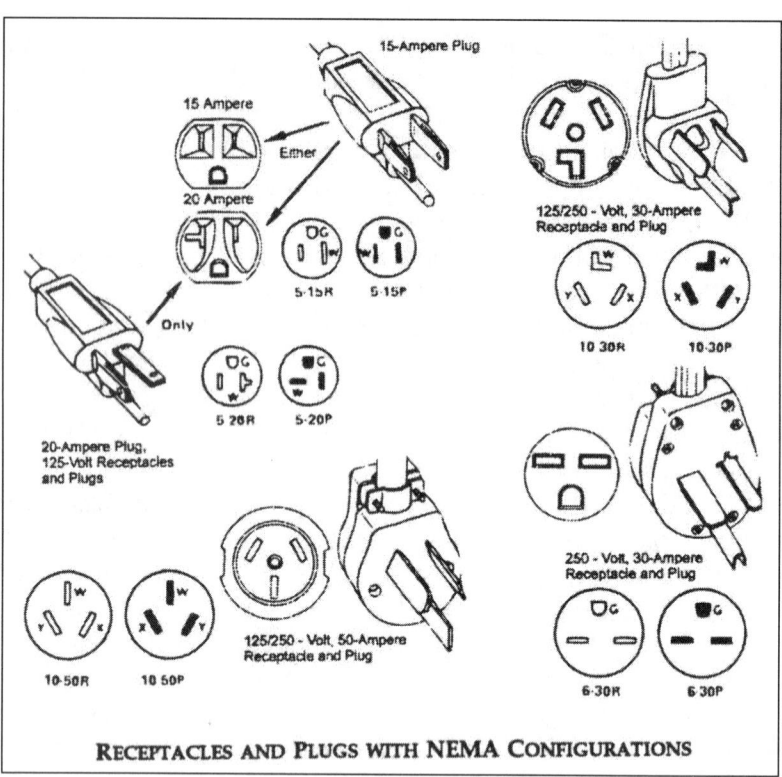

Exhibit 9-18.
Receptacles and plugs with NEMA configurations.

These adapters are not permitted because equipment-grounding connections cannot be made through the two-blade adapters and because the fixture has been designed only for lighting. Only weatherproof lamp holders may be installed in wet or damp areas. Unprotected lamp holders might allow moisture to enter the lamp holder socket, creating an electrical shock hazard.

Receptacles, Cord Connectors and Attachment Plugs

Cord connectors are devices that join two sections of electrical cord together. Attachment plugs are devices that are fastened onto the end of a cord so that electrical contact can be made between the conductors in the cord and the conductors in a receptacle. Connectors, plugs and receptacles are uniquely designed for different voltages and currents so that only matching plugs will fit into the correct receptacle or cord connector. In this way, a piece of equipment rated for one voltage combination cannot be plugged into a power system that is of a different voltage or current capacity.

The only exceptions to this are 125-volt and 250-volt, 20-ampere, T-slot receptacles (See Exhibit 9-18.). A 125-volt and 250-volt, 15-ampere plug will fit into a 20-ampere T-slot receptacle or connector of the same voltage rating as well as in a 120-volt, 15-ampere grounding type receptacle or connector of the same voltage rating. An electrical appliance that is rated for 15 amperes will not overload a 20-ampere circuit, and the 20-ampere breaker will still provide overcurrent protection for 15-ampere equipment. Note that the opposite is not necessarily true and that a 20-ampere plug will not fit into a 15-ampere receptacle or cord connector.

The National Electrical Manufacturer's Association (NEMA) has standard plug and receptacle connector blade configurations. Each has been developed to standardize the use of plugs and receptacles for different voltages, amperages, and phases from 115 volts through 600, from 15 amperes through 60, and for single- and three-phase systems.

A receptacle installed in a wet or damp location shall be suitable for the location.

Appliances

Electrical appliances such as portable air conditioning units, coffee makers and fans must not have any exposed live wires or electrical parts that might create an electrical shock hazard.

Exceptions to this are appliances such as heaters or toasters that must have exposed current-carrying parts that operate at high temperatures to transfer heat (e.g., a space heater). The heat generated by these parts minimizes the possibility of direct contact and resultant electric shock.

A disconnecting means is a switch or plug that can open any electric circuit under load and safely stop the flow of current. All appliances must have a disconnecting means. Each appliance shall be marked with its rating in volts and amperes or volts and watts.

Motors

Disconnecting Means

- A disconnecting means shall be located in sight from the controller (i.e., visible from and located within 50 ft. of the controller). However, a single disconnecting means may be located adjacent to a group of coordinated controllers mounted adjacent to one another on a multi-motor continuous process machine. The controller disconnecting means for motor branch circuits over 600 volts, nominal, may be out of sight of the controller if the controller is marked with a warning label giving the location and identification of the disconnecting means that is to be locked in the open position.
- The disconnecting means shall disconnect the motor and the controller from all ungrounded supply conductors and shall be so designed that no pole can be operated independently.
- If a motor and the driven machinery are not in sight from the controller location, the installation shall comply with one of the following conditions:
  - The controller disconnecting means shall be capable of being locked in the open position.
  - A manually operable switch that will disconnect the motor from its source of supply shall be placed in sight from the motor location.
- The disconnecting means shall plainly indicate whether it is in the open (off) or closed (on) position.
- The disconnecting means shall be readily accessible. If more than one disconnect is provided for the same equipment, only one need be readily accessible.
- An individual disconnecting means shall be provided for each motor, but a single disconnecting means may be used for a group of motors if:
  - A number of motors drive special parts of a single machine or piece of apparatus, such as a metal or woodworking machine, crane or hoist;
  - A group of motors is under the protection of one set of branch-circuit protective devices; or
  - A group of motors is in a single room in sight from the location of the disconnecting means.

Motor Overload, Short-Circuit and Ground Fault Protection

Motors, motor-control apparatus and motor branch-circuit conductors shall be protected against overheating due to motor overloads or failure to start and against short-circuit or ground faults. These provisions shall not require overload protection that will stop a motor where a shutdown is likely to introduce additional or increased hazards, as in the case of fire pumps, or where continued operation of a motor is necessary for a safe shutdown of equipment or process and motor overload-sensing devices are connected to a supervised alarm.

Protection of Live Parts—All Voltages

Stationary motors having commutators, collectors and brush rigging located inside of motor end brackets and not conductively connected to supply circuits operating at more than 150 volts to ground need not have such parts guarded.

Exposed live parts of motors and controllers operating at 50 volts or more between terminals shall be guarded against accidental contact by:

- Installation in a room or enclosure that is accessible only to qualified persons;
- Installation on a suitable balcony, gallery or platform so elevated and arranged as to exclude unqualified persons; or
- Elevation 8 ft. or more above the floor.

Where live parts of motors or controllers operating at over 150 volts to ground are guarded against accidental contact only by location and adjustment or other attendance may be necessary during the operation of the apparatus, suitable insulating mats or platforms shall be provided so the attendant cannot readily touch live parts unless standing on the mats or platforms.

Transformers
These requirements apply to most transformers, with some exceptions noted in the standard.
- Warning signs or visible markings on the equipment or structure shall indicate the operating voltage of exposed live parts of transformer installations.
- Dry-type, high fire point liquid-insulated and asbestos-insulated transformers installed indoors and rated over 35kV shall be in a vault.
- If they present a fire hazard to employees, oil-insulated transformers installed indoors shall be in a vault.
- Combustible material, combustible buildings and parts of buildings, fire escapes, and door and window openings shall be safeguarded from fires that may originate in oil-insulated transformers attached to or adjacent to a building or combustible material.
- Transformer vaults shall be constructed so as to contain fire and combustible liquids within the vault and to prevent unauthorized access. Locks and latches shall be so arranged that a vault door can be readily opened from the inside.
- Any pipe or duct system foreign to the vault installation may not enter or pass through a transformer vault.
- Materials may not be stored in transformer vaults.

Capacitors
Capacitors store electrical charges and can be a source of severe shock unless that charge is drained when the capacitors are disconnected from the power source. Unless some type of automatic discharge is designed into a system, devices such as resistors must be permanently attached across the terminals of the capacitors to drain the charge when the circuit is open (deenergized). Most capacitors are manufactured with this type of discharge resistor already built in. Surge capacitors, which act like lightning rods, do not require an automatic means for draining the charge.

Storage Batteries
Storage batteries, which are usually lead-acid or alkali, produce explosive gases, including hydrogen, if they are overcharged. These explosive gases must not accumulate in quantities that may form an explosive mixture with air. A spark or open flame could ignite the mixture and cause an explosion. Good ventilation must be provided to prevent this accumulation.

Hazardous (Classified) Locations— 1910.307

Hazardous (classified) locations are those areas where a potential for explosion and fire exists because of flammable gases, vapors or finely pulverized dusts in the atmosphere or because of the presence of easily ignitable fibers or flyings. Hazardous locations may result from the normal processing of certain volatile chemicals, gases, grains, etc., or it may result from accidental failure of storage systems for these materials. It also is possible that a hazardous location may be created when volatile solvents or fluids, used in a normal maintenance routine, vaporize to form an explosive atmosphere.

Hazardous (classified) locations may be found in occupancies such as aircraft hangars, gasoline dispensing and service stations, bulk storage plants for gasoline or other volatile flammable liquids, paint-finishing process plants, health care facilities, agricultural or other facilities where excessive combustible dusts may be present, marinas, boat yards, and petroleum and chemical processing plants. Each room, section or area shall be considered individually in determining its classification.

Regardless of the cause of a hazardous location, it is necessary that every precaution be taken to guard against ignition of the atmosphere. Certainly no open flames would be permitted in these locations, but there are other potential sources of ignition, including electrical equipment. The normal operation of switches, circuit breakers, motor starters, contactors and plugs and receptacles release this energy in the form of arcs and sparks as contacts open and close, making and breaking circuits.

Electrical equipment such as lighting fixtures and motors are classified as "heat producing" and they will become a source of ignition if they reach a surface temperature that exceeds the ignition temperature of the particular gas, vapor or dust in the atmosphere.

It is also possible that an abnormality or failure in an electrical system could provide a source of ignition. The failure of insulation from cuts, nicks or aging can also act as an ignition source again from sparking, arcing and heat.

There are several OSHA standards that require the installation of electrical wiring and equipment in hazardous (classified) locations according to the requirements of Subpart S, Electrical. Most of these standards are contained in Subpart H, Hazardous Materials. Some examples include Acetylene, Flammable and Combustible Liquids, Spray Finishing Using Flammable and Combustible Liquids, and Dip Tanks Containing Flammable or Combustible Materials. The basis for OSHA Standard 1910.307 is the National Electrical Code (NEC), NFPA 70. A general overview of the guidelines contained in the NEC for installation of electrical wiring and equipment in hazardous (classified) locations can be found in this document under the section "Hazardous Materials."

Electrical Safety-Related Work Practices

OSHA's Safety-Related Work Practice standards for general industry, 1910.331 to 1910.399 are performance-oriented regulations that complement the existing electrical installation standards. These work-practice standards include requirements for work performed on or near exposed energized and deenergized parts of electric equipment; the use of electrical protective equipment and the safe use of electric equipment.

These rules are intended to protect employees from the electrical hazards that they may be exposed to even though equipment may be in compliance with the installation requirements in Subpart S, Electrical. When employees are working with electric equipment, they must use safe work practices. Such safety-related work practices include keeping a prescribed distance from exposed energized lines, avoiding the use of electric equipment when the employee or the equipment is wet, and locking out and tagging equipment that is deenergized for maintenance.

Another important safety practice involves the use of electrical protective devices, such as rubber gloves and rubber mats, for the purpose of insulation against live parts. Another practice is to have hand tools for purposes of both insulation and manipulation of energized parts from a distance. However, to ensure the protection of the employee, this equipment must be properly manufactured and maintained.

It is important to understand the distinction between these standards and OSHA Standard 1910.147, Control of Hazardous Energy (Lockout/Tagout). The lockout/tagout standard helps safeguard employees from hazardous energy while they are servicing or performing maintenance on machines and equipment. The standard covers electrical energy sources, but it specifically excludes "exposure to electrical hazards from work on, near, or with conductors or equipment in electrical utilization installations," which is covered by Subpart S (Electrical). Thus, the lockout/tagout standard does not cover electrical hazards associated with conductors and equipment, but only covers electrical equipment that relates to machinery and equipment that is covered by the lockout standard.

Covered Work by Both Qualified and Unqualified Persons—1910.331

The provisions of these standards cover electrical safety-related work practices for both qualified persons (those who have training in avoiding the electrical hazards of working on or near exposed energized parts) and unqualified persons (those with little or no such training) working on, near or with the following installations:

- Premises wiring. Installations of electric conductors and equipment within or on buildings or other structures on other premises, such as yards, carnival, parking, and other lots, and industrial substations.

ELECTRICAL

- Wiring for connections to supply. Installations of conductors that connect to the supply of electricity.
- Other wiring. Installations of other outside conductors on the premises.
- Optical fiber cable. Installations of optical fiber cable where such installations are made along with electric conductors.

Other Covered Work by Unqualified Persons

The provisions of these standards also cover work performed by unqualified persons on, near or with the following installations:
- Generation, transmission and distribution installations. Installations for the generation, control, transformation, transmission, and distribution of electric energy (including communication and metering) located in buildings used for such purposes or located outdoors.
- Communications installations. Installations of communications equipment to the extent that the work is covered under 1910.268.
- Installations in vehicles. Installations in ships, watercraft, railway rolling stock, aircraft, or automotive vehicles other than mobile homes and recreational vehicles.
- Railway installations. Installations of railways for generation, transformation, transmission or distribution of power used exclusively for the operation of rolling stock or installation of railways used exclusively for signaling and communication purposes.

Excluded Work by Qualified Persons

The provisions of these standards do not apply to work performed by qualified persons on or directly associated with the four types of installations described above.

Training—1910.332

The training requirements contained in this section apply to employees who face a risk of electric shock that is not reduced to a safe level by the electrical installation requirements of 1910.303 to 1910.308. Employees in occupations listed below face such a risk and are required to be trained:
- Blue-collar supervisors;
- Electrical and electronic engineers;
- Electrical and electronic equipment engineers;
- Electrical and electronic technicians;
- Electricians;
- Industrial machine operators;
- Material handling equipment operators;
- Mechanics and repairers;
- Painters;
- Riggers and roustabouts;
- Stationary engineers; and
- Welders.

With the exception of electricians and welders, workers in the above groups do not need to be trained if their work or the work of those they supervise does not bring them or the employees they supervise close enough to exposed parts of electric circuits operating at 50 volts or more to ground for a hazard to exist. Other employees who also may reasonably be expected to face a comparable risk of injury due to electric shock or other electrical hazards must also be trained.

Content of Training

Employees shall be trained in and be familiar with the safety-related work practices required by 1910.331 to 1910.355 that pertain to their respective job assignments.

Employees who are covered by the scope of this standard but who are not qualified persons shall also be trained in and be familiar with any electrically related safety practices not specifically addressed by 1910.331 to 1910.335 but which are necessary for their safety.

Qualified persons (i.e., those permitted to work on or near exposed energized parts) shall, at a minimum, be trained in and be familiar with the following:
- The skills and techniques necessary to distinguish exposed live parts from other parts of electric equipment;
- The skills and techniques necessary to determine the nominal voltage of exposed live parts; and
- The clearance distances specified in this standard and the corresponding voltages to which the qualified person will be exposed.

Type of Training

The training required by this section shall be of the classroom or on-the-job type. The degree of training provided shall be determined by the risk to the employee.

Selection And Use Of Work Practices—1910.333

Safety-related work practices shall be employed to prevent electric shock or other injuries resulting from either direct or indirect electrical contacts when work is performed near or on equipment or circuits that are or may be energized. The specific safety-related work practices shall be consistent with the nature and extent of the associated electrical hazards.

Deenergized Parts

Live parts to which an employee may be exposed shall be deenergized before the employee works on or near them, unless the employer can demonstrate that deenergizing would introduce additional or increased hazards or is not feasible due to equipment design or operational limitations. Live parts that operate at less than 50 volts to ground need not be deenergized if there will not be increased exposure to electrical burns or explosion due to electric arcs.

Energized Parts

If the exposed live parts are not deenergized, (i.e., for reasons of increased or additional hazards or unfeasibility), other safety-related work practices shall be used to protect employees who may be exposed to the electrical hazards involved. Such work practices shall protect employees from making contact with energized circuit parts—either directly with any part of their body or indirectly through some other conductive object.

Working On or Near Exposed Deenergized Parts

Application

This paragraph applies to work on exposed deenergized parts or near enough to them to expose the employee to any electrical hazard they present.

Conductors and parts of electric equipment that have been deenergized but have not been locked out or tagged shall be treated as *energized* parts.

Lockout and Tagout

While any employee is exposed to contact with parts of fixed electric equipment or circuits that have been deenergized, the circuits energizing the parts shall be locked out, tagged or both in accordance with the requirements of this paragraph in the following order:

1. Procedures shall be in place before equipment may be deenergized.
2. Circuits and equipment to be worked on shall be disconnected from all electrical energy sources.
3. Stored electrical energy that poses a hazard to workers shall be released.
4. Stored nonelectrical energy in devices that could reenergize electric circuit parts shall be blocked or relieved to the extent that the circuit parts could not be accidentally energized by the device.
5. A lock and a tag shall be placed on each disconnecting means used to deenergize circuits and equipment on which work is to be performed, except as provided below.
6. Each tag shall contain a statement prohibiting unauthorized operation of the disconnecting means and removal of the tag.
7. If a lock cannot be applied or if the employer can demonstrate that tagging procedures will provide a level of safety equivalent to that obtained by the use of a lock, a tag may be used without a lock.
8. A tag used without a lock, as permitted above, shall be supplemented by at least one additional safety measure that provides a level of safety equivalent to that obtained by the use of a lock (e.g., the removal of an isolating circuit element, blocking of a controlling switch, opening of an extra disconnecting device).
9. A lock may be placed without a tag only under the following conditions:
   a. Only one circuit or piece of equipment is deenergized;
   b. The lockout period does not extend beyond the work shift; and
   c. Employees exposed to the hazards associated with reenergizing the circuit or equipment are familiar with this procedure.
10. Before any circuits or equipment can be considered and worked as deenergized:

a. A qualified person shall operate the equipment operating controls or otherwise verify that the equipment cannot be restarted.
b. A qualified person shall use test equipment to test the circuit elements and electrical parts of equipment to which employees will be exposed and shall verify that the circuit elements and equipment parts are deenergized.

11. Before circuits and equipment are reenergized, even temporarily, the following requirements shall be met, in the order given:
   a. A qualified person shall conduct tests and visual inspections, as necessary, to verify that all tools, electrical jumpers, shorts, grounds and other such devices have been removed so that the circuits and equipment can be safely energized.
   b. Employees exposed to the hazards associated with reenergizing the circuit or equipment shall be warned to stay clear of such circuits and equipment.
   c. Each lock and tag shall be removed by the employee who applied it or under his or her direct supervision. However, if this employee is absent from the workplace, the lock or tag may be removed by a qualified person designated to perform this task, provided that the employer ensures that the employee who applied the lock or tag is not available at the workplace and is aware that the lock or tag has been removed before he or she resumes work at that workplace.
   d. There shall be a visual determination that all employees are clear of the circuits and equipment.

Lockout/Tagout Injury

The photos in Exhibit 9-19 are the result of an accident that occurred while an apprentice electrician was changing the ballast on a fluorescent light fixture that was not deenergized.

No pre-job briefing or lockout/tagout was performed.

The apprentice electrician stripped the wrong wire and got severely shocked, badly burning his right hand. He was holding a wire nut in his mouth at the time of the accident. He has now undergone three surgeries. The first surgery was to remove the wire nut that he swallowed when he was shocked. The wire nut went all the way to the bottom of his lung. The other surgeries included amputations necessary to try to regain some use of his right hand.

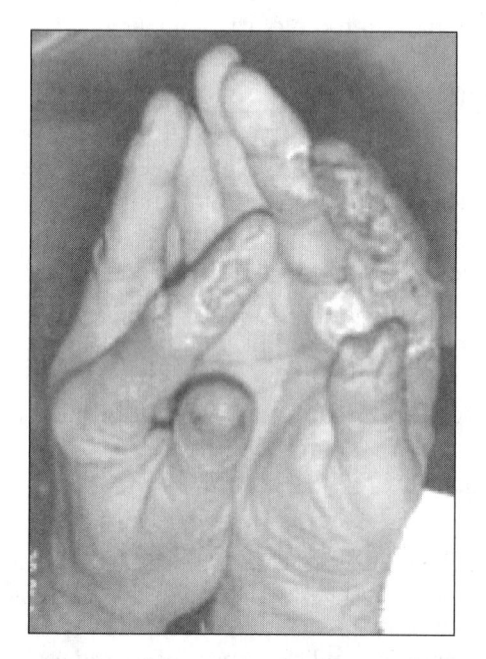

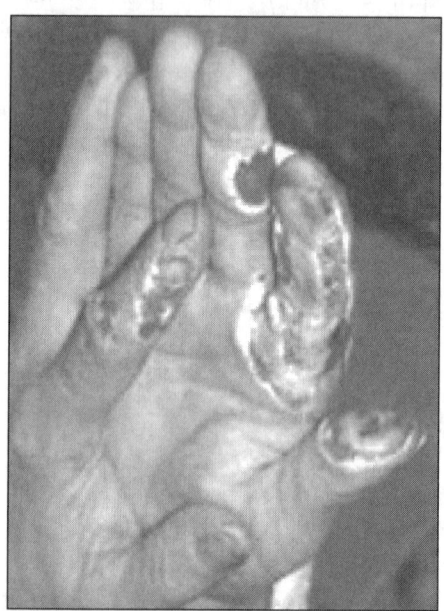

**Exhibit 9-19.
The effects of electrical shock as a result of no lockout/tagout.**

Working On or Near Exposed Energized Parts

Application
This paragraph applies to work performed on exposed live parts (involving either direct contact or contact by means of tools or materials) or near enough to them for employees to be exposed to any hazard they present.

Work on Energized Equipment
Only qualified persons may work on electric circuit parts or equipment that have not been deenergized under the procedures of these standards. Such persons shall be capable of working safely on energized circuits and shall be familiar with the proper use of special precautionary techniques, personal protective equipment, insulating and shielding materials, and insulated tools.

Overhead Lines
If work is to be performed near overhead lines, the lines shall be deenergized and grounded, or other protective measures shall be provided before work is started. If the lines are to be deenergized, arrangements shall be made with the operator or controller of the appropriate electric circuits to have them deenergized or grounded. If protective measures such as guarding, isolating or insulating are provided, these precautions shall prevent employees from contacting such lines directly with any part of their body or indirectly through conductive materials, tools or equipment.

Unqualified Persons
When an unqualified person is working in an elevated position near overhead lines, the location shall be such that the person and the longest conductive object he or she may contact cannot come closer to any unguarded, energized overhead line than the following distances:
- For voltages to ground 50 kV or below, 10 ft.; and
- For voltages to ground over 50 kV, 10 ft. plus 4 in. for every 10 kV over 50 kV.

When an unqualified person is working on the ground in the vicinity of overhead lines, that person may not bring any conductive object closer to unguarded, energized overhead lines than the distances given above.

Qualified Persons
When a qualified person is working in the vicinity of overhead lines, whether in an elevated position or on the ground, the person may not approach or take any conductive object without an approved insulating handle closer to exposed energized parts than shown in Table S-5 of 1910.333(c)(3)(ii), unless certain insulation requirements are met.

Vehicular and Mechanical Equipment
Any vehicle or mechanical equipment capable of having parts of its structure elevated near energized overhead lines shall be operated so that a clearance of 10 ft. is maintained. If the voltage is higher than 50 kV, the clearance shall be increased 4 in. for every 10 kV over that voltage. The standard outlines several conditions under which this clearance may be reduced.

Illumination
Employees may not enter spaces containing exposed energized parts, unless illumination is provided that enables the employees to perform the work safely. Where lack of illumination or an obstruction precludes observation of the work to be performed, employees may not perform tasks near exposed energized parts. Employees may not reach blindly into areas that may contain energized parts.

Confined or Enclosed Work Sites
When an employee works in a confined or enclosed space (e.g., a manhole or vault) that contains exposed energized parts, the employer shall provide, and the employee shall use, protective shields, protective barriers or insulating materials as necessary to avoid inadvertent contact with these parts. Doors, hinged panels and the like shall be secured to prevent their swinging into an employee and causing the employee to contact exposed energized parts.

Conductive Materials and Equipment
Conductive materials and equipment that are in contact with any part of an employee's body shall be handled in a manner that will prevent them from contacting exposed energized conductors

or circuit parts. If an employee must handle long, dimensional conductive objects (e.g., ducts, pipes) in areas with exposed live parts, the employer shall institute work practices (e.g., the use of insulation, guarding, material handling techniques) that will minimize the hazard.

Portable Ladders
Portable ladders shall have nonconductive side rails if they are used where the employee or the ladder could contact exposed energized parts.

Conductive Apparel
Conductive articles of jewelry and clothing may not be worn if they might contact exposed energized parts, unless they are rendered nonconductive by covering, wrapping or other insulating means.

Housekeeping Duties
Where live parts present an electrical contact hazard, employees may not perform housekeeping duties at such close distances to the parts that there is a possibility of contact, unless adequate safeguards (e.g., insulating equipment, barriers) are provided. Electrically conductive cleaning materials may not be used in proximity to energized parts unless procedures are followed that will prevent electrical contact.

Interlocks
Only a qualified person following the requirements of this section may defeat an electrical safety interlock. If the interlock is defeated, it may only be while he or she is working on the equipment; the interlock system must be returned to its operable condition when the work is completed.

Use of Equipment—1910.334

Portable Electric Equipment
This paragraph applies to the use of cord- and plug-connected equipment, including flexible cord sets (extension cords).

Handling
Portable equipment shall be handled in a manner that will not cause damage. Flexible electric cords connected to equipment may not be used for raising or lowering the equipment. Flexible cords may not be fastened with staples or otherwise hung in such a fashion as could damage the outer jacket or insulation.

Visual Inspection
Portable cord- and plug-connected equipment and flexible cord sets (extension cords) shall be visually inspected before use on any shift for external defects and for evidence of possible internal damage. Cord- and plug-connected equipment and extension cords that remain connected once they are put in place and are not exposed to damage need not be visually inspected until they are relocated. Defective or damaged items shall be removed from service until repaired.

Grounding-Type Equipment
A flexible cord used with grounding-type equipment shall contain an equipment-grounding conductor.

Attachment plugs and receptacles may not be connected or altered in a manner that would prevent proper continuity of the equipment-grounding conductor at the point where plugs are attached to receptacles. In addition, these devices may not be altered to allow the grounding pole of a plug to be inserted into slots intended for connection to the current-carrying conductors. Adapters, which interrupt the continuity of the equipment grounding connection, may not be used.

Conductive Work Locations
Portable electric equipment and flexible cords used in highly conductive work locations (e.g., those inundated with water or other conductive liquids) or in job locations where employees are likely to contact water or conductive liquids shall be approved for those locations.

Connecting Attachment Plug
Employee's hands may not be wet when plugging and unplugging flexible cords and cord- and plug-connected equipment if energized equipment is involved.

Energized plug and receptacle connections may be handled only with insulating protective equipment if the condition of the connection could provide a conducting path to the employee's hand. Locking-type connectors shall be properly secured after connection.

Electric Power and Lighting Circuits

Routine Opening and Closing of Circuits
Load-rated switches, circuit breakers or other devices specifically designed as disconnecting means shall be used for the opening, reversing or closing of circuits under load conditions. Cable connectors not of the load-break type, fuses, terminal lugs and cable splice connections may not be used for such purposes, except in an emergency.

Reclosing Circuits after Protective Device Operation
After a circuit is deenergized by a circuit protective device, the circuit may not be manually reenergized until it has been determined that the equipment and circuit can be safely energized. The repetitive manual reclosing of circuit breakers or reenergizing of circuits through replaced fuses is prohibited.

Overcurrent Protection Modification
Overcurrent protection of circuits and conductors may not be modified, even on a temporary basis, beyond that allowed in 1910.304(e), the installation safety requirements for overcurrent protection.

Test Instruments and Equipment
Only qualified persons may perform testing work on electric circuits or equipment.

Visual Inspection
Test instruments and equipment and all associated test leads, cables, power cords, probes and connectors shall be visually inspected for external defects and damage before the equipment is used. If there is a defect or evidence of damage that might expose an employee to injury, the defective or damaged item shall be removed from service and no employee may use it until necessary repairs and tests to render the equipment safe have been made.

Rating of Equipment
Test instruments and equipment and their accessories shall be rated for the circuits and equipment to which they will be connected and shall be designed for the environment in which they will be used.

Occasional Use of Flammable or Ignitable Materials

Where flammable materials are present only occasionally, electric equipment capable of igniting them shall not be used, unless measures are taken to prevent hazardous conditions from developing.

Safeguards For Personnel Protection—1910.335

Use of Protective Equipment

Personal Protective Equipment
Employees working in areas where there are potential electrical hazards shall be provided with, and shall use, electrical protective equipment that is appropriate for the specific parts of the body to be protected and for the work to be performed.

Protective equipment shall be maintained in a safe, reliable condition and shall be periodically inspected or tested, as required by 1910.137. If the insulating capability of protective equipment may be subject to damage during use, the insulating material shall be protected. For example, an outer covering of leather is sometimes used for the protection of rubber insulating material.

Employees shall wear nonconductive head protection wherever there is a danger of head injury from electric shock or burns due to contact with exposed energized parts. Employees shall wear protective equipment for the eyes or face wherever there is danger of injury to the eyes or face from electric arcs or flashes or from flying objects resulting from electrical explosion.

General Protective Equipment and Tools
When working near exposed energized conductors or circuit parts, each employee shall use insulated tools or handling equipment if the tools or handling equipment might make contact with such conductors or parts. If the insulating capability of insulated tools or handling equipment is subject to damage, the insulating material shall be protected.

Fuse-handling equipment, insulated for the circuit voltage, shall be used to remove or install fuses when the fuse terminals are energized. Ropes and hand lines used near exposed energized parts shall be nonconductive.

Protective shields, protective barriers or insulating materials shall be used to protect each employee from shock, burns, or other electrical-related injuries while that employee is working near exposed energized parts that might be accidentally contacted or where dangerous electric heating or arcing might occur. When normally enclosed live parts are exposed for maintenance or repair, they shall be guarded to protect unqualified persons from contact with the live parts.

Alerting Techniques

The following alerting techniques shall be used to warn and protect employees from hazards that could cause injury due to electric shock, burns or failure of electric equipment parts:

- **Safety signs and tags.** Safety signs, safety symbols or accident prevention tags shall be used where necessary to warn employees about electrical hazards that may endanger them, as required by 1910.145.
- **Barricades.** Barricades shall be used in conjunction with safety signs where it is necessary to prevent or limit employee access to work areas that will expose employees to uninsulated energized conductors or circuit parts. Conductive barricades may not be used where they might cause an electrical contact hazard.
- **Attendants.** If signs and barricades do not provide sufficient warning and protection from electrical hazards, an attendant shall be stationed to warn and protect employees.

Hazards of Electricity

Shock. Electric shock occurs when the human body becomes part of a path through which electrons can flow. The resulting effect on the body can be either direct or indirect.

- **Direct.** Injury or death can occur whenever electric current flows through the human body. Currents of less than 30 mA can result in death.
- **Indirect.** Although the electric current through the human body may be well below the values required to cause noticeable injury, human reaction can result in falls from ladders or scaffolds, or movement into operating machinery. Such reaction can result in serious injury or death.

Burns. Burns can result when a person touches electrical wiring or equipment that is improperly used or maintained. Typically, such burn injuries occur on the hands.

Arc-blast. Arc-blasts occur from high-amperage currents arcing through air. This abnormal current flow (arc-blast) is initiated by contact between two energized points. This contact can be caused by persons who have an accident while working on energized components or by equipment failure due to fatigue or abuse. Temperatures as high as 35,000° have been recorded in arc-blast research. The three primary hazards associated with an arc-blast are:

Thermal radiation. In most cases, the radiated thermal energy is only part of the total energy available from the arc. Numerous factors, including skin color, area of skin exposed and type of clothing have an effect on the degree of injury. Proper clothing, work distances and overcurrent protection can improve the chances of curable burns.

Pressure wave. A high-energy arcing fault can produce a considerable pressure wave. Research has shown that a person 2 ft. away from a 25 kA arc would experience a force of approximately 480 lbs. on the front of his or her body. In addition, such a pressure wave can cause serious ear damage and memory loss due to mild concussions.

In some instances, the pressure wave may propel the victim away from the arc-blast, reducing the exposure to the thermal energy. However, such rapid movement could also cause serious physical injury.

Projectiles. The pressure wave can propel relatively large objects over a considerable distance. In some cases, the pressure wave has sufficient force to snap the heads of 3/8 in. steel bolts and knock over ordinary construction walls.

The high-energy arc also causes many of the copper and aluminum components in the electrical equipment to become molten. These "droplets" of molten metal can be propelled great distances by the pressure wave. Although these droplets cool rapidly, they can still be above temperatures capable of causing serious burns or igniting ordinary clothing at distances of 10 ft. or more. In many cases, the burning effect is much worse than the injury from shrapnel effects of the droplets.

Explosions. Explosions occur when electricity provides a source of ignition for an explosive mixture in the atmosphere. Ignition can be due to overheated conductors or equipment or normal arcing (sparking) at switch contacts. OSHA standards, the National Electrical Code and related safety standards have precise requirements for electrical systems and equipment when applied in such areas.

Fires. Electricity is one of the most common causes of fire both in the home and workplace. Defective or misused electrical equipment is a major cause, with high resistance connections being one of the primary sources of ignition. High resistance connections occur where wires are improperly spliced or connected to other components such as receptacle outlets and switches. This was the primary cause of fires associated with the use of aluminum wire in buildings during the 1960s and 1970s.

Heat is developed in an electrical conductor by the flow of current at the rate I^2R. The heat thus released elevates the temperature of the conductor material. A typical use of this formula illustrates a common electrical hazard. If there is a bad connection at a receptacle, resulting in a resistance of 2 ohms, and a current of 10 amperes flows through that resistance, the rate of heat produced W would be:

$$W = I^2R = 10^2 \times 2 = 200 \text{ watts}$$

If you have ever touched an energized 200 watt light bulb, you will realize that this is a lot of heat to be concentrated in the confined space of a receptacle. Situations similar to this can contribute to electrical fires.

Effects of Electricity on the Human Body

The effects of electric shock on the human body depend on several factors. The muscular structure of the body is also a factor in that people having less muscle and more fat typically show similar effects at lesser current values. Major factors are:
- Current and voltage;
- Resistance;
- Path through body; and
- Duration of shock.

Current and Voltage

Although high voltage often produces massive destruction of tissue at contact locations, it is generally believed that the detrimental effects of electric shock are due to the current actually flowing through the body. Any electrical device used on a house wiring circuit can, under certain conditions, transmit a fatal current. Although currents greater than 10 mA are capable of producing painful to severe shock, currents between 100 and 200 mA can be lethal.

With increasing alternating current, the sensations of tingling give way to contractions of the muscles. The muscular contractions and accompanying sensations of heat increase as the current is increased. Sensations of pain develop, and voluntary control of the muscles that lie in the current pathway becomes increasingly difficult. As current approaches 15 mA, the victim cannot let go of the conductive surface being grasped. At this point, the individual is said to "freeze" to the circuit. This is frequently referred to as the "let-go" threshold.

As current approaches 100 mA, ventricular fibrillation of the heart occurs. Ventricular fibrillation is defined as "very rapid uncoordinated contractions of the ventricles of the heart resulting in loss of synchronization between heartbeat and pulse beat." Once ventricular fibrillation occurs, it will continue and death will ensue within a few minutes. Use of a defibrillator is required to save the victim.

Heavy current flow can result in severe burns and heart paralysis. If shock is of short duration, the heart stops during current passage and usually restarts normally on current *interruption*, improving the victim's chances for survival.

Resistance

Studies have shown that the electrical resistance of the human body varies with the amount of moisture on the skin, the pressure applied to the contact point, and the contact area. The outer layer of skin, the epidermis, has very high resistance when dry. Wet conditions, a cut or other break in the skin will drastically reduce resistance.

Shock severity increases with an increase in pressure of contact. Also, the larger the contact area, the lower the resistance. Whatever protection is offered by skin resistance decreases rapidly with increase in voltage. Higher voltages have the capability of "breaking down" the outer layers of the skin, thereby reducing the resistance.

Path Through Body

The path the current takes through the body affects the degree of injury. A small current that passes from one extremity through the heart to the other extremity is capable of causing severe injury or electrocution. There have been many cases where an arm or leg was almost burned off when the extremity came in contact with electrical current and the current only flowed through a portion of the limb before it went out into the other conductor. Had the current gone through the trunk of the body, the person almost surely would have been electrocuted. A large number of serious electrical accidents in industry involve current flow from hands to feet. Because such a path involves both the heart and the lungs, the effects are often fatal.

Current flow through the chest, neck, head, or major nerve centers controlling respiration may result in a failure of the respiratory system. This is usually caused by a disruption of the nerve impulses between the respiratory control center and the respiratory muscles. Such a condition is dangerous since it is possible for the respiratory failure to continue even after the current flow has stopped.

The most dangerous condition can occur when fairly small amounts of current flow through the heart area. Such current flow can cause ventricular fibrillation. This asynchronous movement of the heart causes the heart's usual rhythmic pumping action to cease. Death occurs within minutes.

When relatively large currents flow through the heart area, heart action may be stopped entirely. If the shock duration is short and no physical damage to the heart has occurred, the heart may begin rhythmic pumping automatically when the current ceases. Extensive tissue damage, including internal organ damage due to high temperatures, occurs when very large currents flow through major portions of the body.

Duration of Shock

Current flow greater than the "let-go" threshold of an individual may cause a person to collapse, become unconscious, and even result in death. In most cases, the current flow would have to continue for more than five seconds. Although it may not be possible to determine the exact cause of death with certainty, asphyxiation or heart failure are the prime suspects.

The duration of the shock has a great bearing on the final outcome. If the shock is of short duration, it may only be a painful experience for the person. If the level of current flow reaches the approximate ventricular fibrillation threshold of 100 mA, shock duration of a few seconds could be fatal. This is not much current considering that a small light-duty portable electric drill draws about 30 times as much. At relatively high currents, death is inevitable if the shock is of appreciable duration; however, if the shock is of short duration, and if the heart has not been damaged, interruption of the current may be followed by a spontaneous resumption of its normal rhythmic contractions.

Delayed Effects

There are recorded cases of delayed death after a person has been revived following an electrical shock. This may occur within minutes, hours, or even days after the event has occurred. Several assumptions for such delayed effects are:

- Internal or unseen hemorrhaging;
- Emotional or psychological effects of the shock; and
- Aggravation of a pre-existing condition.

In many accidents, there is a combination of the above effects as well as conditions that develop after the initial accident.

Electrical Protective Devices

As a power source, electricity can create conditions almost certain to result in bodily harm, property damage, or both. It is important for workers to understand the hazards involved when they are working around electrical power tools, maintaining electrical equipment, or installing equipment for electrical operation.

Electrical protective devices, including fuses, circuit breakers, and ground-fault circuit interrupters, are therefore critically important to electrical safety. Overcurrent devices should be installed where required. They should be of the size and type to interrupt current flow when it exceeds the capacity of the conductor. Proper selection takes into account not only the capacity of the conductor, but also the rating of the power supply and potential short circuits.

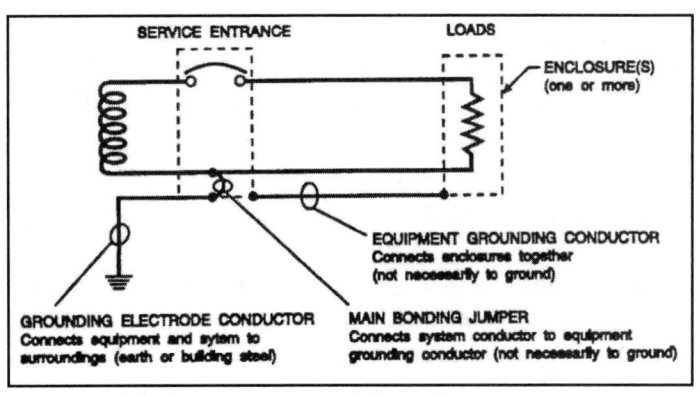

Exhibit 9-20.
The terminology and layout of an electrical system.

Types of Overcurrent

There are two types of overcurrent:
- **Overload** - When you ask a 10 hp motor to do the work of a 12 hp motor, an overload condition exists. The overcurrent may be 150 percent of normal current.
- **Fault** - When insulation fails in a circuit, a fault current can result that may be from 5 times to 50 times that of normal current.

When a circuit is overloaded, the plasticizers in the insulation are vaporized over a long period of time, and the insulation becomes brittle. The brittle insulation has slightly better electrical insulating properties. However, movement of the conductors due to magnetic or other forces can crack the insulation, and a fault can result. Conductors should be protected from overload and the eventual damage that results.

Faults occur in two ways. Most of the time a fault will occur between a conductor and an enclosure. This is called a ground fault. Infrequently, a fault will occur between two conductors. This is called a short circuit.

To predict what will happen in a normal circuit and a ground-fault circuit, it is important to understand the terminology of an electrical system. As illustrated in Exhibit 9-20, the dashed lines represent the enclosures surrounding the electrical system. These enclosures include the service panel, conduit, and boxes enclosing such parts as switches, controllers, and equipment terminals. The conduit bonds all of the enclosures together so there is no electrical potential between them. It also provides an emergency path for ground-fault current to return to the voltage source that, in this case, is shown as secondary windings of a transformer.

Notice that there must be a wire between the grounded conductor and the enclosure to allow the fault current to return to its source. This wire is called the main bonding jumper. If there is no wire, then the electrical system is isolated and requires extra safety features.

The basic idea of an overcurrent protective device is to make a weak link in the circuit. In the case of a fuse, the fuse is destroyed before another part of the system is destroyed. In the case of a circuit breaker, a set of contacts opens the circuit. Unlike a fuse, a circuit breaker can be reused by reclosing the contacts. Fuses and circuit breakers are designed to protect equipment and facilities, and in so doing, they also provide considerable protection against shock in most situations. However, the only electrical protective device whose sole purpose is to protect people is the ground-fault circuit-interrupter.

Fuses

A fuse is an electrical device that opens a circuit when the current flowing through it exceeds the rating of the fuse. The "heart" of a fuse is a special metal strip (or wire) designed to melt and blow out when its rated amperage is exceeded. Overcurrent devices (e.g., fuses, circuit breakers) are always placed in the "hot" side of a circuit (usually a black wire) and in series with the load, so that all the current in the circuit must flow through them.

If the current flowing in the circuit exceeds the rating of the fuse, the metal strip will melt and open the circuit so that no current can flow. A fuse cannot be reused and must be replaced after eliminating the cause of the overcurrent. Fuses are designed to protect equipment and conductors from excessive current. It is important to always replace fuses with the proper type and current rating. Too low a rating will result in unnecessary blowouts, while too high a rating may allow dangerously high currents to pass.

Circuit Breaker

Circuit breakers provide protection for equipment and conductors from excessive current without the inconvenience of changing fuses. Circuit breakers trip (open the circuit) when the current flow is excessive.

There are two primary types of circuit breakers based on the current-sensing mechanism. In a magnetic circuit breaker, a coil that forms an electromagnet senses the current. When the current is excessive, the electromagnet actuates a small armature that pulls the trip mechanism thus opening the circuit breaker. In a thermal-type circuit breaker, the current heats a bi-metallic strip that, when heated sufficiently, bends enough to allow the trip mechanism to operate.

Ground-Fault Circuit Interrupter

A ground-fault circuit interrupter (GFCI) is not an overcurrent device. A GFCI is used to open a circuit if the current flowing to the load does not return by the prescribed route. In a simple 120-volt circuit, the current usually flows through the black (ungrounded) wire to the load and returns to the source through the white (grounded) wire. If it does not return through the grounded wire, then it must go somewhere else, usually to ground. The GFCI is designed to limit electric shock to a current and time-duration level that will not produce serious injury.

The National Electric Code and the NFPA 70E

In 1881 Thomas Edison replied to an article written in the newspaper where the underwriters asked for information about the danger of fire from electric wires. Edison responded with "I beg to say that the system of lighting of the Edison Electric Light company is absolutely free from any possible danger from fire, even in connection with the most inflammable material."

Since electrical systems were a new development, no codes had been developed for safe installation. Installers had to make decisions based on their limited experience of this new technology. It is no wonder why electrical fires became prevalent throughout the United States. Eight years after Edison wrote his safety note about fires there were five different electrical installation codes in use in the United States. Each code was sufficiently different from the others that confusion reigned and the need for a uniform electrical code was recognized.

In 1896 a committee was formed by the American Society of Mechanical Engineers. This committee was given the task of selecting the most suitable rules from the five outstanding codes and developing a consensus through a review of their findings. The code was recognized by the National Board of Fire Underwriters and published as the "National Electrical Code of 1897."

In 1911 the work of periodically revising the National Electric Codes (NEC) was transferred to the Electrical Committee of the National Fire Protection Association (NFPA). This organization had been formed in 1896 to reduce fire and safety hazards. It has published over two hundred articles (codes and standards). These codes cover a variety of topics from NFPA 10 "Standard for Portable Fire Extinguishers" to Article 70 which is the National Electric Code.

When the Williams-Steiger Occupational Safety and Health Act of 1970 created OSHA, the need for industrial regulation of electricity became apparent. OSHA adopted the most widely accepted electrical standard in the world, the National Fire Protection Association's Standard, NFPA 70. However, when the

NEC was updated periodically, OSHA would have to go through an extensive legal process to adopt the new NEC edition and resolve conflicts between the adopted version and the updated version. The NEC only referred to electrical installation and OSHA needed a regulation that addressed operation, maintenance and repair as well.

How did OSHA solve this problem? Employers have a specific duty to follow the rules enacted by OSHA, and a general duty to provide a safe and healthy workplace. This "general duty" requires that each employer furnish to their workers a place that is free from recognized hazards. Generally, if there is a serious hazard that could result in death or serious injury, the hazard must be abated, regardless of whether or not OSHA has specifically addressed the hazard. An important example of recognized hazards that may not be covered by a specific standard drafted by OSHA is the national consensus standard. This type of standard is developed by the same persons it affects, is adopted by a nationally recognized organization and carries the same weight and is as enforceable as the OSHA regulations. NFPA 70E is a national consensus standard.

The most recent edition of the NFPA 70E was adopted in 2004. It reorganizes the article into the NEC format. There are four chapters and thirteen annexes to the article.

- Chapter 1 covers safety related work practices, qualified versus unqualified persons and training. This chapter calls for an electrical safety program that includes electrical hazard analysis for shock and flash, energized electrical work permits and lockout/tagout procedures. Personal protective equipment, protective clothing and approach boundaries are also discussed.
- The second chapter outlines safety related maintenance requirements necessary to ensure that electrical wiring, components and equipment are maintained in a safe condition.
- Chapter 3 discusses special equipment such as batteries, lasers, arc welding equipment, and other select equipment.
- Chapter 4 describes installation safety requirements and is a truncated version of the National Electrical Code.

While parts of the NFPA 70E have been around since 1981, only recently has OSHA begun to refer to the article in their documents and citations. This consensus standard covers electrical safety related work practices and procedures for employees who work on or near exposed energized electrical conductors or circuit parts. Relevant requirements include:

Power must be proven to be off before performing work. This includes
- The safe interruption of the load & opening of the disconnect
- Visual verification/voltage testing to ensure deenergization

The potential electrical hazard must be identified and documented
- Flash hazard analysis must be performed
- Flash protection boundaries must be determined

Appropriate steps must be taken to protect persons working near live parts or within the flash protection boundary.
- Personal Protective Equipment must be provided based on the relevant incident energy exposure levels (cal/cm2)
- Only properly qualified persons shall be allowed to perform work.

NFPA 70 (NEC) differs from NFPA 70E
The National Electrical Code is generally considered an electrical *installation* document and protects employees under normal circumstances. NFPA 70E is intended to provide guidance with respect to electrical *safe work practices*.

NFPA not incorporated by reference
OSHA does not enforce NFPA 70E since it is not incorporated by reference.

However, OSHA has several comparable requirements that are enforceable:
- **29 CFR 1910.132 (d)(1):** Requires employers perform a personal protective equipment (PPE) hazard assessment to determine necessary PPE;
- **29 CFR 1910.269 (l)(6)(iii):** Requires employers ensure each employee working at electric power generation, transmission, and distribution facilities who is exposed to the

hazards of flames or electric arcs does not wear clothing that could increase the extent of injury when exposed to such a hazard;

- **29 CFR 1910.335 (a)(1)(i):** Employees working in areas where there are potential electrical hazards shall use electrical protective equipment appropriate for the specific parts of the body for the work being performed;
- **29 CFR 1910.335 (a)(1)(iv):** Requires employees wear nonconductive head protection whenever exposed to electric shock or burns due to contact with exposed energized parts;
- **29 CFR 1910.335 (a)(1)(v):** Employees shall wear protective equipment for the eyes or face wherever there is danger of injury to the eyes or face from electric arcs or flashes or from flying objects resulting from an electrical explosion;
- **29 CFR 1910.335 (a)(2):** Employees shall use insulated tools or handling equipment when working near exposed energized conductors or circuit parts;
- **29 CFR 1926.28 (a):** Employer shall require employees wear appropriate personal protective equipment (PPE) during construction work.

Boundaries for flash protection

NFPA 70E specifies boundaries within which flash protection is required as a means to reduce the extent of injuries. Protective equipment is also specified including flash resistant clothing and face shielding. Boundary distances vary depending on both the qualifications of the person being exposed and the voltage involved. Unqualified personnel must be accompanied by qualified personnel. All personnel within the defined boundaries must wear specified protective equipment.

The following boundary distances for flash protection have been established in NFPA 70E. For workers within the following approach distances, flash protection is required:

Voltage Flash Protection Boundary

up to 750V	3 feet
750V to 2kV	4 feet
2kV to 15kV	16 feet
15kV to 36kV	19 feet
over 36kV	Must be Calculated

Flash resistant personal protection equipment does not protect you from shock, but it will give substantial protection from the effects of flash, especially burns and eye damage. Flash protective clothing is specified by ASTM F1506 while eye protection must comply with ANSI Z87.1. Clothing coverage must be 100%, i.e., coveralls or shirt and trousers.

NFPA 70E: 130.1 requires employers perform a flash hazard analysis to identify work tasks performed on energized electrical conductors. Appropriate flame retardant clothing (FRC), as well as voltage rated tools, may then be selected.

It is possible to protect an employee exposed to an electric arc incident from experiencing a burn that will cause irreversible tissue damage – a curable burn. This is a 2° burn where the skin temperature does not exceed 175° with a duration no longer than 0.1 second.

Although not specifically required by NFPA 70E, it is recommended that covered employees be provided FRC daily wear with an ATPV (Arc Thermal Performance Value) of at least 8. This satisfies the garment requirements for a Hazard Risk Classification 2 task.

The term "clearing time" is discussed under NFPA 70E: 130.3 (A). This refers to the time necessary for an electrical circuit breaker or disconnect to switch from an energized state to a deenergized state. A faster clearing time reduces the potential for an electric arc incident.

The Hazard Risk Category (HRC) classifications (0 – 4) listed in Table 130.7 (C)(9)(a) do not have a have a direct correlation to voltage. The hazard risk category (HRC) classifications are based upon estimated incident energy. Consider two work tasks with different voltages:

Task #1: Examination of insulated cable in open area ≥ 1,000 volts: **HRC 2 task**.

Task #2: Insertion or removal of 600 volt class individual starter buckets: **HRC 3 task.**

Although task #2 involves less voltage than task #1, it has a higher HRC.

Seventh edition changes

The National Fire Protection Association (NFPA) published the seventh edition of the

70E Standard for Electrical Safety Requirements for Employee Workplaces in February 2004. This edition contains many significant changes from the sixth edition: The edition stresses safe work practices.

The issue of multi-employer relationships (110.4) is addressed. The addition of multi-employer relationships now makes the employer liable whenever outside contractors are engaged in activities covered by the scope and application of this standard. The employer and contractor must inform each other of existing hazards, personal protective equipment/clothing requirements, safe work practices, and emergency evacuation procedures applicable to the work to be performed. This coordination must include a meeting and documentation.

NFPA 70E covers the full range of electrical safety issues, including safety related work practices, maintenance, special equipment requirements, and installation. It focuses on protecting people and identifies requirements that are considered necessary to provide a workplace that is free of electrical hazards. OSHA bases its electrical safety mandates, found in Subpart S part 1910 and Subpart K part 1926, on the comprehensive information found in NFPA 70E. NFPA 70E is recognized as the tool that illustrates how an employer might comply with these OSHA standards. The relationship between the OSHA regulations and NFPA 70E can be described as OSHA is the "shall" and NFPA 70E the "how."

OSHA mandates that all services to electrical equipment be done in a de-energized state. Working live can only be under special circumstances. If it is necessary to work live (>50 volts to ground), the regulations outlined in NFPA 70E, Article 130 should be used as a tool to comply with OSHA mandates Subpart S part 1910.333(a)(1).

- **Shock hazard analysis (paragraph 130.2):** Determines the voltage to which personnel will be exposed, boundary requirements, and PPE necessary. Table 130.2(C) is used to determine boundary distances.
- **Flash hazard analysis (paragraph 130.3):** Determines the flash protection boundary and PPE needed within that boundary. The flash protection boundary is determined by methods found in 130.3(A) or Annex D of the standard. Protective clothing is determined by using tables 130.7(C)(9)(a), 130.7(C)(10), and 130.7(C)(11). See question and answer number 1 for more details.

Remember, OSHA only allows work on live electrical parts under two special circumstances: (1) when continuity of service is required, and (2) when de-energizing equipment would create additional hazards. In all other cases, lockout/tagout is the law.

Employers are also responsible for complying with the 2002 NEC 110.6 labeling requirements. This requires all switchboards, panel boards, industrial control panels, and motor control centers to be field marked. Any equipment installed after 2002 needs to be labeled. For equipment installed before 2002, labeling must be applied if ANY modifications or upgrades take place. Some of the labels listed below require boundary distances calculated in Article 130 of the standard. Examples of labels that meet this requirement are:

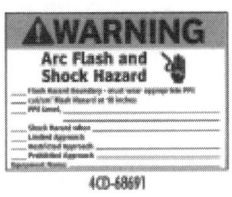

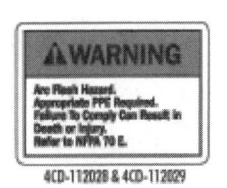

Each FR garment is assigned an ATPV rating by the manufacturer. The ATPV value represents the amount of incident energy that would cause the onset of second-degree burns. It also signifies the amount of protection the clothing affords when an electrical arc comes in contact with the fabric. Most of the industry falls into either Category 1 or 2 protection. Most uniforms already meet Category 1 or 2 requirements, but people who fall into this category typically are not covered by this standard. The employees addressed by this standard fall into Category 3 and 4. The garments must also be designed to withstand a cleaning process to remove soils and then be returned to service without damage to the protective characteristic of the fabric. The label on the garment must contain the following information: tracking ID number, meet ASTM spec F1506, name of manufacturer, size and care instructions, ATPV rating, and must meet ASTM spec f1506.

Sources for More Information
NFPA 70E Standard for Electrical Safety Requirements for Employee Workplaces, National Fire Protection Association, 1-617-770-3000.

Protective Clothing Characteristics

Hazard/Risk Category	Clothing Description Number of Layers ( )	APTV Rating Cal/cm2
0	Untreated Cotton, Wool, Rayon, Silk, or Blend. Fabric weight >4.5oz/Yd2 (1)	N/A
1	FR Shirt and FR Pants or FR Coverall (1)	4
2	Cotton underwear plus FR shirt and FR pants (1 or 2)	8
3	Cotton underwear plus FR shirt and FR pants plus FR coverall, cotton underwear plus two FR Coveralls (2 or 3)	25
4	Cotton underwear plus FR shirt and FR pants plus multilayer flash suit (3 or more)	40

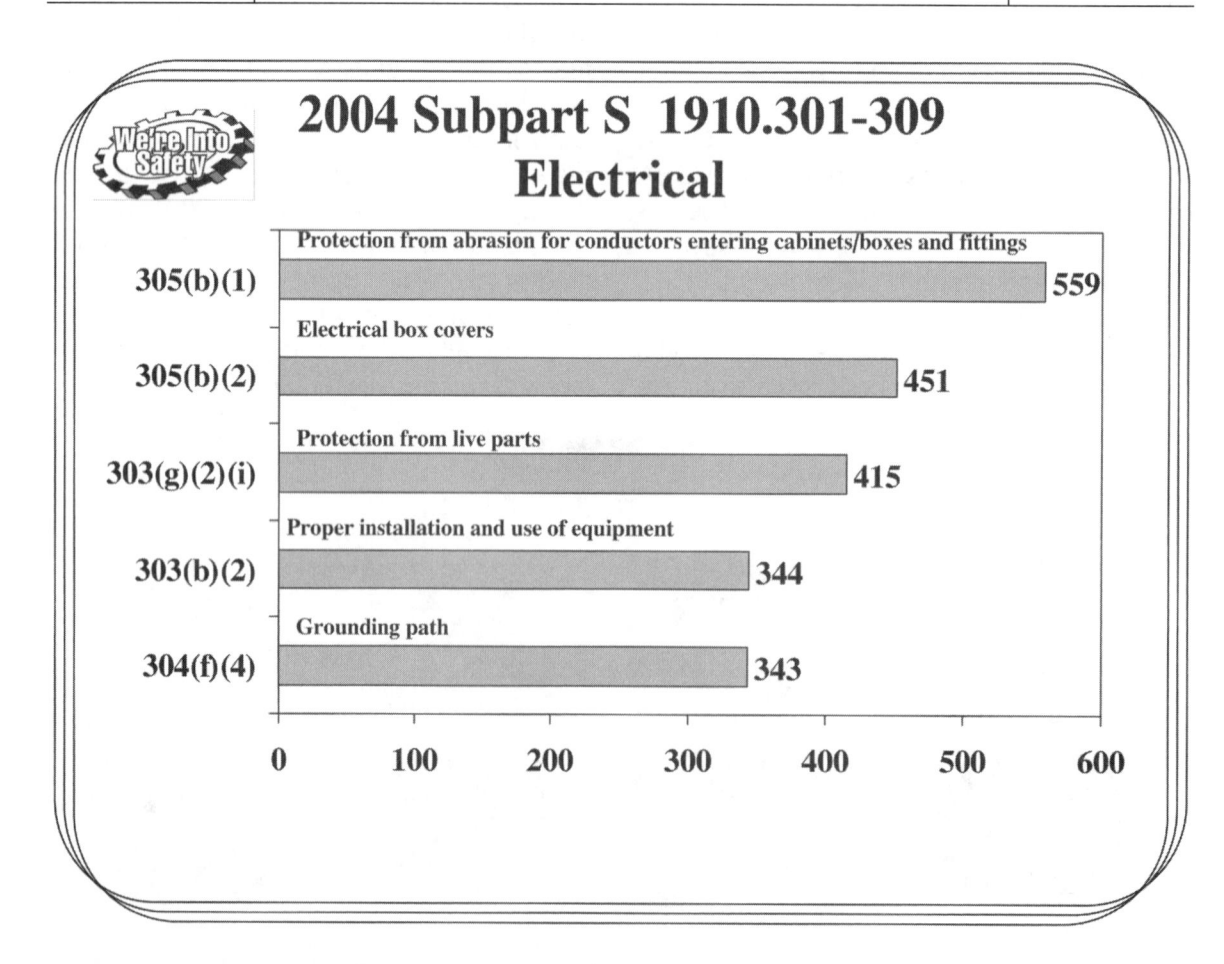

2004 Subpart S 1910.301-309 Electrical

- 305(b)(1) — Protection from abrasion for conductors entering cabinets/boxes and fittings: 559
- 305(b)(2) — Electrical box covers: 451
- 303(g)(2)(i) — Protection from live parts: 415
- 303(b)(2) — Proper installation and use of equipment: 344
- 304(f)(4) — Grounding path: 343

CHAPTER 10

HAZARDOUS MATERIALS

Flammable and Combustible Liquids—1910.106

The primary basis of this standard is the National Fire Protection Association's publication NFPA 30, *Flammable and Combustible Liquids Code*. This standard applies to the handling, storage and use of flammable and combustible liquids with a flash point below 200°F. There are two primary hazards associated with flammable and combustible liquids: explosion and fire. In order to prevent these hazards, this standard addresses the primary concerns of design and construction, ventilation, ignition sources and storage.

Definitions

These definitions were derived from consensus standards and were not uniquely developed for OSHA regulations. Some of the more important definitions are discussed below.

Aerosol shall mean a material that is dispensed from its container as a mist, spray or foam by a propellant under pressure.

Approved shall mean approved or listed by a nationally recognized testing laboratory.

Boiling point shall mean the boiling point of a liquid at a pressure of 14.7 pounds per square inch absolute (psia). This pressure is equivalent to 760 millimeters of mercury (760 mm Hg). At temperatures above the boiling point, the pressure of the atmosphere can no longer hold the liquid in the liquid state and bubbles begin to form. The lower the boiling point, the greater the vapor pressure at normal ambient temperatures and, consequently, the greater the fire risk.

Container shall mean any can, barrel or drum.

Closed container shall mean a container so sealed by means of a lid or other device from which neither liquid nor vapor will escape at ordinary temperatures.

Fire area shall mean an area of a building separated from the remainder of the building by construction, having a fire resistance of at least one hour and having all communicating openings properly protected by an assembly having a fire resistance rating of at least one hour.

Flash point means the minimum temperature at which a liquid gives off vapor within a test vessel in sufficient concentration to form an ignitable mixture with air near the surface of the liquid. The flash point is normally an indication of susceptibility to ignition.

The flash point is determined by heating the liquid in test equipment and measuring the temperature at which a flash will be obtained when a small flame is introduced in the vapor zone above the surface of the liquid.

A standard closed container is used to determine the closed-cup flash point and a standard open-surface dish for the open-cup flash point temperature, as specified by the American Society for Testing and Materials (ASTM). These methods are referenced in OSHA's 1910.106 standard.

Combustible liquid means any liquid having a flash point at or above 100°F (37.8°C). Combustible liquids shall be divided into two classes as follows:

- **Class II liquids** shall include those with flash points at or above 100°F (37.8°C) and below 140°F (60°C), except any mixture having components with flash points of 200°F (93.3°C) or higher, the volume of which makes up 99 percent or more of the total volume of the mixture (See Exhibit 10-1.).
- **Class III liquids** shall include those with flash points at or above 140°F (60°C) (See Exhibit 10-2.). Class II liquids are subdivided into two subclasses:
- **Class IIIA liquids** shall include those with flash points at or above 140°F (60°C) and below 200°F (93.3°C), except any mixture having components with flash points of 200°F (93.3°C), or higher, the total volume of which makes up 99 percent or more of the total volume of the mixture.
- **Class IIIB liquids** shall include those with flash points at or above 200°F (93.3°C). This section does not regulate Class IIIB liquids. Where the term "Class III liquids" is used in this section, it shall mean only Class IIIA liquids. When a combustible liquid is heated to within 30°F (16.7°C) of its flash point, it shall be handled in accordance with the requirements for the next lower class of liquids.

Flammable liquid means any liquid having a flash point below 100°F (37.8°C) or higher, the total of which makes up 99 percent or more of the total volume of the mixture. Flammable liquids shall be known as Class I liquids. Class I liquids are divided into three classes as follows:

- Class IA shall include liquids having flash points below 73°F (22.8°C) and having boiling points below 100°F (37.8°C) (See Exhibit 10-3.).
- Class IB shall include liquids having flash points below 73°F (22.8°C) and having boiling points at or above 100°F (37.8°C) (See Exhibit 10-4.).
- Class IC shall include liquids having flash points at or above 73°F (22.8°C) and below 100°F (37.8°C) (See Exhibit 10-5.).

It should be mentioned that flash point was selected as the basis for classification of flammable and combustible liquids because it is directly related to a liquid's ability to generate vapor (i.e., its volatility). Since it is the vapor of the liquid not the liquid itself that burns, vapor generation becomes the primary factor in determining the fire hazard. The expression "low flash-high hazard" applies. Liquids having flash points below ambient storage temperatures generally display a rapid rate of flame spread over the surface of the liquid because it is not necessary for the heat of the fire to expend its energy in heating the liquid to generate more vapor (See Exhibit 10-7).

Exhibit 10-1 further illustrates these classifications.

Exhibit 10-1.
Examples of Class II material.

Liquid		Flash Point (°F)	Boiling Point (°F)	Flammable Limits		Vapor Density Air =1	PEL (ppm)
Common Name	Other Names			LEL	UEL		
Isoamyl Alcohol		109	268	1.2	-	3.0	100
Cellosolve Acetate	2-Ethooxyethyl Acetate	117	313	1.7	-	4.7	100
Cyclohexanone		111	313	-	-	3.4	50
Fuel Oil #1 & #2		100+	-	-	-	-	-
Fuel Oil #4		110+	-	-	-	-	-
Fuel Oil #5		130+	-	-	-	-	-
Kerosene		110-150	180-300	0.7	5.0	4.5	-
Naphtha (coal tar)	100-110	300-400	-	-	4.3	100	
Naphtha (High Flash)	100° Naphtha Safety Solvent, Stoddard Solvent	100-110	300-400	0.8	6.0	>4.2	500
Methyl Cellosolve	2-Methoxyethanol	115	255	2.5	14.0	-	25

Exhibit 10-2.
Examples of Class III material.

Liquid		Flash Point (°F)	Boiling Point (°F)	Flammable Limits		Vapor Density Air = 1	PEL (ppm)
Common Name	Other Names			LEL	UEL		
Aniline		158	363	1.3	-	3.2	5
Butyl Cellosolve	2-Butoxyethanol	160	340	1.1	10.6	4.1	50
Cellosolve Solvent	2-Ethoxyethanol Cellosolve Solvent	202	275	1.8	14.0	3.1	200
Cyclohexanol		162	322	-	-	2.5	50
Ethylene Glycol	Glycol	232	387	3.2	-	-	-
Furfural		140	324	2.1	19.3	3.3	5
Glycerine	Glycerol	320	554	-	-	3.2	-
Isophorone		184	419	0.8	3.8	-	25
Nitrobenzene		190	412	-	-	4.3	1

Exhibit 10-3.
Examples of Class IA material.

Liquid		Flash Point (°F)	Boiling Point (°F)	Flammable Limits		Vapor Density Air = 1	PEL (ppm)
Common Name	Other Names			LEL	UEL		
1-1 Dichloroethylene	Vinylidene chloride	0	x99	7.3	10.0	3.4	-
Ethulamine		<0	63	3.5	14.0	1.6	10
Ethyl Chloride	Chloroethane	-58	54	3.8	15.4	2.2	1000
Ethyl Ether	Ether	-49	95	1.9	36.0	2.6	400
Isopentane		<-60	82	1.4	7.6	2.5	-
Isopropyl Chloride	2-Chloropropane	-26	97	2.8	10.7	2.7	-
Methyl Formate		-2	90	5.0	23.0	2.1	100
Pentane		<-40	97	1.5	7.8	2.5	1000
Propylene Oxide		-35	93	2.8	37.0	2.0	100

Exhibit 10-4.
Examples of Class IB material.

Liquid		Flash Point (°F)	Boiling Point (°F)	Flammable Limits		Vapor Density Air = 1	PEL (ppm)
Common Name	Other Names			LEL	UEL		
Acetone		0	134	2.6	12.8	2.0	1000
Benzene	Benzol	12	176	1.3	7.1	2.8	1
Carbon Disulfide	Carbon bisulfide	-22	115	1.3	50.0	2.6	20
1.2-Dichloroethylene	Acetylene dichloride	43	140	9.7	12.8	3.4	200
Ethyl Acetate		24	171	2.2	11.0	3.0	400
Ethyl Alcohol	Ethanol, Grain alcohol	55	173	3.3	19	1.6	1000
Ethyl Benzene		59	277	1.0	6.7	3.7	100
Gasoline	x	-45	100-399	1.4	7.6	3-4	-
Hexane	x	-7	156	1.1	7.5	3.0	500
Methyl Acetate		14	135	3.1	16	2.6	200
Methyl Alcohol	Wood alcohol, Methanol	52	147	6.7	36	1.1	200
Methyl Ethyl Ketone	MEK, 2-Butanone	21	176	1.8	10	2.5	200
Methyl Propyl Ketone	2-Pentanone	45	216	1.5	8.2	2.9	200
VM&P Naphtha	76° Naphtha	20-45	212-320	0.9	6.0	4.2	-
Octane		56	257	1.0	6.5	3.9	500
Propyl Acetate		58	215	2.0	8.0	3.5	200
Isopropyl Acetate		58	215	2.0	8.0	3.5	200
Isopropyl Alcohol	IPA, 2-Propanol	53	180	2.0	12	2.1	400
Toulene	Toluol	40	232	1.2	7.1	3.1	200
Butyl Acetate		72	260	1.7	7.6	4.0	150

Exhibit 10-5.
Examples of Class IC material.

Liquid		Flash Point (°F)	Boiling Point (°F)	Flammable Limits		Vapor Density Air = 1	PEL (ppm)
Common Name	Other Names			LEL	UEL		
Isoamyl Acetate	Banana Oil	77	288	1.0	7.5	4.5	100
Amyl Alcohol	Pentanol	91	281	1.2	10	3.0	-
Butyl Alcohol	Butanol	84	243	1.4	11.2	2.6	100
Methyl Isobutyl Ketone	MIBK, Hexone	73	246	1.4	7.5	3.5	100
Naphtha (Petroleum)	Mineral Spirits, Petroleum Ether	85-110	302-399	0.8	6.0	4.2	-
Propyl Alcohol	Propanol	77	208	2.1	13.5	2.1	200
Styrene (Monomer)	Vinyl Benzene	90	295	1.1	6.1	3.6	100
Turpentine		95	307-347	0.8	-	-	100
Xylene	Xylol	81-115	281-291	1.1	7.0	3.7	100

Exhibit 10-6.
Examples of nonflammable material.

Liquid		Boiling Point (°F)	PEL (ppm)
Common Name	Other Names		
Carbon Tetrachloride		171	10
Chloroform	Trichloromethane	142	50
Ethylene Dibromide	1,2-Dibromoethane	270	20
Methyl Chloroform	1,1,1-Trichloroethane	165	350
Methylene Chloride	Dichloromethane	104	500
Perchloroethylene	Tetrachloroethylene	248	100
Trichloroethylene	TCE, Trichlor	190	100

* Non-Flammable under normal conditions. Unstablizied trichloroethylene can decompose violently inpresence of fine aluminum powder.

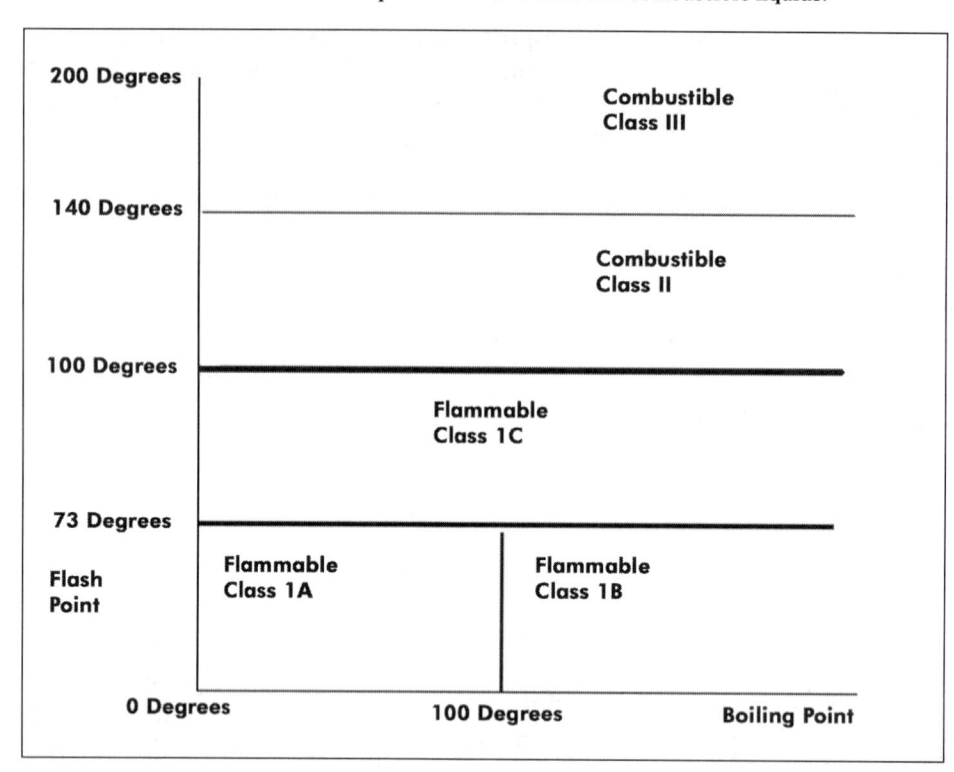

Exhibit 10-7.
Classifications and flashpoints for flammable and combustible liquids.

```
200 Degrees |                          Combustible
            |                          Class III
            |
140 Degrees |─────────────────────────────────────
            |
            |                          Combustible
            |                          Class II
            |
100 Degrees |═════════════════════════════════════
            |
            |                   Flammable
            |                   Class 1C
            |
 73 Degrees |─────────────────────────────────────
Flash       |  Flammable        |  Flammable
Point       |  Class 1A         |  Class 1B
            |                   |
              0 Degrees    100 Degrees    Boiling Point
```

Portable tank shall mean a closed container having a liquid capacity of more than 60 U.S. gallons and not intended for fixed installation.

Safety can shall mean an approved container of not more than 5-gallon capacity, having a spring-closing lid and spout cover and so designed that it will safely relieve internal pressure when subjected to fire exposure.

Vapor pressure shall mean the pressure, measured in pounds per square inch (absolute), exerted by a volatile liquid as determined by the American Society for Testing and Materials' Standard Method of Test for Vapor Pressure of Petroleum Products (Reid Method), ASTM D323-68.

Vapor pressure is a measure of a liquid's propensity to evaporate. The higher the vapor pressure, the more volatile the liquid and thus the more readily the liquid gives off vapors.

Ventilation as specified in this section is for the prevention of fire and explosion. It is considered adequate if it is sufficient to prevent accumulation of significant quantities of vapor-air mixtures in concentration over one-fourth of the lower flammable limit.

Flammable (Explosive) Limits

When vapors of a flammable or combustible liquid are mixed with air in the proper proportions in the presence of a source of ignition, rapid combustion or an explosion can occur. The proper proportion is called the *flammable range* and is also often referred to as the *explosive range*. The flammable range includes all concentrations of flammable vapor or gas in air, in which a flash will occur or a flame will travel if the mixture is ignited. There is a minimum concentration of vapor or gas in air below which propagation of flame does not occur on contact with a source of ignition. There is also a maximum proportion of vapor in air above which propagation of flame does not occur. These boundary-line mixtures of vapor with air are known as the *lower* and *upper flammable* or *explosive limits* (LEL or UEL) respectively, and they are usually expressed in terms of percentage by volume of vapor in air (See Exhibit 10-8.).

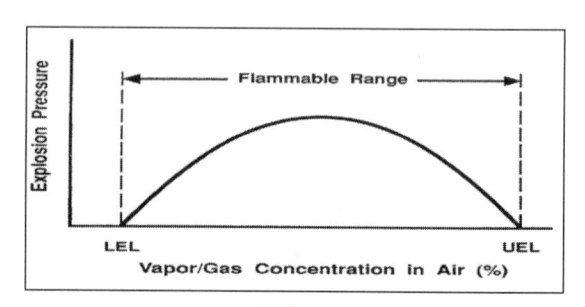

Exhibit 10-8.
When mixed with air, vapors and gases have minimum and maximum concentrations at which propagation of flame will not occur.

In popular jargon, a vapor-air mixture below the flammable limit is too "lean" to burn or explode, and a mixture above the upper flammable limit is too "rich" to burn or explode. No attempt is made to differentiate between the terms *flammable* and *explosive* as applied to the lower and upper limits of flammability.

Container and Portable Tank Storage

This section applies only to the storage of flammable or combustible liquids in drums or other containers (including flammable aerosols) not exceeding 60-gallon individual capacity and portable tanks of less than 660-gallon individual capacity. A portable tank is a closed container that has a liquid capacity of more than 60 gallons and is not intended for fixed installations.

This section does not apply to the following:
- Storage of containers in bulk plants, service stations, refineries, chemical plants and distilleries;
- Class I or Class II liquids in the fuel tanks of a motor vehicle, aircraft, boat, or portable or stationary engine;
- Flammable or combustible paints, oils, varnishes and similar mixtures used for painting or maintenance when not kept for a period in excess of 30 days; and
- Beverages when packed in individual containers not exceeding 1 gallon in size.

Design, Construction and Capacity of Containers

Only approved containers and portable tanks may be used to store flammable and combustible liquids. Metal containers and portable tanks meeting the requirements of the Department of Transportation (DOT) (49 CFR 178) are deemed acceptable when containing products authorized by the DOT (49 CFR 173).

The latest version of NFPA 30, Flammable and Combustible Liquids Code, indicates that certain petroleum products may be safely stored within plastic containers if the terms and conditions of the following specifications are met:
- Plastic Containers (Jerry Cans) for Petroleum Products, ANSI/ASTM D 3435-80;
- Standard for Portable Gasoline Containers for Consumer Use, ASTM F852-86;
- Standard for Portable Kerosene Containers for Consumer Use, ASTM F 976-86; and
- Nonmetallic Safety Cans for Petroleum Products, ANSI/UL 1313-83.

This standard also requires portable tanks to have provision for emergency venting. Top-mounted emergency vents must be capable of limiting internal pressure under fire exposure conditions to 10 psig or 30 percent of the bursting pressure of the tank, whichever is greater. Portable tanks are also required to have at least one pressure-activated vent with a minimum capacity of 6,000 cu. ft. of free air at 14.7 psia and 60°F. These vents must be set to open at not less than 5 psig. If fusible vents are used, they shall be actuated by elements that operate at a temperature not exceeding 300°F.

Maximum allowable sizes of various types of containers and portable tanks are specified based on the class of flammable and combustible liquid they contain.

Design, Construction and Capacity of Storage Cabinets

Not more than 60 gallons of Class I and/or Class II liquids, or not more than 120 gallons of Class III liquids may be stored in an individual cabinet.

This standard permits both metal and wooden storage cabinets. Storage cabinets shall be designed and constructed to limit the internal temperature to not more than 325°F when subjected to a standardized 10-minute fire test. All joints and seams shall remain tight and the door shall remain securely closed during the fire test. Storage cabinets shall be conspicuously labeled "Flammable—Keep Fire Away."

The bottom, top, door and sides of metal cabinets shall be at least No. 18 gage sheet metal and double walled with ½-in. air space. The door shall be provided with a three-point lock, and the doorsill shall be raised at least 2 in. above the bottom of the cabinet.

Design and Construction of Inside Storage Rooms

Construction is to comply with the test specifications included in Standard Methods of Fire Tests of Building Construction and Materials, NFPA 2511969.

Openings to other rooms or buildings shall be provided with non-combustible liquid-tight raised sills or ramps at least 4 inches in height, or the floor in the storage area shall be at least 4 in. below the surrounding floor. Openings shall be provided with approved self-closing fire doors. The room shall be liquid-tight where the walls join the floor. A permissible alternate to the sill or ramp is an open-grated trench inside of the room that drains to a safe location. This method may be preferred if there is an extensive need to transfer flammable liquids into and out of the room by means of hand trucks.

Rating and capacity. Storage in inside storage rooms shall comply with Table H-13 (See Exhibit 10-9.):

Exhibit 10-9.
Table H-13. Storage in Inside Rooms

Fire protection provided	Fire resistance	Maximum size	Total allowable quantities (gals./sq. ft./floor area)
Yes	2 hours	500 sq. ft.	10
No	2 hours	500 sq. ft.	5
Yes	1 hour	150 sq. ft.	4
No	1 hour	150 sq. ft.	2

Wiring. Electrical wiring and equipment located in inside storage rooms used for Class I liquids shall be approved under Subpart S, Electrical, for Class I, Division 2 Hazardous Locations; for Class III and Class III liquids, shall be approved for general use.

Ventilation. Every inside storage room shall be provided with either a gravity or a mechanical exhaust ventilation system designed to provide for a complete change of air within the room at least six times per hour. Ventilation is vital to the prevention of flammable liquid fires and explosions. It is important to ensure that airflow through the system is constant and prevents the accumulation of any flammable vapors.

Storage. In every inside storage room, there shall be maintained an aisle at least 3 ft. wide. Easy movement within the room is necessary in order to reduce the potential for spilling or damaging the containers and to provide both access for fire fighting and a ready escape path for occupants of the room should a fire occur.

Containers with more than a 30-gallon capacity shall not be stacked one upon the other. Such containers are built to DOT specifications and are not required to withstand a drop test greater than 3 ft. when full. Dispensing shall be only by approved pump or self-closing faucet.

Storage Inside Building

Egress. Flammable or combustible liquids, including stock for sale, shall not be stored so as to limit use of exits, stairways or areas normally used for the safe egress of people.

Office occupancies. Storage shall be prohibited except that which is required for maintenance and operation of equipment. Such storage shall be kept in closed metal containers stored in a storage cabinet or in safety cans or in an inside storage room not having a door that opens into that portion of the building used by the public.

General purpose public warehouses. There are tables in the standard summarizing the storage requirements applicable to "General Purpose Public Warehouses." These tables refer to indoor storage of flammable and combustible liquids that are confined in containers and portable tanks. Storage of incompatible materials that create a fire exposure (e.g., oxidizers, water-reactive chemicals, certain acids) is not permitted.

Warehouses or storage buildings. The last type of inside storage covered by this paragraph addresses storage in "warehouses or storage buildings." These structures are sometimes referred to as outside storage rooms. Any quantity of flammable and combustible liquid can be stored in these buildings, provided that they are stored in a configuration consistent with the tables in this paragraph.

Pallets or dunnage shall separate containers in piles, where necessary, to provide stability and to prevent excessive stress on container walls.

Stored material shall not be piled within 3 ft. of beams or girders and shall be at least 3 ft. below sprinkler deflectors, discharge orifices of water spray or other fire protection equipment.

Aisles of at least 3 ft. in width shall be maintained to access doors, windows or standpipe connections.

Storage Outside Building

Requirements covering "storage outside buildings" are summarized in tables in this paragraph. Associated requirements are given for storage adjacent to buildings. Also included are requirements involving controls for diversion of spills away from buildings and security measures for protection against trespassing and tampering. Certain housekeeping requirements are given that relate to control of weeds, debris and accumulation of unnecessary combustibles.

Fire control

Suitable fire control devices, such as small hose or portable fire extinguishers, shall be available at locations where flammable or combustible liquids are stored. At least one portable fire extinguisher having a rating of not less than 12-B units shall be located:

- Outside of, but not more than 10 ft. from, the door opening into any room used for storage; and
- Not less than 10 ft., nor more than 25 ft., from any Class I or Class II liquid storage area located outside of a storage room, but inside a building.

The reason for requiring that portable fire extinguishers be located a distance away from the storage room is that fires involving Class I and Class II flammable liquids are likely to escalate rapidly. If the fire is too close to the storage area, it may be impossible to get to it once the fire has started. Open flames and smoking shall not be permitted in flammable or combustible liquid storage areas.

Materials that react with water shall not be stored in the same room with flammable or combustible liquids. Automatic sprinkler or water spray systems and hose lines protect many flammable and combustible liquid storage areas. Consequently, any storage of water-reactive material in the storage area creates an unreasonable risk.

Industrial Plants

This paragraph shall not apply to chemical plants, refineries or distilleries.

This paragraph applies to those industrial plants where:

- The use of flammable or combustible liquids is incidental to the principal business; or
- Where flammable or combustible liquids are handled or used only in unit physical operations such as mixing, drying, evaporating, filtering, distillation and similar operations that do not involve chemical reaction.

Incidental Storage or Use of Flammable or Combustible Liquids

This subparagraph is applicable to those portions of an industrial plant where the use and handling of flammable or combustible liquids is only incidental to the principal business, such as paint thinner storage in an automobile assembly plant, solvents used in the construction of electronic equipment and flammable finishing materials used in furniture manufacturing.

Containers. Flammable or combustible liquids shall be stored in tanks or closed containers.

The quantity of liquid that may be located outside of an inside storage room or storage cabinet in a building or in any one fire area of a building shall not exceed:

- 25 gallons of Class IA liquids in containers;
- 120 gallons of Class IB, IC, II, or III liquids in containers; or
- 660 gallons of Class IB, IC, II, or III liquids in a single portable tank.

Handling liquids at point of final use. Flammable liquids shall be kept in covered containers when not actually in use. Where flammable or combustible liquids are used or handled, except in closed containers, means shall be provided to dispose promptly and safely of leakage or spills. Flammable or combustible liquids shall be drawn from or transferred into vessels, containers or portable tanks within a building only in the following manner:

- Through a closed piping system;
- From safety cans;
- By means of a device drawing through the top; or
- From containers or portable tanks by gravity through an approved self-closing valve.

Transfer operations must be provided with adequate ventilation. Sources of ignition are not permitted in areas where flammable vapors may travel. Transferring liquids by means of air pressure on the container or portable tanks is prohibited. This may result in an overpressure that could exceed what the container or tank could withstand. In addition, a flammable atmosphere could be created within the container or tank. This atmosphere would be particularly sensitive to ignition because of the increased pressure.

HAZARDOUS MATERIALS

Unit Physical Operations

This subparagraph applies to those portions of industrial plants where flammable or combustible liquids are handled or used in unit physical operations such as mixing, drying, evaporating, filtering, distillation and similar operations that do not involve chemical change. Examples are plants compounding cosmetics, pharmaceuticals, solvents, cleaning fluids, insecticides and similar activities.

Location. Industrial plants shall be located so that each building or unit of equipment is accessible from at least one side for firefighting and fire control purposes.

Drainage. Emergency drainage systems shall be provided to direct flammable or combustible liquid leakage and fire protection water to a safe location.

Ventilation. The standard requires that adequate ventilation be provided in operating areas. Appropriate measures must be taken to trap and remove hazardous vapors.

Tank Vehicle and Tank Car Loading and Unloading

Tank vehicle and tank car loading or unloading facilities shall be separated from above-ground tanks, warehouses and similar facilities by a distance of 25 ft. for Class I liquids and 15 ft. for Class II and Class III liquids measured from the nearest position of any fill stem.

Fire Control

These requirements basically state that hazards shall be evaluated and appropriate fire protection provided. Such an evaluation must consider the hazards of the operation, the various materials used, the design of the plant and equipment, materials handling and transfer requirements, any unusual conditions and the available fire protection sprinkler systems and other types of protective systems that may be necessary to protect employees.

Sources of Ignition

Adequate precautions shall be taken to prevent the ignition of flammable vapors. Sources of ignition include, but are not limited to:
- Open flames;
- Lightning;
- Smoking; cutting and welding;
- Hot surfaces;
- Frictional heat;
- Static, electrical and mechanical sparks;
- Spontaneous ignition, including heat-producing chemical reactions; and
- Radiant heat.

These are some examples of common ignition sources, although the list is neither all-inclusive nor applicable in all cases. Again, it is emphasized that control of ignition sources is the second line of defense; minimizing the possibility of a spill or leak is the primary objective of this regulation.

Bonding. Class I liquids shall not be dispensed into containers unless the nozzle and container are electrically interconnected.

Electrical

All electrical wiring and equipment shall be installed according to the requirements of Subpart S, Electrical.

Locations where flammable vapor-air mixtures may exist under normal operations shall be classified Class I, Division 1, according to the requirements of Subpart S. For those pieces of equipment installed in accordance with the above paragraphs on unit physical operations, the Division 1 area shall extend 5 ft. in all directions from all points of vapor liberation. Unventilated pits within any Class I area are classified as Division 1 locations.

Locations where flammable vapor-air mixtures may exist under abnormal conditions and for a distance beyond Division 1 locations shall be classified Division 2 according to the requirements of Subpart S, Electrical. These locations include an area within 20 ft. horizontally, 3 ft. vertically beyond a Division 1 area, and up to 3 ft. above floor or grade level within 25 ft., if indoors, or 10 ft. if outdoors, from any pump, bleeder, withdrawal fitting, meter or similar device handling Class I liquids. Adequately ventilated pits are Class I, Division 2 locations.

Repairs to Equipment

Hot work, such as welding or cutting operations, use of spark-producing power tools and chipping operations shall be permitted only under supervision of an individual in responsible charge.

Housekeeping

Maintenance and operating practices shall follow established procedures that control leakage and prevent accidental escape of flammable or combustible liquids. Spills shall be cleaned up promptly. Combustible waste material and residues in a building or unit operating area shall be kept to a minimum, stored in covered metal receptacles and disposed of daily.

Bulk Plants

Storage

Class I liquids shall be stored in closed containers, in storage tanks above ground outside of buildings or underground in accordance with the requirements of this section.

Class II and III liquids shall be stored in containers, in tanks within buildings, above ground outside of buildings or underground in accordance with the requirements of this section.

Building

Rooms in which flammable or combustible liquids are stored or handled by pumps shall have exit facilities arranged to prevent occupants from being trapped in the event of fire.

Rooms in which Class I liquids are stored or handled shall be heated only by means not constituting a source of ignition, such as steam or hot water. Adequate ventilation shall be provided for all rooms, buildings or enclosures in which Class I liquids are pumped or dispensed.

Loading and Unloading Facilities

Tank vehicle and tank car loading or unloading facilities shall be separated from aboveground tanks, warehouses and similar facilities a distance of 25 ft. for Class I liquids and 15 ft. for Class II and Class III liquids measured from the nearest position of any fill spout. (Buildings for pumps or shelters for personnel may be considered a part of the loading and unloading facilities).

Equipment such as piping, pumps and meters used for the transfer of Class I liquids between storage tanks and the fill stem of the loading rack shall not be used for the transfer of Class II or Class III liquids.

Static Protection. Bonding facilities for protection against static sparks during the loading of tank vehicles through open domes shall be provided:

- Where Class I liquids are loaded; or
- Where Class II or Class III liquids are loaded into vehicles that may contain vapors from previous cargoes of Class I liquids.

The standard requires appropriate bonding equipment and procedures. Facilities for materials that do not have a static electricity hazard are not required to be bonded.

Container Filling Facilities. Class I liquids shall not be dispensed into containers unless the nozzle and container are electrically interconnected.

Electrical Equipment

This subparagraph applies to areas where Class I liquids are stored or handled. For areas where only Class II or Class III liquids are stored or handled, the electrical equipment may be installed in accordance with requirements for ordinary (i.e., nonhazardous) locations.

Sources of Ignition

Class I liquids shall not be handled, drawn or dispensed where flammable vapors may reach a source of ignition. Smoking shall be prohibited except in designated locations. "No Smoking" signs shall be conspicuously posted where a hazard from flammable liquid vapors is normally present.

Drainage and Waste Disposal

Provision shall be made to prevent flammable or combustible liquids that may be spilled at loading or unloading points from entering public sewers, drainage systems or natural waterways. Connection to such sewers, drains or waterways by which flammable or combustible liquids might enter shall be provided with separator boxes or other approved means whereby such entry is precluded. Crankcase drainings and flammable or combustible liquids shall not be dumped into sewers, but shall be stored in tanks or tight drums outside of any building until removed from the premises.

Fire Control

Suitable fire-control devices, such as a small hose or portable fire extinguishers, shall be available in locations where fires are likely to occur.

Service Stations

Liquids shall be stored in approved closed containers not exceeding 60 gallons in capacity, in underground tanks or in tanks in special enclosures or above ground as provided for in this section.

No Class I liquids may be dispensed into portable containers unless the metal container has a tight closure with screwed or spring cover and is fitted with a spout or so designed that the contents can be poured without spilling.

A clearly identified and easily accessible switch(es) or circuit breaker(s) shall be provided at a location remote from dispensing devices, including remote pumping systems, to shut off the power to all dispensing devices in the event of an emergency.

Processing Plants

This paragraph applies to those plants or buildings that house chemical operations such as oxidation, reduction, halogenation, hydrogenation, alkylation, polymerization and other chemical processes, but does not apply to chemical plants, refineries or distilleries.

Processing Building

This section requires that appropriate facilities be provided for flammable and combustible liquid processing within buildings. Buildings must be safely constructed with appropriate drainage, ventilation and explosion relief. Emergency drainage systems shall be provided to direct flammable or combustible liquid leakage and fire protection water to a safe location. If connected to public sewers or discharged into public waterways, these systems shall be equipped with traps or separators.

Liquid Handling

The storage of flammable or combustible liquids in tanks shall be in accordance with the provisions of this section. Piping must be identified and meet safety requirements.

The transfer of large quantities of flammable or combustible liquids shall be through piping by means of pumps or water displacement. Except as required in process equipment, gravity flow shall not be used. The use of compressed air as a transferring medium is prohibited. Equipment must be designed to ensure containment. Where the vapor space is usually within the flammable range or other operational hazards indicate a need, equipment must be protected against explosion by construction or other appropriate measures.

Fire Control

Fire control provisions, including portable extinguishers, water supply, fixed extinguishing systems and alarm systems shall be provided. An analysis similar to that required for plants having unit physical operations must be performed. Appropriate fire control facilities must be provided as indicated by the special hazards of the plant.

Sources of Ignition

As in other paragraphs of this section, sources of ignition shall be prevented from igniting flammable vapors.

Waste and Residues

Combustible waste material and residues in a building or operating area shall be kept to a minimum, stored in closed metal waste cans and disposed of daily.

Refineries, Chemical Plants and Distilleries

Plants must be protected from catastrophic fire, explosion and/or release of toxic materials. Refer to 29 CFR 1910.119, *Process Safety Management of Highly Hazardous Chemicals*, which requires programmatic controls regarding these hazards. Recognized safe-practice documents from the affected industries identify necessary controls.

Storage Tanks

Flammable and combustible liquids shall be stored in tanks, containers or portable tanks in accordance with the requirements of this section.

Wharves

Wharves handling flammable or combustible liquids shall be in accordance with the requirements of this section.

Location of Process Units

Process units shall be located so that they are accessible from at least one side for the purpose of fire control.

Fire Control

Fire control provisions, including portable fire extinguishers, water supply and fixed extinguishing systems, shall be provided.

Spray Finishing Using Flammable and Combustible Materials—1910.107

This regulation is based on the National Fire Protection Association's standard NFPA 33, Spray Finishing Using Flammable and Combustible Materials, 1969 edition. Many technological changes have occurred in the field of spray finishing since this edition was published. The current edition of NFPA 33 is called Spray Application Using Flammable and Combustible Materials to recognize the hazards inherent to operations such as spray-up molding (e.g., fiberglass).

This section applies to flammable and combustible finishing materials when applied as a spray by compressed air, "airless" or "hydraulic atomization," steam, electrostatic methods or by any other means in continuous or intermittent processes. This section also covers the application of combustible powders by powder spray guns, electrostatic powder spray guns, fluidized beds or electrostatic fluidized beds. This section does not apply to outdoor spray application of buildings, tanks or other similar structures, nor to small portable spraying apparatus not used repeatedly in the same location. In these situations, there would be lesser chance of combustible residue buildup and greater chance of atmospheric dilution of flammable vapors.

Authority having jurisdiction. As defined in NFPA 33, the organization, office or individual responsible for "approving" equipment, an installation or a procedure. Where public safety is primary, the "authority having jurisdiction" may be a federal, state, local or other regional department or individual such as a fire chief, fire marshal, chief of a fire prevention bureau, labor department, health department, building official, electrical inspector or others having statutory authority.

Dry spray booth. A spray booth not equipped with a water-washing system. A dry spray booth may be equipped with (1) distribution or baffle plates to promote an even flow of air through the booth or cause deposit of over spray before it enters exhaust ducts; or (2) over spray dry filters to minimize dust or residue entering exhaust ducts; or (3) overspray dry filter rolls designed to minimize dust or residue entering exhaust ducts; or (4) where dry powders are being sprayed, with powder collection systems so arranged in the exhaust to capture oversprayed material (See Exhibit 10-10.).

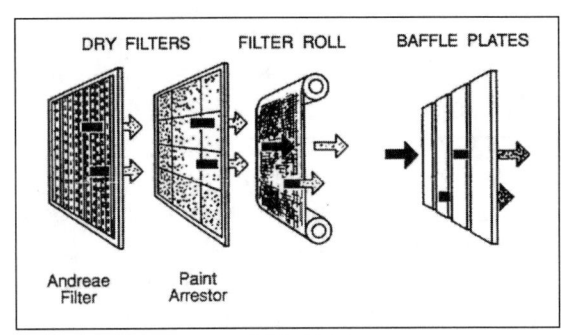

**Exhibit 10-10.
Types of filters for dry spray booths.**

Spray Area. Any area in which dangerous quantities of flammable vapors or mists or combustible residues, dusts or deposits are present due to the operation of spraying processes. The current edition of NFPA 33 offers greater insight into the meaning of the spray area.

According to OSHA, a spray area shall include:
- The interior of spray booths or rooms;
- The interior of ducts exhausting from spraying processes; and
- Any area in the direct path of spraying operations.

The "authority having jurisdiction" may define the limits of the spray area in any specific case. When spraying operations are strictly confined to predetermined spaces that are provided with adequate and reliable ventilation, such as a properly constructed spray booth, the "spray area" will ordinarily not extend beyond the booth enclosure. When the spray operations are not confined to adequately ventilated spaces, the spray area may extend to the entire room that contains spraying operations.

The implications of the above definition are as follows. A spray area is, by definition, a Class I, Division 1 hazardous location. As discussed later in this section, electrical devices and wiring within the spray area must comply with Subpart S, Electrical. The authority having jurisdiction may define the entire room to be a spray area if operations are not confined. Installation of expensive electrical devices and wiring would then be required throughout the entire room.

It is therefore in the employer's interest to conduct spray operations in a spray booth or room in order to confine the vapors, mist and residue and limit the area requiring the use of Class I, Division 1 electrical equipment. Confining spray operations to a spray booth or room also increases the safety of the operation, facilitates maintenance and clean up and provides a healthier working environment. The employer also may then be eligible fir preferred insurance rates.

Spray Room. A power-ventilated fully enclosed room used exclusively for the open spraying of flammable or combustible materials. The entire spray room is a spray area. A spray booth is not a spray room.

Spray Booths

Construction
Spray booths shall be substantially constructed of securely and rigidly supported steel, concrete or masonry; however, aluminum or other substantial non-combustible material may be used for intermittent or low-volume spraying. Spray booths shall be designed to sweep air currents toward the exhaust outlet.

Interiors
The interior surfaces of spray booths shall be smooth and continuous, without edges and otherwise designed to prevent the pocketing of residue and facilitate cleaning and washing without injury (See Exhibit 10-11.).

**Exhibit 10-11.
Anatomy of a spray booth.**

FRONTAL AREA
Each spray booth having a frontal area larger than 9 square feet shall have a metal deflector or curtain not less than 2 1/2 inches deep installed at the outer edge of the booth over the opening.

CONVEYORS
Where conveyors are arranged to carry work into and out of spray booths, the conveyor openings shall be as small as practical.

Space within the spray booth on the downstream and upstream sides of filters shall be protected with an automatic sprinkler, dry chemical, or carbon dioxide extinguishing system.

Visible gauges or audible alarm or pressure-activated devices shall be installed to indicate or insure that the required air velocity is maintained.

ACCESS DOORS
When necessary to facilitate cleaning, exhaust ducts shall be provided with an ample number of access doors.

ARRESTOR BANK
All discarded filter pads and filter rolls shall be immediately removed to a safe, well-detached location or placed in a water-filled metal container or disposed of at the close of the day's operation unless maintained completely in water.

The average air velocity over the open face of the booth (or booth cross section during spraying operations) shall be not less than 100 fpm; for electrostatic spraying, the minimum velocity is 60 fpm.

MINIMUM SEPARATION
There shall be no open flame or spark producing equipment in any spraying area nor within 20 feet thereof, unless separated by a partition.

CONSTRUCTION
Spray booths shall be substantially constructed of steel, securely and rigidly supported, or of concrete or masonry except that aluminum or other substantial noncombustible material may be used for intermittent or low volume spraying.

INTERIORS
The interior surfaces of spray booths shall be smooth and continuous without edges and otherwise designed to prevent pocketing of residues and facilitate cleaning and washing without injury.

DISCHARGE CLEARANCE
Unless the spray booth exhaust duct terminal is from a water-wash spray booth, the terminal discharge point shall be not less than 6 feet from any combustible exterior wall or roof nor discharge in the direction of any combustible construction or unprotected opening in any noncombustible exterior wall within 25 feet.

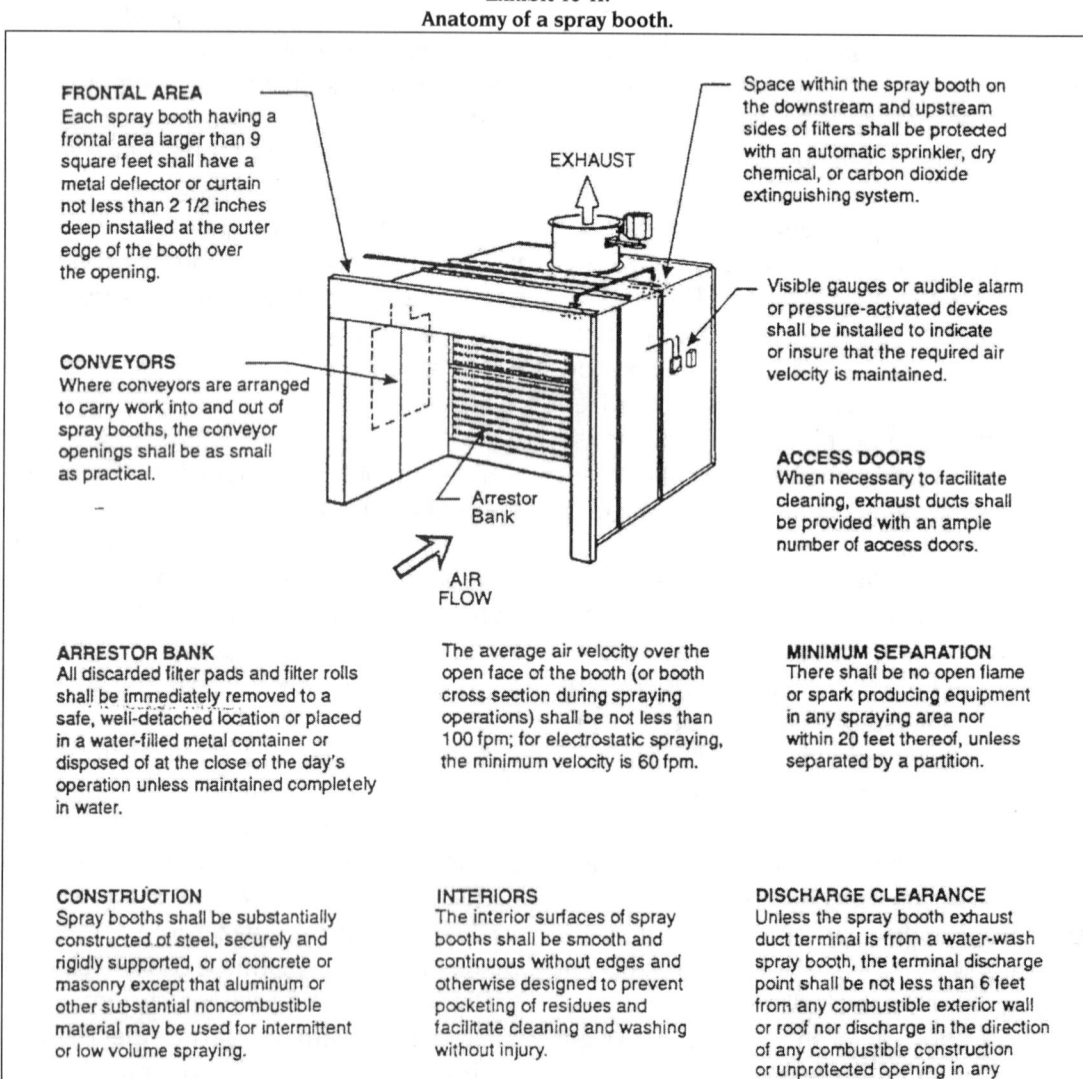

Floors

The floor surface of a spray booth and operator's working area, if combustible, shall be covered with non-combustible material of such character as to facilitate the safe cleaning and removal of residue.

Distribution or Baffle Plates

Distribution or baffle plates, if installed to promote an even flow of air through the booth or cause the deposit of overspray before it enters the exhaust duct, shall be of non-combustible material and readily removable or accessible on both sides for cleaning. Such plates shall not be located in exhaust ducts.

Dry-Type Overspray Collectors

In conventional dry-type spray booths, overspray dry filters or filter rolls, if installed, shall conform to the following:

- The spraying operations except electrostatic spraying operations shall be so designed, installed and maintained that the average air velocity over the open face of the booth (or booth cross section during spraying operations) shall be not less than 100 linear feet per minute (fpm). Electrostatic spraying operations may be conducted with an air velocity over the open face of the booth of not less than 60 fpm.

NOTE: These requirements were taken from NFPA 33-1969. They pertain to those hazards associated with fire protection or the removal of flammable vapor accumulation. These requirements apply to maintaining the concentration of flammable vapors below the lower explosive limit in a spray booth but do not apply to maintaining operator exposure to within the permissible exposure limits. Where a health hazard has been established, controls and modifications may be required and could include increasing the air velocity beyond that stated here.

- Visible gauges or audible alarms or pressure-activated devices shall be installed to indicate or ensure that the required air velocity is maintained. Filter rolls shall be inspected to insure proper replacement of filter media.
- All discarded filter pads and filter rolls shall be immediately removed to a safe, well-detached location or placed in a water-filled metal container and disposed of at the close of the day's operation, unless maintained completely in water.
- Space within the spray booth on the downstream and upstream sides of filters shall be protected with an automatic sprinkler, dry chemical or carbon dioxide extinguishing system.
- Filters or filter rolls shall not be used when applying a spray material known to be highly susceptible to spontaneous heating and ignition.
- Clean filters or filter rolls shall be non-combustible or a type having a combustibility not in excess of class 2 filters as listed by Underwriter's Laboratories, Inc.
- Filters and filter rolls shall not be alternately used for different types of coating materials, where the combination of materials may be conducive to spontaneous ignition.

Frontal Area

Each spray booth having a frontal area larger than 9 sq. ft. shall have a metal deflector or curtain not less than 2½ in. deep installed at the upper outer edge of the booth over the opening.

Conveyors

Where conveyors are arranged to carry work into and out of spray booths, the conveyor openings shall be as small as practical.

Separation of Operations

Each spray booth shall be separated from other operations by not less than 3 ft., or by a greater distance, or by such partition or wall as to reduce the danger from juxtaposition of hazardous operations (See Exhibit 10-12.).

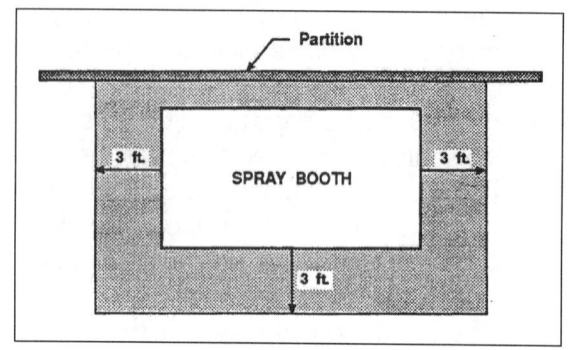

Exhibit 10-12.
Spray booths must be separated from other operations by at least 3 ft. on all sides.

Cleaning
Spray booths shall be so installed that all portions are readily accessible for cleaning. A clear space of not less than 3 ft. on all sides shall be kept free from storage or combustible construction.

Illumination
When spraying areas are illuminated through glass panels or other transparent materials, only fixed lighting units shall be used as a source of illumination. Panels shall effectively isolate the spraying area from the area in which the lighting unit is located and shall be of a noncombustible material of such a nature or so protected that breakage will be unlikely. Panels shall be so arranged that normal accumulation of residue on the exposed surface of the panel will not be raised to a dangerous temperature by radiation or conduction from the source of illumination.

Electrical and Other Sources of Ignition

Minimum Separation
There shall be no open flame or spark-producing equipment in any spraying area nor within 20 ft. thereof, unless separated by a partition.

Hot Surfaces
Space-heating appliances, steam pipes or hot surfaces shall not be located in a spraying area where deposits of combustible residues may readily accumulate.

Wiring Conformance
Electrical wiring and equipment shall conform to the provisions of this section and shall otherwise be in accordance with Subpart S, Electrical.

Combustible Residue, Areas
Unless specifically approved for locations containing both deposits of readily ignitable residue and explosive vapors, there shall be no electrical equipment in any spraying area, whereon deposits of combustible residue may readily accumulate, except wiring in rigid conduit or in boxes or fittings containing no taps, splices or terminal connections.

Wiring Type Approved
Electrical wiring and equipment not subject to deposits of combustible residue but located in a spraying area as herein defined shall be of explosion-proof type approved for Class I, Group D locations and shall otherwise conform to the provisions of Subpart S, Electrical, for Class I, Division 1 hazardous locations. Electrical wiring, motors and other equipment outside of but within 20 ft. of any spraying area, and not separated there from by partitions, shall not produce sparks under normal operating conditions and shall otherwise conform to the provisions of Subpart S, Electrical, for Class I, Division 2 hazardous locations.

The current edition of the NEC is more specific in defining the extent of Class I or Class II, Division 1 and Division 2 locations for spraying operations using flammable and combustible materials. The current definitions are briefly summarized below.

Class I or Class II, Division 1 Locations
- The interiors of spray booths or rooms.
- The interior of exhaust ducts.
- Any area in the direct path of spray operations.

Class I or Class II, Division 2 Locations
The following spaces shall be considered Class I or Class II, Division 2, as applicable:
- For open spraying, all space outside of but within 20 ft. horizontally and 10 ft. vertically of the Class I, Division 1 location and not separated from it by partitions (See Exhibit 10-13.).

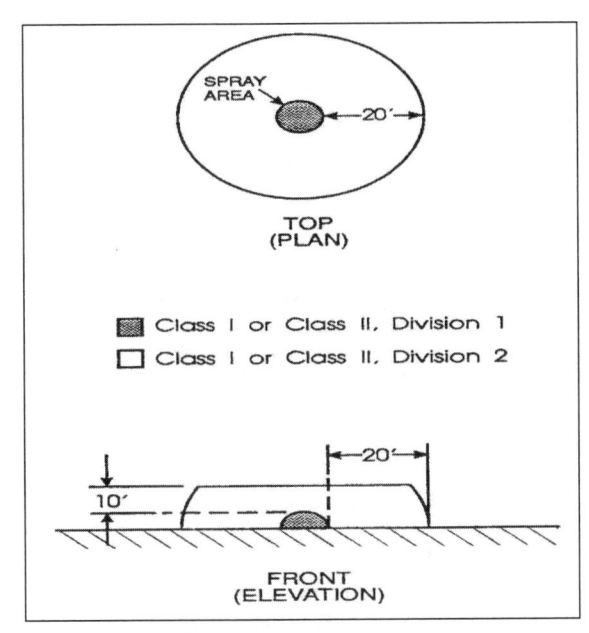

Exhibit 10-13.
Open spraying parameters in a Class, Division 2 location.

- For spraying operations conducted within a closed top, open face or front spray booth, the space within 3 ft. in all directions from openings other than the open face or front.

The Class II, Division 2 location shown in Exhibit 10-14 shall extend from the open face or front of the spray booth in accordance with the following:

- If the ventilation system is interlocked with the spraying equipment so as to make the spraying equipment inoperable when the ventilation system is not in operation, the space shall extend 5 ft. from the open face or front of the spray booth, and as otherwise shown in Figure "A" of Exhibit 10-14.
- If the ventilation system is not interlocked with the spraying equipment so as to make the spraying equipment inoperable when the ventilation system is not in operation, the space shall extend 10 ft. from the open face or front of the spray booth, and as otherwise shown in Figure "B" of Exhibit 10-14.

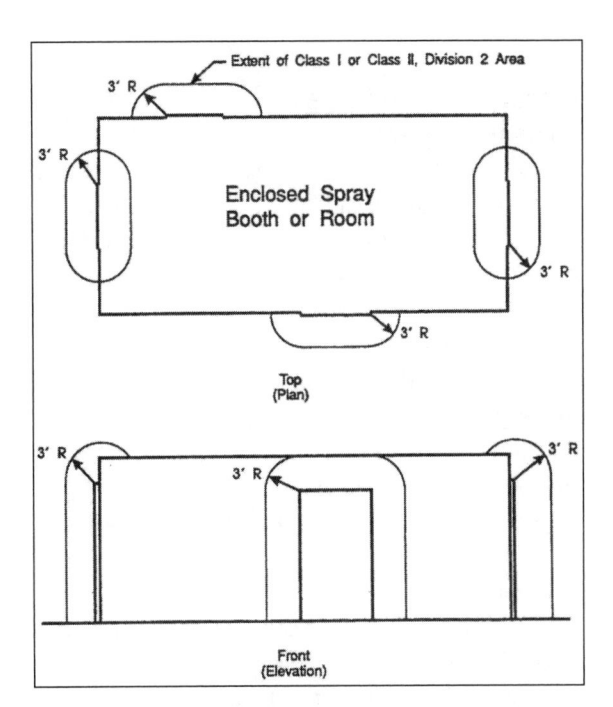

Exhibit 10-14.
Class I or Class II, Division 2 locations adjacent to a closed top, open faced or open front spray booth.

- For spraying operations confined to an enclosed spray booth or room, the space within 3 ft. in all directions from any openings shall be considered Class I or Class II, Division 2 locations, as shown in Exhibit 10-15.

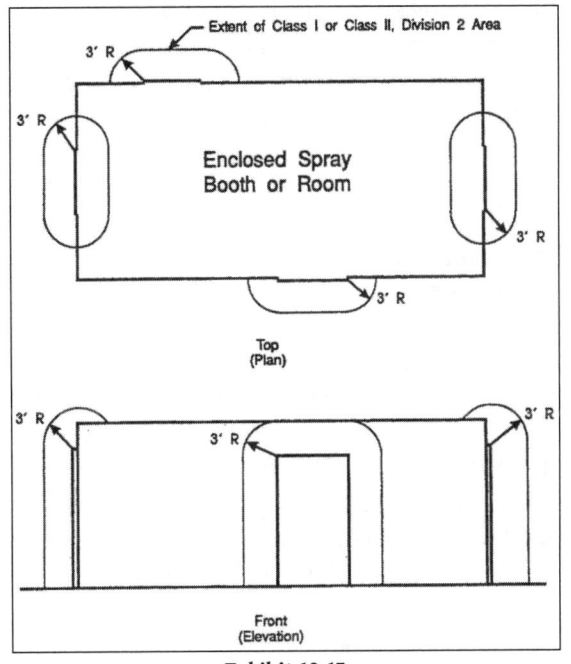

Exhibit 10-15.
In an enclosed spray area, the space within 3 ft. on all sides is a Class I or Class II, Division 2 location.

Lamps
Electric lamps outside of, but within 20 ft. of any spraying area, and not separated therefrom by a partition, shall be totally enclosed to prevent the falling of hot particles and shall be protected from mechanical injury by suitable guards or by location.

Portable Lamps
Portable electric lamps shall not be used in any spraying area during spraying operations. Portable electric lamps, if used during cleaning or repairing operations, shall be of the type approved for hazardous Class I locations.

Grounding
All metal parts of spray booths, exhaust ducts and piping systems conveying flammable or combustible liquids or aerated solids shall be properly electrically grounded in an effective and permanent manner.

Ventilation
All spraying areas shall be provided with mechanical ventilation adequate to remove flammable vapors, mists or powders to a safe location and to confine and control combustible residue so that life is not endangered. Mechanical ventilation shall be kept in operation at all times while spraying operations are being conducted and for a sufficient time thereafter to allow vapors from drying coated articles and finishing material residue to be exhausted.

Independent Exhaust
Each spray booth shall have an independent exhaust duct system discharging to the exterior of the building. However, multiple cabinet spray booths in which identical spray finishing material is used that have a combined frontal area of not more than 18 sq. ft. may have a common exhaust. If more than one fan serves one booth, all fans shall be so interconnected that one fan cannot operate without all fans being operated.

Fan-Rotating Element
The fan-rotating element shall be non-ferrous or non-sparking or the casing shall consist of or be lined with such material. Electric motors driving exhaust fans shall not be placed inside booths or ducts.

Belts
Belts shall not enter the duct or booth unless the belt and pulley within the duct or booth are thoroughly enclosed.

Exhaust Ducts
Exhaust ducts shall be constructed of steel and shall be substantially supported. Exhaust ducts without dampers are preferred; however, if dampers are installed, they shall be maintained so that they will be in a full-open position at all times the ventilating system is in operation.

Discharge Clearance
Unless the spray booth exhaust duct terminal is from a water-wash spray booth, the terminal discharge point shall be not less than 6 ft. from any combustible exterior wall or roof nor discharge in the direction of any combustible construction or unprotected opening in any non-combustible exterior wall within 25 ft.

Air Exhaust
Air exhaust from spray operations shall not be directed so that it will contaminate makeup air being introduced into the spraying area or other ventilating intakes. Air exhausted from spray operations shall not be recirculated.

Access Doors
When necessary to facilitate cleaning, exhaust ducts shall be provided with an ample number of access doors.

Drying Spaces
Freshly sprayed articles shall be dried only in spaces provided with adequate ventilation to prevent the formation of explosive vapors. In the event adequate and reliable ventilation is not provided, such drying spaces shall be considered a spraying area.

Flammable and Combustible Liquids— Storage and Handling

Conformance
The storage of flammable or combustible liquids in connection with spraying operations shall conform to the requirements of 1910.106, where applicable.

Quantity

The quantity of flammable or combustible liquids kept in the vicinity of spraying operations shall be the minimum required for operations and should ordinarily not exceed a supply for one day or one shift.

Containers

Original closed containers, approved portable tanks, approved safety cans or a properly arranged system of piping shall be used to bring flammable or combustible liquids into spray finishing rooms. Open or glass containers shall not be used.

Transferring Liquids

Except as discussed in the paragraph, "Spraying Containers," the withdrawal of flammable and combustible liquids from containers having a capacity of greater than 60 gallons shall be by approved pumps. The withdrawal of flammable or combustible liquids from containers and the filling of containers, including portable mixing tanks, shall be done only in a suitable mixing room or in a spraying area when the ventilating system is in operation. Adequate precautions shall be taken to protect against liquid spillage and from sources of ignition.

Spraying Containers

Containers supplying spray nozzles shall be of closed type or provided with metal covers that are kept closed. Containers supplying spray nozzles by gravity flow shall not exceed 10 gallons in capacity.

Original shipping containers shall not be subject to air pressure for supplying air nozzles. Containers under air pressure supplying spray nozzles shall be of limited capacity, not exceeding that necessary for one day's operation; shall be designed and approved for such use; shall be provided with a visible pressure gauge; and shall be provided with a relief valve; all in conformance with the ASME Code for Unfired Pressure Vessels.

Containers under pressure supplying spray nozzles, air storage tanks and coolers shall conform to the standards of the ASME Code for Unfired Pressure Vessels for construction, tests and maintenance.

Pipes and Hoses

All containers or piping to which a hose or flexible connection is attached shall be provided with a shut-off valve at the connection. Such valves shall be kept shut when spraying operations are not being conducted. When a pump is used to deliver products, automatic means shall be provided to prevent pressure in excess of the design working pressure of accessories, piping and hose.

All pressure hoses and couplings shall be inspected at regular intervals appropriate to this service. The hoses and couplings shall be tested with the hose extended and using the "in-service maximum operating pressures." Any hose showing material deterioration, signs of leakage or weakness in its carcass or at the couplings shall be withdrawn from service and repaired or discarded.

Piping systems conveying flammable or combustible liquids shall be of steel or other material having comparable properties of resistance to heat and physical damage. Piping systems shall be properly bonded and grounded.

Spray Liquid Heaters

Electrical-powered spray liquid heaters shall be approved and fisted for the specific location in which they are used. Heaters shall not be located in spray booths or other locations subject to the accumulation of deposits or combustible residue.

Pump Relief

If flammable or combustible liquids are supplied to spray nozzles by positive displacement pumps, means shall be provided to prevent the discharge pressure from exceeding the safe operating pressure of the system. Any discharge shall be to a safe location.

Grounding

Whenever flammable or combustible liquids are transferred from one container to another, both containers shall be effectively bonded and grounded to prevent discharge sparks of static electricity.

Protection

Conformance
In sprinklered buildings, the automatic sprinkler system in rooms containing spray-finishing operations shall conform to the requirements of 1910.159. In unsprinklered buildings where sprinklers are installed only to protect spraying areas, the installation shall conform to such standards insofar as they are applicable. Sprinkler heads shall be located so as to provide water distribution throughout the entire booth.

Valve Access
Automatic sprinklers protecting each spray booth (together with its connecting exhaust) shall be under an accessibly located separate outside stem and yoke sub-control valve.

Cleaning of Heads
Sprinklers protecting spraying areas shall be kept as free from deposits as practical by cleaning, daily if necessary.

Portable Extinguishers
An adequate supply of suitable portable fire extinguishers shall be installed near all spraying areas.

Operations and Maintenance

Spraying
Spraying shall not be conducted outside of predetermined spraying areas.

Cleaning
All spraying areas shall be kept as free from the accumulation of deposits of combustible residue as practical, with cleaning conducted daily if necessary. Scrapers, spuds, or other such tools used for cleaning shall be constructed of nonsparking material.

Residue Disposal
Residue scrapings and debris contaminated with residue shall be immediately removed from the premises and properly disposed of. Approved metal waste cans shall be provided wherever rags or waste are impregnated with finishing material, and all such rags or waste deposited therein immediately after use. The contents of waste cans shall be properly disposed of at least once daily or at the end of each shift.

Clothing Storage
Spray-finishing employees' clothing shall not be left on the premises overnight unless kept in metal lockers.

Cleaning Solvents
The use of solvents for cleaning operations shall be restricted to those having flash points not less than 100°F; however, for cleaning spray nozzles and auxiliary equipment, solvents having flash points not less than those normally used in spray operations may be used. Such cleaning shall be conducted inside spray booths and ventilating equipment operated during cleaning.

Hazardous Materials Combinations
Spray booths shall not be alternately used for different types of coating materials, where the combination of the materials may be conducive to spontaneous ignition, unless all deposits of the first used material are removed from the booth and exhaust ducts prior to spraying with the second used material. Examples of dangerous combinations are:
- Deposits of lacquers containing nitrocellulose combined with finishes containing drying oils, such as varnishes, oil-based stains, air-drying enamels and primers.
- Bleaching compounds based on hydrogen peroxide, hypochlorites, perchlorates, or other oxidizing compounds combined with any organic finishing materials.

"No Smoking" Signs
"No Smoking" signs in large letters on contrasting color background shall be conspicuously posted at all spraying areas and paint storage rooms.

Fixed Electrostatic Apparatus

Location
Transformers, power packs, control apparatus, and all other electrical portions of the equipment, with the exception of high-voltage grids, electrodes and electrostatic atomizing heads

and their connections, shall be located outside of the spraying area or shall otherwise conform to the requirements for electrical contained earlier in this section.

Insulators, Grounding
High-voltage leads to electrodes shall be properly insulated and protected from mechanical injury or exposure to destructive chemicals. Electrostatic atomizing heads shall be effectively and permanently supported on suitable insulators and shall be effectively guarded against accidental contact or grounding. An automatic means shall be provided for grounding the electrode system when it is electrically de-energized for any reason. All insulators shall be kept clean and dry.

Safe Distance
A safe distance shall be maintained between goods being painted and electrodes or electrostatic atomizing heads or conductors of at least twice the sparking distance. A suitable sign indicating this safe distance shall be conspicuously posted near the assembly.

Conveyors Required
Goods being painted using this process are to be supported on conveyors. The conveyors shall be so arranged as to maintain safe distances between the goods and the electrodes or electrostatic atomizing heads at all times.

Fail-Safe Controls
Electrostatic apparatus shall be equipped with automatic controls that will operate without time delay to disconnect the power supply to the high voltage transformer and to signal the operator under any of the following conditions:
- Stoppage of ventilating fans or failure of ventilating equipment from any cause;
- Stoppage of the conveyor carrying goods through the high voltage field;
- Occurrence of a ground or of an imminent ground at any point on the high voltage system; and
- Reduction of clearance below that specified in "Safe Distance" above.

Guarding
Adequate booths, fencing, railings or guards shall be so placed about the equipment that they, either by their location, character or both, ensure that a safe isolation of the process is maintained from plant storage or personnel. Such railings, fencing and guards shall be of conducting material, adequately grounded.

Ventilation
Where electrostatic atomization is used, the spraying area shall be so ventilated as to ensure safe conditions from a fire and health standpoint.

Fire Protection
Automatic sprinklers shall protect all areas used for spraying, including the interior of the booth, where this protection is available. Where this protection is not available, other approved automatic extinguishing equipment shall be provided.

Electrostatic Hand-Spraying Equipment

Application
This paragraph shall apply to any equipment using electrostatically charged elements for the atomization and/or precipitation of materials for coatings on articles or for other similar purposes in which the atomizing device is hand held and manipulated during the spraying operation.

Spray Gun Ground
The handle of the spraying gun shall be electrically connected to ground by a metallic connection and to be so constructed that the operator in normal operating position is in intimate electrical contact with the grounded handle.

Grounding—General
All electrically conductive objects in the spraying area shall be adequately grounded. This requirement shall apply to paint containers, wash cans and any other objects or devices in the area. The equipment shall carry a prominent, permanently installed warning regarding the necessity for this grounding feature.

Maintenance of Grounds
Objects being painted or coated shall be maintained in metallic contact with the conveyor or other grounded support. Hooks shall be regularly cleaned to ensure this contact and areas of contact shall be sharp points or knife-edges where possible. Points of support of the object shall be concealed from random spray where feasible and where the objects being sprayed are supported from a conveyor; the point of attachment to the conveyor shall be so located as to not collect spray material during normal operation.

Interlocks
The electrical equipment shall be so interlocked with the ventilation of the spraying area that the equipment cannot be operated unless the ventilation fans are in operation.

Ventilation
The spraying operation shall take place within a spray area that is adequately ventilated to remove solvent vapors released from the operation.

Drying, Curing or Fusion Apparatus

Conformance
This standard adopts the provisions of the Standard for Ovens and Furnaces, NFPA 86A-1969, which is incorporated by reference, where process ovens and similar items are used in connection with spray finishing. To prevent oven explosions, the workspace must be preventilated before starting the oven. A safe atmosphere must be maintained at any source of ignition. In addition, the heating system must shut down in the event of a failure of the ventilation system.

Alternate Use Prohibited
Spray booths, rooms or other enclosures used for spraying operations shall not alternately be used for the purpose of drying by any arrangement that will cause a material increase in the surface temperature of the spray booth, room or enclosure.

Alternate Use Permitted
Automobile refinishing spray booths or enclosures, otherwise installed and maintained in full conformity with this section, may alternately be used for drying with portable electrical infrared drying apparatus when conforming with the following:

- Interior (especially floors) of spray enclosures shall be kept free of overspray deposits.
- During spray operations, the drying apparatus and electrical connections and wiring thereto shall not be located within spray enclosures nor in any other location where spray residue may be deposited thereon.
- The spraying apparatus, the drying apparatus and the ventilating system of the spray enclosure shall be equipped with suitable interlocks so arranged that:
    - The spraying apparatus cannot be operated while the drying apparatus is inside the spray enclosure.
    - The spray enclosure will be purged of spray vapors for a period of not less than three minutes before the drying apparatus can be energized.
    - The ventilating system will maintain a safe atmosphere within the enclosure during the drying process and the drying apparatus will automatically shut off in the event of failure of the ventilating system.
- All electrical wiring and equipment of the drying apparatus shall conform with the applicable sections of Subpart S, Electrical.
- The drying apparatus shall contain a prominently located, permanently attached warning sign indicating that ventilation should be maintained during the drying period and that spraying should not be conducted in the vicinity that spray will deposit on apparatus.

Powder Coating

This finishing process is increasing in industrial importance. It generally invokes application of plastic particles to pre-warmed parts. The coating is then fused in place by heating the part in an oven. Electrostatic forces are frequently used to apply the coating.

Electrical and Other Sources of Ignition
Powder-coating operations must conform to the general spray-finishing requirements that provide protection against open flames, spark-producing equipment and hot surfaces. Wiring must conform to OSHA electrical requirements for Class II hazardous locations regarding combustible dusts. Powder-coating equipment and booths must be grounded. Portable lamps may not be used during spraying operations. Only approved portable lamps may be used during cleaning and repair.

Ventilation
This standard requires appropriate exhaust ventilation. Powders must be safely removed to recovery equipment and must not be released to the outside atmosphere.

Drying, Curing, or Fusion Equipment
NFPA 86A-1969 is adopted by reference to safeguard fusion ovens. See description above regarding such ovens.

Operation and Maintenance
Accumulation of combustible dust must be prevented on ledges and similar surfaces. Good housekeeping must be practiced and hazardous dust clouds must not be created during cleaning operations.

Electrostatics
The use of electrostatic energy to apply powder must not create an ignition hazard. The controls for electrostatic application of liquid droplets also apply to powders. In addition, the standard requires that equipment be maintained at less than 150°F. (Note that warmed parts are often coated). There are specific requirements to prevent spark ignition of powder and to provide appropriate bonding and grounding.

Organic Peroxides and Dual Component Coatings

Organic peroxides are a group of chemicals in the high hazard class that have become increasingly useful as chain reaction initiators or catalysts in the manufacture of plastics and other materials. These chemicals, which were formerly consumed by chemical process industries in limited quantities, are now being stored in greater volume and processed in more dangerous concentrations in a wider variety of industries. A number of fires and explosions have been attributed to these chemicals. The reinforced plastics manufacturing industry is one of the larger consumers of organic peroxides. The plastic is frequently applied to the reinforcing material by spraying automatically proportioned mixtures of a resin monomer and an organic peroxide catalyst.

Organic peroxides are marketed in a large number of commercial peroxide preparations and in the form of solids, liquids and pastes. Some of the materials are diluted in order to decrease their hazard potential. The most widely used peroxides are benzyl peroxide and methyl ethyl ketone (MEK) peroxide. These per-oxygen materials contain a large amount of active oxygen; this is an important factor in determining hazard characteristics as it supports combustion even though air is excluded. Organic peroxides burn much more rapidly than ordinary flammable liquids or combustible solids.

Conformance. All spraying operations involving the use of organic peroxides and other dual-component coatings shall be conducted in approved sprinklered spray booths meeting the requirements of this section.

Smoking. Smoking shall be prohibited and "No Smoking" signs shall be prominently displayed and only non-sparking tools shall be used in any area where organic peroxides are stored, mixed or applied.

NOTE: The use of organic peroxides and monomers in spray finishing involves possible hazardous chemical reactions. Reference should be made to various industry-recognized documents that address control methods relating to these processes and materials. The following are examples of relevant standards and guidance documents in this area:

- NFPA 43A, Liquid, Solid Oxidizing Materials;
- NFPA 43B, Organic Peroxide Formulations; and
- NFPA 49, Hazardous Chemicals Data.

Hazardous (Classified) Locations

The National Electrical Code (NEC) defines hazardous locations as those areas "where fire or explosion hazards may exist due to flammable gases or vapors, flammable liquids, combustible dust, or ignitable fibers or flyings."

A substantial part of the NEC is devoted to the discussion of hazardous locations. That's because electrical equipment can become a source of ignition in these volatile areas. The writers of the NEC developed a shorthand method of describing areas classified as hazardous locations. One of the purposes of this discussion is to explain this classification system. Hazardous locations are classified in three ways by the National Electrical Code: Type, Condition and Nature.

Hazardous Location Types

Class I Locations
According to the NEC, there are three types of hazardous locations. The first type of hazard is one that is created by the presence of flammable gases or vapors in the air, such as natural gas or gasoline vapor. When these materials are found in the atmosphere, a potential for explosion exists, which could be ignited if an electrical or other source of ignition is present. The Code writers have referred to this first type of hazard as Class I. So, a *Class I hazardous location* is one in which *flammable gases or vapors* may be present in the air in sufficient quantities to be explosive or ignitable. Some typical Class I locations are:
- Petroleum refineries and gasoline storage and dispensing areas;
- Dry cleaning plants where vapors from cleaning fluids can be present;
- Spray finishing areas;
- Aircraft hangars and fuel servicing areas; and
- Utility gas plants and operations involving storage and handling of liquefied petroleum gas or natural gas.

The above Class I hazardous locations require special Class I hazardous location equipment.

Class II Locations
The second types of hazard listed by the National Electrical Code are those areas made hazardous by the presence of combustible dust. These are referred to in the Code as "Class II Locations." Finely pulverized material, suspended in the atmosphere, can cause as powerful an explosion as one occurring at a petroleum refinery. Some typical Class II locations are:
- Grain elevators;
- Flour and feed mills;
- Plants that manufacture, use or store magnesium or aluminum powders;
- Producers of plastics, medicines and fireworks;
- Producers of starch or candies;
- Spice-grinding plants, sugar plants and cocoa plants; and
- Coal preparation plants and other carbon-handling or processing areas.

Class III Locations
Class III hazardous locations, according to the NEC, are areas where there are easily ignitable fibers or flyings present, due to the types of materials being handled, stored or processed. The fibers and flyings are not likely to be suspended in the air, but can collect around machinery, on lighting fixtures and where heat, a spark or hot metal can ignite them. Some typical Class III locations are:
- Textile mills and cotton gins;
- Cotton seed mills and flax processing plants; and
- Plants that shape, pulverize or cut wood and create sawdust or flyings.

Hazardous Location Conditions
In addition to the types of hazardous locations, the National Electrical Code also concerns itself with the kinds of conditions under which these hazards are present. The code specifies that hazardous material may exist in several different kinds of conditions that, for simplicity, can be described as, first, normal conditions and, second, abnormal conditions.

In normal conditions, the hazard would be expected to be present in everyday production operations or during frequent repair and maintenance activity. When the hazardous material is expected to be confined within closed containers or closed systems and will be present only through accidents rupture, breakage or unusual faulty operation, the situation would be considered abnormal.

The code writers have designated these two kinds of conditions very simply as Division 1—normal and Division 2—abnormal. Class I, Class II and Class III hazardous locations can be either Division 1 or Division 2. Good examples of Class I, Division 1 locations would be the areas near open-dome loading facilities or adjacent to relief valves in a petroleum refinery because the hazardous material would be present during normal plan operations.

Closed storage drums containing flammable liquids in an inside storage room would not normally allow the hazardous vapors to escape into the atmosphere. However, if one of the containers is leaking, it would be considered a Class I, Division 2 (abnormal) hazardous location.

In summary, there are three types of hazardous locations:
- Class I—gas or vapor;
- Class II—dust; and
- Class III—fibers and flyings; and

Two types of conditions:
- Division 1—normal conditions, and
- Division 2—abnormal conditions.

Each of the above categories can be broken down further.

Nature of Hazardous Substances

The gases and vapors of Class I locations are broken into four groups by the code: A, B, C and D. These materials are grouped according to the ignition temperature of the substance, its explosion pressure and other flammable characteristics.

The only substance in Group A is acetylene. Acetylene makes up only a very small percentage of hazardous locations. Consequently, little equipment is available for this type of location. Acetylene is a gas with extremely high explosion pressures.

Group B is another relatively small segment of classified areas. This group includes hydrogen and other materials with similar characteristics. If certain specific restrictions in the code are followed, some of these Group B locations, other than hydrogen, can actually be satisfied with Group C and Group D equipment.

Group C and Group D are by far the most usual Class I groups. They comprise the greatest percentage of all Class I hazardous locations. Found in Group D are many of the most common flammable substances, including butane, gasoline, natural gas and propane.

Class II, dust locations, includes the hazardous materials in Groups E, F and G. These groups are classified according to the ignition temperature and the conductivity of the hazardous substance. Conductivity is an important consideration in Class II locations, especially with metal dusts.

Metal dusts are categorized in the code as Group E. Included here are aluminum and magnesium dusts and other metal dusts of similar nature. Group F atmospheres contain such materials as carbon black, charcoal dust, coal and coke dust. Group G includes grain dusts, flour, starch, cocoa and similar types of materials.

The National Electric Code classifies hazardous locations the same way.

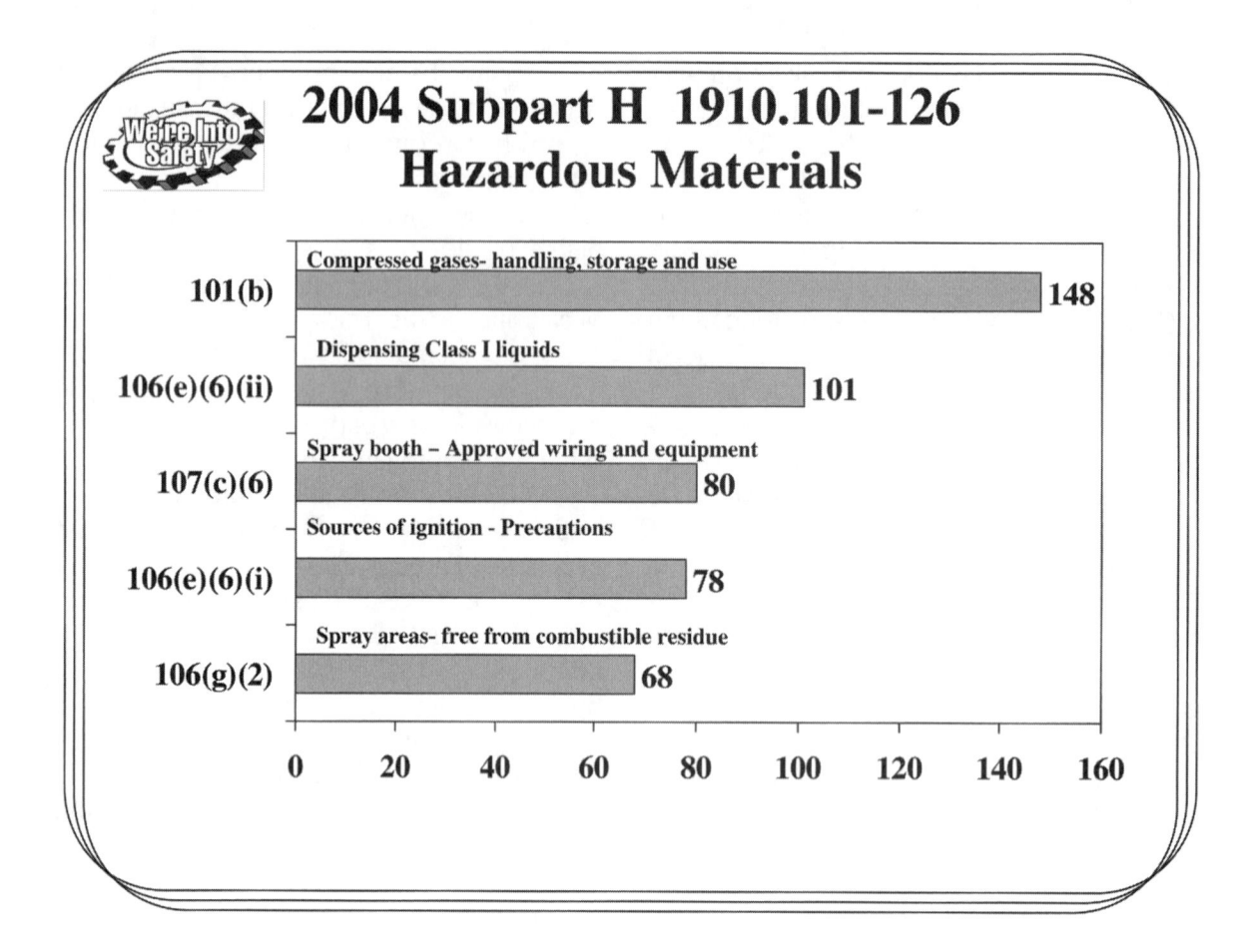

2004 Subpart H 1910.101-126
Hazardous Materials

Citation	Topic	Count
101(b)	Compressed gases- handling, storage and use	148
106(e)(6)(ii)	Dispensing Class I liquids	101
107(c)(6)	Spray booth – Approved wiring and equipment	80
106(e)(6)(i)	Sources of ignition - Precautions	78
106(g)(2)	Spray areas- free from combustible residue	68

WHAT EVERY SUPERVISOR MUST KNOW ABOUT OSHA-GENERAL

Fire Protection

Scope, Application and Definitions—1910.155
This subpart contains requirements for fire brigades and all portable and fixed fire suppression equipment, fire detection systems, and fire and employee alarm systems installed to meet the fire protection requirements of 29 CFR 1910. It applies to employment other than maritime, construction and agriculture.

Definitions
Class A fire. A fire involving ordinary combustible materials such as paper, wood, cloth, and some rubber and plastic materials.

Class B fire. A fire involving flammable or combustible liquids; flammable gases, greases and similar materials; and some rubber and plastic materials.

Class C fire. A fire involving energized electrical equipment where safety to the employee requires the use of electrically nonconductive extinguishing media.

Class D fire. A fire involving combustible metals such as magnesium, titanium, zirconium, sodium, lithium or potassium.

Dry chemical. An extinguishing agent primarily composed of very small particles of chemicals (e.g., sodium bicarbonate, potassium bicarbonate, monoammonium phosphate).

Dry powder. A compound used to extinguish or control Class D fires.

Extinguisher rating. The numerical rating given to an extinguisher, which indicates the extinguishing potential of the unit, based on standardized tests developed by Underwriters' Laboratories Inc.

Fire brigade. An organized group of employees, who are knowledgeable, trained and skilled in at least basic firefighting operations.

Halon 1211. A colorless, faintly sweet-smelling, electrically nonconductive liquefied gas ($CBrClF_2$) that is a medium for extinguishing fires by inhibiting the chemical chain reaction of fuel and oxygen. It is also known as bromochlorodifluoromethane.

Halon 1301. A colorless, odorless, electrically nonconductive gas ($CBrF$) that is a medium for extinguishing fires by inhibiting the chemical chain reaction of fuel and oxygen. It is also known as bromotrifluoromethane.

Incipient stage fire. A fire that is in the initial or beginning stage and that can be controlled or extinguished by portable fire extinguishers, Class II standpipe or small hose systems without the need for protective clothing or breathing apparatus.

Interior structural firefighting. The physical activity of fire suppression, rescue or both, inside of buildings or enclosed structures that are involved in a fire situation beyond the incipient stage.

Multi-purpose dry chemical. A dry chemical that is approved for use on Class A, Class B and Class C fires.

Standpipe systems
- **Class I** system means a 2½ in. hose connection for use by fire departments and those trained in handling heavy fire streams.
- **Class II** system means a 1½ in. hose system that provides a means for the control or extinguishment of incipient stage fires.
- **Class III** system means a combined system of hose that is for the use of employees trained in hose operations and is capable of furnishing effective water discharge during the more advanced stages of fire (beyond the incipient stage) in the interior of workplaces. Hose outlets are available for both 1½ in. and 2½ in. hose.
- Small hose system means a system of hose (5/8 in. to 1½ in. diameter) that is for the use of employees for the control or extinguishment of incipient stage fires.

Fire Brigades—1910.156
This section contains requirements for the organization, training and personal protective equipment of fire brigades whenever they are established by an employer. It should be noted that this regulation does not require an employer to establish fire brigades. If they are established, however, the requirements of this section must be met. The requirements of this section apply to fire brigades, industrial fire departments and private or contractual fire departments. This section does not apply to airport crash rescue or forest firefighting operations.

Organization
The employer must ensure that employees who are expected to do interior structural firefighting are physically capable of performing duties that may be assigned to them during emergencies. The employer shall prepare and maintain a written policy statement that:
- Establishes the fire brigade and its organizational structure;
- Defines the functions to be performed; and
- States training program requirements.

Training and Education
- Training shall be conducted prior to assignment to the fire brigade
- Training shall take place at least annually for all fire brigade members.
- Quarterly training or education sessions are required for those fire brigade members expected to perform interior structural firefighting.
- Training must be provided by qualified training instructors who are available from:
  - Local municipal fire department;
  - State Fire Marshal's office;
  - State University extension service;
  - International Society of Fire Service Instructors; or
  - Community College Fire Science programs.

Firefighting Equipment
The employer shall maintain and inspect firefighting equipment to ensure safe operational condition of the equipment. This inspection shall take place at least annually. Portable fire extinguishers and respirators must be inspected at least monthly.

Protective Clothing
These requirements apply to those employees who perform interior structural firefighting. The requirements do not apply to employees who use fire extinguishers or standpipe systems to control or extinguish fires only in the incipient stage. These requirements are for members of the fire brigade:
- Foot and leg protection;
- Body protection;
- Hand protection;
- Head, eye and face protection; and
- Respiratory protection:
  - Must meet requirements of 1910.134; and
  - Must be certified under 30 CFR Part 11.

Portable Fire Extinguishers—1910.157

Exemptions
The standard does not require the employees to use extinguishers. Where the employer has

a total evacuation policy, an emergency action plan and a fire prevention plan that meet the requirements of 1910.38, and extinguishers are not available in the workplace, the employer is exempt from all requirements of this section unless a specific standard in Part 1910 requires that a portable extinguisher be provided. Where the employer has an emergency action plan meeting the requirements of 1910.38, which establishes fire brigades and requires all other employees to evacuate, the employer is exempt from the distribution requirements of this section.

General Requirements

General requirements regarding portable fire extinguishers include:

- The employer shall provide portable fire extinguishers and shall mount, locate and identify them so they are readily accessible to employees but do not subject them to possible injury.
- Only approved portable fire extinguishers shall be used.
- The employer shall not provide or make available in the workplace portable fire extinguishers that use carbon tetrachloride or chlorobromomethane extinguishing agents.
- The employer shall ensure portable fire extinguishers are maintained in a fully charged and operable condition and kept in their designated places at all times, except during use.
- The employer shall remove from service all soldered or riveted shell inverting type of extinguishers.

Selection and Distribution

Extinguishers shall be provided for employee use and selected and distributed based on the classes of anticipated workplace fires and on the size and degree of hazard that would affect their use. Extinguishers shall be distributed so that the following maximum travel distances apply:

- Class A   75 ft.
- Class B   50 ft.
- Class C   Based on appropriate pattern for existing Class A or B hazards.
- Class D   75 ft.

Inspection, Maintenance and Testing

Extinguishers shall be visually inspected monthly, maintained annually and hydrostatically tested periodically as per Table L-1 of this standard (See Exhibit 11-1.).

Exhibit 11-1. Fire extinguishers must be tested at the intervals listed in the table below.
Table L-1

Type of extinguishers	Time interval (years)
Soda acid (soldered brass shells) (until 1/1/82)	(\1\)
Soda acid (stainless steel shell)	5
Cartridge operated water and/or antifreeze	5
Stored pressure water and/or antifreeze	5
Wetting agent	5
Foam (soldered brass shells) (until 1/1/82)	(\1\)
Foam (stainless steel shell)	5
Aqueous Film Forming foam (AFFF)	5
Loaded stream	5
Dry chemical with stainless steel	5
Carbon dioxide	5
Dry chemical, stored pressure, with mild steel, brazed brass or aluminum shells	12
Dry chemical, cartridge or cylinder operated, with mild steel shells	12
Halon 1211	12
Halon 1301	12
Dry powder, cartridge or cylinder operated with mild steel shells	12

\1\ Extinguishers having shells constructed of copper or brass joined by soft solder or rivets shall not be hydrostatically tested and shall be removed from service by January 1, 1982. (Not permitted)

Training and Education

Employees shall be educated in the use of extinguishers and associated hazards upon initial employment and at least annually thereafter. In addition, The employees who have been designated to use firefighting equipment as part of an emergency action plan shall be trained in the use of the appropriate equipment.

Standpipe and Hose Systems—1910.158

This section applies to all small hose, Class II and Class III standpipe systems required by other OSHA standards. It does not apply to Class I standpipe systems.

Protection of Standpipes

Standpipes shall be located or otherwise protected against mechanical damage. Damaged standpipes shall be repaired promptly.

Equipment

Hose Reels and Cabinets

Where reels or cabinets are provided to contain fire hoses, the employer shall ensure they are designed to facilitate prompt use at the time of an emergency.

Hose Outlets and Connections

Hose outlets and connections must be located high enough above the floor to avoid being obstructed and to be accessible to employees.

Hose

Each required hose outlet shall be equipped with a hose that is connected and ready for use.

Where the hose may be damaged by extreme cold, it may be kept in a protected location as long as it is readily available to be connected for use.

Nozzles

The employer shall ensure the standpipe hose is equipped with shut-off type nozzles. There are two basic nozzle types: straight stream and fog (also referred to as variable stream, spray or combination). While fog is generally preferred, straight stream is acceptable.

Water Supply

The minimum water supply for standpipe and hose systems, which are provided for the use of employees, shall be sufficient to provide 100 gallons per minute for at least 30 minutes.

Tests and Maintenance

Acceptance Tests

Piping and hose of Class II and Class III systems shall be hydrostatically tested before being placed in service.

Maintenance

The following maintenance items are required for standpipe and hose systems:

- Water supply tanks are to be kept filled, except during repairs.
- Valves in the main piping connections to the automatic sources of water supply must always be kept fully open except during repairs.
- Hose systems must be inspected at least annually and after each use.
- Any unserviceable portion of the system must be removed immediately and replaced with equivalent protection during repair.
- Hemp or linen hoses shall be unpacked, inspected for deterioration and repacked using a different fold pattern at least annually. Defective hoses shall be replaced.
- Trained persons shall be designated to conduct all these required inspections.

Automatic Sprinkler Systems—1910.159

This section applies to all automatic sprinkler systems installed to meet a particular OSHA standard. Systems installed solely for property protection are not covered.

All automatic sprinkler designs must provide the necessary discharge patterns, densities, and water flow characteristics for complete coverage. Only approved equipment and devices shall be used.

Maintenance

Systems shall be properly maintained. A main drain flow test must be performed on each system annually. The inspector's test valve shall be opened at least every two years to ensure proper operation of the system.

Acceptance Tests

New systems shall have proper acceptance tests conducted, including:

- Flushing of underground connections;
- Hydrostatic tests of system piping;
- Air tests in dry-pipe systems; and
- Test of drainage facilities.

Water Supplies

Every automatic sprinkler system must be provided with at least one automatic water supply capable of providing design water flow for at least 30 minutes.

Sprinkler Spacing

In order to provide a maximum protection area per sprinkler and a minimum of interference to the discharge pattern, the vertical clearance between sprinklers and material below shall be at least 18 in.

Fixed Extinguishing Systems, General—1910.160

This section applies to all fixed extinguishing systems installed to meet a particular OSHA standard except for automatic sprinkler systems covered by 1910.159. Certain paragraphs of this section also apply to fixed systems not installed to meet a particular OSHA standard, but which, by their operation, may expose employees to possible injury, death or adverse health consequences caused by the extinguishing agent. Specific fixed extinguishing systems using dry chemical, gaseous agents, and water spray and foam are covered in later sections.

Fixed extinguishing system components and agents must be designed and approved for use on the specific fire hazards they are expected to control. If the system becomes inoperable, the employer shall notify employees and take the necessary temporary precautions to ensure their safety until the system is restored to operating order.

Except where discharge is immediately recognizable, a distinctive alarm or signaling system that complies with 1910.165 and is capable of being perceived above ambient noise or light levels shall be provided on all extinguishing systems in those areas covered by the system.

Effective safeguards shall be provided to warn employees against entry into discharge areas where the atmosphere remains hazardous to employee safety or health. Hazard warning or caution signs shall be posted at the entrance to, and inside of, areas protected by systems that use agents in hazardous concentrations. A person knowledgeable in the design and function of the system shall inspect fixed systems annually. The weight and pressure of refillable containers and the weight of non-refillable containers shall be checked at least semi-annually.

Total Flooding Systems with Potential Health and Safety Hazards to Employees

The employer shall provide an emergency action plan per 1910.38 for each area protected by a total flooding system that provides agent concentrations exceeding the maximum safe levels specified in 1910.162(b)(5) and (b)(6).

All systems must have a pre-discharge alarm that complies with 1910.165 and is capable of being perceived above ambient light or noise levels, which will give the employees time to safely exit from the discharge area prior to discharge. An approved fire detection device interconnected with the pre-discharge employee alarm system shall provide automatic actuation of the system.

Fixed Extinguishing Systems, Dry Chemical—1910.161

This section applies to all fixed systems using dry chemical as the extinguishing agent, installed to meet a particular OSHA standard. These systems must also comply with 1910.160.

Dry chemical agents must be compatible with any foams or wetting agents with which they are used. When dry chemical discharge may obscure vision, a pre-discharge employee alarm is required which complies with 1910.165 and which will give employees time to safely exit from the discharge area prior to system discharge.

The rate of application of dry chemicals must be such that the designed concentration of the system will be reached within 30 seconds of initial discharge.

Fixed Extinguishing Systems, Gaseous Agent—1910.162

This section applies to all fixed extinguishing systems, using a gas as the extinguishing agent, installed to meet a particular OSHA standard. These systems shall also comply with 1910.160. For total flooding systems, the designed extinguishing concentration must be reached within 30 seconds of initial discharge except for Halon systems that must achieve design concentration within 10 seconds.

For total flooding systems, a pre-discharge alarm is required on Halon 1211 and carbon dioxide systems with a design concentration of 4 percent or greater and for Halon 1301 systems with a design concentration of 10 percent or greater. The alarm must provide employees time to safely exit the discharge area prior to system discharge.

For total flooding systems using Halon 1301:
- Agent concentrations of more than 7 percent shall not be used where egress from an area takes more than one minute.
- Agent concentrations are limited to 10 percent where egress takes longer than 30 seconds, but less than one minute.
- Agent concentrations greater than 10 percent are only permitted in areas not normally occupied, provided that any employee in the area can escape within 30 seconds. The employer shall assure that no unprotected employees enter the area during agent discharge.

Fixed Extinguishing Systems, Water Spray and Foam—1910.163

This section applies to all fixed extinguishing systems using water or foam solution as the extinguishing agent, installed to meet a particular OSHA standard. These systems must also comply with 1910.160. This section does not apply to automatic sprinkler systems, which are covered under 1910.159.

The foam and water spray systems must be designed to be effective in at least controlling fire in the protected area or on protected equipment. Drainage of water spray systems must be directed away from areas where employees are working and no emergency egress is permitted through the drainage path.

Fire Detection Systems—1910.164

Installation and Restoration
Only approved devices and equipment may be used. All fire detection systems and components shall be restored to normal operating condition as soon as possible after each test or alarm.

Maintenance and Testing
All systems must be maintained in an operable condition, except during repairs or maintenance. Fire detectors and fire detection systems (unless factory calibrated) must be tested and adjusted as often as needed to maintain proper reliability and operating condition. Servicing, maintenance and testing of fire detection systems must be performed by a trained person knowledgeable in the operations and functions of the system.

Protection of Fire Detectors
Fire detection equipment installed outdoors or in the presence of corrosive atmospheres shall be protected from corrosion. Detection equipment must be located and/or protected from mechanical or physical impact.

Response Time
Fire detection systems installed for the purpose of actuating fire extinguishment or suppression systems shall be designed to operate in time to control or extinguish a fire. Detection systems installed for the purpose of employee alarm and evacuation must be designed and installed to provide a warning for emergency action and safe escape of employees.

Number, Location and Spacing of Detecting Devices

The number, location and spacing of fire detectors must be based upon design data obtained from field experience, tests, engineering surveys, manufacturer's recommendations or a recognized testing laboratory listing.

Employee Alarm Systems—1910.165

This section applies to all emergency employee alarms installed to meet a particular OSHA standard.

The employee alarm system shall provide warning for necessary emergency action as called for in the emergency action plan or for the reaction time for safe escape of employees. The employee alarm shall be capable of being perceived above ambient noise or light levels by all employees in the affected portions of the workplace. The alarm must be distinctive and recognizable as a signal to evacuate the work area or to perform actions designated under the emergency action plan. The employer shall explain to each employee the preferred means of reporting emergencies, such as manual pull-box alarms, public address systems, radio or telephones.

Installation and Restoration

All devices, components and systems installed to comply with this standard must be approved. All employee alarm systems must be restored to normal operating condition as promptly as possible after each test or alarm.

Maintenance and Testing

All employee alarm systems shall be maintained in operating condition, except when undergoing repairs or maintenance. A test of the reliability and adequacy of non-supervised employee alarm systems must be made every two months. A different actuation device shall be used in each test of a multi-actuation device system so that no individual device is used for two consecutive tests.

All supervised employee alarm systems must be tested at least annually for reliability and adequacy. Servicing, maintenance and testing of systems must be done by persons trained in the designed operation and functions necessary for reliable and safe operation of the system.

Manual Operation

Manually operated actuation devices for use in conjunction with employee alarms shall be unobstructed, conspicuous and readily accessible.

Classification of Portable Fire Extinguishers

Portable fire extinguishers are classified to indicate their ability to handle specific classes and sizes of fires. Labels on extinguishers indicate the class and relative size of fire that they can be expected to handle.

Class A extinguishers are used on fires involving ordinary combustibles, such as wood, cloth and paper.

Class B extinguishers are used on fires involving liquids, greases and gases.

Class C extinguishers are used on fires involving energized electrical equipment.

Class D extinguishers are used on fires involving metals, such as magnesium, titanium, zirconium, sodium and potassium.

The recommended marking system to indicate the extinguisher suitability according to class of fire is a pictorial concept that combines the uses and non-uses of extinguishers on a single label (See Exhibit 11-2.)

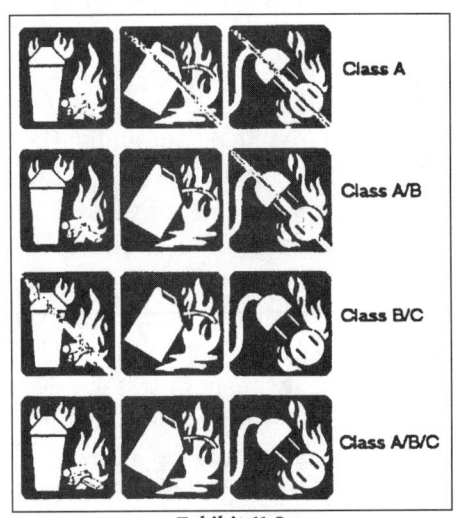

Exhibit 11-2.
Portable fire extinguishers should be labeled to indicate the class and size of fire for which they are appropriate.

The first row of symbols illustrated in Exhibit 11-2 is a label for uses on a Class A extinguisher. The symbol at the left, which depicts a Class A fire, is blue. Since the extinguisher is not recommended for use on Class B or C fires, the remaining two symbols, which depict Class B and Class C fires, are black, with a diagonal red line through them.

The second row of symbols is a label for use on a Class A/B extinguisher. The two left symbols are blue. Since the extinguisher is not recommended for use on Class C fires, the symbol on the far right, which depicts a Class C fire, is black, with a diagonal red line through it.

The third set of symbols is a label for use on Class B/C extinguishers. The two right symbols are blue. Since the extinguisher is not recommended for use on Class A fires, this symbol is black, with a diagonal red line through it. The fourth set of symbols is for use on Class A/B/C extinguishers. All symbols on this label are blue.

Letter-shaped symbol markings are also used to indicate extinguisher suitability according to class of fire (See Exhibit 11-3.). Extinguishers suitable for Class A fires should be identified by a triangle containing the letter "A." If colored, the triangle should be green.

Extinguishers suitable for Class B fires should be identified by a square containing the letter "B." If colored, the square shall be red.

A circle containing the letter "C" should identify extinguishers suitable for Class C fires. If colored, the circle should be blue.

A five-pointed star containing the letter "D" should identify extinguishers suitable for fires involving metals. If colored, the star should be yellow.

Extinguishers suitable for more than one class of fire should be identified by multiple symbols placed in a horizontal sequence.

Class A and Class B extinguishers carry a numerical rating to indicate how large a fire an experienced person can put out with the extinguisher. The ratings are based on reproducible physical tests conducted by Underwriters' Laboratories, Inc. Class C extinguishers have only a letter rating because there is no readily measurable quantity for Class C fires, which are essentially Class A or B fires involving energized electrical equipment. Class D extinguishers likewise do not have a numerical rating. Their effectiveness is described on the faceplate.

Exhibit 11-3. Labels with letter-shaped symbols also may be used to identify the class of fire for which the extinguisher is suitable.

Class A Rating
An extinguisher for Class A fires could have any one of the following ratings: 1-A, 2-A, 3-A, 4-A, 6-A, 10-A, 20-A, 30-A, and 40-A. A 4-A extinguisher, for example, should extinguish about twice as much fire as a 2-A extinguisher.

Class B Rating
An extinguisher for Class B fires could have any one of the following ratings: 1-B, 2-B, 5-B, 10-B, 20-B, 30-B, 40-B, and up to 640-B.

Class C Ratings
Extinguishers rated for Class C fires are tested only for electrical conductivity. However, no extinguisher gets a Class C rating without a Class A and/or Class B rating.

Class D Ratings
Class D extinguishers are tested on metal fires. The agent used depends on the metal for which the extinguisher was designed. Check the extinguisher faceplate for the unit's effectiveness on specific metals.

Common Fire Extinguishing Agents

Water
- Removes heat
- Effective on Class A fires
- Inexpensive
- Plentiful
- Non-toxic
- Disadvantages:
  - Conducts electricity
  - May spread Class B fires
  - Freezes in cold climates
  - May carry pollutants as run-off water

Carbon Dioxide (CO_2)
- Reduces oxygen to less than 15 percent
- Effective on Class B and C fires
- No residue
- Relatively inert
- Disadvantages:
  - Generally greater than 35 percent concentration by volume required for total flooding systems
  - Toxic to humans at greater than 4 percent by volume
  - Not the best agent for smoldering deep-seated fires (maintain concentration for more than 20 minutes)
  - Dissipates rapidly; allows reflash
  - Has a cooling/chilling effect on some electronic components
  - Vapor density = 1.5 (collects in pits and low areas)

Dry Chemical
- Interrupts chemical reactions
- Sodium bicarbonate (baking soda)
- Very effective on Class B and C fires
- Not considered toxic

- Disadvantages:
  - Leaves a residue
  - Obscures vision
  - Not good on deep-seated Class A fires
  - Absorbs moisture and may "cake" within container
  - May be irritating
  - Nozzle pressure may cause burning liquids to splash

Multipurpose Dry Chemical
- Interrupts chemical reactions
- Ammonium phosphate
- Effective on Class A, B, and C fires
- Nonconductive
- Disadvantages:
  - Obscures vision
  - More irritating than ordinary dry chemical
  - Nozzle pressure may cause burning liquids to splash

Halon
- Halon 104: Carbon tetrachloride (CC_4)
- Halon 1211: Bromochlorodifluoromethane ($CBrClF_2$)
- Halon 1301: Bromotrifluoromethane ($CBrF_3$)

Halon 1211
- Interrupts chemical reactions
- Bromochlorodifluoromethane
- Effective on Class A, B and C fires
- No residue
- May be sprayed (Boiling Point = 25°F)
- Used in portable fire extinguishers
- Disadvantages:
  - Acutely toxic at greater than 4 percent by volume (dizziness, impaired coordination and cardiac effects)
  - Must be used at greater than 5 percent by volume
  - Toxic decomposition products are generated by fire
  - Vapor density = 5.7 (collects in pits and low areas)
  - Production restricted per Montreal Protocol due to depletion of ozone layer
- Halon 1211 decomposition products:
  - Hydrogen bromide (HBr)
  - Hydrogen chloride (HCl)
  - Hydrogen fluoride (HF)
  - Bromine (Br_2)
  - Chlorine (C_2)
  - Fluorine (F_2)
  - Carbonyl bromide ($COBr_2$)
  - Carbonyl chloride ($COCl_2$)
  - Carbonyl fluoride (COF_2)

Halon 1301
- Interrupts chemical reactions
- Bromotrifluoromethane
- Effective on Class A, B, and C fires
- Not acutely toxic at less than 10 percent by volume
- Generally used at less than 7 percent by volume
- No residue
- No chilling effect on electronic parts and components
- Disadvantages:
  - Acutely toxic at greater than 10 percent by volume (anesthetic and cardiac effects)
  - Delayed effects and effects of chronic exposure not well known
  - Toxic decomposition products are generated by fire
  - Vapor density = 5 (collects in pits and low areas)
  - Production restricted per Montreal Protocol due to depletion of ozone layer
- Halon 1301 decomposition products:
  - Hydrogen fluoride (HF)
  - Hydrogen bromide (HBr)
  - Bromine (Br_2)
  - Carbonyl Fluoride (COF_2)
  - Carbonyl Bromide ($COBr_2$)

Fire Protection
- Fire brigades
- Portable and fixed fire suppression equipment
- Fire alarm and detection systems

Fire Brigades
- Organization:
  - Written policy statement establishing duties, organization and program requirements (physically capable of duties)
- Training:
  - Initial and annually
  - Quarterly for interior structural firefighting
- Equipment:
  - Firefighting and protective clothing
  - Respiratory devices

Fire Protection

Portable Fire Extinguishers

General requirements
- Mount, locate and identify extinguishers for ready access
- Approved equipment that is maintained fully charged and operable at all times

Selection and distribution
- Class A – Paper, wood, cloth
- Class B – Flammable/combustible liquids, grease or gases
- Class C – Electrical equipment
- Class D – Combustible metals

Maximum travel distances
- Class A – 75 ft.
- Class B – 50 ft.
- Class C – Based on appropriate pattern for existing Class A or B
- Class D – 75 ft.

Inspection, maintenance and testing
- Visual monthly inspection
- Maintained annually
- Periodic hydrostatic testing

Training and education
- Employees educated in use on initial employment and at least annually
- Employees designated to use firefighting equipment shall be trained

Fixed fire-suppression equipment
- Standpipes and hoses
- Automatic sprinkler systems
- Fixed extinguishing systems

General, dry chemical, gaseous, water, foam
- Other fire protection systems
- Fire detection systems
- Employee alarm systems

Fixed Fire-Suppression Equipment
- Protection of standpipes from damage
- Equipment requirements
  - Hose reels, cabinet, connections nozzles—fog, stream
  - Water supply—100 gpm/30min
  - Test and maintenance
  - Filled water supplies, valves open, hose inspected annually

Automatic sprinkler systems
Necessary discharge patterns, densities, and water flow for complete coverage, 18 inch clearance between sprinklers and materials
- Dry chemical—compatible and warning time to exit prior to discharge
- Gaseous agents—total flood time 30 seconds, but specific exit times given
- Foam/water —effective and drain away from exits

Fire Alarm and Detection System
- Installation and restoration
- Maintenance and testing
- Protection of fire detectors
- Corrosive atmospheres
- Mechanical/physical impact
- Response time; time to control fires
- Number, location and spacing of devices

Employee Alarm Systems
- Provide warning and perceived above ambient noise or light levels, distinctive and recognizable and explained to employees
- Installation and restoration
- Maintenance and testing
- Manual operation—unobstructed, conspicuous

2004 Subpart L 1910.155-165
Fire Protection

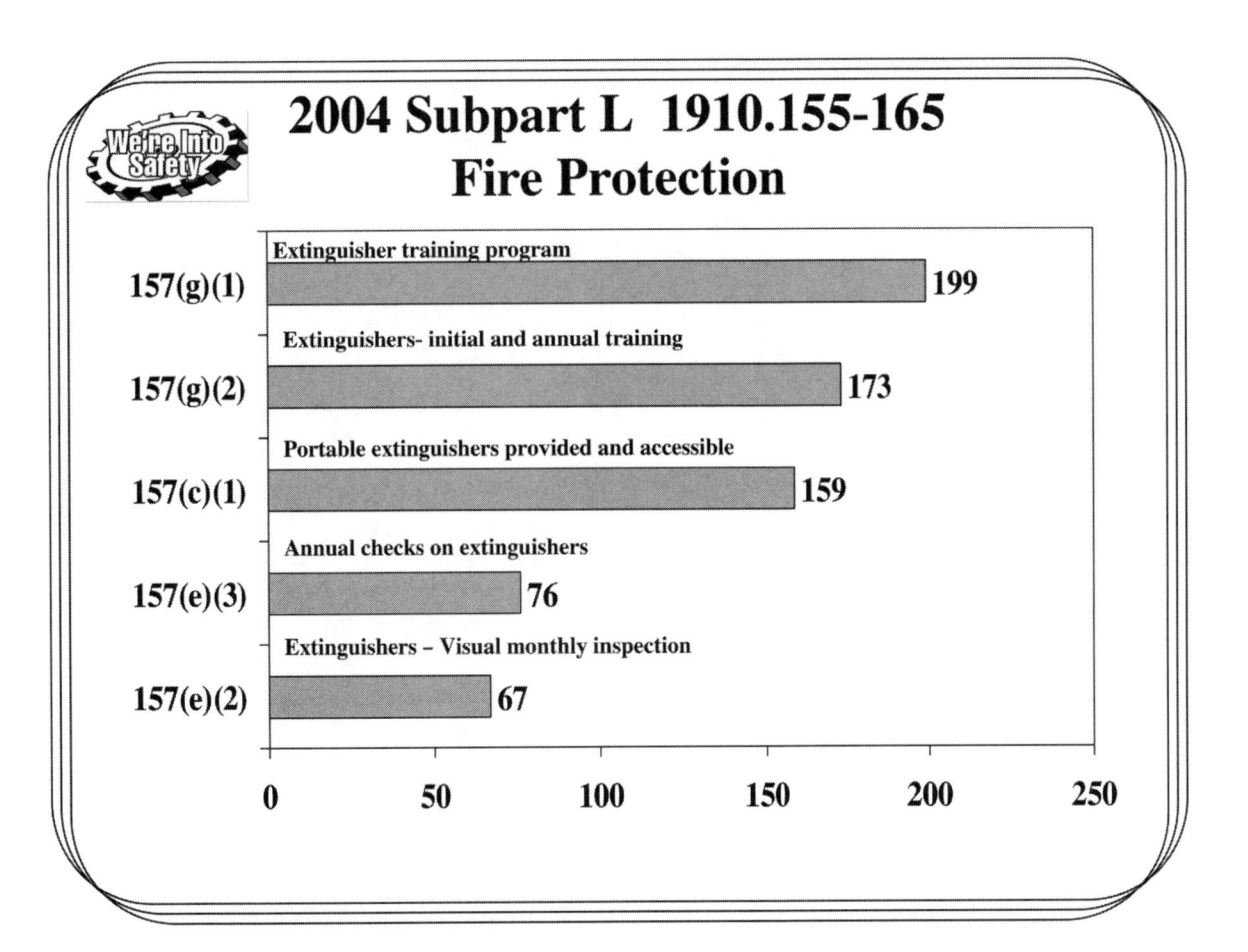

Citation	Description	Count
157(g)(1)	Extinguisher training program	199
157(g)(2)	Extinguishers - initial and annual training	173
157(c)(1)	Portable extinguishers provided and accessible	159
157(e)(3)	Annual checks on extinguishers	76
157(e)(2)	Extinguishers – Visual monthly inspection	67

CHAPTER 12

Exit Routes, Emergency Action Plans, and Fire Prevention Plans

This subpart deals with a subject that has been familiar to all of us since our early childhood days. Many of us remember the fire drills at school, with the primary classes trying for an orderliness award and the older students welcoming a break in the routine. In school, or elsewhere, the exit sign represents a sight so familiar that we hardly take note of it. We see them in stores, factories, theaters, office buildings, hotels, apartment buildings, practically everywhere. Yet we rarely notice them until we look for them.

Egress, as defined by Webster means "a place or means of going out."

Subpart E was rewritten and finalized on December 9, 2002. OSHA kept the intent of the original regulation so that companies already in compliance with the old rule will automatically be in compliance with the revised rule. The goal of the revision was to rewrite the existing requirements in clearer language and make them performance-oriented to the extent possible. The name of the subpart was changed from "Means of Egress" to "Exit Routes, Emergency Action Plans and Fire Prevention Plans" to describe the contents more accurately.

During the review of the regulation, OSHA discovered that some provisions were outdated and not consistent with contemporary life safety options in the NPFA 101, Life Safety Code, 1994 Edition (the current edition at the time of the review). In an effort to expand the employer compliance options without lessening employee safety, the revision included some of the options from NFPA 101. For example, the Life Safety Code allows for the exit to lead to a place of refuge rather than to an outside public way and allows the use of self-luminous and electroluminescent exit signs.

The revised regulation contains definitions of terms related to the topic and was reorganized around the three aspects of exit routes:

1. Design and construction requirements;
2. Maintenance, safety guards and operational requirements; and
3. Requirements for warning employees of the need to escape.

Because this regulation is referenced in other subparts, OSHA referred to those affected subparts in the regulation to ensure that there was continuity throughout the workplace with regard to emergency exits, emergency action plans and fire prevention plans. For example, the latter part of the rule refers to subpart H, L, R and Z where those rules require an Emergency Action Plan or Fire Prevention Plan.

While this discussion is devoted to the subject of emergency exit and subsequent planning, the emphasis may appear to be on escaping from fires. While this is certainly a primary reason for emergency egress from a building, there are a multitude of events that may result in exercising this sort of plan.

Hazards that must be considered include:
- Terrorist activity
- Workplace violence
- Explosion
- Earthquake
- Smoke (without fire)
- Toxic vapors
- Bomb threat
- Storms (e.g., tornado, hurricane)
- Flash floods
- Nuclear radiation exposure and
- Actions or threatened actions of mentally ill persons or political radicals

Each of these hazards to the occupants of a building can occur individually or in combination with others. Depending on the hazard, the people involved, the characteristics of the building and the quality of the means of egress provided, each hazard can be compounded by:
- Panic and confusion
- Poor visibility and
- Lack of information or misinformation.

These compounding factors frequently cause more injuries and fatalities than the hazard itself. Providing the proper means of egress can enable persons to successfully escape from the primary hazard.

Coverage and Definitions—1910.34

This section applies to workplaces in general industry and every employer is covered. This section does not cover mobile workplaces such as vehicles or vessels. OSHA set forth the minimum requirements for exit routes that employers must provide in order to evacuate the workers safely during an emergency as well as cover the minimum requirements for emergency action plans and fire prevention plans.

Definitions

Electroluminescent means a light-emitting capacitor. Alternating current excites phosphor atoms when placed between the electrically conductive surfaces to produce light. This light source is typically contained inside the device.

Exit means that portion of an exit route that is generally separated from other areas to provide a protected way of travel to the exit discharge. An example of an exit is a two-hour fire resistance-rated enclosed stairway that leads from the fifth floor of an office building to the outside of the building.

Exit access means that portion of an exit route that leads to an exit. An example of an exit access is a corridor on the fifth floor of an office building that leads to a two-hour fire resistance-rated enclosed stairway (the exit).

Exit discharge means the part of the exit route that leads directly outside or to a street, walkway, refuge area, public way or open space with access to the outside. An example of an exit discharge is a door at the bottom of a two-hour fire resistance-rated enclosed stairway that discharges to a place of safety outside the building.

Exit route means a continuous and unobstructed path of exit travel from any point within a workplace to a place of safety (including refuge areas). An exit route consists of three parts: the exit access, the exit and the exit discharge. (An exit route includes all vertical and horizontal areas along the route.)

High hazard area means an area inside a workplace in which operations include high hazard materials, processes or contents.

Occupant load means the total number of persons that may occupy a workplace or portion of a workplace at any one time. The occupant load of a workplace is calculated by dividing the gross floor area of the workplace or portion of a workplace by the occupant load factor for that particular type of workplace occupancy. Information regarding "occupant load" is located in NFPA 101-2000, Life Safety Code.

Refuge area means (1) A space along an exit route that is protected from the effects of fire by separation from other spaces within the building by a barrier with at least a one-hour fire resistance-rating; or (2) A floor with at least two spaces, separated from each other by smoke-resistant partitions, in a building protected throughout by an automatic sprinkler system that complies with §1910.159 of this part.

Self-luminous means a light source that is illuminated by a self-contained power source (e.g., tritium) and that operates independently from external power sources. Batteries are not acceptable self-contained power sources. The light source is typically contained inside the device.

Compliance with NFPA 101-2000, Life Safety Code—1910.35

In order to enhance the opportunity for employers to comply with the intent of the regulation, OSHA has provided owners a choice of complying with the OSHA regulation or the most recent NFPA standard on Life Safety. The regulation was revised using the 1996 NFPA standard, which has been replaced with a 2000 standard.

Subpart E is promulgated from NFPA 101, Life Safety Code. This document is prepared, maintained and published by the National Fire Protection Association. Copies of the NFPA standard can be obtained for a small fee from them by writing to them at 11 Tracy Drive, Avon, MA 02322.

Because this code is used as the basis for most local fire codes, it is written for general applicability. Keep in mind that your concern is its application primarily for the protection of employees, not the preservation of facilities.

The Life Safety Code, formerly the Building Exits Code, originated from work performed by the Committee on Safety to Life, The National Fire Protection Association, in 1913. At first, the committee devoted its attention to a study of the notable fires involving loss of life. That led to the preparation of standards for the construction of stairways and fire escapes, for fire drills in various occupancies and for the construction and arrangement of exit facilities for factories, schools, etc. These form the basis of the present code.

In 1921, the committee was enlarged to provide a comprehensive guide to exits and related features of life safety from fire in all classes of occupancy. This guide was to be known as the Building Exits Code.

During the Coconut Grove nightclub disaster in Boston in 1942, 492 lives were lost, focusing national attention upon the importance of adequate exits and related fire safety features. Public attention to exit matters was further stimulated by a series of hotel fires in 1946. The Building Exits Code thereafter was used to an increasing extent for legal regulatory purposes. The code, however, was not in suitable form for adoption into law and the entire code was edited to limit the body of the text to requirements suitable for mandatory application. This involved adding provisions to the code of many features that had not been previously covered in order to produce a complete document.

In 1966, the code title was changed from "Building Exits Code" to "The Code for Safety to Life from Fire in Buildings and Structures," known as the "Life Safety Code."

The regulation reflects that an employer who demonstrates compliance with the exit route provisions of NFPA 101-2000, the Life Safety Code, will be deemed to be in compliance with the corresponding requirements in 1910.34, 1910.36 and 1910.37.

Design and Construction Requirements for Exit Routes—1910.36

This subpart contains general fundamental requirements essential to providing a safe means of egress from fire and other emergencies. The requirements in Subpart E are minimum requirements. The standards do not prohibit better construction, more exits or safer conditions that these minimums set forth.

These requirements are not intended to apply to exits from vehicles, vessels or other mobile structures.

Fundamental Requirements for Exit Routes

This subparagraph contains requirements that apply to all buildings, new or old, that are intended for human occupancy. They may be summarized as follows:

- Each exit route must be a permanent part of the workplace.
- Exits must be separated from other parts of the workplace by fire resistant materials.
  - One-hour rating for less than four story buildings.
  - Two-hour rating for four story buildings or higher.
- Openings to exits must be limited to those needed to allow exit discharge or access to the exit from occupied areas of the workplace. Openings into an exit must be protected by an approved self-closing fire door.
- Number of exit routes must be adequate.
  - Two exit routes located as far away as practical from each other.

EXIT ROUTES, EMERGENCY ACTION PLANS, AND FIRE PREVENTION PLANS

- Single exit route is allowed if the size of the building, its occupancy or the workplace is arranged so that all employees would be able to evacuate safely through one exit.
- NFPA 101-2000, Life Safety Code, provides assistance in determining the required number of exits.
- Exit discharge must lead directly outside to a street, public way, walkway, open space or a refuge area that is large enough to hold all the building occupants.
- Exit stairs that lead beyond the exit discharge must have doors and partitions to clearly indicate the direction to lead to the actual exit.
- Exit doors must be unlocked.
  - Employees must be able to open the exit route door without keys, tools or special knowledge.
  - Panic bars are permitted on exit discharge doors.
  - Exit route doors may be locked from the inside only in mental, penal or correctional facilities only if supervisory personnel are continuously on duty and the employer has a written plan to remove those occupants during an emergency.
- Side-hinged exit doors must be used.
  - If the room is designed to hold more than 50 people or is a high hazard area, the door must swing out in the direction of exit travel.
- Exit routes must support the maximum permitted occupant load of each floor.
- The capacity of an exit route may not get smaller in the direction of exit route travel.
- The ceiling of an exit route must be at least 7 ft. 6 in. high and items reaching down from the ceiling must be at least 6 ft. 8 in. from the floor.
- Exit access must be at least 28 in. wide, but must accommodate the maximum permitted occupant load of each floor served by that route. No object may stick out into the path of the exit route in such a way as to reduce the minimum width requirement.

Exterior Routes of Exit Access

Under certain conditions it is permissible to plan an exterior route as a way of exit access from one interior part of a building to another or to an exterior exit. Such routes may include flat rooftops, enclosed courtyards and balconies.

Normally, when thinking of an exit, people visualize a door or doorway through which one can pass from the inside to the outside. As used in this subpart, an exit can be such a doorway and it can also be an interior stairwell or anteroom. It also can be an exterior fire escape.

Requirements for exit routes that are outside include:
- The ceiling of an exit route must be at least 7 ft. 6 in. high and items reaching down from the ceiling must be at least 6 ft. 8 in. from the floor.
- Exit access must be at least 28 in. wide, but must accommodate the maximum permitted occupant load of each floor served by that route. No object may stick out into the path of the exit route in such a way as to reduce the minimum width requirement.
- If a fall hazard exists, the route must have guardrails.
- If there is the likelihood of snow or ice accumulation, the route must be covered unless the employer can demonstrate that snow and ice is removed before it becomes a slipping hazard.
- The outdoor route must be reasonably straight with smooth, solid and substantially level walkways.
- No route more than 20 ft. long may lead to a dead end.

Width and Capacity of Means of Egress

The capacity in number of persons per unit of exit width for approved components of means of egress shall be as follows:
- Level Egress Components (including Class A ramps): 100 persons/unit
  - A Class A ramp is one that has a slope no greater than 1 3/16 in. in a 12-in. length, a width at least 44 in., and no limit on the maximum height between landings.
- Inclined Egress Components (including Class B ramps): 60 persons/unit
  - A Class B ramp is one that has a slope of 1 3/16 to 2 in. in a 12-in. length, a width of 30 to 44 in., and a maximum height between landings of 12 ft.

Maintenance, Safeguards and Operational Features for Exit Routes—1910.37

Minimizing the danger to employees

This section is established to minimize the danger to employees. During the Coconut Grove fire disaster in Boston, flames ignited wall decorations and wall coverings, which emitted toxic smoke as they burned. Fire exits were locked, forcing patrons to exit through dangerous areas. As a result of the lessons learned at that disaster, this section requires that all exit routes be kept free of any explosive or flammable furnishings or decorations and that the routes to exits must be set up so that employees do not have to travel towards any high hazard area. If the exit requires the employees to move towards this type of hazard, the exit path must be shielded by partitions or barriers that effectively protect the employee.

Nothing may be placed in the path of the exit route and these routes must not be locked. They must not be blocked by materials, even temporarily. Exit access cannot go through a room that may be locked, such as a bathroom, because if that room was locked it would be eliminated as an exit route. If there is a change of elevation along the exit route, stairs or a ramp must be provided.

Any fire-retardant material applied to the surfaces must be renewed as often as necessary. This material is not permanent and may only last for five years. Follow the manufacturers' recommendations for retaining the fire-retardant properties of the materials used in the structure.

Exit Marking and Lighting

Exit markings fall into two categories:
- Signs or markings that clearly identify an exit or the way to an exit.
- Signs or markings that clearly identify doors or areas that are not means of egress.

All exits must have enough light so that the employees can see along the exit route. The exits must be clearly visible and marked by a sign that says "Exit" in legible letters not less than 6 in. high and ¾ in. wide.

Exit signs must be illuminated to a surface value of at least five foot candles and be a distinctive color so that it stands out in an emergency. Glow in the dark signs (Self-luminous or electroluminescent) are permitted as long as they have a minimum luminance surface value of 0.06 footlamberts or more.

Decorations or signs that make it difficult to see or understand the exit route doors are prohibited. Any doorway or passage that is not an exit, but may be mistaken for an exit, must be marked so workers do not attempt to escape through that route. Placing a sign on the door that says "Not an Exit" or a sign that indicates the actual use of the room such as "Store Room" or "Linen Closet" meets the requirements of the standard.

Any time the employee cannot see the exit or the way to the exit is not immediately apparent, signs must be installed indicating the direction of travel to the nearest exit. A line of sight to an exit or an exit sign must be visible to workers at all time.

All workers are not the same height. It is important to realize that shorter workers may be a workstation previously occupied by a tall employee. The shorter worker may not be able to see the exit signs due to his or her height. To ensure compliance and employee safety, the employer must relocate the signs so that all employees can maintain a line of sight to an exit or exit sign.

Construction, Repairs or Alterations

Facilities are constantly being built, upgraded, altered and repaired. When construction crews are working on new buildings, they are focused on the construction job; facility managers are usually focused on occupying the building as soon as possible. However, the facility manager must ensure that workers do not occupy the workplace of a new building until all of the exit routes are completed and ready for use.

During the construction process itself, employees must not be exposed to any hazardous equipment or flammable substances used by the construction workers that would impede their exit from the work area during an emergency. Employees may not work in or occupy the workplace unless the exit routes are available or fire protection is provided that gives an equivalent level of safety.

EXIT ROUTES, EMERGENCY ACTION PLANS, AND FIRE PREVENTION PLANS

Employee Alarm System

An alarm system must be provided that has a distinctive signal to warn employees of fire dangers or other emergencies. Some firms have one sound or signal for a fire and another for the accidental release of chemicals. For example, one long blast on the horn could signal a fire, while two short blasts on the horn could signal a chemical release. Another long, uninterrupted blast may indicate that the facility must be immediately evacuated for unspecified reasons. The alarm system must meet the requirements of 1910.165 and be maintained in an operable manner.

In some small shops, the work area is so small that workers could immediately see the danger and exit the facility. If the employees can see or small the hazard in a sufficient amount of time to provide adequate warning, no alarm system is required.

Emergency Action Plans—1910.38

This section applies to all emergency action and fire prevention plans required by a particular OSHA standard.

For companies with more than 10 employees, the emergency action plan must be in writing. The plan must be kept at the workplace and be available for employees to review. If the company has less than 10 employees, the plan may be communicated to the workers orally.

The emergency action plan should address all potential emergencies that can be expected in the workplace. It may be helpful to perform a hazard audit to determine potentially toxic materials and unsafe conditions. For information on chemicals, the manufacturer or supplier can be contacted to obtain material safety data sheets. These sheets describe the hazards that a chemical may present; list precautions to take when handling, storing, or using the substance; and outline emergency and first aid procedures.

The employer should list in detail the procedures to be taken by those employees who must remain behind for essential plant operations until their evacuation becomes absolutely necessary. This may include monitoring plant power supplies, water supplies, and other essential services that cannot be shut down for every emergency alarm.

For emergency evacuation, floor plans or workplace maps that clearly show emergency escape routes and safe, or refuge, areas should be included in the plan. All employees must be told what actions they are to take in an emergency situation.

Alarm System

Employers shall establish an employee alarm system that complies with 1910.165. Alarms should be audible or able to be seen by all people in the plant and should have an auxiliary power supply in the event electricity is affected. The alarm should be distinctive and recognizable as a signal to evacuate the work area or perform actions designated under the emergency action plan.

Evacuation

The employer shall establish in the emergency action plan the types of evacuation to be used in emergency circumstances. At the time of an emergency, employees should know what type of evacuation is necessary and what their role is in carrying out the plan. In instances where the emergency is very grave, total and immediate evacuation of all employees is necessary. In other emergencies, a partial evacuation of nonessential employees (with a delayed evacuation of others) may be necessary for continued plant operation.

In some cases, only those employees in the immediate area of the fire may be expected to evacuate or move to a safe area, such as when a local application fire suppression system discharge employee alarm is sounded. Employees must be sure that they know what is expected of them in all such foreseeable emergency possibilities to provide assurance of their safety from fire or other emergency.

The designation of refuge or safe areas for evacuation should be identified in the plan. In a building divided into fire zones by firewalls, the refuge area may still be within the same building, but in a different zone from where the emergency occurs.

Exterior refuge or safe areas may include parking lots, open fields or streets that are located away from the site of the emergency and provide sufficient space to accommodate the employees. Employees should be instructed to move away from the exit discharge doors of the building and to avoid congregating close to the building where they may hamper emergency operations.

The emergency action plan must include, at a minimum, the following elements:

- Procedures for reporting fires or emergencies;
- Procedures for emergency evacuation, including:
  - Exit route assignments; and
  - Type of evacuation (partial or full evacuation);
- Procedures to be followed by employees who remain in place to operate critical plant operations before they evacuate;
- Procedures to account for all employees after evacuation;
- Procedures to be followed by employees performing rescue, medical or assistance duties; and
- The name or job title of every employee who may be contacted to learn more about the plan.

Training

Training is important to the effectiveness of an emergency plan. Before implementing an emergency action plan, a sufficient number of persons must be trained to assist in the safe and orderly evacuation of employees. Training for each type of disaster response is necessary so those employees know what actions are required. The employer shall review with each employee upon initial assignment those parts of the plan that the employee must know to protect himself or herself in the event of an emergency. In addition, the employer shall review the plan with each employee covered by the plan:

- Initially when the plan is developed;
- Whenever the employee's responsibilities or designated actions change; or
- Whenever the plan is changed.

The employer should ensure that an adequate number of employees are available at all times during working hours to act as evacuation wardens so that employees can be swiftly moved from the dangerous location to a safe area. Generally, one warden for each 20 employees in the workplace should be able to provide adequate guidance and instruction at the time of a fire emergency.

The employees selected or who volunteer to serve as wardens should be trained in the complete workplace layout and the various alternative escape routes from the workplace. All wardens and fellow employees should be made aware of handicapped employees who may need extra assistance, such as using the buddy system and of hazardous areas to be avoided during emergencies. Before leaving, wardens should check rooms and other enclosed spaces in the workplace for employees who may be trapped or otherwise unable to evacuate the area. After the desired degree of evacuation is completed, the wardens should be able to account for or otherwise verify that all employees are in the safe area.

Medical Assistance

In a major emergency, time is a critical factor in minimizing injuries. Most small businesses do not have a formal medical program, but they are required to have the following medical and first aid services:

- In the absence of an infirmary, clinic or hospital in close proximity to the workplace that can be used for the treatment of all injured employees, the employer must ensure that a person or persons are adequately trained to administer first aid.
- Where the eyes or body of any employee may be exposed to injurious corrosive materials, eye washes or suitable equipment for quick drenching and flushing must be provided in the work area for immediate emergency use. Employees must be trained to use the equipment.
- The employer must ensure the ready availability of medical personnel for advice and consultation on matters of employee health. This does not mean that health care must be provided, but rather that if health problems develop in the workplace, medical help will be available to resolve them.

Fire Prevention Plan—1910.39

Although Subpart E is devoted to the provisions for ensuring that personnel can exit from a building under emergency conditions, it also contains several provisions for preventing or reducing the risk of such an emergency. Detailed requirements for fire protection are in Subpart L.

Subpart E requires that where protection (e.g., automatic sprinklers, fire-retardant paints) are required and/or installed, they shall be regularly inspected or tested, maintained, and replenished or renewed as necessary to keep them in good operating condition.

Employers must inform employees of the fire hazards of the materials and processes to which they are exposed. Upon initial assignment, the employer shall review those parts of the fire prevention plan that each employee must know to protect him or herself in the event of an emergency. The written plan shall be kept in the workplace and available to the employee. The plan may be communicated orally in establishments with 10 or fewer employees.

The following elements, at a minimum, shall be included in a fire prevention plan:

- A list of all major workplace hazards and their proper handling and storage procedures, potential ignition sources and the type of fire equipment or systems used to control a fire involving them;
- Procedures to control the accumulation of flammable and combustible waste materials and residue;
- Procedures for the regular maintenance of safeguards installed on heat-producing equipment to prevent accidental ignition;
- Names or job titles of people responsible for the maintenance of equipment used to prevent or control sources of ignition or fires; and
- Names or job titles of people responsible for the control of fuel source hazards.

The employer shall control accumulations of flammable and combustible waste materials and residues so they do not contribute to a fire emergency. These procedures shall be included in the fire prevention plan.

Upon initial assignment, the employer shall review those parts of the fire prevention plan that each employee must know to protect him or herself in the event of an emergency.

The written plan shall be kept in the workplace and available to all employees. The plan may be communicated orally in establishments with 10 or fewer employees. Equipment and systems installed on heat-producing equipment shall be maintained in order to prevent accidental ignition of combustible materials.

It can be expected that egress will:
- Exist
- Provide convenient escape
- Have redundancy in design of exits and safeguards
- Not complicate escape by the building structure and design
- Not be locked or cluttered
- Require people to be trained for emergencies

Fundamental Requirements
- Shall have exists sufficient for prompt escape
- Designed with redundancy
- Building structures will not cause danger
- No locks to prevent escape
- Clearly visible and appropriately marked
- Illuminated
- Alarm to warn of emergencies
- Side-hinged doors

Exit Components
- Protective enclosure of exit (one-hour rating for three stories, two-hour rating for four or more stories)
- Width (22 in. units) and capacity
- Capacity and occupant load
- Sufficient for occupant load
- Occupant load is the maximum number of persons that may be in the room
- Exit capacity must not constrict or get smaller in direction of travel

Arrangement of and Access to Exits
- Separated from other exits to minimize chance of more than one being blocked
- Readily accessible at all times
- Doors swing with exit travel in rooms designed to hold 50 or more persons
- Clearly recognizable
- Safe discharge

Exit Marking
- Signs or markings that clearly identify exits
- Signs or markings that clearly identify doors and areas as not exits

Fire Prevention and Emergency Action Plans
- Housekeeping—Control accumulations
- Training—Written plan kept in workplace
- Maintenance—Heat producing equipment is maintained to prevent fires

Exhibit 12-1. Mean of Egress Expectations.

These procedures shall be included in the written fire prevention plan.

Emergency Escape Route and Procedures

Employees should be taught not to stop for clothing, food or other possessions when ordered to evacuate and to proceed directly to the evacuation point or exit. Designate emergency escape routes and post them in conspicuous places around the plant. Conduct evacuation drills a number of times each year to ensure that the workers remember where they are supposed to go. Depending on the facility, indicate a secondary escape route in case the primary route is blocked.

Emergency action plan. Emergency action plan shall cover those designated actions employers and employees must take to ensure employee safety from fire and other emergencies.

Fire prevention plan. Fire prevention plan must include a list of the major workplace fire hazards and their proper handling and storage procedures, potential ignition sources and their control procedures, and the type of fire-protection equipment or systems that can control a fire involving them.

Alarm system. The employer shall establish an employee alarm system.

Evacuation. The employer shall establish in the emergency action plan the types of evacuation to be used in emergency circumstances.

Training. Before implementing the emergency action plan, the employer shall designate and train sufficient number of persons to assist in the safe and orderly emergency evacuation of employees.

Housekeeping. The employer shall control the accumulations of flammable or combustible waste materials and residues so that they do not contribute to a fire emergency.

Maintenance. The employer shall regularly and properly maintain equipment and systems installed to prevent accidental ignition of combustible materials.

Fact Sheet No. OSHA 92-19

Responding to Workplace Emergencies

Employers should establish effective safety and health programs and prepare their workers to handle emergencies before they arise.

Planning

Where required by the Occupational Safety and Health Administration (OSHA), firms with more than 10 employees must have a written emergency action plan; smaller companies may communicate their plans orally. [See 29 Code of Federal Regulation (CFR) Part 1910.38(a) for further information.] Essential to an effective emergency action plan are top management support and commitment and the involvement of all employees. Management should review plans with employees initially and whenever the plan itself, or employees responsibilities under it, change. Plans should be re-evaluated and updated periodically. Emergency procedures, including the handling of any toxic chemicals, should include:

- Escape procedures and escape route assignments.
- Special procedures for employees who perform or shut down critical plant operations.
- A system to account for all employees after evacuation.
- Rescue and medical duties for employees who perform them.
- Means for reporting fires and other emergencies.
- Contacts for information about the plan.

Chain of Command

An emergency response coordinator and a back-up coordinator must be designated. The coordinator may be responsible for plant-wide operations, public information and ensuring that outside aid is called in. A back-up coordinator ensures that a trained person is always available. Duties of the coordinator include:

- Determining what emergencies may occur and seeing that emergency procedures are developed to address them.
- Directing all emergency activities including evacuation of personnel.
- Ensuring that outside emergency services such as medical aid and local fire departments are called when necessary.
- Directing the shutdown of plant operations when necessary.

EXIT ROUTES, EMERGENCY ACTION PLANS, AND FIRE PREVENTION PLANS

Emergency Response Teams

Members of emergency response teams should be thoroughly trained for potential emergencies and physically capable of carrying out their duties; know about toxic hazards in the workplace and be able to judge when to evacuate personnel or depend on outside help (e.g. when a fire is too large for them to handle). One or more teams must be trained in:

- Use of various types of fire extinguishers.
- First aid, including cardiopulmonary resuscitation (CPR).
- The requirements of the OSHA bloodborne pathogens standard.
- Shutdown procedures.
- Chemical spill control procedures.
- Use of self-contained breathing apparatus (SCBA).
- Search and emergency rescue procedures.
- Hazardous materials emergency response in accordance with 29 CFR 1910.120.

Response Activities

Effective emergency communication is vital. An alternate area for a communications center other than management offices should be established in the plans and the emergency response coordinator should operate from this center. Management should provide emergency alarms and ensure that employees know how to report emergencies. An updated list of key personnel and off-duty telephone numbers should be maintained. A system should be established for accounting for personnel once workers have been evacuated with a person in the control center responsible for notifying police or emergency response team members of persons believed missing.

Effective security procedures, such as cordoned off areas, can prevent unauthorized access and protect vital records and equipment. Duplicate records can be kept in off-site locations for essential accounting files, legal documents and lists of employees relatives to be notified in case of emergency.

Training

Every employee needs to know details of the emergency action plan, including evacuation plans, alarm systems, reporting procedures for personnel, shutdown procedures, and types of potential emergencies. Drills should be held at random intervals, at least annually, and include, if possible, outside police and fire authorities.

Training must be conducted initially, when new employees are hired, and at least annually. Additional training is needed when new equipment, materials, or processes are introduced, introduced, when procedures have been updated or revised, or when exercises show that employee performance is inadequate.

Personal Protection

Employees exposed to accidental chemical splashes, falling objects, flying particles, unknown atmospheres with inadequate oxygen or toxic gases, fires, live electrical wiring, or similar emergencies need personal protective equipment, including:

- Safety glasses, goggles, or face shields for eye protection.
- Hard hats and safety shoes.
- Properly selected and fitted respirators.
- Whole body coverings, gloves, hoods, and boots.
- Body protection for abnormal environmental conditions such as extreme temperatures.

Medical Assistance

Employers not near an infirmary, clinic, or hospital should have someone on-site trained in first aid, have medical personnel readily available for advice and consultation, and develop written emergency medical procedures.

It is essential that first aid supplies are available to the trained medical personnel, that emergency phone numbers are placed in conspicuous places near or on telephones, and prearranged ambulance services for any emergency are available.

Emergency Preparedness and Response

The following is an excerpt from an Instructional Guide developed for the OSHA's Small Business Outreach Training Program.

Introduction

The importance of an effective workplace safety and health program cannot be overemphasized.

There are many benefits from such a program including increased productivity, improved employee morale, reduced absenteeism and illness, and reduced workers' compensation rates; however, incidents still occur in spite of efforts to prevent them. Therefore, proper planning for emergencies is necessary to minimize employee injury and property damage.

Purpose

This discussion details the basic steps to handle emergencies in the workplace. These emergencies include accidental releases of toxic gases, chemical spills, fires, explosions, and bodily harm and trauma caused by workplace violence. This discussion is intended to assist small businesses that do not have safety and health professionals. It is not intended as an all inclusive safety program but rather to provide guidelines for planning for emergencies.

Planning

The effectiveness of response during emergencies depends on the amount of planning and training performed. Management must show its support of plant safety programs and the importance of emergency planning. If management is not interested in employee protection and minimizing property loss, little can be done to promote a safe workplace. It is therefore management's responsibility to see that a program is instituted and that it is frequently reviewed and updated. The input and support of all employees must be obtained to ensure an effective program. The emergency response plan should be developed locally and be comprehensive enough to deal with all types of emergencies. When emergency action plans are required by a particular OSHA standard, the plan must be in writing. Although for firms with 10 or fewer employees, the plan may be communicated orally to employees. The plan must include the following elements:

- Emergency escape procedures and emergency escape route assignments;
- Procedures to be followed by employees who remain to perform (or shut down) critical plant operations before they evacuate;
- Procedures to account for all employees after emergency evacuation has been completed;
- Rescue and medical duties for those employees who are to perform them;
- The preferred means for reporting fires and other emergencies; and
- Names or regular job titles of persons or departments to be contacted for further information or explanation of duties under the plan.

The emergency action plan should address all potential emergencies that can be expected in the workplace. Therefore, it will be necessary to perform a hazard audit to determine potentially toxic materials and unsafe conditions. For information on chemicals, the manufacturer or supplier can be contacted to obtain material safety data sheets. These sheets describe the hazards that a chemical may present; list precautions to take when handling, storing, or using the substances and outline emergency and first-aid procedures.

The employer should list in detail the procedures to be taken by those employees who must remain behind to care for essential plant operations until their evacuation becomes absolutely necessary. This may include monitoring plant power supplies, water supplies, and other essential services that cannot be shut down for every emergency alarm.

For emergency evacuation, floor plans or workplace maps that clearly show the emergency escape routes and safe or refuge areas should be included in the plan. All employees must be told what actions they are to take in the emergency situations that may occur in the workplace.

This plan should be reviewed with employees initially when the plan is developed, whenever the employees' responsibilities under the plan change and whenever the plan is changed. A copy should be kept where employees can refer to it at convenient times. In fact, to go a step further, the employer could provide the employees with a copy of the plan, particularly all new employees.

Chain of Command

A chain of command should be established to minimize confusion so those employees will have no doubt about who has authority for making decisions. Responsible individuals should be selected to coordinate the work of the emergency response team. In larger organizations, there may be a plant coordinator in charge of plant-wide operations, public relations and ensuring that outside aid is called. Because of the importance of these functions, adequate backup must be arranged so that trained personnel are always available. The duties of the Emergency Response Team Coordinator should include the following:

- Assessing the situation and determining whether an emergency exists that requires activating the emergency procedures;
- Directing all efforts in the area including evacuating personnel and minimizing property loss;
- Ensuring outside emergency services such as medical aid and local fire departments are called when necessary; and
- Directing the shutdown of plant operations when necessary.

Communications

During a major emergency involving a fire or explosion, it may be necessary to evacuate offices in addition to manufacturing areas. Also, normal services, such as electricity, water and telephones may be nonexistent. Under these conditions, it may be necessary to have an alternate area to which employees can report or that can act as a focal point for incoming and outgoing calls. Because time is an essential element for adequate response, the person designated as being in charge should make this the alternate headquarters so that he or she can be easily reached.

Emergency communications equipment, such as amateur radio systems, public address systems or portable radio units, should be present to notify employees of the emergency and contact local authorities, such as law enforcement officials, the fire department and Red Cross.

A method of communication also is needed to alert employees to evacuate or to take other action as required in the plan. Alarms should be audible or seen by all people in the plant and should have an auxiliary power supply in the event electricity is affected. The alarm should be distinctive and recognizable as a signal to evacuate the work area or perform actions designated under the emergency action plan. The employer should explain to each employee the means for reporting emergencies, such as manual pull-box alarms, public address systems or telephones. Emergency phone numbers should be posted on or near telephones, on employees' notice boards or in other conspicuous locations. The warning plan should be in writing and management must be sure each employee knows what it means and what action is to be taken.

It may be necessary to notify other key personnel such as the plant manager or physician during off-duty hours. An updated written list should be kept of key personnel listed in order of priority.

Accounting for Personnel

Management will need to know when all personnel have been accounted for. This can be difficult during shift changes or if contractors are on site. A responsible person in the control center should be appointed to account for personnel and to inform police or Emergency Response Team members of those persons believed missing.

Emergency Response Teams

Emergency Response Teams are the first line of defense in emergencies. Before assigning personnel to these teams, the employer must ensure that employees are physically capable of performing the duties that may be assigned to them. Depending on the size of the plant, there may be one or several teams trained in the following areas:

- Use of various types of fire extinguishers;
- First aid, including cardiopulmonary resuscitation (CPR);
- Shutdown procedures;
- Evacuation procedures;
- Chemical spill control procedures;
- Use of self-contained breathing apparatus (SCBA);
- Search and emergency rescue procedures; and
- Incipient and advanced stage firefighting.

The type and extent of the emergency will depend on the plant operations and the response will vary according to the type of process, the material handled, the number of employees and the availability of outside resources. Emergency Response Teams should be trained in the types of possible emergencies and the emergency actions to be performed. They should be informed of special hazards to which they may be exposed during fire and other emergencies, such as storage and use of flammable materials, toxic chemicals, radioactive sources and water-reactive substances. It is important to determine when not to intervene. For example, team members must be able to determine if the fire is too large for them to handle or whether search and emergency rescue procedures should be performed. If there is a possibility of members of the Emergency Response Team receiving fatal or incapacitating injuries, they should wait for professional firefighters or emergency response groups.

Training

Training is important to the effectiveness of an emergency plan. Before implementing an emergency action plan, a sufficient number of persons must be trained to assist in the safe and orderly evacuation of employees. Training for each type of disaster response is necessary so those employees know what actions are required.

In addition to the specialized training for Emergency Response Team members, all employees should be trained on the firms' evacuation plans, alarm systems, shutdown procedures, types of potential emergency and reporting procedures for all personnel.

These training programs should be provided for all employees as follows:
- Initially when the plan is developed;
- When new equipment, materials or processes are introduced;
- When procedures have been updated or revised;
- When exercises show that employee performance must be improved; and
- At least annually.

The emergency control procedures should be written in concise terms and made available to all personnel. A drill should be held for all personnel at random intervals at least annually, and an evaluation of performance made immediately by management and employees. When possible, drills should include groups supplying outside services such as fire and police departments. In buildings with several employers, the emergency plans should be coordinated with other companies and employees in the building. Finally, the emergency plan should be reviewed periodically and updated to maintain adequate response personnel and program efficiency.

Personal Protection

Personal protection is essential for any person exposed to potentially hazardous substances. In emergency situations, employees may be exposed to a variety of hazardous circumstances, including:
- Chemical splashes or contact with toxic materials;
- Falling objects and flying particles;
- Unknown atmospheres that may contain toxic gases vapors or mists, or inadequate oxygen to sustain life; and
- Fires and electrical hazards.

It is extremely important for employees to be adequately protected in these situations. Some of the safety equipment that may be used includes:
- Safety glasses, goggles or face shields for eye protection;
- Hard hats and safety shoes for head and foot protection;
- Proper respirators for breathing protection;
- Whole-body coverings, gloves, hoods and boots for body protection from chemicals; and
- Body protection for abnormal environmental conditions (e.g., extreme temperatures).

The equipment selected must meet the criteria contained in the OSHA standards. The choice of proper equipment is not a simple matter; consultation should be made with health and safety professionals before making any purchases. Manufacturers and distributors of health and safety products may be able to answer questions if they have enough information about the potential hazards involved. Professional consultation will most likely be needed when providing adequate respiratory protection.

Respiratory protection is necessary for toxic atmospheres of dust, mists, gases, or vapors and for oxygen-deficient atmospheres. There are four basic categories of respirators:
- Air-purifying devices (e.g., filters, gas masks, chemical cartridges), which remove contaminants from the air but cannot be used in oxygen-deficient atmospheres;

- Air-supplied respirators (e.g., hose masks, air line respirators), which should not be used in atmospheres that are immediately dangerous to life or health;
- Self-contained breathing apparatus, which are required for unknown atmospheres, oxygen-deficient atmospheres or atmospheres immediately dangerous to life or health (positive-pressure type only); and
- Escape masks.

Before assigning or using respiratory equipment the following conditions must be met:
- A medical evaluation should be made to determine if the employees are physically able to use the respirator.
- Written procedures must be prepared covering safe use and proper care of the equipment and employees must be trained in these procedures and the use and maintenance of respirators.
- A fit test must be made to determine a proper match between the face piece of the respirator and the face of the wearer. This testing must be repeated periodically. Training must provide the employee an opportunity to handle the respirator have it fitted properly, test its face-to-face seal, wear it in normal air for a familiarity period and wear it in a test atmosphere.
- A regular maintenance program must be instituted, including cleaning, inspecting and testing of all respiratory equipment. Respirators used for the emergency response must be inspected after each use and at least monthly to ensure that they are in satisfactory working condition. A written record of inspection must be maintained.
- Distribution areas for equipment used in emergencies must be readily accessible to employees.

Self-contained breathing apparatus (SCBA) offers the best protection to employees involved in controlling emergency situations. It should have a minimum service life rating of 30 minutes. Conditions that require a SCBA include the following:
- Leaking cylinders or containers, smoke from chemical fires or chemical spills that indicate high potential for exposure to toxic substances.
- Atmospheres with unknown contaminants or unknown contaminant concentrations in confined spaces that may contain toxic substances or oxygen-deficient atmospheres.

Emergency situations may involve entering confined spaces to rescue employees who are overcome by toxic compounds or who lack oxygen. These confined spaces include tanks, vaults, pits, sewers, pipelines and vessels. Entry into confined spaces can expose the employee to a variety of hazards, including toxic gases, explosive atmospheres, oxygen deficiency, electrical hazards and hazards created by mixers and impellers that have not been deactivated and locked out. Personnel should never enter a confined space under normal circumstances unless the atmosphere has been tested for adequate oxygen, combustibility and toxic substances. Conditions in a confined space must be considered immediately dangerous to life and health unless shown otherwise.

If a confined space must be entered in an emergency, the following precautions must be taken:
- All lines containing inert, toxic, flammable or corrosive materials must be disconnected or blocked off before entry.
- All impellers, agitators or other moving equipment inside the vessel must be locked out.
- Employees must wear appropriate personal protective equipment before entering the vessel. Mandatory use of safety belts and harnesses should be stressed.
- Rescue procedures must be specifically designed for each entry. A trained stand-by person must be present. This person should be assigned a fully charged, positive-pressure self-contained breathing apparatus with a full face. The stand-by person must maintain unobstructed lifelines and communications to all workers within the confined space and be prepared to summon rescue personnel if necessary. The stand-by person should not enter the confined space until adequate assistance is present. While awaiting rescue personnel, the stand-by person may make a rescue attempt using lifelines from outside the confined space.

Medical Assistance

In a major emergency, time is a critical factor in minimizing injuries. Most small businesses do not have a formal medical program, but they are required to have the following medical and first-aid services:

- In the absence of an infirmary, clinic or hospital in close proximity to the workplace that can be used for treatment of all injured employees, the employer must ensure that a person or persons are adequately trained to administer first aid.
- Where the eyes or body of any employee may be exposed to injurious corrosive materials, eye washes or suitable equipment for quick drenching or flushing must be provided in the work area for immediate emergency use. Employees must be trained to use the equipment.
- The employer must ensure the ready availability of medical personnel for advice and consultation on matters of employees' health. This does not mean that health care must be provided. Rather, if health problems develop in the workplace, medical help will be available to resolve them.

To fulfill the above requirements, the following actions should be considered:

- Survey the medical facilities near the place of business and make arrangements to handle routine and emergency cases. A written emergency medical procedure should then be prepared for handling accidents with minimum confusion.
- If the business is located far from medical facilities, at least one and preferably more employees on each shift must be adequately trained to render first aid. The American Red Cross, some insurance carriers, local safety councils, fire departments and others may be contacted for this training.
- First-aid supplies should be provided for emergency use. This equipment should be ordered through consultation with a physician.
- Emergency phone numbers should be posted in conspicuous places near or on telephones.
- Sufficient ambulance service should be available to handle any emergency. This requires advance contact with ambulance services to ensure they become familiar with the plant location, access routes and hospital locations.

Security

During an emergency, it is often necessary to secure the area to prevent unauthorized access and to protect vital records and equipment. An off-limits area must be established by cordoning off the area with ropes and signs. It may be necessary to notify local law enforcement personnel or to employ private security personnel to secure the area and prevent the entry of unauthorized personnel.

Certain records also may need to be protected, such as essential accounting files, legal documents and lists of employees' relatives to be notified in case of emergency. These records may be stored in duplicate outside the plant or in protected secure locations within the plant.

Final Rule 29 CFR Part 1910 Subpart E Exit Routes, Emergency Action Plans, and Fire Prevention Plans (67 Fed. Reg. 67949)

SUMMARY: The Occupational Safety and Health Administration (OSHA) is revising its standards for means of egress. The purpose of this revision is to rewrite the existing requirements in clearer language so they will be easier to understand by employers, employees, and others who use them.

The revisions reorganize the text, remove inconsistencies among sections, and eliminate duplicative requirements. The rules are performance-oriented to the extent possible, and more concise than the original, with fewer subparagraphs, and fewer cross-references to other OSHA standards. Additionally, a table of contents has been added that is intended to make the standards easier to use.

Also, OSHA is changing the name of the subpart from "Means of Egress" to **"Exit Routes, Emergency Action Plans, and Fire Prevention Plans"** to better describe the contents.

Finally, OSHA has evaluated the National Fire Protection Association's Standard 101, Life Safety Code, 2000 Edition (NFPA 101-2000), and has concluded that the standard provides comparable safety to the **Exit Routes** Standard. Therefore, employers who wish to comply with the NFPA 101-2000 instead of the OSHA standards for **Exit Routes** may do so.

DATES: The final rule becomes effective December 9, 2002.

ADDRESSES: In accordance with 28 U.S.C. 2112(a), the Agency designates the Associate Solicitor of Labor for Occupational Safety and Health, Office of the Solicitor of Labor, Room S-4004, U.S. Department of Labor, 200 Constitution Avenue, NW., Washington, DC 20210 to receive petitions for review of the final rule.

FOR FURTHER INFORMATION CONTACT: OSHA, Ms. Bonnie Friedman, Director, Office of Public Affairs, N-3647, Occupational Safety and Health Administration, U.S. Department of Labor, 200 Constitution Avenue, NW., Washington, DC 20210; telephone: (202) 693-1999. For additional copies of this Federal Register document, contact: OSHA, Office of Publications, U.S. Department of Labor, Room N-3103, 200 Constitution Avenue, NW., Washington, DC 20210; telephone: (202) 693-1888. For electronic copies of this Federal Register document, as well as news releases, fact sheets, and other relevant documents, visit OSHA's homepage at http://www.osha.gov.

SUPPLEMENTARY INFORMATION: References to comments and testimony in the rulemaking record (Docket S-052) are found throughout the text of the preamble. In the preamble comments are identified by an assigned exhibit number as follows: "Ex. 5-1" means Exhibit 5-1 in Docket S-052. For quoted material in the preamble, the page number where the quote can be located is included if other than page one. The transcript of the public hearing is cited by the page number as follows: Tr. 37. A list of the exhibits, copies of the exhibits, and transcripts are available in the OSHA Docket Office.

I. Background

In 1971 and 1972, OSHA adopted hundreds of national consensus and established Federal standards under section 6(a) of the Occupational Safety and Health Act of 1970. Section 6(a) allowed the Agency to adopt these standards for a limited period of time without going through traditional rulemaking. Many of these "start-up standards" have been criticized for being overly wordy, difficult to understand, repetitive and internally inconsistent.

On September 10, 1996, OSHA published a proposed rule in the Federal Register (61 FR 47712) proposing to revise subpart E of part 1910. OSHA proposed to rewrite the existing requirements of subpart E in plain language so that the requirements would be easier to understand by employers, employees, and others who use them. The proposal did not intend to change the regulatory obligations of employers or the safety and health protection provided to employees by the original standard.

OSHA proposed two versions of the revision of subpart E. The first version was organized in the traditional regulatory format characteristic of most OSHA standards. The second version was in a question and answer format. OSHA invited interested parties to comment on the content and effectiveness of the proposed changes and to indicate which version they preferred. Both versions left unchanged the regulatory obligations placed on employers and the safety and health protection provided to employees. Based on the majority of comments (e.g., Exs. 5-13, 17, 24-26, 45-47, 58-60) OSHA has decided to use its traditional regulatory text format for this final rule. OSHA believes that the revised subpart E is more performance-oriented and more compliance options will be available to employers.

In the proposal, OSHA stated what it expected to achieve by revising subpart E: (1) To maintain the safety and health protection provided to employees without increasing the regulatory burden on employers; (2) to create a regulation that is easily understood and; (3) to state employers' obligations in performance-oriented language to the extent possible.

The proposal attempted to simplify, rather than to substantively revise, OSHA's means of egress standards. In finalizing this proposal, the Agency has been careful to ensure that the protections afforded employees were not weakened. Employers who are in compliance with the original subpart E will continue to be in compliance with the revised subpart E that is being promulgated in this rule.

In developing the proposal, OSHA reviewed relevant OSHA decisions of the Federal courts, the Occupational Safety and Health Review Commission, and Agency letters of interpretation (Ex. 2) to determine how each provision of subpart E has been interpreted. Also, OSHA reviewed comparable State regulations, training materials and current consensus standards including the National Fire Protection Association's Life Safety Code, NFPA 101 (at that time the 1994 Edition). This review enabled OSHA to reorganize subpart E, eliminate duplicative provisions, and have confidence that the revisions did not diminish the safety and health protection afforded by existing rules.

OSHA discovered during the review process that some provisions of subpart E were outdated and not consistent with contemporary fire safety options in then current NFPA 101, Life Safety Code, 1994 Edition. Where it was possible to expand permissible employer compliance options without lessening employee safety, the proposal included these expanded options. For example, OSHA incorporated NFPA 101, 1994 Edition, the Life Safety Code's option to exit to a refuge area rather than to the outside (proposed paragraph 1910.36(f)(3)). The proposal also permitted the use of self-luminous and electroluminescent exit signs (proposed paragraph 1910.37(c)(6)). (E.g., Exs. 5-18, 40, 45, 54.) The proposal enabled employers to avail themselves of these newer options or continue with current compliance methods. In this way OSHA increased compliance flexibility without reducing safety.

OSHA did not substitute performance-oriented language for current language where doing so would either eliminate a requirement that protects employee safety and health, or expand an employer's compliance obligation. For example, the proposal continued the existing requirement that a means of egress must be at least 28 inches wide (proposed paragraph 1910.37(j)). The Agency chose not to substitute performance-oriented criteria for this provision (such as "means of egress be of adequate width to support building occupants") because this change would eliminate the existing minimum width specification and might not provide adequate protection to employees leaving the workplace in an **emergency**. For this reason, OSHA decided not to revise the minimum clearance requirement.

OSHA noted in the proposal that for some employers, reliance on performance-oriented standards might create confusion as to the specific precautions necessary in a variety of situations. In the past, OSHA has used NFPA 101 as an aid in interpreting subpart E. OSHA intends to continue to rely on NFPA 101 as guidance in implementing performance-oriented provisions of revised subpart E.

In addition to organizing the requirements of the revised subpart E in a logical and understandable manner, OSHA has organized the requirements around three aspects of **exit routes:**
(1) Design and construction requirements;
(2) maintenance, safeguards, and operational requirements; and

EXIT ROUTES, EMERGENCY ACTION PLANS, AND FIRE PREVENTION PLANS

(3) requirements for warning employees of the need to escape. Reorganizing subpart E in this manner has enabled OSHA to eliminate many duplicative provisions.

For example, in existing subpart E, both paragraph 1910.36(b)(8) and paragraph 1910.37(e) contain the design requirements that where workplaces are required to have two means of egress, these means of egress must be located as far away as practical (remote) from one another.

Other significant revisions to subpart E include: Removal of obligations that are not related to employee protection but pertain to the protection of the general public, and the deletion of any recommended as opposed to required actions (i.e., provisions that use "should" or "may").

II. Regulatory Format

As noted above, OSHA proposed two versions of subpart E; a traditional regulatory text version and a question and answer version. The traditional regulatory text version was preceded by a descriptive section heading that told the reader what information could be found in that section. The question and answer version was written in a form by which an employer might ask a question about the rule, and this question was then followed by an answer that told the employer about the requirement.

Other efforts to make subpart E more user-friendly included: removal of unused terms and ordinary terms from the definitions; elimination of cross-references to other standards; removal of overly technical terms in favor of more common words; use of the active voice; and, the use of positive as opposed to negative sentences.

The Agency invited public comment and requests for a hearing on the proposed revision to subpart E. An informal public hearing was requested by the National Fire Protection Association (Ex. 5-18) and Hallmark Cards (Ex. 5-51).

On March 3, 1997, OSHA published a notice in the Federal Register (62 FR 9402) announcing an informal public hearing and a reopening of the written comment period. Written comments on the proposed standard were to be postmarked by April 19, 1997. The hearing was held in Washington, DC on April 29-30, 1997.

In the hearing notice, OSHA invited comment on ten issues that will be discussed below in more detail. In summary, OSHA asked: (1) How OSHA should use the Life Safety Code in the final rule; (2) how or if OSHA should use model building codes; (3) whether the use of performance language creates new enforcement problems;

(4) how OSHA should address the issues of exit capacity and the number of required exits;
(5) whether or not the exit sign provisions were too general;
(6) whether or not the revised requirements for exit illumination were too general;
(7) whether or not there were still provisions or terms in the proposed revision that were too technical or difficult to understand;
(8) whether OSHA achieved in the proposed revision its goal of not changing employers obligations;
(9) whether any of the proposed provisions provided greater protection than in the original subpart E; and
(10) whether any of the requirements presented technological feasibility problems for affected employers.

The subpart E rulemaking record contains 23 exhibits, 69 comments, 170 pages of testimony and four post-hearing comments.

III. Summary and Explanation of the Final Rule

This section contains an analysis of the record evidence and policy decisions pertaining to the various provisions of revised subpart E.

As stated previously, OSHA's goals in revising subpart E were to maintain the safety and health protection provided to employees in subpart E without increasing the regulatory burden on employers, create a regulation that is easily understood, and, to the extent possible, express employers' obligations in performance-oriented language.

The majority of commenters supported OSHA's use of plain language. Owens Manufacturing, Inc. (Ex. 5-1) stated they were "in favor of this change as it allows the production people in our manufacturing area to understand the scope and meaning of this regulation much easier."

United Refining Company (Ex. 5-2) remarked "For those individuals who occasionally reference a standard the Plain English version will be beneficial." The commenter from Medical Environment, Inc. (Ex. 5-7) stated "I commend your actions in correcting the highly technical language into wording that is understandable to the average person. I have read your proposed changes, and find them to be significantly improved." The Institute for Interconnecting and Packaging Electronic Circuits (IPC) (Ex. 5-25) observed that:

> * * * Because IPC members are predominantly small companies, they have limited resources to track down, read, understand, and comply with the substantial volume of federal, state, and local regulations. In many firms, the company president, plant manager, or production supervisor is responsible for facility-wide health and safety compliance in addition to running production and perhaps running the company.
>
> Given IPC members' commitment to advancing employee health and safety, IPC applauds OSHA's proposed Means of Egress rule. The proposed changes are designed to make the standard more understandable and, therefore promote industry compliance. "Translating" OSHA's current regulations into "plain English" is an outstanding activity that should be aggressively applied to ALL federal regulations—not just OSHA regulations, and IPC supports OSHA's actions to effect such change.

The International Brotherhood of Teamsters (Ex. 5-31) commended OSHA for undertaking the revision effort and stated that the International:

> [I]s pleased to see the Occupational Safety and Health Administration attempt to develop plain English standards. This International Union feels that this approach to safety and health standards will enable our members and other workers across the country to better understand their OSHA rights and their employer's obligations.

The National Institute for Occupational Safety and Health (NIOSH, Ex. 5-42) also supported the effort observing that "By revising the Means of Egress rule in easy to understand terms as part of a shorter, performance-oriented standard, the standard will be easier to use and provide more compliance options for employers."

Schirmer Engineering Corporation (Ex. 5-57) stated:

> Review of the revisions introduced in the proposed rule indicates an effort to provide language which is more condensed and clear, with the removal of verbose wording. The sections that were deleted from the original version did not greatly affect the overall life safety concept as it pertains to egress from a building. In addition, the reorganization helps to clarify some of the requirements of the code which, in turn, facilitates overall compliance.

(See also Exs. 5-5, 12, 13, 15-17, 20-24, 26, 27, 29, 30, 34, 35, 37, 39, 43, 45, 47, 51, 52, 54-56, 58, 59, 60, 61, 70.)

On the other hand, some commenters did object to the revision of subpart E on the grounds either that it was not productive for OSHA to re-write these standards, or that the revised language actually changed the requirements. For example, James R. Hutton, a fire protection engineer (Ex. 5-9), believed the "proposed revisions will complicate and cause more difficulties, not less, for smaller businesses who do not have the resources to undergo the time or expense required to develop "custom solutions" to "plain English" requirements." OSHA disagrees. The revised subpart E only makes compliance requirements clearer and it refers employers and employees to NFPA 101 for added details, when necessary.

It was also suggested by some commenters that instead of finalizing the proposed revision, OSHA should adopt NFPA 101, the Life Safety Code, or that OSHA should rely on building codes, instead of revising subpart E. (See e.g., Exs. 5-10, 15, 18, 19, 26, 41, 46, 48, 61, 68; Tr. 14, 23; Ex. 10.)

The National Fire Protection Association (NFPA, Ex. 5-18) remarked:

> NFPA agrees with several of the goals as contained in the OSHA/NPRM but find serious flaws in the methodology being

proposed to attain these goals. Specifically, NFPA applauds OSHA's goal "to maintain the safety and health protection provided to employees by subpart E * * *" and "to create a regulation that is easily understood." We also applaud OSHA's desire "to allow employers the flexibility of relying on more contemporary compliance approaches."

However, we do not believe these goals can be achieved by either "plain English" alternative taken together or separately as being proposed by OSHA in the NPRM. Specifically, NFPA recommends OSHA abandon its attempt to rewrite a 25-year old standard as represented in the first alternative of the NPRM * * *.

Further, NFPA asserted that OSHA's rewrite would make enforcement more difficult especially when performance-oriented language is substituted for specifications; that the proposal drops all references to the NFPA Life Safety Code even though the proposal indicated OSHA would continue to rely on that Code; and, that the proposed rewrite did not specifically allow for contemporary compliance options as contemplated by OSHA and as set forth in the current edition of NFPA 101 (1994). NFPA recommended that:

[T]he first alternative be abandoned [traditional regulatory text] and that OSHA instead adopt by reference the 1994 edition of NFPA 101 * * * Further, NFPA believes the adoption of the 1994 edition of NFPA 101, together with a supplemental Q&A (question and answer) format as proposed in the second NPRM alternative, would be the best approach to achieve the desired goals as stated by OSHA in the NPRM.

At the time of the proposal, the latest version of NFPA 101 was the 1994 Edition. NFPA subsequently issued a 1997 edition and then a 2000 edition. OSHA has reviewed the NFPA 101-2000 edition carefully and found that compliance with its provisions would protect employees as well as the parallel provisions of subpart E. Adopting NFPA 101 as an OSHA standard would require OSHA to conduct a full rulemaking under section 6(b) of the OSH Act, scrutinizing each provision, accounting for each cost impact on employers, justifying why the new standard is reasonably necessary and appropriate, and showing that the adoption would reduce significant risk to employees. This would be inconsistent with the goal of this project which was to clarify employer obligations without increasing compliance burdens. However, OSHA has been convinced by commenters that consideration should be given to compliance with NFPA 101.

The 2000 Life Safety Code goes far beyond the requirements of OSHA's standard, both in details of compliance and flexibility for unique workplace conditions. If an employer complies with NFPA 101-2000, OSHA will deem such compliance to be compliance with the OSHA standard. OSHA believes that allowing employers to comply with NFPA 101 as an alternative to the revised Exit Routes standard will provide greater flexibility to employers who want to go beyond OSHA's basic provisions. Additionally, the National Technology Transfer and Advancement Act of 1995 (15 U.S.C. 3701 (1996)) directs Federal agencies to use voluntary consensus standards to the extent practicable. Under section 6(b)(8) of the OSH Act, the Agency must consider using national consensus standards as the basis for its safety and health standards wherever possible. By allowing employers to comply with the exit route provisions of NFPA 101-2000, OSHA has struck a balance that is consistent with its goals for this rulemaking as well as the spirit of the National Technology Transfer and Advancement Act.

OSHA has evaluated NFPA 101-2000 and has concluded that an employer who complies with the provisions of that code for means of egress will provide employees with safety that is comparable with compliance with OSHA's revised Exit Routes standard. OSHA is adding a new Sec. 1910.35 to the final rule to recognize NFPA 101-2000 in this regard.

The South Carolina Department of Labor, Licensing & Regulation (Ex. 5-49, p.2) remarked that "It is a shame to spend this amount of time to adjust the wording when the whole standard is in need of repair."

Others criticized the proposal, feeling that it did not achieve its stated goal. For example, the American Health Care Association (Ex. 53) indicated that by "Developing new terminology for traditional means of egress requirements, we firmly believe, is a step backward and counter to OSHA's stated goal of creating a regulation that is easily understood." The United Steelworkers of America (Ex. 5-69) objected "to the very general performance language of this proposal. The language gives little, if any direction to employers and employees on how to comply with this proposed standard * * * Further, the proposed standard is somewhat confusing." (See also Exs. 5-33, 38, 40, 62, 66-68, 71).

OSHA does not agree with commenters who have concluded that OSHA has failed to meet its goals of (1) maintaining the safety and health protection provided to employees by subpart E without increasing the regulatory burden; (2) creating a regulation that is easily understood; and, (3) stating employers' obligations in performance-oriented language to the extent possible. Many commenters suggested improvements and language changes. Unfortunately in some cases the recommendations would have made substantive changes in the requirements of subpart E (e.g., Exs. 5-4, 11, 18, 21, 24, 40, 47, 49, 63). OSHA has considered and incorporated many comments that improve the clarity of the text, without making substantive changes in the obligations and protections offered by existing subpart E. The final rule as revised and reorganized, incorporates many commenter suggestions. OSHA strongly believes the final rule fulfills its goal of providing employers and employees with much clearer standards in subpart E. In addition, as already discussed, employers may take advantage of a more recent version of NFPA 101 under Sec. 1910.35 which recognizes compliance with the 2000 Edition of the Life Safety Code.

In response to comments, OSHA has changed the name of subpart E to better reflect the contents of the final rule. OSHA proposed to call the subpart "Exit Routes," but several commenters (Exs. 5-24, 40, 45) noted that the subpart contains provisions not only for exit routes but also for emergency action plans, and fire prevention plans. OSHA agrees with these commenters and has therefore changed the name of subpart E to reflect its coverage of Exit Routes, Emergency Action Plans, and Fire Prevention Plans.

In the preamble to the proposal OSHA stated that it included a table of contents to make it easier to access the provisions. The table was inadvertently left out of the proposed regulatory language in the Federal Register notice. OSHA believes that a table of contents will be helpful to employers and employees in locating provisions in the subpart and therefore, is including a table of contents in Sec. 1910.33.

As indicated in the Regulatory Format section above, the proposed rule offered two versions of a revised subpart E. The first version was written in the traditional format of OSHA standards. The second version was written in a question and answer format.

Commenters who addressed this issue indicated a preference for the traditional regulatory format as opposed to the question and answer format. For example, Medical Environment, Inc. (Ex. 5-7) supported the traditional "regulatory format, because this is what everyone is used to seeing. The question/answer format seemed too "loose" to find an answer to a specific question." Similarly, the International Dairy Foods Association (IDFA) (Ex. 5-22) believed "that the "traditional" plain English version is the preferred version. In contrast, we find that the question and answer format quickly becomes condescending, and to a degree, annoying."

The American Petroleum Institute (API) (Ex. 5-29, p.2) supported the traditional format because of perceived pitfalls in the question and answer format.

While the Q/A version has some appeal in terms of better first-impression, API believes that the traditional format makes it easier to understand the rule in total, and to locate specific requirements.

Another API concern is that of confusion. The Q/A format could be associated with OSHA's Field Directives, in which questions and answers are sometimes used to explain requirements. The questions and answers in Field Directives, however, do not hold the same weight as regulatory language. As a result, confusion could be caused by the use of questions and answers in both the OSHA standards and in Field Directives.

API is also concerned that the potential for inadvertent change of requirements is greater during a Q/A conversion. This is because more structural revision and reorganization is required to accommodate the Q/A approach, as demonstrated by comparison of the two approaches in this pilot conversion. It follows that the Q/A approach would face even greater conversion problems for other, more complicated safety and health regulations.

In addition, the International Brotherhood of Teamsters recommended that OSHA not adopt the question and answer format because the union believed that the format is neither well organized nor easy to read. (See also Exs. 5-2, 3, 12, 13, 14, 15, 16, 17, 20, 21, 24, 25, 26, 27, 30, 31, 34, 36, 37, 40, 41, 43, 45, 46, 47, 49.)

Several commenters stated that either version would be acceptable (Exs. 5-12, 17, 25). Other commenters supported the question and answer version (Exs. 5-16, 23, 32, 42, 48). Some suggested that the question and answer version be included in an appendix or some other OSHA publication (Exs. 5-20, 24, 26, 45, 54, 59). The Agency, after considering the comments, has decided to use the traditional format in the final rule. The Agency believes that including the question and answer version in an appendix might result in confusion. OSHA does use the question and answer format for other, non-regulatory documents, and will consider that format for future guidance in this area.

Additional comments ranged from remarks that OSHA should do nothing, revise subpart E and reference NFPA 101, or adopt NFPA 101 entirely (Exs. 5-10, 18, 28, 38, 41, 47, 53, 62, 66, 68, 71). The subject of how to address NFPA 101 in the plain language revision was also issue 1 in the hearing notice (at 62 FR 9403). Liberty Mutual Insurance Group (Ex. 5-19) recommended that OSHA "include a provision that compliance with a national consensus standard such as NFPA 101, Life Safety Code * * *would be recognized as compliance with the OSHA standard." The Building Owners and Managers Association (BOMA) stated that it believed that "it is essential for OSHA to add appendix language stating that compliance with the Life Safety Codes NFPA 101, constitutes compliance with subpart E. Current OSHA practices essentially recognize this now (Tr. 23)."

OSHA's intention in the proposed rule was to simplify subpart E, not to replace it. First, OSHA could not simply adopt "NFPA 101" as an OSHA standard, because it can only consider versions of that standard that are currently in existence. To do otherwise (i.e., attempting to approve a future edition) would result in an illegal delegation of agency authority. Second, adoption of NFPA 101-2000 as the OSHA standard goes beyond the limited purpose of this rulemaking. Such action would involve substantive rulemaking, including detailed analysis of the differences between OSHA current rules and NFPA 101-2000, including costs to employers and benefits to employees.

As discussed earlier, OSHA has reviewed NFPA 101-2000 and has determined that compliance with that standard will provide comparable protection to subpart E. Although the Agency is not adopting NFPA 101-2000, an employer who demonstrates compliance with that standard will be deemed to be in compliance with Sec. Sec. 1910.34, 1910.36, and 1910.37 of subpart E. Many commenters (e.g., Exs. 5-10, 18, 19, 41, 46, 48, 61) supported language that would allow employers to comply with the NFPA 101 standard as an alternative to the OSHA standard for Exit Routes. OSHA has incorporated such language into Sec. 1910.35 of the final rule.

Some commenters also asserted that OSHA should base its standard on the model building codes or allow compliance with the various national building codes (Exs. 5-19, 27, 47, 67; Tr. 23, 26, 32, 43). At the time of the rulemaking, there were three different national building codes in the United States: The Building Officials and Code Administrators' (BOCA) National Building Code, the International Conference of Building Officials' (ICBO) Uniform Building Code, and the Southern Building Code Congress International's (SBCCI) Standard Building Code.

OSHA emphasizes again that it did not propose to substantively revise subpart E, nor did it propose to allow the use of building codes to comply with subpart E. OSHA is not familiar enough with the detailed requirements of the various building codes to determine unequivocally whether compliance with any or all of them could be considered to fulfill employer obligations imposed by subpart E. Moreover the contents of these building codes were not analyzed, evaluated or considered as part of this rulemaking. The BOCA, ICBO, and SBCCI Codes vary considerably in their requirements and coverage relating to areas covered by subpart E. This rulemaking was not designed to address these differences, nor was it intended to expand the coverage of subpart E. Accordingly, OSHA declines to extend recognition to building codes as a means of determining compliance with subpart E. This decision only involves the narrow issue of whether compliance with a given building code demonstrates compliance with subpart E. OSHA recognizes and acknowledges the importance and the value of building codes in assuring that buildings are constructed safely.

Final Rule

Section 1910.34, Coverage and Definitions

In the proposal, Sec. 1910.35 was entitled "Coverage." It noted that all general industry employers were covered by subpart E, and that "exits" and "exit routes" were covered. The section went on to define these unique terms in the proposal. OSHA has retitled this section as "coverage and definitions," and has moved it to Sec. 1910.34 of the final rule. The "coverage" paragraph, Sec. 1910.34(a), specifies that the standard covers all workplaces in general industry except mobile workplaces. Paragraph (b) sets forth the "coverage" of the subpart: The minimum requirements for exit routes, emergency action plans, and fire prevention plans. Paragraph (c) of Sec. 1910.34 includes the definitions pertinent to the subpart.

In the proposal, OSHA included definitions for "Exit" and "Exit Route," eliminating all other definitions, believing they were unnecessary. However, commenters thought that OSHA went too far by not defining other terms or inappropriately failed to define other important terms (e.g., Exs. 5-18, 21, 24, 28, 41, 45, 47, 49.) After due consideration, OSHA agrees with these commenters and in the final rule (now paragraph 1910.34(c)) has added and clarified definitions for words used in the proposal that commenters found unclear. OSHA has clarified the terms "exit" and "exit route" and has added definitions for electroluminescent, exit access, exit discharge, high hazard area, occupant load, refuge area, and self-luminous.

Section 1910.35, Compliance With NFPA 101-2000, Life Safety Code

As discussed previously in this preamble, this section provides that an employer who complies with corresponding provisions of NFPA 101-2000 is deemed to be in compliance with subpart E, sections 1910.34-1910.37.

Section 1910.36, Design and Construction Requirements for Exit Routes

Section 1910.36 contains requirements for the design and construction of exit routes. It includes a requirement that exit routes be permanent, addresses fire resistance-ratings of construction materials used in exit stairways (exits), describes openings into exits, defines the minimum number of exit routes in workplaces, addresses exit discharges, and discusses locked exit route doors, and exit route doors. It also addresses the capacity, height and width of exit routes, and finally, it sets forth requirements for exit routes that are outside a building.

Many of these requirements are identical or nearly the same as those proposed, but have been rearranged in a more logical order or reworded so that the requirements are clearer and easier to understand and follow.

Paragraph (a)(1) of 1910.36 (proposed paragraph 1910.36(a)), requires that exit routes be a permanent part of the workplace. This provision remains as proposed. OSHA believes that exit routes must be a permanent part of a structure and that employees must know the route to safety. Otherwise, during an emergency, employees may become confused and take the wrong path to safety.

Paragraph (a)(2) of 1910.36 (proposed paragraph 1901.36(d)), specifies the fire resistance-rating of construction materials used to separate exits from other parts of the workplace (e.g., stairways). For example, where an exit stairway connects three or fewer stories, it must be constructed of materials having a 1-hour fire resistance-rating. If the exit stairway connects four or more stories, it must be constructed of materials having a 2-hour fire resistance-rating.

One commenter, IMC Global, Inc. (Ex. 5-54), suggested that OSHA include information in the standard or the appendix that would specify what construction materials or combination of materials would meet the fire resistance-ratings required by the standard. They explained that the information would be used by in-house personnel who make alterations or repairs to the building. OSHA believes that the reference to NFPA 101 in Sec. 1910.35 will assist employers and employees in answering these questions.

IMC Global, Inc. also recommended that OSHA define the term "story," suggesting that OSHA use the definition used in the NFPA 101, Life Safety Code, but did not provide any rationale or support to demonstrate that the failure to include a definition would have a negative impact on worker safety or health. OSHA notes that the NPFA 101-2000, defines the term "story" to mean "That portion of a building between the upper surface of a floor and the upper surface of the floor or roof next above." OSHA believes this definition to be generally understood and has determined not to include a definition of "story" in the regulatory text of the final rule.

Another commenter, the American Trucking Association (Ex. 5-52), suggested that OSHA reword proposed paragraph 1910.36(d), to make it similar to the wording in the existing subpart E concerning fire resistant-materials (paragraphs 1910.37(b)(1) and (b)(2)). That wording requires that for exits protected by separation from other parts of the building, the separation shall meet certain construction requirements. The commenter noted that the proposed wording appears to require all exits to be separated by fire resistant-materials. OSHA agrees that the provision was not clearly worded and has revised the language of the final rule to specify the required fire resistance-rating of materials used to construct separations, i.e., enclosed stairways. The revised language reflects the concerns raised by the commenter.

Paragraph (a)(3) of 1910.36 (proposed paragraph 1910.36(c)), restricts the number of openings into exits to those openings necessary to allow access to the exit from occupied areas of the workplace, or from the exit to the exit discharge. It also specifies that openings must be protected by a self-closing fire door that remains closed unless the fire door automatically closes in an emergency when the fire alarm or employee alarm system is sounded.

The final rule differs from the proposal in that it permits fire doors to remain open as long as they close automatically during an emergency. This change was made in response to comments from H. M. Bucci and the NFPA (Exs. 5-10, 18). Both pointed out that NFPA 101, Life Safety Code, permits the exception. OSHA notes that the additional flexibility provided from this provision is in keeping with the Agency's intent in rewriting subpart E, i.e., to add flexibility if it does not detract from employee safety or health and does not impose additional costs or compliance obligations.

A commenter, Dennis Kirson (Ex. 5-4), noted that the proposed provision did not provide guidance on the fire rating for fire doors opening into an exit. Such ratings are based on the purpose of the door. To be listed or approved as a fire door, the door would have to meet the fire rating set by a nationally recognized testing laboratory (see next paragraph).

Paragraph 1910.36(a)(3) (proposed paragraph 1910.36(c)), requires that each fire door, including its frame and hardware, be listed or approved by a nationally recognized testing laboratory. The International Dairy Foods Association (Ex. 5-22), suggested that OSHA include the definition of the terms "listed," "approved," and "nationally recognized testing laboratory" in the regulatory language of the final rule instead of giving a cross-reference to another section of the standards. Section 1910.7 contains what employers need to know about "listed," "approved," and "nationally recognized testing laboratory." OSHA does not agree that adding additional definitions, which are duplicated elsewhere in part 1910, to the standard would be particularly helpful. Therefore, OSHA has retained in the final rule the cross-reference to the standard containing the terms.

Two commenters (Exs. 5-10, 11) commented on OSHA's failure to address other openings in exits made for electrical and mechanical systems. One commenter (Ex. 5-11) suggested that OSHA delete the provision because it precludes the use of protected openings when such openings are necessary for certain mechanical or electrical penetrations. The other commenter (Ex. 5-10) asked OSHA to address such openings by requiring that they be sealed with an approved fire barrier sealant or fire stop. The existing rule does not contain requirements addressing such openings and, as discussed above, the purpose of the revision is not to add new requirements that would impose new obligations on employers. If an employer has these openings, OSHA notes that such openings into exits are addressed in NFPA 101. The employer may use NFPA 101-2000 for guidance even though the final rule does not address this issue.

Paragraph 1910.36(b) of the final rule, the proposal, and issue 4 in the hearing notice (at 62 FR 9403), all address the general requirement that all workplaces have at least two exit routes, as far away as practical from each other, to ensure that all employees and other building occupants can promptly and safely evacuate the workplace during an emergency. Where two are insufficient, the employer must have additional exit routes (see NFPA 101-2000 for guidance). The number of exit routes can be reduced to one where the number of employees, the size of the building, its occupancy, or the arrangement of the workplace is such that all employees would be able to evacuate safely during an emergency.

Although OSHA does not have direct authority to regulate non-employee occupants of a building, in assuring the safe evacuation of employees, the impact of other occupants in a building must be taken into consideration to assure a safe evacuation of all employees. Thus, OSHA refers to "other building occupants" generally as it does in the existing subpart E.

"As far away as practical" ("remote" in the proposal) means that exit routes must be located far enough apart so that if one exit route is blocked by fire or smoke, employees can evacuate using the second exit route. The paragraph also provides a note that employers must consider the number of employees, the size of the building, its occupancy, and the arrangement of the workplace to determine the correct number of exit routes, recommending that employers consult the NFPA 101-2000 for the number of exit routes appropriate to their particular workplace.

The provision in the final rule differs from the proposed rule in that it has been reworded to state specifically that an employer must have at least two exits (final paragraph 1910.36(b)(1)), or a sufficient number of exit routes (final paragraph 1910.36(b)(2)) to ensure that all occupants can safely and promptly leave the workplace during an emergency. An exception to the two-exit route rule is provided in those circumstances where an employer can demonstrate that the number of employees, size of the building or arrangement of the workplace is such that one exit route alone is sufficient (final paragraph 1910.36(b)(3)).

There were a number of comments on the required number of exit routes provision in the proposal (e.g., Exs. 5-4, 5, 8, 11, 18, 24, 26, 40, 41, 43, 45, 47, 49, 54, 63) with many commenters suggesting that the provision be rewritten to state clearly that two exit routes are required. Commenters also suggested that OSHA more fully explain how to determine when one exit route would be permitted or suggested that this exception be eliminated (Exs. 5-4, 5, 8, 26, 40, 41, 43, 45, 49, 54, 63).

OSHA agrees with some of the commenters in part, and has made it clear that employers must have at least two exit routes, except where one exit route would be sufficient to allow all employees to evacuate the workplace safely and promptly. OSHA has added a note to provision stating that employers may consult NFPA 101-2000 for guidance on how to determine the appropriate number of exit routes.

Other commenters suggested that the expression in proposed paragraph 1910.36(b)(2), "other means of escape * * * should be available," invited confusion, made the provision vague, and was unenforceable, and that OSHA should remove it in the final rule (Exs. 5-4, 11, 24, 40). OSHA agrees with the commenters and has eliminated the advisory wording in the final provision.

Paragraph 1910.36(c)(1) of the final rule (proposed paragraph 1910.36(f)) requires that each exit discharge lead directly outside or to a street, walkway, refuge area, public way, or open space with access to the outside. Paragraph 1910.36(c)(2) requires that the street, walkway, refuge area, public way, or open space to which an exit discharge leads must be large enough to accommodate the building occupants likely to use the exit.

Lastly, paragraph 1910.36(c)(3) (proposed paragraph 1910.36(f)(4)) requires that exit stairs that continue beyond the level on which the exit discharge is located must be interrupted at that level by doors, partitions, or other effective means to make clear the direction to go to the exit discharge. This paragraph differs from the proposed provision. It has been reworded to make it clear that where exit stairs continue beyond the level of the exit discharge, there must be some effective way to direct occupants to the exit discharge. This rewording responds to comments questioning the clarity of the provision as proposed (Exs. 5-22, 41).

A number of commenters indicated their support for allowing exit discharges to lead to a refuge area as proposed in paragraph 1910.36(f)(3) (Exs. 5-24, 29, 40, 45); they also suggested that the paragraph heading and the definition of exit route needed to be reworded to reflect the acceptability of refuge areas. The American Petroleum Institute remarked:

Section 1910.35(b)(2) should be revised to clarify that an exit route does not necessarily lead to the outside but could lead to a refuge area * * *.

As currently written, section 1910.35(b)(2) incorrectly defines an `exit route' as a means of travel to safety `outside' and further states that one part of an `exit route' is the way from the exit to the `outside.' is incorrectly misleads users into thinking that the only endpoint for an exit route is outside.

Similarly, the heading of section 1910.36(f) incorrectly states that an exit must lead to the outside. This heading should be amended to include the endpoint of a refuge area. Organization Resources Counselors, Inc. (5-45, p. 3) stated that it "agrees that the concept of refuge areas is one that should be adopted by OSHA."

In response to the comments, OSHA has revised the definition of exit route (paragraph 1910.34(c) of the final rule) to reflect the acceptability of refuge areas. Also, the heading to paragraph 1910.36(f) of the proposal, "An Exit Must Lead Outside," has been changed to "Exit Discharge" in final rule paragraph 1910.36(c).

Paragraphs 1910.36(d)(1), (2), and (3) of the final rule (proposed as paragraph 1910.36(g)), address locking exit route doors. Paragraph 1910.36(d)(1) specifies that employees must be able to open an exit route door from the inside at all times without keys, tools, or special knowledge. Devices that only lock from the outside at the exit discharge door, such as panic bars, are permitted. Paragraph 1910.36(d)(2) specifies that exit route doors must be free of any device or alarm that could restrict emergency use of the exit route if the device or alarm fails. Finally, paragraph 1910.36(d)(3) of the final rule states that in mental, penal or correctional facilities, an exit route door may be locked from the inside if supervisory personnel are continuously on duty and the employer has a plan to remove occupants from the facility during an emergency.

The final rule requirements on locking exit doors are essentially those in the proposal, except that the provisions are now located in paragraph 1910.36(d) in the final rule (instead of paragraph 1910.36(g) in the proposal). There were three comments on the proposal addressing locking exit doors. Commenter Dennis Kirson (Ex. 5-4) suggested that OSHA delete the sentence "A device that locks from the outside such as a panic bar is permitted because," he said, "it deals with ingress (to be locked out) rather than egress (to be locked in), it serves no purpose." Mr. Kirson further noted that this sentence did not modify the first sentence. OSHA has not made the suggested change because to avoid any misunderstandings it believes that the rule should include specific language to indicate what is acceptable. The Agency believes it is necessary in this context to state what is permitted along with what is not permitted, because of the widespread use of panic bars. The commenter also suggested OSHA delete the reference to mental, penal, or correctional institutions because they did not appear to fit the definition of general industry worksites. OSHA has not made the suggested change because such institutions are indeed "general industry" establishments and employees in these establishments are afforded the same protections as employees in other general industry workplaces. In recognition of the unique problems these institutions have with regard to the need to ensure occupants remain inside the facilities, OSHA is providing specific language to indicate clearly the performance to be achieved at these worksites.

Another commenter, the Department of Energy (Ex. 5-11), suggested that this last provision should also reflect national security at Federal locations and that OSHA should add "or other facility requiring security from unauthorized access." While OSHA does not disagree with the commenter, it has not made the suggested change because the inclusion of this additional language is beyond the stated scope of this proceeding. However the Agency will consider adding the suggested language in the future when substantive revisions are made to this subpart.

Paragraph 1910.36(e) (proposed paragraph 1910.36(h)), sets out requirements for doors leading to an exit route. The paragraph requires that a side-hinged door must be used to connect any room to an exit route and that the door that connects any room to an exit route must swing out in the direction of exit travel if the room is designed to be occupied by more than 50 people or if the room is used as a high hazard area (i.e., contains contents that are likely to burn with extreme rapidity or explode).

The final rule provision in paragraph 1910.36(e) is essentially the same as the proposed provision (paragraph 1910.36(h) in the proposal) with minor reorganizing to emphasize the requirements of the provisions. OSHA has divided the paragraph into two concise paragraphs in the final rule, paragraphs 1910.36(e)(1) and (2). Two commenters recommended changing the language of the proposed provision that required exit doors "swing out." Mr. Dennis Kirson (Ex. 5-4) suggested adding an exception to the provision that doors swing out, to allow for containment of hazardous materials, because of the greater hazard (to the public) of loss of containment of such materials. Such a change is beyond the scope of this project but the Agency may consider such a change as part of a future rulemaking. Tenneco (Ex. 5-41) suggested the phrase be changed to "swing with the exit travel" for further clarity. OSHA has revised the provision to incorporate the recommended change.

Eastman Kodak Company (Ex. 5-21) asked if security pass-through gates/turnstiles that free wheel when an alarm goes off would be considered an exit. Another commenter (Ex. 5-18) suggested that sliding doors be acceptable to OSHA if their operation is maintained to NFPA 101 specifications. The commenter noted that the current code (at that time NFPA 101-1994) allows vertical and sliding doors. OSHA has not modified the provision to address sliding doors or turnstiles because it would be a substantive change to the Exit Routes standard. However, these configurations are addressed in NFPA 101-2000. Employers who comply with that standard for the requirements concerning gates, turnstiles, and vertical or sliding doors, will be deemed to comply with this provision of subpart E.

Final rule paragraph 1910.36(f) (proposed paragraph 1910.36(i)) and issue 4 in the hearing notice (at 62 FR 9403)), address the required capacity for exit routes. The paragraph requires that exit routes be able to support the maximum permitted occupant load for each floor served by the exit routes, and that the capacity of exit routes may not decrease in the direction of exit route travel to the exit discharge.

OSHA has divided this proposed provision into two provisions in the final rule. The Agency has also made an editorial change in response to a concern raised by the Tennessee Valley Authority (TVA) (Ex. 5-47). TVA pointed out that in the existing standard, each exit route does not have to support the maximum permitted occupant load; rather, the existing standard requires that the combined capacity of the exits must support the maximum permitted occupant load for that floor. OSHA agrees with the commenter and has revised final paragraph 1910.36(f) accordingly.

Several commenters (Exs. 5-14, 36) expressed concerns about how to determine adequate capacity or the expected occupancy load for each floor. Argonne National Laboratory (Ex. 5-14) suggested that OSHA adopt the latest NFPA 101 to determine "whether or not adequate exiting capacity is provided from an area." Another commenter, Mr. Donald R. Delano (Ex. 5-36), suggested that OSHA define "maximum permitted occupant load" and "expected occupant load." IMC Global, Inc. (Ex. 5-54) asked that OSHA define "occupant load." In response to these comments OSHA has added a definition for the term "occupant load" and explained generally how to calculate the occupant load in the definition. The calculation can be done in accordance with NFPA 101-2000, since there are a wide variety of general industry occupancies which may be subject to different considerations.

Final rule paragraph 1910.36(g) (proposed paragraph 1910.36(j)) addresses the height and width requirements for exit routes and specifies that the ceiling of an exit route must be at least seven feet six inches (2.3 m) high. The paragraph specifies that any projection from the ceiling cannot decrease the space between the projection and the floor to less than six feet eight inches (2.0 m). Paragraph 1910.36(g) also specifies that the width of an exit access must be at least 28 inches (71.1 cm) wide at all points and that where a single way of exit access leads to an exit, its width must be at least equal to the width of the exit to which it leads.

Final paragraph 1910.36(g) also specifies that the width of an exit route must be sufficient to accommodate the maximum permitted occupant load of each floor served by the exit route. Lastly, the paragraph specifies that any objects that project into the exit route must not reduce the width of the exit route to less than the minimum width requirements for exit routes.

Paragraphs 1910.36(h)(1) through (4) (proposed paragraphs 1910.36(k)(1)(i) through (iv)), set out special requirements for exit routes that are outside of a building. The paragraphs require that each outdoor exit route must meet the minimum height and width requirements for indoor exit routes and must also meet certain other requirements. Specifically, (1) an outdoor exit route must have guardrails to protect unenclosed sides if a fall hazard exists; (2) an outdoor exit route must be covered if snow or ice is likely to accumulate along the route, unless the employer can demonstrate that any snow or ice accumulation will be removed before it presents a slipping hazard; (3) an outdoor exit route must be reasonably straight and have smooth, solid, substantially level walkways; and (4) an outdoor exit route must not have a dead-end that is longer than 20 feet (6.2 m).

Several commenters addressed this paragraph. Two commenters (Exs. 5-29, 40) suggested adding the wording "if a fall hazard exists" to the requirement for guardrails. OSHA agrees that guardrails only need to protect unenclosed sides if a fall hazard exists. One commenter (Ex. 5-10) suggested that the Agency use a 50 foot dead-end rather than a 20 foot dead-end. This would be a significant change and appears to be a decrease in safety to employees during emergencies and therefore OSHA has not changed the length of a dead-end. Other changes to these provisions are editorial only.

Section 1910.37, Maintenance, Safeguards, and Operational Features for Exit Routes

OSHA proposed in Sec. 1910.37 to include provisions covering the operation and maintenance of exit routes. OSHA has expanded the name from the proposal's "Operation and Maintenance Requirements for Exit Routes" to better reflect its contents. In the final rule, Sec. 1910.37 is entitled "Maintenance, safeguards, and operational features for exit routes." Provisions of this section include the safe use of exit routes during an emergency, lighting and marking exit routes, fire retardant paints, exit routes during construction, repairs, or alterations, and employee alarm systems.

OSHA has made several changes to paragraph 1910.37(a) of the proposed rule, by combining related provisions. In the final rule, paragraph 1910.37(a) remains titled "The Danger To Employees Must Be Minimized" and addresses furnishings and decorations (proposed paragraph 1910.37(a)(2)), travel toward a high hazard area (proposed paragraph 1910.37(a)(3)), unobstructed access to exit routes (proposed paragraph 1910.36(e)), and properly operating safeguards designed to protect employees (proposed paragraphs 1910.37(a) and 1910.37(e)).

Minor editorial changes have been made to these paragraphs, with the exception that final paragraph 1910.37(a)(2) has been modified because commenters found the requirement confusing (Exs. 5-5, 18, 26, 63). This confusion resulted from OSHA's use of the terminology "An exit route must not require employees to travel toward materials that burn very quickly, emit poisonous fumes, or are explosive." OSHA has modified the language to more closely reflect the current subpart E language: "Exit routes must be arranged so that employees will not have to travel toward a high hazard area, unless the path of travel is effectively shielded from the high hazard area by suitable partitions or other physical barriers." In addition, OSHA added a definition for "high hazard area" to the final rule's definition section, 1910.34. The new definition is from NFPA-101 with slight editorial changes.

In the proposal, paragraph 1910.37(b) required that exit route lighting be adequate, and paragraph 1910.37(c) required that exits be marked appropriately. OSHA has combined these paragraphs into paragraph 1910.37(b) in the final rule, in part because the provisions are closely related and the Agency believes that the standard will be easier to understand and use if all the requirements covering lighting and marking of exit routes are arranged together. The content of these paragraphs remains virtually the same in the final rule except for editorial clarifications (e.g., "lighted" instead of "illuminated") and the addition of specifications (issue 5 in the hearing notice at 62 FR 9403) for exit signs in response to comments (e.g., Exs. 5-4, 14, 18, 21, 43, 54). OSHA believes that these changes will enable employers and employees to have better and clearer information concerning the requirements for exit routes.

Issue 6 in the hearing notice (62 FR at 9403) asked whether the proposed requirements for exit lighting were too general. Some commenters objected to OSHA's use of the word "adequate" to describe the required amount of lighting in exit routes (Exs. 5-4, 18, 19, 22, 54, 57, 63, 64). (Issue 6 in the hearing notice at 62 FR 9403.) OSHA's current subpart E uses the term "adequate" (existing paragraph 1910.36(b)(6)); OSHA did not revise the word "adequate" in the proposal because specifying a level of lighting could be viewed as a substantive change. However, OSHA has clarified in the final rule (paragraph 1910.37(b)(1)), to make it clear and performance-oriented. The revised provision requires that employees with normal vision be able to see their way along an exit route. Therefore, OSHA has retained the word "adequate" but clarified its meaning in the final rule. Employers and employees can refer to NFPA 101-2000 for more detailed guidance.

Final paragraph 1910.37(b)(4) (proposed paragraphs 1910.37(c)(3) and (c)(4)), addresses the marking of the direction of travel to an exit. Signs would be redundant where the direction of travel is apparent. Therefore, OSHA has added the existing subpart E language to the final rule "where the direction of travel to the nearest exit is not immediately apparent" because such signs are needed only in that situation (Exs. 5-4, 14, 21, 64).

Final paragraph 1910.37(b)(5) (proposed paragraph 1910.37(c)(5)), requires that doors that could be mistaken for exit doors must be marked to indicate the actual use of the door. In the proposal, OSHA required the use of the term "Not an Exit" on such doors. Doing so eliminated the provision's performance nature. In the final rule OSHA has added the language currently found in subpart E (paragraph 1910.37(q)(2)) ("Not an Exit" or similar designation"). This change allows employers to comply with the current OSHA language or the NFPA language. (E.g., Exs. 5-14, 36).

In final paragraph 1910.37(b)(6) (proposed paragraph 1910.37(c)(6)), OSHA has restored the language from subpart E referring to the color of exit signs. In the proposal OSHA stated "An exit sign must show a designated color." OSHA has changed the language back to the current subpart E language, "distinctive in color" (paragraph 1910.37(q)(4)) at the request of several commenters (Exs. 5-30, 41). OSHA does not believe that the proposed language improved the provision and has accordingly changed it back to existing subpart E as recommended by commenters. This paragraph also retains the use of "electroluminescent" and "self-luminous" signs and has defined the terms in the definition section (Sec. 1910.34).

Paragraph 1910.37(b)(7) of the final rule was not in the proposed rule. OSHA proposed to delete the following requirement from current subpart E (paragraph 1910.37(q)(8)) "Every exit sign shall have the word `Exit' in plainly legible letters not less than 6 inches high, with the principal strokes of letters not less than three-fourths-inch wide." The Agency believed that this requirement could be handled without specifications (issue 5 in the hearing notice at 62 FR 9403).

Commenters disagreed and suggested that the current exit sign dimensions also be included in the final rule. For example, Donald R. Delano, P.E., (Ex. 5-36, p. 3) remarked:

> Deletion of reference to design parameters for exit signs leaves no adequate frame of reference. Exit signs need to be of a minimum size and design, just as a national standard exists for a highway STOP sign.

Further, Tenneco Newport News Shipbuilding (NNS, Ex. 5-41, p.2) stated:

> The exit signs as dictated by the current standard have become traditional and easily recognized by the general public. An employer's interpretation of `clearly visible' may not create an easily recognized sign. Therefore, in an emergency the lack of the traditional and consistent format may be detrimental. NNS suggests that the text from the current standard stay in effect.

(See also Exs. 5-5, 14, 18, 31, 39, 63.) OSHA agrees with these commenters and has included in the final rule new paragraph 1910.37(b)(7) specifying the height and stroke width of exit signs (as it appears in the existing subpart E, paragraph 1910.37(q)(8)).

Final paragraph 1910.37(c) (proposed paragraph 1910.37(d)), addresses the upkeep of fire-retardant properties of paints or solutions used in the workplace that might impact the safety of an exit route. In the proposal, OSHA stated that an employer must maintain the fire retardant properties of paints or other coatings used in the workplace. Commenters suggested that OSHA return to the existing subpart E language because the proposed language is vague and harder to understand than the existing language (e.g., Exs. 5-4, 18, 21, 43, 54). OSHA believes the language in the final rule has been made clearer by returning to the subpart E language fire-retardant paints or "solutions," rather than "coatings." OSHA has further clarified the requirement by specifying that paints or solutions used in an exit route must be renewed as often as necessary to maintain the necessary flame retardant properties.

Final paragraph 1910.37(d) (proposed paragraph 1910.37(f)) addresses the maintenance of exit routes during construction, repairs, or alterations. "Alterations" were not included in the heading of the proposed provision; however, in the final rule, the heading has been modified to include "alterations." Both the proposal and final rule include the word "alterations" in the regulatory text.

The first paragraph concerning new construction remains the same as proposed and is now paragraph 1910.37(d)(1). Minor editorial changes have been made to final paragraph 1910.37(d)(2) that address repairs and alterations. Final paragraph 1910.37(d)(3) concerning flammable and explosive substances or equipment used during construction, repairs, or alterations, remains the same as proposed except for some minor changes. As discussed above OSHA has added the word "alterations" to the proposed language. In addition, the Agency returned to the use of "substances" instead of "materials." Finally, OSHA has added "equipment" to the paragraph. The words "substances" and "equipment" are in the present subpart E requirement (paragraph 1910.37(c)(3)) but were inadvertently left out of the proposal. OSHA has changed the proposed language "flammable or explosive materials used during construction or repair must not expose employees to hazards * * *" to "Employees must not be exposed to hazards of flammable or explosive substances or equipment used during construction, repairs, or alterations, that are beyond the normal permissible conditions in the workplace * * *."

Final rule paragraph 1910.37(e) (proposed paragraph 1910.37(g)), requires the installation and maintenance of an employee alarm system meeting Sec. 1910.165, unless employees can promptly see or smell a fire or other hazard. This requirement remains unchanged from the proposed rule.

Section 1910.38, Emergency Action Plans, and Section 1910.39, Fire Prevention Plans

In the final rule, OSHA has retained the separate sections for emergency action plans and fire prevention plans, Sec. Sec. 1910.38 and 1910.39 respectively. OSHA believes it is clearer for the plans and their requirements to be contained in separate sections. Because commenters tended to address both plans at the same time in their comments or their comments were quite similar about the plans, OSHA is discussing them together.

Final paragraph 1910.38(a) states that an emergency action plan is required, and final paragraph 1910.39(a) states that a fire prevention plan is required, when an OSHA standard requires such a plan. A number of commenters (Exs. 5-14, 20, 21, 23, 40, 49) recommended that OSHA include a listing of all OSHA standards that require an emergency action plan or a fire prevention plan. The Agency considered modifying the appendix to add a list of such standards. Instead, OSHA has issued a Compliance Directive that contains a list of current OSHA standards that require emergency action plans or fire prevention plans. The Agency has included this information in a Compliance Directive instead of an appendix to the standard because it is easier to amend the Compliance Directive as needed to keep it current.

For informational purposes, OSHA has identified the following general industry standards that require an emergency action plan or a fire prevention plan.

1. Process Safety Management of Highly Hazardous Chemicals, paragraph 1910.119(n), emergency action plan.
2. Hazardous Waste Operations and Emergency Response, paragraphs 1910.120(l)(1)(ii), (p)(8)(i), (q)(1), and (q)(11)(ii), emergency action plan.
3. Portable Fire Extinguishers, paragraphs 1910.157(a) and (b)(1), emergency action plan and fire prevention plan.
4. Grain Handling Facilities, paragraph 1910.272(d), emergency action plan.
5. Ethylene Oxide, paragraph 1910.1047(h)(1)(iii), emergency action plan and fire prevention plan.
6. Methylenedianiline, paragraph 1910.1050 (d)(1)(iii), emergency action plan and fire prevention plan.
7. 1,3-Butadiene, paragraph 1910.1051(j), emergency action plan and fire prevention plan.

Final paragraph 1910.38(b) and paragraph 1910.39(b) address written emergency action plans and fire prevention plans respectively. They require that the plans must be in writing and available; and for employers with 10 or fewer employees the plan may be transmitted orally rather than in writing. In the final rule, proposed paragraphs 1910.38(a)(2) and (a)(3) are combined into one paragraph, 1910.38(b), and proposed paragraphs 1910.39(a)(2) and (a)(3) become final paragraph 1910.39(b). Combining these paragraphs involved some minor editorial changes.

The Department of Energy (Ex. 5-11, p. 2) suggested that plans should be communicated orally to a "limited number" of employees rather than the 10 or fewer required by OSHA because the intent would be better served by not using an arbitrary number. OSHA disagrees with this suggestion. Since their promulgation in 1980, the emergency action plan and the fire prevention plan have used 10 as a reasonable number of employees for a plan to be communicated orally.

The International Brotherhood of Teamsters (IBT) (Ex. 5-31, p. 6) did not agree with the language in proposed paragraph 1910.38(a)(2) and paragraph 1910.39(a)(2), which stated that "the plan must be made available to employees on request." IBT asked the Agency to use the current language of subpart E, requiring the plans "be available for employees to review." The IBT believed the proposed language added an obstacle to employees by making them request to see the plan. OSHA agrees; in the proposal it had inadvertently changed the language from the current subpart E. OSHA fully believes that the plan should be available for employee review and in the final rule the language reflects this intent.

OSHA has reordered final paragraph 1910.38(c), containing the elements of an emergency action plan, to better reflect the order of an emergency response. Final paragraph 1910.38(c)(1) (proposed paragraph 1910.38(b)(3)) requires that the plan include procedures for reporting a fire or other emergency. OSHA believes reporting a fire or other emergency should be the first thing done in an emergency. The rest of the elements remain in the same order.

Final paragraphs 1910.38(c)(2), (3), and (4) remain for the most part the same as the proposed paragraphs—procedures for evacuation and exit route assignments, procedures to be followed by employees who remain to operate critical plant operations before they evacuate, and procedures to account for all employees after evacuation.

Final paragraph 1910.38(c)(3) concerning emergency operations or shutdown of plant equipment during an emergency has been changed back to the current subpart E language. This was done to clarify that this element of the plan does not apply to all employees and all plants, only to those plants that use employees for these emergency or shutdown procedures (Exs. 5-4, 18, 54).

Eastman Kodak Company (Ex. 5-21, p.3) suggested that OSHA delete the wording that addresses accounting for employees (final paragraph 1910.38(c)(4)):

[sbull] Procedures to assure that the fire area is clear of employees, visitors and contractors. Expectations to track employees such as maintenance personnel, service providers, or engineers is very burdensome. In today's work environment many transient employees work in multiple locations making it difficult to track who will be in any work area in an emergency. Hence, many emergency plans require the use of trained searchers to assure that the area being evacuated is clear of all personnel regardless of their normal work locations.

OSHA disagrees with this commenter and believes that accounting for employees after an emergency is critically important information to rescuers. Employees could, for example, be assigned designated locations away from the facility at which to meet.

In final paragraph 1910.38(c)(5), which requires that the plan include procedures for rescue or medical duties, OSHA has added language to clarify that the requirements only apply to those employees who will be performing such duties. This language parallels more closely the current subpart E language (paragraph 1910.38(a)(2)(iv)). The Agency has also changed "rescue and medical duties" in the proposal to "rescue or medical duties" (emphasis added) since employees may do one or the other but not necessarily both.

Final paragraph 1910.38(c)(6), which addresses names or job titles of employees to be contacted for more information or for an explanation of duties, has been revised from the proposal and is closer to the current language in subpart E (paragraph 1910.38(a)(2)(vi)). The change clarifies the requirement.

A few commenters (e.g., Ex. 5-4) contended that proposed paragraphs 1910.38(d) and 1910.37(g), are redundant. However, while both paragraphs require alarm systems, the two provisions are different. Proposed paragraph 1910.37(g) (paragraph 1910.37(e) in the final rule) requires that an employee alarm system be installed and maintained, unless employees can promptly see or smell a fire or other hazard. It applies regardless of whether the employer must have an emergency action plan. Paragraph 1910.38(d) requires that employers have and maintain an alarm system when an employer is required to have an emergency action plan by another OSHA standard. That alarm system must be provided even if employees can promptly see or smell a fire or other hazard. These paragraphs remain the same as proposed in the final rule.

Final paragraph 1910.38(e), regarding training of designated employees to assist in a safe and orderly evacuation of other employees, remains as proposed except for minor reorganization.

Final paragraph 1910.38(f) (proposed paragraph 1910.38(e)) requires that employers review the emergency action plan with each employee when the plan is developed or the employee is assigned initially to a job, when responsibility under the plan changes or the plan changes. Only minor editorial changes have been made to the final provision.

With regard to 29 CFR 1910.39, fire prevention plans, final paragraph 1910.39(c) (proposed paragraph 1910.39(b)) remains the same as proposed. Few comments were received with respect to the elements of the fire prevention plan.

Final rule paragraph 1910.39(d) (proposed rule paragraph 1910.39(c)) requires employers to inform employees of workplace fire hazards and review those parts of the fire prevention plan necessary for the employee's self-protection. Only minor editorial changes were made to this paragraph.

Miscellaneous Changes

OSHA is also amending the sections listed in the preamble's discussion of 1910.38 and 1910.39 above (e.g., 29 CFR 1910.120, 1910.157, etc.). These changes are necessary to conform with new section and paragraph designations for Emergency Action Plans and Fire Protection Plans found in this revised subpart E.

Other Hearing Issues

As discussed earlier in this preamble, OSHA asked a series of questions in its hearing notice (62 FR 9402). To the extent possible, OSHA has included the questions with the pertinent discussions in the preamble. For example, the use of performance-oriented language in the proposal was discussed earlier in this preamble (issue 3). "Are terms too technical" (issue 7) was discussed by commenters addressing the definitions of the standard or when commenters identified unclear language. However, some of the issues raised in the questions were more general and the vast majority of commenters did not definitively respond to these questions. These issues were numbered 3, 7, 8, 9, and 10 in the hearing notice (62 FR at 9403), and they asked: Would performance-oriented standards create compliance problems; are there terms that might be too technical; whether the revision imposes additional obligations; whether any requirements result in greater safety; and whether any requirements present technical feasibility problems. The questions raised in the hearing notice were intended to assure that various aspects of the proposal were fully considered. Some commenters addressed the issues through their comments regarding specific provisions of the proposal and did not respond to the questions specifically set forth in the hearing notice. To the extent that interested persons commented on these issues, OSHA has responded to these comments in the context of specific provisions of the proposed rule.

III. Legal Considerations

Because the final rule is only a plain language redrafting of a former Agency subpart, it is not necessary to determine significant risk or the extent to which the final rule reduces that risk. As noted above, most of the provisions of subpart E were adopted under section 6(a) of the Occupational Safety and Health Act, which gave the Secretary of Labor the authority, for a limited period of time, to adopt as occupational safety and health standards any established Federal Standard or national consensus standards unless the promulgation of such a standard would not result in improved safety and health for designated employees. By including section 6(a) in the OSH Act, Congress implicitly found that the promulgation of occupational safety and health standards was reasonably necessary or appropriate to provide safe or healthful employment and places of employment. In Industrial Union Department, AFL-CIO v. American Petroleum Institute, 448 U.S. 607 (1980), the Supreme Court ruled that before OSHA can increase the protection afforded by a standard, the Agency must find that the hazard being regulated poses a significant risk to employees and that a new, more protective standard is "reasonably necessary and appropriate" to reduce that risk. The final rule that replaces the Agency's former rules regulating means of egress, emergency action plans, and fire prevention plans does not directly increase or decrease the protection afforded to employees, nor does it increase employers' compliance obligations. Therefore, no finding of significant risk is necessary.

The Agency believes, however, that improved employee protection is likely to result from promulgation of the final rule because employers and employees who clearly understand a rule's requirements are more likely to comply with that rule. In addition, employers may find it easier to comply with the final rule because the final rule is more performance-oriented than the former rule.

IV. Economic Analysis

This final rule has been designated as significant and reviewed by the Office of Management and Budget under Executive Order 12866. It is not an economically significant rule under Executive Order 12866 or a major rule under the Unfunded Mandates Reform Act or section 801 of the Small Business Regulatory

Enforcement Fairness Act (SBREFA). The final rule imposes no additional costs on any private or public sector entity and does not meet any of the criteria for an economically significant or major rule specified by the Executive Order or the other statutes.

Certain provisions of the rule that add flexibility, such as permitting fire doors to remain open as long as they close automatically during an emergency and modifying the definition of exit route to reflect the acceptability of refuge areas, may even reduce costs for employers. Because the rule does not impose any additional costs on employers for exit routes, emergency action plans, and fire prevention plans, no economic or regulatory flexibility analysis of the final rule is required.

V. Regulatory Flexibility Certification

In accord with the Regulatory Flexibility act, 5 U.S.C. 601 et seq. (as amended), OSHA has examined the regulatory requirements of the final rule to determine if it will have a significant economic effect on a substantial number of small entities. As indicated in the previous section of this preamble, the final rule does not increase employers' compliance costs, and may even reduce the regulatory burden on all affected employers, both large and small. Accordingly, the Agency certifies that the final rule does not have a significant economic effect on a substantial number of small entities.

VI. Environmental Impact Assessment

OSHA has reviewed the final rule in accordance with the requirements of the National Environmental Policy Act (NEPA) of 1969 (42 U.S.C. 4321 et seq.), of the Council on Environmental Quality regulations (40 U.S.C. part 1500 et seq.), and the Department of Labor's NEPA regulations (29 CFR part 11). As noted earlier in this preamble, the final rule imposes the same requirements on employers as the standards it replaces. Consequently, the final rule has no additional impact beyond the impact imposed by OSHA's former standards for means of egress on the environment, including no impact on the release of materials that contaminate natural resources or the environment.

VII. Paperwork Reduction Act

The final rule contains no information collection requirements (paperwork) that are subject to the Paperwork Reduction Act. Therefore, approval under the Paperwork Reduction Act is unnecessary.

VIII. Unfunded Mandates

For the purposes of the Unfunded Mandates Reform Act of 1995, this rule does not include any Federal mandate that may result in increased expenditures by State, local, and tribal governments, or increased expenditures by the private sector of more than $100 million in any year.

IX. Federalism

OSHA has reviewed this final rule in accordance with the Executive Order on Federalism (Executive Order 13132, 64 FR 43255) which requires that agencies, to the extent possible, refrain from limiting state policy options, consult with states prior to taking any actions that would restrict state policy options, and take such actions only when there is clear constitutional authority and the presence of a problem of national scope. The Order provides for preemption of State law only if there is a clear Congressional intent for the Agency to do so. Any such preemption is to be limited to the extent possible.

Section 18 of the Occupational Safety and Health (OSH) Act (29 U.S.C. 651 et seq.) expresses Congress' intent to preempt state laws where OSHA has promulgated occupational safety and health standards. Under the OSH Act, a state can avoid preemption on issues covered by Federal standards only if it submits, and obtains Federal approval of, a plan for the development of such standards and their enforcement (State-Plan state). 29 U.S.C. 667. Occupational safety and health standards developed by such State-Plan states must, among other things, be at least as effective in providing safe and healthful employment and places of employment as the Federal standards.

Subject to these requirements, State-Plan states are free to develop and enforce their own requirements for exit routes, emergency action plans, and fire prevention plans. Having already adopted OSHA's former standards on means of egress, emergency action plans, and fire prevention plans, (or having developed alternative standards acceptable to OSHA), State-Plan states are not obligated to adopt the final rule; they may, however, choose to adopt the final rule, and OSHA encourages them to do so.

Although Congress has expressed a clear intent for OSHA standards to preempt State job safety and health rules in areas involving the safety and health rules of employees, this rule nevertheless limits State policy options to a minimal extent.

OSHA concludes that this action does not significantly limit State policy options.

X. State Plan States

OSHA encourages the 26 States and Territories with their own OSHA-approved occupational safety and health plans to revise their standards regulating means of egress, emergency action plans, and fire prevention plans according to the final rule that resulted from this rulemaking.

These states include Alaska, Arizona, California, Connecticut (state and local government employees only), Hawaii, Indiana, Iowa, Kentucky, Maryland, Michigan, Minnesota, Nevada, New Jersey (state and local government employees only), New Mexico, New York (state and local government employees only), North Carolina, Oregon, Puerto Rico, South Carolina, Tennessee, Utah, Vermont, Virginia, Virgin Islands, Washington, and Wyoming.

XI. Authority and Signature

This document was prepared under the direction of John L. Henshaw, Assistant Secretary of Labor for Occupational Safety and Health, U.S. Department of Labor, 200 Constitution Avenue, NW., Washington, DC 20210.

Signed in Washington, DC, this 21st day of October, 2002.

John L. Henshaw,
Assistant Secretary of Labor.

OSHA amends 29 CFR part 1910 as follows:

Part 1910—Occupational Safety and Health Standards

Subpart E—Exit Routes, Emergency Action Plans, and Fire Prevention Plans

Sec. 1910.33 Table of contents.

This section lists the sections and paragraph headings contained in Sec. Sec. 1910.34 through 1910.39.

Sec. 1910.34 Coverage and definitions.

(a) Every employer is covered.
(b) Exit routes are covered.
(c) Definitions.

Sec. 1910.35 Compliance with NFPA 101-2000, Life Safety Code.

Sec. 1910.36 Design and construction requirements for exit routes.

(a) Basic requirements.
(b) The number of exit routes must be adequate.
(c) Exit discharge.
(d) An exit door must be unlocked.
(e) A side-hinged exit door must be used.
(f) The capacity of an exit route must be adequate.
(g) An exit route must meet minimum height and width requirements.
(h) An outdoor exit route is permitted.

Sec. 1910.37 Maintenance, safeguards, and operational features for exit routes.

(a) The danger to employees must be minimized.
(b) Lighting and marking must be adequate and appropriate.
(c) The fire retardant properties of paints or solutions must be maintained.
(d) Exit routes must be maintained during construction, repairs, or alterations.
(e) An employee alarm system must be operable.

Sec. 1910.38 Emergency action plans.

(a) Application.
(b) Written and oral emergency action plans.

(c) Minimum elements of an emergency action plan.
(d) Employee alarm system.
(e) Training.
(f) Review of emergency action plan.

Sec. 1910.39 Fire prevention plans.
(a) Application.
(b) Written and oral fire prevention plans.
(c) Minimum elements of a fire prevention plan.
(d) Employee information.

Sec. 1910.34 Coverage and definitions.
(a) Every employer is covered. Sections 1910.34 through 1910.39 apply to workplaces in general industry except mobile workplaces such as vehicles or vessels.
(b) Exits routes are covered. The rules in Sec. Sec. 1910.34 through 1910.39 cover the minimum requirements for exit routes that employers must provide in their workplace so that employees may evacuate the workplace safely during an emergency. Sections 1910.34 through 1910.39 also cover the minimum requirements for emergency action plans and fire prevention plans.
(c) Definitions.

Electroluminescent means a light-emitting capacitor. Alternating current excites phosphor atoms when placed between the electrically conductive surfaces to produce light. This light source is typically contained inside the device.

Exit means that portion of an exit route that is generally separated from other areas to provide a protected way of travel to the exit discharge. An example of an exit is a two-hour fire resistance-rated enclosed stairway that leads from the fifth floor of an office building to the outside of the building.

Exit access means that portion of an exit route that leads to an exit. An example of an exit access is a corridor on the fifth floor of an office building that leads to a two-hour fire resistance-rated enclosed stairway (the Exit).

Exit discharge means the part of the exit route that leads directly outside or to a street, walkway, refuge area, public way, or open space with access to the outside. An example of an exit discharge is a door at the bottom of a two-hour fire resistance-rated enclosed stairway that discharges to a place of safety outside the building.

Exit route means a continuous and unobstructed path of exit travel from any point within a workplace to a place of safety (including refuge areas). An exit route consists of three parts: The exit access; the exit; and, the exit discharge. (An exit route includes all vertical and horizontal areas along the route.)

High hazard area means an area inside a workplace in which operations include high hazard materials, processes, or contents.

Occupant load means the total number of persons that may occupy a workplace or portion of a workplace at any one time. The occupant load of a workplace is calculated by dividing the gross floor area of the workplace or portion of a workplace by the occupant load factor for that particular type of workplace occupancy. Information regarding "Occupant load" is located in NFPA 101-2000, Life Safety Code.

Refuge area means either:
(1) A space along an exit route that is protected from the effects of fire by separation from other spaces within the building by a barrier with at least a one-hour fire resistance-rating; or
(2) A floor with at least two spaces, separated from each other by smoke-resistant partitions, in a building protected throughout by an automatic sprinkler system that complies with Sec. 1910.159 of this part.

Self-luminous means a light source that is illuminated by a self-contained power source (e.g., tritium) and that operates independently from external power sources. Batteries are not acceptable self-contained power sources. The light source is typically contained inside the device.

Sec. 1910.35 Compliance with NFPA 101-2000, Life Safety Code.

An employer who demonstrates compliance with the exit route provisions of NFPA 101-2000, the Life Safety Code, will be deemed to be in compliance with the corresponding requirements in Sec. Sec. 1910.34, 1910.36, and 1910.37.

Sec. 1910.36 Design and construction requirements for exit routes.

(a) **Basic requirements.** Exit routes must meet the following design and construction requirements:
  (1) An exit route must be permanent. Each exit route must be a permanent part of the workplace.
  (2) An exit must be separated by fire resistant materials. Construction materials used to separate an exit from other parts of the workplace must have a one-hour fire resistance-rating if the exit connects three or fewer stories and a two-hour fire resistance-rating if the exit connects four or more stories.
  (3) Openings into an exit must be limited. An exit is permitted to have only those openings necessary to allow access to the exit from occupied areas of the workplace, or to the exit discharge. An opening into an exit must be protected by a self-closing fire door that remains closed or automatically closes in an emergency upon the sounding of a fire alarm or employee alarm system. Each fire door, including its frame and hardware, must be listed or approved by a nationally recognized testing laboratory. Section 1910.155(c)(3)(iv)(A) of this part defines "listed" and Sec. 1910.7 of this part defines a "nationally recognized testing laboratory."

(b) The number of exit routes must be adequate.
  (1) **Two exit routes.** At least two exit routes must be available in a workplace to permit prompt evacuation of employees and other building occupants during an emergency, except as allowed in paragraph (b)(3) of this section. The exit routes must be located as far away as practical from each other so that if one exit route is blocked by fire or smoke, employees can evacuate using the second exit route.
  (2) **More than two exit routes.** More than two exit routes must be available in a workplace if the number of employees, the size of the building, its occupancy, or the arrangement of the workplace is such that all employees would not be able to evacuate safely during an emergency.
  (3) **A single exit route.** A single exit route is permitted where the number of employees, the size of the building, its occupancy, or the arrangement of the workplace is such that all employees would be able to evacuate safely during an emergency.

Note to paragraph 1910.36(b): For assistance in determining the number of exit routes necessary for your workplace, consult NFPA 101-2000, Life Safety Code.

(c) Exit discharge.
  (1) Each exit discharge must lead directly outside or to a street, walkway, refuge area, public way, or open space with access to the outside.
  (2) The street, walkway, refuge area, public way, or open space to which an exit discharge leads must be large enough to accommodate the building occupants likely to use the exit route.
  (3) Exit stairs that continue beyond the level on which the exit discharge is located must be interrupted at that level by doors, partitions, or other effective means that clearly indicate the direction of travel leading to the exit discharge.

(d) An exit door must be unlocked.
  (1) Employees must be able to open an exit route door from the inside at all times without keys, tools, or special knowledge. A device such as a panic bar that locks only from the outside is permitted on exit discharge doors.
  (2) Exit route doors must be free of any device or alarm that could restrict emergency use of the exit route if the device or alarm fails.
  (3) An exit route door may be locked from the inside only in mental, penal, or correctional facilities and then only if supervisory personnel are continuously on duty and the employer has a plan to remove occupants from the facility during an emergency.

(e) A side-hinged exit door must be used.
  (1) A side-hinged door must be used to connect any room to an exit route.
  (2) The door that connects any room to an exit route must swing out in the direc-

tion of exit travel if the room is designed to be occupied by more than 50 people or if the room is a high hazard area (i.e., contains contents that are likely to burn with extreme rapidity or explode).
(f) The capacity of an exit route must be adequate.
   (1) Exit routes must support the maximum permitted occupant load for each floor served.
   (2) The capacity of an exit route may not decrease in the direction of exit route travel to the exit discharge.

Note to paragraph 1910.36(f): Information regarding "Occupant load" is located in NFPA 101-2000, Life Safety Code.

(g) An exit route must meet minimum height and width requirements.
   (1) The ceiling of an exit route must be at least seven feet six inches (2.3 m) high. Any projection from the ceiling must not reach a point less than six feet eight inches (2.0 m) from the floor.
   (2) An exit access must be at least 28 inches (71.1 cm) wide at all points. Where there is only one exit access leading to an exit or exit discharge, the width of the exit and exit discharge must be at least equal to the width of the exit access.
   (3) The width of an exit route must be sufficient to accommodate the maximum permitted occupant load of each floor served by the exit route.
   (4) Objects that project into the exit route must not reduce the width of the exit route to less than the minimum width requirements for exit routes.
(h) An outdoor exit route is permitted. Each outdoor exit route must meet the minimum height and width requirements for indoor exit routes and must also meet the following requirements:
   (1) The outdoor exit route must have guardrails to protect unenclosed sides if a fall hazard exists;
   (2) The outdoor exit route must be covered if snow or ice is likely to accumulate along the route, unless the employer can demonstrate that any snow or ice accumulation will be removed before it presents a slipping hazard;
   (3) The outdoor exit route must be reasonably straight and have smooth, solid, substantially level walkways; and
   (4) The outdoor exit route must not have a dead-end that is longer than 20 feet (6.2 m).

Sec. 1910.37 Maintenance, safeguards, and operational features for exit routes.

(a) The danger to employees must be minimized.
   (1) Exit routes must be kept free of explosive or highly flammable furnishings or other decorations.
   (2) Exit routes must be arranged so that employees will not have to travel toward a high hazard area, unless the path of travel is effectively shielded from the high hazard area by suitable partitions or other physical barriers.
   (3) Exit routes must be free and unobstructed. No materials or equipment may be placed, either permanently or temporarily, within the exit route. The exit access must not go through a room that can be locked, such as a bathroom, to reach an exit or exit discharge, nor may it lead into a dead-end corridor. Stairs or a ramp must be provided where the exit route is not substantially level.
   (4) Safeguards designed to protect employees during an emergency (e.g., sprinkler systems, alarm systems, fire doors, exit lighting) must be in proper working order at all times.
(b) Lighting and marking must be adequate and appropriate.
   (1) Each exit route must be adequately lighted so that an employee with normal vision can see along the exit route.
   (2) Each exit must be clearly visible and marked by a sign reading "Exit."
   (3) Each exit route door must be free of decorations or signs that obscure the visibility of the exit route door.
   (4) If the direction of travel to the exit or exit discharge is not immediately apparent, signs must be posted along the exit ac-

cess indicating the direction of travel to the nearest exit and exit discharge. Additionally, the line-of-sight to an exit sign must clearly be visible at all times.

(5) Each doorway or passage along an exit access that could be mistaken for an exit must be marked "Not an Exit" or similar designation, or be identified by a sign indicating its actual use (e.g., closet).

(6) Each exit sign must be illuminated to a surface value of at least five foot-candles (54 lux) by a reliable light source and be distinctive in color. Self-luminous or electroluminescent signs that have a minimum luminance surface value of at least .06 footlamberts (0.21 cd/m2) are permitted.

(7) Each exit sign must have the word "Exit" in plainly legible letters not less than six inches (15.2 cm) high, with the principal strokes of the letters in the word "Exit" not less than three-fourths of an inch (1.9 cm) wide.

(c) The fire retardant properties of paints or solutions must be maintained. Fire retardant paints or solutions must be renewed as often as necessary to maintain their fire retardant properties.

(d) Exit routes must be maintained during construction, repairs, or alterations.
 (1) During new construction, employees must not occupy a workplace until the exit routes required by this subpart are completed and ready for employee use for the portion of the workplace they occupy.
 (2) During repairs or alterations, employees must not occupy a workplace unless the exit routes required by this subpart are available and existing fire protections are maintained, or until alternate fire protection is furnished that provides an equivalent level of safety.
 (3) Employees must not be exposed to hazards of flammable or explosive substances or equipment used during construction, repairs, or alterations, that are beyond the normal permissible conditions in the workplace, or that would impede exiting the workplace.

(e) An employee alarm system must be operable. Employers must install and maintain an operable employee alarm system that has a distinctive signal to warn employees of fire or other emergencies, unless employees can promptly see or smell a fire or other hazard in time to provide adequate warning to them. The employee alarm system must comply with Sec. 1910.165.

Sec. 1910.38 Emergency action plans.

(a) **Application.** An employer must have an emergency action plan whenever an OSHA standard in this part requires one. The requirements in this section apply to each such emergency action plan.

(b) **Written and oral emergency action plans.** An emergency action plan must be in writing, kept in the workplace, and available to employees for review. However, an employer with 10 or fewer employees may communicate the plan orally to employees.

(c) **Minimum elements of an emergency action plan.** An emergency action plan must include at a minimum:
 (1) Procedures for reporting a fire or other emergency;
 (2) Procedures for emergency evacuation, including type of evacuation and exit route assignments;
 (3) Procedures to be followed by employees who remain to operate critical plant operations before they evacuate;
 (4) Procedures to account for all employees after evacuation;
 (5) Procedures to be followed by employees performing rescue or medical duties; and
 (6) The name or job title of every employee who may be contacted by employees who need more information about the plan or an explanation of their duties under the plan.

(d) **Employee alarm system.** An employer must have and maintain an employee alarm system. The employee alarm system must use a distinctive signal for each purpose and comply with the requirements in Sec. 1910.165.

(e) **Training.** An employer must designate and train employees to assist in a safe and orderly evacuation of other employees.

(f) **Review of emergency action plan.** An employer must review the emergency action

plan with each employee covered by the plan:
(1) When the plan is developed or the employee is assigned initially to a job;
(2) When the employee's responsibilities under the plan change; and
(3) When the plan is changed.

Sec. 1910.39 Fire prevention plans.

(a) **Application.** An employer must have a fire prevention plan when an OSHA standard in this part requires one. The requirements in this section apply to each such fire prevention plan.

(b) **Written and oral fire prevention plans.** A fire prevention plan must be in writing, be kept in the workplace, and be made available to employees for review. However, an employer with 10 or fewer employees may communicate the plan orally to employees.

(c) **Minimum elements of a fire prevention plan.** A fire prevention plan must include:
(1) A list of all major fire hazards, proper handling and storage procedures for hazardous materials, potential ignition sources and their control, and the type of fire protection equipment necessary to control each major hazard;
(2) Procedures to control accumulations of flammable and combustible waste materials;
(3) Procedures for regular maintenance of safeguards installed on heat-producing equipment to prevent the accidental ignition of combustible materials;
(4) The name or job title of employees responsible for maintaining equipment to prevent or control sources of ignition or fires; and
(5) The name or job title of employees responsible for the control of fuel source hazards.

(d) **Employee information.** An employer must inform employees upon initial assignment to a job of the fire hazards to which they are exposed. An employer must also review with each employee those parts of the fire prevention plan necessary for self-protection.

"Appendix E To Part 1910—Exit Routes, Emergency Action Plans, and Fire Prevention Plans."

* * * * *

Subpart H—Hazardous Materials

3. The authority citation for subpart H of part 1910 is revised to read as follows:
Sections 1910.103, 1910.106 through 1910.111, and 1910.119, 1910.120, and 190.122 through 126 also issued under 29 CFR part 1911.

Section 1910.119 also issued under section 304, Clean Air Act Amendments of 1990 (Pub. L. 101-549), reprinted at 29 U.S.C. 655 Note.

Section 1910.120 also issued under section 126, Superfund Amendments and Reauthorization Act of 1986 as amended (29 U.S.C. 655 Note), and 5 U.S.C. 553.

4. In Sec. 1910.119, the first sentence of paragraph (n) is revised to read as follows:
Sec. 1910.119 Process safety management of highly hazardous chemicals.
(n) Emergency planning and response. The employer shall establish and implement an emergency action plan for the entire plant in accordance with the provisions of 29 CFR 1910.38.* * * * * * *

5. In Sec. 1910.120, paragraphs (l)(1)(ii), (p)(8)(i), (q)(1), and the first sentence of paragraph (q)(11)(ii) are revised to read as follows:
Sec. 1910.120 Hazardous waste operations and emergency response.
(l) * * *
(1)(i) * * *
(ii) Employers who will evacuate their employees from the danger area when an emergency occurs, and who do not permit any of their employees to assist in handling the emergency, are exempt from the requirements of this paragraph if they provide an emergency action plan complying with 29 CFR 1910.38. * * * * * * *
(p) * * *
(8) * * *
(i) Emergency response plan. An emergency response plan shall be developed and implemented by all employers. Such plans need not duplicate any of the subjects fully

addressed in the employer's contingency planning required by permits, such as those issued by the U.S. Environmental Protection Agency, provided that the contingency plan is made part of the emergency response plan. The emergency response plan shall be a written portion of the employer's safety and health program required in paragraph (p)(1) of this section. Employers who will evacuate their employees from the worksite location when an emergency occurs and who do not permit any of their employees to assist in handling the emergency are exempt from the requirements of paragraph (p)(8) if they provide an emergency action plan complying with 29 CFR 1910.38.* * * * *

(q) * * *

(1) **Emergency response plan.** An emergency response plan shall be developed and implemented to handle anticipated emergencies prior to the commencement of emergency response operations. The plan shall be in writing and available for inspection and copying by employees, their representatives and OSHA personnel. Employers who will evacuate their employees from the danger area when an emergency occurs, and who do not permit any of their employees to assist in handling the emergency, are exempt from the requirements of this paragraph if they provide an emergency action plan in accordance with 29 CFR 1910.38.* * * * *

(11) * * *

(i) * * *

(ii) Where the clean-up is done on plant property using plant or workplace employees, such employees shall have completed the training requirements of the following: 29 CFR 1910.38, 1910.134, 1910.1200, and other appropriate safety and health training made necessary by the tasks they are expected to perform such as personal protective equipment and decontamination procedures. * * * * * * * *

Subpart L—Fire Protection

6. The authority citation for subpart L of part 1910 is revised to read as follows:

7. In Sec. 1910.157, paragraphs (a) and (b)(1) are revised to read as follows:

Sec. 1910.157 Portable fire extinguishers.

(a) **Scope and application.** The requirements of this section apply to the placement, use, maintenance, and testing of portable fire extinguishers provided for the use of employees. Paragraph (d) of this section does not apply to extinguishers provided for employee use on the outside of workplace buildings or structures. Where extinguishers are provided but are not intended for employee use and the employer has an emergency action plan and a fire prevention plan that meet the requirements of 29 CFR 1910.38 and 29 CFR 1910.39 respectively, then only the requirements of paragraphs (e) and (f) of this section apply.

(b) **Exemptions.** (1) Where the employer has established and implemented a written fire safety policy which requires the immediate and total evacuation of employees from the workplace upon the sounding of a fire alarm signal and which includes an emergency action plan and a fire prevention plan which meet the requirements of 29 CFR 1910.38 and 29 CFR 1910.39 respectively, and when extinguishers are not available in the workplace, the employer is exempt from all requirements of this section unless a specific standard in part 1910 requires that a portable fire extinguisher be provided. * * * *

Subpart R—Special Industries

8. The authority citation for subpart R of part 1910 is revised to read as follows:

9. In Sec. 1910.268, paragraph (b)(1)(iii) is revised to read as follows:

Sec. 1910.268 Telecommunications.

(b) * * *

(1) * * *

(i) * * *

(ii) * * *

(iii) **Working spaces.** Maintenance aisles, or wiring aisles, between equipment frame

lineups are working spaces and are not an exit route for purposes of 29 CFR 1910.34.*
* * * *

10. a. In Sec. 1910.272, paragraph (d) is revised.
b. In Appendix A to Sec. 1910.272, under the heading "2. Emergency Action Plans" the second sentence is revised. The revised text is set forth as follows:

Sec. 1910.272 Grain handling facilities.
(d) Emergency action plan. The employer shall develop and implement an emergency action plan meeting the requirements contained in 29 CFR 1910.38.

* * * * *

Appendix A to Sec. 1910.272
Grain Handling Facilities

2. Emergency Action Plan

* * * The emergency action plan (Sec. 1910.38) covers those designated actions employers and employees are to take to ensure employee safety from fire and other emergencies. * * *
* * * * *

Subpart Z—Toxic and Hazardous Substances

11. The authority citation for subpart Z of part 1910 is revised to read as follows:

All of subpart Z issued under section 6(b) of the Occupational Safety and Health Act of 1970 (29 U.S.C 653), except those substances that have exposure limits in Tables Z-1, Z-2, and Z-3 of 29 CFR 1910.1000. Section 1910.1000 also issued under section (6)(a) of the Act (29 U.S.C. 655(a)). Section 1910.1000, Tables Z-1, Z-2, and Z-3 also issued under 5 U.S.C. 553, but not under 29 CFR part 1911, except for the inorganic arsenic, benzene, and cotton dust listings.
Section 1910.1001 also issued under section 107 of the Contract Work Hours and Safety Standards Act (40 U.S.C. 333) and 5 U.S.C. 553.

Section 1910.1002 also issued under 5 U.S.C. 553, but not under 29 U.S.C. 655 or 29 CFR part 1911. Sections 1910.1018, 1910.1029, and 1910.1200 also issued under 29 U.S.C. 653.

12. In Sec. 1910.1047, paragraph (h)(1)(iii) is revised to read as follows:

Sec. 1910.1047 Ethylene oxide.
(h) * * *
(1) * * *
(i) * * *
(ii) * * *
(iii) The plan shall include the elements prescribed in 29 CFR 1910.38 and 29 CFR 1910.39, "Emergency action plans" and "Fire prevention plans," respectively. * * * * *

13. In Sec. 1910.1050, paragraph (d)(1)(iii) is revised to read as follows:

Sec. 1910.1050 Methylenedianiline
(d) * * *
(1) * * *
(i) * * *
(ii) * * *
(iii) The plan shall specifically include provisions for alerting and evacuating affected employees as well as the elements prescribed in 29 CFR 1910.38 and 29 CFR 1910.39, "Emergency action plans" and "Fire prevention plans," respectively. * * * * *

14. In Sec. 1910.1051, paragraph (j) is revised to read as follows:

Sec. 1910.1051 1,3-Butadiene

(j) **Emergency situations.** Written plan. A written plan for emergency situations shall be developed, or an existing plan shall be modified, to contain the applicable elements specified in 29 CFR 1910.38 and 29 CFR 1910.39, "Emergency action plans" and "Fire prevention plans," respectively, and in 29 CFR 1910.120, "Hazardous Waste Operations and Emergency Response," for each workplace where there is the possibility of an emergency.

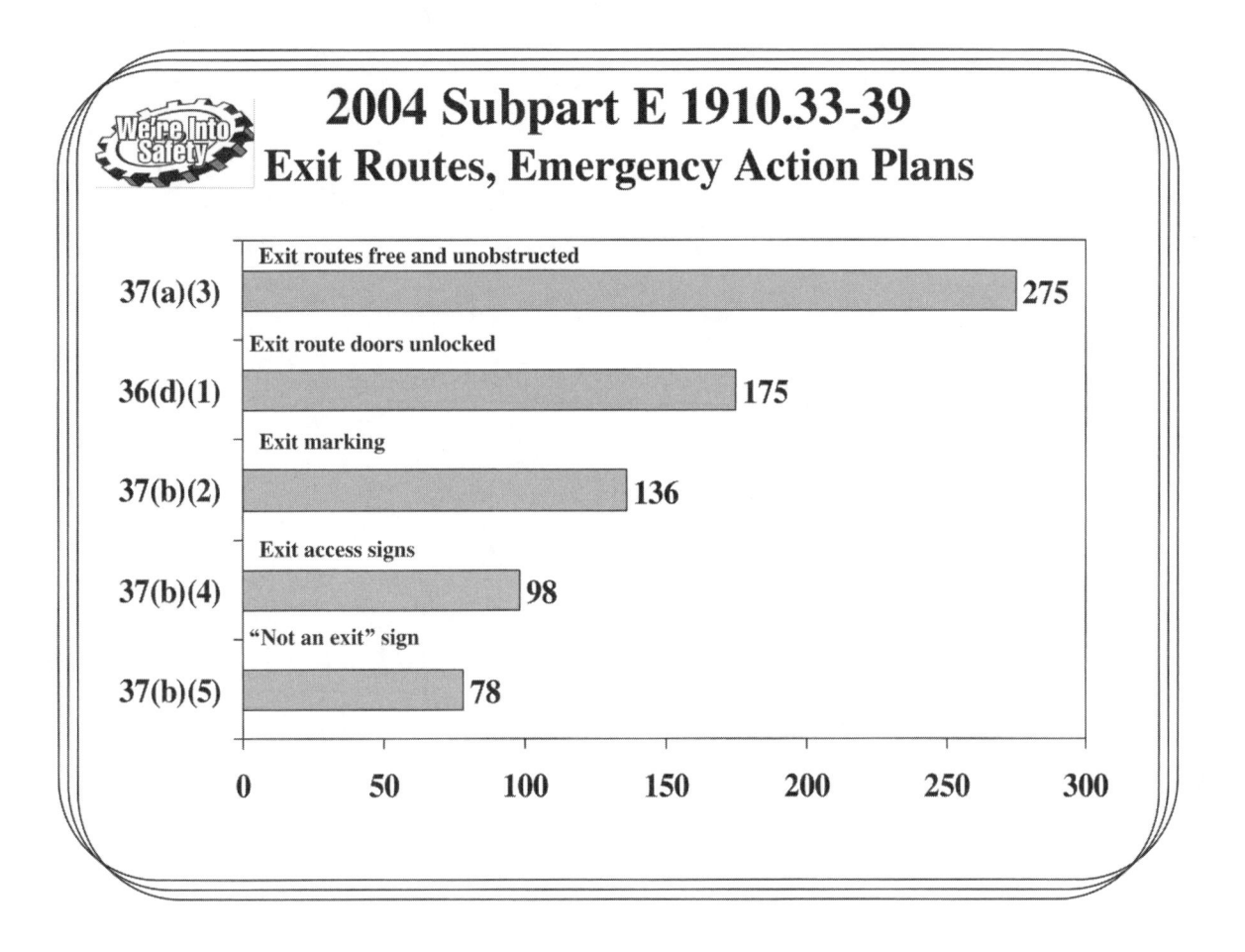

2004 Subpart E 1910.33-39
Exit Routes, Emergency Action Plans

Citation	Description	Count
37(a)(3)	Exit routes free and unobstructed	275
36(d)(1)	Exit route doors unlocked	175
37(b)(2)	Exit marking	136
37(b)(4)	Exit access signs	98
37(b)(5)	"Not an exit" sign	78

CHAPTER 13

MACHINE GUARDING

Crushed hands and arms, severed fingers, blindness—the list of possible machinery-related injuries is as long as it is horrifying. There seem to be as many hazards created by moving machine parts as there are types of machines. Safeguards are essential for protecting workers from needless, preventable injuries. A good rule to remember is that any machine part, function or process that may cause injury must be safeguarded. Where the operation of a machine or accidental contact with it can injure the operator or others in the vicinity, the hazard must either be controlled or eliminated.

Where Mechanical Hazards Occur
Dangerous moving parts in these three basic areas need safeguarding:
- The point of operation: That point where work is performed on the material, such as cutting, shaping, boring or forming of stock.
- Power transmission apparatus: All components of the mechanical system that transmit energy to the part of the machine performing the work. These components include flywheels, pulleys, belts, connecting rods, couplings, cams, spindles, chains, cranks and gears.
- Other moving parts: All parts of the machine that move while the machine is working. These can include reciprocating, rotating and transverse moving parts as well as feed mechanisms and auxiliary parts of the machine.

Hazardous Mechanical Motions and Actions
A variety of mechanical motions and actions may present hazards to the worker. These can include the movement of rotating members, reciprocating arms, moving belts, meshing gears, cutting teeth and any parts that impact or shear. These types of hazardous mechanical motions and actions are basic to nearly all machines, and recognizing them is the first step toward protecting workers from the danger they present.

The basic types of hazardous mechanical motions and actions are:
- Rotating motions (including in-running nip points);
- Reciprocating motions
- Transverse actions
- Cutting
- Punching
- Shearing and
- Bending.

Motions

Rotating motion can be dangerous; even smooth, slowly rotating shafts can grip clothing and through mere skin contact force an arm or hand into a dangerous position. Injuries due to contact with rotating parts can be severe.

Collars, couplings, cams, clutches, flywheels, shaft ends, spindles and horizontal or vertical shafting are some examples of common rotating mechanisms that may be hazardous. There is added danger when bolts, nicks, abrasions and projecting keys or setscrews are exposed on rotating parts on machinery (See Exhibit 13-1.).

The rotating parts on machinery cause in-running nip point hazards. There are three main in-running types.

Parts can rotate in opposite directions while their axes are parallel to each other. These parts may be in contact (producing a nip point) or in close proximity to each other (See Exhibit 13-2.). In the latter case, the stock fed between the rolls produces the nip points. This danger is common on machinery with intermeshing gears, rolling mills and calendars.

Another type of nip point is created between rotating and tangentially moving parts. Examples include:
- The point of contact between a power transmission belt and its pulley;
- A chain and a sprocket; and
- A rack and pinion. (See Exhibit 13-3.)

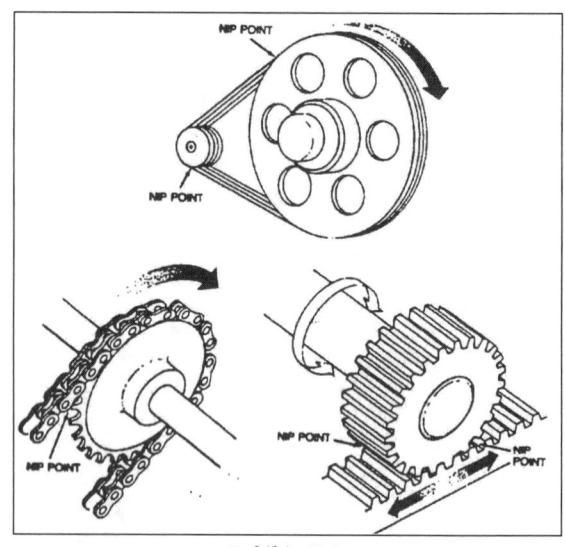

Exhibit 13-2.
Nip points occur where rotating parts come in contact.

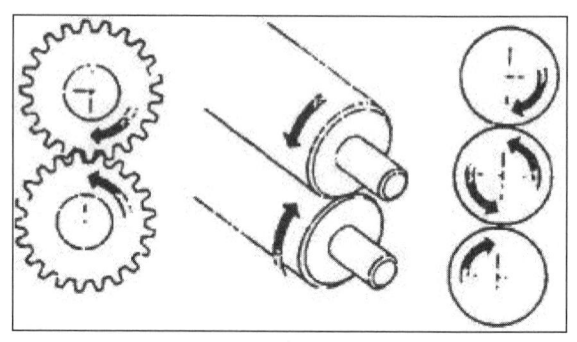

Exhibit 13-3.
Nip points can occur between rotating and tangentially moving parts.

Nip points also can occur between rotating and fixed parts that create a shearing, crushing, or abrading action (See Exhibit 13-4.). Examples include:
- Spoked hand wheels or flywheels;
- Screw conveyors;
- The periphery of an abrasive wheel; and
- An incorrectly adjusted work rest.

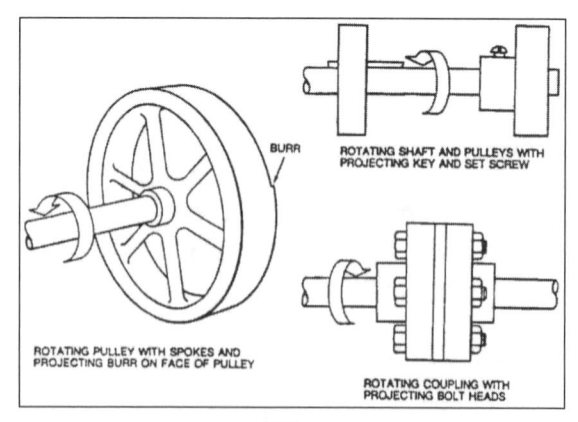

Exhibit 13-1.
Examples of rotating motion.

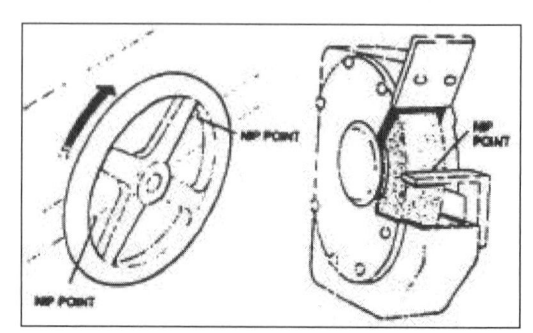

Exhibit 13-4.
Nip points can occur between rotating and fixed parts.

Reciprocating motions may be hazardous because during the back-and-forth or up-and-down motion, a worker may be struck by or caught between a moving and a stationary part (See Exhibit 13-5.).

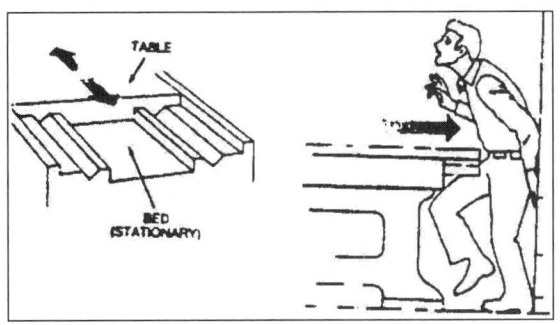

Exhibit 13-5.
Example of a worker being struck by machine with reciprocating motion.

Transverse motion (movement in a straight, continuous line) creates a hazard because a worker may be struck or caught in a pinch or shear point by a moving part (See Exhibit 13-6.).

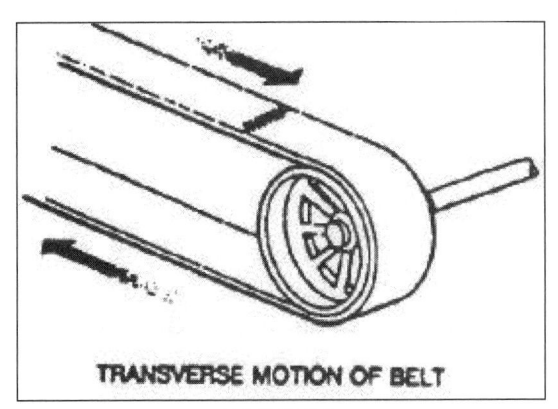

Exhibit 13-6.
Transverse Motion of Belt.

Actions

Cutting action involves rotating, reciprocating or transverse motion. The danger of cutting action exists at the point of operation where finger, head, and arm injuries can occur and flying chips or scrap material can strike the eyes or face. Such hazards are present at the point of operation in cutting wood, metal or other materials. Typical examples of mechanisms involving cutting hazards include bandsaws, circular saws, boring or drilling machines and turning machines, lathes or milling machines.

Punching action results when power is applied to a slide (ram) for the purpose of blanking, drawing or stamping metal or other materials. The danger of this type of action occurs at the point of operation where stock is inserted, held and withdrawn by hand. Typical machinery used for punching operations are power presses and ironworkers.

Shearing action involves applying power to a slide or knife in order to trim or shear metal or other materials. A hazard occurs at the point of operation where stock is actually inserted, held and withdrawn. Typical examples of machinery used for shearing operations are mechanically, hydraulically, or pneumatically powered shears.

Bending action results when power is applied to a slide in order to draw or stamp metal or other materials. A hazard occurs at the point of operation where stock is inserted, held and withdrawn. Equipment that uses bending actions include power presses, press brakes and tube benders.

Requirements for Safeguards

What must a safeguard do to protect workers against mechanical hazards? Safeguards must, at a minimum, meet the following general requirements:

Prevent contact. The safeguard must prevent hands, arms or any part of a worker's body or clothing from making contact with dangerous moving parts. A good safeguarding system eliminates the possibility of the operator or other workers placing parts of their bodies near hazardous moving parts.

Secure. Workers should not be able to easily remove or tamper with the safeguard because a safeguard that can easily be made ineffective is not much of a safeguard. Guards and safety devices should be made of durable material that will withstand the conditions of normal use. They must be firmly secured to the machine.

Protect from falling objects. The safeguard should ensure that no objects can fall into moving parts. A small tool that is dropped into a cycling machine could easily become a projectile that could strike and injure someone.

Create no new hazards. A safeguard defeats its own purpose if it creates a hazard of its own such as a shear point, jagged edge, or unfinished surface which can cause a laceration. The edges of guards, for instance, should be rolled or bolted in such a way that they eliminate sharp edges.

Create no interference. Any safeguard that impedes a worker from performing the job quickly and comfortably might soon be overridden or disregarded. Proper safeguarding can actually enhance efficiency since it can relieve the worker's apprehensions about injury.

Allow safe lubrication. If possible, one should be able to lubricate the machine without removing the safeguards. Locating oil reservoirs outside the guard, with a line leading to the lubrication point, will reduce the need for the operator or maintenance worker to enter the hazardous area.

Training

Even the most elaborate safeguarding system cannot offer effective protection unless the worker knows how to use it and why. Specific and detailed training is therefore a crucial part of any effort to provide safeguarding against machine-related hazards. This kind of safety training is necessary for new operators and maintenance or set-up personnel when any new or altered safeguards are put in service or when workers are assigned to a new machine or operation. Thorough operator training should involve instruction or hands-on training in the following:

- A description and identification of the hazards associated with particular machines;
- The safeguards themselves, how they provide protection and the hazards for which they are intended;
- How to use the safeguards and why;
- How and under what circumstances safeguards can be removed and by whom (in most cases, repair or maintenance personnel only); and
- What to do if a safeguard is damaged, missing, or unable to provide adequate protection (e.g., contact the supervisor).

Methods of Machine Safeguarding

There are many ways to safeguard machinery. The type of operation, size or shape of stock, method of handling, physical layout of the work area, type of material, and production requirements or limitations will help to determine the appropriate safeguarding method for the individual machine.

As a general rule, power transmission apparatus is best protected by fixed guards that enclose the danger area. For hazards at the point of operation, where moving parts actually perform work on stock, several kinds of safeguarding are possible. One must always choose the most effective and practical means available.

Safeguards are classified under five general categories:

1. **Guards**
    - Fixed
    - Interlocked
    - Adjustable
    - Self-adjusting
2. **Devices**
    - Presence Sensing
        - Photoelectrical (optical)
        - Radio frequency (capacitance)
        - Electromechanical
    - Pullback
    - Restraint
    - Safety Controls
        - Safety trip control
        - Pressure-sensitive body bar
        - Safety triprod
        - Safety tripwire cable
    - Two-hand control
    - Two-hand trip
    - Gates
        - Interlocked
        - Other
3. **Location/Distance**
4. **Potential Feeding and Ejection Methods to Improve Safety for the Operator**
    - Automatic feed
    - Semi-automatic feed
    - Automatic ejection
    - Semi-automatic ejection
    - Robot
5. **Miscellaneous Aids**
    - Awareness barriers
    - Miscellaneous protective shields
    - Hand-feeding tools and holding fixtures

Guards

Guards are barriers that prevent access to danger areas. There are four general types of guards.

Fixed

As its name implies, a fixed guard is a permanent part of the machine. It is not dependent upon moving parts to perform its intended function. It may be constructed of sheet metal, screen, wire cloth, bars, plastic or any other material that is substantial enough to withstand whatever impact it may receive and to endure prolonged use. This type of guard is usually preferable to all others because of its relative simplicity and permanence.

For example, a fixed guard on a power press completely encloses the point of operation. The stock is fed through the side of the guard into the die area, with the scrap stock exiting on the opposite side.

Exhibit 13-7 shows a fixed enclosure guard shielding the belt and pulley of a power transmission unit. An inspection panel is provided on top to minimize the need for removing the guard.

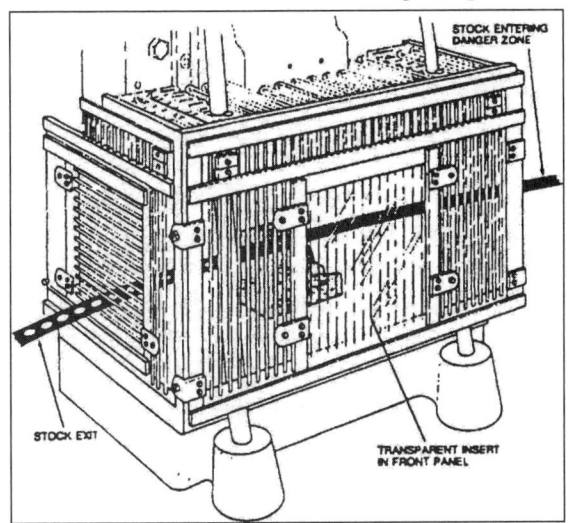

Exhibit 13-7
A fixed guard on a power transmission unit.

Fixed enclosure guards also may be seen on a bandsaw. These guards protect the operator from the turning wheels and moving saw blade. Normally, the only time for the guards to be opened or removed would be for a blade change or maintenance. It is very important that they be securely fastened while the saw is in use.

Interlocked
When this type of guard is opened or removed, the tripping mechanism and/or power automatically shuts off or disengages, and the machine cannot cycle or be started until the guard is back in place. An interlocked guard may use electrical, mechanical, hydraulic, or pneumatic power or any combination of these. Interlocks should not prevent "inching" by remote control, if required. Replacing the guard should not automatically restart the machine.

As illustrated in Exhibit 13-8, the beater mechanism of a picker machine (used in the textile industry) is covered by an interlocked barrier guard. This guard cannot be raised while the machine is running, nor can the machine be restarted with the guard in the raised position.

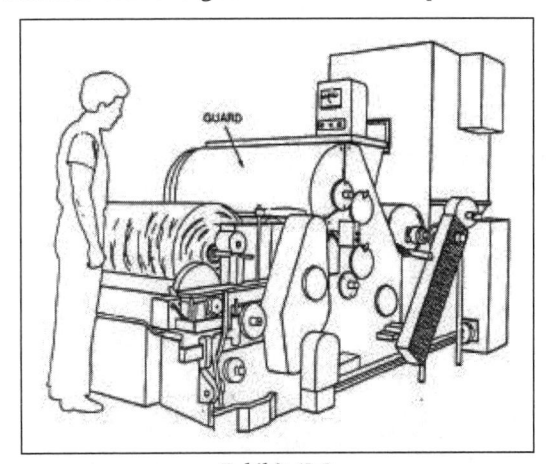

Exhibit 13-8.
A picker machine with an interlocked barrier guard.

Adjustable
Adjustable guards are useful because they allow flexibility in accommodating various sizes of stock (See Exhibit 13-9.).

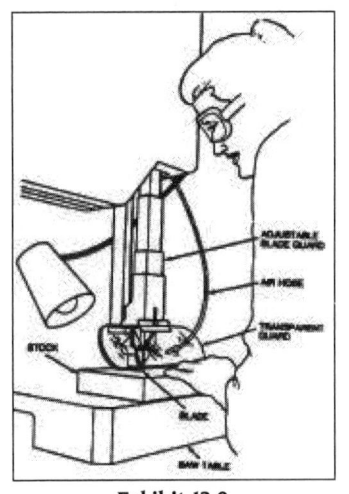

Exhibit 13-9.
An example of an adjustable blade guard.

Self-Adjusting
The movement of the stock determines the openings of these barriers. As the operator moves the stock into the danger area, the guard is pushed away, providing an opening that is only large enough to admit the stock. After the stock is removed, the guard returns to the rest position. This guard protects the operator by placing a barrier between the danger area and

MACHINE GUARDING

the operator. The guards may be constructed of plastic, metal or other substantial material. Self-adjusting guards offer different degrees of protection.

Exhibit 13-10 shows a radial arm saw with a self-adjusting guard. As the blade is pulled across the stock, the guard moves up, staying in contact with the stock.

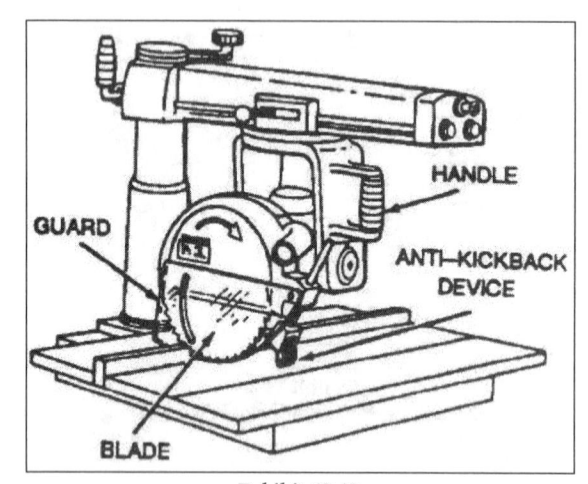

Exhibit 13-10.
A radial arm saw with a self-adjusting guard.

Devices

A safety device may perform one of several functions. It may:

- Stop the machine if a hand or any part of the body is inadvertently placed in the danger area;
- Restrain or withdraw the operator's hands from the danger area during operation;
- Require the operator to use both hands on machine controls thus keeping both hands and body out of danger; or
- Provide a barrier that is synchronized with the operating cycle of the machine to prevent entry to the danger area during the hazardous part of the cycle.

There are several types of devices available.

Presence-Sensing

The **photoelectric** (optical) presence-sensing device uses a system of light sources and controls that interrupt the machine's operating cycle (See Exhibit 13-11.). If the light field is broken, the machine stops and will not cycle. This device must be used only on machines that can be stopped before the worker can reach the danger area.

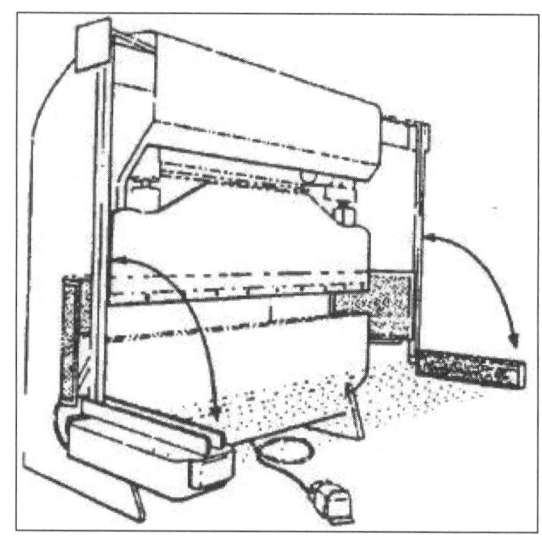

Exhibit 13-11.
A photoelectric presence-sensing device with a press brake.

The **radio frequency** (capacitance) presence-sensing device uses a radio beam that is part of the machine's control circuit (See Exhibit 13-12.). When the capacitance field is broken, the machine will stop or will not activate. Like the photoelectric device, this device shall only be used on machines that can be stopped before the worker can reach the danger area. This requires the machine to have a friction clutch or other reliable means for stopping.

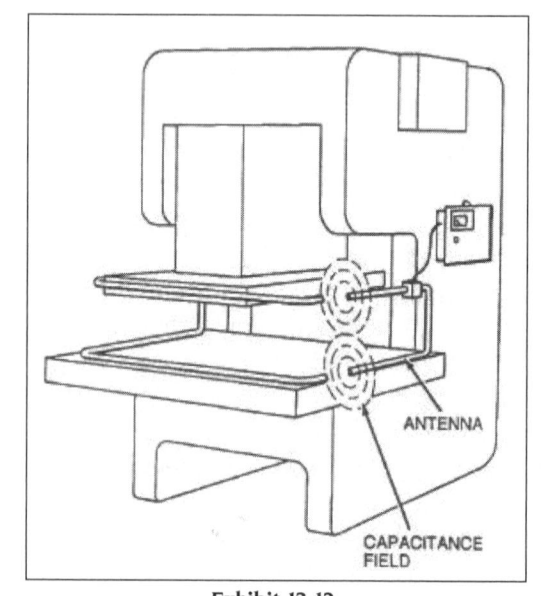

Exhibit 13-12.
A radio frequency presence-sensing device mounted on a part-revolution power press.

The **electromechanical** sensing device has a probe or contact bar that descends to a predetermined distance when the operator initiates the machine cycle (See Exhibit 13-13.). If there is an obstruction preventing it from descending its full predetermined distance, the control circuit will not actuate the machine cycle.

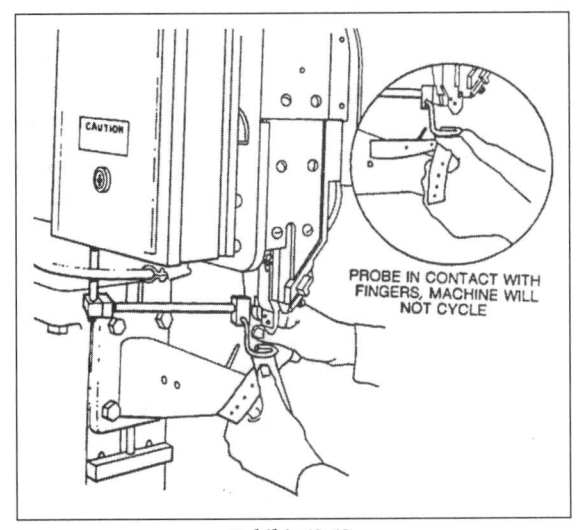

Exhibit 13-13.
An electromechanical sensing device on an eye-letting machine

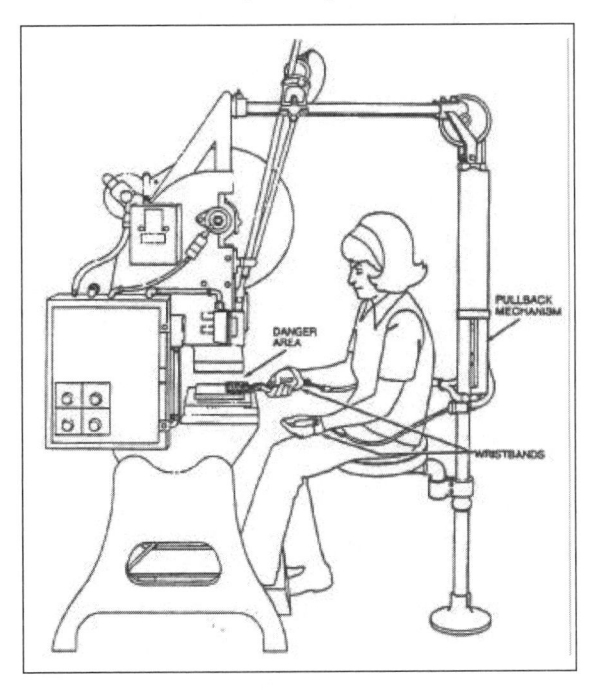

Exhibit 13-14.
A pullback device on a small press.

Pullback

Pullback devices use a series of cables attached to the operator's hands, wrists and/or arms (See Exhibit 13-14.). This type of device is primarily used on machines with stroking action. When the side/ram is up, the operator is allowed access to the point of operation. When the side/ram begins to descend, a mechanical linkage automatically ensures withdrawal of the hands from the point of operation.

Restraint

The restraint (holdout) device in Exhibit 13-15 uses cables or straps that are attached to the operator's hands and a fixed point. The cables or straps must be adjusted to let the operator's hands travel within a predetermined safe area. There is no extending or retracting action involved. Consequently, hand-feeding tools are often necessary if the operation involves placing material into the danger area.

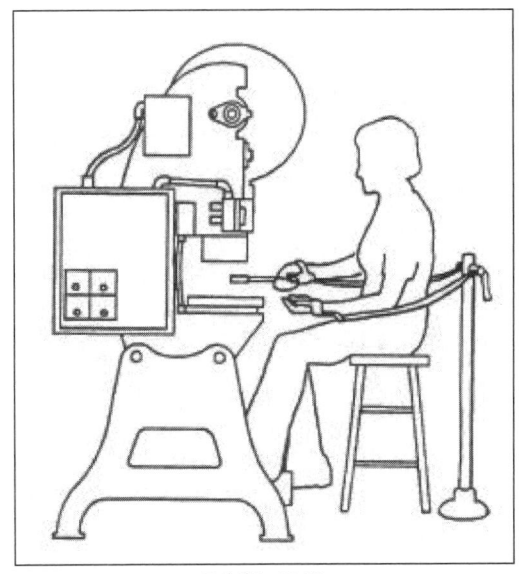

Exhibit 13-15.
A restraint device with wrist straps.

Safety Trip Control

Safety trip controls provide a quick means for deactivating the machine in an emergency situation (See Exhibit 13-16.).

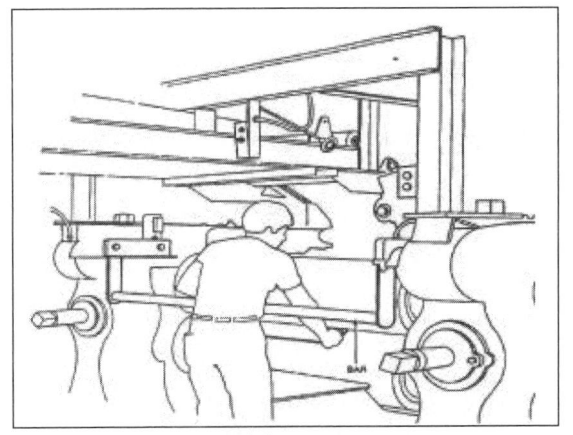

Exhibit 13-16.
A pressure-sensitive body bar located on the front of a rubber mat.

MACHINE GUARDING

A pressure-sensitive body bar, when depressed, will deactivate the machine. If the operator or anyone trips, loses balance or is drawn toward the machine, applying pressure to the bar will stop the operation. The positioning of the bar, therefore, is critical. It must stop the machine before a part of the employee's body reaches the danger area.

All tripwires, rods or other safety devices must be manually reset to restart the machine. When pressed by hand, the safety trip rod deactivates the machine. Because it has to be actuated by the operator during an emergency situation, its proper position is also critical. Exhibit 13-17 shows a trip-rod located above the rubber mill. Exhibit 13-18 shows safety tripwire cables are located around the perimeter of or near the danger area. The operator must be able to reach the cable with either hand to stop the machine.

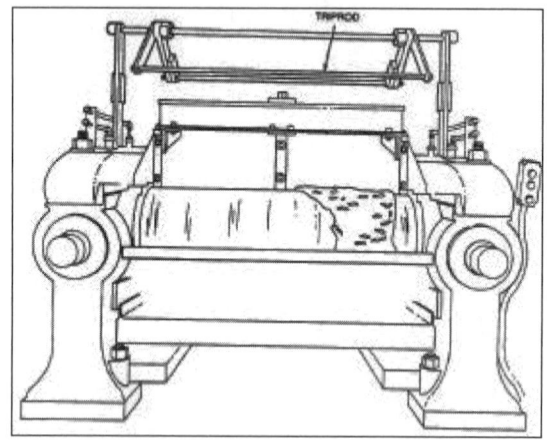

Exhibit 13-17.
A triprod located above the rubber mill.

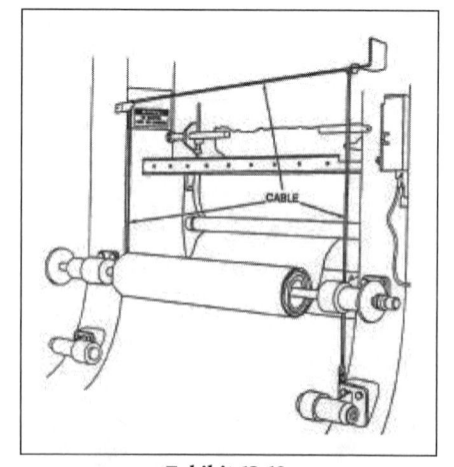

Exhibit 13-18.
Safety tripwire cables located on the perimeter of the danger area.

Two-Hand Control

The two-hand control requires constant, concurrent pressure by the operator to activate the machine. This kind of control requires a part-revolution clutch, brake and brake monitor if used on a power press as illustrated in Exhibit 13-19. With this type of device, the operator's hands are required to be at a safe location (on control buttons) and at a safe distance from the danger area while the machine completes its closing cycle.

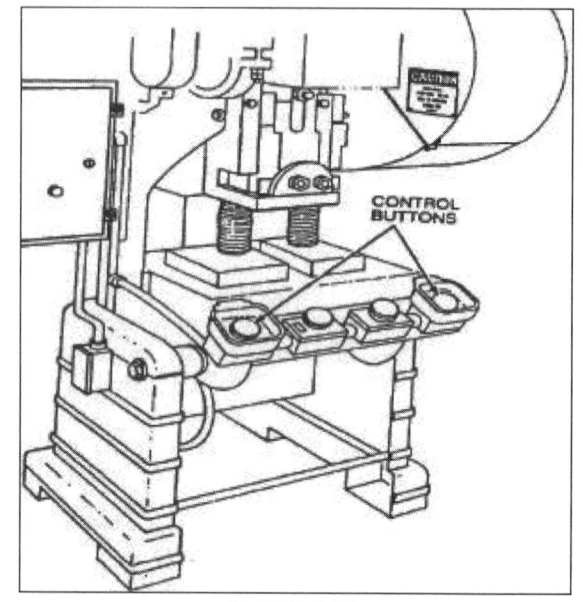

Exhibit 13-19.
The two control buttons require the operator's hands to be a safe distance from the danger area.

Two-Hand Trip

The two-hand trip requires concurrent application of both of the operator's control buttons to activate the machine cycle, after which the hands are free. This device is usually used with machines equipped with full-revolution clutches. The trips must be placed far enough from the point of operation to make it impossible for the operator to move his or her hands from the trip buttons or handles into the point of operation before the first half of the cycle is completed. Thus, the operator's hands are kept far enough away to prevent them from being accidentally placed in the danger area prior to the slide/ram or blade reaching the full "down" position (See Exhibit 13-20.).

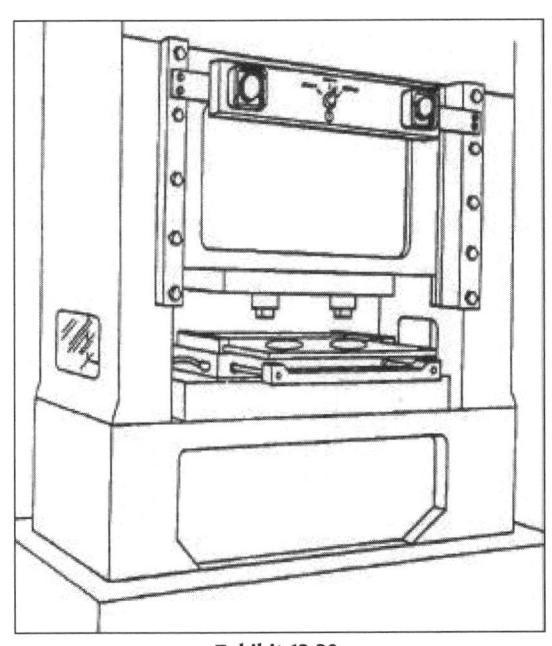

Exhibit 13-20.
The two-hand trip requires both hands to activate the machine cycle.

Gate

A gate is a movable barrier that protects the operator at the point of operation before the machine cycle can be started. Gates are usually designed to be operated with each machine cycle.

Exhibit 13-21 shows a gate on a power press. If the gate is not permitted to descend to the fully closed position, the press will not function.

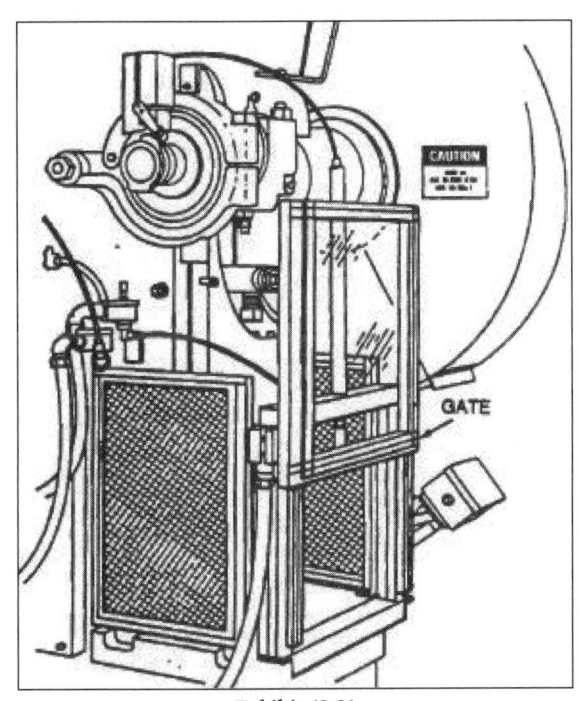

Exhibit 13-21.
A gate on a power press.

Another potential application of this type of guard is where the gate is a component of a perimeter safeguarding system. In this case, the gate would provide protection to the operator as well as to pedestrian traffic.

Safeguarding by Location or Distance

The examples mentioned below are a few of the numerous applications of the principle of safeguarding by location or distance. A thorough hazard analysis of each machine and particular situation is absolutely essential before attempting this safeguarding technique.

To safeguard a machine by location, the machine or its dangerous moving parts must be so positioned that hazardous areas are not accessible or do not present a hazard to a worker during the normal operation of the machine. This may be accomplished by locating a machine so that a plant design feature, such as a wall, protects the worker and other personnel. Additionally, enclosure walls or fences can restrict access to machines. Another possible solution is to have dangerous parts located high enough to be out of the normal reach of any worker.

The feeding process can be safeguarded by location if a safe distance can be maintained to protect the worker's hands. The dimensions of the stock being worked on may provide adequate safety. For instance, if the stock is several feet long and only one end of the stock is being worked on, the operator may be able to hold the opposite end while the work is being performed. An example would be a single-end punching machine. However, depending upon the machine, protection might still be required for other personnel.

The positioning of the operator's control station provides another potential approach to safeguarding by location. Operator controls may be located at a safe distance from the machine if there is no reason for the operator to tend the controls during operation.

Feeding and Ejection Methods to Improve Operator Safety

Many feeding and ejection methods do not require the operator to place his or her hands in the danger area. In some cases, no operator involvement is necessary after the machine is set up. In other situations, operators can manually feed the stock with the assistance of a feeding mechanism. Properly designed ejection methods do not require any operator involvement after the machine starts to function.

However, some feeding and ejection methods may create hazards themselves. For instance, a robot may eliminate the need for an operator to be near the machine, but may create a new hazard when its arm moves.

Using feeding and ejection methods does not necessarily eliminate the need for guards and devices. Guards and devices must be used wherever practical to provide protection from exposure to hazards.

Automatic feed

Automatic feeds reduce the exposure of the operator during the work process and sometimes do not require any effort by the operator after the machine is set up and running.

In Exhibit 13-22, the power press has an automatic feeding mechanism. Notice the transparent fixed enclosure guard at the danger area.

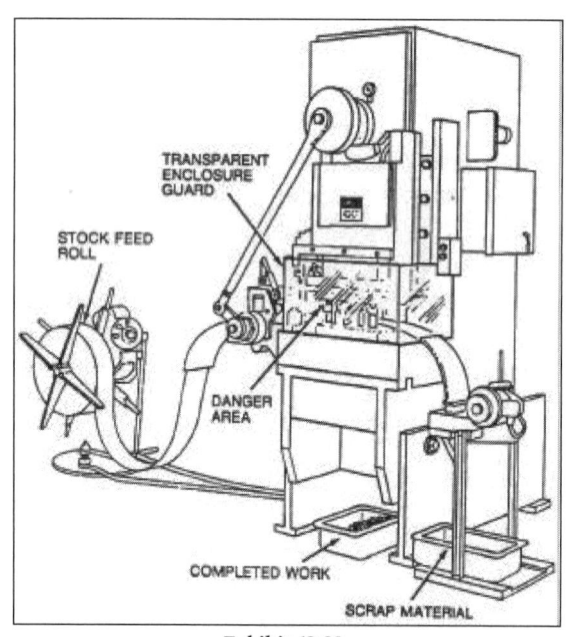

Exhibit 13-22.
A power press with an automatic feeding mechanism.

Semi-automatic feed

With semi-automatic feeding, as in the case of a power press, the operator uses a mechanism to place the piece being processed under the ram at each stroke. The operator does not need to reach into the danger area, and the danger area is completely enclosed.

Exhibit 13-23 shows a chute feed. It may be either a horizontal or an inclined chute into which each piece is placed by hand. Using a chute feed on an inclined press not only helps center the piece as it slides into the die, but may also simplify the problem of ejection.

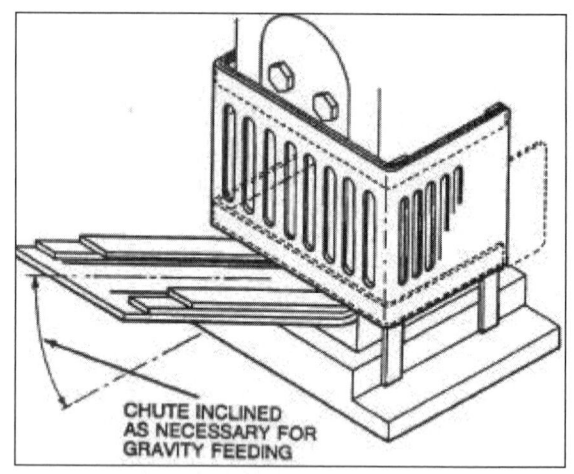

Exhibit 13-23.
A chute feed is one example of a semi-automatic feed.

Automatic Ejection

Automatic ejection may employ either an air-pressure or a mechanical apparatus to remove the completed part from a press. It may be interlocked with the operating controls to prevent operation until the part is ejected. This method requires additional safeguards for operator protection.

In Exhibit 13-24, the pan shuttle mechanism moves under the finished part as the slide moves toward the "up" position. The shuttle then catches the part stripped from the slide by the knockout pins and deflects it into a chute. When the ram moves down toward the next blank, the pan shuttle moves away from the die area.

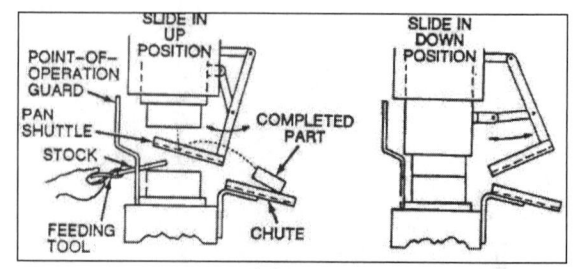

Exhibit 13-24.
The pan shuttle mechanism automatically ejects the part once finished.

WHAT EVERY SUPERVISOR MUST KNOW ABOUT OSHA-GENERAL

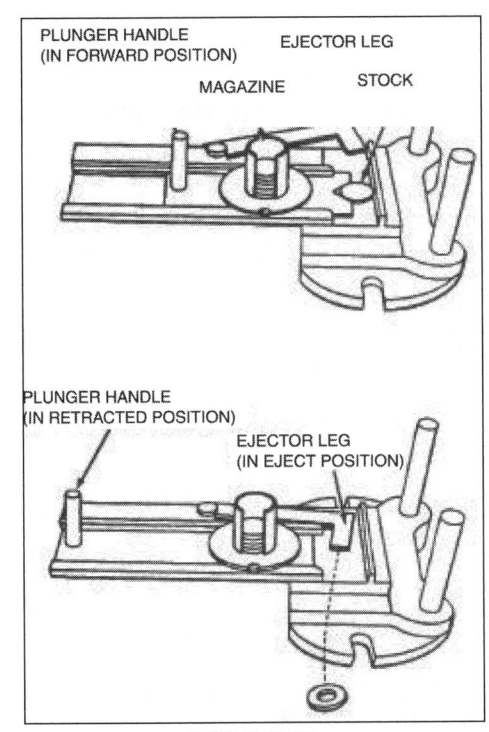

Exhibit 13-25.
The plunger handle and ejector leg work together to eject the part once finished.

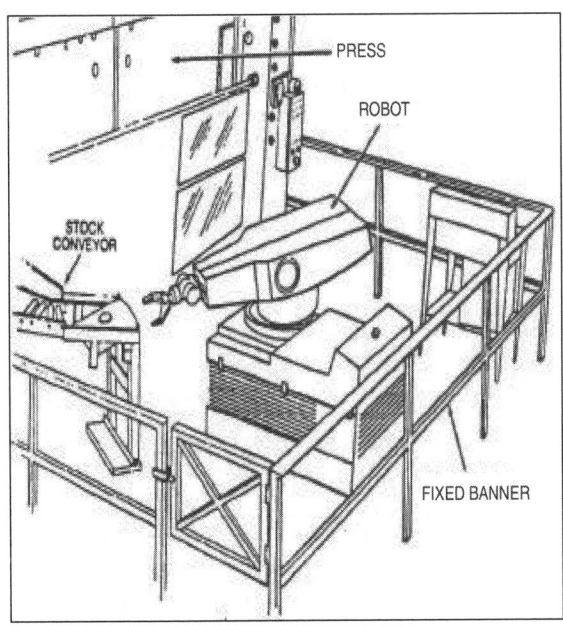

Exhibit 13-26.
Robots may present new hazards to employees thus guards such as perimeter barriers should be used.

Semi-automatic Ejection
Exhibit 13-25 shows a semi-automatic ejection mechanism used on a power press. When the plunger is withdrawn from the die area, the ejector leg, which is mechanically coupled to the plunger, kicks out the completed work.

Robots
Robots are machines that load and unload stock, assemble parts, transfer objects or perform other tasks. Essentially, they perform work otherwise done by an operator. They are best used in high-production processes requiring repeated routines that present hazards to employees. However, robots may create hazards themselves, and, if they do, appropriate guards must be used.

Various techniques are available to prevent employee exposure to the hazards that can be imposed by robots. The most common technique is through the installation of perimeter guarding with interlocked gates (See Exhibit 13-26.). A critical parameter relates to the manner in which the interlocks function. Of major concern is whether the computer program, control circuit, or the primary power circuit, is interrupted when an interlock is activated. The various industry standards should be investigated for guidance; however, it is generally accepted that the interlock should interrupt the primary motive power to the robot.

The American National Standards Institute (ANSI) safety standard for industrial robots, ANSI/RIA R15.06-1986, is informative and presents certain basic requirements for protecting the worker. However, when a robot is to be used in a workplace, the employer should conduct a comprehensive operational safety/health hazard analysis and then devise and implement an effective safeguarding system that is fully responsive to the situation. (Various effective safeguarding techniques are described in ANSI B11.19-1990.)

Studies in Sweden and Japan indicate that many robot accidents have not occurred under normal operating conditions, but rather during programming, program touch-up, maintenance, repair, testing, setup or adjustment. During many of these operations, the operator, programmer or corrective maintenance worker may temporarily be within the robot's working envelope where unintended operations could result in injuries.

All industrial robots are either servo or non-servo controlled. Servo robots are controlled through the use of sensors that are employed to continually monitor the robot's axes for positional and velocity feedback information. This

MACHINE GUARDING

feedback information is compared on an ongoing basis to pre-taught information that has been programmed and stored in the robot's memory.

Non-servo robots do not have the feedback capability of monitoring its axes and velocity and comparing them with a pre-taught program. Their axes are controlled through a system of mechanical stops and limit switches to control the robot's movement.

Types of hazards associated with robots include:

Impact or collision hazards. Impact with the robot's arm or peripheral equipment as a result of unpredicted movements, component malfunctions or unpredicted program changes.

Crushing or trapping hazards. A hazard resulting in some part of a person's body being trapped between the robot's arm and other peripheral equipment or being crushed by peripheral equipment as a result of being impacted by the robot into the equipment.

Mechanical components hazards. Mechanical hazards result from the mechanical failure of components associated with the robot or its power source; drive components, tooling or end-effector and/or peripheral equipment. The failure of gripper mechanisms with resultant release of parts or the failure of end-effector power tools, such as grinding wheels, buffing wheels, de-burring tools, power screwdrivers and nut runners.

Human error. Human error results in hazards both to personnel and equipment. Errors in programming, interfacing peripheral equipment and connecting input/output sensors can result in unpredicted movement or action by the robot that can result in personnel injury or equipment breakage. Human error in judgment frequently results from incorrectly activating the teach pendant or control panel. The greatest human judgment error results from becoming so familiar with the robot's redundant motions that personnel are too trusting in assuming the nature of these motions and place themselves in hazardous positions while programming or performing maintenance within the robot's work envelope.

Miscellaneous Aids

While these aids do not give complete protection from machine hazards, they may provide the operator with an extra margin of safety. Sound judgment is needed in their application and use.

Awareness barriers do not provide physical protection, but serve only to remind a person that he or she is approaching the danger area. Generally, awareness barriers are not considered adequate when continual exposure to the hazard exists.

For example, a rope may be used as an awareness barrier on the rear of a power-squaring shear, as shown in Exhibit 13-27. Although the barrier does not physically prevent a person from entering the danger area, it calls attention to it.

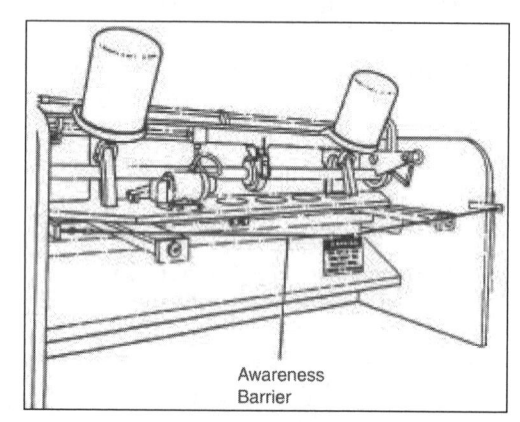

Exhibit 13-27.
A rope being used as an awareness barrier on a power-squaring shear.

Shields may be used to provide protection from flying particles, splashing cutting oils or coolants (See Exhibit 13-28).

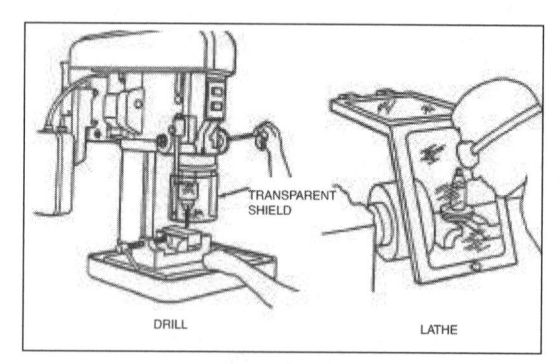

Exhibit 13-28.
Transparent shields provide protection from flying particles.

Holding tools can place and remove stock. A typical use would be for reaching into the danger area of a press or press brake. Holding tools should not be used instead of other machine safeguards; they are merely a supplement to the protection that other guards provide (See Exhibit 13-29). For example, a push stick or block may be used when feeding stock into a saw blade. When it becomes necessary for hands to be in close proximity to the blade, the push stick or block may provide a few inches of safety and prevent a severe injury. In the illustration, the push block fits over the fence.

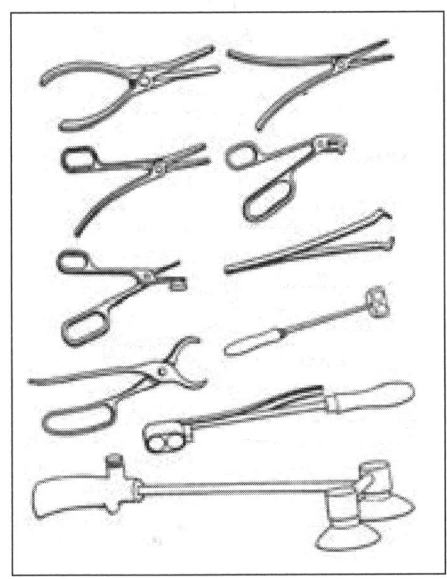

Exhibit 13-29.
Holding tools can protect employees' hands when inserting or removing stock.

Machinery and Machine Guarding

It is important to understand how Subpart 0 applies to machinery in the workplace. Section 212 is a general (or a horizontal) standard that applies to all machines not specifically mentioned elsewhere in other sections of Subpart 0. The other sections are specific (vertical) standards that apply to particular types of machines (e.g., Section 213 applies only to woodworking machinery).

General Requirements for All Machines—1910.212

Machine Guarding

One or more methods of machine guarding shall be provided to protect employees in the machine area from hazards, such as those created by point of operation, ingoing nip points, rotating parts, flying chips and sparks.

Guards shall be affixed to the machine where possible and secured elsewhere if not possible. A guard shall not offer an accident hazard in itself. The point of operation of machines whose operation exposes an employee to inquiry shall be guarded.

Revolving drums, barrels and containers shall be guarded by an enclosure that is interlocked with the drive mechanism

When the periphery of the blades of a fan is less than 7 ft. above the floor or working level, the blades shall be guarded with a guard having openings no larger than 1 in.

Anchoring Fixed Machinery

Machines designed for a fixed location shall be securely anchored to prevent walking or moving.

Woodworking Machinery Requirements—1910.213

Machine Construction, General

Each machine shall be so constructed as to be free from sensible (able to be felt) vibration when the largest size tool is mounted and run idle (no cutting load) at full speed.

Machine Controls and Equipment

A mechanical or electrical power control shall be provided on each machine to make it possible for operators to cut off the power from each machine without leaving their position at the point of operation.

On applications where injury to the operator might result if motors were to restart after power failures, provision shall be made to prevent machines from automatically restarting upon restoration of power.

Power controls and operating controls should be located within easy reach of the operators while they are at their regular work location, making it unnecessary for them to reach over the cutter to make adjustments. This does not apply to constant pressure controls used only for set-up purposes.

Specific Machine Requirements

The remaining paragraphs of Section 213 contain guarding requirements for specific woodworking machines. A discussion of some of these requirements follows.

All woodworking machinery such as table saws, swing saws, radial saws, band saws, jointers, tenoning machines, boring and mortising machines, shapers, planers, lathes, sanders, veneer cutters and other miscellaneous woodworking machinery shall be effectively guarded to protect the operator and other employees from hazards inherent to their operation.

Table Saws
Circular table saws shall have a hood over the portion of the saw above the table, so mounted that the hood will automatically adjust itself to the thickness of and remain in contact with the material being cut.

Circular table saws shall have a spreader aligned with the blade, spaced no more than ½ in. behind the largest blade mounted in the saw. The provision of a spreader in connection with grooving, dadoing or rabbeting is not required.

Circular table saws used for ripping shall have nonkick-back fingers or dogs.

Feed rolls and blades of self-feed circular saws shall be protected by a hood or guard to prevent the hand of the operator from coming into contact with the in-running rolls at any point.

Swing or Sliding Cut-Off Saws
All swing or sliding cut-off saws shall be provided with a guard that will completely enclose the upper half of the saw.

Limit stops shall be provided to prevent swing or sliding cut-off saws from extending beyond the front or back edges of the table.

Each swing or sliding cut-off saw shall be provided with an effective device to return the saw automatically to the back of the table when released at any point of its travel.

Inverted sawing or swing cut-off saws shall be provided with a hood that will cover the part of the saw that protrudes above the top of the table or material being cut.

Radial Saws
The upper hood shall completely enclose the upper portion of the blade down to a point that will include the end of the saw arbor.

The sides of the lower exposed portion of the blade shall be guarded to the full diameter of the blade by a device that will automatically adjust itself to the thickness of the stock and remain in contact with stock being cut (See Exhibit 13-30.). Radial saws used for ripping shall have nonkickback fingers or dogs.

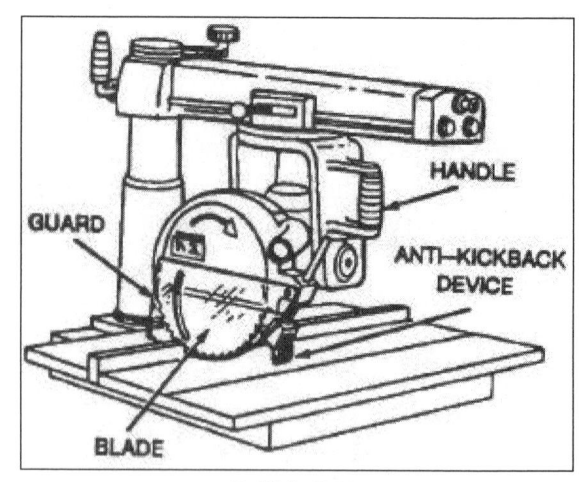

Exhibit 13-30.
The guard extends over the full diameter of the radial saw.

An adjustable stop shall be provided to prevent the forward travel of the blade beyond the position necessary to complete the cut in repetitive operations.

Installation shall be in such a manner that the front end of the unit will be slightly higher than the rear, so as to cause the cutting head to return gently to the starting position when released by the operator.

Bandsaws and Band Resaws
All portions of the saw blade shall be enclosed or guarded, except for the working portion of the blade between the bottom of the guide rolls and the table. Bandsaw wheels shall be fully encased. The outside periphery of the enclosure shall be solid. The front and back shall be either solid or wire mesh or perforated metal.

Jointers
Each hand-fed jointer with horizontal cutting head shall be equipped with an automatic guard that will cover all the sections of the head on the working side of the fence or gauge (See Exhibit 13-31.).

Abrasive Wheel Machinery—1910.215
This section regulates only abrasive wheel machinery. It does not cover wire, buffing or similar wheels. An abrasive wheel is made up of individual particles that are bonded together to form a wheel. The hazard here is that if not properly mounted and used, the wheel can literally explode. Sections of the wheel may fly out at high speeds and strike the operator, causing serious injury or death.

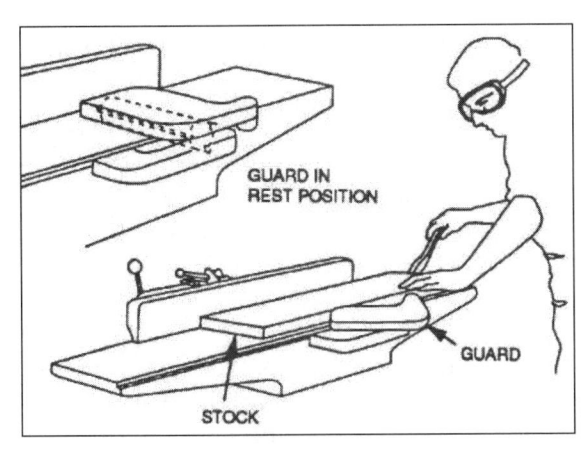

Exhibit 13-31.
A Hand-fed jointer with automatic guard.

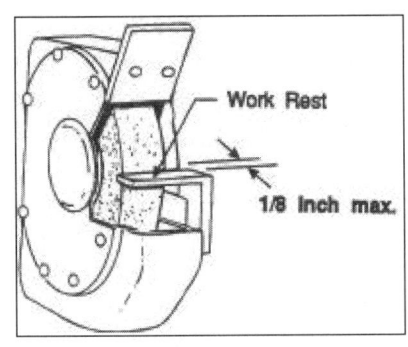

Exhibit 13-32.
Work rests on grinding machines must not open more than 1/8 in.

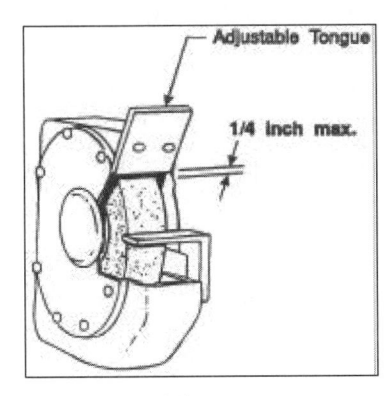

Exhibit 13-33.
Adjustable tongues should not extend beyond ¼ in. from the wheel periphery.

Machine Guarding

Abrasive wheels shall be used only on machines provided with safety guards with the following exceptions:
- Wheels used for internal work while within the work being ground;
- Mounted wheels, used in portable operations, 2 in. and smaller in diameter; and
- Type 16, 17, 18, 18R, and 19 cones, plugs and threaded hole pot balls where the work offers protection.

Guard Design

Abrasive wheel safety guards shall cover the spindle end, nut and flange projections, except:
- Safety guards on all operations where the work provides a suitable measure of protection to the operator may be so constructed that the spindle end, nut and outer flange are exposed;
- Where the nature of the work is such as to entirely cover the side of the wheel, the side covers of the guard may be omitted; and
- The spindle end, nut and outer flange may be exposed on machines designed as portable saws.

Work Rests

On off-hand grinding machines, adjustable work rests of rigid construction shall be used to support the work. Work rests shall be kept adjusted closely to the wheel with a maximum opening 1/8 in. to prevent the work from being jammed between the wheel and the rest, which may cause breakage (See Exhibit 13-32.).

Tongue Adjustment

The protecting member of the abrasive wheel safety guard shall be adjustable for variations in wheel size so that the distance between the wheel periphery and the adjustable tongue or the end of the peripheral member at the top shall never exceed ¼ in. (See Exhibit 13-33.).

Angular Exposure

Abrasive wheel safety guards for bench and floor stands and cylindrical grinders shall not expose the grinding wheel periphery for more than 65° above the horizontal plane of the wheel spindle.

Immediately before mounting, all wheels shall be closely inspected and sounded by the user (ring test) to make sure they have not been damaged.

The spindle speed of the machine shall be checked before mounting of the wheel to be certain that it does not exceed the maximum operating speed marked on the wheel.

Mills and Calendars—1910.216

This section regulates mills and calendars in the rubber and plastics industries. Due to the highly specialized nature of these rules, they are beyond the scope of this manual.

Mechanical Power Presses—1910.217

This section deals with mechanical power presses. This specialized topic is beyond the scope of this manual. However, there are some basic rules that provide insight into this potentially deadly operation.

- The employer shall provide and ensure the use of point-of-operation guards or properly applied and adjusted point-of-operation devices to prevent entry of hands or fingers into the point of operation by reaching through, over, under or around the guard on every operation performed on a mechanical power press. This requirement shall not apply when the point of operation opening is ¼ in. or smaller.
- A substantial guard shall be placed over the treadle on foot-operated presses.
- If the press has pedal counterweights, the path of travel of the weight shall be enclosed.
- Machines using full revolution clutches shall incorporate a single stroke mechanism, except where automatically fed in continuous operation and the points of operation are fully safeguarded by an affixed barrier guard.
- Employers shall establish a program of regular inspections of their power presses to ensure safe operating condition. They shall maintain a record of these inspections and maintenance work.
- All point-of-operation injuries involving a mechanical power press must be reported to OSHA or the state agency (in states with state plans) within 30 days of the injury.

Forging Machines—1910.218

This section regulates forging machines. Due to the highly specialized nature of these rules, they are beyond the scope of this manual.

Mechanical Power-Transmission Apparatus—1910.219

Mechanical power transmission apparatus refers to all components of the mechanical system that transmit energy from the prime source to the part of the machine performing the work. Examples of these components include pulleys, belts, connecting rods, flywheels, shafting, coupling, cams, spindles, chains, cranks and gears. The primary thrust of these regulations is to ensure that employees cannot be "caught" and injured by rotating members, in-running nip points, sprockets, pulleys and similar apparatus.

This section contains detailed requirements for safeguarding all of these components, including:

- Guards for mechanical power transmission equipment shall be made of metal or other suitable material. Wood guards may be used in the woodworking and chemical industries. In industries where atmospheric conditions would rapidly deteriorate metal guards or where extremes in temperature would make metal guards undesirable, wood guards may be used.
- All pulleys, belts, sprockets and chains, flywheels, shaft projections and shafting, gears, couplings or other rotating or reciprocating parts, or any portion thereof within 7 ft. of the floor or working platform shall be guarded effectively.
- All guards for inclined belts shall conform to the standards for construction of horizontal belts. They shall be arranged in such a manner that a minimum clearance of 7 ft. is maintained between the belt and floor at any point outside the guard.
- Flywheels protruding through a working floor shall be guarded.
- Where either runs of horizontal belts are 7 ft. or lower from the floor or working surface, the guard shall extend at least 15 in. above the belt.
- Where gears require a guard, the guard shall extend 6 in. above the mesh point by a bank guard covering the face or be completely enclosed.
- Couplings with bolts, nuts or setscrews that extend beyond the flange of the coupling shall be guarded by a safety sleeve.
- Belts, pulleys and shafting located in rooms used exclusively for power transmission apparatus need not be guarded when the following requirements are met:
    - The basement, tower or room occupied by transmission equipment is locked against unauthorized entrance.

- The vertical clearance in passageways between the floor and power transmission beams, ceiling or other objects is not less than 5 ft. 6 in.
- The intensity of illumination conforms to the requirements of ANSI A11.1-1965
- The route followed by the oiler is protected in such a manner as to prevent accidents.

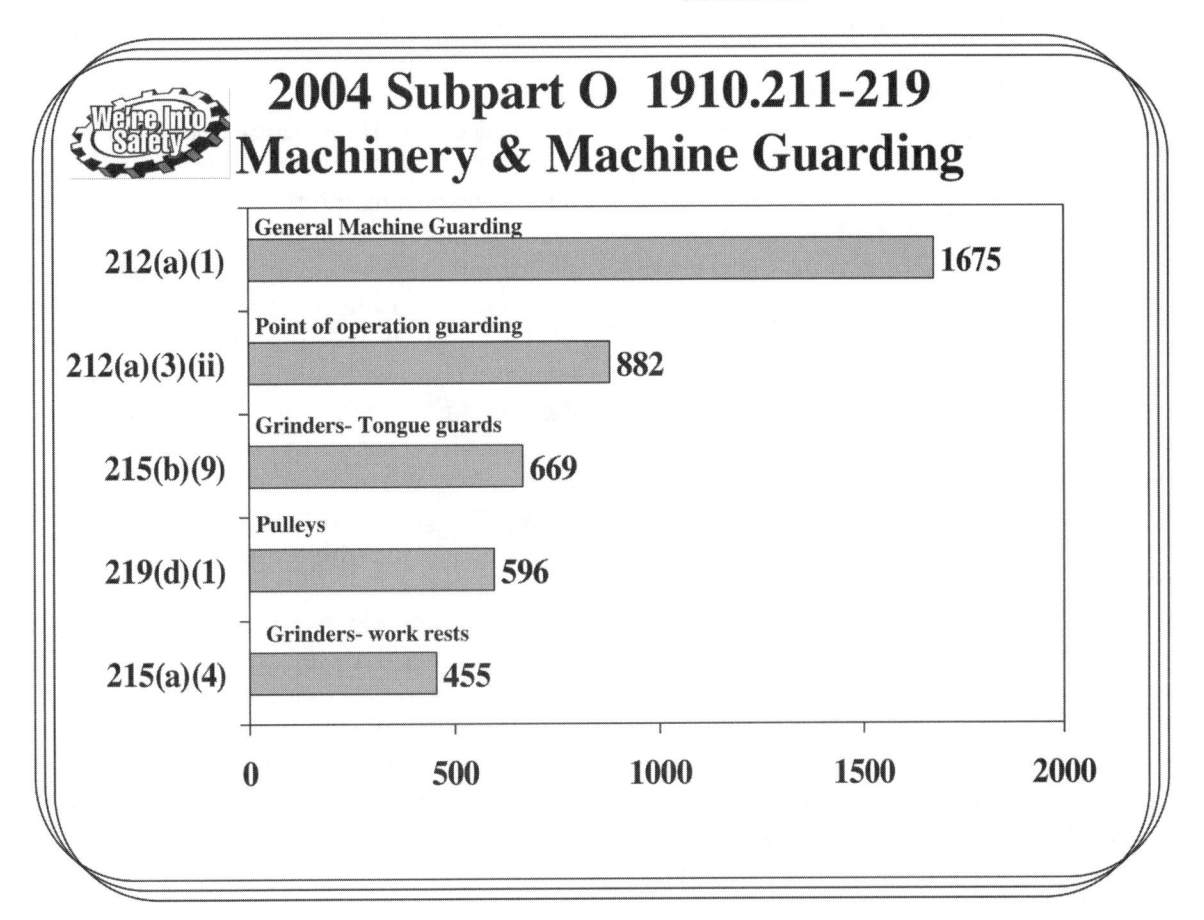

2004 Subpart O 1910.211-219 Machinery & Machine Guarding

Standard	Description	Count
212(a)(1)	General Machine Guarding	1675
212(a)(3)(ii)	Point of operation guarding	882
215(b)(9)	Grinders- Tongue guards	669
219(d)(1)	Pulleys	596
215(a)(4)	Grinders- work rests	455

Control of Hazardous Energy (Lockout/Tagout)

This standard, which went into effect on January 2, 1990, helps safeguard employees from hazardous energy while they are performing service or maintenance on machines and equipment. The standard identifies the practices and procedures necessary to shut down and lock out or tag out machines and equipment, requires that employees receive training in their role in the lockout/tagout program, and mandates that periodic inspections be conducted to maintain or enhance the energy control program.

In the early 1970s, OSHA adopted various lockout-related provisions of the existing national consensus standards and federal standards that were developed for specific types of equipment or industries. When the existing standards require lockout, the new rule supplements these existing standards' by requiring the development and utilization of written procedures, the training of employees, and periodic inspections of the use of the procedures. OSHA has determined that lockout is a more reliable means of deenergizing equipment than tagout and that it is the preferred method for employee use. OSHA contends that, except for limited situations, the use of lockout devices will provide a more secure and more effective means of protecting employees from the unexpected release of hazardous energy or startup of machines and equipment.

The following OSHA standards currently contain lockout/tagout related requirements:
- 1910.178–Powered Industrial Trucks;
- 1910.179–Overhead and Gantry Cranes;
- 1910.181–Derricks;

- 1910.213—Woodworking Machinery;
- 1910.217—Mechanical Power Presses;
- 1910.218—Forging Machines;
- 1910.252—Welding, Cutting and Brazing;
- 1910.262—Textiles;
- 1910.263—Bakery Equipment;
- 1910.265—Sawmills;
- 1910.272—Grain Handling; and
- 1910.305—Electrical.

Scope and Application

The lockout/tagout standard applies to general industry employment and covers the servicing and maintenance of machines and equipment in which the unexpected start-up or release of stored energy could cause injury to employees. The standard establishes minimum performance requirements for the control of hazardous energy.

If employees are performing service or maintenance tasks that do not expose them to the unexpected release of hazardous energy, the standard does not apply. For example, the standard does not apply:

- While servicing or maintaining cord- and plug-connected electrical equipment. (Unplugging the equipment from the energy source must control the hazards; the plug must be under the exclusive control of the employee performing the service and/or maintenance.)
- During hot tap operations that involve transmission and distribution systems for gas, steam, water or petroleum products when:
  - They are performed on pressurized pipelines;
  - Continuity of service is essential and shutdown of the system is impractical; and
  - Employees are provided with an alternative type of protection that is equally effective.

Normal Production Operations

OSHA recognizes that machines and equipment present many hazardous situations during normal production operations (i.e., whenever machines and equipment are used to perform their usual production function). These production hazards are covered by rules in other general industry standards, such as the requirements for general machine guarding and guarding power transmission apparatus in 1910.212 and 1910.219. In certain circumstances, however, some hazards encountered during normal production operations may be covered by the lockout/tagout rule. The following paragraphs illustrate some of these instances.

Servicing and/or Maintenance Operations

If a servicing activity such as lubricating, cleaning or unjamming the production equipment takes place during production, the employee performing the servicing may be subjected to hazards that are not encountered as part of the production operation itself. Workers engaged in these operations are covered by lockout/tagout when any of the following conditions occurs:

- The employee must either remove or bypass machine guards or other safety devices, resulting in exposure to hazards at the point of operation;
- The employee is required to place any part of his or her body in contact with the point of operation of the operational machine or piece of equipment; or
- The employee is required to place any part of his or her body into a danger zone associated with a machine operating cycle.

In the above situations, the equipment must be deenergized and locks or tags must be applied to the energy-isolation devices.

In addition, when normal servicing tasks, such as setting equipment up and making significant adjustments to machines, do not occur during normal production operations, employees performing such tasks are required to lock out or tag out if they can be injured by unexpected energization of the equipment.

OSHA recognizes that some servicing operations must be performed with the power on. Making fine adjustments, such as centering the belt on conveyors or troubleshooting, such as identifying the source of the problem as well as checking to ensure that it has been corrected are examples. OSHA requires the employer to provide effective protection for employees performing such operations.

Minor Servicing Tasks

Employees performing minor tool changes, adjustments and other minor service activities during normal production operations that are routine, repetitive and integral to the use of the production equipment are not covered by the lockout/tagout standard, provided the work is performed using alternative measures that give effective protection.

Provisions of the Standard

The standard requires employers to establish procedures for isolating machines or equipment from the input of energy and affixing appropriate locks or tags to energy-isolating devices to prevent any unexpected energization, start-up or release of stored energy that could injure workers. When tags are used on energy-isolating devices capable of being locked out, the employer must provide additional means to ensure a level of protection equivalent to that of locks. The standard also requires employees be trained and periodic procedures periodically be inspected to maintain or improve their effectiveness.

Energy Control Program

The lockout/tagout rule requires that the employer establish an energy control program that includes:
1. Documented energy control procedures;
2. An employee-training program; and
3. Periodic inspections of the procedures.

The standard requires employers to establish a program to ensure that machines and equipment are isolated and inoperative before any employee performs service or maintenance where the unexpected energization, start up, or release of stored energy could occur and cause injury.

The purpose of the energy control program is to ensure that whenever the possibility of unexpected machine or equipment start-up exists or when the unexpected release of stored energy could occur and cause injury, the equipment is isolated from its energy source(s) and rendered inoperative prior to servicing or maintenance.

Employers have the flexibility to develop a program and procedures that meet the needs of their particular workplace and the particular types of machines and equipment being maintained or serviced.

Energy Control Procedure

This standard requires that energy control procedures be developed, documented and used to control potentially hazardous energy sources whenever workers perform activities covered by the standard.

The written procedures must identify the information that authorized employees must know in order to control hazardous energy during service or maintenance. If this information is the same for various machines or equipment or if other means of logical grouping exists, then a single energy control procedure may be sufficient. If there are other conditions, such as multiple energy sources, different connecting means or a particular sequence that must be followed to shut down the machine or equipment, the employer must develop separate energy control procedures.

The energy control procedure must outline the scope, purpose, authorization, rules and techniques that will be used to control hazardous energy sources as well as the means that will be used to enforce compliance. At a minimum, the procedure must include:

- A statement on how the procedure will be used;
- The steps needed to shut down, isolate, block and secure machines or equipment;
- The steps designating the safe placement, removal and transfer of lockout/tagout devices and who has the responsibility for them; and
- The specific requirements for testing machines or equipment to determine and verify the effectiveness of locks, tags and other energy control measures.

The procedure also must outline the following steps:
- Preparing for shutdown;
- Shutting down the machine(s) or equipment;
- Isolating the machine or equipment from the energy source(s);
- Applying the lockout or tagout device(s) to the energy-isolating device(s);
- Safely releasing all potentially hazardous stored or residual energy; and
- Verifying the isolation of the machine(s) or equipment prior to the start of service or maintenance work.

MACHINE GUARDING

In addition, before lockout or tagout devices are removed and energy is restored to the machines or equipment, certain steps must be taken to reenergize equipment after service is completed, including:
- Ensuring that machines or equipment components are operationally intact;
- Notifying affected employees that lockout or tagout devices have been removed from each energy-isolating device by the employee who applied the device.

Energy-Isolating Devices

The employer's primary tool for providing protection under the standard is the energy-isolating device, which is the mechanism that prevents the transmission or release of energy and to which all locks or tags are attached. This device guards against accidental machine or equipment start-up or the unexpected reenergization of equipment during servicing or maintenance. There are two types of energy-isolating devices: those capable of being locked and those that are not. The standard differentiates between the existence of these two conditions and the employer and employee responsibilities in each case.

When the energy-isolating device cannot be locked out, the employer must use tagout. (The employer may, however, choose to modify or replace the device to make it capable of being locked.) When using tagout, the employer must comply with all tagout-related provisions of the standard and in addition to the normal training required for all employees, train employees in the following limitations of tags:
- Tags are essentially warning devices affixed to energy-isolating devices and do not provide the physical restraint of a lock.
- When a tag is attached to an isolating means, it is not to be removed except by the person who applied it and never to be bypassed, ignored or otherwise defeated.
- Tags must be legible and understandable by all employees.
- Tags and their means of attachment must be made of materials that will withstand the environmental conditions encountered in the work area.
- Tags may evoke a false sense of security. They are only one part of an overall energy control program.
- Tags must be securely attached to the energy-isolating devices so they cannot be detached accidentally during use.

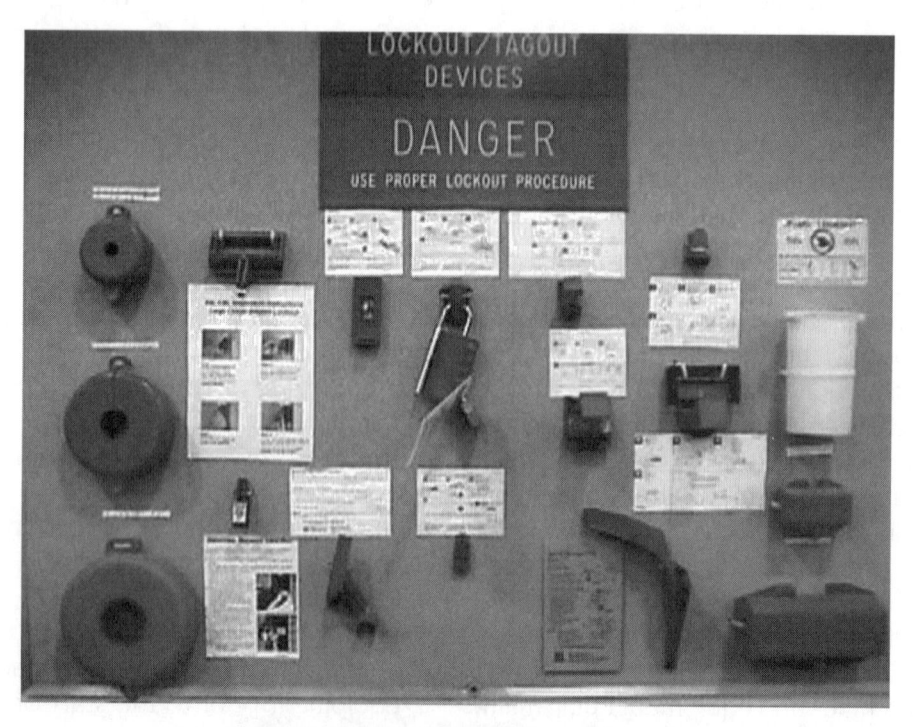

Exhibit 13-34. A lockout/tagout board.

If the energy-isolating device is lockable, the employer shall use locks unless it can prove that the use of tags would provide protection at least as effective as locks and would ensure "full employee protection."

Full employee protection includes complying with all tagout-related provisions and implementing additional safety measures that can provide the level of safety equivalent to that obtained by using lockout. This might include removing and isolating a circuit element, blocking a controlling switch, opening an extra disconnecting device or removing a valve handle to reduce the potential for any inadvertent energization.

Although OSHA acknowledges the existence of energy-isolating devices that cannot be locked out, the standard clearly states that whenever major replacement, repair, renovation or modification of machines or equipment is performed and new machines or equipment are installed, the employer must ensure the energy-isolating devices for such machines or equipment are lockable. Such modifications and/or new purchases are most effectively and efficiently made as part of the normal equipment replacement cycle. All newly purchased equipment must be lockable.

Requirements for Lockout/Tagout Devices

When attached to an energy-isolating device, both lockout and tagout devices are tools that the employer can use in accordance with the requirements of the standard to help protect employees from hazardous energy. The lockout device provides protection by holding the energy-isolating device in the safe position thus preventing the machine or equipment from becoming energized. The tagout device provides protection by identifying the energy-isolating device as a source of potential danger and indicating that the energy-isolating device and the equipment being controlled may not be operated until the tagout device is removed. Whichever devices are used, they must be singularly identified and be the only devices used for controlling hazardous energy. They also must meet the following requirements.

Durable. Lockout and tagout devices must withstand the environment to which they are exposed for the maximum duration of the expected exposure. Tagout devices must be constructed and printed so that they do not deteriorate or become illegible, especially when used in corrosive (acid and alkali chemicals) or wet environments.

Standardized. Both lockout and tagout devices must be standardized according to either color, shape or size. Tagout devices must also be standardized according to print and format.

Substantial. Lockout and tagout devices must be substantial enough to minimize early or accidental removal. Locks must be substantial to prevent removal except by excessive force of special tools, such as bolt cutters or other metal cutting tools. Tag means of attachment must be non-reusable, attachable by hand, self-locking and non-releasable, with a minimum unlocking strength of no less than 50 lbs. The device for attaching the tag also must have the general design and basic characteristics equivalent to a one-piece nylon cable tie that can withstand all environments and conditions.

Identifiable. Locks and tags must clearly identify the employee who applies them. Tags must also warn against hazardous conditions if the machine or equipment is energized and must include a legend such as:
- DO NOT START;
- DO NOT OPEN;
- DO NOT CLOSE;
- DO NOT ENERGIZE; and
- DO NOT OPERATE.

Employee Training

The employer must provide effective initial training and retraining as necessary and certify that such training has been given to all employees covered by the standard. The certification must contain each employee's name and dates of training.

For the purposes of the standard, there are three types of employees: authorized, affected and other. The amount and kind of training that each employee receives is based upon the relationship of that employee's job to the machine or equipment being locked or tagged out and the degree of knowledge relevant to hazardous energy that he or she must possess.

For example, the employer's training program for authorized employees (those who are charged with the responsibility for implementing the energy control procedures and performing the service and maintenance) must cover, at minimum, details about the type and magnitude of the hazardous energy sources present in the workplace and the methods and means necessary to isolate and control those energy sources (i.e., the elements of the energy control procedure(s).)

By contrast, affected employees (usually the machine operators or users) and all other employees need only be able to recognize when the control procedure is being implemented and understand the purpose of the procedure and the importance of not attempting to start up or use the equipment that has been locked or tagged out.

Because an affected employee is not one who is performing the service of maintenance, that employee's responsibilities under the energy control program are simple: Whenever there is a lockout or tagout device in place on an energy isolating device, the affected employee must leave it alone and not attempt to operate the equipment.

Every training program must ensure that all employees understand the purpose, function and restrictions of the energy control program and that authorized employees possess the knowledge and skills necessary for the safe application, use and removal of energy controls. Training programs designed to comply with this standard, which is performance-oriented, should deal with the equipment, type(s) of energy and hazard(s) specific to the work area being covered.

Retraining must be provided whenever there is a change in:
- Job assignments;
- Machines, equipment or processes that present a new hazard; or
- Energy control procedures.

Additional retraining must be conducted when a periodic inspection reveals or the employer has reason to believe that there are deviations from or inadequacies in the employee's knowledge or use of the energy control procedure.

Periodic Inspections
Periodic inspections must be performed at least annually to ensure that the energy control procedures (locks and tags) continue to be implemented properly and that the employees are familiar with their responsibilities under those procedures. In addition, the employer must certify that the periodic inspections have been performed. The certification must identify the machine or equipment on which the energy control procedure was used, the date of the inspection, the employees included in the inspection and the name of the person who performed the inspection. For lockout procedures, the periodic inspection must include a review, between the inspector and each authorized employee, of that employee's responsibilities under the energy control procedure being inspected. When a tagout procedure is inspected, a review on the limitation of tags, in addition to the above requirements, must also be included with each affected and authorized employee.

Application of Controls and Lockout/Tagout Devices
The established procedure of applying energy controls includes the specific elements and actions that must be implemented in sequence.
- Prepare for shut down;
- Shut down the machine or equipment;
- Apply the lockout or tagout device;
- Render safe all stored or residual energy; and
- Verify the isolation and deenergization of the machine or equipment.

Removal of Locks and Tags
Before lockout or tagout devices are removed and energy is restored to the machine or equipment, the authorized employee(s) must take the following actions or observe the following procedures:
- Inspect the work area to ensure that non-essential items have been removed and that machine or equipment components are intact and capable of operating properly;
- Check the area around the machine or equipment to ensure that all employees have been safely positioned or removed;
- Notify affected employees immediately after removing locks or tags and before starting the equipment or machines; and

- Make sure that only those employees who attached the locks or tags remove them. (In the very few instances when this is not possible, the device may be removed under the direction of the employer, provided that he or she strictly adheres to the specific procedures outlined in the standard.)

Additional Safety Requirements

Special circumstances exist when machines need to be tested or repositioned during servicing, outside (contractor) personnel are at the worksite, servicing or maintenance is performed by a group (rather than one specific person) and shift or personnel changes occur.

Testing or positioning of machines. OSHA allows the temporary removal of locks or tags and the reenergization of the machine or equipment only when necessary under special conditions (e.g., when power is needed for the testing or positioning of machines, equipment or components). The reenergization must be conducted in accordance with the sequence of steps listed below:
- Clear the machines or equipment of tools and materials;
- Remove employees from the machines or equipment area;
- Remove the lockout or tagout devices as specified in the standard;
- Energize and proceed with testing or positioning; and
- Deenergize all systems, isolate the machine or equipment from the energy source, and reapply lockout or tagout devices as specified.

Outside personnel (e.g., contractors). The onsite employer and the outside employer must inform each other of their respective lockout or tagout procedures. Each employer must ensure that his or her personnel understand and comply with all restrictions and/or prohibitions of the other employer's energy control program.

Group lockout or tagout. During all group lockout/tagout operations where the release of hazardous energy is possible, each authorized employee performing service or maintenance shall be protected by his or her personal lockout or tagout device or a comparable mechanism that provides equivalent protection.

Shift or personnel changes. Specific procedures must ensure the continuity of lockout or tagout protection during shift or personnel changes.

Lockout/Tagout Glossary

Affected employee. An employee who performs the duties of his or her job in an area in which the energy control procedure is implemented and servicing or maintenance operations are performed. An affected employee does not perform servicing or maintenance on machines or equipment and, consequently, is not responsible for implementing the energy control procedure. An affected employee becomes an "authorized" employee when he or she performs servicing or maintenance functions on machines or equipment that must be locked or tagged.

Authorized employee. An employee who performs servicing or maintenance on machines and equipment. Lockout or tagout is used by these employees for their own protection.

Capable of being locked out. An energy-isolating device is considered capable of being locked out if it meets one of the following requirements:
- It is designed with a clasp to which a lock can be attached;
- It is designed with any other integral part through which a lock can be affixed;
- It has a locking mechanism built into it; or
- It can be locked without dismantling, rebuilding, or replacing the energy isolating device or permanently altering its energy control capability.

Energized. Machines and equipment are energized when (1) they are connected to an energy source or (2) they contain residual or stored energy.

Energy-isolating device. Any mechanical device that physically prevents the transmission or release of energy. These include, but are not limited to, manually operated electrical circuit breakers, disconnect switches, line valves and blocks.

Energy source. Any source of electrical, mechanical, hydraulic, pneumatic, chemical, thermal or other energy.

Energy control procedure. A written document that contains those items of information an authorized employee needs to know to safely control hazardous energy during servicing or maintenance of machines or equipment.

Energy control program. A program intended to prevent the unexpected energizing or the release of stored energy in machines or equipment on which servicing and maintenance is being performed by employees. The program consists of energy control procedures, an employee-training program and periodic inspections.

Lockout. The placement of a lockout device on an energy-isolating device, in accordance with an established procedure, ensuring that the energy-isolating device and the equipment being controlled cannot be operated until the lockout device is removed.

Lockout device. Any device that uses positive means such as a lock, either key or combination, to hold an energy-isolating device in a safe position, thereby preventing the energizing of machinery or equipment. When properly installed, a blank flange or bolted slip blind are considered equivalent to lockout devices.

Tagout. The placement of a tagout device on an energy-isolating device, in accordance with an established procedure, to indicate that the energy-isolating device and the equipment being controlled may not be operated until the tagout device is removed.

Tagout device. Any prominent warning device, such as a tag and a means of attachment that can be securely fastened to an energy-isolating device in accordance with an established procedure. The tag indicates that the machine or equipment to which it is attached is not to be operated until the tagout device is removed in accordance with the energy control procedure.

OSHA Instruction STD 1-7.3 SEP 11, 1990 Directorate of Compliance Programs
SUBJECT: 29 CFR 1910.147, the Control of Hazardous Energy (Lockout/Tagout)—Inspection Procedures and Interpretive Guidance

A. **Purpose.** This instruction establishes policies and provides clarification to ensure uniform enforcement of the Lockout/Tagout Standards.
B. **Scope.** This instruction applies OSHA-wide.
C. **References.**
  1. General Industry Standards, 29 CFR 1910, Subpart O, Subpart S, and other specific subparts.
  2. OSHA Instruction CPL 2.45B, June 15, 1989, the Revised Field Operations Manual (FOM).
D. **Effective Date of Requirements.** All requirements of 29 CFR 1910.147 have an effective date of January 2, 1990. The information collection requirements contained in this section have been approved by the Office of Management and Budget (OMB) and listed under OMB control number 1218-0150, as announced at Federal Register, Volume, 54, No. 199, October 1989.
E. **Action.** Regional Administrators and Area Directors shall ensure that the guidelines and interpretive guidance in this instruction are followed and that compliance officers are familiar with the contents of the standard.
F. **Federal Program Change.** This instruction describes a Federal program change that affects State programs. Each Regional Administrator shall:
  1. Ensure that this change is forwarded to each State designee.
  2. Explain the technical content of this change to the State designee as requested.
  3. Ensure that State designees acknowledge receipt of this Federal program change in writing, within 30 days of notification, to the Regional Administrator. This acknowledgment should include the State's intention to follow the inspection guidelines described in this instruction, or a description of the State's alternative guidelines which are "at least as effective" as the Federal guidelines.
     a. If a State intends to follow the revised inspection guidelines described in this instruction, the State must submit either a revised version of this instruction, adapted as appropriate to reference State law, regulations and administrative structure, or a cover sheet describing how references in this instruction correspond to the State's structure. The State's acknowledgment letter may fulfill the plan supplement requirement if the appropriate documentation is provided.

b. Any alternative State inspection guidelines must be submitted as a State plan supplement within 6 months. If the State adopts an alternative to Federal guidelines, the State's submission must identify and provide a rationale for all substantial differences from Federal guidelines in order for OSHA to judge whether a different State guideline is as effective as a comparable Federal guideline.
4. After Regional review of the State plan supplement and resolution of any comments thereon, forward the State submission to the National Office in accordance with established procedures. The Regional Administrator shall provide a judgment on the relative effectiveness of each substantial difference in the State plan change and an overall assessment thereof with a recommendation for approval or disapproval by the Assistant Secretary.
5. Review policies, instructions and guidelines issued by the State to determine that this change has been communicated to State personnel.

G. **Background.** The Standard for Control of Hazardous Energy (Lockout/Tagout), 29 CFR 1910.147, was promulgated on September 1, 1989, at Federal Register, Volume 54, No. 169 (pages 36644-36696), and was effective on January 2, 1990, as announced at Federal Register, Volume 54, No. 213, November 6, 1989 (page 46610). Previously existing section 29 CFR 1910.147 was redesignated as 29 CFR 1910.150, Sources of Standards.
1. Since the inception of its enforcement program, OSHA has relied on the "General Duty Clause". (Section 5(a)(1) of the OSH Act) to ensure that employers safeguarded their maintenance and service employees through the use of lockout/tagout from the hazards involving the unintentional release of hazardous energy. Such violations reached a level so significant that the development and promulgation of a lockout/tagout standard was required.
2. The new rule addresses practices and procedures that are necessary to disable machinery or equipment and to prevent the release of potentially hazardous energy while maintenance and servicing activities are being performed.
3. The lockout/tagout provisions of this standard are for the protection of general industry workers while performing servicing and maintenance functions and augment the safeguards specified at Subparts O, S, and other applicable portions of 29 CFR 1910.

H. **Inspection Guidelines.** The standard incorporates performance requirements which allow employers flexibility in developing lockout/tagout programs suitable for their particular facilities.
1. The compliance officer shall determine whether servicing and maintenance operations are performed by the employees. If so, the compliance officers shall further determine whether the servicing and maintenance operations are covered by 29 CFR 1910.147 or by the requirements of employee safeguarding specified by other standards as discussed in I.1.
2. Evaluations of compliance with 29 CFR 1910.147 shall be conducted during all general industry inspections within the scope of the standard in accordance with the FOM, Chapter III, D.7. and 8., Additional Information or Supplement Records Review. The review of records shall include special attention to injuries related to maintenance and servicing operations.
3. The compliance officer shall evaluate the employer's compliance with the specific requirements of the standard. The following guidance provides a general framework to assist the compliance officer during inspections:
   a. Ask the employer for any hazard analysis or other basis on which the program related to the standard was developed. Although this is not a specific requirement of the standard, such information, when provided, will aid in determining the adequacy of the program. It should be noted that the absence of a hazard analysis does not indicate non-compliance with the standard.
   b. Ask the employer for the documentation including: procedures for the control of hazardous energy including shutdown, equipment isolation, lockout/tagout application, release of stored energy, verification of isolation; certification of periodic inspections;

MACHINE GUARDING

and certification of training. The documented procedure must identify the specific types of energy to be controlled and, in instances where a common procedure is to be used, the specific equipment covered by the common procedure must be identified at least by type and location. The identification of the energy to be controlled may be by magnitude and type of energy. Note the exception to documentation requirements at paragraph 1910.147(c)(4)(i), "Note". The employer need not document the required procedure for a particular machine or equipment when all eight(8) elements listed in the "Note" exist.

    c. Evaluate the employer's training programs for "authorized", "affected", and "other" employees. Interview a representative sampling of selected employees as a part of this evaluation (29 CFR 1910.147(c)(7)(i)).

        (1) Verify that the training of authorized employees includes:
           (a) Recognition of hazardous energy;
           (b) Type and magnitude of energy found in the workplace;
           (c) The means and methods of isolating and/or controlling energy; and
           (d) The means of verification of effective energy control, and the purpose of the procedures to be used.
        (2) Verify that affected employees have been instructed in the purpose and use of the energy control procedures.
        (3) Verify that all other employees who may be affected by the energy control procedures are instructed about the procedure and the prohibition relating to attempts to restart or reenergize such machines or equipment.
        (4) When the employer's procedures permit the use of tagout, the training of authorized, affected, and other employees shall include the provisions of 29 CFR 1910.147(c)(7)(ii) and (d)(4)(iii).

    d. Evaluate the employer's manner of enforcing the program (29 CFR 1910.147(c)(4)(ii)).

4. In the event that deficiencies are identified by following the guidelines in H.3. of this instruction, the compliance officer shall evaluate the employer's compliance with specific requirements of the standard, with particular attention to the interpretive guidance provided in section I. and to the following:

  a. Evaluate compliance with the requirements for periodic inspection of procedures.

  b. Ensure that the person performing the periodic inspection is an authorized employee other than the one(s) utilizing the procedure being inspected.

  c. Evaluate compliance with retraining requirements which result from the periodic inspection of procedures and practices, or from changes in equipment/processes.

  d. Evaluate the employer's procedures for assessment, and correction of deviations or inadequacies identified during periodic inspections of the energy control procedure.

  e. Identify the procedures for release from lockout/tagout, including:
      (1) Replacement of safeguards, machine or equipment inspection, and removal of non-essential tools and equipment;
      (2) Safe positioning of employees;
      (3) Removal of lockout/tagout device(s); and
      (4) Notification of affected employees that servicing and maintenance is completed.

  f. Ensure that when group lockout or tagout is used, it affords a level of protection equivalent to individual lockout or tagout as amplified in I.7. through I.9. of this instruction

5. The lockout/tagout standard is a performance standard; therefore, additional guidance is provided in Appendix C of this instruction to assist in effective implementation by employers and for uniform enforcement by OSHA field staff.

I. **Interpretive Guidance.** The following guidance relative to specific provisions of 29 CFR 1910.147 is provided to assist compliance officers in conducting inspections where the standard may be applicable:

1. **Scope of the Standard.**
   a. The standard as specified in 29 CFR 1910.147(b), applies to any source of mechanical, hydraulic, pneumatic, chemical, thermal, or other energy.
      (1) The standard applies to piping systems, and requires, at 29 CFR 1910.147(d)(5), that all potentially hazardous stored or residual energy be relieved, disconnected, restrained, and otherwise rendered safe. If there is a possibility of reaccumulation of stored energy to a hazardous level, continued monitoring shall be performed while a potential hazard exists.
      (2) The standard also applies to high intensity electromagnetic fields regulated at 29 CFR 1910.97, nonionizing radiation. Such electromagnetic devices shall be deenergized and held off whenever workers are present within a high intensity ambient field.
      (3) Servicing/maintenance of fire alarm and extinguishing systems and their components, upon which other employees are dependent for fire safety, are not required to meet the requirements of this standard if the workers performing servicing/maintenance upon fire extinguishing systems are protected from hazards related to the unexpected release of hazardous energy by appropriate alternative measures. (See 29 CFR 1910, Subpart L.)
   b. The standard does not apply to servicing and maintenance when employees are not exposed to the unexpected released of hazardous energy.
   c. Safeguarding workers from the hazards of contacting electrically live parts (exposure to electric current) continues to be regulated at Subpart S.
   d. Servicing and maintenance functions conducted during normal production operations are not regulated at 29 CFR 1910.147 if the safeguarding provisions of Subpart 0 or other applicable portions of 29 CFR 1910 prevent worker exposure to hazards created by the unexpected energization or start-up of the machine or equipment. However, lockout/tagout procedures are required if the production safeguards are rendered ineffective while an employee is exposed to hazardous portions of the machines or equipment.
   e. Generally, activities such as lubrication, cleaning or unjamming, servicing of machines or equipment, and making adjustments or tool changes, where the employee may be exposed to the UNEXPECTED energization of start-up of the equipment or release of hazardous energy, are covered by this standard. However, minor tool changes and adjustments, and other minor servicing activities, which take place during normal production operations, are not covered by this standard if they are routine, repetitive, and integral to the use of equipment for production, and if work is performed using alternative protective measures which provide effective employee protection. Thus, lockout or tagout is not required by this standard if the alternative protective measures enable the servicing employee to clean or unjam, or otherwise service the machine without being exposed to unexpected energization or activation of the equipment, or the release of stored energy.
   f. The exclusion of plug and cord connected electric equipment, at 29 CFR 1910.147(a)(2)(iii)(A), applies only when the equipment is unplugged and the plug is under the exclusive control of the employee performing the servicing and/or maintenance.
      (1) The plug is under the exclusive control of the employee if it is physically in the possession of the employee, or in arm's reach and in the line of sight of the employee, or if the employee has affixed a lockout/tagout device on the plug.
      (2) The company lockout/tagout procedures required by the standard at 29 CFR 1910.147(c)(4) shall specify the acceptable procedure for handling cord and plug connected equipment.
2. **Procedures.**
   a. The employer must develop and document procedures and techniques to be used for the control of hazardous energy. The standard, at 29 CFR 1910.147(c)(4)(i) "Note," identifies eight (8) conditions that must exist in order to excuse the employer's obligation to maintain a written procedure for a specific machine or piece of equipment.

MACHINE GUARDING

b. 29 CFR 1910.147(d)(3) and (d)(5) provide that energy isolation be a mandatory part of employer's control procedure where either a lockout system or a tagout system is used.
c. Similar machines and/or equipment (such as those using the same type and magnitude of energy and the same or similar types of controls) can be covered with a single written procedure.

3. **Lockout vs. Tagout.**
   a. OSHA has determined that lockout is a surer means of ensuring deenergization of equipment than tagout, and that it is the preferred method.
   b. 29 CFR 1910.147(c)(3)(ii) provides that: When using a tagout program in those instances where the equipment is capable of being locked out, the employer shall demonstrate that the tagout program will provide a level of safety equivalent to the obtained when using a lockout program. Additional means beyond those necessary for lockout are required. (Additional means include: additional safety measures such as the removal of an isolating circuit element, blocking of a controlling switch, opening of an extra disconnecting device, or the removal of a valve handle to reduce the likelihood of inadvertent energization.)
   c. 29 CFR 1910.147(c)(4)(ii) provides that: Where lockout/tagout programs are used, the employer is required to implement an effective means of enforcing the program.
   d. 29 CFR 1910.147(c)(7)(ii)(A-F) provide that: Additional training of authorized, affected and other employees is required when tagout programs are used.
   e. 29 CFR 1910.147(c)(5)(ii)(A) requires that lockout and tagout devices be capable of withstanding the environment to which they are exposed. Devices which are not exposed to harsh environments need not be capable of withstanding such exposure.
   f. 29 CFR 1910.147(c)(5)(ii)(C)(2) requires that tagout devices having reusable, non-locking, easily detachable means of attachment (such as string, cord, or adhesive) are not permitted.

4. **Employees and Training.**
   a. The standard recognized three types of employees: (1) "authorized" and (2) "affected", defined in 1910.147(b), and (3) "other", defined in 1910.147(c)(7)(ii)(C). Different levels of training are required based upon the respective roles of employees in the control of energy and the knowledge which they must possess to accomplish their tasks safely and to ensure the safety of fellow workers as related to the lockout/tagout procedures (1910.147(c)(7)(i)).
   b. Employees who exclusively perform functions related to normal production operations, and who perform servicing and/or maintenance under the protection of normal machine safeguarding, need only be trained as "affected" (rather than "authorized") employees even if tagout procedures are used. (See I.1.d. and I.1.e. of this instruction.)
   c. The employer's training program must cover, at a minimum, the following three areas: energy control program, elements of energy control procedures relevant to employee duties, and the pertinent requirements of the standard (1910.147(c)(7) and (d) through (f)).
   d. The employer must provide:
      (1) Effective initial training;
      (2) Effective retraining as needed; and
      (3) Certification of training. The certification shall contain each employee's, name and dates of training (1910.147(c)(7)(iv)).
   e. Retraining of authorized and affected employees is required:
      (1) Whenever there is a change in employee job assignments;
      (2) Whenever a new hazard is introduced due to a change in machines, equipment or process;
      (3) Whenever there is a change in the energy control procedures; or

(4) Whenever a periodic inspection by the employer reveals inadequacies in the company procedures or in the knowledge of the employees.

5.
   a. At least annually, the employer shall ensure that an authorized employee other than the one(s) utilizing the energy control procedure being inspected, is required to inspect and verify the effectiveness of the company energy control procedures. These inspections shall at least provide for a demonstration of the procedures and may be implemented through random audits and planned visual observations. These inspections are intended to ensure that the energy control procedures are being properly implemented and to provide an essential check on the continued utilization of the procedures (29 CFR 1910.147(c)(6)(i)).
      (1) When lockout is used, the employer's inspection shall include a review of the responsibilities of each authorized employee implementing the procedure with that employee. Group meetings between the authorized employee who is performing the inspection and all authorized employees who implement the procedure would constitute compliance with this requirement.
      (2) When tagout is used, the employer shall conduct this review with each affected and authorized employee.
      (3) Energy control procedures used less frequently than once a year need be inspected only when used.
   b. The periodic inspection must provide for and ensure effective correction of identified deficiencies (29 CFR 1910.147(c)(6)(i)(B)).
   c. The employer is required to certify that the prescribed periodic inspections have been performed (29 CFR 1910.147(c)(6)(ii)).
6. **Equipment Testing or Positioning.** Under 29 CFR 1910.147(f)(1), OSHA allows the temporary removal of lockout or tagout devices and the reenergization of the machine or equipment ONLY during the limited time necessary for the testing or positioning of machines, equipment or components. After the completion of the temporary reenergization, the authorized employees shall again deenergize the equipment and resume lockout/tagout procedures.
7. **Group Lockout/Tagout.** Group lockout/tagout procedures shall be tailored to the specific industrial operation and may be unique in the manner that employee protection from the release of hazardous energy is achieved. Irrespective of the situation, the requirements of this generic standard specify that each employee performing maintenance or servicing activities shall be in control of hazardous energy during his/her period of exposure.
   a. Group operations normally require that a lockout/tagout program be implemented which ensures that each authorized employee is protected from the unexpected release of hazardous energy by his/her personal lockout/tagout device(s). No employee may affix the personal lockout/tagout device of another employee. Various group lockout/tagout procedures discussed in Appendix C provide for each authorized employee's use of his/her personal lockout/tagout device(s).
   b. One of the most difficult problems addressed by the standard involves the servicing and maintenance of complex equipment. Such equipment is frequently used in the petrochemical and chemical industries. Acceptable group lockout/tagout procedures for complex equipment are discussed further at Appendix C.
8. **Compliance with Group Lockout/Tagout.** These operations shall, at a minimum, provide for the following:
   a. Before the machine or equipment is shut down, each authorized employee who is to be involved during the servicing/maintenance operation shall be made aware by the employer of the type, magnitude, and hazards related to the energy to be controlled and of the method or means to control the energy. In the event that the machine or equipment is already shut down, the authorized employee shall be made aware of these elements

before beginning his/her work (29 CFR 1910.147(d)(1)). Verification shall be performed as noted at I.8.f. of this instruction.
b. An orderly shutdown of the machine or equipment shall be conducted which conforms to the documented company procedure and which will not create hazards (29 CFR 1910.147(d)(2)).
c. All energy isolating devices needed to isolate the machine or equipment shall be effectively positioned and/or installed (29 CFR 1910.147(d)(3)).
d. The authorized employee(s) performing the servicing or maintenance (following the company procedure) shall personally affix a lock or tag upon each energy isolating device (29 CFR 1910.147(d)(4)(i)). The company procedure must ensure that no employee affixes a personal lockout/tagout device for another employee.
    (1) A single lock upon each energy isolating device, together with the use of a lockbox for retention of the keys and to which each authorized employee affixes his/her personal lock or tag, also satisfies the requirements (29 CFR 1910.147(f)(3)(i)).
    (2) Locks shall be affixed in a manner that will hold the energy isolating device in a safe (off) position (29 CFR 1910.147(d)(4)(ii)).
    (3) Tagout devices, where used, shall be affixed at the same location as would a lock if such fittings are provided, or shall be affixed in a manner that will clearly indicate that movement of the isolating device is prohibited (29 CFR 1910.147(d)(4)(iii)).
e. Following the application of locks or tags, all potentially hazardous stored energy or residual energy shall be relieved, disconnected, restrained, and otherwise rendered safe (29 CFR 1910.147(d)(5)(i)).
    (1) Verification of energy isolation shall be monitored as frequently as necessary if there is a possibility of reaccumulation of stored energy (29 CFR 1910.147(d)(5)(ii).
    (2) Monitoring may be accomplished, for example, by observation or with the aid of a monitoring device which will sound an alarm if a hazardous energy level is being approached.
     Authorized employees shall verify that isolation and deenergization have been effectively accomplished before starting servicing/maintenance work. Verification is also necessary by each group of workers before starting work at shift changes.
g. Release from lockout/tagout shall be accomplished in compliance with the requirements of 29 CFR 1910.147(e).
    (1) The machine or equipment area shall be cleared of nonessential items to prevent malfunctions which could result in employee injuries (29 CFR 1910.147(e)(1)).
    (2) The authorized employees shall remove their respective locks or tags from the energy isolating devices or from the group lock-box(s) following the procedure established by the company (29 CFR 1910.147(e)(3)).
    (3) In all instances, the company procedure must provide a system which identifies each authorized employee involved in the servicing/maintenance operation.
    (4) Before reenergization, all employees in the machine or equipment area shall be safely positioned or moved from the area, and the affected employees shall be notified that the lockout/tagout devices have been removed (29 CFR 1910.147(e)(2)).
h. During all group lockout/tagout operations where the release of hazardous energy is possible, each authorized employee performing servicing or maintenance shall be protected by his/her personal lockout or tagout device and by the company procedure. As described at Appendix C, B.1.g., a master tag is a personal tagout device if each employee personally signs on and signs off on it and if the tag clearly identifies each authorized employee who is being protected by it.

9. **Compliance of Outside Personnel.** Outside servicing and maintenance personnel (contractors, etc.) engaged in activities regulated under 29 CFR 1910.147 are subject to the requirements of that standard.

a. The CSHO shall verify that the outside employer and the on-site employer have exchanged information regarding the lockout/tagout energy control procedures used by each employer's workers (29 CFR 1910.147(f)(2)(i)).
   b. The CSHO shall verify that the on-site employer has effectively informed his/her personnel of the restrictions and prohibitions associated with the outside employer's energy control procedures (29 CFR 1910.147(f)(2)(ii)).
   c. When an outside employer's is engaged in servicing and maintenance activities within an on-site employer's facility and if that contractor's activities are subject to the requirements of 29 CFR 1910.147, the CSHO shall coordinate with the Area Director to obtain permission to initiate an independent inspection of the outside contractor's activities.
10. Appendix B contains an example of a functional flow diagram to implement safe lockout/tagout procedures. This flow diagram is presented solely as an aid and does not constitute the exclusive or definitive means of complying with the standard in any particular situation.

J. **Classification of Violations.**
   1. A deficiency in the employer's energy control program and/or procedure that could contribute to a potential exposure capable of producing serious physical harm or death shall be cited as a serious violation.
   2. The failure to train "authorized", "affected", and "other" employees as required for their respective classification should normally be cited as a serious violation
   3. Paperwork deficiencies in lockout/tagout programs where effective lockout/tagout work procedures are in place shall be cited as other-than-serious.

K. **Evaluation.** In keeping with agency policy, each Region shall evaluate the effectiveness of the guidance in this instruction annually. Each Regional Administrator shall submit a written evaluation report to the Directorate of Compliance Program within 30 day of the close of the fiscal year.

Gerard F. Scannell Assistant Secretary
DISTRIBUTION: National, Regional, and Area Offices All Compliance Officers State Designees NIOSH Regional Program Directors (7)(c)(1) Consultation Project Managers OSHA Training Institute

Appendix A

The following listing indicates a number of OSHA standards which currently impose lockout/tagout related requirements. The list does not necessarily include all lockout/tagout related OSHA 29 CFR 1910 standards.

Powered Industrial Trucks	1910.178(q)(4)
Overhead and Gantry Cranes	1910.179(g)(5)(i), (ii), (iii), 1910.179(l)(2)(i)(c), (d)
Derricks	1910.181(f)(2)(i)(c), (d)
Woodworking Machinery	1910.213(a)(10), 1910.213(b)(5)
Mechanical Power Presses	1910.217(b)(8)(i), 1910.217(d)(9)(iv)
Forging Machines	1910.218(a)(3)(iii), (v), 1910.218(d)(2)
	1910.218(e)(1)(ii), (iii), 1910.218(f)(1)(i), (ii), (iii), 1910.218(f)(2)(i), (ii),
	1910.218(h)(2), (5), 1910.218(i)(1), (2), 1910.218(j)(1)
Welding, Cutting and Brazing	1910.252(c)(1)(i)
Pulp, Paper and Paperboard Mills	1910.261(b)(4), 1910.261(f)(6)(i), 1910.261(g)(15)(i)
	1910.261(g)(19)(iii), 1910.261(j)(4)(iii), 1910.261(j)(5)(iii),
	1910.261(k)(2)(ii)
Textiles	1910.262(c)(1), 1910.262(n)(2), 1910.262(p)(1), 1910.262(q)(2)
Bakery Equipment	1910.263(l)(3)(iii)(b), 1910.263(l)(8)(iii)
Sawmills	1910.265(c)(12)(v), 1910.265(c)(13), 1910.265(c)(26)(v)
Grain Handling	1910.272(e)(1)(ii), 1910.272(g)(1)(ii), 1910.272(l)(4)
Electrical	1910.305(j)(4)(ii)(A), 1910.305(j)(4)(ii)(c)(1)

[Appendix B excluded.]

Appendix C

This appendix provides guidelines to assist the compliance officer during evaluations of employer operations.

A. **Normal Production Operations.** The lockout/tagout standard, 29 CFR 1910.147, addresses the safety of employees engaged in servicing and maintenance activities in general industry workplaces. The standard complements the requirements for machine and process operator safety prescribed by the various general industry standards in 29 CFR Part 1910. Subpart 0 of 29 CFR 1910 provides the principal, though not exclusive, machine guarding requirements.
  1. Safeguarding of servicing and maintenance workers can be ensured either by:
     a. Effective machine safeguarding in compliance with Subpart 0, or
     b. Compliance with 29 CFR 1910.147 in situations where the normal production operations safeguards are rendered ineffective or do not protect the servicing/maintenance worker.
  2. Activities which are routine, repetitive, and integral to the use of equipment for production are not covered by this standard if alternative measures provide effective worker protection from hazards associated with unexpected energization. Compliance with the machine guarding requirements of Subpart 0 is an example of such alternative measures. In addition, supplemental personal protective equipment may be necessary during a servicing or maintenance operation when a toxic substance is to be isolated. Under such circumstances, the requirements of applicable standards, such as 29 CFR 1910.134 and Subpart Z, also must be met.
  3. An employer who requires employees to perform routine maintenance and/or servicing while a machine or process is operating in the production mode, must provide employee safeguarding under the applicable requirements of Subpart 0. (Ref. 29 CFR 1910.212(a)(1)). Operations such as lubricating, draining sumps, servicing of filters, and inspection for leaks and/or mechanical malfunction are examples of routine operations which can be

accomplished with effective production-mode safeguards. However, the replacement of machine or process equipment components such as valves, gauges, linkages, support structure, etc., is not considered to be normal routine maintenance function which can safely be accomplished during machine or process equipment operation. Such maintenance requires energy isolation and should be evaluated by OSHA field staff. They also may be an appropriate subject of a variance request.
  4. Several alternative means of safeguarding the hazardous portions of machines and equipment are presented by the national consensus standard, ANSI B11.19-1990. Although that standard is not all inclusive, it describes effective safeguarding alternatives for the protection of employees. The safeguards described include: interlocked barrier guards, presence sensing devices and various devices under the exclusive control of the employee. Such devices or guards, properly applied, may be used in clearing minor jams and performing other minor servicing functions which occur during normal production operations and which meet the criteria described in paragraph A.2. of this appendix.
B. **Group Lockout/Tagout.** The group lockout/tagout procedures described in this instruction at paragraph I.8. require each authorized employee to be in control of potentially hazardous energy release during their servicing/maintenance work assignments. Under most circumstances, where servicing/maintenance is to be conducted during only one shift by an individual or a small number of persons working together, the installation of each individual's lockout/tagout device upon each energy isolating device would not be a burdensome procedure. However, when many energy sources or many persons are involved, and/or the procedure is to extend over more than one shift, (possibly several days, or weeks) consideration must be given to the implementation of a lockout/tagout procedure that will ensure the safety of the employees involved and will provide for each individual's control of the energy hazards. The following procedures are presented as examples to illustrate the implementation of a group lockout/tagout procedure involving many energy isolating devices and/or many servicing/maintenance personnel. They illustrate several alternatives for having authorized employees affix personal lockout/tagout devices in a group lockout/tagout setting. Theses examples are not intended to represent the only acceptable procedures for conducting group operations.
  1. **Definitions.** Various terms used in the examples are defined below.
    a. PRIMARY AUTHORIZED EMPLOYEE is the authorized employee who exercises overall responsibility for adherence to the company lockout/tagout procedures. (See 29 CFR 1910.147(f)(3)(ii)(A).)
    b. PRINCIPAL AUTHORIZED EMPLOYEE is an authorized employee who oversees or leads a group of servicing/maintenance workers (e.g., plumbers, carpenters, electricians, metal workers, mechanics).
    c. JOB-LOCK is a device used to ensure the continuity of energy isolation during a multi-shift operation. It is placed upon a lock-box. A key to the job-lock is controlled by each assigned primary authorized employee from each shift.
    d. JOB-TAG with TAB is a special tag for tagout of energy isolating devices during group lockout/tagout procedures. The tab of the tag is removed for insertion into the lock-box. The company procedure would require that the tagout job-tag cannot be removed until the tab is rejoined to it.
    e. MASTER LOCKBOX is the lockbox into which all keys and tabs from the lockout or, tagout devices securing the machine or equipment are inserted and which would be secured by a "job-lock" during multi-shift operations.
    f. SATELLITE LOCKBOX is a secondary lockbox or lock-boxes to which each authorized employee affixes his/her personal lock or tag.
    g. MASTER TAG is a document used as an administrative control and accountability device. This device is normally controlled by the operations department personnel and is

a personal tagout device if each employee personally signs on and signs off on it and if the tag clearly identifies each authorized employee who is being protected by it.
   h. WORK PERMIT is a control document which authorizes specific tasks and procedures to be accomplished.
2. **Organization.** A group lockout/tagout procedure might provide the following basic organizational structure:
   a. A primary authorized employee would be designated. This employee would exercise primary responsibility for implementation and coordination of the lockout/tagout of hazardous energy sources, for the equipment to be serviced.
   b. The primary authorized employee would coordinate with equipment operators before and after completion of servicing and maintenance operations which require lockout/tagout.
   c. A verification system would be implemented to ensure the continued isolation and deenergization of hazardous energy sources during maintenance and servicing operations.
   d. Each authorized employee would be assured of his/her right to verify individually that the hazardous energy has been isolated and/or deenergized.
   e. When more than one crew, craft, department, etc., is involved, each separate group of servicing/maintenance personnel would be accounted for by a principal authorized employee from each group. Each principal employee is responsible to the primary authorized employee for maintaining accountability of each worker in that specific group in conformance with the company procedure. No person may sign on or sign off for another person, or attach or remove another person's lockout/tagout device, unless the provisions of the exception to 29 CFR 1910.147(e)(3) are met.
3. **Examples of Procedures for Group Lockout/Tagout.** Examples are presented for the various methods of lockout/tagout using lockbox procedures. An example of an applicable method for complex process equipment is also presented.
   a. The following procedures address circumstances ranging from a small group of servicing/maintenance employees during a one-shift operation to a comprehensive operation involving many over a longer period.
      (1) **Type A.** Each authorized employee places his/her personal lock or tag upon each energy isolating device and removes it upon departure from that assignment. Each authorized employee verifies or observes the deenergization of the equipment.
      (2) **Type B.** Under a lockbox procedure, a lock or job-tag with tab is placed upon each energy isolation device after deenergization. The key(s) and removed tab(s) are then placed into a lockbox. Each authorized employee assigned to the job then affixes his/her personal lock or tag to the lockbox. As a member of a group, each assigned authorized employee verifies that all hazardous energy has been rendered safe. The lockout/tagout devices cannot be removed or the energy isolating device turned on until the appropriate key or tab is matched to its lock or tag.
      (3) **Type C.** After each energy isolating device is locked/tagged out and the keys/tabs placed into a master lockbox, each servicing/maintenance group "principal" authorized employee places his/her personal lock or tag upon the master lockbox. Then each principal authorized employee inserts his/her key into a satellite lockbox to which each authorized employee in that specific group affixes his/her personal lock or tag. As a member of a group, each assigned authorized employee verifies that all hazardous energy has been rendered safe. Only after the servicing/maintenance functions of the specific subgroup have been concluded and the personal locks or tags of the respective employees have been removed from the satellite lockbox can the principal authorized employee remove his/her lock from the master lockbox.
      (4) **Type D.** During operations to be conducted over more than one shift or even many days or weeks) a system such as described here might be used. Single locks/tags are affixed upon a lockbox by each authorized employee as described at Type B or Type C above. The master lockbox is first secured with a job-lock before subsequent

locks by the principal authorized employees are put in place on the master lockbox. The job-lock may have multiple keys if they are in the sole possession of the various primary authorized employees (one on each shift). As a member of a group, each assigned authorized employee verifies that all hazardous energy has been rendered safe. In this manner, the security provisions of the energy control system are maintained across shift changes while permitting reenergization of the equipment at any appropriate time or shift.

b. Normal group lockout/tagout procedures require the affixing of individual lockout/tagout devices by each authorized employee to a group lockout device, as discussed in paragraph B.3.a. of this appendix. However, in the servicing and maintenance of sophisticated and complex equipment, such as process equipment in petroleum refining, petroleum production, and chemical production, there may be a need for adaptation and modification of normal group lockout/tagout procedures in order to ensure the safety of the employees performing the servicing and maintenance. To provide greater worker safety through implementation of a more feasible system, and to accommodate the special constraints of the standard's requirement for ensuring employees a level of protection equivalent to that provided by the use of a personal lockout or tagout device, an alternative procedure may be implemented if the company documentation justifies it. Lockout/tagout, blanking, blocking, etc., is often supplemented in these situations by the use of work permits and a system of continuous worker accountability. In evaluating whether the equipment being serviced or maintained is so complex as to necessitate a departure from the normal group lockout/tagout procedures (discussed in paragraph B.3.a.), to the use of an alternative procedures as set forth below, the following (often occurring simultaneously) are some of those which must be evaluated: physical size and extent of the equipment being serviced/maintained; the relative inaccessibility of the energy isolating devices; the number of employees performing the servicing/maintenance; the number of energy isolating devices to be locked/tagged out; and the interdependence and interrelationship of the components in the system or between different systems.

   (1) Once the equipment is shut down and the hazardous energy has been controlled, maintenance/servicing personnel, together with operations personnel, must verify that the isolation of the equipment is effective. The workers may walk through the affected work area to verify isolation. If there is a potential for the release or re-accumulation of hazardous energy, verification of isolation must be continued. The servicing/maintenance workers may further verify the effectiveness of the isolation by the procedures that are used in doing the work (e.g., using a bleeder valve to verify depressurization, flange-breaking techniques, etc.). Throughout the maintenance and/or servicing activity, operations personnel normally maintain control of the equipment. The use of the work permit or "master tag" system (with each employee personally signing on and signing off the job to ensure continual employee accountability and control), combined with verification of hazardous energy control, work procedures, and walk-through, is an acceptable approach to compliance with the group lockout/tagout and shift transfer provisions of the standard. (Note, B.1.g. of this appendix.)

   (2) Specific issues related to the control of hazardous energy in complex process equipment are described below in a typical situation which could be found at any facility. This discussion is intended only as an example and is not anticipated to reflect operations at any specific facility.

      (a) Complex process equipment which is scheduled for servicing/maintenance operations is generally identified by plant supervision. Plant supervision would issue specific work orders regarding the operations to be performed.

      (b) In most instances where complex process equipment is to be serviced or maintained, the process equipment operators can be expected to conduct the shutdown procedure. This is generally due to their in-depth knowledge of the equipment and the need to conduct the shut-down procedure in a safe, economic and specific sequence.

(c) The operations personnel will normally prepare the equipment for lockout/tagout as they proceed and will identify the locations for blanks, blocks, etc., by placing "operations locks and/or tags" on the equipment. The operations personnel can be expected to isolate the hazardous energy, and drain and flush fluids from the process equipment following a standard procedure or a specific work permit procedure.

(d) Upon completion of shutdown, the operations personnel would review the intended job with the servicing and maintenance crew(s) and would ensure their full comprehension of the energy controls necessary to conduct the servicing or maintenance safely. During or immediately after the review of the job, the servicing and maintenance crews would install locks, tags and/or special isolating devices at previously identified equipment locations following the specified work permit procedure.

(e) Line openings necessary for the isolation of the equipment would normally be permitted only by special work permits issued by operations personnel. (Such line openings should be monitored by operations personnel as an added safety measure.)

(f) All of the previous steps should have been documented by a master system of accountability and retained at the primary equipment control station for the duration of the job. The master system of accountability may manifest itself as a Master Tag which is subsequently signed by all of the maintenance/servicing workers if they fully comprehend the details of the job and the energy isolation devices actuated or put in place. This signing by the respective workers further verifies that energy isolation training relative to this operation has been conducted.

(g) After the system has been rendered safe, the authorized employees verify energy controls as described in B.3.b.(1) of the appendix.

(h) Specific work functions are controlled by work permits which are issued for each shift. Each day each authorized employee assigned must sign in on the work permit at the time of arrival to the job and sign out at departure. Signature, date, and time for sign-in and sign-out would be recorded and retained by the applicable crew supervisor who upon completion of the permit requirements would return the permit to the operations supervisor. Work permits could extended beyond a single shift and may subsequently be the responsibility of several supervisors.

(i) Upon completion of the tasks required by the work permit, the authorized employees' names can be signed off the Master Tag by this supervisor once all employees have signed off the work permit. The work permit is then attached to the Master Tag. (Accountability of exposed workers is maintained.)

(j) As the work is completed by the various crews, the work permits and the accountability of personnel are reconciled jointly by the primary authorized employee and the operations supervisor.

(k) During the progress of the work, inspection audits are conducted.

(l) Upon completion of all work, the equipment is returned to the operation personnel after the maintenance and servicing crews have removed their locks, tags, and/or special isolating devices following the company procedure.

(m) At this time all authorized employees who were assigned to the tasks are again accounted for and verified to be clear from the equipment area.

(n) After the completion of the servicing/maintenance work, operations personnel remove the tags originally placed to identify energy isolation.

(o) Operations personnel then begin check-out, verification and testing of the equipment prior to being returned to production service.

c. It should be noted that the purpose of the lockout/tagout standard is to reduce the likelihood of worker injuries and fatalities during servicing/maintenance operations. Therefore, when compliance officers inspect workplaces, they should evaluate the potential for employee exposure to the unexpected release of hazardous energy during servicing/maintenance operations. When a hazard is noted, the various requirements of the standard should be applied in a manner which will result in abatement of the hazardous circumstances.

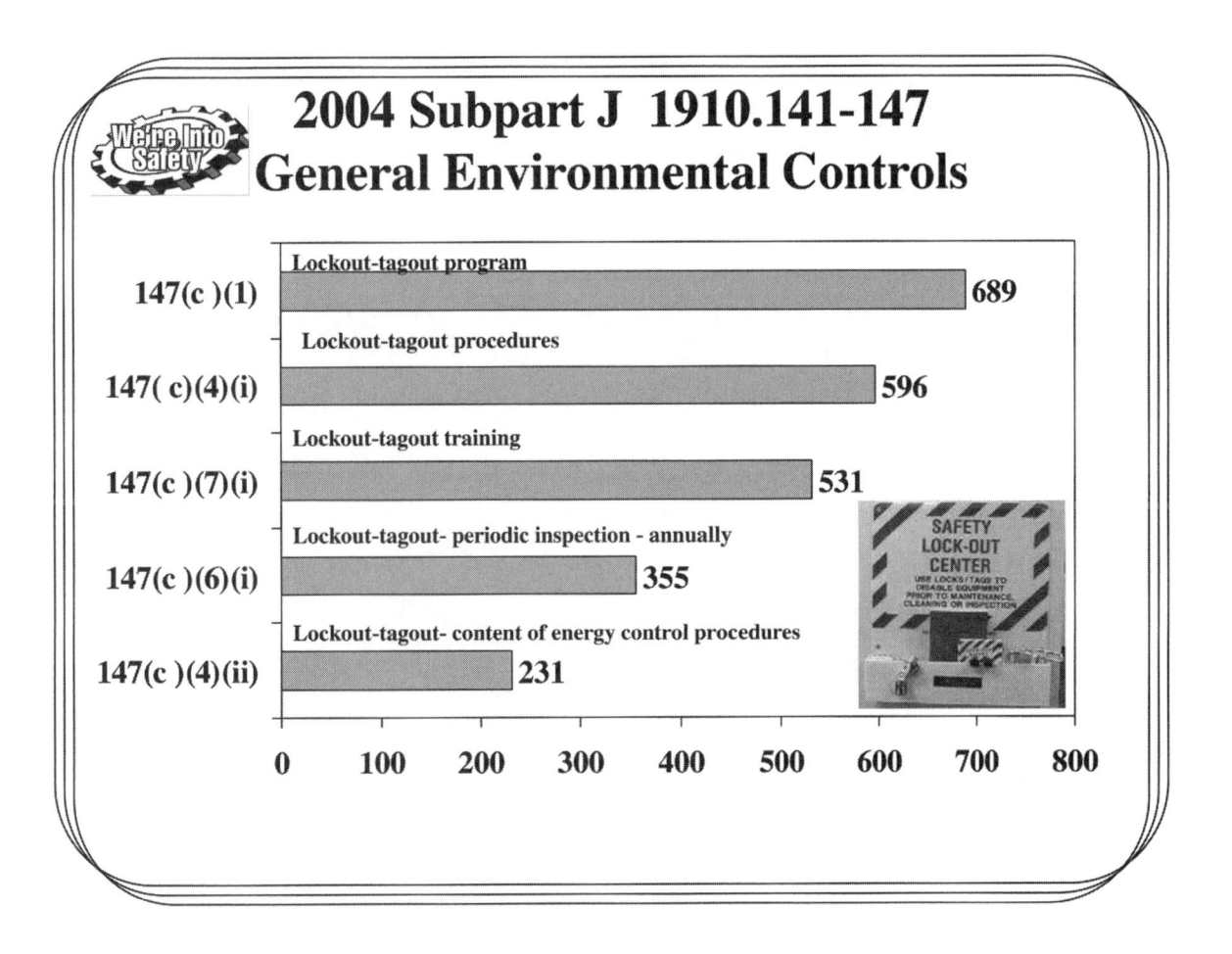

2004 Subpart J 1910.141-147
General Environmental Controls

Standard	Description	Count
147(c)(1)	Lockout-tagout program	689
147(c)(4)(i)	Lockout-tagout procedures	596
147(c)(7)(i)	Lockout-tagout training	531
147(c)(6)(i)	Lockout-tagout- periodic inspection - annually	355
147(c)(4)(ii)	Lockout-tagout- content of energy control procedures	231

CHAPTER 14

HAND AND PORTABLE POWERED TOOLS

Each employer shall be responsible for the safe condition of tools and equipment used by employees, including tools and equipment that may be furnished by employees. All tools shall be restricted to the use for which they are intended and unsafe hand tools shall not be used.

Employees who use hand and power tools and who are exposed to the hazards of falling, flying, abrasive and splashing objects, or exposed to harmful dusts, fumes, mists, vapors or gases must be provided with the particular personal protective equipment necessary to protect them from the hazard.

Following five basic safety rules can prevent all hazards involved in the use of power tools:
- Keep all tools in good condition with regular maintenance;
- Use the right tool for the job;
- Examine each tool for damage before use;
- Operate according to the manufacturer's instructions; and
- Provide and use the proper protective equipment.

Employees and employers have a responsibility to work together to establish safe working, procedures. If a hazardous situation is encountered, it should be brought to the attention of the proper individual immediately.

Hand Tools

Hand tools are non-powered. They include anything from axes to wrenches. The greatest hazards posed by hand tools result from misuse and improper maintenance.

Some examples of misuse and improper maintenance include:
- Using a screwdriver as a chisel may cause the tip of the screwdriver to break and fly, hitting the user or other employees.
- If a wooden handle on a tool such as a hammer or an ax is loose, splintered or cracked, the head of the tool may fly off and strike the user or another worker.
- A wrench must not be used if its jaws are sprung because it might slip.
- Impact tools such as chisels, wedges or drift pins are unsafe if they have mushroomed heads. The heads might shatter on impact, sending sharp fragments flying.

The employer is responsible for the safe condition of tools and equipment used by employees but the employees are responsible for the proper use and maintenance of the tools.

Employers should caution employees to direct saw blades, knives or other tools away from aisle areas and away from other employees working in close proximity. Knives and scissors must be sharp; dull tools can be more hazardous than sharp ones.

Appropriate personal protective equipment (e.g., safety goggles, gloves) should be worn due to hazards that may be encountered while using portable power tools and hand tools. Floors should be kept as clean and dry as possible to prevent accidental slips with or around dangerous hand tools. Sparks produced by iron and steel hand tools can be a dangerous ignition source around flammable substances. Where this hazard exists, spark-resistant tools made from brass, plastic, aluminum or wood will provide safety.

Power Tools

Power tools can be hazardous when improperly used. There are several types of power tools, based on the power source they use: electric, pneumatic, liquid fuel, hydraulic, and powder-actuated.

Employees should be trained in the use of all tools—not just power tools. They should understand the potential hazards as well as the safety precautions to prevent those hazards from occurring.

Power tool users should observe the following general precautions:

- Never carry a tool by its cord or hose.
- Never yank the cord or the hose to disconnect it from the receptacle.
- Keep cords and hoses away from heat, oil and sharp edges.
- Disconnect tools when not in use, before servicing and when changing accessories such as blades, bits and cutters.
- All observers should be kept at a safe distance from the work area.
- Secure work with clamps or a vise, freeing both hands to operate the tool.
- Avoid accidental starting. The worker should not hold a finger on the switch button while carrying a plugged-in tool.
- Tools should be maintained with care. They should be kept sharp and clean for the best performance.
- Follow instructions in the user's manual for lubricating and changing accessories.
- Be sure to keep good footing and maintain good balance.
- The proper apparel should be worn. Loose clothing, ties or jewelry can become caught in moving parts.
- All portable electric tools that are damaged shall be removed from use and tagged "Do Not Use."

Guards

Hazardous moving parts of a power tool need to be safeguarded. For example, belts, gears, shafts, pulleys, sprockets, spindles, drums, flywheels, chains or other reciprocating, rotating or moving parts of equipment must be guarded if such parts are exposed to contact by employees.

Guards, as necessary, should be provided to protect the operator and others from:

- Point of operation
- In-running nip points
- Rotating parts
- Flying chips and sparks.

Safety Switches

All hand-held powered circular saws having a blade diameter greater than 2 in., electric, hydraulic or pneumatic chain saws, and percussion tools without positive accessory holding means shall be equipped with a constant pressure switch or control that will shut off the power when the pressure is released. All hand-held gasoline-powered chain saws shall be equipped with a constant pressure throttle control that will shut off the power to the saw chain when the pressure is released.

All other hand-held powered drills, tappers, fastener drivers, horizontal, vertical and angle grinders with wheels greater than 2 in. in diameter; disc sanders with discs greater than 2 in. in diameter; belt sanders, reciprocating saws, saber, scroll, and jig saws with blade shanks greater than a nominal ¼ in.; and other similarly operated powered tools shall be equipped with a constant pressure switch or control and may have a lock-on control, provided that turn-off can be accomplished with a single motion of the same finger or fingers that turn it on.

All other hand-held powered tools, such as, but not limited to, platen sanders, grinders with wheels 2 in. in diameter or smaller, disc sanders with discs 2 in. in diameter or smaller; routers, planers, laminate trimmers, nibblers, shears, saber, scroll, and jigsaws with blade shanks a nominal ¼ in. wide or smaller, may be equipped with either a positive "on-off" control or other controls as described in this section.

The operating control on hand-held power tools shall be so located as to minimize the possibility of its accidental operation if such accidental operation would constitute a hazard to employees.

The following hand-held powered tools must be equipped with a momentary contact "on-off" control switch:
- Drills
- Tappers
- Fastener drivers;
- Horizontal, vertical and angle grinders with wheels larger than 2 in. in diameter and
- Disc and belt sanders, reciprocating saws, saber saws and other similar tools.

These tools also may be equipped with a lock-on control, provided that turn-off can be accomplished with a single motion of the same finger or fingers that turn it on.

The following hand-held powered tools may be equipped with only a positive "on-off" control: switch-platen sanders; disc sanders with discs 2 in. or less in diameter; grinders with wheels 2 in. or smaller in diameter; and routers, planers, laminate trimmers, nibblers, shears, scroll saws and jigsaws with blade shanks ¼ in. wide or smaller.

Other hand-held power tools, such as circular saws having a blade diameter of more than 2 in., chain saws, and percussion tools without positive accessory holding means, must be equipped with a constant pressure switch that will shut off the power when the pressure is released.

Electric Tools

Among the chief hazards of electric-powered tools are burns and electric shocks, which can lead to injuries or even heart failure. Under certain conditions, even a small amount of current can result in fibrillation of the heart and eventual death. A shock also can cause the user to fall off a ladder or other elevated work surface.

To protect the user from shock, tools must either have a three-wire cord with ground and be grounded, be double insulated, or be powered by a low-voltage isolation transformer.

Three-wire cords contain two current-carrying conductors and a grounding conductor. One end of the grounding conductor connects to the tool's metal housing. The other end is grounded through a prong on the plug. Anytime an adapter is used to accommodate a two-hole receptacle, the adapter wire must be attached to a known ground. The third prong should never be removed from the plug.

Double insulation is more convenient. The user and the tools are protected in two ways: by normal insulation on the wires inside and by a housing that cannot conduct electricity to the operator in the event of a malfunction.

A ground-fault circuit interrupter (GFCI) can provide protection against fatal electrical shocks. A GFCI is not an overcurrent device like a fuse or circuit breaker. GFCIs are designed to sense an imbalance in current flow over the normal path and turn off power within 1/40 second. GFCIs can provide electrical shock protection beyond what a grounded circuit or double insulated tools can provide.

These general practices should be followed when using electric tools:
- Electric tools should be operated within their design limitations;
- Gloves and safety footwear are recommended during use of electric tools;
- When not in use, tools should be stored in a dry place;
- Electric tools should not be used in damp or wet locations;
- Work areas should be well lighted;
- Use a GFCI whenever possible; and
- Never use a tool with damaged cord or missing ground prong.

Powered Abrasive Wheel Tools

Powered abrasive grinding, cutting, polishing, and wire buffing wheels create special safety problems because they may throw off flying fragments. Before an abrasive wheel is mounted, it should be inspected closely and sound, or ring, tested to ensure that it is free from cracks or defects. To test, wheels should be tapped gently with a light non-metallic instrument. If they sound cracked or dead, they could fly apart in operation and must not be used. A sound and undamaged wheel will give a clear metallic tone or ring.

To prevent the wheel from cracking, the user should be sure it fits freely on the spindle. The spindle nut must be tightened enough to hold the wheel in place, but not enough to distort the flange. Follow the manufacturer's recommendations. Care must be taken to ensure that the spindle wheel will not exceed the abrasive wheel specifications.

Due to the possibility of a wheel disintegrating (exploding) during start-up, the employee should never stand directly in front of the wheel as it accelerates to full operating speed.

Portable grinding tools need to be equipped with safety guards to protect workers not only from the moving wheel surface, but also from flying fragments in case of breakage. In addition, when using a grinder, employees should:
- Always use eye protection;
- Turn off power when not in use; and
- Never clamp a hand-held grinder in a vise.

Pneumatic Tools

Pneumatic tools are powered by compressed air. Common pneumatic tools include chippers, drills, hammers and sanders. There are several dangers encountered in the use of pneumatic tools. The main one is the danger of getting hit by one of the tool's attachments or by some kind of fastener the worker is using with the tool.

Eye protection is required and face protection is recommended for employees who work with pneumatic tools. Noise is another hazard. Working with noisy tools such as jackhammers requires proper, effective use of hearing protection.

When using pneumatic tools, employees must check to see that they are fastened securely to the hose to prevent them from becoming disconnected. A short wire or positive locking device that attaches the air hose to the tool will serve as an added safeguard.

A safety clip or retainer must be installed to prevent attachments, (e.g., chisels on chipping hammers) from being unintentionally shot from the barrel.

Screens must be set up to protect nearby workers from being struck by flying fragments around chippers, riveting guns, staplers or air drills. Compressed air guns should never be pointed toward anyone. Users should never "dead-end" it against themselves or anyone else.

Powder-Actuated Tools

Powder-actuated tools operate like a loaded gun and should be treated with the same respect and precautions. In fact, they are so dangerous that only specially trained employees must operate them.

Safety precautions to remember include the following:
- These tools should not be used in an explosive or flammable atmosphere.
- Before using the tool, the worker should inspect it to determine that it is clean, all moving parts operate freely and the barrel is free from obstructions.
- The tool should not be loaded unless it is to be used immediately. A loaded tool should not be left unattended, especially where it would be available to unauthorized persons.
- Hands should be kept clear of the barrel end. To prevent the tool from firing accidentally, two separate motions are required for firing: one to bring the tool into position and another to pull the trigger. The tools must not be able to operate until they are pressed against the work surface with a force of at least 5 lbs. greater than the total weight of the tool.

If a powder-actuated tool misfires, the employee should wait at least 30 seconds before trying to fire it again. If it still will not fire, the user should wait another 30 seconds so that the faulty cartridge is less likely to explode, then carefully remove the load. The bad cartridge should be put in water.

Suitable eye and face protection are essential when using a powder-actuated tool. The muzzle end of the tool must have a protective shield or guard centered perpendicularly on the barrel to confine any flying fragments or particles that might otherwise create a hazard when the tool is fired. The tool must be designed so that it will not fire unless it has this kind of safety device.

All powder-actuated tools must be designed for varying powder charges so the user can select a powder level necessary to do the work without excessive force. If the tool develops a defect during use, it should be tagged and taken out of service immediately until it is properly repaired.

Fasteners

When using powder-actuated tools to apply fasteners, there are some precautions to consider. Fasteners must not be fired into material that would let them pass through to the other side.

In steel, the fastener must not come any closer than ½ in. from a corner or edge. Fasteners must not be driven into very hard or brittle materials that might chip, splatter or make the fastener ricochet.

An alignment guide must be used when shooting a fastener into an existing hole. A fastener must not be driven into a spalled area caused by an unsatisfactory fastening.

Hydraulic Power Tools

The fluid used in hydraulic power tools must be an approved fire-resistant fluid and must retain its operating characteristics at the most extreme temperatures to which it will be exposed. The manufacturer's recommended safe operating pressure for hoses, valves, pipe fitters and other fittings must not be exceeded.

Jacks

All jacks (i.e., lever and ratchet jacks, screw jacks, hydraulic jacks) must have a device that stops them from jacking up too high. Also, the manufacturer's load limit must be permanently marked in a prominent place on the jack and should not be exceeded. A jack should never be used to support a lifted load. Once the load has been lifted, it must immediately be blocked up.

Use wooden blocking under the base if necessary to make the jack level and secure. If the lift surface is metal, place a 1-in. thick hardwood block or equivalent between the lift surface and the metal jack head to reduce the danger of slippage.

When setting up a jack, make certain of the following:
- The base rests on a firm level surface;
- The jack is correctly centered;
- The jack head bears against a level surface; and
- The lift force is applied evenly.

Proper maintenance of jacks is essential for safety. All jacks must be inspected before each use and lubricated regularly. If a jack is subjected to an abnormal load or shock, it should be thoroughly examined to make sure it has not been damaged. Hydraulic jacks exposed to freezing temperatures must be filled with an adequate antifreeze liquid.

OSHA INSTRUCTION STD 1-13.4 AUGUST 5, 1981 OFFICE OF COMPLIANCE PROGRAMMING

SUBJECT: Portable Belt Sanding Machines as Covered by 29 CFR 1910.243(a)(3) and 29 CFR 1926.304(f)

A. Purpose. This instruction provides guidance to allow equitable enforcement of 29 CFR 1910.243 (a)(3) and 29 CFR 1926.304(f) as they pertain to the guarding of portable belt sanders.

B. Scope. This instruction applies OSHA-wide.

C. Action. OSHA Regional Administrators/Area Directors shall classify violations of 29 CFR 1910.243(a)(3) and 29 CFR 1926.304(f) as de minimis violations where the employer has complied with appropriate safeguarding precautions as follows:

1. Portable belt sanding machines shall be provided with guarding on one side at the nip point where the sanding belt runs into a pulley,
2. Handles shall be so located that a barrier interrupts any straight line path between the gripping surface of a handle and the nip point of a pulley, and
3. The unused run of the sanding belt shall be guarded on one side and at the rear.

D. Federal Program Change. This instruction describes a Federal program change which affects State Programs. Each Regional Administrator shall:
1. Ensure that this change is forwarded to each State designee.
2. Explain the technical content of the change to the State designee as requested.
3. Ensure that State designees are asked to acknowledge receipt of this Federal program change in writing, within 30 days of notification, to the Regional Administrator. This acknowledgment should include a description either of the State's plan to implement the change or of the reasons why the change should not apply to that State.
4. Review policies, instructions and guidelines issued by the State to determine if this change has been communicated to State program personnel. Routine monitoring activities (accompanied inspections and case file reviews) shall also be used to determine if this change has been implemented in actual performance.

E. Background. Revision of 29 CFR 1910.243(a)(3) and 29 CFR 1926.304(f) are under consideration by the Mechanical Engineering and Construction Standards Divisions. The standards presently in effect prescribe guarding requirements which essentially prevent flush sanding at corners, and important feature and an application of portable belt sanders. The modified form of guarding , as presented by this instruction, permits flush sanding at corners and provides for operator safety.

Thorne G. Auchter Assistant Secretary

DISTRIBUTION: National, Regional and Area Offices All Compliance Officers State Designees NIOSH Regional Program Directors

OSHA INSTRUCTION STD 1-13.2A DEC 9 1985 DIRECTORATE OF FIELD OPERATIONS

SUBJECT: Explosive Actuated Fastening Tools

A. Purpose. This instruction provides specific interpretation as to when magazine-fed, explosive power operated hand tools are loaded.
B. Scope. This Instruction applies OSHA-wide.
C. Cancellation. OSHA Instruction STD 1-13.2, October 30, 1978, is canceled.
D. Action. OSHA Regional Administrators and Area Directors shall ensure that when magazine-fed tools covered by these standards are inspected, the tool is not considered loaded until the magazine feeds the tool, even though the magazine is in the tool. The operator's instructions shall provide for removing the magazine when the tool is left unattended.
E. Federal Program Change. This instruction describes a Federal program change which affects State programs. Each Regional Administrator shall:
 1. Ensure that this change is promptly forwarded to each State designee.
 2. Explain the technical content of this change to the State designee as requested.
 3. Ensure that State designees are asked to acknowledge receipt of this Federal program change in writing, within 30 days of notification, to the Regional Administrator. This acknowledgment should include a description either of the State's plan to implement the change or of the reasons why the change should not apply to that State.
 4. Review policies, instructions, and guidelines issued by the State to determine that this change has been communicated to State program personnel. Routine monitoring shall also be used to determine if this change has been implemented in actual performance.
F. Application. For the General Industry, Construction and Maritime industries, the following standards shall apply:

General Industry, 29 CFR 1910.243(d)(4)(iv).
Construction, 29 CFR 1926.302(e)(5) and (6).
Maritime, 29 CFR 1915.135(c)(3).

G. Explanation. In developing and promulgating the standards, the magazine or clip-fed explosive power load was not considered. The magazine contains several explosive power loads and single loads are fed into the ram (firing chamber) as needed. Until such time as single loads are fed into the ram, explosive power loads, are not a part of the firing cycle. A separate operation on the part of the operator has to be made to place the load into the firing position.

John B. Miles, Office Director Directorate of Field Operations

DISTRIBUTION National, Regional and Area Offices Compliance Officers State Designees 7(c)(1) Project Managers NIOSH Regional Program Directors

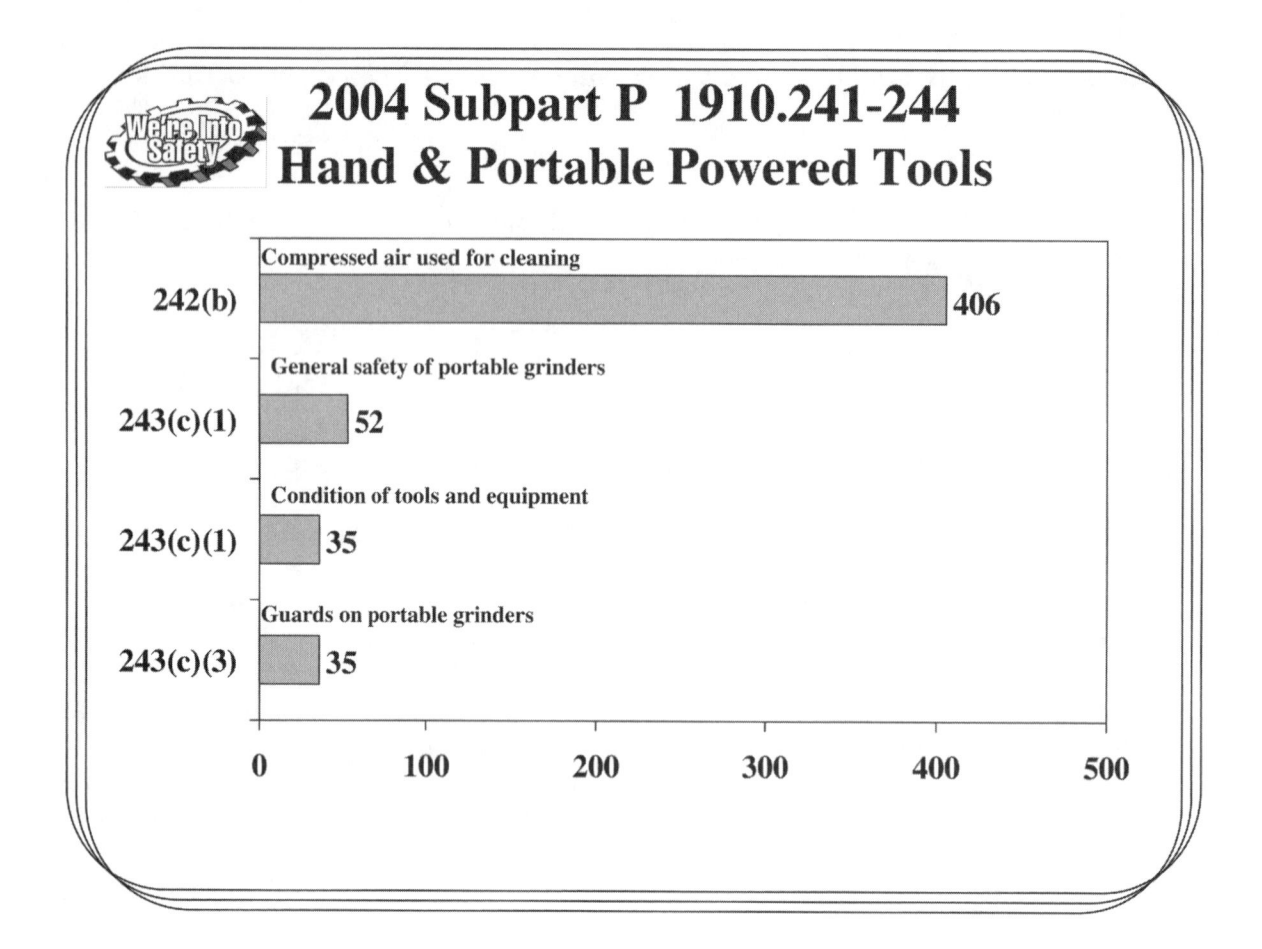

2004 Subpart P 1910.241-244
Hand & Portable Powered Tools

Citation	Description	Count
242(b)	Compressed air used for cleaning	406
243(c)(1)	General safety of portable grinders	52
243(c)(1)	Condition of tools and equipment	35
243(c)(3)	Guards on portable grinders	35

CHAPTER 15

MATERIALS HANDLING

More employees are injured in industry while moving materials than while performing any other single function. In everyday operations, workers handle, transport and store materials. They may do so by hand, by manually operated materials handling equipment or by power operated equipment.

Handling Material—General—1910.176

Where mechanical handling equipment is used, sufficient, safe clearance shall be allowed for aisles, at loading docks, through doorways and wherever turns or passage must be made. Permanent aisles and passageways shall be appropriately marked. Storage of material shall not create a hazard. All materials stored in tiers shall be stacked, blocked, interlocked and limited in height so that they are secure against sliding or collapse. Storage areas shall be kept free from accumulation of materials that constitute hazards from tripping, fire, explosion or pest harborage. Vegetation control will be exercised when necessary.

Servicing Multi-Piece and Single-Piece Rim Wheels—1910.177

Approximately 322,000 employees in more than 100,000 workplaces service large vehicle tires that are mounted on multi-piece or single-piece wheels. OSHA requires:
- Training for all tire-servicing employees;
- The use of industry-accepted procedures that minimize the potential for employee injury;
- The use of proper equipment, such as clip-on chucks, restraining devices or barriers, to retain the wheel components in the event of an incident during the inflation of tires; and
- The use of compatible components.

Types of Wheels/Tires

A rim wheel is the component assembly of wheel (either multi-piece or single-piece), tire and tube and other components. A single-piece wheel is the component of the assembly used to hold the tire, form part of the air chamber (with tubeless wheels) and provide the assembly's means of attachment to the vehicle axle. A multi-piece wheel is a vehicle wheel consisting of two or more parts, one of which is a side- or locking-ring that holds the tire and other components on the rim wheel by interlocking the components when the tire is inflated.

The standard does not apply to the servicing of rim wheels using automobile tires or truck tires designated "LT."

Hazards

The principal difference between accidents involving single-piece rim wheels and those involving multi-piece rim wheels is the effect of the sudden release of the pressurized air contained in a single-piece rim wheel.

In a multi-piece rim wheel accident, the wheel components separate and are released from the rim wheel with violent force. The severity of the hazard is related not only to the air pressure, but also to the air volume.

Single-rim wheel accidents occur when the pressurized air contained in the tire is suddenly released, whether by the bead breaking or slipping over the rim flange. The principal hazards involve pressurized air, which, once released, can either pick up and hurl an employee across the shop if he or she is in close proximity to the rim wheel and within the trajectory; propel the rim wheel in any potential path or route that a rim wheel component may travel during an explosive separation.

Employee Training

The employer must provide a program to train all employees who service rim wheels about the hazards involved and the safety procedures to be followed. The employer must ensure that no employee services any rim wheel unless the worker has been instructed in the correct procedures for mounting, demounting and servicing the wheel as well as the safe operating precautions for the type of wheel being serviced. The employer must regularly evaluate each employee's performance and provide additional training as necessary to ensure that each employee maintains his or her proficiency.

At a minimum, the training program must include the contents of the OSHA standard and the information in the manufacturer's rim manuals or the OSHA charts. Charts are available from OSHA regional, area or national offices.

The instruction must be conducted in an intelligible way. Employees who are unable to read the charts or rim manuals must be trained in the subject matter. The employer must regularly evaluate each employee's performance and provide additional training, as necessary, to ensure each employee maintains his or her proficiency. The employer must ensure each worker demonstrates and maintains the ability to service rim wheels safely by correctly performing the following tasks:
- Demounting tires, including deflation;
- Inspecting and identifying rim wheel components;
- Mounting tires, including inflating them with a restraining device or other safeguard
- Handling rim wheels;
- Inflating tires when single-piece rim wheels are mounted on a vehicle; and
- Understanding the need to stand outside the trajectory during the inflation of the tire and inspection of the rim wheels following inflation.

The Servicing Equipment

The employer must furnish a restraining device for inflating a tire on a multi-piece wheel and must provide a restraining device or barrier for inflating a tire on a single-piece wheel, unless the single-piece rim wheel is bolted onto a vehicle during inflation. The restraining device can be a cage, rack or an assemblage of bars and other parts that will constrain all rim wheel components during an explosive separation of the multi-piece rim wheel or the sudden release of the contained air of a single-piece rim wheel.

A barrier can be a fence, wall, or other structure or object placed between a single-piece rim wheel and an employee during tire inflation to contain the rim wheel components in the event of the sudden release of contained air. Each barrier or restraining device must be able to withstand the maximum force of an explosive rim wheel separation or release of the pressurized air occurring at 150 percent of the maximum tire specification pressure for the rim wheel being serviced.

Restraining devices showing any of the following defects must be immediately removed from service:
- Cracks at welds;
- Cracked or broken components;
- Bent or sprung components caused by mishandling, abuse, tire explosion or rim wheel separation; or
- Component pitted due to corrosion or other structural damage that would decrease its effectiveness.

Restraining devices or barriers removed from service must not be returned to service until they are repaired and inspected. Restraining devices or barriers requiring structural repair such as component replacement or re-welding must not be returned to service until they are certified by either the manufacturer or a Registered Professional Engineer as meeting the strength requirements as stated above (the force of 150 percent of the maximum tire specification pressure).

Current charts or rim manuals containing instructions for the types of wheels being serviced must be available in the service area, including a mobile service unit. Only tools that are recommended in the rim manual may be used for the type of wheel being serviced. The employer must also supply airline equipment with a clip-on chuck with sufficient length of hose between the chuck and in-line valve or regulator to allow the employee to stand outside the trajectory, as well as an in-line valve with a pressure gauge or a pre-settable regulator.

The size (bead diameter and tire/wheel width) and type of both the tire and wheel must be checked for compatibility prior to assembly of the rim wheel. Mismatching of half sizes such as 16-inch and 16.5-inch tires and wheel must be avoided.

Multi-piece wheel components must not be interchanged except as indicated in the applicable charts or rim manuals. Multi-piece wheel components and single-piece wheels must be inspected prior to assembly. Any wheel or wheel component that is bent out of shape, pitted from corrosion, broken or cracked must be marked or tagged "unserviceable" and removed from the service area. Damaged or leaky valves must be replaced.

Rim flanges, rim gutters, rings, and the bead-seating areas of wheels must be free of any dirt, surface rust, scale, or loose or flaked rubber build-up prior to tire mounting and inflation.

Safe Operating Procedures: Multi-Piece Rim Wheels

Employers must instruct employees to use the following steps for safe operating procedures:
1. The tire must be completely deflated by removing the valve core before a rim wheel is removed from the axle in the following situations:
   - When the tire has been driven under inflated at 80 percent or less of its recommended pressure, or
   - When there is obvious or suspected damage to the tire or wheel components.
2. The tire must be completely deflated by removing the valve core before demounting.
3. A rubber lubricant must be applied to the bead and rim mating surfaces when assembling the wheel and inflating the tire, unless the tire or wheel manufacturer recommends against its use.
4. If a tire on a vehicle is underinflated, but has more than 80 percent of the recommended pressure, the tire may be inflated while the rim wheel is on the vehicle, provided remote control inflation equipment is used and no employee remains in the trajectory during inflation.
5. The tire shall be inflated outside a restraining device only to a pressure sufficient to force the tire bead onto the rim ledge and create an airtight seal with the tire and bead.
6. Whenever a rim wheel is in a restraining device, the employee must not rest or lean any part of his or her body or equipment on or against the restraining device.
7. After tire inflation, the tire and wheel must be inspected while still within the restraining device to make sure that they are properly seated and locked. If further adjustment to the tire or wheel components is necessary, the tire shall be deflated by removal of the valve core before the adjustment is made.
8. An attempt must not be made to correct the seating of side and lock rings by hammering, shielding or forcing the components while the tire is pressurized.
9. Cracked, broken, bent or otherwise damaged wheel components must not be reworked, welded, brazed or otherwise heated. Heat must not be applied to a multi-piece wheel.
10. Whenever multi-piece rim wheels are being handled, employees must stay out of the trajectory unless the employer can show that performance of the servicing makes the employee's presence in the trajectory necessary.

MATERIALS HANDLING

Safe Operating Procedures: Single-Piece Rim Wheels

Employees must be instructed in and use the following steps for safe operating procedures with single-piece wheels:

1. The tire must be completely deflated by removing the valve core before demounting.
2. Mounting and demounting of the tire must be performed only from the narrow ledge side of the wheel. Care must be taken to avoid damaging the tire beads, and the tire must be mounted only on a compatible wheel of mating bead diameter and width.
3. A nonflammable rubber lubricant must be applied to bead and wheel mating surfaces before assembling the rim wheel, unless the tire or wheel manufacturer recommends against the use of any rubber lubricant.
4. If a tire-changing machine is used, the tire may be inflated only to the minimum pressure necessary to force the tire bead onto the rim ledge and create an airtight seal before it can be removed from the tire-changing machine.
5. If a bead expander is used, it must be removed before the valve core is installed and as soon as the rim wheel becomes airtight (when the bead slips onto the bead seat).
6. The tire may be inflated only when contained within a restraining device, positioned behind a barrier or bolted on the vehicle with the lug nuts fully tightened.
7. The tire must not be inflated when any flat, solid surface is in the trajectory and within 1 ft. of the sidewall.
8. The tire must not be inflated to more than the inflation pressure stamped in the sidewall, unless the manufacturer recommends a higher pressure.
9. Employees must stay out of the trajectory when a tire is being inflated.
10. Heat must not be applied to a single-piece wheel.
11. Cracked, broken, bent or otherwise damaged wheels must not be reworked, welded, brazed or otherwise heated.

Powered Industrial Trucks—1910.178

This section contains safety requirements relating to fire protection, design, maintenance and use of fork trucks, tractors, platform lift trucks, motorized hand trucks and other specialized industrial trucks powered by electric motors or internal combustion engines. This section does not apply to compressed air or nonflammable compressed gas-operated industrial trucks, farm vehicles or vehicles intended primarily for earth moving or over-the-road hauling.

Approved powered industrial trucks shall bear a label or some other identifying mark indicating approval by the testing laboratory. The user shall not perform modifications and additions that affect capacity and safe operation of trucks without the manufacturer's prior written approval. The term "approved truck" or "approved industrial truck" means a truck that is listed or approved for fire safety purposes for the intended use by a nationally recognized testing laboratory, using nationally recognized testing standards.

Designations

For this standard, there are 11 designations of industrial trucks or tractors:

1. D. Diesel engine-powered units having minimum acceptable safeguards against inherent fire hazards.
2. DS. Diesel-powered units that are provided with additional safeguards to the exhaust, fuel and electrical systems.
3. DY. Diesel-powered units that have all the safeguards of the DS, but do not have any electrical equipment, including the ignition and are equipped with temperature limitation features.
4. E. Electrically powered units that have minimum acceptable safeguards against inherent fire hazards.
5. ES. Electrically powered units that, in addition to all the requirements for E units, are provided with additional safeguards to the electrical system to prevent emission of hazardous sparks and limit surface temperatures.
6. EE. Electrically powered units that have, in addition to all of the requirements for the E and ES units, the electric motors and all other electrical equipment completely enclosed.

7. EX. Electrically powered units that differ from the E, ES, and EE units in that the electrical fittings and equipment are so designed, constructed and assembled that the units may be used in certain atmospheres containing flammable vapors or dusts.
8. G. Gasoline-powered units that have minimum acceptable safeguards against inherent fire hazards.
9. GS. Gasoline-powered units that are provided with additional safeguards to the exhaust, fuel and electrical systems.
10. LP. Liquefied petroleum gas-powered units that are similar to G units, except that liquefied petroleum gas is used for fuel instead of gasoline.
11. LPS. Liquefied petroleum gas-powered units that are provided with additional safeguards to the exhaust, fuel and electrical systems.

Atmospheres or locations throughout the plant must be classified hazardous or nonhazardous before the use of industrial trucks can even be considered. Refer to Table N-1 of 1910.178(c)(2), which is a summary table on the use of industrial trucks in various locations.

Safety Guards

All high-lift rider trucks shall be fitted with overhead guards where overhead lifting is performed unless operating conditions do not permit. In those cases where a high-lift rider truck must enter a structure and the overhead guard will not permit this entry (e.g., truck trailer), the guard may be removed or a powered industrial truck without a guard may be used.

If a powered industrial fork truck carries a load that presents a hazard of falling back onto the operator, the fork truck shall be equipped with a vertical load backrest extension.

Changing and Charging Storage Batteries

Workplaces using electrically powered industrial trucks will have battery-charging areas somewhere in the plant. In many cases, depending on the number of electrically powered industrial trucks, there will be more than one changing and charging area. This section only applies to storage battery changing and charging areas associated with powered industrial trucks. It does not apply to areas where other batteries, such as those used in motor vehicles (cars or trucks), are charged, although some of the same hazardous conditions may exist.

Some of the requirements specified in the regulation include:

- Battery-charging installations shall be located in areas designated for that purpose;
- Facilities shall be provided for:
    - Flushing and neutralizing spilled electrolyte;
    - Fire protection;
    - Protecting charging apparatus from damage by trucks; and
    - Adequate ventilation for dispersal of air contaminants from gassing batteries.
- A conveyor, overhead hoist or equivalent material handling equipment shall be provided for handling batteries.
- Smoking shall be prohibited in the charging area.
- Precautions shall be taken to prevent open flames, sparks or electric arcs in battery charging areas.

Exhibit 15-1.
A battery-charging area.

Trucks and Railroad Cars

In plant receiving and shipping areas, powered industrial trucks are often used to load and unload materials from trucks and railroad cars. The brakes of highway trucks shall be set and wheel chocks placed under the rear wheels to prevent trucks from rolling while they are boarded with powered industrial trucks (See Exhibit 15-2.).

Exhibit 15-2.
Wheel chucks must be used, in addition to truck brakes, to prevent trucks from rolling while being boarded.

Wheel stops or other positive protection shall be provided to prevent railroad cars from moving during loading or unloading operations.

Fixed jacks may be necessary to support a semi trailer and prevent unending during the loading or unloading when the trailer is not coupled to a tractor.

Operator Training

No employee, including supervisory personnel, is permitted to operate a powered industrial truck unless properly trained and authorized to do so.

Operators must pass a written test in a classroom setting as well as demonstrate hands-on capability.

This training must be certified with the name of the operator, the trainer and the date of training. Training is good for three years or until the operator demonstrates a lack of competency.

Truck Operations

Some of the requirements regarding industrial truck operations include:
- No person shall be allowed to stand or pass under the elevated portion of any truck, whether loaded or empty.
- Unauthorized personnel shall not be permitted to ride on powered industrial trucks. A safe place to ride shall be provided where riding of trucks is authorized.
- When a powered industrial truck is left unattended, load-engaging means shall be fully lowered, controls shall be neutralized, power shall be shut off, and brakes set. (A powered industrial truck is "unattended" when the operator is 25 ft. or more away from a vehicle that remains in his or her view or whenever the operator leaves the vehicle and it is not in his or her view.)

Traveling

This section contains requirements for traveling in powered industrial trucks. Some of these requirements include:
- All traffic regulations shall be observed, including authorized plant speed limits.
- The driver shall be required to slow down and sound the horn at cross aisles and other locations where vision is obstructed. If the load being carried obstructs forward view, the driver shall be required to travel with the load trailing.
- Railroad tracks shall be crossed diagonally whenever possible. Parking closer than 8 ft. from the center of railroad tracks is prohibited.
- When ascending or descending grades in excess of 10 percent, loaded trucks shall be driven with the load upgrade.
- Dockboards or bridge plates shall be properly secured before they are driven over. Dockboards or bridge plates shall be driven over carefully, slowly and their rated capacity never exceeded.

Loading

Only stable or safely arranged loads shall be handled. Caution shall be exercised when handling off-center loads that cannot be centered. Only loads within the rated capacity of the truck shall be handled.

Operation of the Truck

If at any time a powered industrial truck is found to be in need of repair, defective or in any way unsafe, the truck shall be taken out of service until it has been restored to safe operating condition. Fuel tanks shall not be filled while the engine is running. Spillage shall be avoided. Any spillage of oil or fuel shall be carefully washed away or completely evaporated and the fuel cap replaced before restarting the engine. Open flames shall not be used for checking electrolyte levels in storage batteries or gasoline levels in fuel tanks.

Maintenance of Industrial Trucks

Any powered industrial truck not in safe operating condition shall be removed from service. Authorized personnel shall make all repairs. No repairs shall be made in Class I, II or III locations. Those repairs to the fuel and ignition systems that involve fire hazards shall be conducted only in locations designated for such repairs.

Industrial trucks shall be examined before being placed in service and shall not be placed in service if the examination shows any condition adversely affecting the safety of the vehicle. Examinations shall be made at least daily. Where trucks are used on a round-the-clock basis, they shall be examined after each shift.

Overhead and Gantry Cranes—1910.179

General Requirements

This section applies to overhead and gantry cranes, including semi-gantry, cantilever gantry, wall cranes, storage bridge cranes and others having the same fundamental characteristics.

Overhead and gantry cranes may not be modified and re-rated unless a qualified engineer or the equipment manufacturer checks the modifications and the supporting structure thoroughly for the new rated load. It is not unusual to find overhead or gantry cranes where it is claimed that the lifting capacity was increased simply by installing a new rated load sign on the bridge of the crane.

The rated loads of the crane shall be plainly marked on each side of the crane. If the crane has more than one hoisting unit, each hoist shall have its rated load marked on it or on its load block.

The potential for overloading the crane increases if the hook-up man or the operator does not know the rated capacity. Only employees selected or assigned by the employer or the employer's representative as being qualified to operate a crane shall be permitted to do so.

Cabs

A cab-operated crane is an overhead or gantry crane controlled by an operator in a cab located on the bridge or trolley. The general arrangement of the cab and the location of control and protective equipment shall be such that all operating handles are within convenient reach of the operator when facing the area to be served by the load hook or while facing the direction of travel of the cab. The arrangement shall allow a view of the load hook in all positions.

The access to all cab-operated cranes shall be checked thoroughly. Serious injuries have occurred because of the following three conditions:
1. There was no conveniently placed fixed ladder, stairs or platform provided to reach the cab or bridge footwalk. It is unacceptable and poses a significant hazard to allow employees to board a crane or climb over guardrails or over, under and around building structures, energized hot rails, portable ladders or movable platforms.
2. There was a gap exceeding 12 in. between a fixed ladder, stairs or platform and access to the cab or bridge footwalk.
3. The fixed ladder used as access to the crane did not meet the American National Standard Safety Code for Fixed Ladders, ANSI A14.3-1956. Common citations include that there were no cages provided for ladders more than 20 ft. in unbroken length, offset platforms were not provided and the ladders themselves were not maintained in a safe condition.

Footwalks and Ladders

Where sufficient headroom is available on cab-operated cranes, a footwalk shall be provided on the drive side along the end length of the bridge of all cranes having the trolley running on the top of the girders.

Significant hazards exist for maintenance and inspection personnel if no footwalk is provided, including work being performed:
- From portable ladders;
- Off the main bridge girder without protection against falling to the floor; or
- From the trolley platform with the same potential for falling to the floor.

Maintenance managers and supervisors shall check very thoroughly the maintenance procedures followed in those cases where cab-operated cranes are not provided with bridge footwalks. This, of course, also applies to all other types of cranes where no footwalk is provided and those cases where bridge footwalks cannot be provided because sufficient headroom is not available.

Bridge footwalks, where provided, shall be of rigid construction and designed to sustain a distributed load of at least 50 lbs. per square foot. In many older workplaces, serious hazards are associated with the bridge footwalk itself. This area cannot be inspected from the floor, and safety people must climb onto the crane to properly document the hazardous conditions. It is quite common to find bridge footwalks not continuous or permanently secured. Although no employees may be on the bridge footwalks or cranes during an inspection, maintenance employees, inspection personnel and the crane operator will be required to go on the bridge footwalk at various times. A common and serious hazard is one where standard railings have not been provided on all open sides of the bridge footwalk. In addition, toe boards shall be installed. The standard railing provisions apply to all sides of the bridge footwalk, including the inside edge next to the bridge girders if a fall potential exists.

All gantry cranes shall be provided with a ladder or a stair that extends from the ground to the footwalk or the cab platform. It is not permitted to board a gantry crane via portable ladders, the structure of the crane or any other method.

Any ladder provided on an overhead or gantry crane shall be permanently and securely fastened in place and also shall be in compliance with 29 CFR 1910.27. Damaged, loose, improperly maintained or unguarded fixed ladders are common.

Stops, Bumpers, Rail Sweeps and Guards

A "stop" is a device to limit travel of a trolley. This device normally is attached to a fixed structure and does not have energy-absorbing ability. Every overhead or gantry crane where the trolley runs on top of the bridge girder shall be provided with stops at either end of the limits of the travel of the trolley. These stops shall be fastened to resist forces applied when contacted, and if the stop engages the tread of the wheel of the trolley, it shall be of a height at least equal to the radius of the wheel.

The obvious hazard concerned with improperly applied trolley stops or no trolley stops at all is that the trolley could be rolled off the trolley runway. The hazards associated with this condition are numerous and present serious injury potential to the employees on the floor below. The trolley itself could fall to the floor, parts of the trolley could come off the crane structure and hit employees below, the load itself could be dumped, or at a minimum, cause unexpected movement of the load, and finally, if the trolley contacted the bridge runway conductors, the entire crane could itself be energized. Again, the only practical method to inspect for this condition is the boarding of the overhead or gantry crane and walking out on the bridge footwalk to look and see if trolley stops are present and installed properly.

Modern cranes must meet or exceed the design specifications of the American National Standard Safety Code for Overhead and Gantry Cranes, ANSI B30.2.0-1967. Another similar hazardous condition is the failure to reinstall crane runway stops at the ends of the limits of travel of the runway. A common problem with overhead crane installations is that the controllers have malfunctioned and become stuck in the open position, after which the crane has run off the ends of the bridge runway—many times through the building wall.

Bridge and Trolley Bumpers

A "bumper"(buffer) is an energy-absorbing device for reducing impact when a moving crane or trolley reaches the end of its permitted travel or when two moving cranes or trolleys come in contact.

Overhead or gantry crane bridges shall be provided with bumpers unless the crane:

- Travels at a slow rate of speed and has a faster deceleration rate due to the use of sleeve bearings;
- Is not operated near the end of bridge travel; or
- Is restricted to a limited distance by the nature of the crane operation and there is no hazard of striking any object in this limited distance.

A common condition that will be observed on many overhead or gantry cranes is that bumpers were not provided where required. However, many times bumpers will be provided that do not have sufficient energy-absorbing capacity to stop the crane when traveling at a speed of at least 40 percent of the rated load speed or the bumpers are not designed and installed as to minimize parts falling from the crane in case of breakage. The hazards of not providing bridge bumpers with energy-absorbing capacities are that when an overhead or gantry crane contacts another crane on the same runway or contacts the bridge stops at the ends of the runway, a shock load is transmitted to the lifting mechanisms that could cause the load to be dropped. In addition, the constant striking of a crane against another object without energy-absorption buffers causes the bridge and end structures to weaken, which could cause cracks in the webbing and lead to the failure of the crane structure.

Trolleys shall also be provided with bumpers, unless the trolley:
- Travels at a slow rate of speed;
- Is not operated near the ends of the trolley travel; or
- Is restricted to a limited distance on the trolley runway and there is no hazard of striking any object in this limited distance.

If there is more than one trolley operated on the same trolley runway, each trolley shall be provided with bumpers on its adjacent ends. If bumpers are installed on the trolley, they shall be designed and installed to minimize parts falling from the trolley in case of breakage and shall be energy absorbing. It must be emphasized that both trolley stops and bumpers shall be provided where required.

Rail Sweeps

Bridge end truck wheels shall be provided with sweeps that extend below the top of the rail and project in front of the truck wheels. Their omission is a common violation. This requirement does not apply to the trolley end truck wheels. By not having rail sweeps, maintenance equipment could be left on the bridge runway rails and as the crane travels into the area, the equipment could derail, causing the movement of the load, a shock load, or dropping of the load. This also applies to gantry and semi-gantry cranes where the truck wheels run on a rail usually located on the floor or working surface.

Guards

Hoisting ropes on overhead and gantry cranes must be closely inspected to ensure they do not run near enough to other parts to make fouling or chafing possible. If a hoisting rope is chafing over a long period of time, the rope will eventually wear through or break and the load will drop to the floor. If the ropes do not run near other parts, guards shall be installed to prevent fouling and chafing. Guards also shall be provided to prevent contact between bridge conductors and hoisting ropes if they could come into contact. Bridge conductors are almost always located on the inside flange of the bridge girders and provide the power to the trolley.

Another very common hazard on overhead or gantry cranes is that exposed moving parts are not properly guarded. For example, gears on or near the bridge footwalk shaft ends on bridge motors usually located on the bridge footwalks, and chain and chain sprockets. There have been several reported fatalities where maintenance employees working on bridge footwalks have been drawn into open gears, projecting shaft ends, and chain and sprocket drives.

Brakes

Each independent hoisting unit of all cranes shall be provided with a holding brake applied directly to the motor shaft or some part of the gear grain. On cab-operated cranes with the cab on the bridge, a bridge brake shall be provided. On occasion, operators of cab-operated overhead cranes that are not equipped with bridge brakes will stop the bridge by plugging—where the operator uses the controller to reverse the direction of the motor. This can be a hazardous practice, especially if it is the only method of stopping cranes. If power to the crane is lost for any reason, the crane will be unstoppable.

All floor, remote and pulpit-operated crane bridge devices shall be provided with a brake or noncoasting mechanical drive (i.e., the crane must be able to stop quickly). It should be noted that overhead or gantry cranes with a cab on the trolley should also be provided with a trolley brake. However, under most conditions trolleys will not be required to have a brake.

Electrical Equipment

All wiring and equipment on overhead or gantry cranes shall comply with the applicable electrical sections of Subpart S.

On floor-operated cranes where a multiple conductor cable is used with a suspended push-button station, the station shall be supported in some manner that will protect the electrical conductor against strain. Simply installing a chair or cable from an upper support to the push-button station to take the strain off the conductor can abate this condition.

Pendant control boxes also shall be clearly marked for identification of functions. Lack of clear labeling is quite common. The hazard is that inexperienced operators or supervisory personnel operating the pendant crane may not know the various functions of the push-button station and could cause an unexpected movement of the crane. Only designated personnel are permitted to operate a crane per § 1910.179(b)(8).

One of the most serious hazards associated with cranes concerns the provision that the hoisting motion of all electric traveling cranes shall be provided with an over-travel limit switch in the hoisting direction. A "limit switch" is a switch that disconnects the power to the drive motor and stops the load if that load is raised above a certain point. Many fatalities and serious injuries have occurred because a crane was not provided with a limit switch or the limit switch malfunctioned.

If a limit switch is not provided or if the limit switch malfunctions, the hoist block could run up into the lifting beam or rope drum, severing the cable and dropping the entire assembly and any load on the hook to the floor below. All inspections of an overhead crane shall include the verification of the presence of a functioning limit switch.

Hoisting Equipment

Sheaves are grooved pulleys that carry the hoisting ropes on overhead cranes. Sheave grooves shall be smooth and free from surface defects.

Sheaves in the bottom blocks shall be equipped with close-fitting guards that will prevent ropes from becoming fouled when the block is lying on the ground with the rope loose. This common condition can be readily observed while watching the crane in operation. The hazard, if the guards are not installed, is that the hoisting rope can come off the sheave groove and become entangled on the shaft, creating binding or shaving of the hoisting rope.

Rope fouling can also occur when there is a slack cable condition, such as when a load block rests on top of a load. It is required that the hoisting ropes have no less than two wraps remaining on the hoist drum when the hook is in its lowest position.

Inspection

Prior to initial use, all new and altered cranes shall be inspected to ensure they are in compliance with the provisions of this section. The inspection procedure for cranes in regular service is divided into two classifications: frequent inspection (daily to monthly intervals) and periodic inspection (one- to 12-month intervals).

Equipment Inspection

All functional operating mechanisms, air and hydraulic systems, chains, rope slings, hooks and other lifting equipment shall be visually inspected daily. Chains, cables, ropes, hooks, etc. on overhead and gantry cranes shall be visually inspected for problems, including deformation, cracks, excessive wear, twists and stretching. Defective gear shall be replaced or repaired. In addition, hooks and chains shall be visually inspected monthly and a certification record kept that includes the date of inspection, the signature of the person who performed the inspection and the serial number or other identifier of the equipment. Running ropes shall be inspected monthly with a certification record that includes the date of inspection the signature of the person who performed the inspection, and the serial number or other identifier of each rope.

Periodic Inspection

Complete inspection of the crane shall be performed at one-month to 12-month intervals, depending on its activity, severity of service and environmental conditions. The inspection shall include the following areas:

- Deformed, cracked, corroded, worn, or loose members or parts;

- The brake system;
- Limit indicators (e.g., wind, load);
- Power plant; and
- Electrical apparatus.

Testing

Prior to initial use, all new and altered cranes shall be tested, including the following functions:
- Hoisting and lowering;
- Trolley travel;
- Bridge travel; and
- Limit switches, locking and safety devices

The trip setting of hoist limit switches shall be determined by tests with an empty hook traveling in increasing speeds up to the maximum speed. The actuating mechanism of the limit switch shall be located so that it will trip the switch, under all conditions, in sufficient time to prevent contact of the hook or hook block with any part of the trolley. A preventive maintenance program based on the crane manufacturer's recommendations shall be established.

Handling the Load

One of the most significant hazards associated with cranes is overloading. A crane shall not be loaded beyond its rated load capacity for any reason except test purposes. "Rated load" means the maximum load for which a crane or individual hoist is designed and built by the manufacturer as shown on the equipment nameplate.

A common misconception is that a safety factor is built in and that an employer may exceed the rated load up to this safety factor. *This is not true.* A load means the total superimposed weight on the load block or hooks and includes lifting devices, such as magnets, spreader bars, chains and slings.

Every load lifted by a crane shall be well secured and properly balanced in the sling or lifting device before it is lifted more than a few inches. In some instances, workers may have balanced, but not secured the load. This is a violation.

To prevent swinging of a load, the hook shall be brought directly over the load when the attachment is made. In addition, no employee is permitted on the load, hook or lifting device while the load is being hoisted, lowered or traveling.

The operator of a crane shall avoid carrying loads over other personnel. This hazard is increased significantly when using a magnet or a vacuum device to lift scrap material. Operators of cranes are not permitted to leave their position at the controls while loads are suspended. This includes suspended lifting devices such as magnets or vacuum lifts.

At the beginning of each operator's shift, the upper limit switch of each hoist shall be tried out under no-load conditions. Additionally, the hoist limit switch, which controls the upper limit of travel of the load block, shall never be used as an operating control.

Crawler Locomotive and Truck Cranes—1910.180

This section applies to crawler cranes, locomotive cranes, wheel-mounted cranes of both truck and self-propelled wheel type, and any variations thereof that retain the same fundamental characteristics. It includes only cranes of the above types, which are powered by internal combustion engines or electric motors and use drums and ropes. Cranes designed for railway and automobile wreck clearances are exceptions. The requirements of this section are applicable only to machines when used as lifting cranes.

All modern crawler, locomotive and truck cranes shall meet the design specifications of the American National Standard Safety Code for Crawler, Locomotive, and Truck Cranes, ANSI B30-1-1968. Only employees selected or assigned by the employer or the employer's representative as being qualified shall be permitted to operate a crane covered by this section.

Load Ratings

Where stability governs lifting performance, load ratings have been established for various types of mounting and are given in a table in 1910.180. A substantial and durable rating chart with clearly legible letters and figures shall be securely fixed in each crane cab in a location easily visible to the operator while seated at the control station.

Inspection

Prior to initial use, all new and altered cranes shall be inspected to ensure compliance with the provisions of this section. The inspection procedure for cranes in regular service is divided into two classifications: frequent inspection (daily to monthly intervals) and periodic inspection (one- to 12-month intervals).

Equipment Inspection

All functional operating mechanisms, control systems, safety devices, air and hydraulic systems, chains, rope slings, hooks, and other lifting equipment shall be visually inspected daily. Items such as cables, ropes and hooks shall be visually inspected daily for problems, including deformation, cracks, excessive wear, twists and stretches. Defective gear shall be replaced or repaired. Running ropes shall be inspected monthly and a certification record kept that includes the date of inspection, the signature of the person who performed the inspection, and the serial number or other identifier of each rope.

Periodic Inspection

Complete inspection of the crane shall be performed at one-month to 12-month intervals, depending on its activity, severity of service and environmental conditions. The inspection shall include the following areas:
- Deformed, cracked, corroded, worn or loose members or parts;
- The brake system;
- Limit indicators (e.g., wind, load);
- Power plant;
- Electrical apparatus; and
- Travel steering, braking and locking devices.

Inspection Records

Certification records that include the date of inspection, signature of the person who performed the inspection, and the serial number or other identifier of the crane that was inspected shall be made monthly on critical items in use, such as brakes, crane hooks and ropes. This certification record shall be kept on file where readily available to appointed personnel.

Testing

Prior to initial use, all new production cranes shall be tested to ensure compliance with the provisions of this section, including the following functions:
- Hoisting and lowering mechanisms;
- Swinging mechanism;
- Travel mechanism; and
- Safety devices.

Maintenance

After adjustments and repairs have been made, the crane shall not be operated until all guards have been reinstalled, safety devices reactivated and maintenance equipment removed.

Handling the Load

One of the most significant hazards associated with cranes is overloading. A crane shall not be loaded beyond its rated load capacity for any reason except test purposes. "Rated load" means the maximum load for which a crane or individual hoist is designed and built by the manufacturer as shown on the equipment nameplate. A load means the total superimposed weight on the load block or hooks and includes lifting devices such as magnets, spreader bars, chains and slings. The hoist rope shall not be wrapped around the load. The load shall be attached to the hook by means of slings or other approved devices.

The employer shall ensure:
- The crane is level and, where necessary, blocked properly;
- The load is well secured and properly balanced in the sling or lifting device before it is lifted more than a few inches;
- Before starting to hoist, the following conditions shall be noted:
  - Hoist rope shall not be kinked;
  - Multiple part lines shall not be twisted around one another; and
  - The hook should be brought over the load in such a manner as to prevent swinging;
- During hoisting, care shall be taken that there is no sudden acceleration or deceleration of the moving load and that the load does not contact any obstructions.

Side loading of booms shall be limited to freely suspended loads. Cranes shall not be used for dragging loads sideways. The operator shall test the brakes each time a load approaching the rated load is handled by raising it a few inches and applying the brakes.

Outriggers shall be used when the load to be handled at that particular radius exceeds the rated load without outriggers as given by the crane manufacturer. Neither the load nor the boom shall be lowered below the point where less than two full wraps of rope remain on their respective drums.

Before traveling a crane with a load, a designated person shall be responsible for determining and controlling safety. When rotating the crane, sudden starts and stops shall be avoided.

When a crane is to be operated at a fixed radius, the boom-hoist pawl or other positive locking device shall be engaged.

Holding the Load
Operators shall not be permitted to leave their position at the controls while the load is suspended. No person shall be permitted to pass under a load on the hook. If the load must remain suspended for considerable time, the operator shall hold the drum from rotating in the lowering direction by activating the positive controllable means of the operator's station.

Operating Near Electric Power Lines

Clearances
Except where the electrical distribution and transmission lines have been deenergized and visibly grounded at the point of work or when insulating barriers not part of the crane have been erected to prevent physical contact with the lines, the minimum clearance between the lines and any part of the crane or load shall be:
- For lines rated 50 kV or below, 10 ft.;
- For lines rated over 50 kV, 10 ft. plus 0.4 inch for every kV over 50 kV or twice the length of the fine insulator, but never less than 10 ft.; and
- In transit with no load and boom lowered, a minimum of 4 ft.

Notification
Before operations near electrical lines begin, the owners of the lines or their authorized representatives shall be notified and provided with all pertinent information.

Overhead Wire
Any overhead wire shall be considered to be an energized line unless and until the owner of the line or the electrical utility authorities indicate that it is not an energized line.

Derricks—1910.181
A derrick is an apparatus consisting of a mast or equivalent member held at the head by guys or braces, with or without a boom, for use with a hoisting mechanism and operating ropes.

This section applies to guy, stiffleg, basket, breast, gin pole, Chicago boom, and A-frame derricks of the stationary type, capable of handling loads at variable reaches and powered by hoists through systems of rope reeling, used to perform lifting hook work, single or multiple line bucket work, grab, grapple and magnet work. Derricks may be permanently installed for temporary use as in construction work. This section also applies to any modification of these types that retain their fundamental features, except for floating derricks.

All modern derricks shall meet the design specifications of the American National Standard Safety Code for Derricks, ANSI B30.6-1969. Only employees designated by the employer or employee's representatives as being qualified shall be permitted to operate these derricks.

Load Ratings

Permanent Installations
For permanently installed derricks with fixed lengths of boom, guy and mast, a durable and clearly legible rating chart shall be securely affixed where it is visible to the operators. The chart shall include:
- Manufacturer's approval load ratings at corresponding ranges of boom angle or operating radii;
- Specific lengths of components on which the load ratings are based; and
- Required parts for hoist reeling.

Nonpermanent Installations
For nonpermanent installations, the manufacturer shall provide sufficient information from which capacity charts can be prepared for the particular installation. The capacity charts shall be located at the derricks or the job site office.

Inspection

Prior to initial use, all new and altered derricks shall be inspected to ensure compliance with the provisions of this section. The inspection procedures for derricks in regular service is divided into two general classifications: frequent inspection (daily to monthly intervals) and periodic inspection (one- to 12-month intervals).

Frequent Inspection

All functional operating systems, control systems, safety devices, cords and lacing, tension in guys, plumb of the mast, air and hydraulic systems, rope reeling, hooks and electrical apparatus shall be visually inspected daily. Running ropes shall be inspected monthly and a certification record shall be kept on file where readily available.

Periodic Inspection

Complete inspection of the derrick shall be performed at one-month to 12-month intervals, depending on its activity, severity of service and environmental conditions. The inspection shall include the following areas:
- Deformed, cracked, corroded, worn or loose members or parts;
- Power plant; and
- Foundation or supports.

Testing

The appointed person shall approve all anchorage. Rock and hairpin anchorage may require special testing. Prior to initial use, all new and altered derricks shall be tested by hoisting and lowering the load and moving the boom up and down. The operator also must inspect the swing and operation of all clutches and brakes of the hoist.

Maintenance

A preventive maintenance program based on the derrick manufacturer's recommendations shall be established.

Handling the Load

No derrick shall be loaded beyond the rated load and the hoist rope shall not be wrapped around the load. The load shall be attached to the hook by means of slings or other suitable devices.

Moving the Load

Some of the requirements for moving loads are stated below:
- The load shall be well secured and properly balanced in the sling or lifting device before it is lifted more than a few inches.
- Before starting to hoist, the following conditions shall be noted:
  - Hoist rope shall not be kinked;
  - Multiple part lines shall not be twisted around each other; and
  - The hook shall be brought over the load in such a manner as to prevent swinging.
- During hoisting, care shall be taken that:
  - There is no sudden acceleration or deceleration of the moving load;
  - Load does not contact any obstructions; and
  - The operator shall test the brakes each time a load approaching the rated load is handled by raising it a few inches and applying the brakes.
- Neither the load nor the boom shall be lowered below the point where less than two full wraps of rope remain on their respective drums.

The operator shall not be allowed to leave his or her position at the controls while the load is suspended. If the load must remain suspended for any considerable length of time, a dog, or pawl and ratchet, or other equivalent means, rather than the brake alone, shall be used to hold the load.

Dogs, pawls or other positive holding mechanism on the hoist shall be engaged. When not in use, the derrick boom shall be:
- Laid down;
- Secured to a stationary member, as nearly under the head as possible, by attachment of a sling to the load block; or
- Hoisted to a vertical position and secured to the mast.

Guards

Exposed moving parts, such as gears, ropes, set-screws, chains, chain sprockets and reciprocating components that constitute a hazard under normal operating conditions shall be guarded.

Operating Near Electrical Power Lines

Except where the electrical distribution and transmission lines have been deenergized

and visibly grounded at the point of work or when insulating barriers not part of the derrick have been erected to prevent physical contact with the lines, the minimum clearance between the lines and any part of the derrick or load shall be:
- For lines rated 50 kV or below, 10 ft.; and
- For lines rated over 50 kV, 10 ft. plus 0.4 inch for every kV over 50 kV, or twice the length of the fine insulator but never less than 10 ft.

Notification
Before operations near electrical lines begin, the owners of the lines or their authorized representatives shall be notified and provided with all pertinent information.

Overhead Wire
Any overhead wire shall be considered to be an energized line unless and until the owner of the line or the electrical utility indicates that it is not an energized line.

Helicopters—1910.183
Helicopter cranes shall comply with any applicable regulations of the Federal Aviation Administration. Prior to each day's operation, a briefing shall be conducted that sets forth the plan of operation for the pilot and ground personnel.

Slings and Tag Lines
Loads shall be properly slung. Tag lines shall be of a length that will not permit their being drawn up into the rotors.

Cargo Hooks
All electrically operated cargo hooks shall have the electrical activating device designed and installed so as to prevent inadvertent operation. In addition, cargo hooks shall be equipped with an emergency mechanical control for releasing the load.

The cargo hooks must be tested prior to each day's operation by a competent person to determine that the release functions properly, both electrically and mechanically.

Personal Protective Equipment
Personal protective equipment (i.e., complete eye protection and hard hats secured by chin straps) shall be provided by the employer and used by employees receiving the load.

Loose-fitting clothing likely to flap in rotor downwash and be snagged on the hoist lines may not be worn.

Loose Gear and Housekeeping
All loose gear within 100 ft. of the place of lifting the load or depositing the load or within all other areas susceptible to rotor downwash shall be secured or removed.

Good housekeeping shall be maintained in all helicopter loading and unloading areas.

Hooking and Unhooking Loads
Employees are not permitted to perform work under hovering aircraft except when necessary to hook or unhook loads.

Static Charge
Static charge on the suspended load shall be dissipated with a grounding device before ground personnel touch the suspended load, unless all ground personnel who may be required to touch are wearing protective rubber gloves.

Signal Systems
The employer shall instruct the aircrew and ground personnel on the signal systems to be used and shall review the systems with the employees before hoisting the load. This applies to both radio and hand signal systems. Hand signals, where used, are to be in conformance with Figure N-1 of 1910.183.

Approach Distance
No employees shall be permitted to approach within 50 ft. of the helicopter when the rotor blades are turning, unless their work duties require their presence in that area.

Communications
There shall be constant, reliable communication between the pilot and a designated employee of the ground crew who acts as a signalman during the period of loading and unloading. The signalman shall be clearly distinguishable from other ground personnel.

Slings—1910.184
This section applies to slings used in conjunction with other material handling equipment for the movement of material by hoisting as

covered elsewhere by this part. The types of slings covered are those made from alloy steel chain, wire rope, metal mesh, natural or synthetic fiber rope (conventional three-strand construction) and synthetic web (nylon, polyester, polypropylene).

Safe Operating Practices

Whenever any sling is used, the following practices shall be observed:

- Slings that are damaged or defective shall not be used.
- Slings shall not be shortened with knots or bolts or other makeshift devices.
- Sling legs shall not be kinked.
- Slings shall not be loaded in excess of their rated capacities.
- Slings used in a basket hitch shall have the loads balanced to prevent slippage.
- Slings shall be securely attached to their loads and padded to protect sharp edges.
- Suspended loads shall be kept clear of all obstructions.
- All employees shall be kept clear of loads about to be lifted and of suspended loads.
- Hands or fingers shall not be placed between the sling and its load while the sling is being tightened around the load.
- Shock loading is prohibited.
- A sling shall not be pulled from under a load when the load is resting on the sling.

Inspections

Each day before being used, the sling and all fastenings and attachments shall be inspected for damage or defects by a competent person designated by the employer. Additional inspections shall be performed during sling use, where service conditions warrant. Damaged or defective slings shall be immediately removed from service.

Alloy Steel Chain Slings

Alloy steel chain slings shall have permanently affixed, durable identification stating size, grade, rated capacity and reach.

Sling Use

Alloy steel chain slings shall not be used with loads in excess of the rated capacities prescribed in Table N-184-1 of 1910.184. Slings not included in this table shall be used only in accordance with the manufacturer's recommendations.

Safe Operating Temperatures

Alloy steel chain slings shall be permanently removed from service if they are heated above 1000°F. When exposed to temperatures in excess of 600°F, maximum working load limits permitted in Table N-184-1 of 1910.18.4 shall be reduced in accordance with the chain or sling manufacturer's recommendations.

Repairing Slings

Worn or damaged alloy steel chain slings or attachments shall not be used until repaired. If the chain size at any point of any link is less than that stated in Table N-1 84-2 of 1910.184, the sling shall be removed from service. Alloy steel chain slings with cracked or deformed master links, coupling links or other components shall be removed from service.

Wire Rope Slings

Sling Use

Wire rope slings shall not be used with loads in excess of the rated capacities shown in Tables N-184-3 through N-184-14 of § 1910.184. Slings not included in these tables shall be used only in accordance with the manufacturer's recommendations.

Safe Operating Temperatures

Fiber core wire rope slings of all grades shall be permanently removed from service if they are exposed to temperatures in excess of 200°F. When non-fiber core wire rope slings of any kind are used at temperatures above 40°F, or below –60°F, recommendations of the sling manufacturer regarding use at that temperature shall be followed.

Removal from Service

Wire rope slings shall be immediately removed from service if any of the following conditions are present:

- There are 10 randomly distributed broken wires in one rope lay or five broken wires in one strand in one rope lay;
- There is wear or scraping of one-third the original diameter of outside individual wires;

- There is kinking, crushing, bird caging or any other damage;
- There is evidence of heat damage;
- There are end attachments that are cracked, deformed or worn;
- There are hooks that have been opened more than 15 percent of the normal throat opening, measured at the narrowest point, or twisted more than 10° from the plane of the unbent hook; or
- The rope or end attachments are corroded.

Metal Mesh Slings

Each metal mesh sling shall have permanently affixed to it a durable marking that states the rated capacity for vertical basket hitch and choker hitch loadings. Handles shall have a rated capacity at least equal to the metal fabric and exhibit no deformation after proof testing. The fabric and handles shall be joined so that:

- The rated capacity of the sling is not reduced.
- The load is evenly distributed across the width of the fabric.
- Sharp edges will not damage the fabric.

Coatings that diminish the rated capacity of a sling shall not be applied.

Sling Test
All new and repaired metal mesh slings, including handles, shall not be used unless proof tested by the manufacturer or equivalent.

Sling Use
Metal mesh slings shall not be used to lift loads in excess of their rated capacities as prescribed in Table N-184-15 of 1910.184. Slings not included in this table shall be used only in accordance with the manufacturer's recommendations.

Safe Operating Temperatures
Metal mesh slings that are not impregnated with elastomers may be used in a temperature range from –20°F to 550°F without decreasing the working load limit. Metal mesh slings impregnated with polyvinyl chloride or neoprene may be used only in a temperature range from 0° to 200°F. For operations outside these temperature ranges or for metal mesh slings impregnated with other materials, the manufacturer's recommendations shall be followed.

Repair
Metal mesh slings that have been repaired, shall not be used unless they were repaired by a metal mesh sling manufacturer or an equivalent entity. Once repaired, records shall be maintained to indicate the date and nature of repairs and the person or organization that performed the repairs.

Removal from Service
Metal mesh slings shall be immediately removed from service if there is a:

- Broken weld or brazed joint along the sling edge;
- Reduction in wire diameter of 25 percent due to abrasion or 15 percent due to corrosion;
- Lack of flexibility due to distortion of the fabric;
- Distortion of the female handle so that the depth of the slot is increased more than 10 percent;
- Distortion of either handle so that the width of the eye is decreased more than 10 percent;
- Fifteen percent reduction of the original cross sectional area of metal at any point around the handle eye; or
- Distortion of either handle out of its plane.

Natural and Synthetic Fiber Rope Slings

Sling Use
Fiber rope slings made from conventional three-strand construction fiber rope shall not be used with loads in excess of the rated capacities prescribed in Tables N-184-16 through N-184-19 of 1910.184. Fiber rope slings shall have a diameter of curvature meeting at least the minimums specified in Figures N-184-4 and N-184-5 of 1910.184.

Slings not included in these tables shall be used only in accordance with the manufacturer's recommendations.

Safe Operating Temperature
Natural and synthetic fiber rope slings, except for wet frozen slings, may be used in a temperature range from –20°F to 180°F without

decreasing the working load limit. For operations outside this temperature range and for wet frozen slings, the sling manufacturer's recommendations shall be followed.

Splicing
Spliced fiber rope slings shall not be used unless they have been spliced in accordance with the minimum requirements specified in 1910.184 and with any additional recommendations of the manufacturer.

Removal from Service
Natural and synthetic fiber rope slings shall be immediately removed from service if there is:
- Abnormal wear;
- Powdered fiber between strands;
- Variations in the size or roundness of strands;
- Discoloration or rotting; or
- Distortion of hardware in the sling.

Only fiber rope slings made from new rope shall be used. Use of repaired or reconditioned fiber rope slings is prohibited.

Synthetic Web Slings

Sling Identification
Each sling shall be marked or coded to show the rated capacities for each type of hitch and type of synthetic web material.

Sling Use
Synthetic web slings illustrated in Figure N-184-6 shall not be used with loads in excess of the rated capacities specified in Tables N-184-20 through N-184-22 of §1910.184. Slings not included in these tables shall be used only in accordance with the manufacturer's recommendations.

When synthetic web slings are used, the following precautions shall be taken:
- Nylon web slings shall not be used where fumes, vapors, sprays, dusts or liquids of acids or phenolics are present;
- Polyester and polypropylene web slings shall not be used where fumes, vapors, sprays, dusts or liquids of caustics are present; and
- Web slings with aluminum fittings shall not be used where fumes, vapors, sprays, dusts or liquids of caustics are present.

Safe Operating Temperature
Synthetic web slings of polyester and nylon shall not be used at temperatures in excess of 180°F. Polypropylene web slings shall not be used at temperatures in excess of 200°F.

Repair
Repaired synthetic web slings shall not be used unless repaired by a sling manufacturer or an equivalent entity.

Removal from Service
Synthetic web slings shall be immediately removed from service if any of the following conditions are present:
- Acid or caustic burns;
- Melting or charring of any part of the sling surface;
- Snags, punctures, tears or cuts;
- Broken or worn stitches; or
- Distortion of fittings.

Sling Safety

Whenever possible, mechanical means should be used to move materials in order to avoid employee injuries such as muscle pulls, strains, and sprains. In addition, many loads are too heavy or bulky to be safely moved manually. Therefore, various types of equipment have been designed specifically to aid in the movement of materials. They include cranes, derricks, hoists, powered industrial trucks, and conveyors. Because cranes, derricks, and hoists rely upon slings to hold their suspended loads, slings are the most commonly used piece of materials-handling apparatus.

Sling Types

The dominant characteristics of a sling are determined by the components of that sling. For example, the strengths and weaknesses of a wire rope sling are essentially the same as the strengths and weaknesses of the wire rope of which it is made.

Slings are generally one of six types:
1. Chain,
2. Wire rope,
3. Metal mesh,
4. Natural fiber rope,
5. Synthetic fiber rope, or
6. Synthetic web.

In general, use and inspection procedures tend to place these slings into three groups:
1. Chain;
2. Wire rope and mesh; and
3. Fiber rope web.

Each type has its own particular advantages and disadvantages. Factors that should be taken into consideration when choosing the best sling for the job include the size, weight, shape, temperature and sensitivity of the material to be moved, as well as the environmental conditions under which the sling will be used.

Chains

Chains are commonly used because of their strength and ability to adapt to the shape of the load. Care should be taken, however, when using alloy chain slings because they are subject to damage by sudden shocks. Misuse of chain slings could damage the sling, resulting in sling failure and possible injury to an employee.

Chain slings are the best choice for lifting materials that are very hot. They can be heated to temperatures of up to 1000°F; however, when alloy chain slings are consistently exposed to service temperatures in excess of 600°F, operators must reduce the working load limits in accordance with the manufacturer's recommendations.

All sling types must be visually inspected prior to use. When inspecting alloy steel chain slings, pay special attention to any stretching, wear in excess of the allowances made by the manufacturer, and nicks and gouges. These are all indications that the sling may be unsafe and should immediately be removed from service.

Wire Rope

A second type of sling is made of wire rope. Wire rope is composed of wire individual wires that have been twisted core to form strands (See Exhibit 15-3.). The strands are then twisted to form a wire rope. When wire rope has a fiber core, it is usually more flexible but less resistant to environmental damage. Conversely, a core that is made of a wire rope strand tends to have greater strength and be more resistant to heat damage.

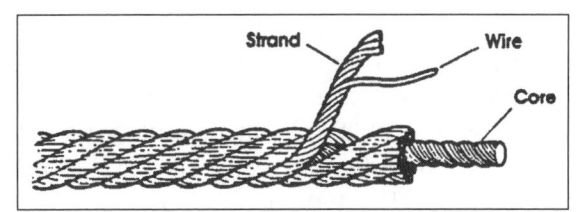

Exhibit 15-3.
Wipe rope is composed of individual wires twisted around a core.

Rope Lay

Wire rope may be further defined by the "lay." The lay of a wire rope can mean any of three things:
1. *One complete wrap of a strand around the core.* One rope lay is one complete wrap of a strand around the core (See Exhibit 15-4.).

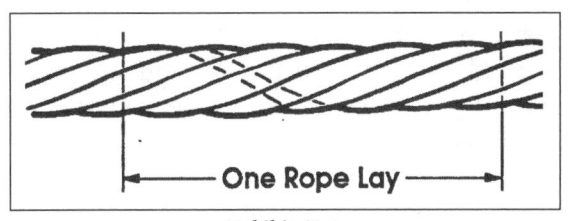

Exhibit 15-4.
An example of one rope lay.

2. The direction the strands are wound around the core. Wire rope is referred to as right lay or left lay. A right lay rope is one in which the strands are wound in a right-hand direction like a conventional screw thread (See Exhibit 15-5.). A left lay rope is just the opposite.

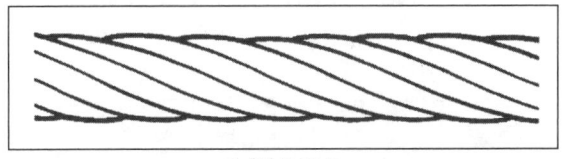

Exhibit 15-5.
An example of right lay rope.

3. *The direction the wires are wound in the strands in relation to the direction of the strands around the core.* In regular lay rope, the wires in the strands are laid in one direction, while the strands in the rope are laid in the opposite direction. In lang lay rope, the wires are twisted in the same direction as the strands. (See Exhibit 15-6.)

In *regular lay ropes*, the wires in the strands are laid in one direction, while the strands in the rope are laid in the opposite direction. The result is that the wire crown runs approximately parallel to the longitudinal axis of the rope. These ropes have good resistance to kinking and twisting and are easy to handle. They are also able to withstand considerable crushing and distortion due to the short length of exposed wires. This type of rope has the widest range of applications.

Lang lay (where the wires are twisted in the *same* direction as the strands) is recommended for such excavating, construction and mining applications, including draglines, hoist lines, dredge lines and other similar lines.

Lang lay ropes are more flexible and have greater wearing surface per wire than regular lay ropes. In addition, because the outside wires in lang lay rope lie at an angle to the rope axis, internal stress due to bending over sheaves and drums is reduced, causing lang lay ropes to be more resistant to bending fatigue.

A *left lay rope* is one in which the strands form a left-hand helix similar to the threads of a left-hand screw thread. Left lay rope has its greatest use in oil fields on rod and tubing lines, blast hole rigs and spudders where rotation of right lay would loosen couplings. The rotation of a left lay rope tightens a standard coupling.

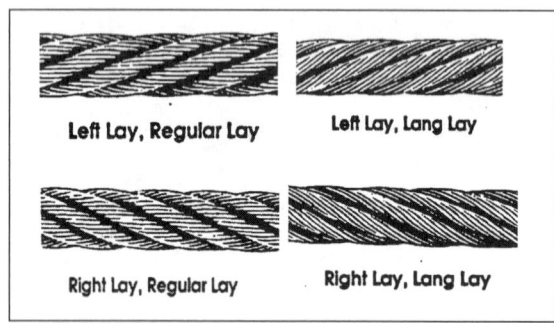

Exhibit 15-6.
Examples of regular lay and lang lay ropes.

Wire Rope Sling Selection

When selecting a wire rope to give the best service, there are four characteristics to consider:
1. Strength;
2. Ability to bend without distortion;
3. Ability to withstand abrasive wear; and
4. Ability to withstand abuse.

Strength. The strength of a wire rope is a function of its size, grade and construction. It must be sufficient to accommodate the maximum load that will be applied. The maximum load limit is determined by means of an appropriate multiplier. This multiplier is the number by which the ultimate strength of a wire rope is divided to determine the working load limit. Thus, a wire rope sling with the strength of 10,000 lbs. and a total working load of 2,000 lbs. has a design factor (multiplier) of 5. New wire rope slings have a design factor of 5. As a sling suffers from the rigors of continued service; however, both the design factor and the sling's ultimate strength will be proportionately reduced. If a sling is loaded beyond its ultimate strength, it will fail. For this reason, older slings must be more rigorously inspected to ensure that rope conditions adversely affecting the strength of the sling are considered when determining whether a wire rope sling should be allowed to continue in service.

Fatigue. A wire rope must have the ability to withstand repeated bending without the failure of the wires from fatigue. Fatigue failure of the wires in a wire rope is the result of the development of small cracks under repeated applications of bending loads (See Exhibit 15-7.). It occurs when ropes make small radius bends. The best means of preventing fatigue failure of wire rope slings is to use blocking or padding to increase the radius of the bend.

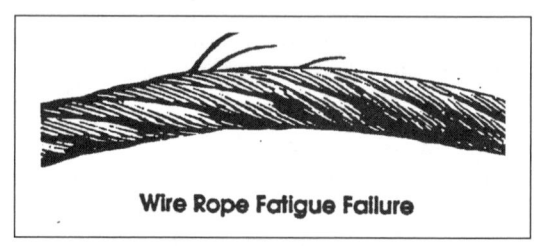

Exhibit 15-7.
Example of wire rope fatigue failure.

Abrasive Wear. The ability of a wire rope to withstand abrasion is determined by the size, number of wires and construction of the rope. Smaller wires bend more readily and therefore offer greater flexibility but are less able to withstand abrasive wear. Conversely, the larger wires of less flexible ropes are better able to withstand abrasion than smaller wires of more flexible ropes.

Abuse. All other factors being equal, misuse or abuse of wire rope will cause a wire rope sling to become unsafe long before any other factor. Abusing a wire rope sling can cause serious structural damage to the wire rope, such as kinking or bird caging, which reduces the strength of the wire rope. (In bird caging, the wire rope strands are forcibly untwisted and become spread outward (See Exhibit 15-8.).) Therefore, to prolong the life of the sling and protect the lives of employees, the manufacturer's suggestion for safe and proper use of wire rope slings must be strictly followed.

Exhibit 15-8.
Example of wire rope "bird cage."

Wire Rope Life

Many operating conditions affect wire rope life, including bending, stresses, loading conditions, speed of load application, jerking, abrasions, corrosion, sling design, materials handled, environmental conditions and history of previous use.

In addition to the above operating conditions, the weight, size and shape of the loads to be handled also affect the service life of a wire rope sling. Flexibility is also a factor. Generally, more flexible ropes are selected when smaller radius bending is required. Less flexible ropes should be used when the rope must move through or over abrasive materials.

Wire Rope Sling Inspection

Wire rope slings must be visually inspected before each use. The operator should check the twists or lay of the sling. If 10 randomly distributed wires in one lay are broken or five wires in one strand of a rope lay are damaged, the sling must not be used. It is not sufficient, however, to check only the condition of the wire rope. End fittings and other components should also be inspected for any damage that could make the sling unsafe.

To ensure safe sling usage between scheduled inspections, all workers must participate in a safety awareness program. Each operator must keep a close watch on the slings he or she is using. If any accident involving the movement of materials occurs, the operator must immediately shut down the equipment and report the accident to a supervisor. The cause of the accident must be determined and corrected before operations can resume.

Field Lubrication

Although every rope sling is lubricated during manufacture, to lengthen its useful service life it must also be lubricated in the field. There is no set rule on how much or how often this should be done. It depends on the conditions under which the sling is used; the more adverse the conditions under which the sling operates, the more frequently lubrication will be required.

Storage

Wire rope slings should be stored in a well-ventilated, dry building or shed. They should never be stored on the ground or continuously exposed to the elements because this will make them vulnerable to corrosion and rust. If it is necessary to store wire rope slings outside, they must be set off the ground and protected.

Note: Using the sling several times a week, even at a light load, is a good practice. Records show that slings that are used frequently or continuously give useful service far longer than those that are idle.

Discarding Slings

Wire rope slings can provide a margin of safety by showing early signs of failure. Factors requiring that a wire sling be discarded include:
- Severe corrosion;
- Localized wear (shiny worn spots) on the outside;
- A one-third reduction in outer wire diameter;
- Damage or displacement of end fittings, hooks, rings, links or collars by overload or misapplication;
- Distortion, kinking, bird caging or other evidence of damage to the wire rope structure; or
- Excessive broken wires.

Fiber Rope and Synthetic Web

Fiber rope and synthetic web slings are used primarily for temporary work, such as construction and painting jobs, and in marine operations. They are also the best choice for use on expensive loads, highly finished parts, fragile parts and delicate equipment.

Fiber Rope

Fiber rope slings are preferred for some applications because they are pliant, grip the load well and do not mar the surface of the load. They should be used only on light loads, however, and must not be used on objects that have sharp edges capable of cutting the rope or in applications where the sling will be exposed to high temperatures, severe abrasion or acids.

The choice of rope type and size will depend upon the application, the weight to be lifted and the sling angle. Before lifting any load with a fiber rope sling, the sling should be inspected carefully because they deteriorate far more rapidly than wire rope slings and their actual strength is very difficult to estimate.

When inspecting a fiber rope sling prior to using it, look first at its surface. Look for dry, brittle, scorched or discolored fibers. If any of these conditions are found, the supervisor must be notified and a determination made regarding the safety of the sling. If the sling is found to be unsafe, it must be discarded.

Next, check the interior of the sling. It should be as clean as when the rope was new. A buildup of powder-like sawdust on the inside of the fiber rope indicates excessive internal wear and is an indication that the sling is unsafe.

Finally, scratch the fibers with a fingernail. If the fibers come apart easily, the fiber sling has suffered some kind of chemical damage and must be discarded.

Synthetic Web Slings

Synthetic web slings offer a number of advantages for rigging purposes. The most commonly used synthetic web slings are made of nylon, dacron and polyester. They have the following properties in common:

- Strength: can handle load of up to 300,000 lbs.;
- Convenience: can conform to any shape;
- Safety: will adjust to the load contour and hold it with a tight, nonslip grip;
- Load protection: will not mar, deface, or scratch highly polished or delicate surfaces;
- Long life: are unaffected by mildew, rot, or bacteria; resist some chemical action and have excellent abrasion resistance;
- Economy: have low initial cost plus long service life;
- Shock absorbency: can absorb heavy shocks without damage; and
- Temperature resistance: are unaffected by temperatures up to 180°F.

Each synthetic material has its own unique properties. Nylon must be used wherever alkaline or greasy conditions exist. It is also preferable when neutral conditions prevail and resistance to chemicals and solvents is important. Dacron must be used where high concentrations of acid solutions, such as sulfuric, hydrochloric, nitric and formic acids, and high-temperature bleach solutions are prevalent. (Nylon will deteriorate under these conditions.) Dacron should not be used in alkaline conditions because it will deteriorate; nylon or polypropylene should be used instead. Polyester must be used where acids or bleaching agents are present and is also ideal for applications where a minimum of stretching is important.

Possible Defects. Synthetic web slings must be removed from service if any of the following defects exist:

- Acid or caustic burns;
- Melting or charring of any part of the surface;
- Snags, punctures, tears or cuts;
- Broken or worn stitches;
- Wear or elongation exceeding the amount recommended by the manufacturer; or
- Distortion of fittings.

Safe Lifting Practices

After the sling has been selected (based upon the characteristics of the load and the environmental conditions surrounding the lift) and inspected prior to use, it can be used—safely. There are four primary factors to take into consideration when safely lifting a load:

- The size, weight and center of gravity of the load;

- The number of legs and the angle the sling makes with the horizontal line;
- The rated capacity of the sling; and
- The history of the care and use of the sling.

Size, Weight and Center of Gravity

The center of gravity of an object is that point at which the entire weight may be considered as concentrated. In order to make a level lift, the crane hook must be directly above this point. While slight variations are usually permissible, if the crane hook is too far to one side of the center of gravity, dangerous tilting will result, causing unequal stresses in the different sling legs. This imbalance must be compensated for at once.

Number of Legs and Angle with the Horizontal

As the angle formed by the sling leg and the horizontal line decreases, the rated capacity of the sling also decreases. In other words, the smaller the angle between the sling leg and the horizontal, the greater the stress on the sling leg and the smaller (lighter) the load the sling can safely support. Larger (heavier) loads can be safely moved if the weight of the load is distributed among more sling legs.

Rated Capacity of the Sling

The rated capacity of a sling varies depending upon the shape of sling, the size of the sling and the type of hitch. Operators must know the capacity of the sling. Charts or tables that contain this information generally are available from sling manufacturers. However, the values given are for *new* slings; older slings must be used with additional caution. Under no circumstances shall a sling's rated capacity be exceeded.

History of Care and Usage

The mishandling and misuse of slings are the leading causes of accidents involving their use. The majority of injuries and accidents, however, can be avoided by becoming familiar with the essentials of proper sling care and use.

Proper care and usage are essential for maximum service and safety. Slings must be protected from sharp bends and cutting edges by means of cover saddles, burlap padding or wood blocking, as well as from unsafe lifting procedures such as overloading.

Before making a lift, check to be certain that the sling is properly secured around the load and that the weight and balance of the load have been accurately determined if the load is on the ground. Do not allow the load to drag along the ground; this could damage the sling if the load is already resting on the sling. Then ensure that there is no sling damage prior to making the lift.

Next, position the hook directly over the load and seat the sling squarely within the hook bowl. This gives the operator maximum lifting efficiency without bending the hook or overstressing the sling.

Wire rope slings are also subject to damage resulting from contact with sharp edges of the loads being lifted. These edges can be blocked or padded to minimize damage to the sling. After the sling is properly attached to the load, there are a number of good lifting techniques that are common to all slings:

- Make sure the load is not clamped or bolted to the floor.
- Guard against shock loading by taking up the slack in the sling slowly.
- Apply power cautiously so as to prevent jerking at the beginning of the lift and accelerate or decelerate slowly.
- Check the tension on the sling.
- Raise the load a few inches, stop, and check for proper balance and that all items are clear of the path of travel.
- Never allow anyone to ride on the hood or load.
- Keep all personnel clear while the load is being raised, moved or lowered.
- Crane or hoist operators should watch the load at all times when it is in motion.
  Obey the following "nevers":
- Never allow more than one person to control a lift;
- Never allow more than on person to give signals to a crane operator;
- Never raise the load more than necessary;
- Never leave the load suspended in the air; and
- Never work under a suspended load or allow anyone else to do so.

Once the lift has been completed, clean the sling, check it for damage and store it in a clean, dry, airy place. It is best to hang it on a rack or wall. Safe and proper use and storage of slings will increase their service life.

Maintenance of Slings

Chains

Chain slings must be cleaned prior to each inspection, as dirt or oil may hide damage. The operator must be certain to inspect the total length of the sling, periodically looking for stretching, binding, wear, or nicks and gouges. If a sling has stretched so that it is now more than 3 percent longer than it was when new, it is unsafe and must be discarded.

Binding is the term used to describe the condition that exists when a sling has become deformed to the extent that its individual links cannot move within each other freely. It is also an indication that the sling is unsafe. Generally, wear occurs on the load-bearing inside ends of the links. Pushing links together so that the inside surface becomes clearly visible is the best way to check for this type of wear. Wear may also occur, however, on the outside of links when the chain is dragged along abrasive surfaces or pulled out from under heavy loads. Either type of wear weakens slings and makes accidents more likely.

Heavy nicks and gouges must be filed smooth, measured with calipers, and then compared with the manufacturer's minimum allowable safe dimensions. When in doubt or in borderline situations, do not use the sling. In addition, never attempt to repair the welded components on a sling. If the sling needs repair of this nature, the supervisor must be notified.

Wire Rope

Wire rope slings, like chain slings, must be cleaned prior to each inspection because they are also subject to damage hidden by dirt or oil. In addition, they must be lubricated according to the manufacturer's instructions. Lubrication prevents or reduces corrosion and wear due to friction and abrasion. Before applying any lubricant, however, the sling user should make certain that the sling is dry. Applying lubricant to a wet or damp sling traps moisture against the metal and hastens corrosion, which deteriorates wire rope. Pitting may indicate it, but it is sometimes hard to detect. Therefore, if a wire rope sling shows any sign of significant deterioration, that sling must be removed until a person who is qualified to determine the extent of the damage can examine it.

Fiber Ropes and Synthetic Webs

In general, fiber ropes and synthetic webs should be discarded rather than serviced or repaired. Operators must always follow manufacturer's recommendations.

Summary

There are good practices to follow to protect employees while using slings to move materials. First, an employee should learn as much as he or she can about the materials with which he or she will be working. Slings come in many different types; analyze the load to be moved—in terms of size, weight, shape, temperature and sensitivity - then choose the sling which best meets those needs. Next, always inspect all the equipment before and after a move. Always be sure to give equipment whatever "in service" maintenance it may need. Finally, use safe lifting practices. Use the proper lifting technique for the type of sling and the type of load.

2004 Subpart N 1910.176-184
Materials Handling & Storage

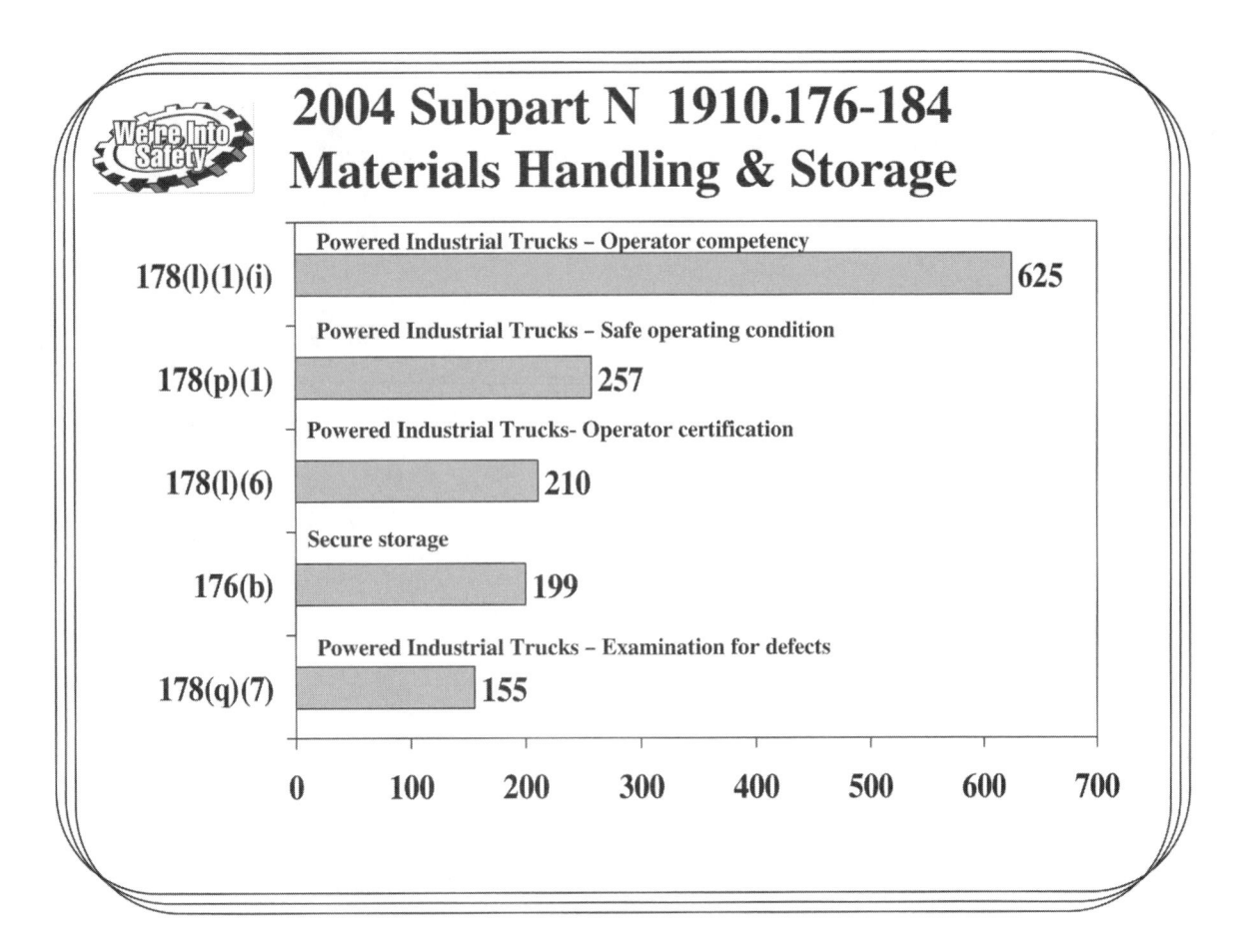

Citation	Description	Count
178(l)(1)(i)	Powered Industrial Trucks – Operator competency	625
178(p)(1)	Powered Industrial Trucks – Safe operating condition	257
178(l)(6)	Powered Industrial Trucks – Operator certification	210
176(b)	Secure storage	199
178(q)(7)	Powered Industrial Trucks – Examination for defects	155

2004 Subpart F 1910.66-68
Powered Platforms

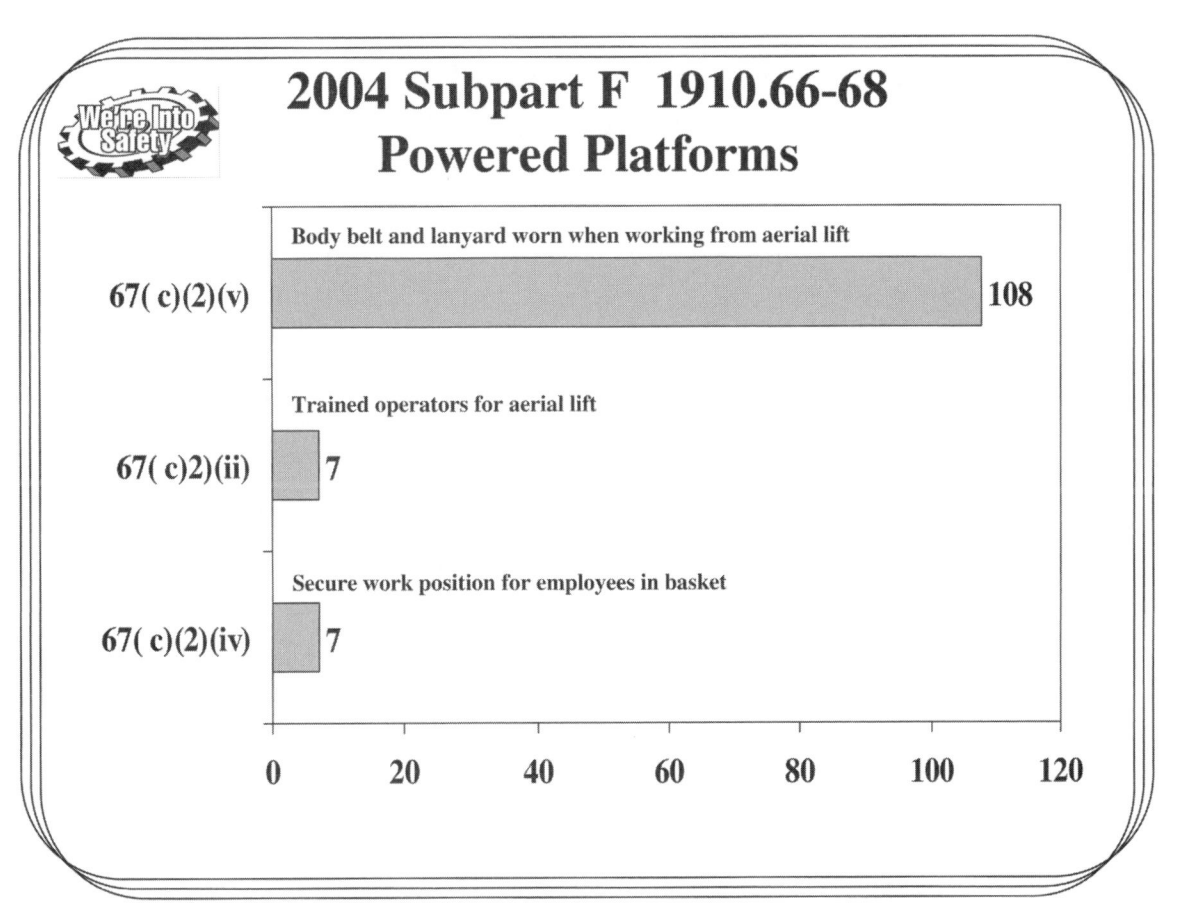

Citation	Description	Count
67(c)(2)(v)	Body belt and lanyard worn when working from aerial lift	108
67(c)2)(ii)	Trained operators for aerial lift	7
67(c)(2)(iv)	Secure work position for employees in basket	7

Chapter 16

Welding, Cutting, and Brazing

Many welding and cutting operations require the use of compressed gases. When compressed gases are consumed in the welding process, such as oxygen fuel gas welding, requirements for their handling, storage and use are contained in Subpart Q.

General requirements for the handling, storage, and use of compressed gases are contained in Subpart H, Hazardous Materials, 1910.101 to 1910.105. Certain welding and cutting operations require the use of compressed gases other than those consumed in the welding process. For example, gas metal arc welding uses compressed gases for shielding. Handling, storage and use of compressed gases in situations such as these require compliance with the requirements contained in Subpart H.

Many hazards are involved in compressed gas handling, storage and use. Compressed gases are stores of potential energy and it takes energy to compress and confine the gas. That energy is stored until purposely released to perform useful work or until accidental release by container failure or other causes.

Some compressed gases (e.g., acetylene) have high flammability characteristics. Flammable compressed gases, therefore, have additional stored energy besides simple compression-release energy. Other compressed gases, such as nitrogen, have simple asphyxiating properties. Some compressed gases, such as oxygen, can augment or compound fire hazards.

Compressed Gases General Requirements—1910.101

Cylinder Inspection

Employers shall determine that compressed gas cylinders under their control are in a safe condition to the extent that this can be determined by visual inspection. Visual and "other" inspections are required, but "other" inspections are not defined. These inspections must be conducted as prescribed in the Hazardous Materials Regulations of the Department of Transportation (DOT) contained in 49 CFR Parts 171 to 179 and 14 CFR Part 103. Where these regulations are not applicable, these inspections shall be conducted in accordance with Compressed Gas Association (CGA) Pamphlets C-6 and C-8. According to DOT regulations:

> A cylinder that leaks, is bulged, has defective valves or safety devices, bears evidence of physical abuse, fire or heat damage, or detrimental rusting or corrosion, must not be used unless it is properly repaired and requalified as prescribed in these regulations.

The term "cylinder" is defined as a pressure vessel designed for pressures higher than 40 psia (pounds per square inch absolute) and having a circular cross section. It does not include a portable tank, multi-unit tank car tank, cargo tank or tank car.

The Department of Transportation (DOT) requires basic information markings on all cylinders. Each required marking on a cylinder must be maintained so that it is legible (See Exhibit 16-1.).

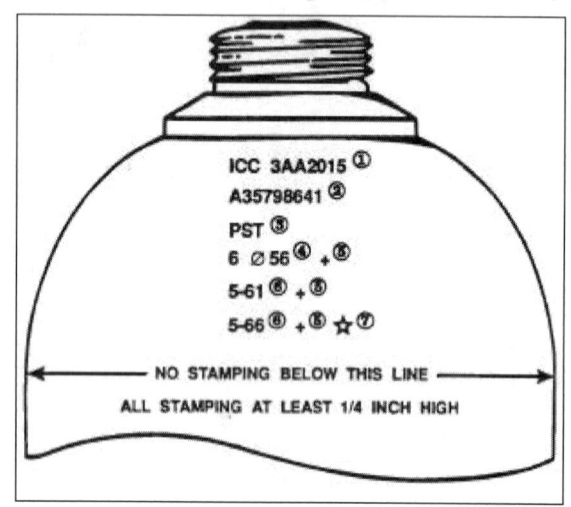

Exhibit 16-1.
The Department of Transportation requires legible markings on all containers.

1. DOT or ICC marking may appear—new manufacture must read "DOT" (49 CFR 171.14). "3AA" indicates specification found in 49 CFR 178.37. "2015" is the marked service pressure.
2. Serial number—no duplications permitted with any particular symbol-serial number combinations.
3. Symbol of manufacturer, user or purchaser.
4. "6 56" date of manufacture. Month and year. "O" is a disinterested inspector's official mark.
5. Plus mark (+) indicates cylinder may be 10% overcharged (49 CFR 173.302(c)).
6. Retest dates.
7. A five-pointed star indicates a 10-year retest interval (49 CFR 173.34(e)(15)).

Handling, Storage and Utilization

The in-plant handling, storage and utilization of all compressed gases in cylinders, portable tanks, rail tank cars or motor vehicle cargo tanks shall be in accordance with Compressed Gas Association (CGA) Pamphlet P-1.

Safety Relief Devices

Compressed gas cylinders, portable tanks and cargo tanks shall have pressure relief devices installed and maintained in accordance with CGA Pamphlets S-1.1 and S-1.2.

Special gases

The OSHA regulations contain some sections regulating specific compressed gases, including acetylene, hydrogen, oxygen, nitrous oxide, anhydrous ammonia and liquefied petroleum gases.

There are many compressed gases that are in common use that fall under regulation 1910.101. Examples include chlorine, vinyl chloride, sulfur dioxide, methyl chloride, hydrogen sulfide, ethane, compressed air and nitrogen.

These compressed gases do not receive explicit coverage by the OSHA regulations, but are covered by the requirements of 1910.101.

Welding, Cutting and Brazing General Requirements—1910.252
Fire Prevention and Protection

Basic Precautions

The basic precautions for fire prevention in welding or cutting work are:
- If the object to be welded or cut cannot readily be moved, all movable fire hazards in the vicinity shall be taken to a safe place.
- If the object to be welded or cut cannot be moved and if all the fire hazards cannot be removed, then guards shall be used to confine the heat, sparks, and slag and to protect the immovable fire hazards.
- If the above requirements cannot be met, then welding and cutting shall not be performed.

Special Precautions

Suitable fire extinguishing equipment shall be maintained in a state of readiness for instant use. Such equipment may consist of pails of water, buckets of sand, hose or portable extinguishers, depending upon the nature and quantity of the combustible material exposed.

Fire watchers are required whenever welding or cutting is performed in locations where other than a minor fire might develop and if the following conditions exist:
- Appreciable combustible materials, in building construction or contents, are closer than 35 ft. to the point of operation; or
- Appreciable combustibles are more than 35 ft. away, but are easily ignited by sparks.

A fire watch also shall be maintained for at least a half hour after the completion of welding or cutting operations to detect and extinguish possible smoldering fires.

Cutting or welding shall not be permitted in the following situations:
- In areas not authorized by management;
- In sprinklered buildings while such protection is impaired;
- In the presence of explosive atmospheres (e.g., mixtures of flammable gases, vapors, liquids, dusts with air), or in explosive atmospheres that may develop inside uncleaned or improperly prepared tanks or equipment that have previously contained such materials or that may develop in areas with an accumulation of combustible dusts.

Welding or Cutting Containers
No welding, cutting or other hot work shall be performed on used drums, barrels, tanks or other containers until they have been cleaned so thoroughly as to make absolutely certain that there are no flammable materials present or any substances such as greases, tars, acids or other materials that when subjected to heat, might produce flammable or toxic vapors. Any pipelines or connections to the drum or vessel shall be disconnected or blanked.

Confined Spaces
When arc welding is to be suspended for any substantial period of time, such as during lunch or overnight, all electrodes shall be removed from the holders and the holders carefully located so that accidental contact cannot occur. The machine also shall be disconnected from the power source.

In order to eliminate the possibility of gas escaping through leaks or improperly closed valves when gas welding or cutting, the torch valves shall be closed and the gas supply to the torch positively shut off at some point outside the confined area whenever the torch is not to be used for a substantial period of time, such as during lunch or overnight. Where practical, the torch and hose shall also be removed from the confined space.

Protection of Personnel
A welder or helper working on platforms, scaffolds or runways shall be protected against falling through the use of railings, safety belts, lifelines or some equally effective safeguard.

Eye Protection
Helmets or hand shields shall be used during all arc welding or arc cutting operations, excluding submerged arc welding. Helpers or attendants shall be provided with proper eye protection.

Helmets and hand shields shall be made of a material that is an insulator for heat and electricity. Helmets, shields and goggles shall not be readily flammable and shall be capable of withstanding sterilization.

Helmets and hand shields shall be arranged to protect the face, neck and ears from direct radiant energy from the arc.

Where the work permits, the welder should be enclosed in an individual booth painted with a finish of low reflectivity such as zinc oxide (an important factor for absorbing ultra-violet radiations) and lamp black or shall be enclosed with non-combustible screens similarly painted. Booths and screens shall permit circulation of air at floor level. Workers or other persons adjacent to the welding areas shall be protected from the rays by noncombustible or flameproof screens or shields or shall be required to wear appropriate goggles.

Protective Clothing
Employees exposed to the hazards created by welding, cutting or brazing operations shall be protected by personal protective equipment in accordance with the requirements of 1910.132. Appropriate protective clothing required for any welding operation will vary with the size, nature and location of the work to be performed (See Exhibit 16-2.).

Exhibit 16-2. Filter Lens Shade Number Guide

Welding Operation	Shade Number
Shielded Metal-Arc Welding, up to 5/32" (4 mm) electrodes	10
Shielded Metal-Arc Welding, 3/16 to ¼" (4.8 to 6.4 mm) electrodes	12
Shielded Metal-Arc Welding, over ¼" (6.4 mm) electrodes	14
Gas Metal-Arc Welding (Nonferrous)	11
Gas Metal-Arc Welding (Ferrous)	12
Gas Tungsten-Arc Welding	12
Atomic Hydrogen Welding	14
Carbon Arc Welding	10-14
Torch Soldering	2
Torch Brazing	3 or 4
Light Cutting, up to 1" (25 mm)	3 or 4
Medium Cutting, 1 to 6" (25 to 150 mm)	4 or 5
Heavy Cutting, more than 6" (150 mm)	5 or 6
Gas Welding (light), up to 1/8" (3.2 mm)	4 or 5
Gas Welding (medium), 1/8 to ½" (3.2 to 12.7 mm)	5 or 6
Gas Welding (heavy), more than ½" (12.7 mm)	6 or 8

Note: In gas welding or oxygen cutting where the torch produces a high yellow light, it is desirable to use a filter lens that absorbs the yellow or sodium line in the visible light of the operation spectrum.

Welders should always select clothing materials that will provide maximum protection from sparks and hot metal. Protective eye wear, safety shoes, clean fire-resistant clothing, and fire-resistant gauntlet gloves are recommended.

In addition, shirts should have full sleeves, no pockets and should be worn outside the trousers with their collars buttoned. Trousers should have no cuffs and should extend well down to the safety shoes.

Work in Confined Spaces

A confined space is defined in this regulation to be a relatively small or restricted space such as a tank, boiler, pressure vessel or small compartment of a ship. Adequate ventilation is a prerequisite to work in confined spaces. Ventilation requirements are discussed later in this section.

When welding or cutting is being performed in any confined space, the gas cylinders and welding machines shall be left outside the space. Where welders must enter a confined space through a manhole or other small opening, means shall be provided for quickly removing them in case of emergency. An attendant with a pre-planned rescue procedure shall be stationed outside to observe the welder at all times and be capable of putting rescue operations into effect.

When arc welding is to be suspended for any substantial period of time, such as during lunch or overnight, all electrodes shall be removed from the holders and the holders carefully located so that accidental contact cannot occur. The machine also shall be disconnected from the power source.

In order to eliminate the possibility of gas escaping through leaks of improperly closed valves when gas welding or cutting, the torch valves shall be closed and the fuel-gas and oxygen supply to the torch positively shut off at some point outside the confined area whenever the torch is not to be used for a substantial period of time, such as during lunch or overnight. Where practical, the torch and hose also shall be removed from the confined space.

Health Protection and Ventilation

Mechanical ventilation is required when welding or cutting is done with materials not specifically mentioned in this section. These materials—fluorine compounds, zinc, lead, beryllium, cadmium, mercury, cleaning compounds and stainless steel—are particularly hazardous and have specific control requirements.

Oxygen-Fuel Gas Welding and Cutting—1910.253

Under no conditions shall acetylene be generated, piped (except in approved cylinder manifolds) or used at a pressure in excess of 15 psig (pounds per square inch gauge) or 30 psia (pounds per square inch absolute). The

30-psia limit is intended to prevent unsafe use of acetylene in pressurized chambers, such as caissons, underground excavations or tunnel construction. (Absolute pressure equal to gauge pressure plus atmospheric pressure, which at sea level averages 14.7 psi. Thus, at sea level a gauge pressure reading of 15 psi is equal to an absolute pressure of 29.7 psi.) This requirement is not intended to apply to storage of acetylene dissolved in a suitable solvent in cylinders manufactured and maintained according to U.S. Department of Transportation requirements or to acetylene for chemical use.

Using acetylene at pressures in excess of 15-psi gauge pressure (or about 30 psi absolute pressure) is a hazardous practice. Free gaseous acetylene is potentially unstable at pressures above 15 psig and could decompose with explosive violence. Experience indicates that 15 psig is generally acceptable as a safe upper pressure limit.

Keeping the gas in liquid solution and storing it in cylinders of unique construction avoids the decomposition characteristics of acetylene gas. Internally, acetylene cylinders are not designed like other kinds of compressed gas cylinders. Acetylene cylinders are never hollow. These cylinders contain a porous, calcium silicate filler and a suitable solvent, usually acetone, because under pressure, acetylene by itself is unstable. Acetone is used because it has the ability to absorb more than 400 times its own volume of acetylene at 70°F.

Millions of microscopic pores make up the calcium silicate filler. Although it appears to fill the steel shell, approximately 90 percent of the filler's volume consists of "pore space" for holding and evenly distributing the acetylene/acetone solution. When absorbed in this filler, the acetylene is divided into such small units that should acetylene decomposition take place in one pore, the heat released is not enough to raise the temperature of the acetylene in surrounding pores to the point where it too will decompose. Acetylene is usually supplied in cylinders that have a capacity of up to 300 cu. ft. of dissolved gas under pressure of 250 psig at 70°F.

Cylinders and Containers

Approval and Marking

All portable cylinders used for the storage and shipment of compressed gases shall be constructed and maintained in accordance with the regulations of the U.S. Department of Transportation (49 CFR Parts 171 to 179).

Compressed gas cylinders shall be legibly marked, for the purpose of identifying the gas content, with either the chemical or trade name of the gas. Such marking shall be by means of stenciling, stamping or labeling and shall not be readily removable. Whenever practical, the marking shall be located on the shoulder of the cylinder.

Storage of Cylinders—General

- Cylinders shall be kept away from radiators and other sources of heat.
- Inside of buildings, cylinders shall be stored in a well-protected, well-ventilated dry location, at least 20 ft. (6.1 m) from highly combustible materials.
- Cylinders should be stored in definitely assigned places away from elevators, stairs, or gangways or other areas where they might be knocked over or damaged by passing or falling objects or subject to tampering.
- Empty cylinders shall have their valves closed.
- Valve protection caps, where the cylinder is designed to accept a cap, shall always be in place, hand-tight, except when cylinders are in use or connected for use. The valve protection cap is designed to take the blow in case the cylinder falls.

Fuel-Gas Cylinder Storage

Inside a building, cylinders, except those in actual use or attached ready for use, shall be limited to a total gas capacity of 2,000 cu. ft. or 300 lbs. of liquefied petroleum gas.

Acetylene cylinders shall be stored valve end up. If the cylinder is on its side, acetone may leak out and create a dangerous condition.

Oxygen Storage

Oxygen cylinders in storage shall be separated from fuel-gas cylinders or combustible materials (especially oil or grease), by a minimum distance of 20 ft. (6.1 m) or by a noncombustible barrier at least 5 ft. (1.5 m) high having a fire resistance rating of at least one-half hour (See Exhibit 16-3.). This requirement is intended to reduce the possibility of any fire support when a fire occurs among the fuel-gas storage.

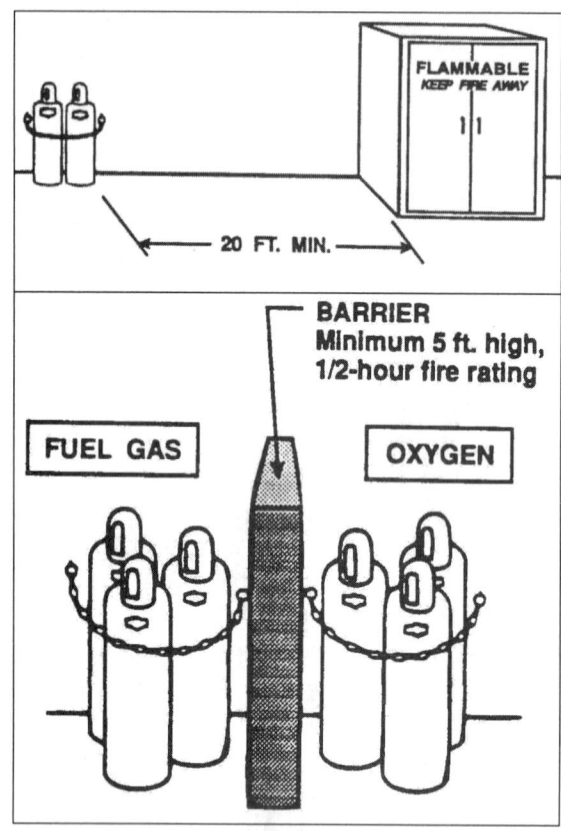

Exhibit 16-3.
Oxygen cylinders in storage shall be separated from fuel-gas cylinders or combustible materials by the distances shown below.

Operating Procedures

Cylinders, cylinder valves, couplings, regulators, hose and apparatus shall be kept free from oily or greasy substances. Oxygen cylinders or apparatus shall not be handled with oily hands or gloves. A jet of oxygen must never be permitted to strike an oily surface, greasy clothes, or enter a fuel oil or other storage tank.

Valve-protection caps shall not be used for lifting cylinders from one vertical position to another. The cap may accidentally and suddenly come loose. Should the cylinder fall or be knocked over, the valve may be damaged or sheared off, causing a sudden release of pressure.

Should the valve outlet of a cylinder become clogged with ice, it should be thawed with warm, not boiling, water.

Unless cylinders are secured on a special truck, regulators shall be removed and valve-protection caps, when provided for, shall be put in place before cylinders are moved.

Cylinders not having fixed hand wheels shall have keys, handles or nonadjustable wrenches on valve stems while they are in service.

Unless connected to a manifold, oxygen from a cylinder shall not be used without first attaching an oxygen regulator to the cylinder valve. In addition, operators must make sure the regulator is proper for the particular gas or service. The regulator also should be clean and have a clean filter installed in its inlet nipple.

Before attaching the regulator, the protective cap should be removed from the cylinder. The valve also should be opened slightly for an instant and then closed. Always stand to one side of the outlet when opening the cylinder valve. This "cracking" of the cylinder valve will clean the valve of dust or dirt that may have accumulated during storage. Dirt can damage critical parts of a regulator and may cause a fire or explosion.

Before a regulator is removed from a cylinder valve, the valve shall be closed and the gas released from the regulator. An acetylene cylinder valve shall not be opened more than one and one-half turns of the spindle. This permits adequate flow of acetylene and allows ready closing of the valve in an emergency situation.

Exhibit 16-4.
Ten Basic Rules for Oxyacetylene Welding

1. Blow out the cylinder valve before you connect the regulator.
2. Release the adjusting screw on the regulator before opening the cylinder valve.
3. Stand to one side of the regulator before you open the cylinder valve.
4. Open the cylinder valve slowly.
5. Do not use or compress acetylene in a free state at pressures more than 15 psig.
6. Purge your acetylene and oxygen passages individually before lighting the torch.
7. Light the acetylene before opening the oxygen on the torch.
8. Never use oil or grease on regulators, tips, etc., in contact with oxygen.
9. Do not use oxygen as a substitute for air.
10. Keep your work area clear of anything that will burn.

Manifolding of Cylinders

Portable Outlet Headers

Portable outlet headers shall not be used indoors except for temporary service where the conditions preclude a direct supply from outlets located on the service piping system.

Each outlet on the service piping from which oxygen or fuel gas is withdrawn to supply a portable outlet header shall be equipped with a readily accessible shut-off valve.

Each service outlet on portable outlet headers shall be provided with a valve assembly that includes a detachable outlet seal cap, chained or otherwise attached to the body of the valve.

The primary reason for using a seal cap is to protect the outlet pipe thread from damage and to prevent the deposit of oil or grease on the threads. However, many times the caps are not used. Damage to threads and ground seals can cause leaky connections.

Service Piping Systems

Materials and Design

Pipe shall be at least Schedule 40 and fittings shall be at least standard weight in sizes up to and including 6-in. nominal.

Schedule 40 pipe is standard black iron pipe that has a working pressure of up to 125 psi and is always tested before use. Problems might arise when line extensions are made with other types of pipe or aluminum tubing. Therefore, a close inspection is necessary.

When oxygen is supplied to a service piping system from a low-pressure oxygen manifold without an intervening pressure regulating device, the piping system shall have a minimum design pressure of 250 psig. A pressure-regulating device shall be used at each station outlet when the connected equipment is for use at pressures less than 250 psig.

Piping for acetylene or acetylenic compounds shall be steel or wrought iron. Unalloyed copper shall not be used for acetylene or acetylenic compounds except in listed equipment. Under certain conditions, acetylene forms explosive compounds with copper (as well as with silver and mercury).

Installation

All piping shall be run as directly as practical and protected against physical damage, with proper allowance being made for expansion and contraction, jarring and vibration. Pipe laid underground shall be located below the frost line and protected against corrosion. After assembly, piping shall be thoroughly blown out with air, nitrogen, or carbon dioxide to remove foreign materials. For oxygen piping, only oil-free air, oil-free nitrogen or oil-free carbon dioxide shall be used.

Low points in piping carrying moist gas shall be drained into drip pots that have been constructed to permit pumping or draining out the condensate at necessary intervals. Drain valves shall be installed for this purpose, having outlets normally closed with screw caps or plugs. No open-end valves or petcocks shall be used. However, when drips are located outdoors or underground, and are not readily accessible, valves may be used at such points if they are equipped with means to secure them in the closed position. Pipes leading to the surface of the ground shall be cased or jacketed where necessary to prevent loosening or breaking.

Piping from overhead lines shall have drip pots at each station. These drip pots either have a plug or petcock on the bottom. Underground installations have no draining system. Pipes leading to the surface from underground lines have to be secured to prevent breaking or to avoid other damage.

Painting

Underground pipe and tubing and outdoor ferrous pipe and tubing shall be covered or painted with a suitable material for protection against corrosion.

Testing

Piping systems shall be tested and proved gas tight at all times during the maximum operating pressure and shall be thoroughly purged of air before being placed in service. The material used for testing oxygen shall be oil free and noncombustible. Flames shall not be used to detect leaks.

Protective Equipment, Hose and Regulators

Pressure Relief Devices

Service piping systems shall be protected by pressure relief devices set to function at not more than the design pressure of the systems and discharge upwards to a safe location. Pressure relief valves are required in fuel-gas piping systems to prevent excessive pressure build-up within the system. Relief valves will vent automatically at preset pressures or may be manually operated to relieve pressure in the system (See Exhibit 16-5.).

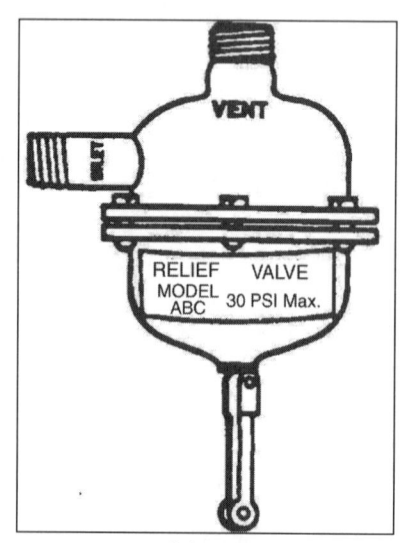

Exhibit 16-5.
Pressure relief valves are required to prevents excessive pressure build-up.

Personal Protective Equipment

Approved protective equipment shall be installed on fuel-gas piping to prevent:
- Backflow of oxygen into the fuel-gas supply system;
- Passage of a flashback into the fuel-gas supply system; and
- Excessive back pressure of oxygen in the fuel-gas supply system.

The three functions of the protective equipment may be combined in one device or may be provided by separate devices.

Exhibit 16-6 illustrates the accepted location of approved protective equipment in fuel-gas piping systems. The protective equipment shall be located in the main supply line, as in Fig. Q-1 or at the head of each branch line as in Fig. Q-2 or withdrawn as shown in Fig. Q-3.

The first system has the most protection; that is, protective equipment is installed in the main supply and one check valve is installed before each outlet. System Q-2 has protective equipment in each branch circuit plus check valves at each outlet. System Q-3 has protective equipment at the fuel gas outlet and check valves at each outlet.

P_F—Protective equipment in fuel gas piping
V_F—Fuel gas station outlet valve
V_O—Oxygen station outlet valve
S_F—Backflow prevention device(s) at fuel gas station outlet
S_O—Backflow prevention device(s) at oxygen station outlet

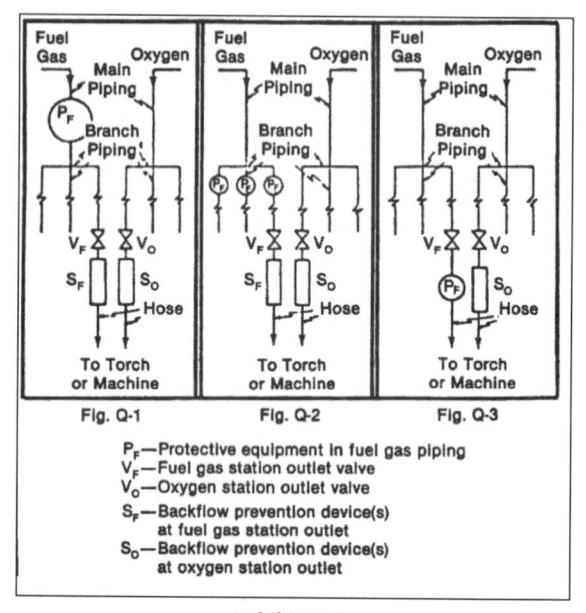

Exhibit 16-6.
Accepted locations of approved protective equipment in fuel-gas systems.

Hose and Hose Connections

Operators must use the proper hose. A fuel-gas hose is usually red (sometimes black) and has a left-hand threaded nut for connecting to the torch. An oxygen hose is green and has a right-hand threaded nut for connecting to the torch (See Exhibit 16-7.).

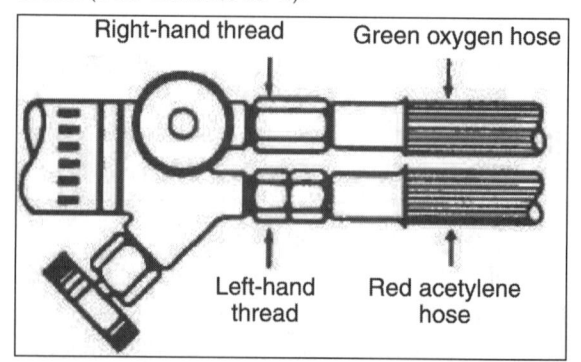

Exhibit 16-7.
Hose and hose connections are color-coded.

Hose and hose connections shall be clamped or otherwise securely fastened in a manner that will withstand, without leakage, twice the pressure to which they are normally subjected in service, but in no case less than a pressure of 300 psi. Oil-free air or an oil-free inert gas shall be used for the test.

Hoses showing leaks, burns, worn places or other defects rendering it unfit for service shall be repaired or replaced. When inspecting hoses, look for charred sections close to the torch. These may have been caused by flashback. Also check that hoses are not taped to cover leaks.

Pressure-Reducing Regulators

Pressure-reducing regulators shall be used only for the gas and pressures for which they are intended. When regulators or parts of regulators, including gauges, need repair, the work shall be performed by skilled mechanics that have been properly instructed. Most production shops do not have the proper equipment to make such repairs. For any equipment repairs or if there are questions about performance reliability, contact the manufacturer.

Gauges on oxygen regulators shall be marked "USE NO OIL."

Arc Welding and Cutting—1910.254

In the arc welding process, an electric current passing through the welding rod, or electrode, is forced to jump, or arc, across a gap. The resulting arc produces the intense heat necessary for the welding or cutting operation. Arc welding is used to fabricate nearly all types of carbon or alloy steels, the common nonferrous metals, and is indispensable in the repair and reclamation of metallic machine parts.

Arc cutting is primarily used for rough cuts or for scrapping because of the unevenness of the cut obtained. It has also been used for underwater cutting in salvaging operations. While most precautions and safe practices are common to oxygen-fuel gas welding, there are some that are unique to either gas or arc welding.

Shielding

It has long been known that welds will have better chemical and physical properties if the air can be kept away from the weld puddle. Such gases as oxygen, hydrogen, nitrogen and water vapor (i.e., moisture) all tend to reduce the quality of the weld. Dirt, dust, and metal oxides (i.e., contaminants) also reduce the weld quality. Shielding of the arc is normally provided in order to preserve the integrity of the weld joint. Shielding is provided either by decomposition of the electrode covering, known as the flux, or by a gas (or gas mixture) that may or may not be inert.

Common Arc Welding and Cutting Processes

Some of the more common arc welding and cutting processes are briefly discussed below.

Shielded metal arc welding (SMAW). This is the most widely used of arc welding, commonly referred to as "stick" welding. In this process, coalescence is achieved by heating with an electric arc between a covered (or coated) electrode and the work surface (See Exhibit 16-8.). Shielding is provided by decomposition of the electrode covering, known as the flux, while filler metal is obtained from the electrode's metal core.

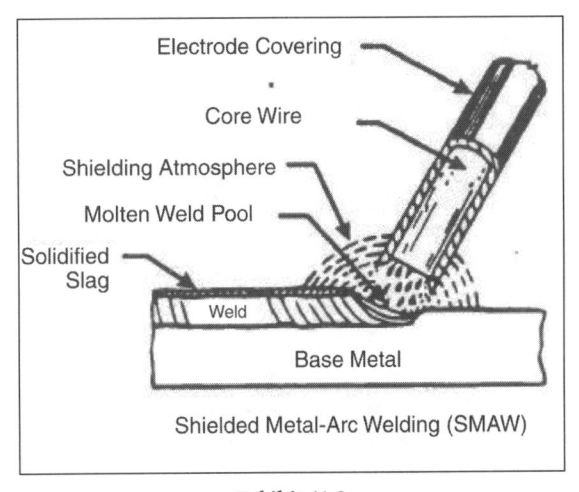

Exhibit 16-8.
Shielded Metal Arc Welding (SMAW).

Gas metal arc welding (GMAW). Commonly known as "MIG" welding. In the gas metal arc welding process, the heat of an electric arc maintained between the end of an electrode and the work surface achieves coalescence (See Exhibit 16-9.). Shielding of the arc is provided by a gas (or gas mixture) that may or may not be inert. The electrode is fed continuously to the weld where it is melted in the intense heat of the arc and deposited as weld metal.

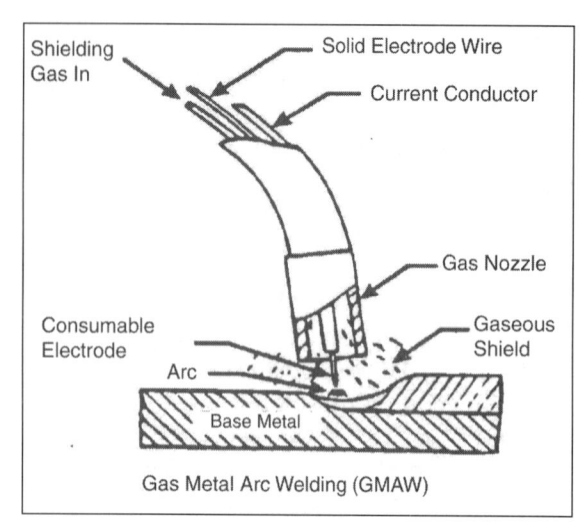

**Exhibit 16-9.
Gas Metal Arc Welding (GMAW).**

Gas tungsten arc welding (GTAW). Commonly known as "TIG" welding.

In gas tungsten arc welding, coalescence is achieved by the same arc and electrode method as in GMAW, except that the tungsten electrode is not consumed (See Exhibit 16-10.). Shielding is provided by an inert gas. This process offers high precision welds.

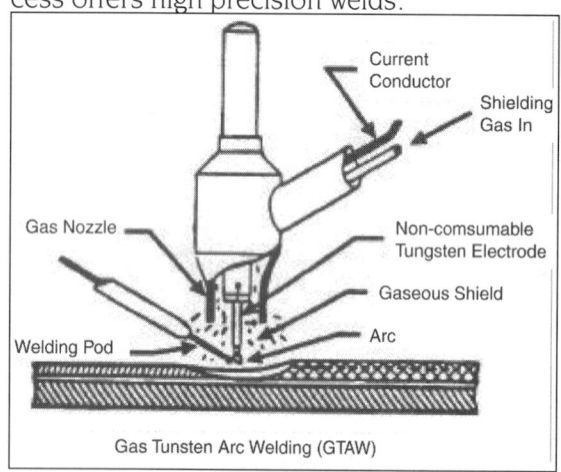

**Exhibit 16-10.
Gas Tungsten Arc Welding.**

Flux cored arc welding (FCAW). Flux cored arc welding is a process that produces coalescence by means of an arc between a continuous consumable electrode and the work surface. Shielding is provided by flux contained within the tubular electrode. Additional shielding may be obtained from a gas or gas mixture.

Submerged arc welding (SAW). In this process, heating produces coalescence with an arc between a bare metal electrode and the work surface. A blanket of granular, fusible flux shields the arc. The tip of the electrode and the welding zone are surrounded and shielded by the molten flux and a layer of unused flux in the granular state. In this process, there is no visible evidence of the passage of current between the electrode and the work surface. This eliminates the sparks, spatter and smoke ordinarily seen in other arc welding processes. Fumes are still produced, but not in quantities generated by other processes.

Arc cutting. Arc cutting is the general process in which the cutting or removal of metals is done by melting with the heat of an arc between an electrode and the base metal.

Plasma arc cutting (PAC). In plasma arc cutting, the metal is cut by melting a localized area with a constricted arc and removing the molten material with a high velocity jet of hot, ionized gas.

Air carbon-arc cutting (AAC). Air carbon-arc is a type of arc cutting in which the metal is cut by melting with the heat of an arc, with use of an air stream to facilitate cutting.

Arc gouging. Arc gouging is an application of arc cutting used to produce a groove or bevel in the metal surface (See Exhibit 16-11.).

Voltage

The following limits shall not be exceeded:

	Alternating Current (AC)	Direct Current (DC)
Manual	80 Volts	100 Volts
Automatic	100 Volts	100 Volts

For AC welding under wet conditions or warm surroundings where perspiration is a factor, the use of reliable automatic controls for reducing no-load voltage is recommended to reduce the hazard of shock. Some of the older AC machines do not have an automatic control and are on load all the time. It is easy to receive an electric shock when the equipment is not handled properly.

Installation

Grounding

The frame or case of the welding machine (except engine-driven machines) shall be grounded under the conditions and according to the methods prescribed in Subpart S, Electrical. Conduits containing electrical conductors shall not be used for completing a work-lead circuit. Pipelines shall not be used as a permanent part of a work-lead circuit, but may be used during construction, extension or repair, providing current is not carried through threaded joints, flanged bolted joints or caulked joints and that special precautions are used to avoid sparking at connection of the work-lead cable.

Operation and Maintenance

Machine Hook Up
Before starting operations, all connections to the machine shall be checked to make certain they are properly made. The work-lead shall be firmly attached to the work; magnetic work clamps shall be freed from adherent metal particles of spatter on contact surfaces. Coiled welding cable shall be spread out before use to avoid serious overheating and damage to insulation.

Electric Shock
Cables with splices within 10 ft. (3 m) of the holder shall not be used. Welders should not coil or loop welding electrode cable around parts of their body.

Maintenance
Cables with damaged insulation or exposed bare conductors shall be replaced. Joining lengths of work and electrode cables shall be done by the use of connecting means specifically intended for the purpose. The connecting means shall have insulation adequate for the service conditions.

Resistance Welding—1910.255
A qualified electrician in conformance with Subpart S, Electrical, shall install all equipment.

Spot and Seam Welding Machines

Interlocks
All doors and access panels of all resistance welding machines and control panels shall be kept locked and interlocked to prevent access by unauthorized persons to live portions of the equipment.

Guarding
All press welding operations, where there is a possibility of the operator's fingers being under the point of operation, shall be effectively guarded by the use of a device such as an electronic eye safety circuit, two hand controls or protection similar to that prescribed for punch press operations.

Shields
Installing a shield guard of safety glass or suitable fire-resistant plastic at the point of operation shall, wherever practical, eliminate the hazard of flying sparks. Additional shields

COMMON WELDING PROCESSES					
Process	**Used For**	**Shielding**	**Smoke/Fume**	**Light/Radiation**	**Specific Hazard**
Stick (SMAW)	General purpose; more than 50% of all welding	Electrode coating	High	moderate-variable	Depends on electrode, e.g., low hydrogen-flourides, stainless steel-nickel
MIG (GMAW)	High production/ automation	Inert gas (argon, helium, CO_2)	Moderate	High-especially with reflective metal and argon shield	Ozone, CO (CO_2 shield), stainless steel-chromium and nickel
TIG (GTAW)	High precision	Inert gas	Low	High	Ozone, light
Plasma (PAW)	Process can be used to weld, cut, metal spray	Gas	Moderate/high	Moderate/high	Noise, electrical shock, potential X-radiation
Sub Arc (SAW)	Horizontal welds, high production	Granular flux	Low	No visible arc unless have "breathrough"	Generally low hazard
Air Arc (AAW)	"Gouging," weld preparation	None	Very high	High	Noise, high fume levels
Flux Core (FCW)	High production/ automation, e.g., MIG with flux filled wire	Wire filling with or without gas	High	Moderate/high	High fume levels
Oxyacetylene (OAW)	Thin to medium thickness metals, steel and non-ferrous in all positions	Filler rod coating	Low/moderate	Low	Compressed gas cylinders; depends on filler rod, e.g., silver brazing-cadmium

WELDING, CUTTING, AND BRAZING

or curtains shall be installed as necessary to protect passing persons from flying sparks.

Foot Switches
All foot switches shall be guarded to prevent accidental operation of the machine.

Stop Buttons
Two or more safety emergency stop buttons shall be provided on all special multi-spot-welding machines, including two-post and four-post weld presses.

Portable Welding Machines

Safety Chains
All portable welding guns, transformers and related equipment suspended from overhead structures, beams, trolleys, etc., shall be equipped with safety chains or cables. Safety chains or cables shall be capable of supporting the total shock load in the event of failure of any component of the supporting system.

Welding Health Hazards

I. CHEMICAL AGENTS

ZINC. Zinc is used in large quantities in the manufacture of brass, galvanized metals, and various other alloys. Inhalation of zinc oxide fumes can occur when welding or cutting on zinc-coated metals. Exposure to these fumes is known to cause metal fume fever. Symptoms of metal fume fever are very similar to those of common influenza. They include fever (rarely exceeding 102°F), chills, nausea, dryness of the throat, cough, fatigue, and general weakness and aching of the head and body. The victim may sweat profusely for a few hours, after which the body temperature begins to return to normal. The symptoms of metal fume fever have rarely, if ever, lasted beyond 24 hours. The subject can then appear to be more susceptible to the onset of this condition on Mondays or on weekdays following a holiday than they are on other days.

CADMIUM. Cadmium is used frequently as a rust-preventive coating on steel and also as an alloying element. Acute exposures to high concentrations of cadmium fumes can produce severe lung irritation, pulmonary edema, and in some cases, death. Long-term exposure to low levels of cadmium in air can result in emphysema (a disease affecting the ability of the lung to absorb oxygen) and can damage the kidneys. Cadmium is classified by OSHA, NIOSH, and EPA as a potential human carcinogen.

BERYLLIUM. Beryllium is sometimes used as an alloying element with copper and other base metals. Acute exposure to high concentrations of beryllium can result in chemical pneumonia. Long-term exposure can result in shortness of breath, chronic cough, and significant weight loss, accompanied by fatigue and general weakness.

IRON OXIDE. Iron is the principal alloying element in steel manufacture. During the welding process, iron oxide fumes arise from both the base metal and the electrode. The primary acute effect of this exposure is irritation of nasal passages, throat, and lungs. Although long-term exposure to iron oxide fumes may result in iron pigmentation of the lungs, most authorities agree that these iron deposits in the lung are not dangerous.

MERCURY. Mercury compounds are used to coat metals to prevent rust or inhibit foliage growth (marine paints). Under the intense heat of the arc or gas flame, mercury vapors will be produced. Exposure to these vapors may produce stomach pain, diarrhea, kidney damage, or respiratory failure. Long-term exposure may produce tremors, emotional instability, and hearing damage.

LEAD. The welding and cutting of lead-bearing alloys or metals whose surfaces have been painted with lead-based paint can generate lead oxide fumes. Inhalation and ingestion of lead oxide fumes and other lead compounds will cause lead poisoning. Symptoms include metallic taste in the mouth, loss of appetite, nausea, abdominal cramps, and insomnia. In time, anemia and general weakness, chiefly in the muscles of the wrists, develop. Lead adversely affects the brain, central nervous system, circulatory system, reproductive system, kidneys, and muscles.

FLUORIDES. Fluoride compounds are found in the coatings of several types of fluxes used in welding. Exposure to these fluxes may irritate the eyes, nose, and throat. Repeated exposure to high concentrations of fluorides in air over a long period may cause pulmonary edema (fluid in the lungs) and bone damage. Exposure to fluoride dusts and fumes has also produced skin rashes.

CHLORINATED HYDROCARBON SOLVENTS. Various chlorinated hydrocarbons are used in degreasing or other cleaning operations. The vapors of these solvents are a concern in welding and cutting because the heat and ultraviolet radiation from the arc will decompose the vapors and form highly toxic and irritating phosgene gas. (See Phosgene.)

PHOSGENE. Phosgene is formed by decomposition of chlorinated hydrocarbon solvents by ultraviolet radiation. It reacts with moisture in the lungs to produce hydrogen chloride, which in turn destroys lung tissue. For this reason, any use of chlorinated solvents should be well away from welding operations or any operation in which ultraviolet radiation or intense heat is generated.

CARBON MONOXIDE. Carbon monoxide is a gas usually formed by the incomplete combustion of various fuels. Welding and cutting may produce significant amounts of carbon monoxide. In addition, welding operations that use carbon dioxide as the inert gas shield may produce hazardous concentrations of carbon monoxide in poorly ventilated areas. This is caused by a "breakdown" of shielding gas. Carbon monoxide is odorless, colorless and tasteless and cannot be readily detected by the senses. Common symptoms of overexposure include pounding of the heart, a dull headache, flashes before the eyes, dizziness, ringing in the ears, and nausea.

OZONE. Ozone (O_3) is produced by ultraviolet light from the welding arc. Ozone is produced in greater quantities by gas metal arc welding (GMAW or short-arc), gas tungsten arc welding (GTAW or heli-arc), and plasma arc cutting. Ozone is a highly active form of oxygen and can cause great irritation to all mucous membranes. Symptoms of ozone exposure include headache, chest pain, and dryness of the upper respiratory tract. Excessive exposure can cause fluid in the lungs (pulmonary edema). Both nitrogen dioxide and ozone are thought to have long-term effects on the lungs.

NITROGEN OXIDES. The ultraviolet light of the arc can produce nitrogen oxides (NO, NO_2), from the nitrogen (N) and oxygen (O_2) in the air. Nitrogen oxides are produced by gas metal arc welding (GMAW or short-arc), gas tungsten arc welding (GTAW or heli-arc), and plasma arc cutting. Even greater quantities are formed if the shielding gas contains nitrogen. Nitrogen dioxide (NO_2), one of the oxides formed, has the greatest health effect. This gas is irritating to the eyes, nose and throat but dangerous concentrations can be inhaled without any immediate discomfort. High concentrations can cause shortness of breath, chest pain, and fluid in the lungs (pulmonary edema).

II. PHYSICAL AGENTS

ULTRAVIOLET RADIATION. Ultraviolet radiation (UV) is generated by the electric arc in the welding process. Skin exposure to UV can result in severe burns, in many cases without prior warning. UV radiation can also damage the lens of the eye. Many arc welders are aware of the condition known as "arc-eye," a sensation of sand in the eyes. This condition is caused by excessive eye exposure to UV. Exposure to ultraviolet rays may also increase the skin effects of some industrial chemicals (coal tar and cresol compounds, for example).

INFRARED RADIATION. Exposure to infrared radiation (IR), produced by the electric arc and other flame cutting equipment may heat the skin surface and the tissues immediately below the surface. Except for this effect, which can progress to thermal burns in some situations, infrared radiation is not dangerous to welders. Most welders protect themselves from IR (and UV) with a welder's helmet (or glasses) and protective clothing.

INTENSE VISIBLE LIGHT. Exposure of the human eye to intense visible light can produce adaptation, pupillary reflex, and shading of the eyes. Such actions are protective mechanisms to prevent excessive light from being focused on the retina. In the arc welding process, eye exposure to intense visible light is prevented for the most part by the welder's helmet. However, some individuals have sustained retinal damage due to careless "viewing" of the arc. At no time should the arc be observed without eye protection.

Care of Gas Cylinders

Acetylene

Acetylene is a compound of the elements carbon and hydrogen. At atmospheric temperatures and pressures, acetylene is a colorless gas that is slightly lighter than air. Acetylene of 100% purity is odorless, but gas of ordinary commercial purity has a distinctive, garlic-like odor. In air at atmospheric pressure, the upper explosive limit is about 80% acetylene and 20% air. The lower explosive limit is 2.5% acetylene in air.

The maximum safe working pressure of acetylene is 15 psig, yet most gauges available in the field permit higher unsafe pressures. The hydrogen and carbon components of acetylene will come apart (dissociate) explosively when pressures are too high. At 1435°F, or at 30 psig, acetylene can explode by itself.

Acetylene cylinders avoid the decomposition characteristics of the gas by providing a porous mass packing material having minute cellular spaces so that no pockets of appreciable size remain where "free" acetylene in gaseous form can collect. This porous filler is saturated with liquid acetone. Acetone has the ability to hold 400 times its own volume of acetylene at the working pressure of 250 psig. The acetylene is dissolved within the acetone, making it stable. A small air space at the top of the cylinder is the only place where gas is permitted to collect. Acetylene cylinders should always be stored, transported and used in a vertical position. If the cylinder is on its side, acetone may leak out and dissolve the rubber hose, permitting a dangerous build-up of acetylene gas in the atmosphere. Another potential hazard exists during heavy cutting or use of large tips. When demands exceed 42 cu. ft. per hour, acetone may be sucked out and burned. Under these conditions, some of the acetylene gas may come free inside the cylinder, creating a potential hazard. Heavy gas volume demands require that more than one tank be manifolded together. Appropriate reverse flow check valves and flashback arresters add safety to any system and become important whenever cylinders are manifolded together.

Oxygen

Oxygen cylinders are pressurized to 2200 psig. If the valve protection cap is not in place and the cylinder is not secured, there is potential for damage to the valve stem. The fastest drag racer can reach 30 mph in 1 to 2 seconds. A compressed oxygen cylinder can reach 30 mph in less than 0.01 second. This is enough momentum to make it rocket through a cinder block wall over 200 ft. away. Consequently, the valve stem must be protected at all times.

An employee's clothing can become saturated with oxygen if he or she uses the torch to blow dust off of his or her clothing. A lit cigarette or any accidental spark will then cause that employee to catch fire. Do not permit oil or grease to come into contact with (or even near) regulators, hoses or torches for oxygen. Oxygen and oils or greases are an explosive combination.

Cylinder Storage For Acetylene

Acetylene gas may be stored in cylinders specifically designed for this purpose. The gas is first passed through filters and purifiers and then compressed into cylinders to a pressure of approximately 250 psig. The storage of acetylene in its gaseous form under pressure is not safe at pressures above 15 psig. The method used to safely store acetylene in cylinders follows:

The cylinders are filled with a substance such as pith from cornstalk, fuller's earth, lime silica or similar substances that absorb acetone. The cylinders are then charged with acetone that absorbs acetylene. The theory is that the acetylene molecules fit in between the acetone molecules. Using both of these techniques prevents the accumulation of a pocket of high-pressure acetylene.

Safety fuse plugs have a metal center that will melt at a temperature of approximately 212°F. If the cylinder should be subjected to a high temperature, the plugs will melt and allow the gas to escape before the pressure builds up enough to burst the cylinder. These precautions are necessary as the pressure in an acetylene cylinder builds up rapidly with an increase of temperature.

Acetylene cylinder valves come in two types. A common type is provided with a 3/8-in. square shank. It is turned by means of a

3/8-in. square box socket wrench. It is recommended that this cylinder valve be opened only 1/4 to 1-1/2 turns. The wrench should be left on the valve stem at all times that the valve is open so that the valve may be closed quickly in emergencies.

Another type of acetylene cylinder valve is fitted with a hand wheel. The regulator fitting is a female fitting.

The pressure in the cylinder cannot estimate the amount of acetylene in a cylinder because the gas discharge pressure will remain fairly constant (depending on the temperature) until most of the gas is consumed.

Oxygen Cylinder—Fuel-Gas Cylinder
Do not drop or strike an electric arc against a cylinder or heat it with an open flame. Recap it whenever gauges and regulators are removed. Never transport a cylinder without safeguards against shifting or falling. Cylinders should not be allowed to stand alone without being secured with lashing or chain to prevent them from toppling over. Avoid laying an acetylene cylinder on its side. Never use choker slings or magnets, nor hook into the protection cap to hoist a cylinder. Instead, use "racks" or "baskets."

Regulator (Oxygen And Fuel Gas)
Inspect to be sure there are no ignition sources in the immediate area; then "crack" the valve to blow out any dust or dirt before connecting the regulator. When setting up a rig, be sure to install a regulator in the line. Screw the regulator stem all the way out before placing pressure on the regulator. Before removing the regulator, close the cylinder valve and vent regulator of any residual gas.

Line And Cylinder Pressure Gauges (Oxygen And Fuel Gas)
Make sure these gauges are made up properly the oxygen gauge to the oxygen cylinder and the fuel-gas gauge to the fuel-gas cylinder. The difference in the thread sizes will prevent exchanging the connections. Observe each gauge frequently to note any trouble, such as unusual movement or sticking of the pointer. When any trouble sign is noted, close the cylinder valve and discontinue use of the rig until the trouble is located and corrected. Never lubricate gauges, regulators, valves, connections or open valves while wearing greasy gloves. Lubricant can react violently with oxygen. Note: The gauge for acetylene line pressure should register less than 15 psi; higher pressures are unstable.

Cylinder Cap Threads
Avoid striking or scoring the cylinder cap threads or distorting the caps or they will not screw on properly. Never lubricate them with oil or grease.

Cylinder Shut-off Valves (Oxygen And Fuel Gas)
If you hear a leak around the stem after opening the valve, try to stop the leak by tightening the gland nut. If you are not immediately successful, close the valve and tag the cylinder. Then have the cylinder removed from service. The wrench used to open a cylinder valve should be left in position on the stem to permit quick shut-off of the gas flow in case of emergency. For manifold or coupled cylinders, at least one such wrench should be available for immediate use.

Torch
Before beginning work, check torches for leaks in shut-off valves, hose couplings or tip connections. Never use a defective torch. Inspect torch tips for clogging. Clear blockages with proper cleaning tools before beginning work. Never light a torch from hot work or with matches; use a friction lighter.

Torch Shut-off Valves (Oxygen And Fuel Gas)
Inspect frequently for leaks. Do not depend on the valves to prevent gas leakage when torch is left unattended at the worksite for extended periods of time. Always remove torch and hose from enclosed spaces when work is interrupted or concluded. As an additional precaution, close shut-off valves on cylinders.

Hose (Oxygen And Fuel Gas)
Hoses, probably the most fragile part of any burning or welding system, should be inspected frequently, especially before being used. Never use a system that does not distinguish (by color or other means) between the hose carrying oxygen from the one carrying fuel-gas. Couplings are designed so as not to

be interchangeable; fuel-gas couplings are left-hand threaded, while oxygen hose couplings are right-hand threaded.

Never use a hose that carries fuel-gas and oxygen simultaneously (having a common wall construction to separate them). Should a breakdown occur in the separation wall, the two gases could mix, with serious results. Connections that can be joined by being pushed or separated by a straight pull motion should not be used. Only couplings that require a rotary motion to be joined should be used. Pressure test any hose that has a flashback or that shows excessive wear upon inspection.

Fuel-gas and oxygen hoses that have been disconnected from the torch or other gas-consuming device should be removed immediately from a confined space.

OSHA Instruction STD 1-13.1 October 30, 1978 Office of Program Operations
February 14, 1972
OSHA PROGRAM DIRECTIVE #100-1

TO: National and Field Offices
SUBJECT: Reduction of Air Pressure Below 30 psi for Cleaning Purposes
ATTACHMENT: Acceptable Methods for Complying with 41 CFR 50.204.8 and 29 CFR 1910.242(b)

1. Purpose. To provide guidance and examples of what alternate systems will meet the requirements of this section, and to clarify its intent.
2. Background. A number of inquiries have been received requesting a clarification of the meaning of 1910.242(b) also known as 41 CFR 50-2048 under the Walsh-Healey Act.
3. Interpretation. The phrase "reduce to less than 30 psi" means that the downstream pressure of the air at the nozzle (nozzle pressure) or opening of a gun, pipe, cleaning lance, etc., used for cleaning purposes will remain at a pressure level below 30 psi for all static conditions. The requirements for dynamic flow are such that in the case when dead ending occurs a static pressure at the main orifice shall not exceed 30 psi. This requirement is necessary in order to prevent a back pressure buildup in case the nozzle is obstructed or dead ended. See enclosure (1) for two acceptable methods of meeting this requirement. Also, there is no intent to restrict the diameter of the nozzle orifice or the volume (CFM) flowing from it.

    "Effective chip guarding" means any method or equipment which will prevent a chip or particle (of whatever size) from being blown into the eyes or unbroken skin of the operator or other workers. Effective chip guarding may be separate from the air nozzle as in the case where screens or barriers are used. The use of protective cone air nozzles are acceptable in general for protection of the operator but barriers, baffles or screens may be required to protect other workers if they are exposed to flying chips or particles.
4. Action. Inquiries about subject section should be handled in accordance with this instruction.
5. Effective Date. This instruction is effective immediately, and will remain in effect until canceled or superseded.

Director, of Program Operations

DISTRIBUTION: National Field Office A/SEC. (3) Regional Administrators (6) Dep. A/Sec (2) Area/District Offices (3) Spec. Asst. (1) Training Institute (1) Directors (3) RAO (2) SOL (1) Professional Staff (1) BLS (1) Review Commission (6)
Originator: OCSDG

2004 Subpart Q 1910.251-255
Welding, Cutting and Brazing

Citation	Description	Count
253(b)(4)(iii)	Oxygen cylinder storage- separation	254
252(b)(2)(ii)	Cylinder storage inside buildings	101
253(b)(2)(Iii)	Protection from arc welding rays	74
253(b)(2)(iv)	Valve Caps in place	52
254(d)(9)(iii)	Arc Welding/cutting- cable maintenance	48

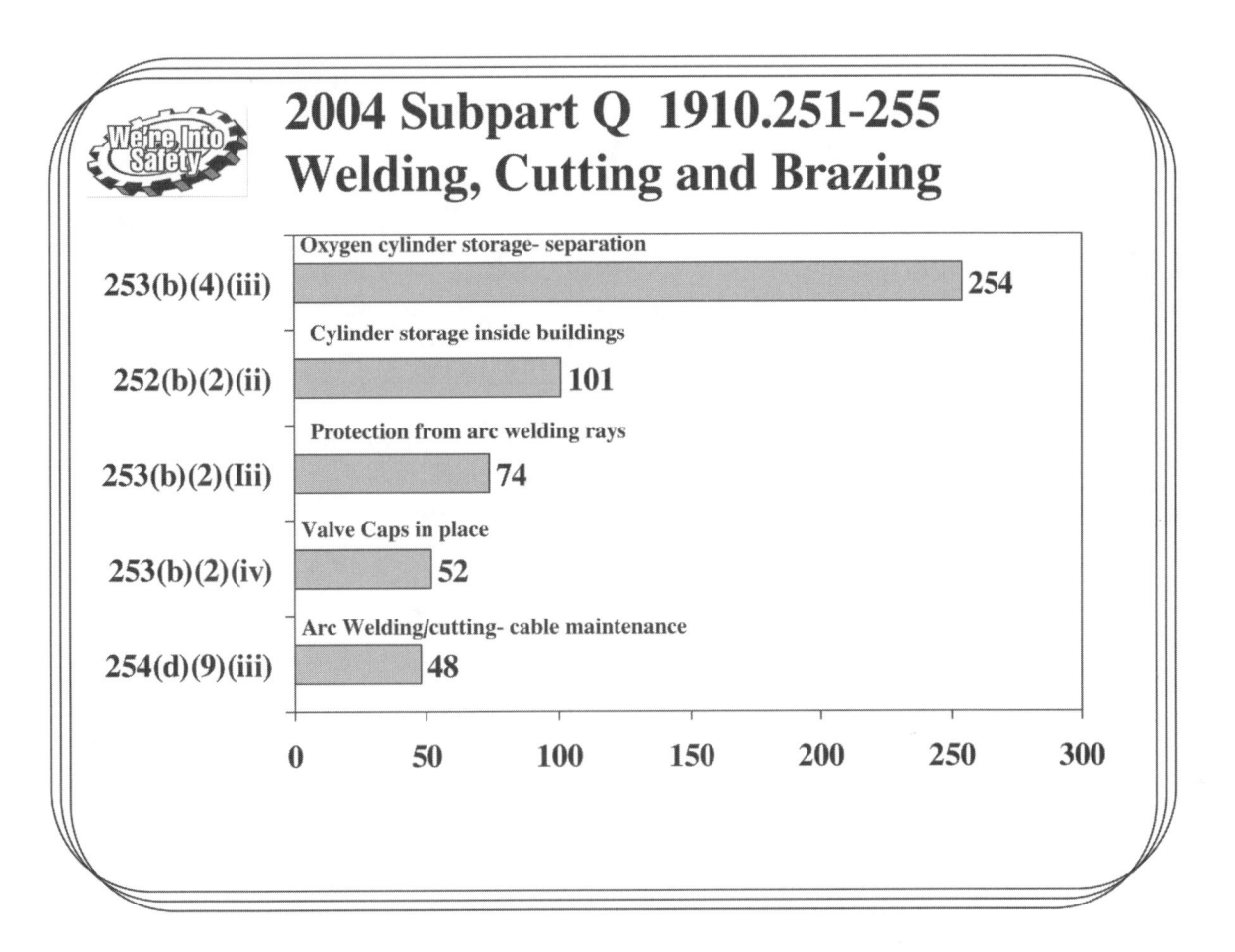

2004 Subpart M 1910.166-169
Compressed Gas & Air Equipment

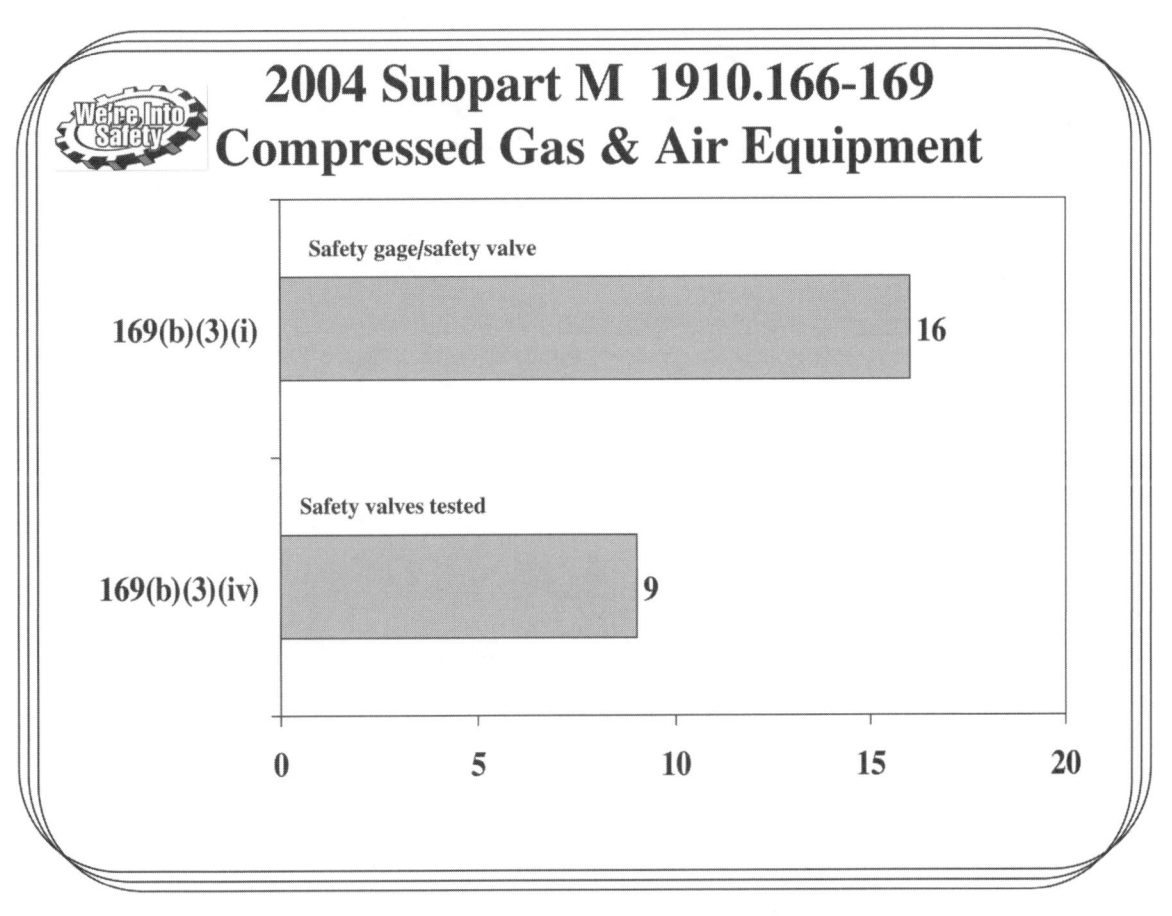

Citation	Description	Count
169(b)(3)(i)	Safety gage/safety valve	16
169(b)(3)(iv)	Safety valves tested	9

WELDING, CUTTING, AND BRAZING

CHAPTER 17

SAFETY PROGRAMS

To effectively control hazards on work sites, companies recognize the need for a safety and health program. This program addresses work-related hazards including *potential* hazards that can result from changes in work conditions or practices whether or not they are regulated by government standards. A good safety and health program makes a big impact on loss prevention, worker productivity and the bottom line of a company.

OSHA has found that those companies with an effective safety and health program are rewarded with fewer and less severe accidents. Employee morale and productivity also are increased, and there is a significant reduction in workers' compensation costs and other costs associated with on the job injuries. The key components of any safety and health program consist of:

- Hazard analysis—Assessment of the hazard;
- Hazard prevention—Steps taken to keep workers safe;
- Polices and procedures—Rules that can be used by workers and subcontractors;
- Employee training—Type and frequency of training for all workers;
- Follow-up inspections—Audits or walk around inspections of the work site; and
- Enforcement—What specific steps are taken when a worker violates a safety rule.

A safety and health program describes the policies, procedures and practices that an organization uses to provide effective safety protection on the job. These guidelines are normally performance oriented. A large number of worksites lack the professional resources to develop this type of program on their own. Most small and medium-sized business do not have a full time safety staff to create such guidelines and this task often falls to a worker who picks up the additional job as a "collateral duty." In these cases it is even more important to have a written document to guide the safety program.

The safety program should spell out the authority, responsibility and accountability of all parties concerned. It also should be documented in writing to prevent confusion and uncertainty. Understanding the practices can be too easily confused considering, personnel turnover, cultural differences, and other factors that influence the work performance on any job. Written guidance is needed to cover basic polices, practices, procedures, emergency plans, posted signs, performance objectives and disciplinary actions.

An effective program includes provisions for the systematic identification and control of hazards in the workplace. The system must go beyond mere regulatory compliance to address all hazards.

Safety programs can be described as a system. Individual parts of the system work together to achieve a common goal. Companies often develop a baseline comprehensive worksite audit and then perform surveys on a routine basis to update that audit. Safety teams or committees work with those audits to correct unsafe conditions. Management and employees must work together to make the system work and they should review the safety program annually to make sure it still fits the needs of the organization.

Major Elements of a Safety Program

OSHA established voluntary guidelines for safety programs in 1989. These rules are guidelines and are not enforceable by OSHA in the general industry sector. However, they have been used by successful companies to develop programs that have reduced their number of accidents significantly. Firms that are involved with the Voluntary Protection Program consistently report that safety programs that follow federal guidelines work very well. There are common elements in any safety program, including:
- Commitment by management;
- Involvement of the workforce;
- Analysis of the work area to identify and control hazardous situations;
- Safety training; and
- Enforcement of the rules.

Management Commitment and Employee Involvement

Safety is a team effort where management is convinced that on-the-job safety is just as important as productivity and cost control. Managers demonstrate their safety leadership by communicating freely with their staff on safety matters. The legal responsibility for safety and health lies with the employer. The employer establishes the program and is responsible for decision making in the facility.

A clearly stated safety policy involves top management and is understood by all workers on a site. An example of demonstrated commitment on the part of management can be found in a company letter from the senior manager explaining that safety is of the utmost importance and is a requirement for the job. The person in charge of a facility often signs a policy statement on safety and health protection.

The role of management in the safety system is to create the needed resources so that responsible workers have the authority and ability to react to safety needs. Management must also hold supervisors accountable for enforcing the safety standards.

Individual workers have a right to a safe and healthy workplace. In successful safety programs, individuals feel that they are involved in the overall safety culture, accept responsibility for their workplace and realize that they have an active role in the safety of others. Safety programs should have committees that meet periodically. Hazard analysis teams or other worker-led groups should assist in inspections or the revision of safety rules as well as accident investigations. This will encourage employees to report workplace conditions that appear to be hazardous.

When workers are involved in decisions that affect safety the results are better management decisions, more effective safety protection and increased employee support for the safety program. A work site that solicits safety suggestions is a better place to work. This type of solicitation should be systematic and workers should feel that they are protected from reprisal for reporting an unsafe condition, and they should be provided with the appropriate response to their suggestion in a timely fashion. Workers will support a company where they know that management wants to be made aware of safety issues and will take action to correct them.

A key part of any program is fixing responsibility at the correct level. Both management and workers share responsibility for safety. Management establishes and enforces the rules and provides financial support for safety training and equipment. When the disciplinary procedures are fairly enforced and clearly understood, and management is held accountable for enforcing those standards, there is little opportunity to push workers into taking short cuts. Management must be held accountable for enforcing safety rules.

Workers are required to follow the rules. They must understand that there is no tolerance for unsafe behavior. Discipline is an indispensable piece of the whole approach to safety and health protection. Safe work procedures must be enforced and must cover all personnel, from site manager to hourly employee.

Accountability can be demonstrated by a written human resource policy that provides for corrective actions to be taken when a safety rule is ignored. This may involve a verbal reprimand, a written reprimand, suspension without pay or even termination. A clearly communicated system of progressive discipline lets all workers and supervisors know what is expected of them as well as the consequences for failing to meet those expectations. By using readily available counseling forms from the Human Resources department to accomplish counseling for safety misconduct, the supervisors reinforce the company's commitment to safety.

Work Site Analysis

Companies should examine the work site to identify existing hazards or conditions that could create hazards. Dangerous operations should be analyzed in such a way that hazards can be anticipated. This involves time and resources that must be approved by management. Safety committees, hazard evaluation teams, insurance carriers or outside consultants can assist in a workplace analysis. The people performing the inspections should be familiar with the safety program and applicable regulations for the work site.

The analysis can take the form of a comprehensive walk-through audit to establish a baseline for future safety audits. Some companies conduct a Job Hazard Analysis that involves reviewing each step of a job in a systematic method to determine the safest way to perform that task. Safety inspections should be performed on a periodic basis. A weekly inspection is recommended, although some companies perform a walk-around inspection every other week. Checklists can be developed that identify regulations that apply to the operation or safety challenges that have developed in the past. A review of accident records or injury reports often provides an insight into hazardous operations. The safety team should review new facilities and processes, materials, and equipment that may be brought into the facilities. The team should also look into accidents and even "near-miss" incidents to find ways to prevent future accidents from occurring.

Workers who report safety problems should be able to notify management or committees about unsafe conditions without fear of reprisal. They should receive feedback in a timely manner and should be encouraged to use the system.

Hazard Prevention and Control

Armed with a list of hazardous operations, the employees can begin a systematic method of controlling the hazards. Effective design or engineering controls often eliminate the hazard at the source. The most effective method of controlling hazards is to establish an engineering control. If that is not feasible, some form of administrative control, personal protective equipment, or another form of work safety practice may be used. Such things as installing a guardrail or electrical insulation often are cost-effective ways of eliminating the hazard. This must be done in a timely manner to minimize the exposure of workers to the danger.

If the hazard cannot be eliminated, workers may have to use some form of hazard control, such as using a spotter or wearing personal protective equipment to continue to work in a safe manner. Sometimes the hazard is the result of poor maintenance or use or abuse of the facility and normal maintenance procedures correct the problem. In other cases, an emergency event such as the release of chlorine gas or reaction to a fire require special training for the workers.

The best preventive measure for controlling unsafe conditions is to create an engineered solution, such as a guardrail or a physical system to protect the worker. Sometimes a feasible physical solution cannot be found. In these cases a procedure for safe work can be established. These procedures may not be as effective at ensuring a safe workplace as building engineered controls, but with the right type of training, positive reinforcement and correction of unsafe worker performance, the workplace can be made significantly safer. Firms should consider the use of a clearly communicated disciplinary system to enforce worker safety performance. In the event that the unsafe act cannot be engineered to a safe degree, or procedures do not provide sufficient safety, the firm may rely on personal protective equipment that can protect the workers. If none of these protective systems are deemed sufficient, the firm may rely on administrative controls to maintain safety. In this event, the facility may rotate workers through the hazardous area in order to minimize the amount of time that the employees are exposed to a hazard.

For example, imagine there is a wastewater treatment facility that has a lot of toxic gases being vented into an area. The first thing to do is try an engineering control, such as installing a ventilation system. If that is not feasible, the second best system is to develop procedures for safe work. In this case, workers may be required to manually vent the area prior to entering the work area. If that procedure is not feasible, the company may elect to use personal protective equipment and require the workers to wear respirators when they are in that area. Finally, if none of those solutions are feasible, the firm may rotate workers through that area so that no single worker is exposed to toxic gases for more than 30 minutes at a time, based on the maximum amount of exposure listed for short-term exposures on the material safety data sheet.

Safety and Health Training

Training is a key element in any business. In order to maximize the human element, workers must be trained in new and productive techniques. Maintaining a safety system is no different than maintaining a production system. Some training is required by federal or state regulations, while other safety training topics enhance the overall safety of the job.

The safety program should identify the types of training required, the personnel who need the training and the frequency of that training. A test should be given to ensure that workers understand the training material. This test can be a written quiz or a hands-on demonstration. It is important to provide sufficient training for safety so that responding to emergencies becomes second nature to the workers. Workers must understand the hazards to which they may be exposed and the training must ensure that they understand how to prevent themselves from getting injured as a result of those hazards.

Training can easily be broken into two areas: initial training and refresher training. When new employees are brought into the job site, they must be taught about the hazards of their new position. If a new machine or process is brought into place, the workers must receive initial training so that they can perform their required tasks safely.

Once they have been on the job for a while, refresher training helps to improve their performance and update the worker regarding any changes in policy or procedures.

Workers need to understand the hazards associated with their jobs, how to identify hazards and how to use the safety equipment that is provided to them. This training should be incorporated into their other training. Contractors and short-term workers must receive orientation training to prepare them to work in the area safely in an environment that may be new to them or hazards that they normally do not encounter. All workers must be trained on how to report incidents and accidents that may require posting in the OSHA 300 log.

Supervisors must understand company polices and procedures, hazard detection and control, how to conduct accident investigations, how to handle emergencies and how to reinforce training. Management training should be different than training that is provided to new hires, contract workers and other employees. It is important to stress to management that their role is to guide the safety program and provide support so that the workers can accomplish the job in a safe and efficient manner.

Elements of a Written Safety Program

These safety programs can be published as a written document and stored in a three-ring binder for easy updating. There are some topics that are common in any safety and health program. The basic plan sets the safety policy of the company and establishes responsibility, accountability and

disciplinary procedures. Some topics included in the basic plan are:
- Safety policy
- Management responsibilities
- Supervisor responsibilities
- Employee responsibilities
- Safety committee or safety team (employee participation)
- Safety meetings
- Hazard recognition
- Incident investigation
- Elimination of workplace hazards
- Basic safety rules
- Job related safety rules
- Disciplinary policy
- Emergency planning
- Reporting accidents and incidents (OSHA 300 log)
- Training
  - Initial (new employee)
  - Refresher
  - Specialized
  - Supervisory

Appendices can then be added to the basic program to provide specific guidance on site-specific hazards that workers will encounter. This allows a company or a facility to personalize the plan to handle situations that may be unique to that area. Some examples of appendices that may be added to the safety program are:
- Personal protective equipment hazard analysis and program
- Respirator program
- Lockout/tagout (control of hazardous energy)
- Permit-required confined space entry
- Electrical safety
- First-aid plan
- Hazard communication program
- Hearing conservation program
- Welding hot work program
- Bloodborne pathogen plan
- Violence in the workplace plan
- Emergency action plan
- Hazardous waste operations plan

State Programs

States often develop regulations that provide safety for their workers that are as effective or even more so than the federal regulations. Two such states are Washington and California. California enacted Senate Bill 198, a worker safety bill that applies to all employers in the state of California. Washington State requires that all companies have an "Accident Prevention Program." Both states require that the document be written and meet certain criteria.

California Requirements

The State of California passed S.B. 198 requiring that all California employers have a safety program. California compares workers' compensation data to target employers for safety inspections. S.B. 198 requires employers to:
- Designate a responsible person;
- Periodically conduct inspections and accident investigations, as necessary;
- Identify workplace hazards;
- Correct safety deficiencies or hazards;
- Train employees;
- Establish a system for employee communication; and
- Establish a disciplinary system.

California requires that a person be designated as a "responsible person" who has the authority to allocate resources of the company to ensure compliance with the rules and regulations. This person should be at the management level in order to be effective. In order to ensure accountability for the program, this responsible person is personally liable and if it is found that he or she concealed danger in a product or business practice, he or she can incur fines and imprisonment.

Any time new equipment is installed or new material or chemicals are introduced to a work area, workers must be trained on those hazards. A list of workplace hazards must be maintained. The list should include such recognized hazards as employees in a company vehicle, operation of dangerous machinery, exposure to electricity, air quality, environmental noise, walking and working surfaces, chemical fumes, mists, vapors, and smoke.

Hazards must be corrected on a timely basis. Employers should identify every possible hazard and establish a priority to correct the deficiency based on the severity of the hazard. Those hazards that can kill, maim, or cause injury to people should be corrected first. If the hazard cannot be eliminated, the area or equipment must be clearly labeled with a sign that informs employees of the danger involved. All employees who might enter the danger area must be given instructions about the hazard. Both the corrective action and the employee training must be documented.

Trained personnel must conduct periodic inspections on a scheduled basis. The person conducting the inspection must be able to identify new potential hazards or investigate workplace safety. Those unsafe conditions and practices that can be corrected on the spot should be immediately corrected. It is a good idea to document these inspections and the corrective actions that result from them.

Before any worker is assigned to a job, he or she must be trained on the general safety procedures for that job. This includes workers on their initial assignment to the work area and workers who are transferred to new positions or the acquisition of new equipment. California requires that this training be accomplished by a trainer and documented. Workers may not be "self-taught" on the job.

Refresher training also should be provided to ensure proficiency in the safety area. Safety training should be incorporated into regular department meetings and routine employee training. Some training topics to regularly cover include:
- The location and operation of firefighting equipment;
- Proper lifting techniques;
- Defensive driving techniques;
- Chemical hazards;
- Electrocution hazards; and
- General housekeeping, which can help reduce slip and fall hazards.

California requires that employers and employees communicate with one another. This communication can be in the form of posters, meetings, training sessions, and labor unions. The intent of the communication is to ensure workers have a way to inform management of unsafe conditions so management can correct them. A company discipline policy for safety violations also must be established, posted, documented, and enforced by management.

Washington State Requirements

The State of Washington requires companies to develop an accident prevention plan (APP) in writing. This plan must be tailored to the workplace. The written plan must contain an orientation and a description of the company safety program. The plan should detail the requirement for a safety committee as well as how to report injuries and unsafe conditions.

Washington's plan requires specific training for employees, such as the initial on-the-job orientation; use and care of personal protective equipment; emergency actions to be taken to evacuate the workplace; and how to identify, store and use hazardous gases, chemicals and materials. Other training topics should include materials-handling equipment, machine tool operations, toxic materials and the operation of utility systems. The plan should require employers to enforce training programs to ensure that the work is performed in a safe manner. Washington companies must establish, supervise and enforce accident prevention programs in a manner that is effective.

A large part of the APP is the establishment of a safety committee and the use of safety meetings. Companies with more than 11 employees at the same location in the same shift are required to have a safety committee. The committee should have an equal number of employee-elected and employer-selected members. It also is important to include union members in the selection of the committee if a union is present. The committee should serve for a maximum of one year. The group must elect a chairperson of the committee and decide how often, when, and where to meet. The committee meetings should not last longer than one hour.

Firms with less than 10 employees may elect to have safety meetings instead of a safety committee. These safety meetings must be held monthly and there must be at least one representative of management at the meetings. The purpose of these meetings should be to review safety inspections, evaluate accident investigations and the overall performance of the APP. A record of the meeting must be made and include the names of the attendees and the subjects discussed.

The safety committee shall:
- Document the attendance at the meeting;
- Record subjects discussed;
- Review safety inspection reports;
- Evaluate accident investigations;
- Evaluate the workplace APP; and
- Maintain meeting records for one year.

The State of Washington allows companies to establish additional programs, plans and other rules that go beyond the APP described in order to develop a total safety and health plan.

Sample Plan

The federal program and the two state programs that have been reviewed provide an insight into how organizations are expected to respond with safety programs. By combining the intent of those programs into one document, a sample safety program that companies can use to meet the requirements of the overall safety program can be created. Companies can use this form by filling in the blank spaces with the pertinent information. It is important to review the plan once it is completed with the management and staff to ensure that the plan is workable for the facility as well as understood by all.

The safety plan provided here is a sample that can be used to develop a facility safety program. The base document describes roles and responsibilities and sets forth guidelines for the effective operation of a safety system. The guidelines may be modified to include more specific details as necessary.

With the overall plan in place, the facility should add appendices as needed to establish site-specific safety operations for detailed tasks. These appendices can be added or deleted as the facility establishes or discontinues procedures.

Sample Safety and Health Policy

It is the policy of _____ to provide a safe and healthy workplace for all of our employees. We feel that all accidents are preventable and that safety is a team effort involving the workforce, supervisors and management.

This company has established a safety and health program that is designed to involve all workers in the identification and reduction of hazards on each of our work sites. This program is designed to protect our most valuable resource, our workers, as well as meet federal and state regulations.

Safety is a key element in all workers' jobs. Unsafe behavior will not be tolerated and supervisors are required to enforce safety rules the same as they would enforce quality and timeliness issues with the employees. Active participation and adherence to our Safety Program is a condition of employment. No worker is allowed to work on a job that he or she knows is not safe. That includes our contractors and subcontractors.

Our goal is to eliminate accidents and injuries. There is no excuse or reason to take a shortcut or cut corners when they could result in an unsafe condition or action.

We're all in this together. If we work together, we will all go home safe and sound.

_____
Senior person

Responsibilities

Safety is a team effort and all employees and subcontractors of _____ are responsible for ensuring that the work area is safe. Different levels of management hold levels of responsibility. The person identified to maintain this program is _____, who is the head of _____ and can be reached at telephone extension _____.

Management is responsible to ensure that a safety committee is formed, staffed and resourced to allow it to complete its task of assisting the safety program. Managers will ensure that workers have sufficient time, resources and support from their supervisors to carry out the program. Managers will include an evaluation of safety performance in the annual review of workers' performance. Managers are responsible to ensure that all incidents (accidents and close calls) are investigated within five days and that corrective action is taken to prevent hazardous conditions from recurring.

Managers will ensure that the OSHA 300 form is maintained in an up-to-date, accurate manner with all recordable incidents being posted within seven calendar days of their occurrence. Managers are expected to report unsafe practices and to take action to correct unsafe behavior when they see it.

Managers of this facility are required to set a good example by following established safety rules, attending required safety training and encouraging all workers to do the same. Managers must enforce the disciplinary rules established in this safety program by ensuring that supervisors counsel workers as necessary.

Supervisors are responsible for the workers entrusted to their care. They must ensure that workers receive initial safety training prior to being allowed to work. Such training should be documented and evaluated to ensure that the worker understands what is expected of him or her. Supervisors are responsible for ensuring workers have the correct personal protective equipment and that the equipment is maintained correctly and used in the correct manner.

Supervisors are required to conduct a daily walk-around safety check of the work area and to direct the correction of any safety deficiencies found. In the event that the safety deficiency is caused by worker misconduct or non-compliance with established safety rules, the supervisor is required to counsel the individual and record that counseling on the employee's work record as outlined in the section on discipline.

Supervisors are not allowed to remove or defeat any safety device or to encourage workers by word or example to take short cuts that could result in an unsafe condition. Supervisors are required to support the safety committee by allowing workers the time needed to participate in safety activities.

Employees are required to comply with the safety rules established by the facility, federal or state law, safety training and any actions directed by their supervisors with regard to safety. When employees identify unsafe conditions, they are to report those to their supervisors or the safety committee promptly. All injuries or accidents are to be reported immediately to supervisors. All personal protective equipment should be maintained in good working condition and used correctly. Workers may not remove or defeat any safety device nor encourage coworkers by word or deed to take short cuts that may lead to an unsafe condition.

The Safety Committee has been formed to help employees and management work together to resolve safety problems and develop solutions identified to ensure that the safety program works smoothly. The safety committee consists of an equal number from management and employees. The union is invited to sit on the safety committee. The leader of the safety committee reports directly to _____ who has responsibility for the safety program. The leader is elected from the group. Members of the safety committee serve for one year and the group meets once a quarter on company time. The committee will review the OSHA 300 log and any accident reports or audits that have occurred during that time. The Safety Committee will conduct a monthly safety audit of the facility and the results of the audit will be posted on the company safety bulletin board.

The Safety Committee meets for one hour on the first Tuesday of the month in the employee lunchroom. The minutes of each meeting will be posted on the company's safety bulletin board and kept on file for two years.

Safety meetings for all employees will be held monthly by the supervisor. All employees are required to attend these meetings to enhance safety-related communication in the facility.

Employees are required to report any injury or work-related illness to their immediate supervisor regardless of how serious. Minor injuries such as cuts and scrapes can be entered on the minor injury log that is posted at _____.

The site supervisor will:
- Investigate a serious injury or illness using procedures in the "Incident Investigation" section below. Complete an "Incident Investigation Report" form. Give the "Employee's Report" and the "Incident Investigation Report" to _____, who is responsible for this form.
- Determine from the Employee's Report, Incident Investigation Report, and any loss and injury claim form associated with the incident, whether it must be recorded on the OSHA Injury and Illness Log and Summary according to the instructions for that form.
- Enter a recordable incident within six days after the company becomes aware of it.
- If the injury is not recorded on the OSHA log, add it to a separate incident report log, which is used to record non-OSHA recordable injuries and near misses.
- Each month before the scheduled safety committee meeting, make any new injury reports and investigations available to the safety committee for review, along with an updated OSHA and incident report log.

The safety committee will review the log for trends and may decide to conduct a separate investigation of any incident.

The management will post a signed copy of the OSHA log summary for the previous year on the safety bulletin board each February 1 until April 30. The log will be kept on file for at least five years. Any employee can view an OSHA log upon request at any time during the year.

Incident Investigation

If an employee dies while working or is not expected to survive, or when three or more employees are admitted to a hospital as a result of the same work-related incident the management will contact the OSHA representative (federal or state) within eight hours after becoming aware of the incident. The scene of the incident should not be disturbed except to aid in rescue or make the scene safe.

Whenever there is an incident that results in death or serious injuries that have immediate symptoms, a preliminary investigation should be conducted by the immediate supervisor of the injured person(s), a person designated by management, an employee representative of the safety committee, and any other persons whose expertise would help the investigation.

The investigation team should take written statements from witnesses and photograph the incident scene and equipment involved. The team also should document as soon as possible after the incident, the condition of the equipment and any anything else in the work area that may be relevant. The team should make a written report of its findings. The report also should include a sequence of events leading up to the incident, conclusions about the incident and any recommendations to prevent a similar incident in the future. The safety committee should review the report at its next regularly scheduled meeting.

When a supervisor becomes aware of an employee injury where the injury was not serious enough to warrant a team investigation as described above, the supervisor should write an Incident Report to accompany the "Employee's Injury/Illness Report Form" and forward them to _____, who maintains the OSHA 300 log.

Whenever there is an incident that did not but could have resulted in serious injury to an employee (a near-miss), the incident should be investigated by the supervisor or a team depending on the seriousness of the injury that would have occurred. The "Incident Investigation Report" form should be used to investigate the near miss. The form should be clearly marked to indicate that it was a near miss and that no actual injury occurred. The report should be forwarded to the bookkeeper to record on the incident log.

Safety Inspection Procedures

This company is committed to aggressively identifying hazardous conditions and practices that are likely to result in injury or illness to employees. We will take prompt action to eliminate any hazards we find. In addition to reviewing injury records and investigating incidents for their causes, management and the safety committee will regularly check the workplace for hazards as described below:

Annual Site Survey—An inspection team made up of members of the safety committee will do a wall-to-wall, walk-through inspection of the entire worksite. They will write down any safety hazards or potential hazards they find. The results of this inspection will be used to eliminate or control obvious hazards, target specific work areas for more intensive investigation, assist in revising the checklists used during regular monthly safety inspections and as part of the annual review of the effectiveness of our accident prevention program.

Weekly Safety Inspection—Each week, safety representatives will inspect their areas for hazards using the standard safety inspection checklist. They will talk to employees about their safety concerns. Committee members will report any hazards or concerns to the whole committee for consideration. The results of the area inspection and any action taken will be posted in the affected area. Occasionally, committee representatives may agree to inspect each other's area rather than their own.

Job Hazard Analysis—As a part of our ongoing safety program, we will use a "Job Hazard Analysis" form to look at each type of job task our employees perform. The supervisor of that job task or a member of the safety committee will conduct this analysis. We will change how the job is performed as needed to eliminate or control any hazards. We will also determine whether the employee needs to use personal protective equipment (PPE) while performing the job. Employees will be trained in the revised operation and to use any required PPE. The results will be reported to the safety committee. Each job task will be analyzed at least once every two years, whenever there is a change in how the task is performed or if a serious injury occurs while the task is being performed.

Hazard Prevention and Control

This company is committed to eliminating or controlling workplace hazards that could cause injury or illness to our employees. We will meet the requirements of federal and state safety standards where there are specific rules about a hazard or potential hazard in our workplace. Whenever possible we will design our facilities and equipment to eliminate employee exposure to hazards. Where these engineering controls are not possible, work rules that effectively prevent employee exposure to the hazard will be established. When the above methods of control are not possible or are not fully effective, we will require employees to use personal protective equipment (PPE) such as safety glasses, hearing protection, and foot and hand protection.

Basic Safety Rules

The following basic safety rules have been established to help make our company a safe and efficient place to work. These rules are in addition to safety rules that must be followed when doing particular jobs or operating certain equipment. Those rules are listed elsewhere in this program. Failure to comply with these rules will result in disciplinary action.

- Never do anything that is unsafe in order to get the job done. If a job is unsafe, report it to your supervisor or safety committee representative. We will find a safer way to do that job.
- Do not remove or disable any safety device! Keep guards in place at all times when machinery is operating.
- Never operate a piece of equipment unless you have been trained on it and are authorized to use it.
- Use your personal protective equipment whenever it is required.
- Obey all safety-warning signs.
- Working under the influence of alcohol or illegal drugs or using them at work is prohibited and will not be tolerated.
- Do not bring firearms or explosives onto company property.
- Smoking is only permitted in designated areas.
- Horseplay, running and fighting are prohibited
- Clean up spills immediately. Replace all tools and supplies after use. Do not allow scraps to accumulate where they will become a hazard. Good housekeeping helps prevent injuries.

Disciplinary Policy

This facility has established a disciplinary policy to provide appropriate consequences for failure to follow safety rules. This policy is designed not so much to punish as to bring unacceptable behavior to the employee's attention in a way that the employee will be motivated to make corrections. The following consequences apply to the violation of the same rule or the same unacceptable behavior:

- First instance—Verbal warning, notation in employee file, and instruction on proper actions.
- Second instance—One-day suspension, written reprimand, and instruction on proper actions.
- Third instance—One-week suspension, written reprimand, and instruction on proper actions.
- Fourth instance—Termination of employment.

An employee may be subject to immediate termination when a safety violation places the employee or his or her coworkers at risk of permanent disability or death.

If an Injury Occurs

A first-aid kit is maintained at _____ _____. Each company vehicle is equipped with a first-aid kit located in the glove box or under the driver's seat. Members of the safety committee check these kits monthly. An inventory of each kit is taped to the inside cover of the box. If you are injured, promptly report it to any supervisor.

All supervisors are required to have first-aid cards. Other employees may have been certified. A list of current first-aid and CPR-certified supervisors and employees is posted on the company safety bulletin board along with the expiration dates of their cards.

In case of serious injury, do not move the injured person unless absolutely necessary. Only provide assistance to the level of your training. Call for help. If there is no response, call 911.

Safety Training

Training is an essential part of our plan to provide a safe workplace at this facility. To ensure that all employees are trained *before* they start a task that requires training, we have a training coordinator whose name is posted on the company safety bulletin board. Our safety-training supervisor is _____ _____.

That person is responsible for verifying that each employee has received an initial orientation by his or her supervisor has received any training needed to do the job safely and that the employee file includes documentation of the training. The coordinator will make sure that an outline and materials list is available for each training course we provide:

Course	Who must attend
Basic Orientation	All employees (given by the employee's supervisor)
Chemical Hazards (General)	All employees
Chemical Hazards (Specific)	An employee who uses or is exposed to a particular chemical
Fire extinguisher Safety	All employees
Respirator Training	Employees who use respirators
Forklift Training	Employees who operate forklifts
Electrical Safety	Employees who work with electricity
Power Tool Safety	Employees who use power tools

How This Safety Program Meets Federal and State Requirements

	Federal Guidelines	California	Washington
Written Plan	X	X	X
Policy Statement by Management	X		
Responsible Person		X	
Accountability	X	X	
Employee Involvement	X	X	X
Hazard Identification	X	X	
Hazard Control	X	X	
Training	X	X	X
Safety Committee	X	X	
Safety Rule Enforcement		X	X
Disciplinary Guides		X	X

Checklist to Assess Your Safety Program

- Title, signature and phone number of Responsible Person
- Policy statement of company safety and health
- Administrative responsibilities for implementing the plan
- Identification and accountability of worksite personnel responsible for accident prevention
- Means for controlling work activities of subcontractors and suppliers
- Responsibilities of subcontractors and vendors
- Plans for safety indoctrination of new employees
- Plans for continued safety training
- Provisions for safety inspections
- Responsibilities for investigation and reporting accidents/exposure
- Responsibilities for maintaining accident data, reports and logs
- Emergency response plan for disasters
- Public safety requirements (e.g., fencing/signs)
- Plans for monthly safety meetings and weekly employee safety meetings.
- Meetings shall be documented, including the date, attendance, subjects, and the names of the individuals who conducted the meeting.
- Documents shall be available for inspection on the job site.

Appendices and Local Requirements as Needed for the Job

- First-aid procedures—local emergency phone numbers posted
- Personal protective equipment—warning signs posted
- Hazard communications—list of chemicals on the job site/container labels
- Confined space entry—identification, entry procedures, rescue procedures
- Hearing conservation
- Respirator program
- Hot work program (welding)
- Bloodborne pathogens
- Violence in the workplace
- Emergency action plan
- Hazardous waste operations
- Fire prevention/protection
- Electrical safety
- Machinery and mechanized equipment
- Hand and power tools
- Fall protection
- Working over water or liquids

CHAPTER 18

RISK MANAGEMENT

INTRODUCTION:

WHY USE A RISK MANAGEMENT APPROACH
The risk management process and the benefits and opportunities it provides have given companies a way to fix responsibility and accountability at the appropriate level. It requires managers and supervisors to accept responsibility for safety actions in a written form that reinforces their responsibility and liability, resulting in more safe decision making.

Risk Management enhances the workplace environment by eliminating or controlling hazards before they result in losses, whether personnel or materiel, losses that can degrade or halt the operation.

Effective use of Risk Management techniques provides worker, manager and leader with a workplace for which the risks are acceptable, in fact, a safe workplace. The Occupational Health and Safety Act mandates that every workplace be free of recognized hazards. This 'general duty' clause is used when there are no specific standards applicable to a particular hazard.

WHAT IS RISK MANAGEMENT?
Risk Management is a disciplined, organized, and logical decision making process to identify, evaluate and control hazards. With training and practice, personnel will be better able to spot hazards, analyze risk and make risk decisions at the appropriate level of control.

All work and daily routines involve risk. All job tasks and daily living require decisions that include assessing and managing hazards. Each supervisor, along with every individual, is responsible for identifying potential risks and adjusting or compensating appropriately. Decisions must then be made at a level of responsibility that corresponds to the degree of risk and the ability to apply resources, taking into consideration the significance of the task and the timeliness of the required decision. The decision-making techniques apply to all operations.

Risk Management is a process that gives an added edge to those with a long history of meeting "minimum" requirements, or, of meeting standards. It is an opportunity to go beyond the minimum, to reach greater effectiveness in any task or operation. Its prerequisite is a firm foundation in standards. One must understand the standards well enough to know the consequences of compromising them, and therefore how to control the task with a new set of 'standards' that provide risk controls or that the decision authority accepts the risk of loss.

Managers, supervisors and individuals must:
- Take ownership of workplace hazards.
- Not allow uncontrolled hazards.

Risk management is not a radical new way of doing business. Each of us manages risk on a daily basis, while working, driving, shopping, investing, etc. Most decisions are automatic, guided by years of experience coping with the same or similar situations. In a sense, we are all experienced "Risk Managers".

Simply put, Risk Management is an organized framework for decision-making. The aim is to minimize losses, whether associated with money, equipment or personnel safety, while maximizing operation success. It is the rational decision process: Weigh expected costs against expected benefits; if benefits outweigh costs—go for it; otherwise don't. The dilemma most often is how to quantify expected costs and benefits.

WHAT ABOUT COMPLIANCE?
Meeting OSHA standards is required in the workplace. It's important to understand there is no conflict between standards and the application of Risk Management techniques. The OSHA standards provide a foundation for hazard identification and control. Knowledge of OSHA standards provides valuable insight into understanding and implementing safety programs and systems. Application of OSHA standards provides safe workplaces from recognized hazards. However OSHA standards don't cover all hazards. Risk Management provides a process to make the workplace safer by reviewing and prioritizing tasks and operations either not covered, or not covered effectively, by OSHA standards.

HOW DOES IT WORK?
The basic decision making principles are applied before any job, action, or operation is executed. As an operation progresses and evolves, risk management controls are continuously reevaluated. Risk Management requires that owners and CEOs accept the process and take an active part in it. This is done by enforcing its use and establishing the amount of risk that workers can accept. For example, a low risk task, such as setting up a ladder and changing a light bulb can be approved by the worker doing the task. On the other hand, if the light bulb has electrical wiring that is shorting out and exposing that worker to an electrical shock, and it is in a wet environment, the task may be defined as a high risk task. The CEO may require that all high risk tasks be approved by the plant manager, thereby shifting the authority and acceptance of liability to the plant manager. This is an important concept that eliminates the situation where a plant manager tells a subordinate to engage in a risky task with impunity. Under the Risk Management system, the worker would be required by company policy to inform the supervisor that as performed, this is a highly dangerous risk that cannot be accepted by the worker or the supervisor. It must be sent up the chain of command to someone who can accept the risk and liability. This protects the company, its officers and the workers by forcing a process to be enacted.

Integrate into planning.
- Risks are more easily assessed and managed in the early planning stages.
- The acceptable plans are in proportion to risks and worth the anticipated cost.
- Operation is accomplished without incurring excessive losses in personnel, equipment, time, or position.

Accept no Unnecessary Risk.
- Unnecessary risk has no payback in terms of real benefits or available opportunities.
- RM provides tools to determine which risk or what level of risk is unnecessary. CEOs

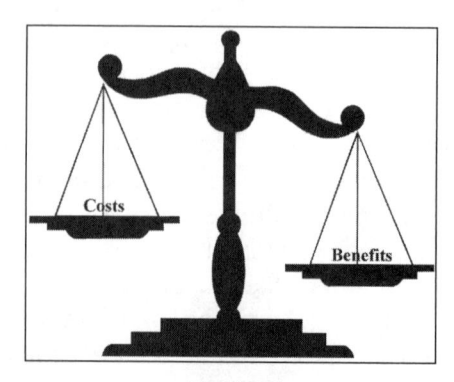

FIGURE I.
COST AND BENEFIT SCALES

must make this decision to protect their company from lawsuits.

Make Risk Decisions at the Appropriate Level.
- Appropriate level for risk decisions is the one that can allocate the resources to reduce the risk or eliminate the hazard and implement controls.

Accept Risk When Benefits Outweigh the Costs.
- All identified benefits should be compared to all identified costs.
- Figure 1 depicts this principle. Balancing costs and benefits may be a subjective process and open to interpretation.

Ultimately, the balance may have to be determined by the appropriate decision authority.

Integrate RM into Company Doctrine and Planning at all Levels.
Risks are more easily assessed and managed in the planning stages of an operation. Integrating risk management into planning as early as possible provides the decision maker the greatest opportunity to apply RM principles.

Five-Step Risk Management Process. RM is a continuous process designed to detect, assess, and control risk while enhancing performance and maximizing capabilities. Figure 2 depicts the five-step process.
- **Identify Hazards.** A hazard is any real or potential condition that can cause (mission degradation) injury, illness, death to personnel or damage to or loss of equipment or property. Experience, common sense, and specific risk management tools help identify real or potential hazards.
- **Assess the Hazards.** Risk is the probability and severity of loss from exposure to the hazard. Assessment is the application of quantitative or qualitative measures to determine the level of risk associated with a specific hazard. The assessment step in the process defines the probability and severity of a mishap that could result from the hazard and determines the exposure of personnel or assets to that hazard.
- **Develop Control Measures and Make Risk Decision.** Investigate specific strategies and tools that reduce, mitigate, or eliminate the risk. Effective control measures reduce one or more of the three components (probability, severity, or exposure) of risk. Decision makers at the appropriate level choose controls based on the analysis of overall costs and benefits.
- **Implement Controls.** Once control strategies have been selected, an implementation strategy needs to be developed and then applied by management and the work force. Implementation requires commitment of time and resources.
- **Supervise and Evaluate.** Risk management is a process that continues throughout the life cycle of the system, mission, or activity. Once controls are in place, the process must be scrutinized and reevaluated to determine its effectiveness. Mission performance is periodically evaluated to determine the effectiveness of risk control measures.

WHAT IS RISK?

The following chart depicts risk: as risk is evaluated some emerge as clearly unacceptable. The unacceptable risk will either be controlled or the task will not be performed as planned. Some risk is not identified until later in the planning process or in the actual operation phase. That risk is evaluated as it's identified and additional controls either put into place or the risk is accepted or eliminated. Risk perception, risk tolerance, and risk acceptance play major roles in an individual's definition of "risk". Diversity of age, experience, and training are among the filters individuals use to consider risk. Level of experience with the specific mission or task adds another dimension to each individual's concept of "risk".

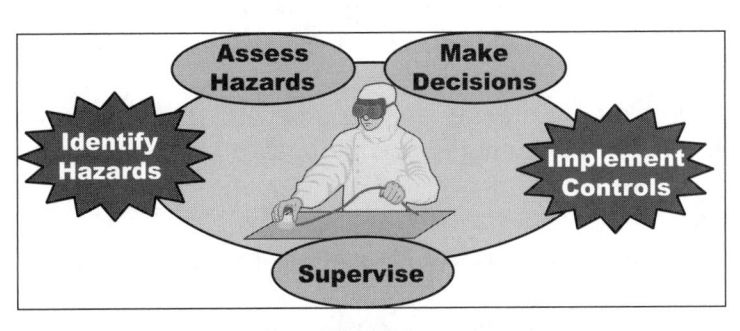

FIGURE 2.
FIVE-STEP PROCESS OF RISK MANAGEMENT

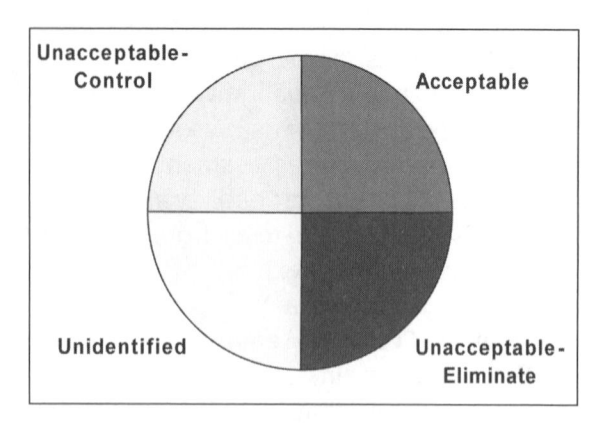

**FIGURE 3.
TYPES OF RISK**

WHAT ARE THE BENEFITS?
Benefits are not limited to reduced accident rates or decreased injuries, but may be actual increases in efficiency or job effectiveness. Examples of potential benefits include:
- Improved ability to protect the workforce by minimizing accidents—analysis of current processes may reduce risks.
- Enhanced job efficiency—reducing hazards by placing controls may make the task flow decisions are based on a reasoned and repeatable process instead of guess work or intuition.
- Improved confidence in capabilities—adequate risk analysis provides a clear picture of workforce strengths and weaknesses.

WHAT IS UNNECESSARY RISK?
Applying risk management requires a clear understanding of what constitutes "unnecessary risk", when benefits actually outweigh costs, and some guidance as to the appropriate level. Accepting risk is a function of both risk assessment and risk management. Risk acceptance is not as simple a matter as it may first appear. Several points must be kept in mind.
- Some risk is a fundamental reality.
- Risk management is a process of tradeoffs.
- Quantifying risk doesn't ensure safety.

General risk management guidelines are:
- Many activities involving a technical device or complex process entail some risk during their execution.
- Weigh risks and make judgments according to knowledge, experience, and mission requirements.
- Encourage other disciplines to adopt risk management principles. Hazard analysis and risk assessment do not free us from reliance on good judgment.
- It is more important to establish clear objectives and parameters for risk assessment than to find a "cookbook" approach and procedure.
- There may be no "best solution." There are a variety of directions to go. Each of these directions may produce some degree of risk reduction.
- Point out improvements to established controls rather than to say their approach will not work.
- Complete risk control is not the goal; total risk elimination is seldom achieved in a practical manner.

WHO DOES WHAT?
CEOs/plant managers:
- Set the example
- Establish the acceptable level of risk for subordinates..
- Establish, endorse, and enforce established standards.
- Reinforce personal accountability.
- Accept responsibility for effective management of risk.
- Select from risk reduction options provided by the staff.
- Accept or reject risk based on the benefit to be derived.
- Train and motivate leaders to use risk management.
- Elevate to the appropriate decision maker; based on clearly defined risk levels.

Supervisors:
- Apply the risk management process and direct personnel to use it both on and off duty.
- Consistently apply effective risk management concepts and methods to operations and tasks.
- Elevate risk issues beyond their control or authority to superiors for resolution.

Individuals:
- Understand, accept, and implement risk management processes.
- Maintain a constant awareness of the hazards associated with a task.

- Make supervisors immediately aware of any unrealistic risk reduction measures or high risk procedures.

HOW MUCH IS ENOUGH?

Somewhere between the back of an envelope and a multi-year, multi-million dollar, contractor-engineered PERT chart lies the right amount of risk management for the task or mission at hand. How much is enough? How do you advise the CEO or manager as to the most appropriate level of detail for a particular task or operation? What criteria would you use to make your recommendation?

- We've talked about "effective application of the Risk Management process". It means that "hands on", not just following the "fine print", defines an efficient and effective process when it comes to Risk Management. It's easy to fall into using some format that soon becomes a thoughtless process. That road may be efficient; but it's rarely effective and it's full of potholes. We've talked about Risk Management providing insights into balancing cost and benefit. Some of the costs and benefits are subjective. The question arises "What 'criteria' would you use to advise the CEO on how much Risk Management is enough?" There is no easy answer. We recognize that the cost benefit of effective Risk Management for a major mobilization would likely be greater than for an everyday delivery of goods and services already in the pipeline. "Enough" Risk Management for continued operations is different than "enough" for an office staff with little change in their day to day operations.

Some suggested criteria:
- Is hasty or is deliberate decision making underway?
- Is time available to plan?
- What's the Importance of operation success—100% OK? 75%, 50%? What if it fails completely?
- Is the operation or task a new one or an old familiar one?
- Are personnel involved new to the operation or task? Unit? Experienced/Inexperienced?
- What's been their operation tempo recently?

HOW DO I EVALUATE OUR PROGRAM?
Use structured internal and external assessments. Leaders and managers take advantage of the entrepreneurial genius of the people within the organization to develop better ways of helping people and getting work done. It is a process which encourages ideas and initiatives to float upward. Embedding the risk management process into organizational processes is one way to help achieve better results. The idea behind risk management is to identify strengths and weaknesses in planning and execution with emphasis on hazard control. The value of self-assessments is the awakening of self awareness. With this self awareness, change is more readily accepted.

Risk management integration and self assessment can help change the thinking from "minimal essential" to "maximum possible" philosophies in providing support to workers. Employees are deserving of nothing less than excellence. Authority and responsibility must be pushed down into the organization.

WHERE DOES THE SAFETY PROFESSIONAL FIT IN?

The safety professional is the link between the compliance world and the world of uncontrolled risk. The first step the safety professional makes is to help leaders recognize that risks can be controlled, that accidental losses need not be the cost of doing business. Whether an individual views a task or operation as having group or individual risk plays a major role in whether the individual believes he can take action to reduce the risk. The safety professional can facilitate a change of attitude from the idea that "warehousemen have back injuries", or group risk about which little can be done; to the reality of individual risk, where worksite redesign, individual work-hardening, or use of manual materials handling equipment and eliminate the hazard. Once leaders believe that losses are unnecessary, the next step is that the safety professional can provide the Risk Management process as a deliberative, logical decision making methodology to control the hazards that lead to unnecessary losses. Now, the safety professional must advise leaders on how much Risk Management is appropriate to the task at hand and facilitate the process.

Professional safety personnel provide knowledge of Occupational Health and Safety standards, application of safety practices to recognize, avoid/eliminate hazards, and for hazard control and evaluation techniques that provide the foundation of a safe workplace. Safety professionals are investigators when hazardous conditions turn into accidents or incidents related to inadequately controlled hazards. Safety offices assist in training, program management and coordinating with health and safety related offices; to include preventive medicine, engineers, security and others.

AUDITS. Measurements are necessary to ensure accurate evaluations of how effectively controls eliminated hazards or reduced risks. After action reports, surveys, and in progress reviews provide great starting places for measurements. To be meaningful, measurements must quantitatively or qualitatively identify reductions of risk, improvements in operation success, or enhancement of capabilities. The Risk Management Evaluation Profile (an adaptation of OSHA's Program Evaluation Profile) in appendix B, provides an audit process of a safety program's overall effectiveness. Benefits of conducting an audit include:
- Formally going through the internal audit process with managers and employees,
- Formally reviewing the elements of a safety and health program,
- Getting managers involved in the audit process,
- Making managers, supervisors and employees aware of the scope and complexity of a formal safety and health program, and of their roles and responsibilities in the programs success.

After finishing your audit share the results with managers and legal personnel since the process of doing an audit creates a paper trail of the program's weaknesses. The audit is a formal tool to uncover weak points in the management system that create unsafe work practices and unsafe conditions that can injure workers, diminish their health, interrupt production, or damage products and property.

CONCLUSION. Risk management provides a logical and systematic means of organizing information for rational decision-making, to identify and control risk. Risk management is a process that requires individuals, supervisors and leaders to support and implement the basic principles, along with the discipline to apply them on a continuing basis. Risk management offers individuals and organizations a powerful tool for eliminating accidents and increasing effectiveness. This process has the unique advantage of being accessible to and usable by everyone in every conceivable setting or scenario.

Appendix A: Risk Management Steps.

Tools to accomplish the various steps are explained and examples are given in, The Risk Manager's *Hazard Identification Tool Box*.

Step 1. IDENTIFY THE HAZARDS.
Hazards lead to risk, so the first step is to identify relevant hazards. Consider all aspects of current and future situations, environment, and known historical problems.

In identifying hazards, experience and training cannot be overemphasized; it is the most effective tool available. Those who have experience must use it, if an organization is to effectively use the RM process. Still, everyone is responsible for, and should be involved in finding potential hazards and informing their supervisor.

Visualization is an effective method to identify hazards. Picture the planned operation, think of what could go wrong—**ask yourself what if**? This can be done by an individual or a group, and can also use quality techniques such as brainstorming, "five whys", mental imaging, affinity diagrams, or cause-effect diagrams. The bottom line is: Honestly assess the planned procedure—**Think of what could go wrong, no matter how unlikely**.

Recognize Hazards: The Activity Hazard Analysis and Job Hazard Analysis (JHA) are both excellent tools to help identify risk as you think through a course of action to be examined. This is accomplished by reviewing current and planned operations describing the task. The supervisor defines what is required to accomplish the tasks and the conditions under which these tasks are to be conducted.

Construct a list or chart depicting the major steps in the job process, normally in time

RESOURCE

STEP 1 - IDENTIFY THE HAZARD

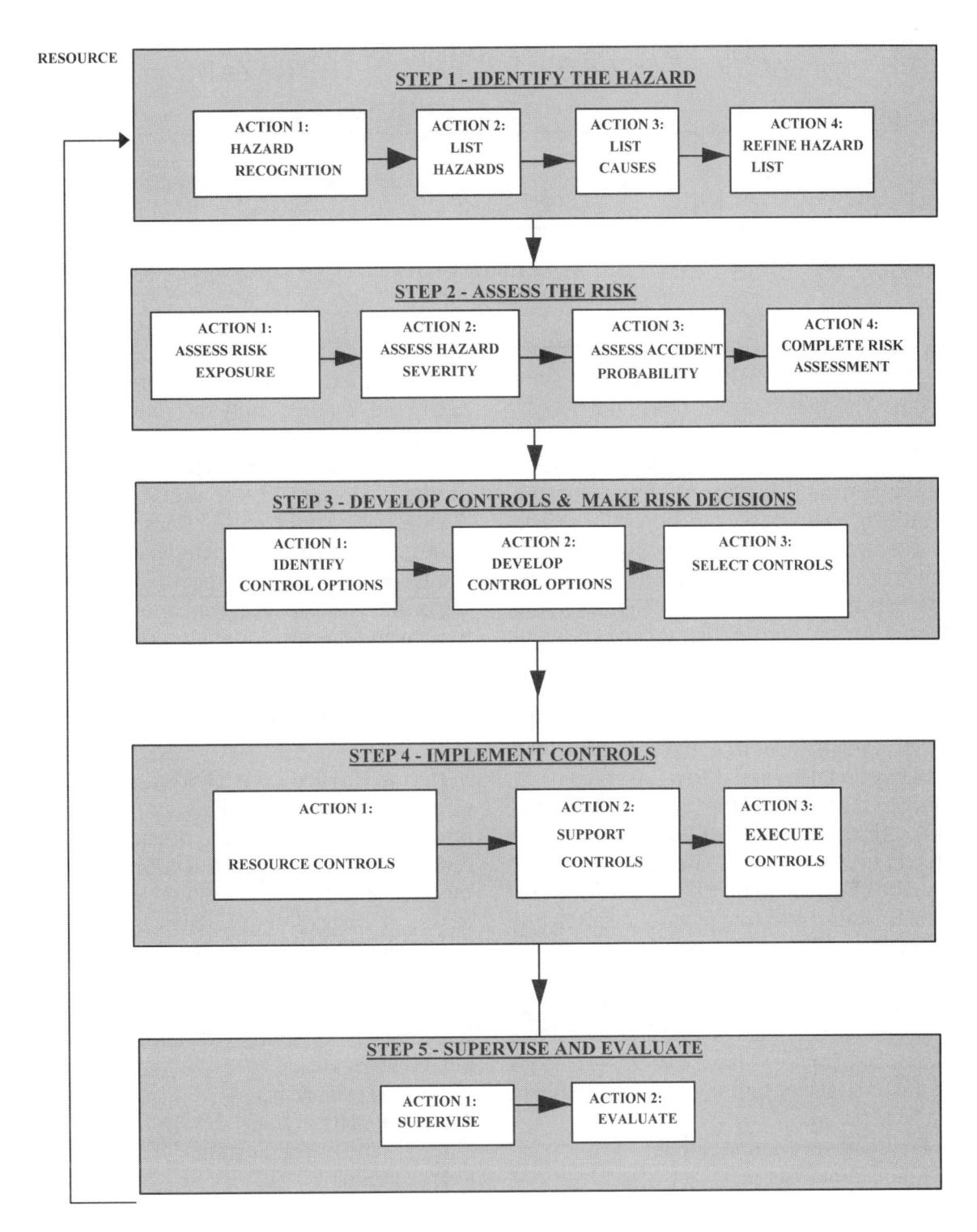

ACTION 1: HAZARD RECOGNITION → ACTION 2: LIST HAZARDS → ACTION 3: LIST CAUSES → ACTION 4: REFINE HAZARD LIST

STEP 2 - ASSESS THE RISK

ACTION 1: ASSESS RISK EXPOSURE → ACTION 2: ASSESS HAZARD SEVERITY → ACTION 3: ASSESS ACCIDENT PROBABILITY → ACTION 4: COMPLETE RISK ASSESSMENT

STEP 3 - DEVELOP CONTROLS & MAKE RISK DECISIONS

ACTION 1: IDENTIFY CONTROL OPTIONS → ACTION 2: DEVELOP CONTROL OPTIONS → ACTION 3: SELECT CONTROLS

STEP 4 - IMPLEMENT CONTROLS

ACTION 1: RESOURCE CONTROLS → ACTION 2: SUPPORT CONTROLS → ACTION 3: EXECUTE CONTROLS

STEP 5 - SUPERVISE AND EVALUATE

ACTION 1: SUPERVISE → ACTION 2: EVALUATE

**FIGURE 4
RISK MANAGEMENT STEPS**

sequence. Break the operation down into 'bite size' chunks.

Some tools that will help perform mission/task analysis are:
- Activity Hazard Analysis
- Job Hazard Analysis
- Flow Diagram
- Multilinear Event Sequence (MES)
- Sequentially Timed Event Plot (STEP)

List Hazards: Hazards, and factors that could generate hazards, are identified based

ACTIONS FOR STEP 1: IDENTIFY THE HAZARD

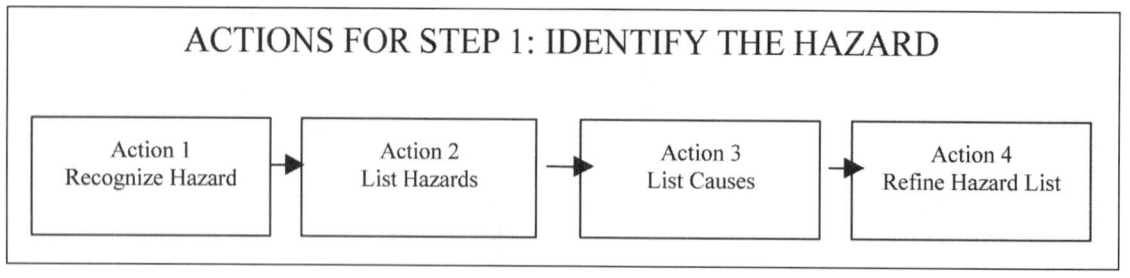

Action 1 — Recognize Hazard → Action 2 — List Hazards → Action 3 — List Causes → Action 4 — Refine Hazard List

FIGURE 5
STEP 1 ACTIONS

on the deficiency to be corrected and the definition of the task and system requirements. The identification phase produces a listing of hazards or adverse conditions and the accidents which could result. Examples of inherent hazards in any one of the elements include fire, explosion, collision with objects, or electrocution. The analysis must also search for factors that can lead to hazards such as alertness, ambiguity, or escape route. In addition to a hazard list for the elements above, interfaces between or among these elements should be investigated for hazards. An individual required to make critical and delicate adjustment to equipment on a cold, dark night may be at risk to a frost-bite injury, maybe an example of the "interface hazards." Make a list of the hazards associated with each step in the task process. Stay focused on the specific steps in the task being analyzed. Try to limit your list to "big picture" hazards (the final link in the chain of events leading to task degradation, personnel injury, death, or property damage). Hazards should be tracked on paper or in a computer spreadsheet/database system to organize ideas and serve as a record of the analysis for future use. Tools that help list hazards are:

- Preliminary Hazard Analysis.
- Change Analysis.
- Brainstorming.
- "What if" Analysis.

Identify hazards associated with these three categories:

- Task Degradation.
- Personal Injury or Death.
- Property Damage.

List Causes: Make a list of the causes associated with each hazard identified in the hazard list. A hazard may have multiple causes related to man, machine and environments. In each case, try to identify the root cause (the first link in the chain of events leading to mission degradation, personnel injury, death, or property damage). Risk controls can be effectively applied to root causes. Causes should be annotated with the associated hazards in the same paper or computer record mentioned in the previous action. Suggested tools are:

- Change Analysis.
- Brainstorming.
- "What if" Analysis.
- Job Hazard Analysis.

Refine Hazard Lists: If time and resources permit, and additional hazard information is required, use strategic hazard analysis techniques. These are normally used for medium and long term planning, complex tasks, or operations in which the hazards are not well understood. The first step of in-depth analysis should be to examine existing databases or available historical and hazard information regarding the operation. Suggested tools are:

- Database analysis.
- Accident History.
- Cause and effect diagrams.
- Tree Diagrams

The following tools are particularly useful for complex, coordinated operations in which multiple units, participants, and system components and simultaneous events are involved: Complex operations risk management tools are:

- Sequentially timed event plot.
- Multilinear event sequence.
- Interface analysis.
- Failure mode and effect analysis.

ACTIONS FOR STEP 2: ASSESS HAZARDS

Action 1: Assess Risk Exposure → Action 2: Assess Hazard Severity → Action 3: Assess Accident Probability → Action 4: Complete Assessment

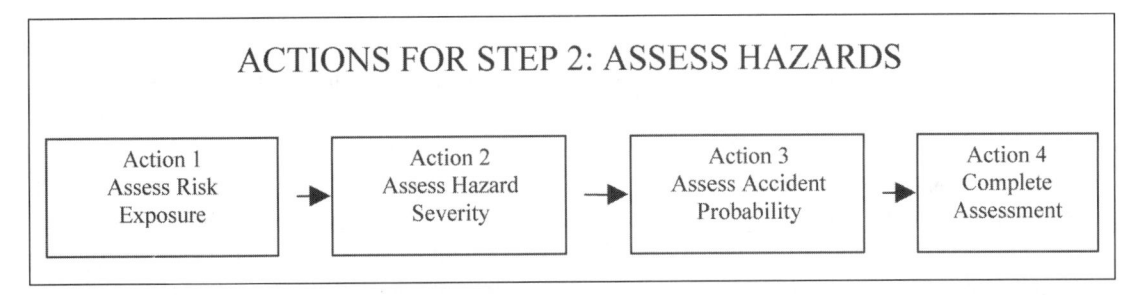

**FIGURE 6
STEP 2 ACTIONS**

There are many additional tools that can help identify hazards. One of the best is through a group process involving representatives directly from the workplace. A simple brainstorming process with a trained facilitator is very productive. The following is a partial list of sources of hazard identification:

- *Accident Reports*: These can come from within the organization, from tenants, within the chain of command, the safety department, etc. Other sources might be medical reports, maintenance records, and fire and police reports.
- *Quality reports*: Quality audits provide important feedback and written documentation on local process management.
- *Accident Databases*: The OSHA 300 and insurance claims are information resources.
- *Surveys*: These can be unit generated. Target an audience and ask some very simple questions related to such topics as: What will your next accident be? Who will have it? What task will cause it? When will it happen? The survey can be a powerful tool because it pinpoints people in the workplace with first hand knowledge of the job. Often, first line supervisors in the same workplace won't have as good an understanding of risk as those who confront it every day.
- *Inspections*: Safety inspections can consist of spot checks, walk-throughs, checklist inspections, site surveys, and mandatory inspections. Use onsite workers to provide input beyond the standard third-party inspection.

Step 2. ASSESS HAZARDS.

For each hazard identified in the previous step:

Assess Hazards. Once hazards are found, the next step is to analyze the associated risk—how likely and how big a loss is possible? **Recognition and assessment is the core of the Risk Management process.** Risk Management process depends on doing good analyses at each step in the process.

Assess Hazard Exposure: Probability is effected by exposure. Repeated exposure to a hazard greatly increases the total likelihood of an accident. This can be expressed in terms of time, proximity, volume, or repetition. Does it happen often, or near personnel or equipment? Does the event happen to a lot of people or equipment? This level can aid in determining the severity or the probability of the event. Additionally, it may serve as a guide for devising control measures to limit exposure. Another important concept is **interaction**. Interaction occurs when two (or more) hazards are present and their total risk is much greater than simply adding their separate risks. It's more like multiplying than adding. Often it is the combination of several factors that make a situation hazardous, rather than any single factor. Experience and clear thinking are the best ways to consistently assess interaction.

Assess Hazard Severity: Determine the severity of the hazard in terms of its potential impact on the people, equipment, or mission. Cause and effect diagrams, scenarios and "What-If" analysis are some of the best tools for assessing the hazard severity. Severity assessment should be based upon the worst possible outcome that can reasonably be expected. Severity categories are defined to provide a qualitative measure of the worst credible accident resulting from personnel error, environmental FIGURE 7 conditions; design inadequacies; procedural deficiencies; or system, subsystem, or component failure or malfunction. Using severity categories provide guidance to a wide variety of missions and systems.

Assess Accident Probability: Determine the probability that the hazard will cause an accident or loss of the severity assessed above. Accident probability is proportional to the cumulative probability of the identified causes for the hazard. Probability may be determined through estimates or actual numbers, if they are available. Assigning a quantitative accident probability to a new mission or system may not be possible early in the planning process.

A qualitative accident probability may be derived from research, analysis, and evaluation of historical safety data from similar missions and systems. The typical accident sequence is much more complicated than a single line of erect dominos where tipping the first domino (hazard) triggers a clearly predictable reaction. Supporting rationale for assigning a accident probability should be documented for future reference.

Complete Risk Assessment: Combine severity and probability estimates to form a risk assessment for each hazard. By combining the probability of occurrence with severity, a matrix is created where intersecting rows and columns define a Risk Assessment Index (RAI), table 3-3, AR 385-10. The Risk Assessment Index forms the basis for judging both the acceptability of a risk and the management level at which the decision of acceptance will be made.

The index may also be used to prioritize resources to resolve risks due to hazards or to standardize hazard notification or response actions. Severity, probability, and risk assessment should be recorded to serve as a record of the analysis for future use. Existing databases, Risk Assessment Index matrix, or a panel of personnel experienced with the mission and hazards can be used to help complete the risk assessment.

The following are some analytical pitfalls that should be avoided in the assessment:

- *Overoptimism*: "It can't happen to us. We're already doing it." This pitfall results from not being totally honest and not looking for root causes of risk.
- *Misrepresentation*: Individual perspectives may distort data.
- *Alarmist*: "The sky's falling" approach, or "worst case" estimates are used.
- *Indiscrimination*: All data is given equal weight.
- *Prejudice*: Subjectivity and/or hidden agendas are used, rather than facts.
- *Inaccuracy*: Bad or misunderstood data nullifies accurate risk assessment.
  - It is difficult to assign a numerical value to human behavior.

SEVERITY	PROBABILITY				
	Frequent	Likely	Occasional	Seldom	Unlikely
Catastrophic	E	E	H	H	M
Critical	E	H	H	M	L
Marginal	H	M	M	L	L
Negligible	M	L	L	L	L

Risk Level: E-Extremely High, H-High, M-Moderate, L-Low

PROBABILITY - The likelihood that an event will occur.
FREQUENT - Occurs often, continuously experienced.
LIKELY - Occurs several times.
OCCASIONAL - Occurs sporadically.
SELDOM - Unlikely, but could occur at some time.
UNLIKELY - Can assume it will not occur.

SEVERITY - The expected consequence of an event in terms of degree of injury, property damage, or other mission-impairing factors.
CATASTROPHIC - Death or permanent total disability, system loss, major property damage, not able to accomplish mission.
CRITICAL - Permanent partial disability, temporary total disability in excess of 3 months, major system damage, significant property damage, significantly degrades mission capability.
MARGINAL - Minor injury, lost workday accident, minor system damage, minor property damage, some degradation of mission capability.
NEGLIGIBLE - First aid or minor medical treatment, minor system impairment, little/no impact on accomplishment of mission.

FIGURE 7

ACTIONS FOR STEP 3: DEVELOP CONTROLS AND MAKE RISK DECISION

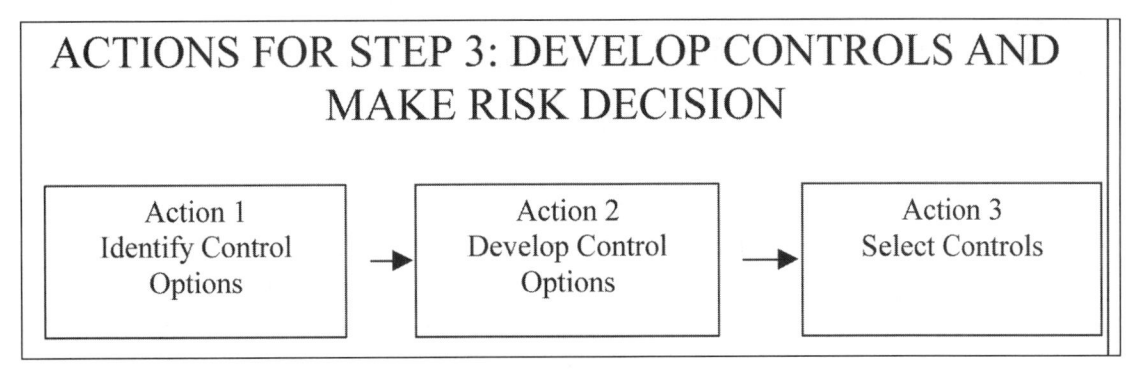

```
┌─────────────────┐     ┌─────────────────┐     ┌─────────────────┐
│    Action 1     │ →   │    Action 2     │ →   │    Action 3     │
│ Identify Control│     │ Develop Control │     │ Select Controls │
│     Options     │     │     Options     │     │                 │
└─────────────────┘     └─────────────────┘     └─────────────────┘
```

FIGURE 8
STEP 3 ACTIONS

- Numbers may oversimplify real life situations.
- It may be difficult to get enough applicable data, which could force inaccurate estimates.
- Oftentimes simple numbers take the place of reasoned judgment.
- Risk can be unrealistically traded off against benefit by relying solely on numbers.

Step 3. DEVELOP CONTROLS & MAKE RISK DECISIONS.

In this area, one must "develop control measures that eliminate the hazard or reduce its risk. As control measures are developed, risks are re-evaluated until all risks are reduced to a level where benefits outweigh potential cost."

Identify Control Options: The process of developing controls starts by taking the risk levels determined in Step 2, then identifying as many risk control options as possible for all hazards which exceed an acceptable level of risk. Refer to the list of possible causes from Step 1 for control ideas. Brainstorming, mission accident analysis and "What-If" analysis are excellent tools to identify control options. Risk control options include: **avoidance, reduction, spreading** and **transference**.

Avoiding risk altogether requires canceling or delaying the job, mission, or operation, but is an option that is rarely exercised due to mission importance. However, it may be possible to avoid specific risks: like wearing proper personal protective equipment can reduce risks in most job areas.

Risk can be **reduced**. The overall goal of risk management is to plan missions or design systems that do not contain uncontrolled hazards. A proven order of precedence for dealing with hazards and reducing the resulting risks is:

- **Plan or Design for Minimum Risk.** From the first, plan the mission or design the system to eliminate hazards. Without a hazard there is no probability, severity or exposure. If an identified hazard cannot be eliminated, reduce the associated risk to an acceptable level.
- **Incorporate Safety Devices.** If identified hazards cannot be eliminated or their associated risk adequately reduced by modifying the mission or system elements or their inputs, that risk should be reduced to an acceptable level through the use of safety design features or devices. Safety devices usually do not effect probability but reduce severity: an automobile seat belt doesn't prevent a collision but reduces the severity of injuries. Nomex gloves and steel toed boots won't prevent the hazardous event, or even change the probability of one occurring, but they prevent, or decrease the severity of, injury. Physical barriers fall into this category.
- **Provide Warning Devices.** When mission planning, system design, and safety devices cannot effectively eliminate identified hazards or adequately reduce associated risk, devices should be used to detect the condition and warn personnel of the hazard. Warning signals and their application should be designed to minimize the probability of the incorrect personnel reaction to the signals and should be standardized. Flashing red lights or sirens are a common warning device that most people understand.

Risk Management

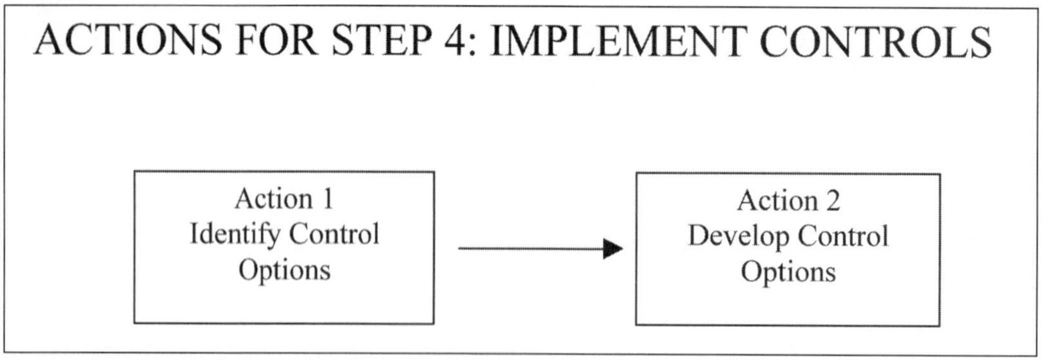

ACTIONS FOR STEP 4: IMPLEMENT CONTROLS

Action 1
Identify Control Options → Action 2
Develop Control Options

FIGURE 9
STEP 4. CONTROL ACTION

- **Develop Procedures and Training.** Where it is impractical to eliminate hazards through design selection or adequately reduce the associated risk with safety and warning devices, procedures and training should be used. If the system is well designed and the mission well planned, the only remaining risk reduction strategies may be procedures and training. *Emergency procedure training and disaster preparedness exercises improve human response to hazardous situations.*

Risk is commonly **spread** out by either increasing the exposure distance or by lengthening the time between exposure events. Administratively controlling exposure events, substitution of less hazardous chemicals or re-engineering an operation to reduce exposures to chemicals or toxic agents are examples.

Risk **transference** does not change probability or severity, however, possible losses or costs are shifted to another entity. An example is locating a remote sensing device in a high risk environment instead of risking personnel.

COMMON WAYS TO CONTROL RISK
- Protective equipment, clothing, or safety devices (PPE)
- Highlight hazards for extra care and handling
- Warnings (signs, color coding, audio/visual alarms)
- Repair hazards or build new facilities
- Limit exposure consistent with mission needs
- Train and educate
- Incorporate firm, fail-safe go/no-go criteria
- Select experienced or specialized personnel
- Increase and/or select more highly qualified and experienced supervision

- New policy—formal/informal, written/unwritten
- Develop new procedures

Develop Control Options: Determine the controls for the risk associated with the hazard. A computer spread sheet or data form (Job Hazard Analysis) may be used to list control ideas and indicate control effects. The estimated value(s) for severity and/or probability after implementation of control measures and the change in overall risk assessed from the Risk Assessment Index should be recorded. Scenario building and next accident assessment provide the greatest ability to determine control effects.

Select Controls: For each hazard, select those risk controls that will reduce the risk to an acceptable level.. The decision maker selects the control options after being briefed on all the possible controls. The best controls will be consistent with mission objectives and optimum use of available resources (manpower, material, equipment, money, time. It is not an ad hoc decision, but rather a logical, sequenced part of the risk management process. Decisions are made with awareness of hazards and how important hazard control is to success or failure of the mission (cost versus benefit).

Step 4. IMPLEMENT CONTROLS.
The decision maker must allocate resources to control risk. Control decisions must be made at the appropriate level. The standard for risk management is leadership at the appropriate level of authority making informed decisions control hazards or accept risks.

Safety advisors and consultants do not control the necessary resources to implement the

ACTIONS FOR STEP 5: SUPERVISE AND EVALUATE

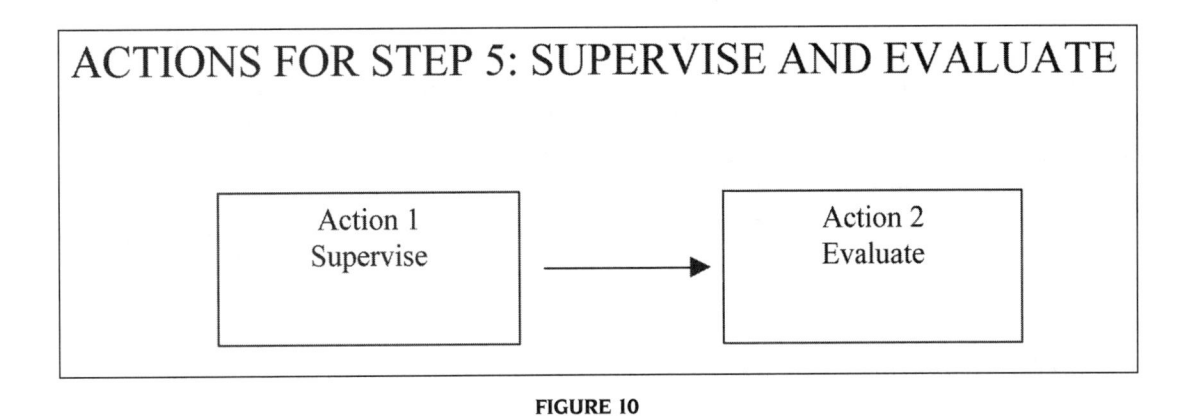

Action 1: Supervise → Action 2: Evaluate

**FIGURE 10
STEP 5 ACTIONS**

control decisions. Appropriate levels of decisions making reflect the ability of the decision maker to resource the controls.

Resource Controls: For each identified hazard, resource controls that will reduce the risk to an acceptable level. The best controls will be consistent with mission objectives and optimum use of available resources (manpower, material, equipment, money, and time). Record implementation decisions for future reference. Should management determine that the controls require resources beyond their authority, they should elevate the risk decision to higher authority.

Order Controls: To be successful, command must support the control measures put in place. Then, explore appropriate ways to demonstrate command commitment. Provide the personnel and resources necessary to implement the control measures. Design in sustainability from the beginning. Deploy the control measure with a feedback mechanism that provides information on whether the control measure is achieving the intended purpose.

Step 5. SUPERVISE AND EVALUATE.

Supervise: Monitor the operation to ensure:
- The controls are effective and remain in place.
- Changes which require further risk management are identified.
- Action is taken when necessary to correct ineffective risk controls and reinitiate the risk management steps in response to new hazards.

Any time personnel, equipment or operations change or new operations are anticipated in an environment not covered in the initial risk management analysis, the risks and control measures should be reevaluated. The goal of measurement is to answer the question of whether the control measure in fact controlled the associated hazard. The best tool for accomplishing this action of supervision is Change Analysis.

Evaluation: The process review must be systematic. After assets are expended to control risks, then a cost benefit analysis must be accomplished to see if risk and cost are in balance. Any changes in the system (the flow charts from the earlier steps provide convenient benchmarks to compare the present system to the original) are recognized and appropriate risk management controls are applied

To accomplish an effective review:
- Identify whether the actual cost is in line with expectations.
- What effect the control measure has had on mission performance.

Provide operation feedback to ensure that the corrective or preventive action taken was effective and that any new hazards identified during the mission are analyzed and corrective action taken.

Audits: Measurements are necessary to ensure accurate evaluations of how effectively controls eliminated hazards or reduced risks. After action reports, surveys, and in progress reviews provide great starting places for measurements. To be meaningful, measurements must quantitatively or qualitatively identify reductions of risk, improvements in operation success, or enhancement of capabilities. The Risk Management Evaluation Profile (OSHA's PEP) in appendix B, provides an audit process of a programs overall effectiveness.

Benefits of conducting an audit includes:
- Formally going through the internal audit process with managers and employees,
- Formally reviewing the elements of a safety and health program,
- Getting managers involved in the audit process,
- Making managers, supervisors and employees aware of the scope and complexity of a formal safety and health program, and of their roles and responsibilities in the programs success.

After finishing your audit share the results with managers and legal personnel. However the process of doing an audit creates a paper trail of the programs weaknesses. The audit is a formal tool to uncover weak points in the management system that create unsafe work practices and unsafe conditions that can injure workers, diminish their health, interrupt production, or damage products and property.

Conclusion. Risk management provides a logical and systematic means of organizing information for rational decision-making, to identify and control risk. Risk management is a process that requires individuals, supervisors and leaders to support and implement the basic principles, along with the discipline to apply them on a continuing basis. Risk management offers individuals and organizations a powerful tool for eliminating accidents and increasing effectiveness. This process has the unique advantage of being accessible to and usable by everyone in every conceivable setting or scenario.

Appendix B: The Risk Management Evaluation Profile (RMEP)

Subject: The Risk Management Evaluation Profile (RMEP)

A. **Purpose.** To establish policies and procedures for the Risk Management Evaluation Profile (RMEP), the RMEP form, which can be used in assessing safety and health programs in general workplaces.
B. **Scope.** The RMEP is applicable to all company workplaces.
C. **References.**
  1. AR 385-10, Army Safety Program.
  2. OSHA Instruction CPL 2.103, September 26, 1994, Field Inspection Reference Manual (FIRM).
D. **Action.** CEOs, directors and managers should ensure that the guidelines and procedures set forth here are followed in using the RMEP.
E. **Background.** Assessment of safety and health conditions in the workplace depends on a clear understanding of the programs and management systems that an employer is using for safety and health compliance. The organization places a high priority on safety and health programs and mandates their implementation.
F. **Application.**
  1. The RMEP should be completed for a general evaluation of workplace safety and health programs.
    a. The RMEP is an educational document for workers and employers, as well as a source of information for use in the inspection process.
    b. In multi-employer workplaces, a RMEP shall be completed for the safety and health program of the host organization. This RMEP will normally apply to all subordinate organizations, and individual RMEPs need not be completed for them.
    c. The RMEP shall be used in experimental programs that require evaluation of an organizations safety and health program, except where other program evaluation methods/tools are specifically approved.
  2. The evaluation of the safety and health program contained in the RMEP shall be shared with the employer and with employee representatives.
G. **Using the RMEP.** The RMEP will be used as a source of safety and health program evaluation for the employer, employees.
  1. **Gathering Information for the RMEP** begins during the opening conference

and continues through the inspection process.
a. The evaluator shall explain the purpose of the RMEP and obtain information about the employer's safety and health program in order to make an initial assessment about the program.
b. This initial assessment shall be verified—or modified—based on information obtained in interviews of an appropriately representative number of employees and by observation of actual safety and health conditions during the inspection process.
2. **Recording the Score.** The program elements in the RMEP correspond generally to the major elements of the Guidelines.
  a. **Elements.** The **six** elements to be scored in the RMEP are:
    (1) Management.
    (2) Leadership.
    (3) Employee Participation.
    (4) Hazard Identification.
    (5) Hazard Control.
    (6) Training.
  b. **Factors.** These elements will also be scored. The score for an element will be determined by the factor scores. The factors are:
    - Management
    - Leadership.
      – Employee participation.
      – Contractor safety.
      – Survey and hazard identification.
      – Reporting.
      – Investigation of accidents and near-miss incidents
      – Data analysis.
      – Hazard control.
      – Maintenance.
    - Medical program.
    - Training (as a whole).
  c. **Scoring.** The evaluator shall objectively score the organization on each of the individual factors and elements after obtaining the necessary information to do so. These shall be given a score of 1, 2, 3, 4, or 5. If the element or factor does not apply to the worksite being inspected, a notation of **"Not Applicable"** shall be made in the space provided. This shall not affect the score.
    (1) The attachment contains the **RMEP Tables**, which provide verbal descriptors of workplace characteristics for each factor for each of the five levels. Evaluators shall refer to these tables as appropriate to ensure that the score they assign to a factor corresponds to the descriptor that best fits the worksite.
    (2) Determine scores for each of the six elements as follows:

NOTE: The factors of "Management", "Leadership" and "Employee Participation" are given greater weight because they are considered the foundation of a safety and health program.

  (a) For each of the other elements, **average** the scores for the factors.
  (b) In **averaging** factor scores, round to the nearest whole number (1, 2, 3, 4, or 5). Round up from one-half (.5) or greater; round down from less than one-half (.5).
3. **Program Levels.** The Overall Score on the RMEP constitutes the "level" at which the establishment's safety and health program is scored. **Remember: This level is a relatively informal assessment.** The following chart summarizes the levels:

Score	Level of Safety and Health Program
5	Outstanding program
4	Superior program
3	Basic program
2	Developmental program
1	No program or ineffective program

4. **Specific Scoring Guidance.** The following shall be taken into account in assessing specific factors:
   a. **Written Programs.** Employer safety and health programs should be in writing in order to be effectively implemented and communicated.
      (1) Nevertheless, a program's **effectiveness** is more important than whether it is in writing.
         (a) An employer's failure to comply with a paperwork requirement is normally penalized only when there is a serious hazard related to this requirement.
         (b) An employer's failure to comply with a written program requirement is normally not penalized if the employer is actually taking the actions that are the subject of the requirement.
      (2) Thus, evaluators should follow the general principle that "performance counts more than paperwork." In using the RMEP, the evaluator is responsible for evaluating the organizations actual management of safety and health in the workplace, not just the organizations documentation of a safety and health program.
   b. **Employee Participation.**
      (1) Employee involvement in an establishment's safety and health program is essential to its effectiveness. Thus, evaluation of safety and health programs must include objective assessment of the ways in which workers' rights and responsibilities are addressed in form and practice.
      (2) Employee involvement should also include participation walk-around inspections, interviews, informal conferences, and formal settlement discussions, as may be appropriate. Many methods of employee involvement may be encountered in individual workplaces.
   c. **Comprehensiveness.** The importance of a safety and health program's comprehensiveness is implicitly addressed in Hazard Identification under both hazard identification and Data analysis. An effective safety and health program shall address all known and potential sources of workplace injuries and illnesses, whether or not they are covered by a specific OSHA standard.
   d. **Consistency with Violations/Hazards Found.** The RMEP evaluation and the scores assigned to the individual elements and factors should be consistent with the types and numbers of violations or hazards found during the inspection.
5. **Scope of the RMEP Review.** The duration of the RMEP review will vary depending on the circumstances of the workplace and the inspection. In all cases, however, this review shall include:
   a. A review of any appropriate employer documentation relating to the safety and health program.
   b. A walk-around inspection of pertinent areas of the workplace.
   c. Interviews with an appropriate number of employer and employee representatives.

The RMEP Tables

- The text in each block provides a description of the program element or factor that corresponds to the level of program that the employer has implemented in the workplace.
- To avoid duplicative language, each level should be understood as containing all positive factors included in the level below it.

MANGEMENT, LEADERSHIP and EMPLOYEE PARTICIPATION	
Management	
Visible management provides the motivating force for an effective safety and health program.	
1	Management demonstrates no policy, goals, objectives, or interest in safety and health issues at this worksite.
2	Management sets and communicates safety and health policy and goals, but remains detached from all other safety and health efforts.
3	Management follows all safety and health rules, and gives | visible support to the safety and health efforts of others.
4	Management participates in significant aspects of the site's safety and health program, such as site inspections, incident reviews, and program reviews. Incentive programs that discourage reporting of accidents, symptoms, injuries, or hazards are absent. Other incentive programs may be present.
5	Site safety and health issues are regularly included on agendas of management operations meetings. Management clearly demonstrates—by involvement, support, and example—the primary importance of safety and health for everyone on the worksite. Performance is consistent and sustained or has improved over time.
Notes:	

Employee Participation
Employee participation provides the means through which workers identify hazards, recommend and monitor abatement, and otherwise participate in their own protection.

1	Worker participation in workplace safety and health concerns is not encouraged. Incentive programs are present which have the effect of discouraging reporting of incidents, injuries, potential hazards or symptoms. Employees/employee representatives are not involved in the safety and health program.
2	Workers and their representatives can participate freely in safety and health activities at the worksite without fear of reprisal. Procedures are in place for communication between employer and workers on safety and health matters. Worker rights under the Occupational Safety and Health Act to refuse or stop work that they reasonably believe involves imminent danger are understood by workers and honored by management. Workers are paid while performing safety activities.
3	Workers and their representatives are involved in the safety and health program, involved in inspection of work area, and are permitted to observe monitoring and receive results. Workers' and representatives' right of access to information is understood by workers and recognized by management. A documented procedure is in place for raising complaints of hazards or discrimination and receiving timely employer responses.
4	Workers and their representatives participate in hazard identification analysis, inspections and investigations, and development of control strategies throughout facility, and have necessary training and education to participate in such activities. Workers and their representatives have access to all pertinent health and safety information, including safety reports and audits. Workers are informed of their right to refuse job assignments that pose serious hazards to themselves pending management response.
5	Workers and their representatives participate fully in development of the safety and health program and conduct of training and education. Workers participate in audits, program reviews conducted by management or third parties, and collection of samples for monitoring purposes, and have necessary training and education to participate in such activities. Employer encourages and authorizes employees to stop activities that present potentially serious safety and health hazards.

Notes:

	LEADERSHIP
	Implementation
	Implementation means tools, provided by management, that include: — budget — information — personnel — assigned responsibility — adequate expertise and authority — means to hold responsible persons accountable (line accountability) — program review procedures.
1	Tools to implement a safety and health program are inadequate or missing.
2	Some tools to implement a safety and health program are adequate and effectively used; others are ineffective or inadequate. Management assigns responsibility for implementing a safety and health program to identified person(s). Management's designated representative has authority to direct abatement of hazards that can be corrected without major capital expenditure.
3	Tools to implement a safety and health program are adequate, but are not all effectively used. Safety representative is knowledgeable in hazard recognition and applicable OSHA requirements. Management keeps or has access to applicable OSHA standards at the facility, and seeks appropriate guidance information for interpretation of OSHA standards. Management representative has authority to order/purchase safety and health equipment.
4	All tools to implement a safety and health program are more than adequate and effectively used. Written safety procedures, policies, and interpretations are updated based on reviews of the safety and health program. Safety and health expenditures, including training costs and personnel, are identified in the facility budget. Hazard abatement is an element in management performance evaluation.
5	All tools necessary to implement a good safety and health program are more than adequate and effectively used. Management safety and health representative has expertise appropriate to facility size and process, and has access to professional advice when needed. Safety and health budgets and funding procedures are reviewed periodically for adequacy.
	Notes:

Hazard Identification

Hazard Recognition and Evaluation

Survey and hazard analysis: An effective, proactive safety and health program will seek to identify and evaluate all hazards. In large or complex workplaces, components of such analysis are the **comprehensive survey** and **evaluations of job hazards and changes in conditions**.

1	No system or requirement exists for hazard review of planned/changed/new operations. There is no evidence of a comprehensive survey for safety or health hazards or for routine job hazard analysis.
2	Surveys for violations of standards are conducted by knowledgeable person(s), but only in response to accidents or complaints. The employer has identified principal OSHA standards which apply to the worksite.
3	Process, task, and environmental surveys are conducted by knowledgeable person(s) and updated as needed and as required by applicable standards. Current hazard analyses are written (where appropriate) for all high-hazard jobs and processes; analyses are communicated to and understood by affected employees. Hazard analyses are conducted for jobs/tasks/workstations where injury or illnesses have been recorded.
4	Methodical surveys are conducted periodically and drive appropriate corrective action. Initial surveys are conducted by a qualified professional. Current hazard analyses are documented for all work areas and are communicated and available to all the workforce; knowledgeable persons review all planned/changed/new facilities, processes, materials, or equipment.
5	Regular surveys including documented comprehensive workplace hazard evaluations are conducted by certified safety and health professional or professional engineer, etc. Corrective action is documented and hazard inventories are updated. Hazard analysis is integrated into the design, development, implementation, and changing of all processes and work practices.

Notes:

WORKPLACE ANALYSIS

Hazard Assessment

Inspection: To identify new or previously missed hazards and failures in hazard controls, an effective safety and health program will include regular **site inspections.**

1	No routine physical inspection of the workplace and equipment is conducted.
2	Supervisors dedicate time to observing work practices and other safety and health conditions in work areas where they have responsibility.
3	Competent personnel conduct inspections with appropriate involvement of employees. Items in need of correction are documented. Inspections include compliance with relevant OSHA standards. Time periods for correction are set.
4	Inspections are conducted by specifically trained employees, and all items are corrected promptly and appropriately. Workplace inspections are planned, with key observations or check points defined and results documented. Persons conducting inspections have specific training in hazard identification applicable to the facility. Corrections are documented through follow-up inspections. Results are available to workers.
5	Inspections are planned and overseen by certified safety or health professionals. Statistically valid random audits of compliance with all elements of the safety and health program are conducted. Observations are analyzed to evaluate progress.

Notes:

WORKPLACE ANALYSIS
Hazard Reporting
A reliable **hazard reporting system** enables employees, without fear of reprisal, to notify management of conditions that appear hazardous and to receive timely and appropriate responses. [Guidelines, (c)(2)(iii)]

1	No formal hazard reporting system exists, or employees are reluctant to report hazards.
2	Employees are instructed to report hazards to management. Supervisors are instructed and are aware of a procedure for evaluating and responding to such reports. Employees use the system with no risk of reprisals.
3	A formal system for hazard reporting exists. Employee reports of hazards are documented, corrective action is scheduled, and records maintained.
4	Employees are periodically instructed in hazard identification and reporting procedures. Management conducts surveys of employee observations of hazards to ensure that the system is working. Results are documented.
5	Management responds to reports of hazards in writing within specified time frames. The workforce readily identifies and self-corrects hazards; they are supported by management when they do so.

Notes:

WORKPLACE ANALYSIS

Accident Investigation

Accident investigation: An effective program will provide for **investigation of accidents and "near miss" incidents**, so that their causes, and the means for their prevention, are identified. [Guidelines, (c)(2)(iv)]

1	No investigation of accidents, injuries, near misses, or other incidents is conducted.
2	Some investigation of incidents takes place, but root cause may not be identified, and correction may be inconsistent. Supervisors prepare injury reports for lost time cases.
3	OSHA-101 is completed for all recordable incidents. Reports are generally prepared with cause identification and corrective measures prescribed.
4	OSHA-recordable incidents are always investigated, and effective prevention is implemented. Reports and recommendations are available to employees. Quality and completeness of investigations are systematically reviewed by trained safety personnel.
5	All loss-producing accidents and "near-misses" are investigated for root causes by teams or individuals that include trained safety personnel and employees.

Notes:

WORKPLACE ANALYSIS
Data Analysis
Data analysis: An effective program will **analyze injury and illness records** for indications of sources and locations of hazards, and jobs that experience higher numbers of injuries. By analyzing injury and illness trends over time, patterns with common causes can be identified and prevented.

1	Little or no analysis of injury/illness records; exposure monitoring) are kept or conducted.
2	Data is collected and analyzed, but not widely used for prevention. Reports are completed for all recordable cases. Exposure records and analyses are organized and are available to safety personnel.
3	Injury/illness logs and exposure records are kept correctly, are audited by facility personnel, and are essentially accurate and complete. Rates are calculated so as to identify high risk areas and jobs. Workers compensation claim records are analyzed and the results used in the program. Significant analytical findings are used for prevention.
4	Employer can identify the frequent and most severe problem areas, the high risk areas and job classifications, and any exposures responsible for OSHA recordable cases. Data are fully analyzed and effectively communicated to employees. Illness/injury data are audited and certified by a responsible person.
5	All levels of management and the workforce are aware of results of data analyses and resulting preventive activity. External audits of accuracy of injury and illness data, including review of all available data sources are conducted. Scientific analysis of health information, including non-occupational data bases is included where appropriate in the program.

Notes:

HAZARD CONTROL

Hazard Control

Hazard Control: Workforce exposure to all current and potential hazards should be prevented or controlled by using **engineering controls** wherever feasible and appropriate, **work practices** and **administrative controls**, and **personal protective equipment** (PPE).

1	Hazard controls are seriously lacking or absent from the facility.
2	Hazard controls are generally in place, but effectiveness and completeness vary. Serious hazards may still exist. Employer has achieved general compliance with applicable OSHA standards regarding hazards with a significant probability of causing serious physical harm. Hazards that have caused past injuries in the facility have been corrected.
3	Appropriate controls (engineering, work practice, and administrative controls, and PPE) are in place for significant hazards. Some serious hazards may exist. Employer is generally in compliance with voluntary standards, industry practices, and manufacturers, and suppliers' safety recommendations. Documented reviews of needs for machine guarding, energy lockout, ergonomics, materials handling, bloodborne pathogens, confined space, hazard communication, and other generally applicable standards have been conducted. The overall program tolerates occasional deviations.
4	Hazard controls are fully in place, and are known and supported by the workforce. Few serious hazards exist. The employer requires strict and complete compliance with all OSHA, consensus, and industry standards and recommendations. All deviations are identified and causes determined.
5	Hazard controls are fully in place and continually improved upon based on workplace experience and general knowledge. Documented reviews of needs are conducted by certified health and safety professionals or professional engineers, etc.

Notes:

HAZARD CONTROL
Maintenance
Maintenance: An effective safety and health program will provide for **facility and equipment maintenance**, so that hazardous breakdowns are prevented.

1	No preventive maintenance program is in place; break-down maintenance is the rule.
2	There is a preventive maintenance schedule, but it does not cover everything and may be allowed to slide or performance is not documented. Safety devices on machinery and equipment are generally checked before each production shift.
3	A preventive maintenance schedule is implemented for areas where it is most needed; it is followed under normal circumstances. Manufacturers' and industry recommendations and consensus standards for maintenance frequency are complied with. Breakdown repairs for safety related items are expedited. Safety device checks are documented. Ventilation system function is observed periodically.
4	The employer has effectively implemented a preventive maintenance schedule that applies to all equipment. Facility experience is used to improve safety-related preventative maintenance scheduling.
5	There is a comprehensive safety and preventive maintenance program that maximizes equipment reliability.

Notes:

HAZARD CONTROL

Medical Program

An effective safety and health program will include a suitable **medical program** where it is appropriate for the size and nature of the workplace and its hazards.

1	Employer is unaware of, or unresponsive to medical needs. Required medical surveillance, monitoring, and reporting are absent or inadequate.
2	Required medical surveillance, monitoring, removal, and reporting responsibilities for applicable standards are assigned and carried out, but results may be incomplete or inadequate.
3	Medical surveillance, removal, monitoring, and reporting comply with applicable standards. Employees report early signs/symptoms of job-related injury or illness and receive appropriate treatment.
4	Health care providers provide follow-up on employee treatment protocols and are involved in hazard identification and control in the workplace. Medical surveillance addresses conditions not covered by specific standards. Employee concerns about medical treatment are documented and responded to.
5	Health care providers are on-site for all production shifts and are involved in hazard identification and training. Health care providers periodically observe the work areas and activities and are fully involved in hazard identification and training.

Notes:

TRAINING
Safety and health training should cover the safety and health responsibilities of all personnel who work at the site or affect its operations. It is most effective when incorporated into other training about performance requirements and job practices. It should include all subjects and areas necessary to address the hazards at the site.

1	Facility depends on experience and peer training to meet needs. Managers/supervisors demonstrate little or no involvement in safety and health training responsibilities.
2	Some orientation training is given to new hires. Some safety training materials (e.g., pamphlets, posters, videotapes) are available or are used periodically at safety meetings, but there is little or no documentation of training or assessment of worker knowledge in this area. Managers generally demonstrate awareness of safety and health responsibilities, but have limited training themselves or involvement in the site's training program.
3	Training includes OSHA rights and access to information. Training required by applicable standards is provided to all site employees. Supervisors and managers attend training in all subjects provided to employees under their direction. Employees can generally demonstrate the skills/knowledge necessary to perform their jobs safely. Records of training are kept and training is evaluated to ensure that it is effective.
4	Knowledgeable persons conduct safety and health training that is scheduled, assessed, and documented, and addresses all necessary technical topics. Employees are trained to recognize hazards, violations of OSHA standards, and facility practices. Employees are trained to report violations to management. All site employees—including supervisors and managers—can generally demonstrate preparedness for participation in the overall safety and health program. There are easily retrievable scheduling and record keeping systems.
5	Knowledgeable persons conduct safety and health training that is scheduled, assessed, and documented. Training covers all necessary topics and situations, and includes all persons working at the site (hourly employees, supervisors, managers, contractors, part-time and temporary employees). Employees participate in creating site-specific training methods and materials. Employees are trained to recognize inadequate responses to reported program violations. Retrievable record keeping system provides for appropriate retraining, makeup training, and modifications to training as the result of evaluations.

Notes:

CHAPTER 19

SAFETY AND HEALTH HAZARDS IN THE OFFICE

Despite common beliefs that the office provides a safe environment to work in, many hazards exist that cause thousands of injuries and health problems each year among office workers. Since one-third of the workforce is in offices, even low rates of work-related injuries and illnesses can have an immense impact on employee safety and health.

Modern offices today are substantially different from the office environment of 20 years ago. Sweeping changes have occurred in the U.S. workplaces as a result of new office technology and the automation of office equipment. Unfortunately, office workers are now faced with many more hazards.

In addition to obvious hazards such as a slippery floor or an open file drawer, an office also may contain hazards such as poor lighting, noise, poorly designed furniture and equipment, and machines that emit noxious gases and fumes. Even the nature of office work itself has produced a host of stress-related symptoms and musculoskeletal strains. For example, long hours at the visual display terminal can cause pains in the neck and back, eyestrain, and a general feeling of tension and irritability.

Leading Types of Disabling Accidents

It is estimated that office workers sustain 76,000 fractures, dislocations, sprains, strains and contusions each year. The leading types of disabling accidents that occur within the office are:
- Falls, strains and overexertion;
- Struck by or striking against objects; and
- Caught in or between objects.

Office workers also are injured as a result of foreign substances in the eye, spilled hot liquids, and burns from fire and electric shock.

In recent years, illness has increased among the office worker population This may be attributed, in part, to the increased presence of environmental toxins within the office and to stress-producing factors associated with the automated office. Resulting illnesses may include respiratory problems, skin diseases and stress-related conditions.

Common Safety and Health Hazards in the Office

There are several types of safety and health hazards common to the office environment; however, there also are control measures that can reduce or eliminate these hazards.

Ventilation

Sources of air pollution in the office include both natural agents (e.g., carbon monoxide, microorganisms, radon) and synthetic chemicals (e.g., formaldehyde, cleaning fluids, cigarette smoke, asbestos).

An adequate ventilation system that delivers quality indoor air and provides for comfortable humidity and temperature is a necessity for the office. Where printing or copying machines are present, an exhaust ventilation system that draws fumes away from the employees' breathing zone should be present. Office machines and ventilation system components should be checked and maintained on a regular basis.

Illumination

Lighting problems in the office include glare, shadows and visual problems (e.g., eyestrain, fatigue, double vision). Poor lighting can also be a contributing factor in accidents.

Controls to prevent poor lighting conditions include:

- Regular maintenance of the lighting system;
- Light-colored matte finish on walls, ceilings and floors to reduce glare;
- Adjustable shades on windows; and
- Indirect lighting.

Noise

In an office, workers are subjected to many noise sources, such as video display terminals, high-speed printers, telephones and human voices. Noise can produce tension and stress as well as damage to hearing. A variety of measures to control unwanted noise is available.

- Noisy machines should be placed in an enclosed space;
- Carpeting, draperies and acoustical ceiling tiles should be used to muffle noise;
- Telephone volume should be adjusted to its lowest level; and
- Traffic routes within the office should be rearranged to reduce traffic within and between work areas.

Physical Layout and Housekeeping

Poor design and/or poor housekeeping can lead to crowding, lack of privacy, and slips, trips and falls. Important factors related to office layout and orderliness include:

- At least 3 ft. distance between desks and at least 50 sq. ft. per employee;
- Keep telephone and electrical cords out of aisles;
- Group employees who use the same machines;
- Office machines should be kept away from edges of desks and tables;
- Regular inspection, repair and replacement of faulty carpets;
- Place mats inside building entrances; and
- Proper placement of electrical and telephone wires.

Exits/Egress

Blocked or improperly planned means of egress can lead to injuries as a result of slips, trips and falls. If during an emergency, employees become trapped due to improper egress, more serious injuries or fatalities may result.

Controls to ensure proper means of egress include providing:

- Access to exits with a minimum width of 28 in.;
- Two exits, if possible;
- Exits and access to exits that are clearly marked;
- Means of egress, including stairways used for emergency exit, that are free of obstructions and adequately lit; and
- Employees with training on the locations of exits and procedures for evacuation.

Fire Hazards

A serious problem associated with office design is the potential for creating fire hazards. Another danger found in modern offices is combustible materials (e.g., furniture, rugs, fibers) that can easily ignite and emit toxic fumes.

A number of steps can be taken to reduce office fire hazards:

- Store unused records or papers in fire-resistant files or vaults;
- Use flame-retardant materials;
- Allow smoking in designated areas only and provide proper ashtrays; and
- Place fire extinguishers and alarms in conspicuous, easily accessible locations.

Handling and Storage Hazards

The improper lifting of materials can cause musculoskeletal disorders such as sprains, strains and inflamed joints. Office materials that are improperly stored can lead to hazards such as objects falling on workers, poor visibility and fires.

There are several controls that can reduce handling and storage hazards:
- An effective control program incorporating employee awareness and training and ergonomic design of work tasks can reduce back injuries.
- Materials should not be stored on top of cabinets.
- Heavy objects should be stored on lower shelves and materials neatly stacked.
- Materials should be stored inside cabinets, files or lockers.
- There should be no storage of materials in aisles, corners or passageways.
- Fire equipment should remain unobstructed.
- Flammable and combustible materials and liquids should be identified and stored properly.
- Material safety data sheets must be provided for each hazardous chemical identified.

Electrical Equipment

Electrical accidents in an office usually occur as a result of faulty or defective equipment, unsafe installation or misuse of equipment. The following guidelines should be adhered to when installing or using electrical equipment:
- Equipment must be properly grounded to prevent shock injuries;
- A sufficient number of outlets will prevent overloading of circuits;
- The use of poorly maintained or nonapproved equipment should be avoided;
- Cords should not be dragged over nails, hooks or other sharp objects;
- Receptacles should be installed and electric equipment maintained so that no live parts are exposed; and
- Machines should be disconnected before cleaning or adjusting. Generally, machines and equipment should be locked or tagged out during maintenance.

Office Furniture

Defective furniture or the misuse of chairs or file cabinets by office workers can lead to serious injuries. Controls related to chairs and cabinets include:

- Chairs should be properly designed and regularly inspected for missing casters, shaky legs and loose parts;
- Employees should not lean back in a chair with feet on a desk;
- Employees should not scoot across the floor while sitting on a chair;
- Employees should never stand on a chair to reach an overhead object;
- Employees should open only one file drawer at a time and use drawer handles to close file drawers; and
- File cabinets should not be located close to doorways or in aisles.

Office Machinery

Machines with ingoing nip points or rotating parts, if not adequately guarded, can cause lacerations, abrasions, fractures and amputations. Machines such as conveyors, electric hole punches and paper shredders with hazardous moving parts must be guarded so that office workers cannot contact the moving parts. Fans must have substantial bases and fan blades must be properly guarded.

Ladders, Stands and Stools

Improper use of ladders, ladder stands and stools can lead to falls. To reduce related injuries include workers should:
- Always face the ladder when climbing up or down;
- Ensure ladders are in good condition and inspected regularly;
- Never use the top of a ladder as a step; and
- Ensure the ladder is fully open and the spreaders are locked before climbing on it.

Office Tools

Misuse of office tools, such as pens, pencils, paper, letter openers, scissors and staplers, can cause cuts, punctures and related infections. Precautions when using these materials can prevent such injuries. Precautions include:
- The blade on a paper cutter should be closed when not in use. A guard also should be provided. Employees should never place their fingers near the blade.
- Employees should always use a staple remover to remove staples. An employee also

should never test a jammed stapler with his or her fingers.
- Employees should store sharp objects (e.g., scissors, pen, pencil) in a drawer or with the point down. Employees should never hand someone a sharp object point first.

Photocopying Machines

Potential health hazards associated with photocopying machines include toxic chemicals, excessive noise and intense light. They can also be a source of indoor air pollution when used in offices that are not well ventilated.

Controls to reduce these hazards include:
- Keeping the document cover closed;
- Reducing noise exposure by isolating the machine;
- Placing machines in well-ventilated rooms away from workers' desks;
- Having machines routinely serviced to prevent chemical emissions;
- Avoiding skin contact with photocopying chemicals; and
- Cleaning all spills immediately and disposing of waste properly.

Video Display Terminals (VDTs)

Health concerns relating to VDT, or computer monitor, use involve radiation, noise, eye irritation, stress and low back, neck and shoulder pain.

Studies have shown that the radiation levels emitted from VDTs are well below those allowed by current standards. However, to minimize any potential exposure, only equipment for which the manufacturer will supply data on emissions should be used.

To minimize noise, VDTs should not be clustered and sound absorbent screens can be used if needed.

Proper ergonomic design, including the relation of the operator to the screen, the operator's posture, lighting and background, should be carefully tailored to prevent discomfort. The keyboard position, document holder, screen design, characters, and color are other factors to consider.

Vision testing should be conducted before office workers operate VDTs and annually thereafter. Work breaks and variation of tasks enable VDT operators to rest their eyes. Postural strain related to VDT use can be relieved by performing simple exercises. Finally, a training program to inform workers of the capabilities of the equipment they are using should be conducted.

Some Do's and Dont's of Office Safety

1. Glass doors should have some conspicuous design, either painted or decal, about 4 ft. above the floor and centered on the door so people will not walk into them.
2. Frosted glass in doors gives a view through for accident prevention, but still preserves privacy. The see-through feature also prevents collisions.
3. If it is necessary that a door be solid, the hazardous area that the door swings over can be marked by yellow and black tape or painted a bright color, or the path of the swinging door can be outlined by colored plastic circles. On carpeted floors, a quarter or half-circle of different colored carpet can be used.
4. Employees should not face windows, unshielded lamps or other sources of glare. Many factors associated with poor illumination are contributing causes of office accidents. Some of these causes are direct glare, reflected glare from the work and harsh shadows. Excessive visual fatigue may be an element leading toward accidents. Accidents may also be prompted by the delayed eye adaptation that a person experiences when moving from bright to dark. Some accidents that are attributed to the individual's "carelessness" actually can be traced to difficulty in seeing.
5. Fans in the office should be placed where they cannot fall on anybody and they should be secured in place.
6. Where possible, outlets (receptacles) should be installed to eliminate extension cords. If cords must cross the floors, they should be covered with rubber channels designed for this purpose.
7. A policy should be adopted to prevent the use of poorly maintained or unsafe, poor quality coffee makers, radios, lamps, hot plates etc., provided by or used by employees, particularly in out-of-the-way locations. Such appliances create fire and shock hazards.

8. Switches should be provided, either in the equipment or in the cords, so that it is not necessary to pull the plugs to shut off the power.
9. Outlets should not only be located under desks to eliminate tripping hazards but they should also be placed where they will not accidentally be kicked or used as a footrest. When they loosen or wear, outlets can become sources of electric shock.
10. Where materials are stored on shelves, the heavy objects should be on the lower shelves.
11. Rolling ladders and stands used for reaching high storage should have brakes that automatically operate when weight is applied to them.
12. Good housekeeping is essential to prevent falls. Wipe up spilled liquid immediately and pick up pieces of paper, paper clips, rubber bands and pencils as soon as they are spotted.
13. Broken glass should be swept up immediately. It should not be placed loose in a waste paper basket but it should be carefully wrapped in heavy paper and marked "Broken Glass." Glass that shatters into fine pieces should be blotted up with damp paper towels.
14. Leaning back in the chair and placing the feet on the desk should not be allowed.
15. Chairs should have at least four legs on the ground whenever they are being occupied.
16. Only one file drawer should be opened at a time to prevent the cabinet from toppling over. File cabinets should be bolted together or otherwise secured to prevent tipping.
17. Desks or files should never be moved by office personnel; they should be moved by maintenance people using special dollies or trucks.
18. Do not pile boxes, papers, books or other heavy objects on top of file cabinets. This could cause the cabinet to tip and the materials to fall on to the employee.
19. Stacks of materials should not be carried on stairs; the elevator should be used instead.
20. Accident records are absolutely necessary if an office safety program is to succeed. Office employees should report every accident, no matter how minor the injury. Near-miss accidents should be reported because they often serve as warnings of worse accidents to come.
21. To develop proper safety attitudes, safety instructions should be properly given to new office employees.
22. A company safety program cannot succeed unless it has the wholehearted backing of top management. The supervisor must know that his or her accident prevention performance is watched and that good performance is appreciated.
23. Supervisors should personally investigate each accidental injury and they should report what caused the accident as well as what steps have been taken to prevent a recurrence.
24. When safety inspections are made and accident hazards are found, the report of these hazards should go to the vice president or other top officers in charge of the department. This top executive should then request a report of corrective measures directly from the supervisor responsible for the hazard. This procedure will go along way in making a safety expert of every office manager and supervisor and reducing the problem of office accidents to a negligible size.

Checklist for Office Safety Concerns

- ☐ Exit safety
- ☐ Condition of furniture and ladders
- ☐ Housekeeping and aisles
- ☐ Grounding
- ☐ Toolbox power tools
- ☐ Floor cleaning equipment
- ☐ Cleaning supplies (corrosives)
- ☐ Guarding of equipment
- ☐ Flammable materials
- ☐ Minimize quantities
- ☐ Proper Cans and Cabinets

Emergency Action Plan

- ☐ Written plan (written for more than 10 employees)
- ☐ Alarm system
- ☐ Evacuation
- ☐ Training

Plan Elements

- ☐ Escape procedure and routes
- ☐ Critical operation (control or shutdown)
- ☐ Accounting for employees
- ☐ Rescue duties
- ☐ Medical duties
- ☐ Preferred method of reporting emergency
- ☐ Authorities

Fire Prevention Plans

- ☐ Hazard lists (written for more than 10 employees)
- ☐ Responsible persons
- ☐ Trash and waste control
- ☐ Training
- ☐ Maintenance of combustion controls

Personal Office Safety and Health Checklist

- ☐ Is there an active safety and health program in your office?
- ☐ Is one person clearly responsible for the overall activities of the safety and health program?
- ☐ Is there a procedure for handling complaints regarding safety and health?
- ☐ Do you know how to locate the nearest doctor or hospital?
- ☐ Are emergency numbers posted?
- ☐ Are first-aid kits easily accessible to each work area?
- ☐ Are you familiar with basic first-aid procedures in case of an emergency?
- ☐ Are all work areas clean and orderly?
- ☐ Are floor surfaces clean, dry, level and in good condition?
- ☐ Are carpets secured to the floor and in good condition?
- ☐ Are aisles and doorways free from obstructions to permit visibility and movement?
- ☐ Are there sufficient exits to permit prompt escape in case of emergency?
- ☐ Are all exits clearly marked and visible?
- ☐ Are emergency exits adequately lighted and free of debris?
- ☐ Do you know where emergency exits are and how to reach them?
- ☐ Are stairways in good condition and covered with skid-resistant materials?
- ☐ Do you know where fire extinguishers are and how to use them?
- ☐ Do you know where fire alarms are?
- ☐ Are you familiar with fire evacuation procedures for your building and what to do in case of fire in your area?
- ☐ Are electrical appliances and equipment in good condition and properly grounded?
- ☐ Are a sufficient number of outlets available to eliminate overloading of circuits?
- ☐ Are file cabinets arranged so that drawers do not open into aisles? Can only one drawer be opened at a time?
- ☐ Are chairs in good condition with no loose casters? Is furniture free from sharp edges, points and splinters?
- ☐ Is your office equipped with a step stool or ladder so that you can safely reach overhead objects? Are you familiar with the correct way to use a ladder?
- ☐ Are photocopying machines placed in well-ventilated rooms away from workers' desks? Are machines serviced routinely?
- ☐ If you work with hazardous substances such as cleaning fluids, are you aware of the related hazards?
- ☐ Are hazardous substances properly stored?
- ☐ Are work areas properly illuminated?
- ☐ Does the ventilation system deliver quality indoor air?
- ☐ Are noise levels within acceptable levels?
- ☐ If you use a video display terminal (VDT), are the keyboard, table, screen and chair adjustable?
- ☐ For VDT users, are work breaks and variation of tasks incorporated into work schedules?
- ☐ Are you trained in proper lifting techniques?

CHAPTER 20

OSHA Programs

Voluntary Protection Programs

The Voluntary Protection Programs (VPP) is designed to recognize and promote effective safety and health management. In the VPP, management, labor, and OSHA establish a cooperative relationship at a workplace that has implemented a strong program. Management agrees to operate an effective program that meets an established set of criteria, while employees agree to participate in the program and work with management to ensure a safe and healthful workplace. OSHA verifies that the program meets VPP criteria and then publicly recognizes the site's exemplary program. OSHA also removes the site from its scheduled routine inspection lists. To confirm the site continues to meet VPP criteria, OSHA reassesses periodically the site program (e.g., every three years or every year).

The VPP concept recognizes that compliance enforcement alone cannot fully achieve the objectives of the Occupational Safety and Health Act. Good safety management programs that go beyond OSHA standards can protect workers more effectively than simple compliance.

VPP participants are a select group of employers that have designed and implemented outstanding health and safety programs. Star Program participants meet all VPP requirements. Merit Program participants have demonstrated the potential and willingness to achieve Star Program status and are implementing planned steps to fully meet all Star requirements.

Participants in the VPP enjoy benefits, such as:
- Improved employee motivation to work safely, which leads to better quality and productivity;
- Reduced workers' compensation costs;
- Recognition in the community;
- Improvement of programs that are already good, through the internal and external review that is part of the VPP application process; and
- Fewer lost-workday injuries (VPP participants generally experience from 60 percent to 80 percent fewer lost-workday injuries than average sites of the same size and industry).

Companies that would like to participate in the VPP must submit a written application to OSHA. The application guideline is included in the VPP information kit. After OSHA has reviewed the written application, an Onsite Review will be scheduled. "What Happens When OSHA Comes Onsite," included in the VPP information kit, describes the onsite review.

Participants provide OSHA with examples of the most effective way to protect workers in their industries, often in ways that exceed the requirements of OSHA standards. Participants from several companies help provide OSHA compliance and program personnel training in safety and health program management. Many participants in the petro-chemical industry provided OSHA standards setters with models of effective process safety management. For example, Mobil Oil Company's Joliet, Ill. refinery provides OSHA compliance officers with hands-on training in process safety management.

The experience with the VPP has led to the issuance of OSHA's Voluntary Safety and Health Program Management Guidelines. CIBA Inc.'s McIntosh, Ala. site and Dow Chemical's Freeport, Texas facility have provided OSHA standards setters with demonstrations of effective 100 percent fall protection. Participation in VPP onsite evaluations provides OSHA with model plants that can be shown to others. As one compliance officer said at an evaluation closing conference, "I've always felt it could be done this way, and now I've seen that it can be done right. So no one will ever be able to tell me again that it can't be done."

The three VPPs—Star, Merit, and Demonstration—are designed to:
- Recognize outstanding achievement of those who have successfully incorporated comprehensive safety and health programs into their total management systems;
- Motivate others to achieve excellent safety and health results in the same outstanding way; and
- Establish a relationship between employers, employees, and OSHA that is based on cooperation rather than coercion.

Star Program. This program is the most demanding and the most prestigious. It is open to employers in any industry who have successfully managed a comprehensive safety and health program to reduce injury rates below the national average for that industry. Specific requirements for the program include:
- Management commitment and employee participation;
- A high quality worksite analysis program;
- Hazard prevention and control programs; and
- Comprehensive safety and health training for all employees.

These requirements must all be in place and operating effectively.

Merit Program. This program is primarily a stepping stone to Star Program. An employer with a basic safety and health program built around the Star requirement who is committed to improving the company's program and has the resources to do so within a specified period of time may work with OSHA to meet Star qualifications.

Demonstration Program. This program is for companies that provide Star-quality worker protection in industries where certain Star requirements may not be appropriate or effective. It allows OSHA both the opportunity to recognize outstanding safety and health programs that would otherwise be unreached by the VPP and to determine if general Star requirements can be changed to include these companies as Star participants.

OSHA reviews an employer's VPP application and conducts an onsite review to verify that the safety and health program described is in operation at the site. Evaluations are conducted on a regular basis: annually for Merit and Demonstration programs and triennially for Star. All participants must send their injury information annually to their OSHA regional office. Sites participating in the VPP are not scheduled for programmed inspection; however, any employee complaints, serious accidents or significant chemical releases that may occur are handled according to routine enforcement procedures.

An employer may make application for any VPP at the nearest OSHA regional office. Once OSHA is satisfied that, on paper, the employer qualifies for the program, an onsite review will be scheduled. The review team will present its findings in a written report for the company's review prior to submission to the Assistant Secretary of Labor, who heads OSHA. If approved, the employer will receive a letter from the Assistant Secretary informing the site of its participation in the VPP. A certificate of approval and flag are presented at a ceremony held at or near the approved worksite. Star sites receiving reapproval after each triennial evaluation receive plaques at similar ceremonies.

The VPPs described are available in states under federal jurisdiction. Some state-plan states have similar programs. Interested employers in these states should contact the appropriate state designee for more information.

Additional information on VPPs is available from OSHA national, regional, and area offices.

OSHA Outreach Program

The OSHA Outreach Training Program is a primary way to train workers in the basics of occupational safety and health. Through the program, individuals who complete a one-week OSHA trainer course are authorized to teach 10-hour or 30-hour courses in construction or general industry safety and health standards. Authorized trainers can receive OSHA course completion cards for their students.

Becoming an Authorized Trainer

To become an authorized trainer, you must complete the required Course 501, Trainer Course in Occupational Safety and Health Standards for General Industry, and pass a test at the end of the course. This course provides an overview of the most hazardous and referenced standards. It is conducted by the OSHA Training Institute and OSHA Training Institute Education Centers located around the country. When you complete the course, you are authorized to train for four years. Before the end of four years, you must take Course 503, Update for General Industry Outreach Trainers, to renew your authorization for another four years. Trainers also may also retake Course 500 to maintain their trainer status. If the trainer authorization status has expired, taking a 500 course available through the OSHA Training Institute Education Centers for course scheduling information can reinstate it.

General industry outreach trainers are authorized to conduct 10- and 30-hour general industry outreach courses and receive OSHA course completion cards to issue to the students.

Outreach Training Program Guidelines

Designated Training Topics
Training must cover the designated topics for the 10-or 30-hour General Industry Outreach Training Program. The 10-hour program is intended to provide a variety of instruction on general industry safety and health standards to entry-level workers. Of the topics listed below, four hours are mandatory and three hours must be chosen from the options provided. For the remainder of the class, you may teach any other general industry standards and policies and/or expand on the required topics.

The following course topics are required:
- Introduction to OSHA, OSH Act/General Duty Clause 5(a)(1), Inspections, Citations, and Penalties (CFR Part 1903);
- Subpart D, Walking and Working Surfaces;
- Subparts E and L, Exit Routes and Fire Protection; and
- Subpart S, Electrical.

Trainers must cover at least three of the following topics:
- Subpart H, Flammable and Combustible Liquids;
- Subpart I, Personal Protective Equipment;
- Subpart O, Machine Guarding;
- Subpart Z, Hazard Communication; or
- Subpart Z, Introduction to Industrial Hygiene, Bloodborne Pathogens and/or Ergonomics
- Safety and Health Programs.

Additional Topics
By tailoring your 10 hours to meet the needs of all industries, you can easily develop two-day seminars for specific industries. For example, when developing separate seminars for maintenance workers and medical workers, the core material (required topics) will remain the same. The medical workers may not need to know about machine guarding or flammable liquid storage, so you can include that information in the maintenance seminar, but substitute it with hazard communication and bloodborne pathogens for the medical seminar.

Additional topics for specific industries could include:
- **Medical/Healthcare**
  - Introduction to Industrial Hygiene, Bloodborne Pathogens (two hours); and
  - Ergonomics, Personal Protective Equipment, Hazard Communication, and Workplace Violence (at least two hours).
- **Maintenance**
  - Ergonomics;
  - Hazard Communication; and
  - Powered Industrial Trucks.

- **Utility**
  - Ergonomics;
  - Power Generation;
  - Confined Spaces; and
  - Ladder Safety.
- **General Office**
  - Ergonomics;
  - Office Safety; and
  - Hazard Communication.

Topic Length
Training sessions for each course topic should last for at least one hour. In most cases, one hour is the least amount of time necessary to cover a topic adequately. The minimum amount of time spent on any topic is 30 minutes.

Trainers must spend at least 10 or 30 class hours covering course topics. Breaks and lunch periods are not counted as class time.

Conducting a class over a period of time. Trainers may teach classes in segments. Each segment must be at least one hour and the course must be completed within six months.

Guest trainers. Trainers may have other trainers help conduct classes. An authorized outreach trainer must design and coordinate the course, teach more of the class than anyone else, and attend all sessions to answer questions and ensure the topics are adequately covered and all students are in attendance.

In-person training. Outreach training must be delivered in person, unless the trainer has received an individual exception. Trainers considering using other training methods, such as online, CDROM, and video conferencing, must contact the OSHA Outreach Coordinator. The basic guidelines for video conferencing are that the trainer be able to ensure the full attendance of all trainees, that off-site locations have a training monitor, and that there is a setup to answer trainee questions quickly and effectively.

Class size. If class will exceed 50 students, contact the OSHA Outreach Coordinator before the class begins to get permission for the class to be considered an outreach training class. Trainers must have a way for students to ask questions apart from the classroom. E-mail is the usual method. It is a good idea to have more than one trainer in large classes and to break the class into work groups whenever possible. Small classes encourage trainee involvement through discussion and group participation and through sharing of knowledge and experiences.

10 + 20 Hours = 30 hours. If a student that was trained in the 10-hour training course is interested in taking the 30-hour course, trainers may supplement the training with 20 additional hours and receive a 30-hour card for the trainee. The limitations are that the same trainer must do the training and all the training must take place within six months. Trainees who do this should only receive the 30-hour card.

Combined 10-hour construction and general industry class. A person who is authorized as a trainer in Construction and General Industry may *not* receive 10-hour student cards for both Construction and General Industry when holding a class for less than 20 full hours.

Outreach Training Tips

Worker Emphasis
The outreach classes are designed to be presented to workers; therefore, they must emphasize hazard identification, avoidance, control, and prevention—not OSHA standards. Trainers must tailor their presentations to the needs and understanding of their audience.

Site-Specific Training
The most rewarding classes for students are the ones they can relate to because the trainer uses examples, pictures, and scenarios that come from the students' workplaces.

Homogenous Class
The ideal class to teach is one in which students have similar needs because they hold comparable positions. Therefore, it is best to conduct separate training for supervisors, managers, and workers. Also, separate workers into like groups (e.g., office personnel, machine operators, and maintenance staff).

Use Objectives
Inform students what is expected of them after receiving training on each topic. Describe the skills and abilities the students should have or exhibit, be specific, and relate the objectives to the students' work if possible.

Presentation Assortment
Students learn in different manners and get tired of one style of training over a lengthy period. Use different trainers, computer presentations, videos, case studies, exercises, and graphics to make the course interesting and enjoyable. By doing this you will be employing the three levels of training techniques:
- Presentation (presenting the material in a variety of ways);
- Discussion (getting the students involved in the learning); and
- Performance (students practice the material they have learned).

Testing
Use quizzes and tests to ensure the students understand key objectives of the training. Students should receive feedback on incorrect answers.

Student Evaluations
Have students evaluate the class. This feedback will help the trainer determine whether the course is accomplishing its goals and provide input that can be used to improve the training.

Monitoring
Staff from the OSHA Office of Training and Education periodically attend outreach training classes to observe training, obtain feedback from the trainer and the students on the training, and to ensure awareness of the outreach guidelines and the materials and assistance that are available to help trainers conduct classes. Through these visits, OSHA aims to help trainers, improve the outreach-training program, ensure consistent program implementation, and assist the trainer in designing the class to meet the needs of the audience.

Advertising
When advertising outreach-training courses, trainers must take the proper care to correctly describe their outreach trainer designation and outreach courses. Trainer authorization is limited to conducting the 10- and 30-hour General Industry outreach training courses.

Outreach advertising restrictions include the following:
- **Certified.** Neither the trainer, the students, nor the curriculum is certified or accredited. The trainer is authorized, the students receive course completion cards, and the OSHA-produced curriculum is approved.
- **OSHA.** Do not make it appear that you are an OSHA employee or that the course is an OSHA course.
- **Course 501.** Do not refer to your outreach course as a 501 course. The 501 course is the train-the-trainer course that is conducted by the OSHA Training Institute or its OSHA Training Institute Education Centers.
- **Department of Labor Logo.** Do not use the logo that you see on outreach cards in your advertising.
- **Train-the-Trainer Course.** Students who complete the 10- or 30-hour training are not entitled to receive cards for students they may train.

If OSHA notifies a trainer that their advertising appears false or misleading and the advertising is not corrected, that trainer will be removed from the program.

Program Administration

Obtaining Student Course Completion Cards
After the class has been conducted, trainers must submit documentation about the course to the OSHA Training Institute to receive OSHA student course completion cards. Prepare a separate packet for each class taught that contains the Outreach Training Program Report, a list of the topics that were taught, and the student roster.
- Submit course documentation within six months of course completion.
- Send one list of legible student names; do not send multiple sign-in sheets.
- You may send more than one class submission in one envelope.
- Do not staple the documentation.

For each outreach class you complete, send the following, in this order, to OSHA (**Note:** OSHA may return submittals that are not complete):
- The OSHA Outreach Training Program Report, which includes information on the course and the trainer. Make sure you include your most recent trainer date and the address to send the cards. (An excerpt of the format is shown on the following pages.)

- A copy of your course certificate/trainer card (note the place you took the training) if:
  - This is your first outreach training class; or
  - You are a trainer with an ID number, but you have taken a more recent trainer course than the one shown on your preprinted Outreach Training Program Report.
- The names of students who completed the course. Keep a copy of the students' names and addresses before sending them to OSHA. You will need them to complete and distribute the cards
- A topic outline that lists the topics taught and the amount of time spent on each. For 10-hour classes, you may complete the topic outline on the bottom of the OSHA Outreach Training Program Report or send a separate outline. For 30-hour classes, you must send a separate outline.

Mail your course documentation to the Education Center that issued you your trainer card. The OSHA Training Institute oversees the program and can be reached at:

Construction (or General Industry)
Outreach Program Coordinator
OSHA Training Institute
2020 S. Arlington Heights Road
Arlington Heights, IL 60005

Student course completion cards are sent to the trainer for completion and distribution.

The trainer completes the card by listing the student's name, the end date of the course, and signing it. If the trainer is unavailable, the trainer's name may be typed in (See Exhibit 18-1.).

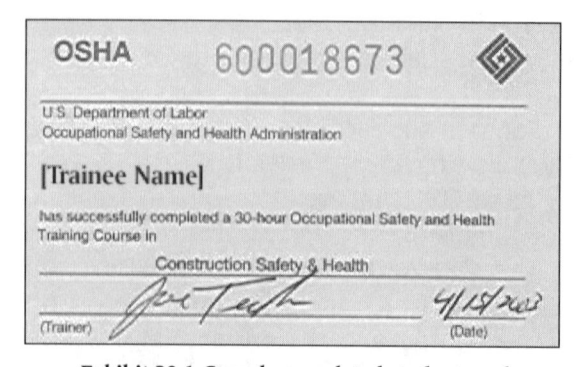

Exhibit 20-1. Sample completed student card.

Cards may be laminated. Trainers may use the back of the cards for other identification or training information purposes, but no other alterations are permitted.

The student course completion cards do not expire.

Trainers may provide students with a certificate of training. This may help students have proof they took the training before they receive their pocket student cards. The advertising restrictions also apply to a certificate.

The OSHA Training Institute will send a few extra cards to trainers for each class completed in case of card errors or to enable trainers to replace lost student cards for students trained.

OSHA recommends Outreach Training Program courses as an orientation to occupational safety and health for workers. Participation is voluntary. Workers must receive additional training on specific hazards of their job.

Replacing Lost or Damaged Trainer or Student Cards
Trainer cards. Contact the organization that issued the card. Inform them what course you took and when you took it. After they validate the class was taken and the test was passed, they will issue a replacement card.

Student cards. Trainers receive a few extra cards when OSHA sends them student cards. Trainers can use these cards to replace student cards, after the proper verification is made. If you do not have an extra card, you or the student may contact OSHA for the replacement. You must provide the following information:
- Student name;
- Trainer name;
- Training date; and
- Type of class (10- or 30-hour, construction or general industry).

OSHA training records are only kept for five years plus the present year. If the training took place prior to this, OSHA will not be able to issue a new card.

OSHA OUTREACH TRAINING PROGRAM REPORT

Construction (or General Industry) Outreach Program Coordinator

Course Conducted (Document each class with a separate Report):
- 10-Hour Construction
- 10-Hour General Industry
- 30-Hour Construction
- 30-Hour General Industry

End Date of Course: _____

Number of Students: _____
(List students' names on back, or on a separate sheet)

Primary Trainer Course Information

ID Number* (see note below): _____

Name: _____

Course (500/501/502/503) Expiration Date: _____
*ID number – new trainers do not have one – this only applies to trainers who have received student cards)

Address – cards will be sent here:
(If you have an ID number and your address is the same, you do not need to complete this)
❏ **Check if this is a new address**
Name: _____
Company / Dept: _____
Address: _____
City /State /Zip: _____

Phone No: _____ extension: _____ Best time(s) to call: _____

Your documentation must include these items:
- OSHA Outreach Training Program Report
- Student names
- List of course topics and the time spent on each

Do not include these items with your documentation: Student evaluation forms, Student sign-in sheets, Long topic outline, stapled pages

OSHA Programs

OSHA OUTREACH TRAINING PROGRAM REPORT (continued)

10 Hour Topics (for a 30-hour class, send in a separate topic list)

Construction

HOURS*
____ Required Introduction to OSHA
____ Required Electrical (K)
____ Required Fall Protection (M)
____ Required Electrical (S)

General Industry

HOURS*
____ Required Introduction to OSHA
____ Required Walking and Working Surfaces (D)
____ Required Egress and Fire Prot. (E & L)

Required – Choose three or more:

____ Personal Protective and Lifesaving Equipment (E)
____ Required – Choose three or more
____ Materials Handling, Storage, Use and Disposal (H)
____ Flammable and Combustible Liquids (H)
____ Tools – Hand and Power (I)
____ Personal Protective Equipment (I)
____ Scaffolds (L)
____ Machine Guarding (O)
____ Cranes, Derricks, Hoists, Elevators, and Conveyors (N)
____ Hazard Communication (Z)
____ Excavations (P)
____ Introduction to Industrial Hygiene/Bloodborne Pathogens (Z)
____ Stairways and Ladders (X)
____ Ergonomics
____ Safety and Health Programs
____ Electives: Any OSHA Construction standard or policy_____
____ Electives: Any OSHA General Industry standard or policy_____

Indicate the amount of time spent on each of the topics in the class.

Shortcut Procedures for Outreach Trainers with ID Numbers

OSHA will accept student card requests by e-mail and fax if the trainer has an ID number.

OSHA also will accept a short mail-in format, which lessens the paperwork required. These procedures are only available to trainers with ID numbers. Trainers should carefully read the instructions below before using these procedures.

1. **Who is Eligible?** Outreach Trainers with ID Numbers.
   - ID Numbers are provided to trainers on preprinted Outreach Training Reports that are returned with the student cards.
   - If you've never received student cards from OSHA, you do not have an ID number.
   - ID numbers are not assigned to each person that takes a trainer class (#500 or #502).
   - If you are sending in your first submission, you must use the standard procedure.
   - ID numbers cannot be provided over the phone.

2. **Sending Requests for Student Cards** – Procedures (also see the following pages)
   a. E-mail to the Education Center that issued the instructor card. OTI email addresses are:
      Construction: diane.uramkin@osha.gov
      General Industry: diana.ward@osha.gov
      Subject: 10- or 30-Hour; Construction or General Industry
      Format your e-mail request based on the format that follows these procedures
      Do not send attachments
   b. Fax to the Education Center that issued the instructor card. OTI FAX number is:
      (847) 297-6636
      Send a cover page and indicate the number of pages you are sending
      Subject: 10- or 30-Hour; Construction or General Industry
      Format your fax request based on the format that follows these procedures
   c. Short Mail-in Format

   Use in place of preprinted Outreach Training Report. Use the short mail-in format that follows these procedures. Mail to the Education Center that issued the instructor card. OTI address is:

   OSHA Training Institute
   2020 S. Arlington Heights Road
   Arlington Heights, IL 60005

3. **Instructions**
   - Use your ID Number when requesting cards using one of these procedures
   - For each class – send a separate e-mail, fax, or mail-in format
   - Keep a file on each course that includes:
     - Topics taught and time spent on each
     - Student names and addresses – the student's work or home address.
     - A copy of the e-mail or fax you sent to request cards.

4. **Monitoring.** At times, OSHA may ask for a copy of your training topics. They conduct monitoring visits for the OSHA Outreach Training Program and may ask to see documentation of the topics and students taught for each outreach training class conducted.

E-Mail Format – Request for Outreach Training Cards – For Trainers with ID Numbers

This is a sample of the format you may send to us. You will have to create the format. Send a separate format for each class. Do not send attachments. Your e-mail should look like this:

TO:

SUBJECT: 10-Hour Construction or 30-Hour Construction
10-Hour General Industry or 30-Hour General Industry

Course End Date: ____/_____/_____

Number of Students_____

ID #: _____
(*IDs are provided on Outreach Training Reports that are sent with student cards to all trainers*)

Trainer Name: _____

Trainer Course Data: *If the Trainer or Update Course information shown on your Outreach Training Report is incorrect, please enter the following information:*

Check one, as applicable:
Construction – 500 – 502   Training Date: _____
General Industry – 501 – 503   Training Date: _____

Education Center that provided/sponsored the training

Trainer Address: (Only needed if your address has changed)
Street: _____
City: _____   State: _____ ZIP: _____

___(Check) I certify *that the topics taught and the time spent on the topics in this class met the requirements of the OSHA Outreach Training Program.*

Student Names (addresses are not necessary): (list names or attach a separate sheet)

FAX Format Request for Outreach Training, Cards – For Trainers with ID Numbers

Fax:
Course: 10-hour Construction   10-hour General Industry
         30-hour Construction   30-hour General Industry

Course End Date: _____

Number of Students: _____

ID Number: _____
(ID is provided on Outreach Training Report that is sent with student cards to trainer)

Trainer Name: _____

Trainer Course Data: *If the Trainer or Update Course information shown on your Outreach Training Report is incorrect, please enter the following information:*

Check one, as applicable:
Construction – 500 – 502   Training Date: _____
General Industry – 501 – 503 Training Date: _____

Education Center that provided/sponsored the training: _____

Trainer Address: (Only needed if your address has changed)
Street: _____
City: _____   State: _____ ZIP: _____

I certify that the topics taught and the time spent on the topics in this class met the requirements of the OSHA Outreach Training Program   (sign) _____

Student Names (addresses are not necessary):

1 _____	11 _____
2 _____	12 _____
3 _____	13 _____
4 _____	14 _____
5 _____	15 _____
6 _____	16 _____
7 _____	17 _____
8 _____	18 _____
9 _____	19 _____
10 _____	20 _____

Continue on another page if more students

Short Mail Format – Request for Outreach Training Cards – For Trainers with ID Numbers

Course: 10-hour Construction   10-hour General Industry
          30-hour Construction   30-hour General Industry

Course End Date: _____

Number of Students: _____

ID Number: _____
(ID is provided on Outreach Training Report sent with student cards to trainer)

Trainer Name: _____

Trainer Course Data: *If the Trainer or Update Course information shown on your Outreach Training Report is incorrect, please enter the following information:*

Check one, as applicable:
Construction   500 – 502   Training Date: _____
General Industry 501 – 503   Training Date: _____
Education Center that provided/sponsored the training: _____

Trainer Address: (Only needed if your address has changed)
Street: _____
City: _____   State: _____ ZIP: _____

I certify that the topics taught and the time spent on the topics in this class met the requirements of the OSHA *Outreach Training Program*   (Sign) _____

Student Names (addresses are not necessary):

1 _____	11 _____
2 _____	12 _____
3 _____	13 _____
4 _____	14 _____
5 _____	15 _____
6 _____	16 _____
7 _____	17 _____
8 _____	18 _____
9 _____	19 _____
10 _____	20 _____

Continue on another page if more students

Further Information

OSHA Web Sites
To keep up on OSHA and to obtain training materials, there are two main Web sites:
- The OSHA home page at www.osha.gov; and
- The OSHA Outreach Training Program page at www.osha.gov/outreach.
- The OSHA Outreach Training Program site includes the outreach program guidelines, teaching aids, and frequently asked questions. All pertinent program changes however are communicated at OSHA's home page.

Additional information can be found online for:
- OSHA Technical Links at www.osha-slc.cov/SLTC;
- Multimedia at www.osha-slc.gov/SLTC/multimedia.html;
- OSHA Video Loan Program at www.osha-sle.gov/Publications/video/video.html;
- Training Materials at www.osha-slc.gov/Training/OutReach.html;
- Industry Profiles at www.osha-slc.gov/SLTC/industries.html; and
- OSHA Small Business Page at www.osha-slc.gov/SmallBusiness/index.html.

OSHA Publications
OSHA has many helpful publications, forms, posters, and fact sheets. You can find information on these at www.osha-slc.gov/OshDoc/Additional.html. Single free copies of some of these are available by calling (202) 693-1888, faxing (202) 6932498, or contacting your nearest OSHA Area or Regional Office.

One helpful publication you can order or download is the OS114 Handbook for Small Businesses, OSHA 2209. This handbook assists small business owners in implementing OSHA's recommended safety and health program management guidelines.

U.S. Government Bookstores
These bookstores sell OSHA standards, publications, and subscriptions to the OSHA CD-ROM and Job Safety and Health Quarterly (OSHA's magazine). The bookstores are located throughout the country. The U.S. Government Printing Office (GPO) in Washington, D.C., is the main office. Materials may be ordered by calling (202) 512-1800 or visiting www.osha.gov/oshpubs/gpopubs.html.

National Technical Information Service (NTIS)
NTIS sells some OSHA informational and training associated materials. NTIS is the federal government's central source for the sale of scientific, technical, engineering, and related business information produced by or for the U.S. government. You can reach them at 1-800-553-6847 or visit their web site at www.ntis.gov/nac/safety.htm.

General Industry Outreach Training Program
If you cannot find the answers for your outreach questions in this guide or at the OSHA Web site, contact:
    Mr. Don Guerra
    Outreach Program Coordinator
    OSHA Office of Training and Education
    OSHA Training Institute
    2020 S. Arlington Heights Rd.
    Arlington Heights, IL 60005
    (847) 759-7735
    e-mail: outreachgosha.gov

Resource Center
The Resource Center Audiovisual Catalog is distributed in all 501 and 503 classes. The catalog is also available at the Outreach Training Program Web site. To receive an updated catalog or for further information contact:
    Ms. Linda Vosburgh
    Librarian
    OSHA Office of Training and Education
    (847) 759-7736
    e-mail: linda-vosburgh@osha.gov

Outreach Trainer State Lists
OSHA distributes lists of active trainers (two or more classes conducted within a year), by state, to persons looking for 10- or 30-hour outreach training and also to trainers who need assistance. To obtain a state list contact:

Diana Ward
OSHA Training Institute
2020 S. Arlington Heights Rd.
Arlington Heights, IL 60005
(847) 759-7737
e-mail: diana.ward@osha.gov

Provide:
- The state(s) you are looking for (up to three);
- Whether you want the list(s) for General Industry or Construction; and
- Your fax number or mailing address.

OSHA Support

Outreach trainers can also receive assistance from OSHA Regional and Area Offices. These offices provide publications, answer questions on OSHA standards, and provide other reference services. Contact information for these offices can be found at www.osha-slc.gov/html/RAmap.html

OSHA Training Institute Education Centers

The OSHA Training Institute (OTI) has authorized other educational institutions to conduct selected OSHA training courses. The education centers conduct all outreach courses, including:
- Course 500 Trainer Course in Occupational Safety and Health Standards for the Construction Industry
- Course 501 Trainer Course in Occupational Safety and Health Standards for General Industry
- Course 502 Update for Construction Industry Outreach Trainers
- Course 503 Update for General Industry Outreach Trainers

Region I
Keene State College
OSHA Education Center
175 Ammon Drive
Manchester, NH 03103-3308
Phone: (800) 449-6742
Fax: (603) 358-2569
www.keene.edu/conted/osha.cfm

Region II
Rochester Institute of Technology
31 Lomb Memorial Dr.
Rochester, NY 14623-5603
Phone: (866) 385-7470 x-2919
Fax: (585) 475-6292
www.rit.edu/osha

Atlantic OSHA Training Center
Univ. of Medicine & Dentistry NJ
683 Hoes Lane West
Piscataway, NJ 08854
Phone: (732) 235-9450
Fax: (732) 235-9460
sph.umdnj.edu/ophp

Atlantic OSHA Training Center
University at Buffalo
3435 Main Street Room 134
Buffalo, NY 14214-3000
Phone: (716) 829-2125
Fax: (716) 829-2806
wings.buffalo.edu/trc

Atlantic OSHA Training Center
Universidad Metropolitana
PO Box 21150
San Juan, PR 00928-1150
Phone: (787) 766-1717 x-6553
Fax: (787) 751-5540
www.suagm.edu/umet

Region III
National Labor College
George Meany Campus
10000 New Hampshire Avenue
Silver Spring, MD 20903-1706
Phone: (800) 367-6724
Fax: (301) 431-5411
www.georgemeany.org/html/nrc_for_osha_training.html

Bldg. Const. Trades Dept. AFL-CIO/
Center to Protect Workers' Rights
815 16th St. NW Ste 600
Washington, D.C. 20006-4101
Phone: (202) 756-4636
Fax: (202) 756-4675
www.cpwr.com/nrccoursecat.html

West Virginia University
Safety and Health Extension
130 Tower Lane
Morgantown, WV 26506-6615
Phone: (800) 626-4748
Fax: (304) 293-5905
www.wvu.edu/~exten/depts/she/osha.htm

Indiana University of Pennsylvania
1010 Oakland Ave
Indiana, PA 15705-1087
Phone: (724) 357-3019
Fax: (724) 357-3992
www.hhs.iup.edu/sa/oti/index.shtm

Region IV

Eastern Kentucky University
521 Lancaster Ave. Room 202
Richmond, KY 40475-3100
Phone: (888) 401-1956
Fax: (859) 622-6205
www.extendedprograms.eku.edu/osha

Georgia Institute of Technology
151 Sixth Street
Atlanta, GA 30332-0837
Phone: (404) 385-3500
Fax: (404) 894-8275
www.conted.gatech.edu/courses/oti/oti.html

University of South Florida
13201 Bruce B. Downs Blvd.
MDC 56
Tampa, FL 33612-3805
Phone: (866) 697-0975
Fax: (813) 974-9972
www.usfoticenter.org

Region V

Mid-America OSHA Trng. Inst.
Sinclair Community College
444 W. 3rd St.
Dayton, OH 45402-1460
Phone: (937) 512-3242
Fax: (937) 512-2279
midamericaosha.org

Mid-America OSHA Trng. Inst.
Ohio Valley Const. Ed. Foundation
33 Greenwood Lane
Springboro, OH 45066-3034
Phone: (866) 444-4412
Fax: (937) 704-9394
midamericaosha.org

National Safety Education Center
Northern Illinois University
590 Garden Rd. RM 318
DeKalb, IL 60115-2854
Phone: (800) 656-5317
Fax: (815) 753-4203
www.earnyourcard.com

National Safety Education Center
Construction Safety Council
4100 Madison Street
Hillside, IL 60162-1768
Phone: (800) 552-7744
Fax: (708) 544-2371
www.buildsafe.org

National Safety Education Center
National Safety Council
1121 Spring Lake Drive
Itasca, IL 60143-3201
Phone: (800) 621-7615
Fax: (630) 285-1613

Great Lakes Regional
OTI Education Center
UAW Health and Safety Dept.
8000 East Jefferson Ave
Detroit, MI 48214-3963
Phone: (800) 932-8689
Fax: (734) 481-0509
tp://www.uaw.org/hs/02/05/hs01.html

Great Lakes Regional
OTI Education Center
Eastern Michigan University
2000 Huron River Drive, Ste. 101
Ypsilanti, MI 48197-1699
Phone: (800) 932-8689
Fax: (734) 481-0509
www.emich.edu/public/osha

Great Lakes Regional
OTI Education Center
MWC for Occup. Safety & Health
2221 Univ. Ave. SE Ste. 350
Minneapolis, MN 55414-3078
Phone: (800) 493-2060
Fax: (612) 626-4525
www1.umn.edu/mcohs

Great Lakes Regional
OTI Education Center
University of Cincinnati
P.O. Box 670567
Cincinnati, OH 45267-0567
Phone: (800) 207-9399
Fax: (513) 558-1756
www.greatlakesosha.org

Region VI

Southwest Education Center
Texas Engineering Ext. Service
15515 IH-20 at Lumley
Mesquite, TX 75181-3710
Phone: (800) 723-3811
Fax: (972) 222-2978
teexweb.tamu.edu/osha

Region VII

Metropolitan Community Colleges/
Business & Technology College
1775 Universal Avenue
Kansas City, MO 64120-1313
Phone: (800) 841-7158
Fax: (816) 482-5408
kcmetro.edu/btcnew/oshatrain.asp

Midwest OSHA Education Centers
Saint Louis University
3545 Lafayette Ste. 300
St. Louis, MO 63104-8150
Phone: (888) 382-3756
Fax: (314) 977-8150
ceet.slu.edu

Midwest OSHA Education Center
National Safety Council
11620 M Circle
Omaha, NE 68137-2231
Phone: (800) 592-9004
Fax: (402) 896-6331
www.safenebraska.org

Midwest OSHA Education Center
Kirkwood Community College
6301 Kirkwood Blvd. SW
Cedar Rapids, IA 52404-5260
Phone: (800) 464-6874
Fax: (319) 398-1250
www.hmtri.org/moec/moec_index.htm

Region VIII

Rocky Mountain Education Center
Red Rocks Community College
13300 West Sixth Avenue
Lakewood, CO 80228-1255
Phone: (800) 933-8394
Fax: (303) 980-8339
www.rrcc.cccoes.edu/rmec

Mountain West
OSHA Training Center
Consortium of Salt Lake C.C./
University of Utah
75 South 2000 East
Salt Lake City, UT 84112-5120
Phone: (801) 581-4055
Fax: (801) 585-5275
www.rmcoeh.utah.edu/ce/otc.html

Region IX

University of California, San Diego
15373 Innovation Drive, Ste. 105
San Diego, CA 92128-3424
Phone: (800) 358-9206
Fax: (858) 485-7390
osha.ucsd.edu

WESTEC
Westside Energy Services
210 East Center Street
Taft, CA 93268-3605
Phone: (866) 493-7832
Fax: (661) 763-5162
www.westec.org/osha

Region X

University of Washington
4225 Roosevelt Way NE #100
Seattle, WA 98105-6099
Phone: (800) 326-7568
Fax: (206) 685-3872
depts.washington.edu/ehce

OSHA Outreach Training Program Fact Sheet—West Coast

As of October 1, 2003, outreach trainers will send their requests for 10- and 30-hour general industry and construction student cards to the OSHA Training Institute Education Center where they last took their trainer course (500, 501, 502, or 503).

WHEN:
Instructors must send requests to the OSHA Education Center that issued your trainer card (OSHA 500)

Note: Trainers who took their last general industry and construction trainer courses at different Education Centers will have to send their requests to different Education Centers.

WHERE:
Effective October 1, 2003, fax, mail or e-mail requests for student cards to:

Region IX OSHA Training Institute
ATTN: Outreach Training Coordinator
15373 Innovation Drive, Ste 105
San Diego, CA 92128
FAX: 858-485-7390
Online: http://osha.ucsd.edu

Region X OSHA Training Institute
ATTN: Registrar
4225 Roosevelt Way NE #100
Seattle, WA 98105
FAX: 206-685-3872
Online: http://depts.washington.edu/ehce

QUESTIONS:
If you have questions about the outreach training program, including the new administration procedures, contact either of the following staff members.

Region IX – UCSD
OSHA Outreach Coordinator
(858) 358-9206

Region X – UW
Lauren Baker, Registrar
(206) 685-3089

HOW THE PROGRAM ADMINISTRATION WORKS FOR EDUCATION CENTERS:
Education Centers will administer the outreach card requests the same way that the OSHA Training Institute does. Refer to the outreach guidelines for details, the guidelines are also available at *www.osha.gov/fso/ote/training/outreach/training_program.html*.

Requests require the same information: Provide a copy of your trainer certificate if this is your first request. We will review requests for: trainer qualification (valid authorization status), compliance to the guidelines (topics, time, etc.), and completeness of information. Discrepancies will be reviewed with the trainer, and complete, valid requests will be filled within three weeks of receipt.

CHAPTER 21

TRAINING TECHNIQUES

As an OSHA Outreach trainer you are a key player in helping OSHA's goal to reduce accidents and injuries. Workers do not want to get hurt and employers want to provide a safe workplace. Unfortunately, many safety training sessions are conducted in a manner that leaves the worker with a less-than-enthusiastic regard for safety training. It is necessary to liven up the topic to keep workers' attention. A good training philosophy is to present the material three different times in three different ways so the student is not bored, and yet the material is emphasized in such a way that he or she cannot forget it. A good trainer will establish the climate for training by introducing him or herself to the class as well as any guest speakers. Group discussions and workshops enhance training. A good rule to follow when setting up a class is to stress the key points of the topic.

- Tell them what you are going to teach them.
- Teach them.
- Tell them what you have taught them.

Presenting the material in a lecture is one way of presenting the material. Then have the students refer to written material that you have provided in a handout. Finally, show them a video of the subject as a third method of instruction. This allows different learners to obtain information in different ways and helps to break up the monotony of a class.

Good trainers realize that they are working with an adult audience that has needs and expectations. These trainers also realize that they are the key element in the company safety training program. Trainers must be confident of their own skills and technical knowledge to provide benefit to the learners. This confidence enhances the importance of the message.

Adult Learners

Human beings have a natural potential for learning. Trainers must understand that people are naturally curious, yet they are simultaneously eager and *not* eager to learn. The tendency to learn is more likely to appear when people see relevance to themselves in the learning process.

Significant learning takes place when the subject matter is perceived by students as being meaningful for their own purposes.

Confronting students with practical, social, and ethical problems is effective in promoting learning. Brief, intensive experiences for individuals facing immediate problems are especially effective.

Learning is facilitated when the student participates responsively in the learning process. When students set their own goals, locate and learn to use their resources, and take responsibility for their actions, they tend to learn quickly. Students should be involved in and participate in the class to enhance their educational experience.

Classroom Training in General

Students will not normally pursue the learning objectives on their own and must be prompted to do so. Tests and quizzes can be used as motivators to get students focused on the learning objectives. These evaluations are part of the education itself as students learn what they need to master and build upon their skill levels "one brick at a time."

Instructors must assign projects and tasks to the students and supervise the work to ensure it is completed. The facilitator must provide guidance to the learners and some form of evaluation.

As a trainer, you are expected to know the subject inside and out, as well as the objective of the training session. Prepare your materials the day before the training session and be prepared to teach an outstanding class. The reason that many instructors carry their own LCD projectors and laptop computers with them is that often different brands are not compatible and simply do not work together. Check out your equipment so that you know how it works. Do a function check of the equipment to ensure your material works with your equipment. Do not wait until the last minute to plan, prepare, and practice.

It is the trainers' responsibility to set the initial mood of the group and establish a climate of learning. Students may be motivated by allowing them to have input into the learning structure. One way of doing this is to have the students introduce themselves at the beginning of a class and identify one particular problem that they seek to have resolved during this class. The trainer can write those concerns on a chalkboard and tailor the session to ensure that those concerns are met. Facilitators can provide a range of resources for the students, including the instructor. Facilitators should keep a close eye on the climate of the class and modify the presentation to maximize class learning.

Confidence comes from being thoroughly prepared. Instructors can expect to be nervous. Some instructors use acting skills and exercises to overcome this natural nervousness. They realize that trainers who *act* enthusiastic will *be* enthusiastic. Keep a positive attitude when you are training and commit yourself to the training session preparation and the class time itself. Speak loudly and clearly. Your clothes say a lot about you, so dress appropriately. Wear comfortable shoes and match your dress to your audience. An executive briefing may require a coat and tie, while the same set of clothes would not be appropriate for forklift operator training.

Understand how long you have to conduct the class and meet that time expectation. Start and finish the class on time. This may require that the class be rehearsed and timed. Understanding the audience is also important. Giving a presentation to new employees is different than giving a presentation to experienced workers or job foremen. Executive briefings and seminars require an entirely different presentation.

What you hear, you forget.
What you hear and see, you remember.
Use PowerPoints, videos, flipcharts
What you do, you understand and remember.
Use field trips, demonstrations, hands-on activities

There are several ways that training can be presented. Each of these methods has its own advantages and disadvantages. The simplest form of education is the lecture, where the instructor talks to the students about the subject. A foreman getting the workers together and talking to them about how to set up and use a ladder can accomplish ladder safety training.

Using a chalkboard or easel and paper to show the workers the 1:4 pitch, how to tie off the ladder, and the 3-ft. extension can enhance a ladder presentation.

The presentation could be further enhanced by using PowerPoint slides to show photographs of ladders being used correctly and incorrectly. Pictures help to maintain focus on the material. Another teaching aid is a videotape or DVD that brings in sound and action. Other things that can add to the class are personal experiences of the instructor or the students, case studies, and "fatal facts" (available from OSHA's web page www.osha.gov) as well as diagrams and sketches. Training is always enhanced when the students receive a handout to refer to after the class. Use a variety of teaching techniques to present your materials to keep the information interesting.

In some cases, students cannot be brought together in one location for the training. In these cases a remote broadcast or teleconference may be used to provide training over long distances. Some companies use computer-based training via CD-ROM or the Internet. Both of these systems provide sound, action, and student interactions so that the training is effective.

When conducting classes in person, set up your class area early. It is a good policy to arrive at least one hour prior to the scheduled class time so that you can locate the light switches, room temperature controls, exits, restrooms, vending machines, and smoking area. Identify where the electrical outlets and extension cords are if you need them. Emergency exit plans should also be reviewed at this time. Set up the seats in the arrangement that you want and organize your equipment so that when students begin to arrive you are prepared to greet them.

About Your Session

It is all about you! You are the instructor and your personal style is what sets you apart from the thousands of other instructors out there. Learn what your strengths and weaknesses are and work to correct and minimize the weak parts of your style. Use your strong points to your advantage.

You may find that you need to be more forceful when leading a group so the students do not get out of control. Or you may find that you need to speak more deliberately, slowly, or loudly. Gestures work for some instructors and are a disaster for others. Stress the point that everyone involved in the session is going to learn from each other, instructor and student alike. Most adult students have experiences to share with the class, so stress the equality of the group as opposed to the superiority of any one person. Focus on solving problems that the group has identified and try to avoid ego trips of individual students...and the instructor.

Involve the students as much as possible using your personal style, attitude, and philosophy. Be open and honest with the students and do not be afraid to display your weaknesses as well as your strengths. This allows the group to open up. When students realize that everyone has foibles and shortcomings, they tend to become more involved. Make eye contact with the students and smile.

Some in the group may initially be a little defensive. They may be a little tense, wondering what they have gotten into and what demands will be made on them. The instructor must allay their fears and lower their tension levels by creating a climate that is open and positive and encourages discussion. Remind the group that their combined experience is far greater than yours and that the instructor is there to serve as a guide to help them use the resources of the group constructively and positively. Mention that quite often they will learn more from one another than they will from the programmed course of instruction.

Handling Questions

Skillfully used questions help guide discussions and extract and develop information. Questions invite participation and test the effectiveness of the communication. Questions should motivate student thought and opinion and should be directed to everyone in the group. The purpose of questions is to get the students to participate and think.

There are several types of questions that instructors can ask:
- Open. Allows great freedom, is less threatening, and the student controls the amount of detail in answer.
- Closed. The instructor limits the response length and allows for re-direction of the discussion.
- Direct. Directed at one student.
- Overhead. Covers the class as a whole.
- Return. Redirects the question back to the person who asked it.
- Relay. Takes a question from one student and redirects it at another.

Good questions require the students to think so after you ask a question, allow adequate time for answer formulation and response. Be patient. As a rule of thumb, count to 12 before asking the class another question.

TRAINING TECHNIQUES

Presenting Material

You have had time to prepare, but most people in the group are hearing the material for the first time. Understanding this new material can be difficult. Presenting it in a way that others understand is even more difficult.

Preparation enables you to organize your thoughts, plan your approach, expand your understanding and develop illustrations, examples and analogies. Only preparation will give you the skill and confidence you need to achieve your goals. Rehearsal is difficult, but it is one of the key tools of good preparation. So run through your act before you go on stage. Concentrate on the things you will say and go over it a few times.

Make sure you allow students time to become familiar with a topic before moving on to the next topic. Stress how the concepts can be applied to add value for the students in their own lives. Increase their understanding by relating concepts presented to their own experiences.

Stories hold a group's interest a lot more than explaining something. Everybody loves a story; they are "the propelling power of persuasion". Good stories are real, vital, timely, specific, and about people. The best stories are drawn from your own experience and tailored to fit the topic. Variety keeps them interested and can make a good program outstanding.

CHAPTER 22

INSPECTION PROCEDURES

Purpose of the Inspection
- Uncover unsafe acts and conditions.
- Reveal the need for specific guards for workers, machines and materials.
- Help "sell" safety program to workers.
- Encourage supervisors to inspect their own areas, tools, equipment, materials, and work practices.
- Bring about a closer liaison between safety personnel and line personnel.
- Provide an additional set of eyes to identify unsafe conditions before an accident occurs.

Preparation for the Inspection
- The person making inspection should acquire and digest all pertinent information available on the type of operation to be inspected.
- He or she should review and acquaint himself or herself with details of these functions, generally and specifically, in relation to one another.
- He or she should determine which standards will apply.
- He or she should define the work area.
- He or she should plan the inspection route.
- He or she should review previous inspections and results, looking for any outstanding work orders.
- He or she should learn what OSHA is inspecting for similar industries (go to www.osha.gov and use the Establishment Search).
- He or she should make or obtain checklists (if they are to be used).

Conduct of the Inspection
- Inspect while employees are working (if possible).
- Use simple forms or notes during the inspection process.
- Be alert for all hazards and conditions.
- Record all unsafe practices and conditions.
- Check for specific items as well as general conditions.
- Check all areas; do not be steered away.
- Do not be a disturbing influence.
- Be constructive.
- Look for "why" conditions exist.

- Advise supervisors; do not argue with them.
- Discuss recommendations with supervisors.
- Try to "sell" your recommendations to ensure early abatement.

Examples of What to Check in a Safety Inspection

Receiving, Shipping, Storage
Equipment, job planning, layout, heights, floor loads, projection of materials, material-handling methods. Dockboards on trucks securely fastened, guardrails in place to keep workers from walking off the docks. Forklift operations require seat belts for operators and 18-wheel trucks must have at least one wheel chocked when forklifts are inside them.

Building Conditions
Floors, walls, ceilings, exits, stairs, walkways, ramps, platforms, driveways, and aisles. Exit signs, markings, and emergency lighting systems.

Housekeeping
Waste disposal, tools, objects, materials, leakage and spillage, methods, schedules, work areas, remote areas, windows, and ledges.

Electricity
Equipment, switches, breakers, fuses, switchboards, junctions, special fixtures, circuits, insulation, extensions, tools, motors, grounding and code compliance. Access to circuit breakers and fuse panels, unguarded electrical wiring. Overused extension cords or extension cords used in place of permanent wiring.

Lighting
Type, intensity, controls, condition, diffusion, location, glare and shadow control, and standards applied.

Heating And Ventilation
Type, effectiveness, temperature, humidity, controls, natural and artificial ventilation and exhausting.

Machines
Points of operation, flywheels, gears, shafts, pulleys, key ways, belts, couplings, sprockets, chains, frames, controls, lighting, tools and equipment, brakes, exhausting, feeding, oiling, adjusting, maintenance, grounding, how attached, workspace, and location.

Personnel
Training, experience, methods of checking machines before use, methods of cleaning, oiling, or adjusting machinery, type of clothing, personal protective equipment, use of guards, tool storage, and work practices.

Hand and Power Tools
Purchasing standards, inspection, storage, repair, types, maintenance, grounding, use and handling, guards in place on tools.

Chemicals
Storage, handling, transportation, amounts used, warning signs, supervision, training, protective clothing and equipment. Hazard communication plan, list of current hazardous chemicals, MSDS for each chemical onsite, all containers labeled.

Fire Prevention
Extinguishers, alarms, sprinklers, smoking rules, exits, personnel assigned, separation of flammable materials and dangerous operations, explosion-proof fixtures, and waste disposal.

Maintenance
Regularity, effectiveness, training of personnel, materials, and equipment used, method of locking out machinery and general methods. Lockout/tagout and confined space entry programs.

Personal Protective Equipment
Type, size, maintenance, repair, storage, assignment of responsibility, purchasing methods, standards observed, rules of use and method of assignment. ANSI standards on equipment. Head, eye, face, and foot protection. Use of respirators and respirator program.

Common Occupational Safety Inconsistencies Found During Inspections

Inconsistency	Section	Risk Code
EGRESS		
Fire exit doors do not operate	29CFR1910.37 (q)(3)	3
Two means of egress as required	29CFR1910.36 (b) (2)	3
Exit routes free of obstruction	29CFR1910.37 (q)(3)	3
Emergency Lighting not working	29CFR1910.37 (b) (1)	3
Adequate number of exits	29CFR1910.36 (b)(1)	3
Egress routes less than 28 inches wide	29CFR1910.36(g)(2)	3
All exits marked with signs	29CFR1910.37 (b) (2)	3
All NON EXIT's labeled as such	29CFR1910.37 (b) (5)	3
Direction signs to exits not marked	29CFR1910.37 (b) (4)	3
Exit lights not illuminated	29CFR1910.37 (b) (6)	3
Exit letters on signs six inches by ¾ inch	29CFR1910.37 (b) (7)	3
Emergency Action Plan	29CFR1910.38 (b) (1)	3
Emergency Alarm system	29CFR1910.38.(d)	3
ELECTRICAL		
Missing ground fault circuit interrupters	NEC 210-8 (b)	3
Circuit breakers not labeled	29CFR1910.303 (f)	3
Grounding plug missing	29CFR1910.304 (f) (5) (v)	3
Ungrounded outlet	29CFR1910.304 (s) (1) (iv)	3
Loose or broken electrical outlet	29CFR1910.305 (b) (2)	3
Exposed live wires	29CFR1910.305 (g)	2
Flexible cords through doors/windows	29CFR1910.305 (g) (1) (iii) (a)	3
Temporary wiring used as permanent	29CFR1910.305 (q) (1)	4
FIRE		
Extinguishers mounted, accessible	29CFR1910.157 (c) (1)	3
Extinguishers fully charged	29CFR1910.157 (c) (4)	3
Extinguishers of the correct type	29CFR1910.157 (d) (1)	3
Extinguishers within 75 feet	29CFR1910.157 (d) (2)	3
Extinguishers monthly inspection	29CFR1910.157 (e) (2)	3
Extinguishers annual inspection	29CFR1910.157 (e) (3)	3
Fire alarms readily accessible	29CFR1910.164 (e)	3
Emergency phone numbers posted	29CFR1910.165 (b) (4)	3
Flammable liquid storage	29CFR1910.106	3
HAZMAT		
Bulk combustibles in storage cabinet	29CFR1910.106 (d) (3)	3
Combustibles in storage room	29CFR1910.106 (d) (3)	3
Flammable liquids in office	29CFR1910.106 (d) (5) (iii)	3
Hazard Communication Program	29CFR1910.1200	3

SANITATION

Proper housekeeping	29CFR1910.141 (a) (3) (i)	3
Adequate toilet facilities	29CFR1910.141 (c) (1) (i)	3
Toilets with privacy	29CFR1910.141 (c) (2) (i)	3
Adequate washing facilities	29CFR1910.141 (d) (2) (iii)	3
Emergency eye wash station	29CFR1910.151 (c)	3

SHOP

Grinding wheel rest not 1/8 inch	29CFR1910.215 (a) (4)	3
Tongue guard adjusted to ¼ inch	29CFR1910.215 (a) (9)	3
All wheels ring tested prior to mounting	29CFR1910.215 (d) (1)	3
Machine guards in place	29CFR1910.212(a)(1)	3
Fixed machinery not anchored to floor	29CFR1910.212(b)	3
Eye and face protection	29CFR1910.133	3
Eyewash stations	29CFR1910.151(c)	3
Lockout Tagout Program	29CFR1910.147	3

WALKING/WORKING

Roof/pipe leaks- wet floor	29CFR1910.22 (a) (2)	3
Slip hazard/broken door	29CFR1910.22 (b) (1)	3
Aisles clear and in good repair	29CFR1910.22 (b) (1)	3
Fixed ladder over 20 feet without cage	29CFR1910.27 (d) (1) (iii)	3
Guarding Floor and wall openings	29CFR1910.23	3

WELDING

Gas cylinder caps in place	29CFR1910.253 (a) (2) (iii)	3
Gas cylinders marked	29CFR1910.253 (b) (1) (ii)	3
Gas cylinders 20 feet from combustibles	29CFR1910.253 (b) (2) (ii)	3
Acetylene cylinders stored upright	29CFR1910.253 (b) (3) (ii)	3
Oxygen cylinders near oil or grease	29CFR1910.253 (4) (i)	3
Oxygen cylinders separated from fuel	29CFR1910.253 (4) (iii)	3

MISC

General Duty Clause	PL 91-156	3
Forklift Seat belt in use	General Duty Clause	3
Respiratory Protection when required	29CFR1910.134	2
Forklift Training	29CFR1910.178(l)	3
Record keeping	29CFR1904	3

Sample Inspection Form

Location	Discrepancy	OSHA Citation	Corrective Action

Comments: How many citations did your team find? _____
What was the single most important item to fix?

CHAPTER 23

ACCIDENT INVESTIGATION

Thousands of workplace accidents occur throughout the United States every day. The failure of people, equipment, supplies or surroundings to behave or react as expected cause most accidents. Accident investigations determine how and why these failures occur. By using the information gained through an investigation, a similar or perhaps more disastrous accident may be prevented. Conduct accident investigations with accident prevention in mind. Investigations are *not* to place blame.

An accident is any unplanned event that results in personal injury or property damage. When the personal injury requires little or no treatment, it is minor. If it results in a fatality or in a permanent total, permanent partial or temporary total (lost-time) disability, it is serious. Similarly, property damage may be minor or serious. Investigate all accidents regardless of the extent of injury or damage.

Accidents are part of a broad group of events that adversely affect the completion of a task. These events are incidents. For simplicity, the procedures discussed in later sections refer only to accidents. They are, however, also applicable to incidents.

Accident Prevention

Accidents are usually complex. An accident may have 10 or more events that can be causes. A detailed analysis of an accident will normally reveal three cause levels: basic, indirect and direct. At the lowest level, an accident results only when a person or object receives an amount of energy or hazardous material that cannot be absorbed safely. This energy or hazardous material is the *direct cause* of the accident. The direct cause is usually the result of one or more unsafe acts, unsafe conditions or both. Unsafe acts and conditions are the *indirect causes* or symptoms. In turn, indirect causes are usually traceable to poor management policies and decisions or to personal or environmental factors. These are the *basic causes*.

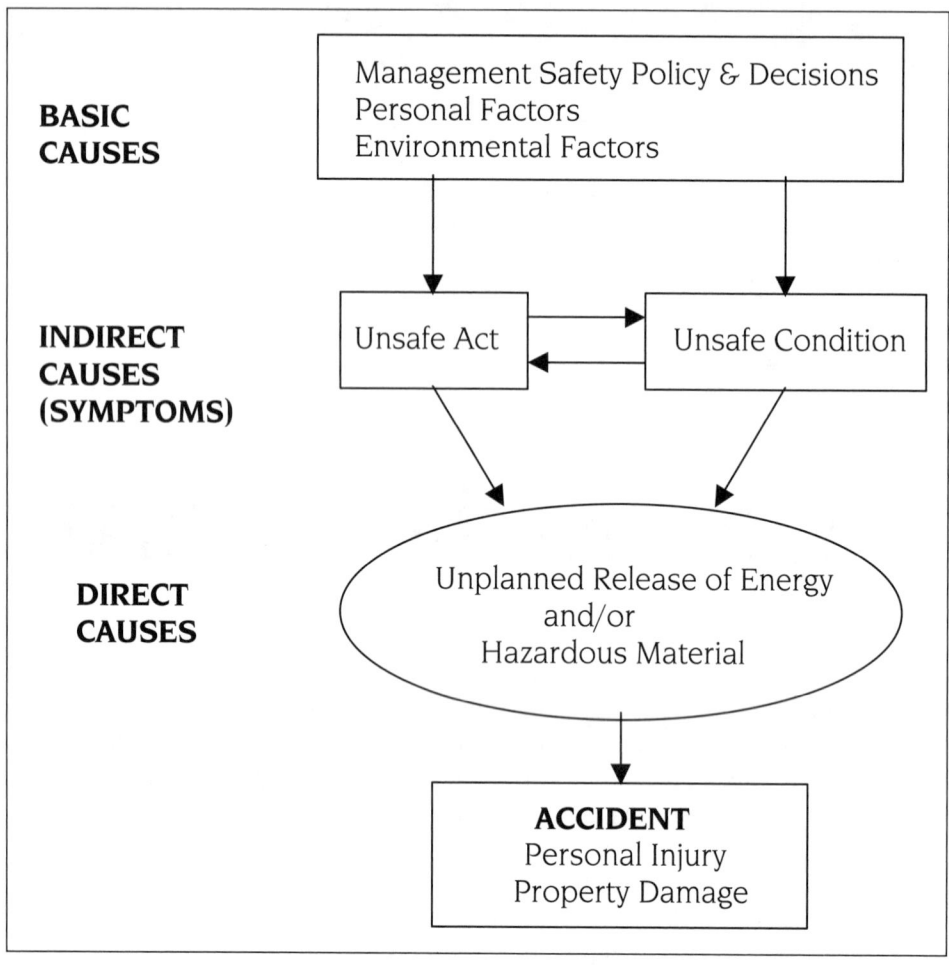

Exhibit 23-1.
A detailed analysis of an accident will normally reveal three cause levels: basic, indirect and direct.

Despite their complexity, most accidents can be prevented by eliminating one or more causes. Accident investigations determine not only what happened, but also how and why. Accident investigators are interested in each event as well as in the sequence of events that led to an accident. The accident type is also important to the investigator. The recurrence of accidents of a particular type or those with common causes shows areas needing special accident-prevention emphasis.

Investigative Procedures

The actual procedures used in a particular investigation depend on the nature and results of the accident. The agency having jurisdiction over the location determines the administrative procedures. In general, responsible officials will appoint an individual to be in charge of the investigation. The investigator uses most of the following steps:

1. Define the scope of the investigation.
2. Select the investigators. Assign specific tasks to each (preferably in writing).
3. Present a preliminary briefing to the investigating team, including:
   a. Description of the accident, with damage estimates.
   b. Normal operating procedures.
   c. Maps (local and general).
   d. Location of the accident site.
   e. List of witnesses.
   f. Events that preceded the accident.
4. Visit the accident site to get updated information.
5. Inspect the accident site:
   a. Secure the area. Do not disturb the scene unless a hazard exists.
   b. Prepare the necessary sketches and photographs. Label each carefully and keep accurate records.
6. Interview each victim and witness. Also interview those who were present before

the accident and those who arrived at the site shortly after the accident. Keep accurate records of each interview. Use a tape recorder if desired and approved.
7. Determine:
   a. What was not normal before the accident.
   b. Where the abnormality occurred.
   c. When it was first noted.
   d. How it occurred.
8. Analyze the data obtained in step 7. Repeat any of the prior steps, if necessary.
9. Determine:
   a. Why the accident occurred.
   b. A likely sequence of events and probable causes (direct, indirect, basic).
   c. Alternative sequences.
10. Check each sequence against the data from step 7.
11. Determine the most likely sequence of events and the most probable causes.
12. Conduct a post-investigation briefing.
13. Prepare a summary report, including the recommended actions to prevent a recurrence. Distribute the report according to applicable instructions.

An investigation is not complete until all data is analyzed and a final report is completed. In practice, the investigative work, data analysis and report preparation often occur simultaneously.

Fact-Finding

The investigator should gather evidence from many sources during an investigation, obtaining information from witnesses and reports as well as by observation. Interview witnesses as soon as possible after an accident. Inspect the accident site before any changes occur. Take photographs and draw sketches of the accident scene. Record all pertinent data on maps. Get copies of all reports. Documents containing normal operating procedures, flow diagrams, maintenance charts, or reports of difficulties or abnormalities are particularly useful. Record pre-accident conditions, the accident sequence, and post-accident conditions. In addition, document the location of victims, witnesses, machinery, energy sources, and hazardous materials. Keep complete and accurate notes in a bound notebook.

In some investigations, a particular physical or chemical law, principle or property may explain a sequence of events. Include laws in the notes taken during the investigation or in the later analysis of data. In addition, gather data during the investigation that may lend itself to analysis by these laws, principles or properties. An appendix in the final report can include an extended discussion.

Interviews

In general, experienced personnel should conduct interviews. If possible, the team assigned to this task should include an individual with a legal background. In conducting interviews, the team should:
1. Appoint a speaker for the group.
2. Get preliminary statements as soon as possible from all witnesses.
3. Locate the position of each witness on a master chart (including the direction of view).
4. Arrange for a convenient time and place to talk to each witness.
5. Explain the purpose of the investigation (accident prevention) and put each witness at ease.
6. Listen. Let each witness speak freely, and be courteous and considerate.
7. Take notes without distracting the witness. Use a tape recorder only with the consent of the witness.
8. Use sketches and diagrams to help the witness.
9. Emphasize areas of direct observation. Label hearsay accordingly.
10. Be sincere and do not argue with the witness.
11. Record the exact words used by the witness to describe each observation. Do not "put words into a witness's mouth."
12. Word each question carefully and be sure the witness understands.
13. Identify the qualifications of each witness (e.g., name, address, occupation, years of experience).
14. Supply each witness with a copy of his or her statements. Have each witness sign his or her statement if possible.

After interviewing all witnesses, the team should analyze each witness statement. They may wish to re-interview one or more witnesses to confirm or clarify key points. While there may be inconsistencies in witnesses' statements, investigators should assemble the available testimony into a logical order. Analyze this information along with data from the accident site.

Not all people react in the same manner to a particular stimulus. For example, a witness within close proximity to the accident may have an entirely different story from one who saw it at a distance. Some witnesses may also change their stories after they have discussed it with others. The reason for the change may be an additional clue.

A witness who has had a traumatic experience may not be able to recall the details of the accident. A witness who has a vested interest in the results of the investigation may offer biased testimony. Finally, eyesight, hearing, reaction time and the general condition of each witness may affect his or her powers of observation. A witness may omit entire sequences because of a failure to observe them or because their importance was not realized.

Problem Solving Techniques

Accidents represent problems that must be solved through investigations. Several formal procedures solve problems of any degree of complexity. This section considers these problem-solving procedures.

The Scientific Method

The scientific method forms the basis of nearly all problem-solving techniques. It is used for conducting research. In its simplest form, it involves the following sequence: making observations, developing hypotheses and testing the hypotheses.

Even a simple research project may involve many observations. A researcher records all observations immediately. A good investigator must do the same thing. Where possible, the observations should involve quantitative measurements. Quantitative data is often important in later development and testing of the hypotheses. Such measurements may require the use of many instruments in the field as well as in the laboratory.

When making observations, the investigator develops one or more hypotheses that explain the observations. The hypothesis may explain only a few of the observations or it may try to explain all of them. At this stage, the hypothesis is merely a preliminary idea. Even if later rejected, the investigator has a goal toward which to proceed.

Test the hypothesis against the original observations. A series of controlled experiments is often useful in performing this evaluation. If the hypothesis explains all of the observations, testing may be a simple process. If not, either make additional observations, change the hypothesis or develop additional hypotheses.

As with scientific research, the most difficult part of any investigation is the formulation of worthwhile hypotheses. Use the following three principles to simplify this step:

- **The principle of agreement.** An investigator uses this principle to find one factor that associates with each observation.
- **The principle of differences.** This principle is based on the idea that variations in observations are due only to differences in one or more factors and no other factors.
- **The principle of concomitant variation.** This principle is the most important because it combines the ideas of both of the preceding principles. In using this principle, the investigator is interested in the factors that are common as well as those that are different in the observations.

In using the scientific method, the investigator must be careful to eliminate personal bias. The investigator must be willing to consider a range of alternatives. Finally, he or she must recognize that accidents often result from the chance occurrence of factors that are too numerous to evaluate fully.

Change Analysis

As its name implies, this technique emphasizes change. To solve a problem, an investigator must look for deviations from the norm. As with the scientific method, change analysis also follows a logical sequence. It is based on the principle of differences described in the discussion of the scientific method. Consider all problems to result from some unanticipated change. Make an analysis of the change to determine its causes. Use the following steps in this method:

- Define the problem (What happened?).
- Establish the norm (What should have happened?).

- Identify, locate and describe the change (What, where, when and to what extent?).
- Specify what was and what was not affected.
- Identify the distinctive features of the change.
- List the possible causes.
- Select the most likely causes.

Sequence Diagrams
Gantt charts are sequence diagrams. Use them for scheduling investigative procedures. They can also aid in the development of the most probable sequence of events that led to the accident. Such a chart is especially useful in depicting events that occurred simultaneously.

Gross Hazard Analysis
Perform a gross hazard analysis (GHA) to get a rough assessment of the risks involved in performing a task. It is "gross" because it requires further study. It is particularly useful in the early stages of an accident investigation in developing hypotheses. A GHA will usually take the form of a logic diagram or table. In either case, it will contain a brief description of the problem or accident and a list of the situations that can lead to the problem. In some cases, analysis goes a step further to determine how the problem could occur. A GHA diagram or table thus shows at a glance the potential causes of an accident. One of the following analysis techniques can then expand upon a GHA.

Job Safety Analysis
Job safety analysis (JSA) is part of many existing accident prevention programs. In general, a JSA breaks a job into basic steps and identifies the hazards associated with each step. The JSA also prescribes controls for each hazard. A JSA is a chart listing these steps, hazards and controls. If available, review the JSA during the investigation. If not available, perform a JSA if one is not available. Perform a JSA as a part of the investigation to determine the events and conditions that led to the accident.

Failure Mode and Effect Analysis
Failure mode and effect analysis (FMEA) determines where failures occurred. Consider all items used in the task involved in the accident (e.g., people, equipment, machine parts, materials). In the usual procedure, FMEA lists each item on a chart. The chart lists the manner or mode in which each item can fail and determines the effects of each failure. Included in the analysis are the effects on other items and on overall task performance. In addition, make evaluations about the risks associated with each failure. That is, project the chance of each failure and the severity of its effects. Determine the most likely failures that led to the accident. Comparing these projected effects and risks with actual accident results does this.

Fault Tree Analysis
Fault tree analysis (FTA) is a logic diagram. It shows all the potential causes of an accident or other undesired event. The undesired event is at the top of the "tree." Reasoning backward from this event, determine the circumstances that can lead to the problem. These circumstances are then broken down into the events that can lead to them and so on. Continue the process until the identification of all events can produce the undesired event. Use a logic tree to describe each of these events and the manner in which they combine. This information determines the most probable sequence of events that led to the accident.

Report of Investigation
As noted earlier, an accident investigation is not complete until a report is prepared and submitted to proper authorities. Special report forms are available in many cases. Other instances may require a more extended report. Such reports are often very elaborate and may include a cover page, title page, abstract, table of contents, commentary or narrative portion, discussion of probable causes and a section on conclusions and recommendations.

The following outline is useful when developing the formal report:
1. Background information
   a. Where and when the accident occurred
   b. Who and what were involved
   c. Operating personnel and other witnesses
2. Account of the accident (What happened?)
   a. Sequence of events
   b. Extent of damage

  c. Accident type
  d. Agency or source (of energy or hazardous material)
3. Discussion (Analysis of the accident—how, why?)
  a. Direct causes (energy sources, hazardous materials)
  b. Indirect causes (unsafe acts, conditions)
  c. Basic causes (management policies, personal or environmental factors)
4. Recommendations (to prevent a recurrence) for immediate and long-range action to remedy
  a. Basic causes
  b. Indirect causes
  c. Direct causes (e.g., reduced quantities, protective equipment, structures)

Summary

A successful accident investigation determines not only what happened, but also finds how and why the accident occurred. Investigations are an effort to prevent a similar or perhaps more disastrous sequence of events.

Accident investigations involve gathering and analyzing facts, and developing hypotheses to explain these facts. Each hypothesis must then be tested against the facts. An investigation is not complete, however, until a final report is written. Responsible officials can then use the resulting information and recommendations to prevent future accidents.

CHAPTER 24

FIELD INSPECTION REFERENCE MANUAL

The following is the official Field Inspection Reference Manual from the Occupational Safety and Health Administration, including change memos. The index has been excluded.

July 13, 1999

MEMORANDUM FOR:	REGIONAL ADMINISTRATORS STATE DESIGNEES
THROUGH:	CHARLES N. JEFFRESS ASSISTANT SECRETARY
FROM:	R. DAVIS LAYNE DEPUTY ASSISTANT SECRETARY
SUBJECT:	Child Labor: Probability Assessment and Good Faith Penalty Adjustment Considerations

This is to advise you of a change to OSHA's probability assessment and "good faith" penalty adjustment considerations for violations involving minor employees which will be effective on the date of this memorandum. This change is being made to provide more effective protection for young workers, who are especially vulnerable to workplace hazards. Young workers in this situation are those employees who are less than 18 years old.

Please make the following pen-and-ink changes at C.2.f.(2)(f) on page IV-10 of the OSHA Field Inspection Reference Manual (FIRM):

C. 2. f. **Probability Assessment...**
    (2)**Violations**. The following circumstances may normally be considered, as appropriate, when violations likely to result in injury or illness are involved.
        (a)Number of workers exposed
        ...
        ~~(f) Other pertinent working conditions~~
        (f) Youth and inexperience of workers, especially those under 18 years old.
        (g)Other pertinent working conditions.

Please make the following pen-and-ink changes at C.2.i.(5)(b) on page IV-14 of the FIRM:

C. 2. i. (5)(b) *Good Faith*. A penalty reduction of up to 25 percent, based on the CSHO's professional judgment, is permitted in recognition of an employer's "good faith".
<add>
5. Where young workers (i.e., less than 18 years old) are employed, the CSHO's evaluation must consider whether the employer's safety and health program appropriately addresses the particular needs of such workers with regard to the types of work they perform and the hazards to which they are exposed.
<add>

State Plan States. OSHA believes that this policy change is an effective tool for dealing with employers who expose children to safety and health violations. We strongly encourage the States to adopt a similar penalty adjustment policy for violations exposing young workers and to notify their Regional Administrator as to their intent.

If you have any questions, please contact Patrick Kapust or William Smith in the Office of General Industry Compliance Assistance at (202) 693-1850.

June 21, 1996

MEMORANDUM FOR: ALL REGIONAL ADMINISTRATORS

FROM: MICHAEL G. CONNORS
Deputy Assistant Secretary

SUBJECT: FIRM Change: Mandatory Collection of OSHA-200 and Lost Workday Injury and Illness (LWDII) Data During Inspections

This is to advise you that effective July 1, 1996 all OSHA Compliance Safety and Health Officers (CSHOs) shall review the employer's injury and illness records for three prior calendar years, record the information on a copy of the OSHA-200 screen, and enter the employer's data using the IMIS Application on the NCR (micro). This shall be done for all general industry, construction, maritime, and agriculture inspections and investigations.

For construction inspections/investigations, only the OSHA-200 information for the prime/general contractor need be recorded. It will be left to the discretion of the area office or the CSHO as to whether OSHA-200 data should also be recorded for any of the subcontractors.

CSHOs will not need to calculate the LWDII rate since it is automatically calculated when the OSHA-200 data is entered into the micro. If one of the three years is a partial year, so indicate, and the software will calculate accordingly. CSHOs shall, however collect the Number of Employee Hours Worked in each of the reference years, or if the information is not available, the Average Number of Full Time Employees (FTEs).

CSHOs can print blank OSHA-200 screens through any form containing establishment processing (1A, 7, 36, 90, 55). The NCR's Systems Administrator can also print blank 200 forms from selection **N. Blankforms Menu** on the Systems Administration Screen.

Select **Q. UnNumbered 200 & PEP**.

It will be necessary for training to be provided, in each area office, for CSHOs who are not using the CSHO Application, to ensure that they can gather and enter the necessary data for calculation of the LWDII. The PC-CSHO Application on the laptops for the OSHA-200 is not yet available.

This memorandum supersedes any policy or procedure to the contrary that is found in the Field Inspection Reference Manual (FIRM) or what might be found in any memorandum issued before the date of this memorandum. This has been cleared with NCFLL.

If you have any questions, contact Art Buchanan at (202) 219-8041 x114.

March 23, 1995
MEMORANDUM FOR: REGIONAL ADMINISTRATORS

FROM: JAMES W. STANLEY
    Deputy Assistant Secretary

SUBJECT: FIRM Change: Minimum Serious Willful Penalty

This is to advise you of a change to OSHA's civil willful penalty which will be effective as of the date of this memorandum. The change concerns the proposed minimum serious willful penalties for the smaller employers. The small employers in this situation are those with 50 or fewer employees. In no case will the proposed penalty be less than the statutory minimum, i.e., $5,000 for these employers.

Please make the following pen-and-ink changes at C.2.b.(1) on page IV-8 of the OSHA Field Inspection Reference Manual (FIRM):

C.2.b.(1) Minimum Penalties. The following guidelines apply:

(1) ~~The proposed penalty for any willful violation shall not be less than $5,000 for other than serious and regulatory violations, and shall not be less than $25,000 for serious violations. The $5,000 penalty is a statutory minimum and not subject to administrative discretion.~~
<add>
(1) The proposed penalty for any willful violation shall not be less than $5,000. The $5,000 penalty is a statutory minimum and not subject to administrative discretion. See C.2.m.(1)(a)1, below, for applicability to small employers.
<add>

Please make the following pen-and-ink changes at C.2.m.(1)(a)1, 3, and 4 on page IV-18 of the FIRM:

C.2.m.(1)(a)   *Serious Violations.* For willful serious violations, a gravity of **high, ~~medium~~,** <add>**moderate**<add>, or **low** shall be assigned based on the GBP of the underlying serious violation, as described at C.2.g.(2).

~~1 The adjustment factor for size shall be applied at **one half** of the values stated at C.2.c(5)(A)1, i.e., a reduction of 30 percent (1-25 employees), 20 percent (26-100 employees), 10 percent (101-250 employees), or no reduction (251 or more employees).~~
<add>
1 The adjustment factor for size shall be applied as shown in the following chart:

Employees	Percent Reduction
10 or less	80
11-20	60
21-30	50
31-40	40
41-50	30
51-100	20
101-250	10
251 or more	0

<add>

2. The adjustment factor for history shall be applied as described at C.2.i.(5)(c); i.e., a reduction of 10 percent shall be given to employers who have not been cited by OSHA for any serious, willful, or repeated violations in the past 3 years. There shall be no adjustment for good faith.
3. The proposed penalty shall then be determined from the table below:

Penalties to be Proposed

Total percentage reduction for size and/or history	High Gravity	Moderate Gravity	Low Gravity
0%	$70,000	$55,000	$40,000
10%	$63,000	$49,500	$36,000
20%	$56,000	$44,000	$32,000
30%	$49,000	$38,500	$28,000
40%	$42,000	$33,000	$24,000
50%	$35,000	$27,500	$20,000
60%	$28,000	$22,000	$16,000
70%	$21,000	$16,500	$12,000
80%	$14,000	$11,000	$8,000
90%	$7,000	$5,500	$5,000

~~Penalties to be proposed~~

~~Total percentage reduction for size and/or history~~	~~0%~~	~~10%~~	~~20%~~	~~30%~~	~~40%~~
~~High Gravity~~	~~$70,000~~	~~$63,000~~	~~$56,000~~	~~$49,000~~	~~$42,000~~
~~Moderate Gravity~~	~~$55,000~~	~~$49,500~~	~~$44,000~~	~~$38,500~~	~~$33,000~~
~~Low Gravity~~	~~$40,000~~	~~$36,000~~	~~$32,000~~	~~$28,000~~	~~$25,000~~[1]

~~[1] See C.2.m.(1)(a)1 below.~~

4. In no case shall the proposed penalty be less than ~~$25,000~~ <add>$5,000<add>.

State Plan States. Regional Administrators shall ensure that this memorandum is promptly forwarded to each State designee and explain its content as requested. This changes the minimum willful penalty policy which was previously transmitted by memorandum on June 14, 1994 and later incorporated in CPL 2.103, the Field Inspection Reference Manual (FIRM), on September 26, 1994. OSHA believes that this policy change is an effective tool for dealing with significant willful safety and health violations especially as modified in the above table which minimizes penalties for employers with 50 or fewer employees. States are encouraged to adopt this or an equivalent policy. State designees shall advise the Regional Administrator of their intention within 30 days.

States that adopt an identical or alternative policy should submit appropriate plan documentation. States wishing to pilot alternatives may do so and should consider entering into limited Performance Agreements on this issue.

If you have any questions, please contact Helen Rogers or William Smith in the Office of General Industry Compliance Assistance at (202) 219-8031.

OSHA Instruction CPL 2.103
September 26, 1994
Office of General Industry Compliance Assistance

SUBJECT: Field Inspection Reference Manual (FIRM)

A. Purpose. The purpose of this instruction is to transmit the Field Inspection Reference Manual (FIRM). The FIRM was developed by the Field Operations Manual (FOM) Revision Team to provide the field offices a reference document for identifying the responsibilities associated with the majority of their inspection duties.

B. Scope. This instruction applies OSHA-wide.

C. References.
1. OSHA Instruction CPL 2.45B, June 15, 1989, Field Operations Manual (FOM).
2. OSHA Instruction CPL 2.51H, March 22, 1993, Exemptions and limitations under the Current Appropriations Act.
3. OSHA Instruction CPL 2.77, December 30, 1986, Critical Fatality/Catastrophe Investigation.
4. OSHA Instruction CPL 2.80, October 21, 1990, Handling of Cases to be Proposed for Violation-by-Violation Penalties.
5. OSHA Instruction CPL 2.90, June 3, 1991, Guidelines for Administration of Corporate-Wide Settlement Agreements.
6. OSHA Instruction CPL 2.94, July 22, 1991, OSHA Response to Significant Events of Potentially Catastrophic Consequences.
7. OSHA Instruction CPL 2.97, January 26, 1993, Fatality/Catastrophe Reports to the National Office ("Flash Reports").
8. OSHA Instruction CPL 2.98, October 12, 1993, Guidelines for Case File Documentation for Use with Videotapes and Audiotapes.
9. OSHA Instruction CPL 2.102, March 28, 1994, Procedures for Approval of Local Emphasis Programs (LEPs) and Experimental Programs.
10. OSHA Instruction CPL 2-2.20B, April 19, 1993, OSHA Technical Manual.
11. OSHA Instruction CPL 2-2.35A, December 19, 1983, 29 CFR 1910.95(b)(1), Guidelines for Noise Enforcement, Appendix A, Noise Control Guidelines.
12. OSHA Instruction CPL 2-2.54, February 10, 1992, Respiratory Protection Program Manual.
13. OSHA Instruction ADM 1-1.31, September 20, 1993, The IMIS Enforcement Data Processing Manual
14. OSHA Instruction ADM 4.4, August 19, 1991, Administrative Subpoenas. 15. OSHA Instruction ADM 11.3A, July 29, 1991, Revision A to the OSHA Mission and Function Statements.
16. Memorandum dated March 31, 1994, to the Regional Administrators from H. Berrien Zettler, Deputy Director, Directorate of Compliance Programs regarding Policy Regarding Voluntary Rescue Activities.

D. Action. The issuance of the FIRM will require a change to the FOM, but until this change is accomplished, if there are any discrepancies between the FIRM and the FOM, the FIRM prevails. Regional Administrators and Area Directors shall ensure that the policies and procedures established in this instruction are transmitted to all Area and District Offices, and to appropriate staff.

E. Effective Date. September 30, 1994.

F. Federal Agencies. This instruction describes a change that affects Federal agencies. Executive Order 12196, Section 1-201, and 29 CFR 1960.16, maintains that Federal agencies must also follow the enforcement policy and procedures contained in this instruction.

G. Federal Program Change. This instruction describes a Federal program change which affects State programs. Each Regional Administrator shall:
1. Ensure that this change is promptly forwarded to each State designee using a format consistent with the Plan Change Two-way Memorandum in Appendix P, OSHA Instruction STP 2.22A, CH-3. Explain the content of this change to the State designee as requested.

2. Encourage the State designees to review the streamlined procedures contained in the Field Inspection Reference Manual (FIRM) and adopt parallel procedures through a State FIRM or changes to the State Field Operations Manual (FOM).
3. Advise the State designees that specific policy changes have been made in the following areas, through the FIRM, which require a State response:
    a. serious willful penalty increase (previously transmitted by June 14, 1994, memorandum "Revised Penalties and Willful Violations") - FIRM Chapter IV, C.2.m.
    b. good faith credit - FIRM Chapter IV, C.2.i.(5)(b)
    c. definitions concerning complaints - FIRM Chapter I, C.2.
    d. nonformal complaint procedures - FIRM Chapter I, C.7.
    e. voluntary rescue operations - FIRM Chapter II, B.2.e.
    f. imminent danger investigations - FIRM Chapter II, B.3.
    g. unobserved exposure - FIRM Chapter III, C.1.b.(4)
    h. reporting of fatalities/catastrophe (1904.8) - FIRM Chapter II, B.2.a.
    i. reporting time for fatalities (1904.8) - FIRM Chapter IV, C.2.n.(3)(c)
    j. notification of employee representatives of citation issuance - FIRM Chapter IV, B.1.b.
    k. economic feasibility - FIRM Chapter IV, A.6.a.(4)(b) and NOTE
    l. employee involvement in informal conferences/settlements - FIRM Chapter IV, D.1.b.
4. Ensure that the State designees are asked to acknowledge receipt of this Federal program change in writing to the Regional Administrator as soon as possible, but not later than 70 calendar days after the date of issuance (10 days for mailing and 60 days for response). The acknowledgment must include a statement indicating the State's general intention with regard to adopting the FIRM or incorporating the FIRM's procedures into its Field Manual and specific intention in regard to the 12 policy changes specified above.
5. Ensure that State designees submit an appropriate State plan supplement within 6 months and that it is reviewed and processed in accordance with paragraphs I.1.a.(3)(a) and (b), Part I of the State Plan Policies and Procedures Manual (SPM).
6. Review policies, instructions and guidelines issued by the State to determine that this change has been communicated to State compliance personnel.

Joseph A. Dear   Ron Yarman
Assistant Secretary   Executive Vice President, NCFLL

DISTRIBUTION: National, Regional, and Area Offices
All Compliance Officers State Designees
NIOSH Regional Program Directors 7(c)(1) Project Managers

OSHA Instruction CPL 2.103
September 26, 1994
Office of General Industry Compliance Assistance

This manual is intended to provide guidance regarding some of the internal operations of the Occupational Safety and Health Administration (OSHA), and is solely for the benefit of the Government. No duties, rights, or benefits, substantive or procedural, are created or implied by this manual. The contents of this manual are not enforceable by any person or entity against the Department of Labor or the United States. Guidelines which reflect current Occupational Safety and Health Review Commission or court precedents do not necessarily indicate acquiescence with those precedents.

TABLE OF CONTENTS
(Page numbers deleted)

CHAPTER I. PRE-INSPECTION PROCEDURES

A. GENERAL RESPONSIBILITIES AND ADMINISTRATIVE PROCEDURES
1. Regional Administrator
2. Area Director
3. Assistant Area Director
4. Compliance Safety and Health Officer
    a. General
    b. Subpoena Served on CSHO
    c. Testifying in Hearings
    d. Release of Inspection Information
    e. Disposition of Inspection Records
5. General Area Office Responsibilities

B. INSPECTION SCHEDULING
1. Program Planning
2. Inspection/Investigation Types
    a. Unprogrammed
    b. Unprogrammed Related
    c. Programmed
    d. Programmed Related
3. Inspection Priorities
    a. Order of Priority
    b. Efficient Use of Resources
    c. Followup Inspections
4. Inspection Selection Criteria
    a. General Requirements
    b. Employer Contacts

C. COMPLAINTS AND OTHER UNPROGRAMMED INSPECTIONS
1. General
2. Definitions
3. Identity of Complaint
4. Formalizing Oral Complaints
5. Imminent Danger Report Received by the Field
6. Formal Complaints
7. Nonformal Complaints
8. Results of Inspection to Complainant
9. Discrimination Complaints
10. Referrals
11. Accidents

D. PROGRAMMED INSPECTIONS

E. INSPECTION PREPARATION
1. General
2. Planning
3. Advance Notice of Inspections
4. Preinspection Compulsory Process
5. Expert Assistance
6. Personal Security Clearance
7. Disclosure of Records
8. Classified and Trade Secret Information

FIELD INSPECTION REFERENCE MANUAL

CHAPTER II. INSPECTIONS PROCEDURES
A. GENERAL INSPECTION PROCEDURES
1. Inspection Scope
   a. Comprehensive
   b. Partial
2. Conduct of the Inspection
   a. Time of Inspection
   b. Presenting Credentials
   c. Refusal to Permit Inspection
      (1) Refusal of Entry or Inspection
      (2) Employer Interference
      (3) Administrative Subpoena
      (4) Obtaining Compulsive Process
      (5) Compulsory Process
      (6) Action to be Taken Upon Receipt of Compulsory Process
      (7) Federal Marshal Assistance
      (8) Refused Entry or Interference with a Compulsory Process
   d. Forcible Interference with Conduct of Inspection or Other Official Duties
   e. Release for Entry
   f. Bankrupt or Out of Business
   g. Strike or Labor Dispute
   h. Employee Participation
3. Opening Conference
   a. Attendance at Opening Conference
   b. Scope
   c. Forms Completion
   d. Employees of Other Employers
   e. Voluntary Compliance Programs
      (1) Section 7(c)(1) and Contract Consultations
      (2) Voluntary Protection Programs (VPP)
   f. Walkaround Representatives
      (1) Employees Represented by a Certified or Recognized Bargaining Agent
      (2) Safety Committee
      (3) No Certified or Recognized Bargaining Agent
   g. Preemption by Another Agency
   h. Disruptive Conduct
   i. Trade Secrets
   j. Classified Areas
   k. Examination of Record Programs and Posting Requirements
      (1) Records
      (2) Lost Workday Injury (LWDI) Rate
      (3) Posting
4. Walkaround Inspection
   a. Evaluation
   b. Record All Facts Pertinent to an Apparent Violation
   c. Collecting Samples
   d. Taking Photographs and/or Videotapes
   e. Interviews
      (1) Purpose
      (2) Employee Right of Complaint
      (3) Time and Location

- (4) Privacy
- (5) Interview Statements
- f. Employer Abatement Assistance
  - (1) Policy
  - (2) Disclaimers
- g. Special Circumstances
  - (1) Trade Secrets
  - (2) Violations of Other Laws
5. Closing Conference

B. SPECIAL INSPECTION PROCEDURES
1. Followup and Monitoring Inspections
   a. Inspection Procedures
   b. Failure to Abate
   c. Reports
   d. Followup Files
2. Fatality/Catastrophe Investigations
   a. Definitions
   b. Selection of CSHO
   c. Families of Victims
   d. Criminal
   e. Rescue Operations
   f. Public Information Policy
3. Imminent Danger Investigations
   a. Definition
   b. Requirements
   c. Inspection
4. Construction Inspections
   a. Standards Applicability
   b. Definition
   c. Employer Worksite
   d. Entry of the Workplace
   e. Closing Conference
5. Federal Agency Inspections

CHAPTER III INSPECTION DOCUMENTATION
A. FOUR STAGE CASE FILE DOCUMENTATION
1. General
2. Case File Stages

B. SPECIFIC FORMS
1. Narrative, Form OSHA-1A
2. Photo Mounting Worksheet, Form OSHA-89
3. Inspection Case File Activity Diary Insert

C. VIOLATIONS
1. Basis of Violations
   a. Standards and Regulations
   b. Employee Exposure
2. Types of Violations
   a. Other-than-serious
   b. Serious
   c. Violations of the General Duty Clause
      (1) Evaluation of Potential Section 5(a)(1) Situations

        (2) Discussion of Section 5(a)(1) Elements
        (3) Limitations on Use of the General Duty Clause
        (4) Pre-Citation Review
     d. Willful
     e. Criminal/Willful
     f. Repeated
     g. De Minimis
  3. Health Standard Violations
     a. Citation of Ventilation Standards
     b. Violations of the Noise Standard
     c. Violations of the Respirator Standard
     d. Additive and Synergistic Effects
     e. Absorption and Ingestion Hazards
     f. Biological Monitoring
  4. Writing Citations
  5. Combining and Grouping of Violations
     a. Combining
     b. Grouping
     c. When Not to Group
  6. Multiemployer Worksites
  7. Employer/Employee Responsibilities
  8. Affirmative Defenses
     a. Definition
     b. Burden of Proof
     c. Explanations

CHAPTER IV. POST-INSPECTION PROCEDURES

A. ABATEMENT
  1. Period
  2. Reasonable Abatement Date
  3. Verification of Abatement
  4. Effect of Contest Upon Abatement Period
  5. Long-Term Abatement Date for Implementation of Feasible Engineering Controls
  6. Feasible Administrative, Work Practice and Engineering Controls
     a. Definitions
     b. Responsibilities
     c. Reducing Employee Exposure

B. CITATIONS
  1. Issuing Citations
  2. Amending or Withdrawing Citation and Notification of Penalty
     a. Citation Revision Justified
     b. Citation Revision Not Justified
     c. Procedures for Amending or Withdrawing Citations

C. PENALTIES
  1. General Policy
  2. Civil Penalties
     a. Statutory Authority
     b. Minimum Penalties
     c. Penalty Factors
     d. Gravity of Violation

       e. Severity Assessment
       f. Probability Assessment
       g. Gravity-Based Penalty
       h. Gravity Calculations for Combined or Grouped Violations
       i. Penalty Adjustment Factors
       j. Effect on Penalties if Employer Immediately Corrects
       k. Failure to Abate
       l. Repeated Violations
       m. Willful Violations
       n. Violation of 29 CFR Parts 1903 and 1904 Regulatory Requirements
    3. Criminal Penalties
D. POST-CITATION PROCESSES
    1. Informal Conferences
       a. General
       b. Procedures
       c. Participation by OSHA Officials
       d. Conduct of the Informal Conference
       e. Decisions
       f. Failure to Abate
    2. Petitions for Modification of Abatement Date (PMA)
       a. Filing Date
       b. Failure to Meet All Requirements
       c. Delayed Decisions
       d. Area Office Position on the PMA
       e. Employee Objections
    3. Services Available to Employers
    4. Settlement of Cases by Area Directors
       a. General
       b. Pre-Contest Settlement (Informal Settlement Agreement)
       c. Procedures for Preparing an Informal Settlement Agreement
       d. Post-Contest Settlement (Formal Settlement Agreement)
       e. Corporate-Wide Settlement Agreement
    5. Guidance for Determining Final Dates of Settlements and Review Commission Orders
E. REVIEW COMMISSION
    1. Transmittal of Notice of Contest and Other Documents to Review Commission
       a. Notice of Contest
       b. Documents to Executive Secretary
       c. Petitions for Modification of Abatement Dates (PMAs)
    2. Transmittal of File to Regional Solicitor
       a. Notification of the Regional Solicitor
       b. Subpoena
    3. Communications with Commission Employees
    4. Dealings with Parties while Proceedings are Pending Before the Commission

CHAPTER I
PRE-INSPECTION PROCEDURES

A. **General Responsibilities and Administrative Procedures.** The following are brief descriptions of the general responsibilities for positions within OSHA. Reference OSHA Instruction ADM 11.3A for complete information on OSHA mission and function statements. Employees should refer to their position descriptions for individual job responsibilities. This document empowers OSHA personnel to make decisions as situations warrant with the ability to act efficiently to accomplish the mission of OSHA and to enforce the Occupational Safety and Health Act.

1. **Regional Administrator.** It is the duty or mission of the Regional Administrator to manage, execute and evaluate all programs of the Occupational Safety and Health Administration (OSHA) in the region. The Regional Administrator reports to the Assistant Secretary through the career Deputy Assistant Secretary.
2. **Area Director (AD).** It is the duty or mission of the Area Director to accomplish OSHA's programs within the delegated area of responsibility of the Area Office. This includes administrative and technical support of the Compliance Safety and Health Officers (CSHOs) assigned to the Area Office, and the issuing of citations.
3. **Assistant Area Director (AAD).** The Assistant Area Director has first level supervisory responsibility over CSHOs in the discharge of their duties and may also conduct compliance inspections. Assistant Area Directors ensure technical adequacy in applying the policies and procedures in effect in the Agency. The Assistant Area Director shall implement a quality assurance system suitable to the work group, including techniques such as random review of selected files, review based on CSHO recommendation/request, and verbal briefing by CSHOs and/or review of higher profile or non-routine cases.
4. **Compliance Safety and Health Officer.**
    a. **General.** The primary responsibility of the Compliance Safety and Health Officer (CSHO) is to carry out the mandate given to the Secretary of Labor, namely, "to assure so far as possible every working man and woman in the Nation safe and healthful working conditions...." To accomplish this mandate the Occupational Safety and Health Administration employs a wide variety of programs and initiatives, one of which is enforcement of standards through the conduct of effective inspections to determine whether employers are:
        (1) Furnishing places of employment free from recognized hazards that are causing or are likely to cause death or serious physical harm to their employees, and
        (2) Complying with safety and health standards and regulations. Through inspections and other employee/employer contact, the CSHO can help ensure that hazards are identified and abated to protect workers. During these processes, the CSHO must use professional judgment to adequately document hazards in the case file, as required by the policies and procedures in effect in the Agency. The CSHO will be responsible for the technical adequacy of each case file.

    A.4.b. **Subpoena Served on CSHO.** If a CSHO is served with a subpoena, the Area Director shall be informed immediately and shall refer the matter to the Regional Solicitor.
    c. **Testifying in Hearings.** The CSHO is required to testify in hearings on OSHA's behalf. The CSHO shall be mindful of this fact when recording observations during inspections. The case file shall reflect conditions observed in the workplace as accurately as possible. If the CSHO is called upon to testify, the case file will be invaluable as a means for recalling actual conditions.
    d. **Release of Inspection Information.** The information obtained during inspections is confidential, but is to be determined as disclosable or nondisclosable on the basis of criteria established in the Freedom of Information Act, as amended, in 29 CFR Part 70, and in Chapter XIV of OSHA Instruction CPL 2.45B or a superseding directive. Requests for release of inspection information shall be directed to the Area Director.
    e. **Disposition of Inspection Records.** "Inspection Records" are any records made by a CSHO that concern, relate to, or are part of any inspection or that concern, relate to, or are part of the performance of any official duty. Such original material and all copies shall be included in the case file. These records are the property of the United States Government and a part of the case file. Inspection records are not the property of the CSHO and under no circumstances are they to be retained or used for any private purpose. Copies of documents, notes or other recorded information not necessary or pertinent, or not suitable for inclusion in the case file shall, with the concurrence and permission of the Area Director, be destroyed in accordance with an approved record disposition schedule.
5. **General Area Office Responsibilities.** The Area Director shall ensure that the Area Office maintains an outreach program appropriate to local conditions and the needs of the service area. The plan may include utilization of Regional Office Support Services, training and education services, referral services, voluntary compliance programs, abatement assistance, and technical services.

B. **Inspection Scheduling.**
1. **Program Planning.** Effective and efficient use of resources requires careful, flexible planning. In this way, the overall goal of hazard abatement and worker protection is best served.
2. **Inspection/Investigation Types.**
    a. **Unprogrammed.** Inspections scheduled in response to alleged hazardous working conditions that have been identified at a specific worksite are unprogrammed. This type of inspection responds to reports of imminent dangers, fatalities/catastrophes, complaints and referrals. It also includes followup and monitoring inspections scheduled by the Area Office. **NOTE:** This category includes all employers directly affected by the subject of the unprogrammed activity, and is especially applicable on multiemployer worksites.
    b. **Unprogrammed Related.** Inspections of employers at multiemployer worksites whose operations are not directly affected by the subject of the conditions identified in the complaint, accident, or referral are unprogrammed related. An example would be a trenching inspection conducted at the unprogrammed worksite, where the trenching hazard was not identified in the complaint, accident report, or referral.
    c. **Programmed.** Inspections of worksites which have been scheduled based upon objective or neutral selection criteria are programmed. The worksites are selected according to national scheduling plans for safety and for health or special emphasis programs.

d. **Programmed Related.** Inspections of employers at multiemployer worksites whose activities were not included in the programmed assignment, such as a low injury rate employer at a worksite where programmed inspections are being conducted for all high injury rate employers. All high hazard employers at the worksite shall normally be included in the programmed inspections. (See Chapter II, F.2. of OSHA Instruction CPL 2.45B or a superseding directive).

3. **Inspection Priorities.**
   a. **Order of Priority.** Generally, priority of accomplishment and assignment of staff resources for inspection categories shall be as follows:

Priority	Category
First	Imminent Danger
Second	Fatality/Catastrophe Investigations
Third	Complaints/Referrals Investigation
Fourth	Programmed Inspections

   b. **Efficient Use of Resources.** Based on the nature of the alleged hazard, unprogrammed inspections normally shall be scheduled and conducted prior to programmed inspections. Deviations from this priority list are allowed so long as they are justifiable, lead to efficient use of resources, **and** contribute to the effective protection of workers. An example of such a deviation would be for Area Directors to commit a certain percentage of IH resources to programmed Local Emphasis Program (LEP) inspections.

   c. **Followup Inspections.** In cases where followup inspections are necessary, they shall be conducted as promptly as resources permit. Except in unusual circumstances, followup inspections shall take priority over all programmed inspections and any unprogrammed inspections with hazards evaluated as other than serious. Followup inspections should not normally be conducted within the 15 working day contest period unless high gravity serious violations were issued. See Chapter IV at A.4. regarding effect of contest upon abatement period.

4. **Inspection Selection Criteria.**
   a. **General Requirements.** OSHA's priority system for conducting inspections is designed to distribute available OSHA resources as effectively as possible to ensure that maximum feasible protection is provided to the working men and women of this country.
      (1) **Scheduling.** The Area Director shall ensure that inspections are scheduled within the framework of this chapter, that they are consistent with the objectives of the Agency, and that appropriate documentation of scheduling practices is maintained. (See OSHA Instruction CPL 2.51H, or most current version, for congressional exemptions and limitations on OSHA inspection activity.)

B. 4. a. (2) **Effective Use of Resources.** The Area Director shall ensure that OSHA resources are effectively distributed during inspection activities. If an inspection is of a complex nature, the Area Director may consider utilizing outside OSHA resources (e.g., the Health Response Team) to more effectively employ the Area or District Office resources. The Area Office will retain control of the inspection.

   (3) **Effect of Contest.** If an employer scheduled for inspection, either programmed or unprogrammed, has contested a citation and/or a penalty received as a result of a previous inspection and the case is still pending before the Review Commission, the following guidelines apply:
      (a) If the employer has contested the penalty only, the inspection shall be scheduled as though there were no contest.
      (b) If the employer has contested the citation itself or any items thereon, then programmed and unprogrammed inspections shall be scheduled in accordance with the guidelines in B.3.a. of this chapter. The scope of unprogrammed inspections normally shall be partial. All items under contest shall be excluded from the inspection unless a potential imminent danger is involved.

   b. **Employer Contacts.** Contacts for information initiated by employers or their representatives shall not trigger an inspection, nor shall such employer inquiries protect them against regular inspections conducted pursuant to guidelines established by the agency. Further, if an employer or its representative indicates that an imminent danger exists or that a fatality or catastrophe has occurred, the Area Director shall act in accordance with established inspection priority procedures.

C. **Complaints & Other Unprogrammed Inspections.**
   1. **General.** This section relates to information received and processed at the area office before an inspection rather than information which is given to the CSHO during an inspection. Complaints will be evaluated according to local Area Office procedures, including using the criteria established in Chapter III, C.2. for classifying the alleged violations as serious or other. When essential information is not provided by the complainant, the complaint is too vague to evaluate, or the Area Office has other specific information that the complaint is not valid, an attempt shall be made to clarify or supplement available information. If a decision is made that the complaint is not valid, a letter will be sent to the complainant advising of the decision and its reasons.
   2. **Definitions.**
      a. **Complaint (OSHA-7).** Notice of an alleged hazard (over which OSHA has jurisdiction), or a violation of the Act, given by a past or present employee, a representative of employees, a concerned citizen, or an 11(c) officer seeking resolution of a discrimination complaint.
      b. **Referral (OSHA-90).** Notice of an alleged hazard or a violation of the Act given by any source not listed under C.2.a., above, including media reports.
      c. **Formal Complaint.** A signed complaint alleging an imminent danger or the existence of a violation threatening physical harm, submitted by a current employee, a representative of employees (such as unions, attorneys, elected representatives, and family members), or present employee of another company if that employee is exposed to the hazards of the complained-about workplace. Reference Section 8(f) of the Act and 29 CFR 1903.11.
      d. **Nonformal Complaint.** Oral or unsigned complaints, or complaints by former employees or non-employees.
   3. **Identity of Complainant.** The identity of the complainant shall be kept confidential unless otherwise requested by the complainant, in accordance with Section 8(f)(1) of the Act. No information shall be given to employers which would allow them to identify the complainant.

FIELD INSPECTION REFERENCE MANUAL

4. **Formalizing Oral Complaints.**
   a. If the complainant meets the criteria in C.2.c., above, for filing formal complaints and wishes to formalize an oral complaint, all pertinent information will be entered on an OSHA-7 form, or equivalent, by the complainant or a member of the Area Office staff. A copy of this completed form can be sent to the complainant for signature, or the complainant shall be asked to sign a letter with the particular details of the complaint to the area office.
   b. The complainant shall be informed that, if the signed complaint is not returned within 10 working days, it shall be treated as a nonformal complaint. If the signed complaint arrives after the 10 working days but prior to OSHA's contact with the employer, it will be treated as a formal complaint.
   c. The following are examples of deficiencies which would result in the failure of an apparent formal complaint to meet the requirements of the definition:
      (1) A thorough evaluation of the complaint does not establish reasonable grounds to believe that the alleged violation can be classified as an imminent danger or that the alleged hazard is covered by a standard or, in the case of an alleged serious condition, by the general duty clause (Section 5(a)(1)).
      (2) The complaint concerns a workplace condition which has no direct relationship to safety or health and does not threaten physical harm; e.g., a violation of a regulation or a violation of a standard that is classified as de minimis.
      (3) The complaint alleges a hazard which violates a standard but describes no actual workplace conditions and gives no particulars which would allow a proper evaluation of the hazard. In such a case the Area Director shall make a reasonable attempt to obtain such information.
5. **Imminent Danger Report Received By the Field.** Any allegation of imminent danger received by an OSHA office shall be handled in accordance with the following procedures:
   a. The Area Director shall immediately determine whether there is a reasonable basis for the allegation.
   b. Imminent danger investigations shall be scheduled with the highest priority.
   c. When an immediate inspection cannot be made, the Area Director or CSHO shall contact the employer (and when known, the employee representative) immediately, obtain as many pertinent details as possible concerning the situation and attempt to have any employees affected by imminent danger voluntarily removed. Such notification shall be considered advance notice and shall be handled in accordance with the procedures given in E.3. of this chapter.
6. **Formal Complaints.** All formal complaints meeting the requirements of Section 8(f)(1) of the Act and 29 CFR 1903.11 shall be scheduled for workplace inspections.
   a. **Determination.** Upon determination by the Assistant Area Director that a complaint is formal, an inspection shall be scheduled in accordance with the priorities in C.6.b.
   b. **Priorities for Responding by Inspections to Formal Complaints.** Inspections resulting from formal complaints shall be conducted according to the following priority:
      (1) Formal complaints, other than imminent danger, shall be given a priority based upon the classification and the gravity of the alleged hazards as defined in Chapters III and IV.
      (2) Formal serious complaints shall be investigated on a priority basis within 30 working days and formal other-than-serious complaints within 120 days.
      (3) If resources do not permit investigations within the time frames given in (2), a letter to the complainant shall explain the delay and shall indicate when an investigation may occur. The complainant shall be asked to confirm the continuation of the alleged hazardous conditions.
      (4) If a late complaint inspection is to be conducted, the Area Director may contact the complainant to ensure that the alleged hazards are still existent.
7. **Nonformal Complaints.**
   a. **Serious.** If a decision is made to handle a serious nonformal complaint by letter, a certified letter shall be sent to the employer advising the employer of the complaint items and the need to respond to OSHA within a specified time. When applicable, the employer shall be informed of Section 11(c) requirements, and that the complainant will be kept informed of the complaint progress. Follow-up contact may be by telephone at the option of the Area Office.
   b. **Other-Than-Serious and De Minimis.** Nonformal complaints about other-than-serious hazards or de minimis conditions may be investigated by telephone if they can be satisfactorily resolved in that manner, with follow-up telephone contact to the complainant with the results of the employer's investigation and corrective actions. Corrective action shall be documented in the case file. If, however, the telephone contact is inadequate, a letter will be sent to the employer.
   c. **Letter to Complainant.** Concurrent with the letter to the employer, a letter to the complainant shall be sent containing a copy of the letter to the employer. The complainant will be asked to notify OSHA if no corrective action is taken within the indicated time frame, or if any adverse or discriminatory action or threats are made due to the complainant's safety and health activities. Copies of subsequent correspondence related to the complaint shall be sent to the complainant, if requested.
   d. **Inspection of Nonformal Complaint.**
      (1) Where the employer fails to respond or submits an inadequate response, the employer may be contacted to find out what corrections will be made, or the nonformal complaint will be activated for inspection. If no action has been taken, the nonformal complaint shall normally be activated for inspection.
      (2) Nonformal complaints, when received by the Area Office, may be activated for inspection if the Area Director or representative judges the hazard to be high gravity serious in nature, and the inspection can be performed with efficient use of resources.
8. **Results of Inspection to Complainant.** After an inspection, the complainant shall be sent a letter addressing each complaint item, with reference to the citations and/or with a sufficiently detailed description of the findings and why they did not result in a violation. The complainant shall also be informed of the appeal rights under 29 CFR 1903.12.
9. **Discrimination Complaints.** The complainant shall be advised of the protection against discrimination afforded by Section 11(c) of the Act and shall be informed of the procedure for filing an 11(c) complaint.
   a. Safety and/or health complaints filed by former employees who allege that they were fired for exercising their rights under the Act are nonformal complaints and will not be scheduled for investigation.

(1) Such complaints shall be recorded on an OSHA-7 Form and handled in accordance with the procedures outlined in OSHA Instruction DIS .4B. They shall be transmitted to the appropriate 11(c) personnel for investigation of the alleged 11(c) discrimination complaint.

(2) No letter shall be sent to the employer until after the Regional Supervisory Investigator has reviewed the case and decided that no recommendation for inspection will be submitted to the Area Director.

(3) This screening process by the Regional Supervisory Investigator is not anticipated to take more than 3 work days and usually less. The Area Director can expect to be informed by telephone of the decision within that time frame.

b. In those instances where the Regional Supervisory Investigator determines that the existence or nature of the alleged hazard is likely to be relevant to the resolution of the 11(c) discrimination complaint, the complaint shall be sent back to the Area Director for an OSHA inspection to be handled as a referral.

c. When the decision is that no inspection is necessary, the Area Director shall ensure that the complaint has been recorded on an OSHA-7 Form and proceed to send a letter to the employer in accordance with procedures for responding to nonformal complaints.

d. Any 11(c) complaint alleging an imminent danger shall be handled in accordance with the instructions in C.5.

10. **Referrals.** Referrals shall be handled in a manner similar to that of complaints.
    a. **Letters.** Referrals shall normally be handled by letter or telephone. For those referrals handled by letter, complaint letters can be revised to fit the particular circumstances of the referral.
    b. **Inspections.** High gravity serious referrals shall normally be handled by inspection. A letter transmitting the results of the investigation shall be sent to any referring agency/department.

11. **Accidents.** Accidents involving significant publicity or any other accident not involving a fatality or a catastrophe, however reported, may be considered as either a complaint or a referral, depending on the source of the report, and shall be handled according to the directions given in Chapter II.

D. Programmed Inspections.

1. Programmed inspections shall be scheduled in accordance with Chapter II, F.2. of OSHA Instruction CPL 2.45B or a superseding directive.
2. Local emphasis programmed inspections shall be conducted in accordance with OSHA Instruction CPL 2.102.

E. Inspection Preparation.

1. **General.** The conduct of effective inspections requires professional judgment in the identification, evaluation and reporting of safety and health conditions and practices. Inspections may vary considerably in scope and detail, depending upon the circumstances in each case.

2. **Planning.** It is most important that the CSHO adequately prepares for an inspection. The CSHO shall also ensure the selection of appropriate inspection materials and equipment, including personal protective equipment, based on anticipated exposures and training received in relation to the uses and limitations of such equipment. Refer to OSHA Instruction CPL 2-2.54 regarding respiratory protection.

    a. 29 CFR 1903.7(c) requires that the CSHO comply with all safety and health rules and practices at the establishment and wear or use the safety clothing or protective equipment required by OSHA standards or by the employer for the protection of employees.

    b. The CSHO shall not enter any area where special entrance restrictions apply until the required precautions have been taken. It shall be the Assistant Area Director's responsibility to procure whatever materials and equipment are needed for the safe conduct of the inspection.

3. **Advance Notice of Inspections.**
    a. **Policy.** Section 17(f) of the Act and 29 CFR 1903.6 contain a general prohibition against the giving of advance notice of inspections, except as authorized by the Secretary or the Secretary's designee.

        (1) The Occupational Safety and Health Act regulates many conditions which are subject to speedy alteration and disguise by employers. To forestall such changes in worksite conditions, the Act, in Section 8(a), prohibits unauthorized advance notice.

        (2) There may be occasions when advance notice is necessary to conduct an effective investigation. These occasions are narrow exceptions to the statutory prohibition against advance notice.

        (3) Advance notice of inspections may be given only with the authorization of the Area Director and only in the following situations:
            (a) In cases of apparent imminent danger to enable the employer to correct the danger as quickly as possible;
            (b) When the inspection can most effectively be conducted after regular business hours or when special preparations are necessary;
            (c) To ensure the presence of employer and employee representatives or other appropriate personnel who are needed to aid in the inspection; and
            (d) When the giving of advance notice would enhance the probability of an effective and thorough inspection; e.g., in complex fatality investigations.

        (4) Advance notice exists whenever the Area Office sets up a specific date or time with the employer for the CSHO to begin an inspection. Any delays in the conduct of the inspection shall be kept to an absolute minimum. Lengthy or unreasonable delays shall be brought to the attention of the Assistant Area Director. Advance notice generally does not include non-specific indications of potential future inspections.

        (5) In unusual circumstances, the Area Director may decide that a delay is necessary. In those cases the employer or the CSHO shall notify affected employee representatives, if any, of the delay and shall keep them informed of the status of the inspection.

    b. **Documentation.** The conditions requiring advance notice and the procedures followed shall be documented in the case file.

4. **Pre-Inspection Compulsory Process.** 29 CFR 1903.4 authorizes the agency to seek a warrant in advance of an attempted inspection if circumstances are such that "pre-inspection process (is) desirable or necessary." The Act authorizes the agency to issue administrative subpoenas to obtain relevant information.

a. Although the agency generally does not seek warrants without evidence that the employer is likely to refuse entry, the Area Director may seek compulsory process in advance of an attempt to inspect or investigate whenever circumstances indicate the desirability of such warrants. **NOTE:** Examples of such circumstances would be evidence of denied entry in previous inspections, or awareness that a job will only last a short time or that job processes will be changing rapidly.
b. Administrative subpoenas may also be issued prior to any attempt to contact the employer or other person for evidence related to an OSHA inspection or investigation. (See OSHA Instruction ADM 4.4. and Chapter II, A.2.c.(3).)
5. **Expert Assistance.**
   a. The Area Director shall arrange for a specialist and/or specialized training, preferably from within OSHA, to assist in an inspection or investigation when the need for such expertise is identified.
   b. OSHA specialists may accompany the CSHO or perform their tasks separately. A CSHO must accompany outside consultants. OSHA specialists and outside consultants shall be briefed on the purpose of the inspection and personal protective equipment to be utilized.
6. **Personal Security Clearance.** Some establishments have areas which contain material or processes which are classified by the U.S. Government in the interest of national security. Whenever an inspection is scheduled for an establishment containing classified areas, the Assistant Area Director shall assign a CSHO who has the appropriate security clearances. The Regional Administrator shall ensure that an adequate number of CSHOs with appropriate security clearances are available within the Region and that the security clearances are current.
7. **Disclosure of Records.** The disclosure of inspection records is governed by the Department's regulations at 29 CFR part 70, implementing the Freedom of Information Act (FOIA).
8. **Classified and Trade Secret Information.** Any classified or trade secret information and/or personal knowledge of such information by OSHA personnel shall be handled in accordance with the regulations of OSHA or of the responsible agency. The collection of such information, and the number of personnel with access to it shall be limited to the minimum necessary for the conduct of compliance activities. The CSHO shall identify classified and trade secret information as such in the case file. Title 18 USC, Section 1905, as referenced by Section 15 of the OSH Act, provides for criminal penalties in the event of improper disclosure.

CHAPTER II
INSPECTION PROCEDURES

A. General Inspection Procedures.
1. **Inspection Scope.** Inspections, either programmed or unprogrammed, fall into one of two categories depending on the scope of the inspection:
   a. **Comprehensive.** A substantially complete inspection of the potentially high hazard areas of the establishment. An inspection may be deemed comprehensive even though, as a result of the exercise of professional judgment, not all potentially hazardous conditions, operations and practices within those areas are inspected.
   b. **Partial.** An inspection whose focus is limited to certain potentially hazardous areas, operations, conditions or practices at the establishment. A partial inspection may be expanded based on information gathered by the CSHO during the inspection process. Consistent with the provisions of Section 8(f)(2) of the Act, and Area Office priorities, the CSHO shall use professional judgment to determine the necessity for expansion of the inspection scope, based on information gathered during records or program review and walkaround inspection.
2. **Conduct of the Inspection.**
   a. **Time of Inspection.** Inspections shall be made during regular working hours of the establishment except when special circumstances indicate otherwise. The Assistant Area Director and CSHO shall confer with regard to entry during other than normal working hours.
   b. **Presenting Credentials.**
      (1) At the beginning of the inspection the CSHO shall locate the owner representative, operator or agent in charge at the workplace and present credentials. On construction sites this will most often be the representative of the general contractor.
      (2) When neither the person in charge nor a management official is present, contact may be made with the employer to request the presence of the owner, operator or management official. The inspection shall not be delayed unreasonably to await the arrival of the employer representative. This delay should normally not exceed one hour. If the person in charge at the workplace cannot be determined, record the extent of the inquiry in the case file and proceed with the physical inspection.
   c. **Refusal to Permit Inspection.** Section 8 of the Act "provides that CSHOs may enter without delay and at reasonable times any establishment covered under the Act for the purpose of conducting an inspection". Unless the circumstances constitute a recognized exception to the warrant requirement (i.e., consent, third party consent, plain view, open field, or exigent circumstances) an employer has a right to require that the CSHO seek an inspection warrant prior to entering an establishment and may refuse entry without such a warrant. **NOTE:** On a military base or other Federal Government facility, the following guidelines do not apply. Instead, a representative of the controlling authority shall be informed of the contractor's refusal and asked to take appropriate action to obtain cooperation.
      (1) **Refusal of Entry or Inspection.** When the employer refuses to permit entry upon being presented proper credentials or allows entry but then refuses to permit or hinders the inspection in some way, a tactful attempt shall be made to obtain as much information as possible about the establishment. (See A.2.c.(4), below, for the information the CSHO shall attempt to obtain.)
         (a) If the employer refuses to allow an inspection of the establishment to proceed, the CSHO shall leave the premises and immediately report the refusal to the Assistant Area Director. The Area Director shall notify the Regional Solicitor.
         (b) If the employer raises no objection to inspection of certain portions of the workplace but objects to inspection of other portions, this shall be documented. Normally, the CSHO shall continue the inspection, confining it only to those certain portions to which the employer has raised no objections.

(c) In either case the CSHO shall advise the employer that the refusal will be reported to the Assistant Area Director and that the agency may take further action, which may include obtaining legal process.

(d) On multiemployer worksites, valid consent can be granted by the owner, or another co-occupier of the space, for site entry.

(2) **Employer Interference.** Where entry has been allowed but the employer interferes with or limits any important aspect of the inspection, the CSHO shall determine whether or not to consider this action as a refusal. Examples of interference are refusals to permit the walkaround, the examination of records essential to the inspection, the taking of essential photographs and/or videotapes, the inspection of a particular part of the premises, indispensable employee interviews, or the refusal to allow attachment of sampling devices.

(3) **Administrative Subpoena.** Whenever there is a reasonable need for records, documents, testimony and/or other supporting evidence necessary for completing an inspection scheduled in accordance with any current and approved inspection scheduling system or an investigation of any matter properly falling within the statutory authority of the agency, the Regional Administrator, or authorized Area Director, may issue an administrative subpoena. (See OSHA Instruction ADM 4.4.)

(4) **Obtaining Compulsory Process.** If it is determined, upon refusal of entry or refusal to produce evidence required by subpoena, that a warrant will be sought, the Area Director shall proceed according to guidelines and procedures established in the Region for warrant applications.

(a) With the approval of the Regional Solicitor, the Area Director may initiate the compulsory process.

(b) The warrant sought when employer consent has been withheld shall normally be limited to the specific working conditions or practices forming the basis of the unprogrammed inspection. A broad scope warrant, however, may be sought when the information available indicates conditions which are pervasive in nature or if the establishment is on the current list of targeted establishments.

(c) If the warrant is to be obtained by the Regional Solicitor, the Area Director shall transmit in writing to the Regional Solicitor, within 48 hours after the determination is made that a warrant is necessary, all information necessary to obtain a warrant, as determined through contact with the Solicitor, which may include the following:

1 Area/District Office, telephone number, and name of Assistant Area Director involved.
2 Name of CSHO attempting inspection and inspection number, if assigned. Identify whether inspection to be conducted included safety items, health items or both.
3 Legal name of establishment and address including City, State and County. Include site location if different from mailing address.
4 Estimated number of employees at inspection site.
5 SIC Code and high hazard ranking for that specific industry within the State, as obtained from statistics provided by the National Office.
6 Summary of all facts leading to the refusal of entry or limitation of inspection, including the following:
  a Date and time of entry.
  b Date and time of denial.
  c Stage of denial (entry, opening conference, walkaround, etc.).
7 Narrative of all actions taken by the CSHO leading up to during and after refusal, including the following information:
  a Full name and title of the person to whom CSHO presented credentials.
  b Full name and title of person(s) who refused entry.
  c Reasons stated for the denial by person(s) refusing entry.
  d Response, if any, by CSHO to c, above.
  e Name and address of witnesses to denial of entry.
8 All previous inspection information, including copies of the previous citations.
9 Previous requests for warrants. Attach details, if applicable.
10 As much of the current inspection report as has been completed.
11 If a construction site involving work under contract from any agency of the Federal Government, the name of the agency, the date of the contract, and the type of work involved.
12 Other pertinent information such as description of the workplace; the work process; machinery, tools and materials used; known hazards and injuries associated with the specific manufacturing process or industry.
13 Investigative techniques which will be required during the proposed inspection; e.g., personal sampling, photographs, audio/ videotapes, examination of records, access to medical records, etc.
14 The specific reasons for the selection of this establishment for the inspection including proposed scope of the inspection and rationale:
  a **Imminent Danger.**
    o Description of alleged imminent danger situation.
    o Date received and source of information.
    o Original allegation and copy of typed report, including basis for reasonable expectation of death or serious physical harm and immediacy of danger.
    o Whether all current imminent danger processing procedures have been strictly followed.
  b **Fatality/Catastrophe.**
    o The OSHA-36F filled out in as much detail as possible.
  c **Complaint or Referral.**
    o Original complaint or referral and copy of typed complaint or referral.
    o Reasonable grounds for believing that a violation that threatens physical harm or imminent danger exists, including standards that could be violated if the complaint or referral is true and accurate.
    o Whether all current complaint or referral processing procedures have been strictly followed.
    o Additional information gathered pertaining to complaint or referral evaluation.

FIELD INSPECTION REFERENCE MANUAL

                    d **Programmed.**
                        - Targeted safety—general industry, maritime, construction.
                        - Targeted health.
                        - Special emphasis program—Special Programs, Local Emphasis Program, Migrant Housing Inspection, etc.
                    e **Followup.**
                        - Date of initial inspection.
                        - Details and reasons followup was to be conducted.
                        - Copies of previous citations on the basis of which the followup was initiated.
                        - Copies of settlement stipulations and final orders, if appropriate.
                        - Previous history of failure to correct, if any.
                    f **Monitoring.**
                        - Date of original inspection.
                        - Details and reasons monitoring inspection was to be conducted.
                        - Copies of previous citations and/or settlement agreements on the basis of which the monitoring inspection was initiated.
                        - PMA request, if applicable.
            (5) **Compulsory Process.** When a court order or warrant is obtained requiring an employer to allow an inspection, the CSHO is authorized to conduct the inspection in accordance with the provisions of the court order or warrant. All questions from the employer concerning reasonableness of any aspect of an inspection conducted pursuant to compulsory process shall be referred to the Area Director.
            (6) **Action to be Taken Upon Receipt of Compulsory Process.** The inspection will normally begin within 24 hours of receipt of a warrant or of the date authorized by the warrant for the initiation of the inspection.
                (a) The CSHO shall serve a copy of the warrant on the employer and make a separate notation as to the time, place, name and job title of the individual served.
                (b) The warrant may have a space for a return of service entry by the CSHO in which the exact dates of the inspection made pursuant to the warrant are to be entered. Upon completion of the inspection, the CSHO will complete the return of service on the original warrant, sign and forward it to the Assistant Area Director for appropriate action.
                (c) Even where the walkaround is limited by a warrant or an employer's consent to specific conditions or practices, a subpoena for production of records shall be normally served, if necessary, in accordance with A.2.c.(3), above. The records specified in the subpoena shall include (as appropriate) injury and illness records, exposure records, the written hazards communication program, the written lockout-tagout program, and records relevant to the employer's safety and health management program, such as safety and health manuals or minutes from safety meetings.
                (d) The Regional Administrator, or Area Director authorized to do so, may issue, for each inspection, an administrative subpoena which seeks production of the above specified categories of documents. The subpoena may call for immediate production of the records with the exception of the documents relevant to the safety and health management program, for which a period of 5 working days normally shall be allowed.
                (e) If circumstances make it appropriate, a second warrant may be sought based on the review of records or on "plain view" observations of other potential violations during a limited scope walkaround.
            (7) **Federal Marshal Assistance.** A U.S. Marshal may accompany the CSHO when the compulsory process is presented.
            (8) **Refused Entry or Interference with a Compulsory Process.**
                (a) When an apparent refusal to permit entry or inspection is encountered upon presenting the warrant, the CSHO shall specifically inquire whether the employer is refusing to comply with the warrant.
                (b) If the employer refuses to comply or if consent is not clearly given, the CSHO shall not attempt to conduct the inspection but shall leave the premises and contact the Assistant Area Director concerning further action. The CSHO shall make notations (including all witnesses to the refusal or interference) and fully report all relevant facts. Under these circumstances the Area Director shall contact the Regional Solicitor and they shall jointly decide what further action shall be taken.
        d. **Forcible Interference with Conduct of Inspection or Other Official Duties.** Whenever an OSHA official or employee encounters forcible resistance, opposition, interference, etc., or is assaulted or threatened with assault while engaged in the performance of official duties, all investigative activity shall cease.
            (1) The Assistant Area Director shall be advised by the most expeditious means.
            (2) Upon receiving a report of such forcible interference, the Area Director or designee shall immediately notify the Regional Administrator.
        e. **Release for Entry.**
            (1) The CSHO shall not sign any form or release or agree to any waiver. This includes any employer forms concerned with trade secret information.
            (2) The CSHO may obtain a pass or sign a visitor's register, or any other book or form used by the establishment to control the entry and movement of persons upon its premises. Such signature shall not constitute any form of a release or waiver of prosecution of liability under the Act.
        f. **Bankrupt or Out of Business.** If the establishment scheduled for inspection is found to have ceased business and there is no known successor, the CSHO shall report the facts to the Assistant Area Director. If an employer, although adjudicated bankrupt, is continuing to operate on the date of the scheduled inspection, the inspection shall proceed. An employer must comply with the Act until the day the business actually ceases to operate.
        g. **Strike or Labor Dispute.** Plants or establishments may be inspected regardless of the existence of labor disputes involving work stoppages, strikes or picketing. If the CSHO identifies an unanticipated labor dispute at a proposed inspection site, the Assistant Area Director shall be consulted before any contact is made.

(1) **Programmed Inspections.** Programmed inspections may be deferred during a strike or labor dispute, either between a recognized union and the employer or between two unions competing for bargaining rights in the establishment.

(2) **Unprogrammed Inspections.** Unprogrammed inspections (complaints, fatalities, etc.) will be performed during strikes or labor disputes. However, the seriousness and reliability of any complaint shall be thoroughly investigated by the supervisor prior to scheduling an inspection to ensure as far as possible that the complaint reflects a good faith belief that a true hazard exists. If there is a picket line at the establishment, the CSHO shall inform the appropriate union official of the reason for the inspection prior to initiating the inspection.

h. **Employee Participation.** The CSHO shall advise the employer that Section 8(e) of the Act and 29 CFR 1903.8 require that an employee representative be given an opportunity to participate in the inspection.

(1) CSHOs shall determine as soon as possible after arrival whether the employees at the worksite to be inspected are represented and, if so, shall ensure that employee representatives are afforded the opportunity to participate in all phases of the workplace inspection.

(2) If an employer resists or interferes with participation by employee representatives in an inspection and this cannot be resolved by the CSHO, the continued resistance shall be construed as a refusal to permit the inspection and the Assistant Area Director shall be contacted in accordance with A.2.c. of this chapter. **NOTE:** For the purpose of this chapter, the term "employee representative" refers to (1) a representative of the certified or recognized bargaining agent, or, if none, (2) an employee member of a safety and health committee who has been chosen by the employees (employee committee members or employees at large) as their OSHA representative, or (3) an individual employee who has been selected as the walkaround representative by the employees of the establishment.

3. **Opening Conference.** The CSHO shall attempt to inform all effected employers of the purpose of the inspection, provide a copy of the complaint if applicable, and shall include employees unless employer objects. The opening conference shall be kept as brief as possible and may be expedited through use of an opening conference handout. Conditions of the worksite shall be noted upon arrival as well as any changes which may occur during the opening conference. **NOTE:** The CSHO shall determine if the employer is covered by any of the exemptions or limitations noted in the current Appropriations Act (see OSHA Instruction CPL 2.51 H, or the most current version), or deletions in OSHA Instruction CPL 2.45B Chapter II, F.2.b.(1)(b)**5 b** or superseding directive.

a. **Attendance at Opening Conference.** OSHA encourages employers and employees to meet together in the spirit of open communication. The CSHO shall conduct a joint opening conference with employer and employee representatives unless either party objects. If there is objection to a joint conference, the CSHO shall conduct separate conferences with employer and employee representatives.

b. **Scope.** The CSHO shall outline in general terms the scope of the inspection, including private employee interviews, physical inspection of the workplace and records, possible referrals, discrimination complaints, and the closing conference(s).

c. **Forms Completion.** The CSHO shall obtain available information for the OSHA-1 and other appropriate forms.

d. **Employees of Other Employers.** During the opening conference, the CSHO shall determine whether the employees of any other employers are working at the establishment. If these employers may be affected by the inspection, the scope may be expanded to include others or a referral made at the discretion of the CSHO. At multiemployer sites, copies of complaint(s), if applicable, shall be provided to all employers affected by the alleged hazard(s), and to the general contractor.

e. **Voluntary Compliance Programs.** Employers who participate in selected voluntary compliance programs may be exempted from programmed inspections. The CSHO shall determine whether the employer falls under such an exemption during the opening conference.

(1) **Section 7(c)(1) and Contract Consultations.** In accordance with 29 CFR 1908.7 and Chapter IX of the Consultation Policies and Procedures Manual (CPPM), the CSHO shall ascertain at the opening conference whether an OSHA-funded consultation is in progress or whether the facility is pursuing or has received an inspection exemption through consultation under current procedures.

(a) An on site consultation visit in progress has priority over programmed inspections except as indicated in 29 CFR 1908.7(b) (2)(iv), which allows for critical inspections as determined by the Assistant Secretary.

(b) If a consultation visit is in progress, the inspection may be rescheduled.

(c) If a followup inspection (including monitoring) or an imminent danger, fatality/catastrophe, complaint or referral investigation is to be conducted, the inspection shall not be deferred, but its scope shall be limited to those areas required to complete the purpose of the investigation. The consultant must interrupt the onsite visit until the compliance inspection shall have been completed (Ref. 29 CFR 1908.7).

(2) **Voluntary Protection Programs (VPP).** In the event a CSHO enters a facility that has been approved for participation in a VPP and is currently under an inspection exemption, the approval letter shall be copied and the inspection either be terminated (if it is a programmed inspection) or limited to the specified items in the complaint or referral (if it is unprogrammed).

f. **Walkaround Representatives.** Those representatives designated to accompany the CSHO during the walkaround are considered walkaround representatives, and will generally include employer designated and employee designated representatives. At establishments where more than one employer is present or in situations where groups of employees have different representatives, it is acceptable to have a different employer/employee representative for different phases of the inspection. More than one employer and/or employee representative may accompany the CSHO throughout or during any phase of an inspection if the CSHO determines that such additional representatives will aid, and not interfere with, the inspection (29 CFR 1903.8(a)).

(1) **Employees Represented by a Certified or Recognized Bargaining Agent.** During the opening conference, the highest ranking union official or union employee representative on-site shall designate who will participate in the walkaround. OSHA regulation 29 CFR 1903.8(b) gives the CSHO the authority to resolve all disputes as to who is the representative authorized by the employer and employees. Title 29 CFR 1903.8(c) states that the representative authorized by the employees shall be an employee of the employer. The CSHO can decide to include others.

FIELD INSPECTION REFERENCE MANUAL

(2) **Safety Committee.** The employee members of an established plant safety committee or the employees at large may have designated an employee representative for OSHA inspection purposes or agreed to accept as their representative the employee designated by the committee to accompany the CSHO during an OSHA inspection.

(3) **No Certified or Recognized Bargaining Agent.** Where employees are not represented by an authorized representative, where there is no established safety committee, or where employees have not chosen or agreed to an employee representative for OSHA inspection purposes whether or not there is a safety committee, the CSHO shall determine if any other employees would suitably represent the interests of employees on the walkaround. If selection of such an employee is impractical, the CSHO shall consult with a reasonable number of employees during the walkaround.

g. **Preemption by Another Agency.** Section 4(b)(1) of the OSH Act states that the OSH Act does not apply to working conditions over which other Federal agencies exercise statutory responsibility. The determination of preemption by another Federal agency is, in many cases, a highly complex matter. Any such situations shall be brought to the attention of the Area Director as soon as they arise, and dealt with on a case by case basis.

h. **Disruptive Conduct.** The CSHO may deny the right of accompaniment to any person whose conduct interferes with a full and orderly inspection (29 CFR 1903.8(d)). If disruption or interference occurs, the CSHO shall use professional judgment as to whether to suspend the walkaround or take other action. The Assistant Area Director shall be consulted if the walkaround is suspended. The employee representative shall be advised that during the inspection matters unrelated to the inspection shall not be discussed with employees.

i. **Trade Secrets.** The CSHO shall ascertain from the employer if the employee representative is authorized to enter any trade secret area(s). If not, the CSHO shall consult with a reasonable number of employees who work in the area (29 CFR 1903.9(d)).

j. **Classified Areas.** In areas containing information classified by an agency of the U.S. Government in the interest of national security, only persons authorized to have access to such information may accompany a CSHO (29 CFR 1903.8(d)).

k. **Examination of Record Programs and Posting Requirements.**

(1) **Records.** As appropriate, the CSHO shall review the injury and illness records to the extent necessary to determine compliance and identify trends. Other OSHA programs and records will be reviewed at the CSHO's professional discretion as necessary.

(2) **Lost Workday Injury (LWDI) Rate.** The LWDI may in the CSHO's discretion be used in determining trends in injuries and illnesses. The LWDI rate is calculated according to the following formula:

If the number of employees hours worked is available from the employer, use:

$$\text{LWDI Rate} = \frac{\text{\# LWDI's} \times 200{,}000}{\text{\# employee hours worked}}$$

Where:
- LWDI's = sum of LWDI's in the reference years.
- employee hours worked = sum of employee hours in the reference years.
- 200,000 = base for 100 full-time workers, working 40 hours per week, 50 weeks per year.

EXAMPLE: An establishment scheduled for inspection in October 1993 employed an average of 54 workers in 1992, 50 workers in 1991, and 50 workers in 1990. Therefore, injury and employment data for the two preceding calendar years will be used.
- LWDI's in 1991 = 5
- LWDI's in 1992 = 3
- Employee hours worked in 1991 = 100,000
- Employee hours worked in 1992 = 108,000

$$\text{LWDI Rate} = \frac{(5 + 3) \times 200{,}000}{100{,}000 + 108{,}000}$$

$$= \frac{1{,}600{,}000}{208{,}000}$$

$$= 7.69 \text{ (rounded to 7.7)}$$

(3) **Posting.** The CSHO shall determine if posting requirements are met in accordance with 29 CFR Parts 1903 and 1904.

4 **Walkaround Inspection.** The main purpose of the walkaround inspection is to identify potential safety and/or health hazards in the workplace. The CSHO shall conduct the inspection in such a manner as to eliminate unnecessary personal exposure to hazards and to minimize unavoidable personal exposure to the extent possible.

a. **Evaluation.** The employer's safety and health program shall be evaluated to determine the employer's good faith. See Chapter IV, C.2.i.(5)(b).

b. **Record All Facts Pertinent to an Apparent Violation.** Apparent violations shall be brought to the attention of employer and employee representatives at the time they are documented. CSHOs shall record at a minimum the identity of the exposed employee, the hazard to which the employee was exposed, the employee's proximity to the hazard, the employer's knowledge of the condition, and the manner in which important measures were obtained.

NOTE: If employee exposure (either to safety or health hazards) is not observed, the CSHO shall document facts on which the determination is made that an employee has been or could be exposed.

c. **Collecting Samples.**

(1) The CSHO shall determine as soon as possible after the start of the inspection whether sampling, such as but not limited to air sampling and surface sampling, is required by utilizing the information collected during the walkaround and from the pre-inspection review.

(2) If either the employer or the employee representative requests sampling results, summaries of the results shall be provided to the requesting representative as soon as practicable.

(3) The CSHO may reference the sampling strategy located in OSHA Instruction CPL 2-2.20B for additional information on sampling techniques.

d. **Taking Photographs and/or Videotapes.** Photographs and/or videotapes shall be taken whenever the CSHO judges there is a need. Photographs that support violations shall be properly labeled, and may be attached to the appropriate OSHA-1B. The CSHO shall ensure that any photographs relating to confidential or trade secret information are identified as such. All film and photographs shall be retained in the case file. Videotapes shall be properly labeled and stored. Refer to OSHA Instruction CPL 2.98 for further information on videotaping.

e. **Interviews.** A free and open exchange of information between the CSHO and employees is essential to an effective inspection. Interviews provide an opportunity for employees or other individuals to point out hazardous conditions and, in general, to provide assistance as to what violations of the Act may exist and what abatement action should be taken. Employee interviews are also an effective means to determine if advance notice of inspection, when given under the guidelines in Chapter I, E.3., has adversely affected the inspection conditions.

   (1) **Purpose.** Section 8(a)(2) of the Act authorizes the CSHO to question any employee privately during regular working hours in the course of an OSHA inspection. The purpose of such interviews is to obtain whatever information the CSHO deems necessary or useful in carrying out the inspection effectively. Such interviews shall be kept as brief as possible. Individual interviews are authorized even when there is an employee representative present.

   (2) **Employee Right of Complaint.** The CSHO may consult with any employee who desires to discuss a possible violation. Upon receipt of such information, the CSHO shall investigate the alleged hazard, where possible, and record the findings. If a written complaint is received, the written response procedures in Chapter I shall be followed.

   (3) **Time and Location.** Interviews shall be conducted in a reasonable manner and normally will be conducted during the walkaround; however, they may be conducted at any time during an inspection. If necessary, interviews may be conducted at locations other than the workplace.

   (4) **Privacy.** Employers shall be informed that the interview is to be in private. Whenever an employee expresses a preference that an employee representative be present for the interview, the CSHO shall make a reasonable effort to honor that request. Any employer objection to private interviews with employees may be construed as a refusal of entry and handled in accordance with the procedures in A.2.c. of this chapter.

   (5) **Interview Statements.** Interview statements of employees or other individuals shall be obtained whenever the CSHO determines that such statements would be useful in documenting adequately an apparent violation.

      (a) Interviews shall normally be reduced to writing, and the individual shall be encouraged to sign and date the statement. The CSHO shall assure the individual that the statement will be held confidential to the extent allowed by law, but they may be used in court/hearings. See OSHA Instruction CPL 2.98 for guidance on videotaping.

      (b) Interview statements shall normally be written in the first person and in the language of the individual.

         1 Any changes or corrections shall be initialed by the individual; otherwise, the statement shall not be changed, added to or altered in any way.

         2 The statements shall end with wording such as: "I have read the above, and it is true to the best of my knowledge." The statement shall also include the following: "I request that my statement be held confidential to the extent allowed by law." The individual, however, may waive confidentiality. The individual shall sign and date the statement and the CSHO shall then sign it as a witness.

         3 If the individual refuses to sign the statement, the CSHO shall note such refusal on the statement. The statement shall, nevertheless, be read to the individual and an attempt made to obtain agreement. A note that this was done shall be entered into the case file.

      (c) A transcription of a recorded statement shall be made if necessary.

f. **Employer Abatement Assistance.**

   (1) **Policy.** CSHOs shall offer appropriate abatement assistance during the walkaround as to how workplace hazards might be eliminated. The information shall provide guidance to the employer in developing acceptable abatement methods or in seeking appropriate professional assistance. CSHO's shall not imply OSHA endorsement of any product through use of specific product names when recommending abatement measures. The issuance of citations shall not be delayed.

   (2) **Disclaimers.** The employer shall be informed that:

      (a) The employer is not limited to the abatement methods suggested by OSHA;

      (b) The methods explained are general and may not be effective in all cases; and

      (c) The employer is responsible for selecting and carrying out an effective abatement method.

g. **Special Circumstances.**

   (1) **Trade Secrets.** Trade secrets are matters that are not of public or general knowledge. A trade secret is any confidential formula, pattern, process, equipment, list, blueprint, device or compilation of information used in the employer's business which gives an advantage over competitors who do not know or use it.

      (a) **Policy.** It is essential to the effective enforcement of the Act that the CSHO and all OSHA personnel preserve the confidentiality of all information and investigations which might reveal a trade secret.

      (b) **Restrictions and Controls.** When the employer identifies an operation or condition as a trade secret, it shall be treated as such. Information obtained in such areas, including all negatives, photographs, videotapes, and OSHA documentation forms, shall be labeled:
      "ADMINISTRATIVELY CONTROLLED INFORMATION"
      "RESTRICTED TRADE INFORMATION"

         1 Under Section 15 of the Act, all information reported to or obtained by a CSHO in connection with any inspection or other activity which contains or which might reveal a trade secret shall be kept confidential. Such information shall not be disclosed except to other OSHA officials concerned with the enforcement of the Act or, when relevant, in any proceeding under the Act.

         2 Title 18 of the United States Code, Section 1905, provides criminal penalties for Federal employees who disclose such information. These penalties include fines of up to $1,000 or imprisonment of up to one year, or both, and removal from office or employment.

         3 Trade secret materials shall not be labeled as "Top Secret," "Secret," or "Confidential," nor shall these security classification designations be used in conjunction with other words unless the trade secrets are also classified by an agency of the U.S. Government in the interest of national security.

(c) **Photographs and Videotapes.** If the employer objects to the taking of photographs and/or videotapes because trade secrets would or may be disclosed, the CSHO should advise the employer of the protection against such disclosure afforded by Section 15 of the Act and 29 CFR 1903.9. If the employer still objects, the CSHO shall contact the Assistant Area Director.

(2) **Violations of Other Laws.** If a CSHO observes apparent violations of laws enforced by other government agencies, such cases shall be referred to the appropriate agency. Referrals shall be made using appropriate Regional procedures (see A.3.g. of this chapter).

5. **Closing Conference.**
   a. At the conclusion of an inspection, the CSHO shall conduct a closing conference with the employer and the employee representatives, jointly or separately, as circumstances dictate. The closing conference may be conducted on site or by telephone as deemed appropriate by the CSHO.
   NOTE: When conducting separate closing conferences for employers and labor representatives (where the employer has declined to have a joint closing conference with employee representatives), the CSHO shall normally hold the conference with employee representatives first, unless the employee representative requests otherwise. This procedure will ensure that worker input, if any, is received—and that any needed changes are made—before employers are informed of violations and proposed citations.
   b. The CSHO shall describe the apparent violations found during the inspection and other pertinent issues as deemed necessary by the CSHO. Both the employer and the employee representatives shall be advised of their rights to participate in any subsequent conferences, meetings or discussions, and their context rights. Any unusual circumstances noted during the closing conference shall be documented in the case file.
      (1) Since the CSHO may not have all pertinent information at the time of the first closing conference, a second closing conference may be held by telephone or in person to inform the employer and the employee representatives whether the establishment is in compliance.
      (2) The CSHO shall advise the employee representatives that:
         (a) Under 29 CFR 2200.20 of the Occupational Safety and Health Review Commission regulations, if the employer contests, the employees have a right to elect "party status" before the Review Commission.
         (b) They must be notified by the employer if a notice of contest or a petition for modification of abatement date is filed.
         (c) They have Section 11(c) rights.
         (d) They have a right to contest the abatement date. Such contest must be in writing and must be filed within 15 working days after receipt of the citation.

B. **Special Inspection Procedures.**
   1. **Followup and Monitoring Inspections.**
      a. **Inspection Procedures.** The primary purpose of a followup inspection is to determine if the previously cited violations have been corrected. Monitoring inspections are conducted to ensure that hazards are being corrected and employees are being protected, whenever a long period of time is needed for an establishment to come into compliance, or to verify compliance with the terms of granted variances. Issuance of willful, repeated and high gravity serious violations, failure to abate notifications, and/or citations related to imminent danger situations are examples of prime candidates for followup or monitoring inspections. Followup or monitoring inspections would not normally be conducted when evidence of abatement is provided by the employer or employee representatives. Normally, there shall be no additional inspection activity unless, in the judgment of the CSHO, there have been significant changes in the workplace which warrant further inspection activity.
      b. **Failure to Abate.**
         (1) A failure to abate exists when the employer has not corrected a violation for which a citation has been issued and abatement date has passed or which is covered under a settlement agreement, or has not complied with interim measures involved in a long-term abatement within the time given.
         (2) If the cited items have not been abated, a Notice of Failure to Abate Alleged Violation shall normally be issued. If a subsequent inspection indicates the condition has still not been abated, the Regional Solicitor shall be consulted for further guidance.
         NOTE: If the employer has exhibited good faith, a late PMA may be considered in accordance with Chapter IV, D.2. where there are extenuating circumstances.
         (3) If it is determined that the originally cited violation was abated but then recurred, a citation for repeated violation may be appropriate.
      c. **Reports.**
         (1) A copy of the previous OSHA-1B, OSHA-1BIH, or citation can be used, and "corrected" written on it, with a brief explanation of the correction if deemed necessary by the CSHO, for those items found to be abated. This information may alternately be included in the narrative or in video/audio documentation.
         (2) In the event that any item has not been abated, complete documentation shall be included on an OSHA-1B.
      d. **Followup Files.** The followup inspection reports shall be included with the original (parent) case file.
   2. **Fatality/Catastrophe Investigations.** For guidance on conducting fatality and catastrophe inspections, refer to OSHA Instruction CPL 2.77 and CPL 2.94.
      a. **Definitions.** The following definitions apply for purposes of this section:
         (1) **Fatality.** An employee death resulting from a work-related incident or exposure; in general, from an accident or illness caused by or related to a workplace hazard.
         (2) **Catastrophe.** The hospitalization of three or more employees resulting from a work-related incident; in general, from an accident or illness caused by a workplace hazard.
      (3) **Hospitalization.** To be admitted as an **inpatient** to a hospital or equivalent medical facility for examination or treatment.
      (4) **Reporting.** Area Directors shall report all job-related fatalities and catastrophes which may result in high media attention or have national implications and that appear to be within OSHA's jurisdiction as soon as they become aware of them to the Regional Administrator. See CPL 2.97.

b. **Selection of CSHO.** A CSHO, preferably with expertise in the particular industry or operation involved in the accident or illness, shall be selected by the Area Director and sent to the establishment as soon as possible. If a potential criminal violation appears possible during the inspection, staff who have received criminal investigation training at the Federal Law Enforcement Training Center shall be assigned, if available.

c. **Families of Victims.**
   (1) Family members of employees involved in fatal occupational accidents or illnesses shall be contacted at an early point in the investigation, given an opportunity to discuss the circumstances of the accident or illness, and provided timely and accurate information at all stages of the investigations as directed in (2), below.
   (2) All of the following require special tact and good judgment on the part of the CSHO. In some situations, these procedures should not be followed to the letter; e.g., in some small businesses, the employer, owner, or supervisor may be a relative of the victim. In such circumstances, such steps as issuance of the form letter may not be appropriate without some editing.
      (a) As soon as practicable after initiating the investigation, the CSHO shall attempt to compile a list of all of the accident victims and their current addresses, along with the names of individual(s) listed in the employer's records as next-of-kin (family member(s)) or person(s) to contact in the event of an emergency.
      (b) The standard information letter should be sent to the family member(s) or the person(s) listed as the emergency contact person(s) indicated on the victims' employment records within 5 working days of the time their identities have been established.
      (c) The compliance officer, when taking a statement from families of victims, shall explain that the interview will be kept confidential to the extent allowed by law and that the interview will be handled following the same procedures as employee interviews. The greatest sensitivity and professionalism is required for such an interview. The information received must be carefully evaluated and corroborated during the investigation.
      (d) Followup contact shall be maintained with a key family member or other contact person, when requested, so that the survivors can be kept up-to-date on the status of the investigation. The victim's family members shall be provided a copy of all citations issued as a result of the accident investigation within 5 working days of issuance.

d. **Criminal.** Section 17(e) of the Act provides criminal penalties for an employer who is convicted of having willfully violated an OSHA standard, rule or order when that violation caused the death of an employee. In an investigation of this type, therefore, the nature of the evidence available is of paramount importance. There shall be early and close liaison between the OSHA investigator, the Area Director, the Regional Administrator and the Regional Solicitor in developing any finding which might involve a violation of Section 17(e) of the Act. An OSHA investigator with criminal investigation training shall be assigned at an early stage to assist in developing the case.

e. **Rescue Operations.** OSHA has no authority to direct rescue operations—this is the responsibility of the employer and/or of local political subdivisions or State agencies. OSHA does have the authority to monitor and inspect the working conditions of covered employees engaged in rescue operations to make certain that all necessary procedures are being taken to protect the lives of the rescuers. See also memorandum on Policy Regarding Voluntary Rescue Activities, dated March 31, 1994, to the Regional Administrators from H. Berrien Zettler, Deputy Director, Directorate of Compliance Programs.

f. **Public Information Policy.** The OSHA public information policy regarding response to fatalities and catastrophes is to explain Federal presence to the news media. It is not to provide a continuing flow of facts nor to issue periodic updates on the progress of the investigation. The Area Director or his/her designee shall normally handle responses to media inquiries.

3. **Imminent Danger Investigations.**
   a. **Definition.** Section 13(a) of the Act defines imminent danger as "... any conditions or practices in any place of employment which are such that a danger exists which could reasonably be expected to cause death or serious physical harm immediately or before the imminence of such danger can be eliminated through the enforcement procedures otherwise provided by this Act."
   b. **Requirements.** The following conditions must be met before a hazard becomes an imminent danger:
      (1) It must be reasonably likely that a serious accident will occur immediately (see B.3.c.(2)(b), below) **or**, if not immediately, then before abatement would otherwise be required (see B.3.c.2.(c), below). If an employer contests a citation, abatement will not be required until there is a final order of the Review Commission affirming the citation.
      (2) The harm threatened must be death or serious physical harm. For a health hazard, exposure to the toxic substance or other health hazard must cause harm to such a degree as to shorten life or cause substantial reduction in physical or mental efficiency even though the resulting harm may not manifest itself immediately.
   c. **Inspection.**
      (1) **Scope.** CSHO may consider expanding the scope of inspection based on the information available during the inspection process.
      (2) **Elimination of the Imminent Danger.** As soon as reasonably practicable after it is concluded that conditions or practices exist which constitute an imminent danger, the employer shall be so advised and requested to notify its employees of the danger and remove them from exposure to the imminent danger. The employer should be encouraged to do whatever is possible to eliminate the danger promptly on a voluntary basis.
         (a) **Voluntary Elimination of the Imminent Danger.** The employer may voluntarily and permanently eliminate the imminent danger as soon as it is pointed out. In such cases, no imminent danger proceeding need be instituted; and, no Notice of Alleged Imminent Danger completed. An appropriate citation and notification of penalty shall be issued.
         (b) **Action Where the Danger is Immediate and Voluntary Elimination Is Not Accomplished.** If the employer either cannot or does not voluntarily eliminate the hazard or remove employees from the exposure and the danger is immediate, the following procedures shall be observed:
            1   The CSHO shall post the OSHA-8 and call the Area Director, who will decide whether to contact the Regional Solicitor to obtain a Temporary Restraining Order (TRO). The Regional Administrator shall be notified of the TRO proceedings.

FIELD INSPECTION REFERENCE MANUAL

NOTE: The CSHO has no authority to order the closing of the operation or to direct employees to leave the area of the imminent danger or the workplace.

2. The CSHO shall notify employees and employee representatives of the posting of the OSHA-8 and shall advise them of their Section 11(c) rights.
3. The employer shall be advised that Section 13 of the Act gives United States District Courts jurisdiction to restrain any condition or practice which is an imminent danger to employees.
4. The Area Director and the Regional Solicitor shall assess the situation and make arrangements for the expedited initiation of court action, if warranted, or instruct the CSHO to remove the OSHA-8.
5. The CSHO's first priority in scheduling activities is to prepare for litigation related to TRO's in imminent danger matters.

(c) **Action Where the Danger is that the Harm will Occur Before Abatement is Required.** If the danger is that the harm will occur before abatement is required, i.e. before a final order of the Commission can be obtained in a contested case, the CSHO shall contact the Area Director and Regional Solicitor.

1. In many cases, the CSHO or the AD may not decide there is such an imminent danger at the time of the physical inspection of the plant. Further evaluation of the file or additional evidence may warrant consultation with the Regional Solicitor.
2. In appropriate cases, the imminent danger notice may be posted at the time citations are delivered or even after the notice of contest is filed.

4. **Construction Inspections.**
   a. **Standards Applicability.** The standards published as 29 CFR Part 1926 have been adopted as occupational safety and health standards under Section 6(a) of the Act and 29 CFR 1910.12. They shall apply to every employment and place of employment of every employee engaged in construction work, including non-contract construction work.
   b. **Definition.** The term "construction work" means work for construction, alteration, and/or repair, including painting and decorating. These terms are discussed in 29 CFR 1926.13. If any question arises as to whether an activity is deemed to be construction for purposes of the Act, the Director of Compliance Programs shall be consulted.
   c. **Employer Worksite.**
      (1) **General.** Inspections of employers in the construction industry are not easily separable into distinct worksites. The worksite is generally the site where the construction is being performed (e.g., the building site, the dam site). Where the construction site extends over a large geographical area (e.g., road building), the entire job will be considered a single worksite. In cases when such large geographical areas overlap between Area Offices, generally only operations of the employer within the jurisdiction of any Area Office will be considered as the worksite of the employer.
      (2) **Beyond Single Area Office.** When a construction worksite extends beyond a single Area Office and the CSHO believes that the inspection should be extended, the affected Area Directors shall consult with each other and take appropriate action.
   d. **Entry of the Workplace.**
      (1) **Other Agency.** The CSHO shall ascertain whether there is a representative of a Federal contracting agency at the worksite. If so, the CSHO shall contact the representative, advise him/her of the inspection and request that he/she attend the opening conference. (For Federal Agencies, see Chapter XIII and following Appendix A, of OSHA Instruction CPL 2.45B or a superseding directive).
      (2) **Complaints.** If the inspection is being conducted as a result of a complaint, a copy of the complaint is to be furnished to the general contractor and any affected sub-contractors.
   e. **Closing Conference.** Upon completion of the inspection, the CSHO shall confer with the general contractors and all appropriate subcontractors or their representatives, together or separately, and advise each one of all the apparent violations disclosed by the inspection to which each one's employees were exposed, or violations which the employer created or controlled. Employee representatives participating in the inspection shall also be afforded the right to participate in the closing conference(s).

5. **Federal Agency Inspections.** Policies and procedures for Federal agencies are to be the same as those followed in the private sector, except as specified in Chapter XIII, and the following Appendix A, of OSHA Instruction CPL 2.45B or a superseding directive.

CHAPTER III
INSPECTION DOCUMENTATION

A. **Four Stage Case File Documentation.**
   1. **General.**
      a. **Guidelines.** These guidelines are developed to assist the CSHO in determining the minimum level of written documentation appropriate for each of four case file stages. **All necessary information relative to violations shall be obtained during the inspection, using any means deemed appropriate by the CSHO (i.e., notes, audio/videotapes, photographs, and employer records).**
      b. **Solicitor Coordination.** Consultation in accordance with regional procedures, including Solicitor procedures, shall be considered when the inspection or investigation could involve important, novel or complex litigation or when consultation is necessary in the CSHO or Area Director's professional judgment. If consultation is deemed necessary, such consultation shall be conducted at the earliest stage possible of the investigation.
   2. **Case File Stages.** The following paragraphs indicate what documentation is required for each of the four case file stages.
      NOTE: The difference between Stage III and Stage IV is one of format and organization only. A Stage III case file is not understood as involving a lesser degree of documentation.
      a. **Stage I.**
         No on-site inspection conducted —
         o OSHA-1 or equivalent, and brief statement expanding upon the reason for not conducting the inspection.
         o If refusal of entry, information necessary to secure a warrant (see Chapter II, A.2.c.).
         o Complainant/referral response, if complaint/referral inspection.

b. **Stage II.**
In-compliance inspection --
- o OSHA-1 or equivalent.
- o OSHA-1A or pertinent information (see B.1. of this chapter).
- o Records obtained during the inspection, based on the CSHO's professional judgment as to what should be obtained.
  NOTE: The CSHO need not document that a condition was in compliance beyond a general statement that no conditions were observed in violation of any standard.
- o Complainant/referral response, if complaint/referral inspection.

c. **Stage III.**
Inspection conducted, citations to be issued --
- o SHA-1 or equivalent.
- o OSHA-1A or equivalent (see B.1. of this chapter).
- o Records obtained during the inspection which, based on the CSHO's professional judgment, are necessary to support the violations.
- o OSHA-1B forms or the equivalent with the following included: Inspection # Instances on page (a,b,/) Type of violation (S,W,R,O,FTA) Citation number and item number Number exposed REC Abatement Period SAVE, AVD, and/or standard reference Photo/video location Severity Rating (H,M,L) and brief justification Probability Rating (G,L) and brief justification GBP and multiplier if applicable % reduction (adjustment) Proposed penalty
  NOTE: Information in relation to exposed employees shall be documented on the OSHA-1B, or referenced on the OSHA-1B as to the specific location of this information.
- o Complainant/referral response, if complaint/referral inspection.

d. **Stage IV.**
Citations are contested --
- o CSHO's will determine after consultation with the Solicitor if the documentation obtained during the inspection needs to be transferred to a different format or location within the file (e.g., transfer of video/audio information to a written format). The information will then be transferred to the appropriate areas as needed. Items which may be considered include transfer of exposed employee information, instance description, employer knowledge, employer's affirmative defenses, employer/ employee comments, and other employer information to the OSHA-1B or equivalent.

B. **Specific Forms.**
1. **Narrative, Form OSHA-1A.**
   a. **General.** The OSHA-1A Form, or its equivalent, shall be used to record information relative to the following, at a minimum:
   ITEM: Establishment Name.
   ITEM: Inspection Number.
   ITEM: Additional Citation Mailing Addresses.
   ITEM: Names and Addresses of all Organized Employee Groups.
   ITEM: Names, Addresses, and Phone Numbers of Authorized Representatives of Employees.
   ITEM: Employer Representatives Contacted and extent of their participation in the inspection.
   ITEM: Comment on S&H program to the extent necessary, based on CSHO's professional judgment, including penalty reduction justifications for good faith.
   ITEM: Document whether closing conference was held, describe any unusual circumstances.
   ITEM: Additional Comments (CSHO's shall use their professional judgment to determine if any additional information shall be added to the case file.)
   b. **Specific.** The following information may be located on the OSHA-1A Form or referenced on the OSHA-1A as to the specific location of this information:
   ITEM: Names, Addresses, and Phone Numbers of Other Persons Contacted.
   ITEM: Accompanied By.
2. **Photo Mounting Worksheet, Form OSHA-89.** This worksheet may be utilized by the CSHO, if mounting is necessary. Other methods of mounting the photograph may be used, such as attaching it to the OSHA-1B. The photograph shall be annotated "trade secret," if applicable.
3. **Inspection Case File Activity Diary Insert.** The Inspection Case File Activity Diary is designed to provide a ready record and summary of all actions relating to a case. The diary sheet will be used to document important events related to the case, especially those not found elsewhere in the case file.

C. **Violations.**
1. **Basis of Violations.**
   a. **Standards and Regulations.** Section 5(a)(2) of the Occupational Safety and Health Act states that each employer has a responsibility to comply with the occupational safety and health standards promulgated under the Act. The specific standards and regulations are found in Title 29 Code of Federal Regulations (CFR) 1900 series. Subparts A and B of 29 CFR 1910 specifically establish the source of all the standards which are the basis of violations.
      NOTE: The most specific subdivision of the standard shall be used for citing violations.
      (1) **Definition and Application of Universal Standards (Horizontal) and Specific Industry Standards (Vertical).** Specific Industry standards are those standards which apply to a particular industry or to particular operations, practices, conditions, processes, means, methods, equipment or installations. Universal standards are those standards which apply when a condition is not covered by a specific industry standard. Within both universal and specific industry standards there are general standards and specific standards.
          (a) When a hazard in a particular industry is covered by both a specific industry (e.g., 29 CFR Part 1915) standard and a universal (e.g., 29 CFR Part 1910) standard, the specific industry standard shall take precedence. **This is true even if the universal standard is more stringent.**

FIELD INSPECTION REFERENCE MANUAL

(b) When determining whether a universal or a specific industry standard is applicable to a work situation, the CSHO shall focus attention on the activity in which the employer is engaged at the establishment being inspected rather than the nature of the employer's general business.

   (2) **Variances.** The employer's requirement to comply with a standard may be modified through granting of a variance, as outlined in Section 6 of the Act.

      (a) An employer will not be subject to citation if the observed condition is in compliance with either the variance or the standard.

      (b) In the event that the employer is not in compliance with the requirements of the variance, a violation of the standard shall be cited with a reference in the citation to the variance provision that has not been met.

  b. **Employee Exposure.**

   (1) **Definition of Employee.** Whether or not exposed persons are employees of an employer depends on several factors, the most important of which is who controls the manner in which the employees perform their assigned work. The question of who pays these employees may not be the determining factor. Determining the employer of an exposed person may be a very complex question, in which case the Area Director may seek the advice of the Regional Solicitor.

   (2) **Proximity to the Hazard.** The proximity of the workers to the point of danger of the operation shall be documented.

   (3) **Observed Exposure.** Employee exposure is established if the CSHO witnesses, observes, or monitors exposure of an employee to the hazardous or suspected hazardous condition during work or work-related activities. Where a standard requires engineering or administrative controls (including work practice controls), employee exposure shall be cited regardless of the use of personal protective equipment.

   (4) **Unobserved Exposure.** Where employee exposure is not observed, witnessed, or monitored by the CSHO, employee exposure is established if it is determined through witness statements or other evidence that exposure to a hazardous condition has occurred, continues to occur, or could recur.

      (a) In fatality/catastrophe (or other "accident") investigations, employee exposure is established if the CSHO determines, through written statements or other evidence, that exposure to a hazardous condition occurred at the time of the accident.

      (b) In other circumstances, based on the CSHO's professional judgment and determination, exposure to hazardous conditions has occurred in the past, and such exposure may serve as the basis for a violation when employee exposure has occurred in the previous six months.

   (5) **Potential Exposure.** A citation may be issued when the possibility exists that an employee could be exposed to a hazardous condition because of work patterns, past circumstances, or anticipated work requirements, and it is reasonably predictable that employee exposure could occur, such as:

      (a) The hazardous condition is an integral part of an employer's recurring operations, but the employer has not established a policy or program to ensure that exposure to the hazardous condition will not recur; or

      (b) The employer has not taken steps to prevent access to unsafe machinery or equipment which employees may have reason to use.

2. **Types of Violations.**

  a. **Other-Than-Serious Violations.** This type of violation shall be cited in situations where the most serious injury or illness that would be likely to result from a hazardous condition cannot reasonably be predicted to cause death or serious physical harm to exposed employees but does have a direct and immediate relationship to their safety and health.

  b. **Serious Violations.**

   (1) Section 17(k) of the Act provides ". . . a serious violation shall be deemed to exist in a place of employment if there is a substantial probability that death or serious physical harm could result from a condition which exists, or from one or more practices, means, methods, operations, or processes which have been adopted or are in use, in such place of employment unless the employer did not, and could not with the exercise of reasonable diligence, know of the presence of the violation."

   (2) The CSHO shall consider four elements to determine if a violation is serious.

      (a) **Step 1.** The **types of accident** or health hazard exposure which the violated standard or the general duty clause is designed to prevent.

      (b) **Step 2.** The most serious **injury or illness** which could reasonably be expected to result from the type of accident or health hazard exposure identified in Step 1.

      (c) **Step 3**. Whether the results of the injury or illness identified in Step 2 could **include death or serious physical harm.** Serious physical harm is defined as:

        1 Impairment of the body in which part of the body is made **functionally useless** or is **substantially reduced in efficiency** on or off the job. Such impairment may be permanent or temporary, chronic or acute. Injuries involving such impairment would usually require treatment by a medical doctor.

        2 Illnesses that could shorten life or significantly reduce physical or mental efficiency by inhibiting the normal function of a part of the body.

      (d) **Step 4.** Whether the **employer knew**, or with the exercise of reasonable diligence, could have known of the presence of the hazardous condition.

        1 In this regard, the supervisor represents the employer and a supervisor's knowledge of the hazardous condition amounts to employer knowledge.

        2 In cases where the employer may contend that the supervisor's own conduct constitutes an isolated event of employee misconduct, the CSHO shall attempt to determine the extent to which the supervisor was trained and supervised so as to prevent such conduct, and how the employer enforces the rule.

        3 If, after reasonable attempts to do so, it cannot be determined that the employer has actual knowledge of the hazardous condition, the knowledge requirement is met if the CSHO is satisfied that the employer could have known through the exercise of reasonable diligence. As a general rule, if the CSHO was able to discover a hazardous condition, and the condition was not transitory in nature, it can be presumed that the employer could have discovered the same condition through the exercise of reasonable diligence.

c. **Violations of the General Duty Clause.** Section 5(a)(1) of the Act requires that "Each employer shall furnish to each of his (sic) employees employment and a place of employment which are free from recognized hazards that are causing or are likely to cause death or serious physical harm to his (sic) employees." The general duty provisions shall be used only where there is no standard that applies to the particular hazard involved, as outlined in 29 CFR _1910.5(f).

(1) **Evaluation of Potential Section 5(a)(1) Situations.** In general, Review Commission and court precedent has established that the following elements are necessary to prove a violation of the general duty clause:

  (a) The employer failed to keep the workplace free of a hazard to which employees of that employer were exposed;
  (b) The hazard was recognized;
  (c) The hazard was causing or was likely to cause death or serious physical harm; and
  (d) There was a feasible and useful method to correct the hazard.

(2) **Discussion of Section 5(a)(1) Elements.** The above four elements of a Section 5(a)(1) violation are discussed in greater detail as follows:

  (a) **A Hazard to Which Employees Were Exposed.** A general duty citation must involve both a serious hazard and exposure of employees.

  1 **Hazard.** A hazard is a danger which threatens physical harm to employees.

  a **Not the Lack of a Particular Abatement Method.** In the past some Section 5(a)(1) citations have incorrectly alleged that the violation is the failure to implement certain precautions, corrective measures or other abatement steps rather than the failure to prevent or remove the particular hazard. It must be emphasized that Section 5(a)(1) does not mandate a particular abatement measure but only requires an employer to render the workplace free of certain hazards by any feasible and effective means which the employer wishes to utilize.

  EXAMPLE: In a hazardous situation involving high pressure gas where the employer has failed to train employees properly, has not installed the proper high pressure equipment, and has improperly installed the equipment that is in place, there are three abatement measures which the employer failed to take; there is only one hazard (that is, exposure to the hazard of explosion due to the presence of high pressure gas) and hence only one general duty clause citation.

  b **The Hazard Is Not a Particular Accident.** The occurrence of an accident does not necessarily mean that the employer has violated Section 5(a)(1) although the accident may be evidence of a hazard. In some cases a Section 5(a)(1) violation may be unrelated to the accident. Although accident facts may be relevant and shall be gathered, the citation shall address the hazard in the workplace, not the particular facts of the accident.

  EXAMPLE: A fire occurred in a workplace where flammable materials were present. No employee was injured by the fire itself but an employee, disregarding the clear instructions of his/her supervisor to use an available exit, jumped out of a window and broke a leg. The danger of fire due to the presence of flammable materials may be a recognized hazard causing or likely to cause death or serious physical harm, but the action of the employee may be an instance of unpreventable employee misconduct. The citation should deal with the fire hazard, not with the accident involving the employee who broke his/her leg.

  c **The Hazard Must Be Reasonably Foreseeable.** The hazard for which a citation is issued must be reasonably foreseeable.

  i. All the factors which could cause a hazard need not be present in the same place at the same time in order to prove foreseeability of the hazard; e.g., an explosion need not be imminent.

  EXAMPLE: If combustible gas and oxygen are present in sufficient quantities in a confined area to cause an explosion if ignited but no ignition source is present or could be present, no Section 5(a)(1) violation would exist. If an ignition source is available at the workplace and the employer has not taken sufficient safety precautions to preclude its use in the confined area, then a foreseeable hazard may exist.

  ii. It is necessary to establish the reasonable foreseeability of the general workplace hazard, rather than the particular hazard which led to the accident.

  EXAMPLE: A titanium dust fire may have spread from one room to another only because an open can of gasoline was in the second room. An employee who usually worked in both rooms was burned in the second room from the gasoline. The presence of gasoline in the second room may be a rare occurrence. It is not necessary to prove that a fire in both rooms was reasonably foreseeable. It is necessary only to prove that the fire hazard, in this case due to the presence of titanium dust, was reasonably foreseeable.

  2 **The Hazard Must Affect the Cited Employer's Employees.** The employees exposed to the Section 5(a)(1) hazard must be the employees of the cited employer.

  (b) **The Hazard Must be Recognized.** Recognition of a hazard can be established on the basis of industry recognition, employer recognition, or "common-sense" recognition. The use of common-sense as the basis for establishing recognition shall be limited to special circumstances. Recognition of the hazard must be supported by satisfactory evidence and adequate documentation in the file as follows:

  1 **Industry Recognition.** A hazard is recognized if the employer's industry recognizes it. Recognition by an industry other than the industry to which the employer belongs is generally insufficient to prove this element of a Section 5(a)(1) violation. Although evidence of recognition by the employer's specific branch within an industry is preferred, evidence that the employer's industry recognizes the hazard may be sufficient.

  a In cases where State and local government agencies not falling under the preemption provisions of Section 4(b)(1) have codes or regulations covering hazards not addressed by OSHA standards, the Area Director shall determine whether the hazard is to be cited under Section 5(a)(1) or referred to the appropriate local agency for enforcement.

FIELD INSPECTION REFERENCE MANUAL

b Regulations of other Federal agencies or of State atomic energy agencies generally shall not be used. They raise substantial difficulties under Section 4(b)(1) of the Act, which provides that OSHA is preempted when such an agency has statutory authority to deal with the working condition in question.

2 **Employer Recognition.** A recognized hazard can be established by evidence of actual employer knowledge. Evidence of such recognition may consist of written or oral statements made by the employer or other management or supervisory personnel during or before the OSHA inspection, or instances where employees have clearly called the hazard to the employer's attention.

3 **Common-Sense Recognition.** If industry or employer recognition of the hazard cannot be established in accordance with (a) and (b), recognition can still be established if it is concluded that any reasonable person would have recognized the hazard. This theory of recognition shall be used only in flagrant cases.

(c) **The Hazard Was Causing or Was likely to Cause Death or Serious Physical Harm.** This element of Section 5(a)(1) violation is identical to the elements of a serious violation, see C.2.b. of this chapter.

(d) **The Hazard Can Be Corrected by a Feasible and Useful Method.**

1 To establish a Section 5(a)(1) violation the agency must identify a method which is feasible, available and likely to correct the hazard. The information shall indicate that the recognized hazard, rather than a particular accident, is preventable.

2 If the proposed abatement method would eliminate or significantly reduce the hazard beyond whatever measures the employer may be taking, a Section 5(a)(1) citation may be issued. A citation shall not be issued merely because the agency knows of an abatement method different from that of the employer, if the agency's method would not reduce the hazard significantly more than the employer's method. It must also be noted that in some cases only a series of abatement methods will alleviate a hazard. In such a case all the abatement methods shall be mentioned.

(3) **limitations on Use of the General Duty Clause.** Section 5(a)(1) is to be used only within the guidelines given in C.2.c. of this chapter.

(a) Section 5(a)(1) may be cited in the alternative when a standard is also cited to cover a situation where there is doubt as to whether the standard applies to the hazard.

(b) Section 5(a)(1) violations shall not be grouped together, but may be grouped with a related violation of a specific standard.

(c) Section 5(a)(1) shall not normally be used to impose a stricter requirement than that required by the standard. For example, if the standard provides for a permissible exposure limit (PEL) of 5 ppm, even if data establishes that a 3 ppm level is a recognized hazard, Section 5(a)(1) shall not be cited to require that the 3 ppm level be achieved unless the limits are based on different health effects. If the standard has only a time-weighted average permissible exposure level and the hazard involves exposure above a recognized ceiling level, the Area Director shall consult with the Regional Solicitor.

NOTE: An exception to this rule may apply if it can be documented that "an employer knows a particular safety or health standard is inadequate to protect his workers against the specific hazard it is intended to address." **International Union, U.A.W. v. General Dynamics Land Systems Div.,** 815 F.2d 1570 (D.C. Cir. 1987). Such cases shall be subject to pre-citation review.

(d) Section 5(a)(1) shall normally not be used to require an abatement method not set forth in a specific standard. If a toxic substance standard covers engineering control requirements but not requirements for medical surveillance, Section 5(a)(1) shall not be cited to require medical surveillance.

(e) Section 5(a)(1) shall not be used to enforce "should" standards.

(f) Section 5(a)(1) shall not normally be used to cover categories of hazards exempted by a standard. If, however, the exemption is in place because the drafters of the standard (or source document) declined to deal with the exempt category for reasons other than the lack of a hazard, the general duty clause may be cited if all the necessary elements for such a citation are present.

(4) **Pre-Citation Review.** Section 5(a)(1) citations shall undergo a pre-citation review following established area office procedures when required by the Area Director or Assistant Area Director.

NOTE: If a standard does not apply and all criteria for issuing a Section 5(a)(1) citation are not met, but it is determined that the hazard warrants some type of notification, a letter shall be sent to the employer and the employee representative describing the hazard and suggesting corrective action.

d. **Willful Violations.** The following definitions and procedures apply whenever the CSHO suspects that a willful violation may exist:

(1) A willful violation exists under the Act where the evidence shows either an intentional violation of the Act or plain indifference to its requirements.

(a) The employer committed an intentional and knowing violation if:

1 An employer representative was aware of the requirements of the Act, or the existence of an applicable standard or regulation, and was also aware of a condition or practice in violation of those requirements, and did not abate the hazard.

2 An employer representative was not aware of the requirements of the Act or standards, but was aware of a comparable legal requirement (e.g., state or local law) and was also aware of a condition or practice in violation of that requirement, and did not abate the hazard.

(b) The employer committed a violation with plain indifference to the law where:

1 Higher management officials were aware of an OSHA requirement applicable to the company's business but made little or no effort to communicate the requirement to lower level supervisors and employees.

2 Company officials were aware of a continuing compliance problem but made little or no effort to avoid violations.

EXAMPLE: Repeated issuance of citations addressing the same or similar conditions.

3 An employer representative was not aware of any legal requirement, but was aware that a condition or practice was hazardous to the safety or health of employees and made little or no effort to determine the extent

of the problem or to take the corrective action. Knowledge of a hazard may be gained from such means as insurance company reports, safety committee or other internal reports, the occurrence of illnesses or injuries, media coverage, or, in some cases, complaints of employees or their representatives.

    4 Finally, in particularly flagrant situations, willfulness can be found despite lack of knowledge of either a legal requirement or the existence of a hazard if the circumstances show that the employer would have placed no importance on such knowledge even if he or she had possessed it, or had no concern for the health or safety of employees.

(2) It is not necessary that the violation be committed with a bad purpose or an evil intent to be deemed "willful." It is sufficient that the violation was deliberate, voluntary or intentional as distinguished from inadvertent, accidental or ordinarily negligent.

(3) The CSHO shall carefully develop and record, during the inspection, all evidence available that indicates employer awareness of and the disregard for statutory obligations or of the hazardous conditions. Willfulness could exist if an employer is advised by employees or employee representatives of an alleged hazardous condition and the employer makes no reasonable effort to verify and correct the condition. Additional factors which can influence a decision as to whether violations are willful include:

  (a) The nature of the employer's business and the knowledge regarding safety and health matters which could reasonably be expected in the industry.
  (b) The precautions taken by the employer to limit the hazardous conditions.
  (c) The employer's awareness of the Act and of the responsibility to provide safe and healthful working conditions.
  (d) Whether similar violations and/or hazardous conditions have been brought to the attention of the employer.
  (e) Whether the nature and extent of the violations disclose a **purposeful disregard** of the employer's responsibility under the Act.

(4) If the Area Office cannot determine whether to issue a citation as a willful or a repeat violation due to the raising of difficult issues of law and policy which will require the evaluation of complex factual situations, the Area Director shall normally consult with the Regional Solicitor.

e. **Criminal/Willful Violations.** Section 17(e) of the Act provides that: "Any employer who willfully violates any standard, rule or order promulgated pursuant to Section 6 of this Act, or of any regulations prescribed pursuant to this Act, and that violation caused death to any employee, shall, upon conviction, be punished by a fine of not more than $10,000 or by imprisonment for not more than six months, or by both; except that if the conviction is for a violation committed after a first conviction of such person, punishment shall be a fine of not more than $20,000 or by imprisonment for not more than one year, or by both."

(1) The Area Director, in coordination with the Regional Solicitor, shall carefully evaluate all willful cases involving worker deaths to determine whether they may involve criminal violations of Section 17(e) of the Act. Because the nature of the evidence available is of paramount importance in an investigation of this type, there shall be early and close liaison between the OSHA investigator, the Area Director, the Regional Administrator, and the Regional Solicitor in developing any finding which might involve a violation of Section 17(e) of the Act.

(2) The following criteria shall be considered in investigating possible criminal/willful violations:

  (a) In order to establish a criminal/willful violation OSHA must prove that:
    1 The employer violated an OSHA standard. A criminal/willful violation cannot be based on violation of Section 5(a)(1).
    2 The violation was willful in nature.
    3 The violation of the standard caused the death of an employee. In order to prove that the violation of the standard caused the death of an employee, there must be evidence in the file which clearly demonstrates that the violation of the standard was the cause of or a contributing factor to an employee's death.

  (b) Although it is generally not necessary to issue "Miranda" warnings to an employer when a criminal/willful investigation is in progress, the Area Director shall seek the advice of the Regional Solicitor on this question.

  (c) Following the investigation, if the Area Director decides to recommend criminal prosecution, a memorandum containing that recommendation shall be forwarded promptly to the Regional Administrator. It shall include an evaluation of the possible criminal charges, taking into consideration the greater burden of proof which requires that the Government's case be proven beyond a reasonable doubt. In addition, if the correction of the hazardous condition appears to be an issue, this shall be noted in the transmittal memorandum because in most cases the prosecution of a criminal/willful case delays the affirmance of the civil citation and its correction requirements.

  (d) The Area Director shall normally issue a civil citation in accordance with current procedures even if the citation involves allegations under consideration for criminal prosecution. The Regional Administrator shall be notified of such cases, and they shall be forwarded to the Regional Solicitor as soon as practicable for possible referral to the U.S. Department of Justice.

(3) When a willful violation is related to a fatality, the Area Director shall ensure the case file contains succinct documentation regarding the decision **not** to make a criminal referral. The documentation should indicate which elements of a criminal violation make the case unsuitable for criminal referral.

f. **Repeated Violations.** An employer may be cited for a repeated violation if that employer has been cited previously for a **substantially similar condition** and the citation has become a final order.

  (1) **Identical Standard.** Generally, similar conditions can be demonstrated by showing that in both situations the identical standard was violated.

    EXCEPTION: Previously a citation was issued for a violation of 29 CFR 1910.132(a) for not requiring the use of safety-toe footwear for employees. A recent inspection of the same establishment revealed a violation of 29 CFR 1910.132(a) for not requiring the use of head protection (hard hats). Although the same standard was involved, the hazardous conditions found were not substantially similar and therefore a repeated violation would not be appropriate.

FIELD INSPECTION REFERENCE MANUAL

(2) **Different Standards.** In some circumstances, similar conditions can be demonstrated when different standards are violated. Although there may be different standards involved, the hazardous conditions found could be substantially similar and therefore a repeated violation would be appropriate.

(3) **Time limitations.** Although there are no statutory limitations upon the length of time that a citation may serve as a basis for a repeated violation, the following policy shall be used in order to ensure uniformity.
   (a) A citation will be issued as a repeated violation if:
      1 The citation is issued within 3 years of the final order of the previous citation, or,
      2 The citation is issued within 3 years of the final abatement date of that citation, whichever is later.
   (b) When a violation is found during an inspection, and a repeated citation has been issued for a substantially similar condition which meets the above time limitations, the violation may be classified as a second instance repeated violation with a corresponding increase in penalty (see Chapter IV, C.2.l.).
   (c) For any further repetition, the Area Director shall be consulted for guidance.

(4) **Obtaining Inspection History.** For purposes of determining whether a violation is repeated, the following criteria shall apply:
   (a) **High Gravity Serious Violations.** When high gravity serious violations are to be cited, the Area Director shall obtain a history of citations previously issued to this employer at all of its identified establishments, nationwide, (Federal enforcement only) within the same two-digit SIC code. If these violations have been previously cited within the time limitations described in C.2.f.(3), above, and have become a final order of the Review Commission, a repeated citation may be issued. Under special circumstances, the Area Director, in consultation with the Regional Solicitor, may also issue citations for repeated violations without regard for the SIC code.
   (b) **Violations of Lesser Gravity.** When violations of lesser gravity than high gravity serious are to be cited, Agency policy is to encourage the Area Director to obtain a national inspection history whenever the circumstances of the current inspection will result in a large number of serious, repeat, or willful citations. This is particularly so if the employer is known to have establishments nationwide and if significant citations have been issued against the employer in other areas, or at other mobile worksites.
   (c) **Geographical limitations.** Where a national inspection history has **not** been obtained, the following criteria regarding geographical limitations shall apply:
      1 **Multifacility Employer.** A multifacility employer shall be cited for a repeated violation if the violation recurred at any worksite within the same OSHA Area Office jurisdiction.
         EXAMPLE: Where the construction site extends over a large area and/or the scope of the job is unclear (such as road building), that portion of the workplace specified in the employer's contract which falls within the Area Office jurisdiction is the establishment. If an employer has several worksites within the same Area Office jurisdiction, a citation of a violation at Site A will serve as the basis for a repeated citation in Area B.
      2 **Longshoring Establishment.** A longshoring establishment will encompass all longshoring activities of a single stevedore within any single port area. Longshoring employers are subject to repeated violation citations based on prior violations occurring anywhere. Other maritime employers covered by OSHA standards (e.g., shipbuilding, ship repairing) are multifacility employers as defined in **a**., above.

(5) **Repeated vs. Willful.** Repeated violations differ from willful violations in that they may result from an inadvertent, accidental or ordinarily negligent act. Where a repeated violation may also meet the criteria for willful but not clearly so, a citation for a repeated violation shall normally be issued.

(6) **Repeated vs. Failure to Abate.** A failure to abate situation exists when an item of equipment or condition previously cited has never been brought into compliance and is noted at a later inspection. If, however, the violation was not continuous (i.e., if it had been corrected and then reoccurred), the subsequent occurrence is a repeated violation.

(7) Alleged Violation Description (AVD). If a repeated citation is issued, the CSHO must ensure that the cited employer is fully informed of the previous violations serving as a basis for the repeated citation, by notation in the AVD portion of the citation, using the following or similar language:
THE (COMPANY NAME) WAS PREVIOUSLY CITED FOR A VIOLATION OF THIS OCCUPATIONAL SAFETY AND HEALTH STANDARD OR ITS EQUIVALENT STANDARD (NAME PREVIOUSLY CITED STANDARD) WHICH WAS CONTAINED IN OSHA INSPECTION NUMBER_____, CITATION NUMBER_____, ITEM NUMBER_____, ISSUED ON (DATE), WITH RESPECT TO A WORKPLACE LOCATED AT _____.

g. **De Minimis Violations.** De Minimis violations are violations of standards which have no direct or immediate relationship to safety or health and shall not be included in citations. An OSHA-1B/1BIH is no longer required to be completed for De Minimis violations. The employer should be verbally notified of the violation and the CSHO should note it in the inspection case file. The criteria for finding a de minimis violation are as follows:

(1) An employer complies with the clear intent of the standard but deviates from its particular requirements in a manner that has no direct or immediate relationship to employee safety or health. These deviations may involve distance specifications, construction material requirements, use of incorrect color, minor variations from recordkeeping, testing, or inspection regulations, or the like.
EXAMPLE #1: 29 CFR 1910.27(b)(1)(ii) allows 12 inches (30 centimeters) as the maximum distance between ladder rungs. Where the rungs are 13 inches (33 centimeters) apart, the condition is de minimis.
EXAMPLE #2: 29 CFR 1910.28(a)(3) requires guarding on all open sides of scaffolds. Where employees are tied off with safety belts in lieu of guarding, often the intent of the standard will be met, and the absence of guarding may be de minimis.
EXAMPLE #3: 29 CFR 1910.217(e)(1)(ii) requires that mechanical power presses be inspected and tested at least weekly. If the machinery is seldom used, inspection and testing prior to each use is adequate to meet the intent of the standard.

(2) An employer complies with a proposed standard or amendment or a consensus standard rather than with the standard in effect at the time of the inspection and the employer's action clearly provides equal or greater employee protection or the employer complies with a written interpretation issued by the OSHA Regional or National Office.

(3) An employer's workplace is at the "state of the art" which is technically beyond the requirements of the applicable standard and provides equivalent or more effective employee safety or health protection.

3. **Health Standard Violations.**
   a. **Citation of Ventilation Standards.** In cases where a citation of a ventilation standard may be appropriate, consideration shall be given to standards intended to control exposure to recognized hazardous levels of air contaminants, to prevent fire or explosions, or to regulate operations which may involve confined space or specific hazardous conditions. In applying these standards, the following guidelines shall be observed:
      (1) **Health-Related Ventilation Standards.** An employer is considered in compliance with a health-related airflow ventilation standard when the employee exposure does not exceed appropriate airborne contaminant standards; e.g., the PELs prescribed in 29 CFR 1910.1000.
         (a) Where an over-exposure to an airborne contaminant is detected, the appropriate air contaminant engineering control requirement shall be cited; e.g., 29 CFR 1910.1000(e). In no case shall citations of this standard be issued for the purpose of requiring specific volumes of air to ventilate such exposures.
         (b) Other requirements contained in health-related ventilation standards shall be evaluated without regard to the concentration of airborne contaminants. Where a specific standard has been violated **and** an actual or potential hazard has been documented, a citation shall be issued.
      (2) **Fire- and Explosion-Related Ventilation Standards.** Although they are not technically health violations, the following guidelines shall be observed when citing fire- and explosion-related ventilation standards:
         (a) **Adequate Ventilation.** In the application of fire- and explosion-related ventilation standards, OSHA considers that an operation has **adequate** ventilation when both of the following criteria are met:
            1 The requirement of the specific standard has been met.
            2 The concentration of flammable vapors is 25 percent or less of the lower explosive limit (LEL).
               EXCEPTION: Certain standards specify violations when 10 percent of the LEL is exceeded. These standards are found in maritime and construction exposures.
         (b) **Citation Policy.** If 25 percent (10 percent when specified for maritime or construction operations) of the LEL has been exceeded and:
            1 The standard requirements have not been met, the standard violation normally shall be cited as serious.
            2 There is no applicable specific ventilation standard, Section 5(a)(1) of the Act shall be cited in accordance with the guidelines given in C.2.c. of this chapter.
   b. **Violations of the Noise Standard.** Current enforcement policy regarding 29 CFR 1910.95(b)(1) allows employers to rely on personal protective equipment and a hearing conservation program rather than engineering and/or administrative controls when hearing protectors will effectively attenuate the noise to which the employee is exposed to acceptable levels as specified in Tables G-16 or G-16a of the standard.
      (1) Citations for violations of 29 CFR 1910.95(b)(1) shall be issued when engineering and/or administrative controls are feasible, both technically and economically; and
         (a) Employee exposure levels are so high that hearing protectors alone may not reliably reduce noise levels received by the employee's ear to the levels specified in Tables G-16 or G-16a of the standard. Given the present state of the art, hearing protectors which offer the greatest attenuation may not reliably be used when employee exposure levels border on 100 dBA (See OSHA Instruction CPL 2-2.35A, Appendix.); or
         (b) The costs of engineering and/or administrative controls are less than the cost of an effective hearing conservation program.
      (2) A control is not reasonably necessary when an employer has an ongoing hearing conservation program and the results of audiometric testing indicate that existing controls and hearing protectors are adequately protecting employees. (In making this decision such factors as the exposure levels in question, the number of employees tested, and the duration of the testing program shall be taken into consideration.)
      (3) When employee noise exposures are less than 100 dBA but the employer does not have an ongoing hearing conservation program or the results of audiometric testing indicate that the employer's existing program is not working, the CSHO shall consider whether:
         (a) Reliance on an effective hearing conservation program would be less costly than engineering and/or administrative controls.
         (b) An effective hearing conservation program can be established or improvements can be made in an existing hearing conservation program which could bring the employer into compliance with Tables G-16 or G-16a.
         (c) Engineering and/or administrative controls are both technically and economically feasible.
      (4) If noise levels received by the employee's ear can be reduced to the levels specified in Tables G-16 or G-16a by means of hearing protectors and an effective hearing conservation program, citations under the hearing conservation shall normally be issued rather than citations requiring engineering controls. If improvements in the hearing conservation program cannot be made or, if made, cannot be expected to reduce exposure sufficiently and feasible controls exist, a citation under 1910.95(b)(1) shall normally be issued.
      (5) When hearing protection is required but not used and employee exposure exceeds the limits of Table G-16, 29 CFR 1910.95(i)(2)(i) shall be cited and classified as serious (see (8), below) whether or not the employer has instituted a hearing conservation program. 29 CFR 1910.95(a) shall no longer be cited except in the case of the oil and gas drilling industry.
         NOTE: Citations of 29 CFR 1910.95(i)(2)(ii)(b) shall also be classified as serious.
      (6) If an employer has instituted a hearing conservation program and a violation of the hearing conservation amendment (other than 1910.95 (i)(2)(i) or (i)(2)(ii)(b)) is found, a citation shall be issued if employee noise exposures equal or exceed an 8-hour time-weighted average of 85 dB.
      (7) If the employer has not instituted a hearing conservation program and employee noise exposures equal or exceed an 8-hour time-weighted average of 85 dB, a citation for 1910.95(c) only shall be issued.
      (8) Violations of 1910.95(i)(2)(i) from the hearing conservation amendment may be grouped with violations of 29 CFR 1910.95(b)(1) and classified as serious when an employee is exposed to noise levels above the limits of Table G-16 and:

(a) Hearing protection is not utilized or is not adequate to prevent overexposure to an employee; or
(b) There is evidence of hearing loss which could reasonably be considered:
1 To be work-related, and
2 To have been preventable, at least to some degree, if the employer had been in compliance with the cited provisions.
(9) When an employee is overexposed but effective hearing protection is being provided and used, an effective hearing conservation program has been implemented and no feasible engineering or administrative controls exist, a citation shall not be issued.

c. **Violations of the Respirator Standard.** When considering a citation for respirator violations, the following guidelines shall be observed:
(1) **In Situations Where Overexposure Does Not Occur.** Where an overexposure has not been established:
(a) But an improper type of respirator is being used (e.g., a dust respirator being used to reduce exposure to organic vapors), a citation under 29 CFR 1910.134(b)(2) shall be issued, provided the CSHO documents that an overexposure is possible.
(b) And one or more of the other requirements of 29 CFR 1910.134 is not being met; e.g., an unapproved respirator is being used to reduce exposure to toxic dusts, generally a de minimis violation shall be recorded in accordance with OSHA procedures. (Note that this policy does **not** include emergency use respirators.) The CSHO shall advise the employer of the elements of a good respirator program as required under 29 CFR 1910.134.
(c) In **exceptional** circumstances a citation may be warranted if an adverse health condition due to the respirator itself could be supported and documented. Examples may include a dirty respirator that is causing dermatitis, a worker's health being jeopardized by wearing a respirator due to an inadequately evaluated medical condition or a significant ingestion hazard created by an improperly cleaned respirator.
(2) **In Situations Where Overexposure Does Occur.** In cases where an overexposure to an air contaminant has been established, the following principles apply to citations of 1910.134:
(a) 29 CFR 1910.134(a)(2) is the general section requiring employers to provide respirators "... when such equipment is necessary to protect the health of the employee" and requiring the establishment and maintenance of a respiratory protection program which meets the requirements outlined in 29 CFR 1910.134(b). Thus, if no respiratory program at all has been established, 1910.134(a)(2) alone shall be cited; if a program has been established and some, but not all, of the requirements under 1910.134(b) are being met, the specific standards under 1910. 134(b) that are applicable shall be cited.
(b) An acceptable respiratory protection program includes all of the elements of 29 CFR 1910.134; however, the standard is structured such that essentially the same requirement is often specified in more than one section. In these cases, the section which most adequately describes the violation shall be cited.

d. **Additive and Synergistic Effects.**
(1) Substances which have a known additive effect and, therefore, result in a greater probability/severity of risk when found in combination shall be evaluated using the formula found in 29 CFR _ 1910.1000(d)(2). The use of this formula requires that the exposures have an additive effect on the same body organ or system.
(2) If the CSHO suspects that synergistic effects are possible, it shall be brought to the attention of the supervisor, who shall refer the question to the Regional Administrator. If it is decided that there is a synergistic effect of the substances found together, the violations shall be grouped, when appropriate, for purposes of increasing the violation classification severity and/or the penalty.

e. **Absorption and Ingestion Hazards.** The following guidelines apply when citing absorption and ingestion violations. Such citations do **not** depend on measurements of airborne concentrations, but shall normally be supported by wipe sampling.
(1) Citations under 29 CFR 1910.132, 1910.141 and/or Section 5(a)(1) may be issued when there is reasonable probability that employees will be exposed to these hazards.
(2) Where, for any substance, a serious hazard is determined to exist due to the potential of ingestion or absorption of the substance for reasons other than the consumption of contaminated food or drink (e.g., smoking materials contaminated with the toxic substance), a serious citation shall be considered under Section 5(a)(1) of the Act.

f. **Biological Monitoring.** If the employer has been conducting biological monitoring, the CSHO shall evaluate the results of such testing. The results may assist in determining whether a significant quantity of the toxic material is being ingested or absorbed through the skin.

4. **Writing Citations.**
a. **General.** Section 9 of the Act controls the writing of citations.
(1) **Section 9(a).** "... the Secretary or his authorized representative ... shall with reasonable promptness issue a citation to the employer." To facilitate the prompt issuance of citations, the Area Director may issue citations which are unrelated to health inspection air sampling, prior to receipt of sampling results.
(2) **Section 9(c).** "No citation may be issued ... after the expiration of six months following the occurrence of any violation." Accordingly, a citation shall not be issued where any violation alleged therein last occurred 6 months or more prior to the date on which the citation is actually signed and dated. Where the actions or omissions of the employer concealed the existence of the violation, the time limitation is suspended until such time that OSHA learns or could have learned of the violation. The Regional Solicitor shall be consulted in such cases.

b. **Alternative Standards.**
(1) In rare cases, the same factual situation may present a possible violation of more than one standard. For example, the facts which support a violation of 29 CFR 1910.28(a)(1) may also support a violation of 1910.132(a) if no scaffolding is provided when it should be and the use of safety belts is not required by the employer.
(2) Where it appears that more than one standard is applicable to a given factual situation and that compliance with any of the applicable standards would effectively eliminate the hazard, it is permissible to cite alternative standards using the words "in the alternative." A reference in the citation to each of the standards involved shall be accompanied by a separate Alleged Violation Description (AVD) which clearly alleges all of the necessary elements of a violation of that standard. Only one penalty shall be proposed for the violative condition.

5. **Combining and Grouping of Violations.**
    a. **Combining.** Violations of a single standard having the same classification found during the inspection of an establishment or worksite generally shall be combined into one alleged citation item. Different options of the same standard shall normally also be combined. Each instance of the violation shall be separately set out within that item of the citation. Other-than-serious violations of a standard may be combined with serious violations of the same standard when appropriate.

       NOTE: Except for standards which deal with multiple hazards (e.g., Tables Z-1, Z-2 and Z-3 cited under 29 CFR 1910.1000 (a), (b), or (c)), the same standard may not be cited more than once on a single citation. The same standard may be cited on different citations on the same inspection, however.

    b. **Grouping.** When a source of a **hazard** is identified which involves interrelated violations of different standards, the violations may be grouped into a single item. The following situations normally call for grouping violations:
        (1) **Grouping Related Violations.** When the CSHO believes that violations classified either as serious or as other-than-serious are so closely related as to constitute a single hazardous condition.
        (2) **Grouping Other-Than-Serious Violations Where Grouping Results in a Serious Violation.** When two or more individual violations are found which, if considered individually represent other-than-serious violations, but if grouped create a substantial probability of death or serious physical harm.
        (3) **Where Grouping Results in Higher Gravity Other-Than-Serious Violation.** Where the CSHO finds during the course of the inspection that a number of other-than-serious violations are present in the same piece of equipment which, considered in relation to each other affect the overall gravity of possible injury resulting from an accident involving the combined violations.
        (4) **Violations of Posting and Recordkeeping Requirements.** Violations of the posting and recordkeeping requirements which involve the same document; e.g., OSHA-200 Form was not posted or maintained. (See Chapter IV, C.2.n. for penalty amounts.)
        (5) **Penalties for Grouped Violations.** If penalties are to be proposed for grouped violations, the penalty shall be written across from the first violation item appearing on the OSHA-2.

    c. **When Not to Group.** Times when grouping is normally inappropriate.
        (1) **Multiple Inspections.** Violations discovered in multiple inspections of a single establishment or worksite may not be grouped. An inspection in the same establishment or at the same worksite shall be considered a single inspection even if it continues for a period of more than one day or is discontinued with the intention of resuming it after a short period of time if only one OSHA-1 is completed.
        (2) **Separate Establishments of the Same Employer.** Where inspections are conducted, either at the same time or different times, at two establishments of the same employer and instances of the same violation are discovered during each inspection, the employer shall be issued separate citations for each establishment. The violations shall not be grouped.
        (3) **General Duty Clause Violations.** Because Section 5(a)(1) of the Act is cited so as to cover all aspects of a serious hazard for which no standard exists, **no** grouping of separate Section 5(a)(1) violations is permitted. This provision, however, does not prohibit grouping a Section 5(a)(1) violation with a related violation of a specific standard.
        (4) **Egregious Violations.** Violations which are proposed as violation-by-violation citations shall **not** normally be combined or grouped. (See OSHA Instruction CPL 2.80.)

6. This paragraph has been replaced by a revised multi-employer policy contained in OSHA Instruction CPL 2-0.124. ~~Multiemployer Worksites. On multiemployer worksites, both construction and non-construction, citations normally shall be issued to employers whose employees are exposed to hazards (the exposing employer).~~
    a. ~~Additionally, the following employers normally shall be cited, whether or not their own employees are exposed, but see C.2.c.(2)(a)2 of this chapter for Section 5(a)(1) violation guidance:~~
        (1) ~~The employer who actually creates the hazard (the creating employer);~~
        (2) ~~The employer who is responsible, by contract or through actual practice, for safety and health conditions on the worksite; i.e., the employer who has the authority for ensuring that the hazardous condition is corrected (the controlling employer);~~
        (3) ~~The employer who has the responsibility for actually correcting the hazard (the correcting employer).~~
    b. ~~Prior to issuing citations to an exposing employer, it must first be determined whether the available facts indicate that employer has a legitimate defense to the citation, as set forth below:~~
        (1) ~~The employer did not create the hazard;~~
        (2) ~~The employer did not have the responsibility or the authority to have the hazard corrected;~~
        (3) ~~The employer did not have the ability to correct or remove the hazard;~~
        (4) ~~The employer can demonstrate that the creating, the controlling and/or the correcting employers, as appropriate, have been specifically notified of the hazards to which his/her employees are exposed;~~
        (5) ~~The employer has instructed his/her employees to recognize the hazard and, where necessary, informed them how to avoid the dangers associated with it.~~
            (a) ~~Where feasible, an exposing employer must have taken appropriate alternative means of protecting employees from the hazard.~~
            (b) ~~When extreme circumstances justify it, the exposing employer shall have removed his/her employees from the job to avoid citation.~~
    c. ~~If an exposing employer meets all these defenses, that employer shall not be cited. If all employers on a worksite with employees exposed to a hazard meet these conditions, then the citation shall be issued only to the employers who are responsible for creating the hazard and/or who are in the best position to correct the hazard or to ensure its correction. In such circumstances the controlling employer and/or the hazard-creating employer shall be cited even though no employees of those employers are exposed to the violative condition. Penalties for such citations shall be appropriately calculated, using the exposed employees of all employers as the number of employees for probability assessment.~~

7. **Employer/Employee Responsibilities.**
   a. Section 5(b) of the Act states: "Each employee shall comply with occupational safety and health standards and all rules, regulations, and orders issued pursuant to the Act which are applicable to his own actions and conduct." The Act does not provide for the issuance of citations or the proposal of penalties against employees. Employers are responsible for employee compliance with the standards.
   b. In cases where the CSHO determines that employees are systematically refusing to comply with a standard applicable to their own actions and conduct, the matter shall be referred to the Area Director who shall consult with the Regional Administrator.
   c. Under no circumstances is the CSHO to become involved in an **onsite** dispute involving labor-management issues or interpretation of collective-bargaining agreements. The CSHO is expected to obtain enough information to understand whether the employer is using all appropriate authority to ensure compliance with the Act. Concerted refusals to comply will not bar the issuance of an appropriate citation where the employer has failed to exercise full authority to the maximum extent reasonable, including discipline and discharge.

8. **Affirmative Defenses.**
   a. **Definition.** An affirmative defense is any matter which, if established by the employer, will excuse the employer from a violation which has otherwise been proved by the CSHO.
   b. **Burden of Proof.** Although affirmative defenses must be proved by the employer at the time of the hearing, OSHA must be prepared to respond whenever the employer is likely to raise or actually does raise an argument supporting such a defense. The CSHO, therefore, shall keep in mind the potential affirmative defenses that the employer may make and attempt to gather contrary evidence when a statement made during the inspection fairly raises a defense. The CSHO should bring the documentation of the hazards and facts related to possible affirmative defenses to the attention of the Assistant Area Director. Where it appears that each and every element of an affirmative defense is present, the Area Director may decide that a citation is not warranted.
   c. **Explanations.** The following are explanations of the more common affirmative defenses with which the CSHO shall become familiar. There are other affirmative defenses besides these, but they are less frequently raised or are such that the facts which can be gathered during the inspection are minimal.
      (1) **Unpreventable Employee Misconduct or "Isolated Event".** The violative condition was:
         (a) Unknown to the employer; and
         (b) In violation of an adequate work rule which was effectively communicated and uniformly enforced.
            EXAMPLE: An unguarded table saw is observed. The saw, however, has a guard which is reattached while the CSHO watches. Facts which the CSHO shall document may include: Who removed the guard and why? Did the employer know that the guard had been removed? How long or how often had the saw been used without guards? Did the employer have a work rule that the saw guards not be removed? How was the work rule communicated? Was the work rule enforced?
      (2) **Impossibility.** Compliance with the requirements of a standard is:
         (a) Functionally impossible or would prevent performances of required work; and
         (b) There are no alternative means of employee protection.
            EXAMPLE: During the course of the inspection an unguarded table saw is observed. The employer states that the nature of its work makes a guard unworkable. Facts which the CSHO shall document may include: Would a guard make performance of the work impossible or merely more difficult? Could a guard be used part of the time? Has the employer attempted to use guards? Has the employer considered alternative means or methods of avoiding or reducing the hazard?
      (3) **Greater Hazard.** Compliance with a standard would result in greater hazards to employees than noncompliance and:
         (a) There are no alternative means of employee protection; and
         (b) An application of a variance would be inappropriate.
            EXAMPLE: The employer indicates that a saw guard had been removed because it caused particles to be thrown into the operator's face. Facts which the CSHO shall consider may include: Was the guard used properly? Would a different type of guard eliminate the problem? How often was the operator struck by particles and what kind of injuries resulted? Would safety glasses, a face mask, or a transparent shelf attached to the saw prevent injury? Was operator technique at fault and did the employer attempt to correct it? Was a variance sought?
      (4) **Multiemployer Worksites.** Refer to C.6. of this chapter.

CHAPTER IV
POST-INSPECTION PROCEDURES

A. **Abatement.**
   1. **Period.** The abatement period shall be the shortest interval within which the employer can **reasonably** be expected to correct the violation. An abatement date shall be set forth in the citation as a specific date, not a number of days. When the abatement period is very short (i.e., 5 working days or less) and it is uncertain when the employer will receive the citation, the abatement date shall be set so as to allow for a mail delay and the agreed-upon abatement time. When abatement has been witnessed by the CSHO during the inspection, the abatement period shall be "Corrected During Inspection" on the citation.
   2. **Reasonable Abatement Date.** The establishment of the shortest practicable abatement date requires the exercise of professional judgment on the part of the CSHO.
      NOTE: Abatement periods exceeding 30 calendar days should not normally be necessary, particularly for safety violations. Situations may arise, however, especially for health violations, where extensive structural changes are necessary or where new equipment or parts cannot be delivered within 30 calendar days. When an initial abatement date is granted that is in excess of 30 calendar days, the reason, if not self-evident, shall be documented in the case file.
   3. Verification of Abatement. The Area Director is responsible for determining if abatement has been accomplished. When abatement is not accomplished during the inspection or the employer does not notify the Area Director by letter of the abatement, verification shall be determined by telephone and documented in the case file.

NOTE: If the employer's abatement letter indicates that a condition has not been abated, but the date has passed, the Area Director shall contact the employer for an explanation. The Area Director shall explain Petition for Modification of Abatement (PMA) procedures to the employer, if applicable.

4. **Effect of Contest Upon Abatement Period.** In situations where an employer contests either (1) the period set for abatement or (2) the citation itself, the abatement period generally shall be considered not to have begun until there has been an affirmation of the citation and abatement period. In accordance with the Act, the abatement period begins when a final order of the Review Commission is issued, and this abatement period is not tolled while an appeal to the court is ongoing unless the employer has been granted a stay. In situations where there is an employee contest of the abatement date, the abatement requirements of the citation remain unchanged.
    a. Where an employer has contested only the proposed penalty, the abatement period continues to run unaffected by the contest.
    b. Where the employer does not contest, he must abide by the date set forth in the citation even if such date is within the 15-working-day notice of contest period. Therefore, when the abatement period designated in the citation is 15 working days or less and a notice of contest has not been filed, a followup inspection of the worksite may be conducted for purposes of determining whether abatement has been achieved within the time period set forth in the citation. A failure to abate notice may be issued on the basis of the CSHO's findings.
    c. Where the employer has filed a notice of contest to the initial citation within the contest period, the abatement period does not begin to run until the entry of a final Review Commission order. Under these circumstances, any followup inspection within the contest period shall be discontinued and a failure to abate notice shall not be issued.
        NOTE: There is one exception to the above rule. If an early abatement date has been designated in the initial citation and it is the opinion of the CSHO and/or the Area Director that a situation classified as imminent danger is presented by the cited condition, appropriate imminent danger proceedings may be initiated notwithstanding the filing of a notice of contest by the employer.
    d. If an employer contests an abatement date in good faith, a Failure to Abate Notice shall not be issued for the item contested until a final order affirming a date is entered, the new abatement period, if any, has been completed, and the employer has still failed to abate.
5. **Long-Term Abatement Date for Implementation of Feasible Engineering Controls.** Long-term abatement is abatement which will be completed more than one year from the citation issuance date. In situations where it is difficult to set a specific abatement date when the citation is originally issued; e.g., because of extensive redesign requirements consequent upon the employer's decision to implement feasible engineering controls and uncertainty as to when the job can be finished, the CSHO shall discuss the problem with the employer at the closing conference and, in appropriate cases, shall encourage the employer to seek an informal conference with the Area Director.
    a. **Final Abatement Date.** The CSHO and the Assistant Area Director shall make their best judgment as to a reasonable abatement date. A specific date for final abatement shall, in all cases, be included in the citation. The employer shall not be permitted to propose an abatement plan setting its own abatement dates. If necessary, an appropriate petition may be submitted later by the employer to the Area Director to modify the abatement date. (See D.2. of this chapter for PMA's.)
    b. **Employer Abatement Plan.** The employer is required to submit an abatement plan outlining the anticipated long-term abatement procedures.
        NOTE: A statement agreeing to provide the affected Area Offices with written periodic progress reports shall be part of the long-term abatement plan.
6. **Feasible Administrative, Work Practice and Engineering Controls.** Where applicable, the CSHO shall discuss control methodology with the employer during the closing conference.
    a. Definitions.
    (1) **Engineering Controls.** Engineering controls consist of substitution, isolation, ventilation and equipment modification.
    (2) **Administrative Controls.** Any procedure which significantly limits daily exposure by control or manipulation of the work schedule or manner in which work is performed is considered a means of administrative control. The use of personal protective equipment is not considered a means of administrative control.
    (3) **Work Practice Controls.** Work practice controls are a type of administrative controls by which the employer modifies the manner in which the employee performs assigned work. Such modification may result in a reduction of exposure through such methods as changing work habits, improving sanitation and hygiene practices, or making other changes in the way the employee performs the job.
    (4) **Feasibility.** Abatement measures required to correct a citation item are feasible when they can be accomplished by the employer. The CSHO, following current directions and guidelines, shall inform the employer, where appropriate, that a determination will be made as to whether engineering or administrative controls are feasible.
        (a) **Technical Feasibility.** Technical feasibility is the existence of technical know-how as to materials and methods available or adaptable to specific circumstances which can be applied to cited violations with a reasonable possibility that employee exposure to occupational hazards will be reduced.
        (b) **Economic Feasibility.** Economic feasibility means that the employer is financially able to undertake the measures necessary to abate the citations received.
            NOTE: If an employer's level of compliance lags significantly behind that of its industry, allegations of economic infeasibility will not be accepted.
    b. **Responsibilities.**
        (1) The CSHO shall document the underlying facts which give rise to an employer's claim of infeasibility.
            (a) When economic infeasibility is claimed the CSHO shall inform the employer that, although the cost of corrective measures to be taken will generally not be considered as a factor in the issuance of a citation, it may be considered during an informal conference or during settlement negotiations.
            (b) Serious issues of feasibility should be referred to the Area Director for determination.
        (2) The Area Director is responsible for making determinations that engineering or administrative controls are or are not feasible.

FIELD INSPECTION REFERENCE MANUAL

c. **Reducing Employee Exposure.** Whenever feasible engineering, administrative or work practice controls can be instituted even though they are not sufficient to eliminate the hazard (or to reduce exposure to or below the permissible exposure limit (PEL)). Nonetheless, they are required in conjunction with personal protective equipment to reduce exposure to the lowest practical level.

B. **Citations.**
 1. **Issuing Citations.**
    a. **Sending Citations to the Employer.** Citations shall be sent by certified mail; hand delivery of citations to the employer or an appropriate agent of the employer may be substituted for certified mailing if it is believed that this method would be more effective. A signed receipt shall be obtained whenever possible; otherwise the circumstances of delivery shall be documented in the file.
    b. **Sending Citations to the Employee.** Citations shall be mailed to employee representatives no later than one day after the citation is sent to the employer. Citations shall also be mailed to any employee upon request.
    c. **Followup Inspections.** If a followup inspection reveals a failure to abate, the time specified for abatement has passed, and no notice of contest has been filed, a Notification of Failure to Abate Alleged Violation (OSHA-2B) may be issued immediately without regard to the contest period of the initial citation.
 2. **Amending or Withdrawing Citation and Notification of Penalty in Part or In Its Entirety.**
    a. **Citation Revision Justified.** Amendments to or withdrawal of a citation shall be made when information is presented to the Area Director which indicates a need for such revision under certain conditions which may include:
       (1) Administrative or technical error.
          (a) Citation of an incorrect standard.
          (b) Incorrect or incomplete description of the alleged violation.
       (2) Additional facts establish a valid affirmative defense.
       (3) Additional facts establish that there was no employee exposure to the hazard.
       (4) Additional facts establish a need for modification of the correction date, or the penalty, or reclassification of citation items.
    b. **Citation Revision Not Justified.** Amendments to or withdrawal of a citation shall not be made by the Area Director under certain conditions which include:
       (1) Valid notice of contest received.
       (2) The 15 working days for filing a notice of contest has expired and the citation has become a final order.
       (3) Employee representatives have not been given the opportunity to present their views unless the revision involves only an administrative or technical error.
       (4) Editorial and/or stylistic modifications.
    c. **Procedures for Amending or Withdrawing Citations.** The following procedures are to be followed in amending or withdrawing citations. The instructions contained in this section, with appropriate modification, are also applicable to the amendment of the Notification of Failure to Abate Alleged Violation, OSHA-2B Form:
       (1) Withdrawal of or modifications to the citation and notification of penalty, shall normally be accomplished by means of an informal settlement agreement (ISA). (See D.4.b. of this chapter for further information in ISA's).
       (2) Changes initiated by the Area Director without an informal conference are exceptions. In such cases the procedures given below shall be followed:
          (a) If proposed amendments to citation items change the classification of the items; e.g., serious to other-than-serious, the original citation items shall be withdrawn and new, appropriate citation items issued.
          (b) The amended Citation and Notification of Penalty Form (OSHA-2) shall clearly indicate that:
             1 The employer is obligated under the Act to post the amendment to the citation along with the original citation until the amended violation has been corrected or for 3 working days, whichever is longer;
             2 The period of contest of the amended portions of the OSHA-2 will begin from the day following the date of receipt of the amended Citation and Notification of Penalty; and
             3 The contest period is not extended as to the unamended portions of the original citation.
          (c) A copy of the original citation shall be attached to the amended Citation and Notification of Penalty Form when the amended form is forwarded to the employer.
          (d) When circumstances warrant it, a citation may be withdrawn in its entirety by the Area Director. Justifying documentation shall be placed in the case file. If a citation is to be withdrawn, the following procedures apply:
             1 A letter withdrawing the Citation and Notification of Penalty shall be sent to the employer. The letter shall refer to the original citation and penalty, state that they are withdrawn and direct that the letter be posted by the employer for 3 working days in those locations where the original citation was posted.
             2 When applicable to the specific situation (e.g., an employee representative participated in the walkaround inspection, the inspection was in response to a complaint signed by an employee or an employee representative, or the withdrawal resulted from an informal conference or settlement agreement in which an employee representative exercised the right to participate), a copy of the letter shall also be sent to the employee or the employee representative as appropriate.

C. **Penalties.**
 1. **General Policy.** The penalty structure provided under Section 17 of the Act is designed primarily to provide an incentive toward correcting violations voluntarily, not only to the offending employer but, more especially, to other employers who may be guilty of the same infractions of the standards or regulations.
    a. While penalties are not designed primarily as punishment for violations, the Congress has made clear its intent that penalty amounts should be sufficient to serve as an effective deterrent to violations.
    b. Large proposed penalties, therefore, serve the public purpose intended under the Act; and criteria guiding approval of such penalties by the Assistant Secretary are based on meeting this public purpose. (See OSHA Instruction CPL 2.80.)
    c. The penalty structure outlined in this section is designed as a general guideline. The Area Director may deviate from this guideline if warranted, to achieve the appropriate deterrent effect.

2. **Civil Penalties.**
   a. **Statutory Authority.** Section 17 provides the Secretary with the statutory authority to propose civil penalties for violations of the Act.
      (1) Section 17(b) of the Act provides that any employer who has received a citation for an alleged violation of the Act which is determined to be of a serious nature shall be assessed a civil penalty of up to $7,000 for each violation. (See OSHA Instruction CPL 2.51H, or the most current version, for congressional exemptions and limitations placed on penalties by the Appropriations Act.)
      (2) Section 17(c) provides that, when the violation is specifically determined not to be of a serious nature, a proposed civil penalty of up to $7,000 may be assessed for each violation.
      (3) Section 17(i) provides that, when a violation of a posting requirement is cited, a civil penalty of up to $7,000 shall be assessed.
   b. **Minimum Penalties. The following guidelines apply:**
      (1) The proposed penalty for any willful violation shall not be less than $5,000. The $5,000 penalty is a statutory minimum and not subject to administrative discretion. See C.2.m.(1)(a)1, below, for applicability to small employers.
      (2) When the adjusted proposed penalty for an other-than-serious violation (citation item) would amount to less than $100, no penalty shall be proposed for that violation.
      (3) When, however, there is a citation item for a posting violation, this minimum penalty amount does not apply with respect to that item since penalties for such items are mandatory under the Act.
      (4) When the adjusted proposed penalty for a serious violation (citation item) would amount to less than $100, a $100 penalty shall be proposed for that violation.
   c. **Penalty Factors.** Section 17(j) of the Act provides that penalties shall be assessed on the basis of four factors:
      (1) The gravity of the violation,
      (2) The size of the business,
      (3) The good faith of the employer, and
      (4) The employer's history of previous violations.
   d. **Gravity of Violation.** The gravity of the violation is the primary consideration in determining penalty amounts. It shall be the basis for calculating the basic penalty for both serious and other violations. To determine the gravity of a violation the following two assessments shall be made:
   (1) The severity of the injury or illness which could result from the alleged violation.
   (2) The probability that an injury or illness could occur as a result of the alleged violation.
   e. **Severity Assessment.** The classification of the alleged violations as serious or other-than-serious, in accordance with the instructions in Chapter III, C.2., is based on the severity of the injury or illness that could result from the violation. This classification constitutes the first step in determining the gravity of the violation. A severity assessment shall be assigned to a hazard to be cited according to the most serious injury or illness which could reasonably be expected to result from an employee's exposure as follows:
      (1) **High Severity:** Death from injury or illness; injuries involving permanent disability; or chronic, irreversible illnesses.
      (2) **Medium Severity:** Injuries or temporary, reversible illnesses resulting in hospitalization or a variable but limited period of disability.
      (3) **Low Severity:** Injuries or temporary, reversible illnesses not resulting in hospitalization and requiring only minor supportive treatment.
      (4) **Minimal Severity:** Other-than-serious violations. Although such violations reflect conditions which have a direct and immediate relationship to the safety and health of employees, the injury or illness most likely to result would probably not cause death or serious physical harm.
   f. **Probability Assessment.** The probability that an injury or illness will result from a hazard has no role in determining the classification of a violation but does affect the amount of the penalty to be proposed.
      (1) **Categorization.** Probability shall be categorized either as greater or as lesser probability.
         (a) Greater probability results when the likelihood that an injury or illness will occur is judged to be relatively high.
         (b) Lesser probability results when the likelihood that an injury or illness will occur is judged to be relatively low.
      (2) **Violations.** The following circumstances may normally be considered, as appropriate, when violations likely to result in injury or illness are involved:
         (a) Number of workers exposed.
         (b) Frequency of exposure or duration of employee overexposure to contaminants.
         (c) Employee proximity to the hazardous conditions.
         (d) Use of appropriate personal protective equipment (PPE).
         (e) Medical surveillance program.
         (f) Youth and inexperience of workers, especially those under 18 years old.
         (g) Other pertinent working conditions.
      (3) **Final Probability Assessment.** All of the factors outlined above shall be considered together in arriving at a final probability assessment. When strict adherence to the probability assessment procedures would result in an unreasonably high or low gravity, the probability may be adjusted as appropriate based on professional judgment. Such decisions shall be adequately documented in the case file.
   g. **Gravity-Based Penalty.** The gravity-based penalty (GBP) is an unadjusted penalty and is calculated in accordance with the following procedures:
      (1) The GBP for each violation shall be determined based on an appropriate and balanced professional judgment combining the severity assessment and the final probability assessment.
      (2) For serious violations, the GBP shall be assigned on the basis of the following scale:

Severity	Probability	GBP	Gravity
High	Greater	$5,000	high ($5,000+)
Medium	Greater	$3,500	—
Low	Greater	$2,500	⎤ — moderate
High	Lesser	$2,500	⎥
Medium	Lesser	$2,000	—
Low	Lesser	$1,500	low

NOTE: The gravity of a violation is defined by the GBP.
- o A high gravity violation is one with a GBP of $5,000 or greater.
- o A moderate gravity violation is one with GBP of $2,000 to $3,500.
- o A low gravity violation is one with a GBP of $1,500.

(3) The highest gravity classification (high severity and greater probability) shall normally be reserved for the most serious violative conditions, such as those situations involving danger of death or extremely serious injury or illness. If the Area Director determines that it is appropriate to achieve the necessary deterrent effect, a GBP of $7,000 may be proposed. The reasons for this determination shall be documented in the case file.

(4) For other-than-serious safety and health violations, there is no severity assessment.

(5) The Area Director may authorize a penalty between $1,000 and $7,000 for an other-than-serious violation when it is determined to be appropriate to achieve the necessary deterrent effect. The reasons for such a determination shall be documented in the case file.

Probability	GBP
Greater	$1,000 – $7,000
Lesser	$0

(6) A GBP may be assigned in some cases without using the severity and the probability assessment procedures outlined in this section when these procedures cannot appropriately be used.

(7) The Penalty Table (Table IV-1) may be used for determining appropriate adjusted penalties for serious and other-than-serious violations.

h. **Gravity Calculations for Combined or Grouped Violations.** Combined or grouped violations will normally be considered as one violation and shall be assessed one GBP. The following procedures apply to the calculation of penalties for combined and grouped violations:

(1) The severity and the probability assessments for combined violations shall be based on the instance with the highest gravity. It is not necessary to complete the penalty calculations for each instance or subitem of a combined or grouped violation if it is clear which instance will have the highest gravity.

(2) For grouped violations, the following special guidelines shall be adhered to:

(a) **Severity Assessment.** There are two considerations to be kept in mind in calculating the severity of grouped violations:
1 The severity assigned to the grouped violation shall be no less than the severity of the most serious reasonably predictable injury or illness that could result from the violation of any single item.
2 If a more serious injury or illness is reasonably predictable from the grouped items than from any single violation item, the more serious injury or illness shall serve as the basis for the calculation of the severity factor of the grouped violation.

(b) **Probability Assessment.** There are two considerations to be kept in mind in calculating the probability of grouped violations:
1 The probability assigned to the grouped violation shall be no less than the probability of the item which is most likely to result in an injury or illness.
2 If the overall probability of injury or illness is greater with the grouped violation than with any single violation item, the greater probability of injury or illness shall serve as the basis for the calculation of the probability assessment of the grouped violation.

(3) In egregious cases an additional factor of up to the number of violation instances may be applied. Such cases shall be handled in accordance with OSHA Instruction CPL 2.80. Penalties calculated with this additional factor shall not be proposed without the concurrence of the Assistant Secretary. (See also C.2.k.(2)(c)4 of this chapter.)

i. **Penalty Adjustment Factors.** The GBP may be reduced by as much as 95 per cent depending upon the employer's "good faith," "size of business," and "history of previous violations." Up to 60-percent reduction is permitted for size; up to 25-percent reduction for good faith, and 10-percent for history.

(1) Since these adjustment factors are based on the general character of a business and its safety and health performance, the factors generally shall be calculated only once for each employer. After the classification and probability ratings have been determined for each violation, the adjustment factors shall be applied subject to the limitations indicated in the following paragraphs.

(2) Penalties assessed for violations that are classified as high severity and greater probability shall be adjusted only for size and history.

(3) Penalties assessed for violations that are classified as repeated shall be adjusted only for size.

(4) Penalties assessed for regulatory violations, which are classified as willful, shall be adjusted for size. Penalties assessed for serious violations, which are classified as willful, shall be adjusted for size and history.

NOTE: If one violation is classified as willful, no reduction for good faith can be applied to any of the violations found during the same inspection. The employer cannot be willfully in violation of the Act and at the same time, be acting in good faith.

(5) The rate of penalty reduction for size of business, employer's good faith and employer's history of previous violations shall be calculated on the basis of the criteria described in the following paragraphs:

(a) **Size.** A maximum penalty reduction of 60 percent is permitted for small businesses. "Size of business" shall be measured on the basis of the maximum number of employees of an employer at all workplaces at any one time during the previous 12 months.
1 The rates of reduction to be applied are as follows:

Employees	Percent reduction
1-25	60
26-100	40
101-250	20
251 or more	None

2 When a small business (1-25 employees) has one or more serious violations of high gravity or a number of serious violations of moderate gravity, indicating a lack of concern for employee safety and health, the CSHO may recommend that only a partial reduction in penalty shall be permitted for size of business.

(b) **Good Faith.** A penalty reduction of up to 25 percent, based on the CSHO's professional judgment, is permitted in recognition of an employer's "good faith".
1 The 25% credit for "good faith" normally requires a written safety and health program. In exceptional cases, the compliance officer may recommend the full 25% for a smaller employer (1-25 employees) who has implemented an efficient safety and health program, but has not reduced it to writing.
   a Provides for appropriate management commitment and employee involvement; worksite analysis for the purpose of hazard identification; hazard prevention and control measures; and safety and health training.
   NOTE: One example of a framework for such a program is given in OSHA's voluntary "Safety and Health Program Management Guidelines" (Federal Register, Vol. 54, No. 16, January 26, 1989, pp. 3904-3916, or later revisions as published).
   b Has deficiencies that are only incidental.
2 A reduction of 15 percent shall normally be given if the employer has a documentable and effective safety and health program, but with more than only incidental deficiencies.
3 No reduction shall be given to an employer who has no safety and health program or where a willful violation is found.
4 Only these percentages (15% or 25%) may be used to reduce penalties due to the employer's good faith. No intermediate percentages shall be used.
5 Where young workers (i.e., less than 18 years old) are employed, the CSHO's evaluation must consider whether the employer's safety and health program appropriately addresses the particular needs of such workers with regard to the types of work they perform and the hazards to which they are exposed.

(c) **History.** A reduction of 10 percent shall be given to employers who have not been cited by OSHA for any serious, willful, or repeated violations in the past three years.

(d) **Total.** The total reduction will normally be the sum of the reductions for each adjustment factors.

j. **Effect on Penalties If Employer Immediately Corrects or Initiates Corrective Action.** Appropriate penalties will be proposed with respect to an alleged violation even though, after being informed of such alleged violation by the CSHO, the employer immediately corrects or initiates steps to correct the hazard.

k. **Failure to Abate.** A Notification of Failure to Abate an Alleged Violation (OSHA-2B) shall be issued in cases where violations have not been corrected as required.

(1) **Failure to Abate.** Failure to abate penalties shall be applied when an employer has not corrected a previously cited violation which had become a final order of the Commission. Citation items become final order of the Review Commission when the abatement date for that item passes, if the employer has not filed a notice of contest prior to that abatement date. See D.5. of this chapter for guidance on determining final dates of settlements and Review Commission orders.

(2) **Calculation of Additional Penalties.** A GBP or unabated violations is to be calculated for failure to abate a serious or other-than-serious violation on the basis of the facts noted upon reinspection. This recalculated GBP, however, shall not be less than that proposed for the item when originally cited, except as provided in C.2.k.(4), below.
(a) In those instances where no penalty was initially proposed, an appropriate penalty shall be determined after consulting with the Assistant Area Director. In no case shall the unadjusted penalty be less than $1,000 per day.
(b) Only the adjustment factor for size—based upon the circumstances noted during the reinspection—shall then be applied to arrive at the daily proposed penalty.
(c) The daily proposed penalty shall be multiplied by the number of calendar days that the violation has continued unabated, except as provided below:
1 The number of days unabated shall be counted from the day following the abatement date specified in the citation or the final order. It will include all calendar days between that date and the date of reinspection, excluding the date of reinspection.
2 Normally the maximum total proposed penalty for failure to abate a particular violation shall not exceed 30 times the amount of the daily proposed penalty.
3 At the discretion of the Area Director, a lesser penalty may be proposed with the reasons for doing so (e.g., achievement of an appropriate deterrent effect) documented in the case file.
4 If a penalty in excess of the normal maximum amount of 30 times the amount of the daily proposed penalty is deemed appropriate by the Area Director, the case shall be treated under the violation-by-violation (egregious) penalty procedures established in OSHA Instruction CPL 2.80.

(3) Partial Abatement.
(a) When the citation has been partially abated, the Area Director may authorize a reduction of 25 percent to 75 percent to the amount of the proposed penalty calculated as outlined in C.2.k.(2), above.
(b) When a violation consists of a number of instances and the followup inspection reveals that only some instances of the violation have been corrected, the additional daily proposed penalty shall take into consideration the extent that the violation has been abated.

FIELD INSPECTION REFERENCE MANUAL

EXAMPLE: Where 3 out of 5 instances have been corrected, the daily proposed penalty (calculated as outlined in C.2.k.(2), above, without regard to any partial abatement) may be reduced by 60 per cent.

   (4) **Good Faith Effort to Abate.** When the CSHO believes, and so documents in the case file, that the employer has made a good faith effort to correct the violation and had good reason to believe that it was fully abated, the Area Director may reduce or eliminate the daily proposed penalty that would otherwise be justified.

l. **Repeated Violations.** Section 17(a) of the Act provides that an employer who repeatedly violates the Act may be assessed a civil penalty of not more than $70,000 for each violation.

   (1) **Gravity-Based Penalty Factors.** Each violation shall be classified as serious or other-than-serious. A GBP shall then be calculated for repeated violations based on facts noted during the current inspection. Only the adjustment factor for size, appropriate to the facts at the time of the reinspection, shall be applied.

   (2) **Penalty Increase Factors.** The amount of the increased penalty to be assessed for a repeated violation shall be determined by the size of the employer.

      (a) **Smaller Employers.** For employers with 250 or fewer employees, the GBP shall be doubled for the first repeated violation and quintupled if the violation has been cited twice before. If the Area Director determines that it is appropriate to achieve the necessary deterrent effect, the GBP may be multiplied by 10.

      (b) **Larger Employers.** For employers with more than 250 employees, the GBP shall be multiplied by 5 for the first repeated violation and multiplied by 10 for the second repeated violation.

   (3) **Other-Than-Serious, No Initial Penalty.** For a repeated other-than-serious violation that otherwise would have no initial penalty, a GBP penalty of $200 shall be assessed for the first repeated violation, $500 if the violation has been cited twice before, and $1,000 for a third repetition.

   NOTE: This penalty will not have the penalty increase factors applied as discussed under C.2.l.(2).

   (4) **Regulatory Violations.** For repeated instances of regulatory violations, the initial penalty shall be doubled for the first repeated violation and quintupled if the violation has been cited twice before. If the Area Director determines that it is appropriate to achieve the necessary deterrent effect, the initial penalty may be multiplied by 10.

   NOTE: See Chapter III, C.2.f., for additional guidance on citing repeated violations.

m. **Willful Violations.** Section 17(a) of the Act provides that an employer who willfully violates the Act may be assessed a civil penalty of not more than $70,000 but not less than $5,000 for each violation.

   (1) **Gravity-Based Penalty Factors.** Each willful violation shall be classified as serious or other-than-serious.

      (a) Serious Violations. For willful serious violations, a gravity of high, medium moderate, or low shall be assigned based on the GBP of the underlying serious violation, as described at C.2.g.(2).

         1 The adjustment factor for size shall be applied as shown in the following chart:

Employees	Percent Reduction
10 or less	80
11–20	60
21–30	50
31–40	40
41–50	30
51–100	20
101–250	10
251 or more	0

         2 The adjustment factor for history shall be applied as described at C.2.i.(5)(c); i.e., a reduction of 10 percent shall be given to employers who have not been cited by OSHA for any serious, willful, or repeated violations in the past 3 years. There shall be no adjustment for good faith.

         3 The proposed penalty shall then be determined from the table below:

Penalties to be proposed

Total percentage reduction for size and/or history	0%	10%	20%	30%	40%	50%	60%	70%	80%	90%
High Gravity	$70,000	$63,000	$56,000	$49,000	$42,000	$35,000	$28,000	$21,000	$14,000	$7,000
Moderate Gravity	$55,000	$49,000	$44,000	$38,000	$33,000	$27,500	$22,000	$16,500	$11,000	$5,500
Low Gravity	$40,000	$36,000	$32,000	$28,000	$24,000	$20,000	$16,000	$12,000	$8,000	$5,000

         4 In no case shall the proposed penalty be less than $25,000 $5,000.

      (b) **Other-Than-Serious Violations.** For willful other-than-serious violations, the minimum willful penalty of $5,000 shall be assessed.

   (2) **Regulatory Violations.** In the case of regulatory violations (see C.2.n., below) that are determined to be willful, the unadjusted initial penalty shall be multiplied by 10. In no event shall the penalty, after adjustment for size, be less than $5,000.

n. **Violation of 29 CFR Parts 1903 and 1904 Regulatory Requirements.** Except as provided in the Appropriations Act, Section 17 of the Act provides that an employer who violates any of the posting requirements shall be assessed a civil penalty of up to $7,000 for each violation and may be assessed a like penalty for recordkeeping violations.

   (1) **General Application.** Unadjusted penalties for regulatory violations, including posting requirements, shall have

the adjustment factors for size and history applied (excluding willful violations, see C.2.m.(2), above).
(2) **Posting Requirements.** Penalties for violation of posting requirements shall be proposed as follows:
   (a) **OSHA Notice (Poster).** If the employer has not displayed (posted) the notice furnished by the Occupational Safety and Health Administration as prescribed in 29 CFR 1903.2 (a), an other-than-serious citation shall normally be issued. The unadjusted penalty for this alleged violation shall be $1,000 provided that the employer has received a copy of the poster or had knowledge of the requirement.
   (b) **Annual Summary.** If an employer fails to post the summary portion of the OSHA-200 Form during the month of February as required by 29 CFR 1904.5(d)(1), and/or fails to complete the summary prior to February 1, as required by 29 CFR 1904.5(b), even if there have been no injuries, an other-than-serious citation shall be issued. The unadjusted penalty for this violation shall be $1,000.
   (c) **Citation.** If an employer received a citation that has not been posted as prescribed in 29 CFR 1903.16, an other-than-serious citation shall normally be issued. The unadjusted penalty shall be $3,000.
(3) **Reporting and Recordkeeping Requirements.** Section 17(c) of the Act provides that violations of the recordkeeping and reporting requirements may be assessed civil penalties of up to $7,000 for each violation.
   (a) **OSHA-200 Form.** If the employer does not maintain the Log and Summary of Occupational Injuries and Illnesses, OSHA-200 Form, as prescribed in 29 CFR Part 1904, an other-than-serious citation shall be issued. There shall be an unadjusted penalty of $1,000 for each year the form was not maintained, for each of the preceding 3 years.
      1 When no recordable injuries or illnesses have occurred at a workplace during the current calendar year, the OSHA 200 need not be completed until the end of the calendar year for certification of the summary.
      2 An OSHA-200 with significant deficiencies shall be considered as not maintained.
   (b) **OSHA-101 Forms.** If the employer does not maintain the Supplementary Record, OSHA 101 Form (or equivalent), as prescribed in 29 CFR Part 1904, an other-than-serious citation shall be issued. There shall be an unadjusted penalty of $1000 for each OSHA-101 Form not maintained.
      1 A penalty of $1000 for each OSHA-101 Form not maintained at all up to a maximum of $7000.
      2 A penalty of $1,000 for each OSHA-101 Form inaccurately maintained up to a maximum of $3000.
      3 Minor inaccuracies shall be cited, but with no penalties.
      4 If large numbers of violations or other circumstances indicate that the violations are willful, then other penalties including, violation-by-violation, may be applied.
   (c) **Reporting.** Employers are required to report either orally or in writing to the nearest Area Office within 8 hours, any occurrence of an employment accident which is fatal to one or more employees or which results in the hospitalization of three or more employees.
      1 An other-than-serious citation shall be issued for failure to report such an occurrence. The unadjusted penalty shall be $5,000.
      2 If the Area Director determines that it is appropriate to achieve the necessary deterrent effect, an unadjusted penalty of $7,000 may be assessed.
      3 If the Area Director becomes aware of an incident required to be reported under 29 CFR 1904.8 through some means other than an employer report, prior to the elapse of the 8-hour reporting period and an inspection of the incident is made, a citable violation for failure to report does not exist.
(4) **Grouping.** Violations of the posting and recordkeeping requirements which involve the same document (e.g., summary portion of the OSHA-200 Form was neither posted nor maintained) shall be grouped as an other-than-serious violation for penalty purposes. The unadjusted penalty for the grouped violations would then take on the highest dollar value of the individual items, which will normally be $1,000.
(5) **Access to Records.**
   (a) **29 CFR Part 1904.** If the employer fails upon request to provide records required in 1904.2 for inspection and copying by any employee, former employee, or authorized representative of employees, a citation for violation of 29 CFR 1904.7(b)(1) shall normally be issued. The unadjusted penalty shall be $1,000 for each form not made available.
      1 Thus, if the OSHA-200 for the 3 preceding years is not made available, the unadjusted penalty would be $3,000.
      2 If the employer is to be cited for failure to maintain these records, no citation of 1904.7 shall be issued.
   (b) **29 CFR 1910.20.** If the employer is cited for failing to provide records as required under 29 CFR 1910.20 for inspection and copying by any employee, former employee, or authorized representative of employees, an unadjusted penalty of $1,000 shall be proposed for each record; i.e., either medical record or exposure record, on an individual employee basis. A maximum $7,000 may be assessed for such violations. This policy does not preclude the use of violation-by-violation penalties where appropriate. (See OSHA Instruction CPL 2.80.)
      EXAMPLE: If all the necessary evidence is established where an authorized employee representative requested exposure and medical records for 3 employees and the request was denied by the employer, a citation would be issued for 6 instances of violation of 29 CFR 1910.20, with an unadjusted penalty of $6,000.
(6) **Notification Requirements.** When an employer has received advance notice of an inspection and fails to notify the authorized employee representative as required by 29 CFR 1903.6, an other-than-serious citation shall be issued. The violation shall have an unadjusted penalty of $2,000.

TABLE IV-1
PENALTY TABLE

Percent Reduction			PENALTY (in dollars)				
0	1,000	1,500	2,000	2,500	3,000	3,500	5,000	7,000
10	900	1,350	1,800	2,250	2,700	3,150	4,500	6,300
15	850	1,275	1,700	2,125	2,550	2,975	4,250*	5,950
20	800	1,200	1,600	2,000	2,400	2,800	4,000	5,600
25	750	1,125	1,500	1,875	2,250	2,625	3,750*	5,250*
30	700	1,050	1,400	1,750	2,100	2,450	3,500	4,900
35	650	975	1,300	1,625	1,950	2,275	3,250*	4,550*
40	600	900	1,200	1,500	1,800	2,100	3,000	4,200
45	550	825	1,100	1,375	1,650	1,925	2,750*	3,850*
50	500	750	1,000	1,250	1,500	1,750	2,500	3,500
55	450	675	900	1,125	1,350	1,575	2,250*	3,150*
60	400	600	800	1,000	1,200	1,400	2,000	2,800
65	350	525	700	875	1,050	1,225	1,750*	2,450*
70	300	450	600	750	900	1,050	1,500	2,100
75	250	375	500	625	750	875	1,250*	1,750*
85	150	225	300	375	450	525	750*	1,050*
95	100**	100**	100	125	150	175	250*	350*

* Starred figures represent penalty amounts that would not normally be proposed for high gravity serious violations because no adjustment for good faith is made in such cases. They may occasionally be applicable for other-than-serious violations where the Area Director has determined a high unadjusted penalty amount to be warranted.

** Administratively, OSHA will not issue a penalty less than $100 for a serious violation.

3. Criminal Penalties.
   a. The Act and the U.S. Code provide for criminal penalties in the following cases:
      (1) Willful violation of an OSHA standard, rule, or order causing the death of an employee (Section 17(e)).
      (2) Giving unauthorized advance notice. (Section 17(f).)
      (3) Giving false information. (Section 17(g).)
      (4) Killing, assaulting or hampering the work of a CSHO. (Section 17(h)(2).)
   b. Criminal penalties are imposed by the courts after trials and not by the Occupational Safety and Health Administration or the Occupational Safety and Health Review Commission.

D. Post-Citation Processes.
1. Informal Conferences.
   a. **General.** Pursuant to 29 CFR 1903.19, the employer, any affected employee or the employee representative may request an informal conference. When an informal conference is conducted, it shall be conducted within the 15 working day contest period. If the employer's intent to contest is not clear, the Area Director shall contact the employer for clarification.
   b. **Procedures.** Whenever an informal conference is requested by the employer, an affected employee or the employee representative, both parties shall be afforded the opportunity to participate fully. If either party chooses not to participate in the informal conference, a reasonable attempt shall be made to contact that party to solicit their input prior to signing an informal settlement agreement if the adjustments involves more than the penalty. If the requesting party objects to the attendance of the other party, separate informal conferences may be held. During the conduct of a joint informal conference, separate or private discussions shall be permitted if either party so requests. Informal conferences may be held by any means practical.
      (1) The employer shall be requested to complete and post the form found at the end of the informal conference letter until after the informal conference has been held.
      (2) Documentation of the Area Director's actions notifying the parties of the informal conference shall be placed in the case file.
   c. **Participation by OSHA Officials.** The inspecting CSHOs and their Assistant Area Directors shall be notified of an upcoming informal conference and, if practicable, given the opportunity to participate in the informal conference (unless, in the case of the CSHO, the Area Director anticipates that only a penalty adjustment will result).
      (1) At the discretion of the Area Director, one or more additional OSHA employees (in addition to the Area Director) may be present at the informal conference. In cases in which proposed penalties total $100,000 or more, a second OSHA staff member shall attend the informal conference.
      (2) The Area Director shall ensure that notes are made indicating the basis for any decisions taken at or as a result of the informal conference. It is appropriate to tape record the informal conference and to use the tape recording in lieu of written notes.
   d. **Conduct of the Informal Conference.** The Area Director shall conduct the informal conference in accordance with the following guidelines:
      (1) **Opening Remarks.** The opening remarks shall include discussions of the following:
         (a) Purpose of the informal conference.
         (b) Rights of participants.
         (c) Contest rights and time restraints.
         (d) limitations, if any.
         (e) Settlements of cases.
         (f) Other relevant information.
         (g) If the Area Director states any views on the legal merits of the employer's contentions, it should be made clear that those views are personal opinions only.

(2) **Closing.** At the conclusion of the discussion the main issues and potential courses of action shall be summarized. A copy of the summary, together with any other relevant notes or tapes of the discussion made by the Area Director, shall be placed in the case file.

e. **Decisions.** At the end of the informal conference, the Area Director shall make a decision as to what action is appropriate in the light of facts brought up during the conference.

(1) Changes to citations, penalties or abatement dates normally shall be made by means of an informal settlement agreement in accordance with current OSHA procedures; the reasons for such changes shall be documented in the case file. For more detail on settlement agreements, see D.4.b., below.

(2) Employers shall be informed that they are required by 29 CFR 1903.19 to post copies of all amendments to the citation resulting from informal conferences. Employee representatives must also be provided with copies of such documents. This regulation covers amended citations, citation withdrawals and settlement agreements.

f. **Failure to Abate.** If the informal conference involves an alleged failure to abate, the Area Director shall set a new abatement date in the informal settlement agreement, documenting for the case file the time that has passed since the original citation, the steps that the employer has taken to inform the exposed employees of their risk and to protect them from the hazard, and the measures that will have to be taken to correct the condition.

2. **Petitions for Modification of Abatement Date (PMA).** Title 29 CFR 1903.14a governs the disposition of PMAs. If the employer requests additional abatement time after the 15-working-day contest period has passed, the following procedures for PMAs are to be observed:

a. **Filing Date.** A PMA must be filed in writing with the Area Director who issued the citation no later than the close of the next working day following the date on which abatement was originally required.

(1) If a PMA is submitted orally, the employer shall be informed that OSHA cannot accept an oral PMA and that a written petition must be mailed by the end of the next working day after the abatement date. If there is not sufficient time to file a written petition, the employer shall be informed of the requirements below for late filing of the petition.

(2) A late petition may be accepted only if accompanied by the employer's statement of exceptional circumstances explaining the delay.

b. **Failure to Meet All Requirements.** If the employer's letter does not meet all the requirements of 1903.14a(b)(1)-(5), the employer shall be contacted within 10 working days and notified of the missing elements. A reasonable amount of time for the employer to respond shall be specified during this contact with the employer.

(1) If no response is received or if the information returned is still insufficient, a second attempt (by telephone or in writing) shall be made. The employer shall be informed of the consequences of a failure to respond adequately; namely, that the PMA will not be granted and the employer may, consequently, be found in failure to abate.

(2) If the employer responds satisfactorily by telephone and the Area Director determines that the requirements for a PMA have been met, appropriate documentation shall be placed in the case file.

c. **Delayed Decisions.** Although OSHA policy is to handle PMAs as expeditiously as possible, there are cases where the Area Director's decision on the PMA is delayed because of deficiencies in the PMA itself, a decision to conduct a monitoring inspection and/or the need for Regional Office or National Office involvement. Requests for additional time (e.g., 45 days) for the Area Director to formulate a position shall be sent to the Review Commission through the Regional Solicitor. A letter conveying this request shall be sent at the same time to the employer and the employee representatives.

d. **Area Office Position on the PMA.** After 15 working days following the PMA posting, the Area Director shall determine the Area Office position, agreeing with or objecting to the request. This shall be done within 10 working days following the 15 working days (if additional time has not been requested from the Review Commission; in the absence of a timely objection, the PMA is automatically granted even if not explicitly approved). The following action shall be taken:

(1) If the PMA requests an abatement date which is two years or less from the issuance date of the citation, the Area Director has the authority to approve or object to the petition.

(2) Any PMA requesting an abatement date which is more than two years from the issuance date of the citation requires the approval of the Regional Administrator as well as the Area Director.

(3) If the PMA is approved, the Area Director shall notify the employer and the employee representatives by letter.

(4) If supporting evidence justifies it (e.g., employer has taken no meaningful abatement action at all or has otherwise exhibited bad faith), the Area Director or the Regional Administrator, as appropriate and after consultation with the Regional Solicitor, shall object to the PMA. In such a case, all relevant documentation shall be sent to the Review Commission in accordance with 29 CFR 1903.14a(d). Both the employer and the employee representatives shall be notified of this action by letter, with return receipt requested.

(a) The letters of notification of the objection shall be mailed on the same date that the agency objection to the PMA is sent to the Review Commission.

(b) When appropriate, after consultation with the Regional Solicitor, a failure to abate notification may be issued in conjunction with the objection to the PMA.

e. **Employee Objections.** Affected employees or their representatives may file an objection in writing to an employer's PMA with the Area Director within 10 working days of the date of posting of the PMA by the employer or its service upon an authorized employee representative.

(1) Failure to file such a written objection with the 10-working-day period constitutes a waiver of any further right to object to the PMA.

(2) If an employee or an employee representative objects to the extension of the abatement date, all relevant documentation shall be sent to the Review Commission.

(a) Confirmation of this action shall be mailed (return receipt requested) to the objecting party as soon as it is accomplished.

(b) Notification of the employee objection shall be mailed (return receipt requested) to the employer on the same day that the case file is forwarded to the Commission.

3. **Services Available to Employers.** Employers requesting abatement assistance shall be informed that OSHA is willing to work with them even after citations have been issued.

FIELD INSPECTION REFERENCE MANUAL

4. **Settlement of Cases By Area Directors.**
   a. **General. Area** Directors are granted settlement authority, using the following policy guidelines to negotiate settlement agreements.
      (1) Except for egregious cases, or cases which affect other jurisdictions, Area Directors are authorized to enter into Informal Settlement Agreements with an employer before the employer files a written notice of contest.
         NOTE: After the employer has filed a written notice of contest, the Area Director may proceed toward a Formal Settlement Agreement with the concurrence of the Regional Solicitor in cases where a settlement appears probable without the need for active participation by an attorney.
      (2) Area Directors are authorized to change abatement dates, to reclassify violations (e.g., willful to serious, serious to other-than-serious), and to modify or withdraw a penalty, a citation or a citation item if the employer presents evidence during the informal conference which convinces the Area Director that the changes are justified.
         (a) If an employer, having been cited as willfully or repeatedly violating the Act, decides to correct all violations, but wishes to purge himself or herself of the adverse public perception attached to a willful or repeated violation classification and is willing to pay all or almost all of the penalty and is willing to make significant additional concessions, then a Section 17 designation may be applicable. Decisions to make a Section 17 designation shall be based on whether the employer is willing to make significant concessions.
            NOTE: Significant concessions may include the company entering into a corporate-wide settlement agreement subject to OSHA Instruction CPL 2.90, providing employee training of a specified type and frequency, hiring a qualified safety and health consultant and implementing the recommendations, effecting a comprehensive safety and health program, reporting new construction jobs or other worksites to OSHA, or waiving warrants for specified inspections/periods.
         (b) A Section 17 designation also may be considered if the employer has advanced substantial reasons why the original classification is questionable but is willing to pay the penalty as proposed.
            NOTE: Where the original classification clearly was excessive, Section 17 is not appropriate. Instead, the citation shall be amended to the appropriate classification.
      (3) The Area Director has authority to actively negotiate the amount of penalty reduction, depending on the circumstances of the case and what improvements in employee safety and health can be obtained in return.
      (4) Employers shall be informed that they are required by 29 CFR 1903.19 to post copies of all amendments or changes resulting from informal conferences. Employee representatives must also be provided with copies of such documents. This regulation covers amended citations, citation withdrawals and settlement agreements.
   b. **Pre-Contest Settlement (Informal Settlement Agreement).** Pre-contest settlements generally will occur during, or immediately following, the informal conference and prior to the completion of the 15 working day contest period.
      (1) If a settlement is reached during the informal conference, an Informal Settlement Agreement shall be prepared and the employer representative shall be invited to sign it. The Informal Settlement Agreement shall be effective upon signature by both the Area Director and the employer representative so long as the contest period has not expired. Both shall date the document as of the day of actual signature.
         (a) If the employer representative requests more time to consider the agreement and if there is sufficient time remaining of the 15-working-day period, the Area Director shall sign and date the agreement and provide the signed original for the employer to study while considering whether to sign it. A letter explaining the conditions under which the agreement will become effective shall be given (or mailed by certified mail, return receipt requested) to the employer and a record kept in the case file.
         (b) The Area Director shall sign and date the agreement and provide the original (in person, or by certified mail, return receipt requested) to the employer if any other circumstances warrant such action; the agreement may also be sent to the employer for signature, and returned to the Area Director, via facsimile if circumstances warrant.
            1 If the signed agreement is provided to the employer, a copy shall be kept in the case file and the employer informed in writing that no changes are to be made to the original by the employer without explicit prior authorization for such changes from the Area Director.
            2 In every case the Area Director shall give formal notice in writing to the employer that the citation will become final and unreviewable at the end of the contest period unless the employer either signs the agreement or files a written notice of contest.
            3 If the employer representative wishes to make any changes to the text of the agreement, the Area Director must agree to and authorize the proposed changes prior to the expiration of the contest period.
               a If the changes proposed by the employer are acceptable to the Area Director, they shall be authorized and the exact language to be written into the agreement shall be worked out mutually. The employer shall be instructed to incorporate the agreed-upon language into the agreement, sign it and return it to the Area Office as soon as practicable by telefacsimile, if possible.
               b Annotations incorporating the exact language of any changes authorized by the Area Director shall be made to the retained copy of the agreement, and a dated record of the authorization shall be signed by the Area Director and placed in the case file.
            4 Upon receipt of the Informal Settlement Agreement signed by the employer, the Area Director shall ensure that any modified text of the agreement is in accordance with the notations made in the case file.
               a If so, the citation record shall be updated in IMIS in accordance with current procedures.
               b If not, and if the variations substantially change the terms of the agreement, the agreement signed by the employer shall be considered as a notice of intent to contest and handled accordingly. The employer shall be so informed as soon as possible.
            5 A reasonable time shall be allowed for return of the agreement from the employer.
               a After that time, if the agreement has still not been received, the Area Director shall presume that the employer is not going to sign the agreement; and the citation shall be treated as a final order until such time as the agreement is received, properly signed prior to the expiration of the contest period.

    b The employer shall be required to certify that the informal settlement agreement was signed prior to the expiration of the contest period.
   (2) If the Area Director's settlement efforts are unsuccessful and the employer contests the citation, the Area Director shall state the terms of the final settlement offer in the case file.
  c. **Procedures for Preparing the Informal Settlement Agreement.** The Informal Settlement Agreement shall be prepared and processed in accordance with current OSHA policies and practices. For guidance for determining final dates of settlements and Review Commission orders see D.5., below.
  d. **Post-Contest Settlement (Formal Settlement Agreement).** Post-contest settlements will generally occur before the complaint is filed with the Review Commission.
   (1) Following the filing of a notice of contest, the Area Director shall, unless other procedures have been agreed upon, notify the Regional Solicitor when it appears that negotiations with the employer may produce a settlement. This shall normally be done at the time when the notice of contest transmittal memorandum is sent to the Regional Solicitor.
   (2) If a settlement is later requested by the employer with the Area Director, the Area Director shall communicate the terms of the settlement to the Regional Solicitor who will then draft the settlement agreement.
  e. **Corporate-Wide Settlement Agreements.** Corporate-wide Settlement Agreements (CSAs) may be entered into under special circumstances to obtain formal recognition by the employer of cited hazards and formal acceptance of the obligation to seek out and abate those hazards throughout all workplaces under its control. Guidelines, policies and procedures for entering into CSA negotiations are found in OSHA Instruction CPL 2.90.
5. **Guidance for Determining Final Dates of Settlements and Review Commission Orders.**
  a. **Citation/Notice of Penalty Not Contested.** The Citation/Notice of Penalty and abatement date becomes a final order of the Commission on the date the 15-working-day contest period expires.
  b. **Citation/Notice of Penalty Resolved by Informal Settlement Agreement (ISA).** The ISA becomes final, with penalties due and payable, 15 working days after the date of the last signature.
   NOTE: A later due date for payment of penalties may be set by the terms of the ISA.
   NOTE: The Review Commission does NOT review the ISA.
  c. **Citation/Notice of Penalty Resolved by Formal Settlement Agreement (FSA).** The Citation/Notice of Penalty becomes final 30 days after docketing of the Administrative Law Judge's (ALJ's) Order "approving" the parties' stipulation and settlement agreement, assuming there is no direction for review. The Commission's Notice of Docketing specifies the date upon which the decision becomes a final order. If the FSA is "approved" by a Commission's Order, it will become final after 60 days.
   NOTE: A later due date for payment of penalties may be set by the terms of the FSA.
   NOTE: Settlement is permitted and encouraged by the Commission at any stage of the proceedings. (See 29 CFR 2200.100(a).)
  d. **Citation/Notice of Penalty Resolved by an ALJ Decision.** The ALJ decision/report becomes a final order of the Commission 30 days after docketing unless the Commission directs a review of the case. The Commission's Notice of Docketing specifies the date upon which the decision becomes a final order.
  e. **ALJ Decision is Reviewed by Commission.** According to Section 11 of the OSH Act, the Commission decision becomes final 60 days after the Notice of Commission Decision if no appeal has been filed with the U.S. Court of Appeals. The Notice of Commission Decision specifies the date the Commission decision is issued.
  f. **Commission Decision Reviewed by the U.S. Court of Appeals.** The U.S. Court of Appeals' decision becomes final 90 days after the entry of the judgment, if no appeal has been filed with the U.S. Supreme Court.

E. **Review Commission.**
1. **Transmittal of Notice of Contest and Other Documents to Commission.**
  a. **Notice of Contest.** In accordance with the Occupational Safety and Health Review Commission (OSHRC) revised Rules of Procedure (51 F.R. 32020, No. 173, September 8, 1986), the original notice of contest, together with copies of all relevant documents (all contested Citations and Notifications of Penalty and Notifications of Failure to Abate Alleged Violation, and proposed additional penalty) shall be transmitted by the Area Director to the OSHRC post-marked prior to the expiration of 15 working days after receipt of the notice of contest (29 CFR 2200.33). The Regional Solicitor shall be consulted in questionable cases.
   (1) The envelope that contained the notice of contest shall be retained in the case file with the postmark intact.
   (2) Where the Area Director is certain that the notice of contest was not mailed; i.e., postmarked, within the 15-working-day period allowed for contest, the notice of contest shall be returned to the employer who shall be advised of the statutory time limitation. The employer shall be informed that OSHRC has no jurisdiction to hear the case because the notice of contest was not filed within the 15 working days allowed and, therefore, that the notice of contest will not be forwarded to the OSHRC. A copy of all untimely notices of contest shall be retained in the case file.
   (3) If the notice of contest is submitted to the Area Director after the 15-working-day period, but the notice contests only the reasonableness of the abatement period, it shall be treated as a Petition for Modification of Abatement and handled in accordance with the instructions in D.2. of this chapter.
   (4) If written communication is received from an employer containing objection, criticism or other adverse comment as to a citation or proposed penalty, which does not clearly appear to be a notice of contest, the Area Director shall contact the employer as soon as possible to clarify the intent of the communication. Such clarification must be obtained within 15 working days after receipt of the communication so that if, in fact, it is a notice of contest, the file may be forwarded to the Review Commission within the allowed time. The Area Director shall make a memorandum for the case file regarding the substance of this communication.
   (5) If the Area Director determines that the employer intends the document to be a notice of contest, it shall be transmitted to the OSHRC in accordance with E.1.a., above. If the employer did not intend the document to be a notice of contest, it shall be retained in the case file with the memorandum of the contact with the employer. If no contact can be made with the employer, communications of the kind referred to in E.1.a.(4), above, shall be timely transmitted to the OSHRC.

FIELD INSPECTION REFERENCE MANUAL

(6) If the Area Director's contact with the employer reveals a desire for an informal conference, the employer shall be informed that an informal conference does not stay the running of the 15-working-day period for contest.

b. **Documents to Executive Secretary.** The following documents are to be transmitted within the 15-working day time limit to the Executive Secretary, Occupational Safety and Health Review Commission, 1825 K Street, N.W., Washington, D.C. 20006:

NOTE: In order to give the Regional Solicitor the maximum amount of time to prepare the information needed in filing a complaint with the Review Commission, the notice of contest and other documents shall not be forwarded to the Review Commission until the final day of the 15-working-day period.

(1) **All Notices of Contest.** The originals are to be transmitted to the Commission and a copy of each retained in the case file.

(2) **All Contested Citations and Notices of Proposed Penalty or Notice of Failure to Abate Issued in the Case.** The signed copy of each of these documents shall be taken from the case file and sent to the Commission after a copy of each is made and placed in the case file.

(3) **Certification Form.** The certification form shall be used for all contested cases and a copy retained in the case file. It is essential that the original of the certification form, properly executed, be transmitted to the Commission.

(a) When listing the Region number in the heading of the form, do not use Roman numerals. Use 1, 2, 3, 4, 5, 6, 7, 8, 9, or 10. Insert "C" in the CSHO Job Title block if a safety CSHO or "I," if a health CSHO.

(b) Item 3 on the certification form shall be filled in by inserting only the word "employer" or "employee" in the space provided. This holds true even when the notice of contest is filed by an attorney for the party contesting the action. An item "4" shall be added where other documents, such as additional notices of contest, are sent to the Commission.

(c) Use a date stamp with the correct date for each item in the document list under the column headed "Date".

(d) Be sure to have the name and address of the Regional Solicitor or attorney who will handle the case inserted in the box containing the printed words "FOR THE SECRETARY OF LABOR." The Commission notifies this person of the hearing date and other official actions on the case. If this box is not filled in by the Area Director, delay in receipt of such notifications by the appropriate Regional Solicitor or attorney could result.

(4) **Documents Sent to OSHRC.** In most cases, the envelope sent to the OSHRC Executive Secretary will contain only four documents—the certification form, the employer's letter contesting OSHA's action, and a copy of the Citation and Notification of Penalty Form (OSHA-2) or of the Notice of Failure to Abate Form (OSHA-2B).

c. **Petitions for Modification of Abatement Dates (PMAs).**

(1) In accordance with the OSHRC Rules of Procedure the Secretary or duly authorized agent shall have the authority to approve petitions for modification of abatement filed pursuant to 29 CFR 2200.37(b) and (c).

(2) The purpose of this transfer of responsibility is to facilitate the handling and to expedite the processing of PMAs to which neither the Secretary nor any other affected party objects. The Area Director who issued the citation is the authorized agent of the Secretary and shall receive, process, approve, disapprove or otherwise administer the petitions in accordance with 29 CFR 2200.37 and 2200.38, 29 CFR 1903.14a, and D.2. of this chapter. In general, the Area Director shall:

(a) Ensure that the formal requirements of 2200.37(b) and (c) and 1903.14a are met.

(b) Approve or disapprove uncontested PMA's within 15 working days from the date the petition was posted where all affected employees could have notice of the petition.

(c) Forward to the Review Commission within 10 working days after the 15-working-day approval period all petitions objected to by the Area Director or affected employees.

(d) File a response setting forth the reasons for opposing granting of the PMA within 10 working days after receipt of the docketing by the Commission.

2. **Transmittal of File to Regional Solicitor.**

a. **Notification of the Regional Solicitor.** Under the Commission's Rules of Procedure the Secretary of Labor is required to file a complaint with the Commission within 20 calendar days after the Secretary's receipt of a notice of contest.

b. **Subpoena.** The Commission's rules provide that any person served with a subpoena, whether merely to testify in any Commission hearing or to produce records and testify in such hearing, shall, within 5 days after the serving of the subpoena, move to revoke the subpoena if the person does not intend to comply with the subpoena. These time limitations must be complied with, and expeditious handling of any subpoena served on OSHA employees is necessary. In addition, OSHA personnel may be subpoenaed to participate in nonthird-party OSHA actions. In both types of cases, the Solicitor will move to revoke the subpoena on OSHA personnel. Therefore, when any such subpoena is served on OSHA personnel, the Regional Solicitor shall immediately be notified by telephone.

3. **Communications with Commission Employees.** There shall be no ex parte communication, with respect to the merits of any case not concluded, between the Commission, including any member, officer, employee, or agent of the Commission who is employed in the decisional process, and any of the parties or interveners. Thus, CSHOs, Area Directors, Regional Administrators, or other field personnel shall refrain from any direct or indirect communication relevant to the merits of the case with Administrative Law Judges or any members or employees of the Commission. All inquiries and communications shall be handled through the Regional Solicitor.

4. **Dealings With Parties While Proceedings Are Pending Before the Commission.**

a. **Clearance with Regional Solicitor.** After the notice of contest is filed and the case is within the jurisdiction of the Commission, there shall be no investigations of or conferences with the employer without clearance from the appropriate Regional Solicitor. Such requests shall be referred promptly to the Regional Solicitor for a determination of the advisability, scope and timing of any investigation, and the advisability of and participation in any conference. To the maximum extent possible, there shall be consultation with the Solicitor on questions of this nature so as to insure no procedural or legal improprieties.

b. **Inquiries.** Once a notice of contest has been filed, all inquiries relating to the general subject matter of the Citation and Notification of Penalty raised by any of the parties of the proceedings, including the employer and affected employees or authorized employee representative, shall be referred promptly to the Regional Solicitor. Similarly, all other inquiries, such as from prospective witnesses, insurance carriers, other Government agencies, attorneys, etc., shall be referred to the Regional Solicitor.

Notes

Appendix

ANSWERS TO TEST AND WORKSHOPS

Answers to General Pre-Test

1. **What are the two basic purposes of the OSH Act of 1970?**
   a. f and h
   b. g and i
   c. g and h
   d. f and I
   f. To eliminate or minimize occupational hazards in the workplace.
   g. To provide a safe and healthful workplace.
   h. To recognize, avoid and control hazards in the workplace.
   i. To conserve human resources.

2. **What is one important effect that Executive Order 12196 has on Federal Agencies?**
   a. Mandates compliance with all standards applicable to private industry, except where alternative standards are approved by the Secretary of Labor.
   b. Mandates that Federal Agencies' programs be consistent with private industry.
   c. Provides for OSHA response to reports of hazardous conditions from employees of any Federal Agency.
   d. Mandates the establishment of Occupational Safety and Health Committees at the national and other appropriate levels.

3. **A confined space supervisor must ensure the following before allowing entry into a confined space.**
   a. Employee's are trained and qualified
   b. The atmosphere has been tested and is safe for entry
   c. A standby attendant is present
   d. All of the above

4. **OSHA regulates the use of compressed air for cleaning purposes. What is the main reason for this and how is it regulated?**
   a. Compressed air can cause increased noise exposure, and this is controlled by the use of sound-dampening nozzles.
   b. Air hoses can rupture causing the hose to "whip" and strike workers. Airlines must be equipped with pressure loss closure valves.
   c. Compressed air blows particles and dust causing eye and pressure with guarding.
   d. Gases in compressed air can be toxic. Filters and gas absorbents must be used on air compressors.

5. **Air purifying respirators are:**
   a. Permitted in oxygen deficient atmospheres
   b. Not permitted in oxygen deficient atmospheres

c. Worn with full beards if the employee will not shave
  d. Designed to cool the respirator user

6. Arc welding refers to the use of an electrical current to produce the weld.
  a. True
  b. False

7. What is the purpose of ring-testing abrasive grinding wheels?
  a. A method of detecting cracks in the wheels.
  b. Checks the wheel diameter to assure proper size and mounting.
  c. Tests the rated capacity (in rpm) for the compatibility of the wheel and grinder.
  d. Checks the abrasive action to assure that older wheels are not overworked.

8. The 29 CFR 19 10, Subpart N addresses storage and handling of materials. Which of the following does Subpart N not address?
  a. Flammable storage
  b. Fork lift trucks
  c. Hoisting equipment
  d. Railroad cars

9. Doors, passageways, and stairways, which are not a means of egress, shall be:
  a. Clearly marked "NOT AN EXIT"
  b. Locked to prevent entry
  c. Removed to prevent confusion
  d. No action required

10. The most important factor in the use of hearing protectors is:
  a. The Noise Reduction Factor published by the vendor
  b. The case hardened shell
  c. The polymeric expanded foam adhesive
  d. The fitting to an employee

11. The permissible exposure limit set by OSHA for lead is:
  a. 200 ugh/m3
  b. 20 ugh/m3
  c. 50 ugh/m3
  d. 100 ugh/m3

12. What is a safety can?
  a. A tightly sealed metal container used to store corrosive material.
  b. A spring-loaded, self-closing container of 10 gallons or less which will not allow the escape of vapor when heated.
  c. A glass container used to store materials that are corrosive to metal containers.
  d. A spring-loaded, self-closing metal container, 5 gallons or less, that will allow the escape of vapor when heated.

13. Rolling scaffolds must have guardrails at what height?
  a. 4 feet
  b. 6 feet
  c. 10 feet
  d. 12 feet

14. Flammable materials have a flash point of less than _____ degrees F.
  a. 200
  b. 110
  c. 160
  d. 100

15. Guardrails must be 42" high with a midrail between the top rail and floor.
  a. True
  b. False

16. A spreader must be used whenever cross cutting stock.
  a. True
  b. False

17. High pressure compressed gas cylinders can be handled by their protective caps:
  a. Anytime, that's what the caps are made for
  b. Only when lifting vertically
  c. During transportation and storage
  d. Never, the cap is only designed to protect the valve.

18. Which of the following is not one of the storage requirement flammable and combustible liquids?
  a. The quantity of material stored
  b. The types of containers

c. The stacking and arrangement of containers
d. **Continuous monitoring for vapor concentration**

19. **OSHA makes only three basic requirements of employers with regard to Medical and First Aid. What are they?**
    a. f, g, and h
    b. f, g, and j
    c. g, h, and j
    d. **g, h, and I**
    e. h, 1, and j
    f. Availability of ambulance service
    g. First Aid
    h. Availability of medical personnel
    i. Availability of drenching & flushing
    j. Personal Protection Equipment

20. **A chemical with a skin designation indicates:**
    a. **The chemical may be absorbed into the bloodstream through the skin, the mucous membranes, and/or the eye. This absorption contributes to overall exposure.**
    b. Skin absorption is the only significant route of exposure; inhalation potential is minimal.
    c. All chemicals are readily absorbed through the skin; those chemicals with a skin designation are corrosive or irritating to the skin.
    d. All of the above

21. **Reverse polarity would be a serious OSHA citation.**
    a. **True**
    b. False

22. **A class "A" GFCI trips at 5 mA plus or minus 1 mA.**
    a. **True**
    b. False

23. **"Medical personnel" as defined by OSHA, may be physicians, nurses, or trained (Red Cross or equivalent) First Aid personnel.**
    a. **True**
    b. False

24. **Which of the following is not a function of foot protection?**
    a. Protect the foot from crushing
    b. Protect the sole of the foot from puncture by nails
    c. Electrical protection
    d. **High visibility with reflective footwear**

25. **A good disinfectant mixture is 1 part chlorine bleach to _ part(s) water:**
    a. 1
    b. 7
    c. 5
    d. **10**

26. **A "Point of Operation" can be described as:**
    a. The area where a belt and pulley create a nip point
    b. The location where the operator stands.
    c. **The point at which cutting, shaping, boring, or forming is accomplished upon the stock**
    d. The area of the press where material is actually positioned and work is being performed during any process such as shearing, punching, forming, or assembling.

27. **Which of the following devices may be used for overcurrent protection?**
    a. Circuit Breaker
    b. Main Disconnect
    c. Fuse
    d. GFCI
    e. **Both A and C**

28. **Which of the following best describes the term egress as applied by OSHA?**
    a. **The way to, through, and away from an exit.**
    b. A specially designated exit or escape.
    c. A function of the fire protection requirements of a building.
    d. Non-portable fire protection, such as a sprinkling system.

29. Water is a good extinguishing agent but should never be used on a Class___ fire.
   a. A
   b. B
   c. C
   d. It can be used on all fires

30. 29 CFR 1910, Subpart D, covers the safe use of walking and working surfaces. Which of the following is not considered under Subpart D?
   a. Stairways
   b. Ladders
   c. Handrails
   d. Slippery floors
   e. Workbenches

31. Egress (emergency exiting) fundamentals include:
   a. Sufficient exits for occupant load
   b. Illuminated exit markings
   c. Exits not locked or obstructed
   d. All of the above

32. The Hazard Communication standard requires which of the following:
   a. Employees must be provided with a comprehensive medical examination.
   b. Employees must receive training on the hazardous chemicals in their work area.
   c. Each employee must be provided with his/her own set of MSDSs.
   d. All of the above

33. Tongue guards on bench grinders must be adjusted within:
   a. 1/8"
   b. 1/4"
   c. 3/8"
   d. 1/2"

34. Which of the following are not points of operation guards or devices
   a. Photoelectric light curtain
   b. Pullbacks
   c. Emergency pull cord
   d. Two hand controls

35. A worker is using a ladder improperly. Which of the following situations constitutes improper use of a ladder?
   a. An extension ladder is being used in place of a single ladder.
   b. An employee is using a ten-foot stepladder and standing on the next to-the-top step,
   c. A single ladder is leaning against a roof, extending one foot above the roof surface.
   d. An extension ladder is extended to a length of forty feet and the footing is about ten feet from the supporting wall.

Answers to Hazard Violation Search Workshop #1

Description of Hazard **Standard**

1. Exit access blocked by pallet. Less than 20 inches width. .... 1910.36(g)(2)
2. Fire extinguisher (charged) found lying on floor. .... 1910.157(c)(1)
3. Type BC fire extinguisher used near bales of shredded paper - no type A available. .... 1910.157(d)(1)
4. At some locations in large facility, directions of travel to exit are not apparent. .... 1910.37(b)(4)
5. No portable fire extinguisher inspection/maintenance program. .... 1910.157(e)(1)
6. Fixed carbon dioxide extinguishing system not being inspected and tested annually. .... 1910.160(b)(6)
7. No exit signs anywhere in plant employing 50 employees. .... 1910.37(b)(2)
8. Exit door blocked from outside. .... 1910.37(a)(3)

9. Sprinkler heads in storage area -blocked by stacked bags .... 1910.159(c)(10)
10. Fire extinguishers obstructed by containers. .... 1910,157(c)(1)
11. Exit access door swings against exit travel; .... 1910.37(e)(2)
12. No eyewash facility in battery charging area- .... 1910.151(c)
13. Employee observed smoking in a flammable storage area. .... 1910.106(d)(7)(iii)
14. Eye protection -items do not meet ANSI requirements. .... 1910.133(a)(6)
15. Respirators required - no training program for users. .... 1910.134(e)(5)(i-iii)
16. Baseball caps being worn by maintenance employees replacing take-up
    roller bearing on conveyor while standing on a maintenance catwalk. .... 1910.132(a)
17. Equipment ground prong broken off. .... 1910.304(0(5)(v)(C)(3)
18. Forklift being used as a man lift using a standard pallet. 1910.178(m)(12)(i)
19. Permanent aisles in container storage area not marked. .... 1910.176(a)
20. Containers being used for storing flammable liquids -
    not of the approved type. .... 1910.106(d)(2)(i)
21. Paint spray booth with electric motor and fan placed inside exhaust duct. ...... 1910.107(d)(5)
22. End attachments on a wire rope sling are cracked and deformed. 1910.184(f)(5)(v)
23. General industrial plant: 150 gallons of toluene (Flashpoint = 45 degrees F, B.P. =
    232 degrees F) stored outside of an inside storage room. .... 1910.106(e)(2)(ii)(b)(2)
24. Welding operation being done in paint spray booth. .... 1910.107(c)(2)
25. Forklift purchased in 1966, without overhead guard. .... 1910.178(e)(1)
26. LP gas dispensing area without "No Smoking " signs. .... 1910. 110(h)(1) 2)
27. 150 gallons of Class I liquid stored in a storage cabinet. .... 1910.106(d)(3)(i)
28. Forklift operator giving a ride to two personnel on a pallet. .... 1910,178(m)(3)
29. The dry filter spray booth had an exhaust of less than 50 L.F.M .... 1910.107(b)(5)(i)
30. Containers used to transfer flammables were not
    bonded to the dispensing drum. .... 1910.106(e)(6)(iii)
31. Safety cans containing flammable liquids not
    in red can nor contents identified. .... 1910.144(a)(1)(ii); 1910.1200(f)(5)
32. Stairway - 6 risers - no handrail. .... 1910.23(d)(1)
33. Hospital 20 miles away, no employee trained in first aid. .... 1910.151(b)
34. Open-sided floor - 6-foot drop; no guardrails. .... 1910.23(c)(1)
35. Fixed ladder - 30 feet high with no cage guard .... 1910.27(d)(1)(ii)
36. Wooden extension ladder – ladder safety shoes broken. .... 1910.25 (d)(1)(iv)
37. Stairway more than 44 inches wide does not
    have stair rails on open sides. .... 1910.23(d)(1)(iv)
38. Opened-sided work platform (less than adjacent to dangerous equipment, not
    guarded with standard railing and toeboard. .... 1910.23 (c)(3)
39. Elevated area used for storage has no approved floor loading signs posted
    in a conspicuous place. .... 1910.22(d)(1)
40. Portable dockboard without any securing device to prevent slipping. .... 1910.30(a)(2)
41. Stairways with steam pipe 6 feet above stair tread causing overhead obstruction. ..... 1910.24(i)
42. Defective portable wooden ladder with split side rail being used
    in building maintenance operation. .... 1910.25(d)(2)(viii)
43. Compressed gas cylinders in storage -no valve protection caps. .... 1910.253(b)(2)(iv)
44. Oxygen and fuel-gas cylinders stored together. .... 1910.253(b)(4)(i)
45. Respirators not stored in a clean, sanitary location. .... 1910.134(b)(6)
46. Arc welding area - work area personnel exposed to rays.   1910.252(b)(2)(iii)
47. Welding electrode cable lead is exposing bare wire.   .... 1910.254(d)(9)(iii)
48. Air receiver not equipped with indicating pressure gage. .... 1910.169(b)(3)(i)
49. There is a loose conduit feeding into a fuse box. .... 1910.305(a)(1)(i)

50. A powered industrial truck is sitting and idling while its
    operator drinks coffee 200 feet away. .... 1910.178(m)(5)(i)
51. Steel frame of sorter machine not grounded.
    Machine located 3 feet from water pipe.
    1910.304(f)(5)(iv)(A)
52. Hand-held portable electric grinder with no guard. .... 1910.243(c)(3) & (4)
53. Hand-held electric circular saw equipped with a lock-on control switch. .... 1910.243(a)(2)(1)
54. 150 psi air being used for cleaning purposes in machine shop. .... 1910.242(b)
55. Hand-held circular saw being used -no lower blade guard. .... 1910.243(a)(1)(i)
56. Portable electric drill owned by employee had defective switch. .... 1910.242(a)
57. Radial arm saw has lower portion of blade unguarded. .... 1910.213(h)(1)
58. Explosive-actuated fastening tool without a protective shield. .... 1910.243(d)(2)(i)(a)
59. Work rest on bench grinder has ½ inch space between wheel and rest .... 1910.215(a)(4)
60. Employee working in a mechanical equipment room daily
    for eight hours is exposed to 92 dBA. .... 1910.95(a-e)
61. Employer had no energy control (lockout/tagout) procedures. .... 1910.147(c)(1)
62. No guarding provided to protect employees
    from flying chips of a metal lathe... 1910.133(a)(1) 1910.212(a)(1)

Answers to Hazard Violation Search Workshop #2

(The purpose of the workshop is to stimulate discussion among the students. There is more than one correct answer and some answers are provided here. The student must determine which solution best fits the needs of their organization)

Photo #1

Discrepancy: Workers are exposed to a hazard of falling. In this case one of the companies workers and one of the contractors were conducting a weekly inspection of the cables and antenna system. They would then climb down the ladder, walk across the top of the breezeway, then climb up the other side and inspect on the other building. Temporary ladders are installed incorrectly.

Cited Regulation: Subpart D, Walking and Working Surfaces (see 1910.26(c) (use of ladders) 1910.23(c)(1) protection of open sided floors

Corrective Action: Employee should be retrained on fall hazards Contractor should be advised by the contracting officer that they are in violation of OSHA rules and the terms of the contract which requires them to comply with Federal, state and local regulations and to retrain their workers.

Install guardrails on the roof and a permanent ladder system by using a work order. OR- Simply retrain the workers and have them use each buildings rooftop access system, instead of taking a shortcut across the roof of the breezeway.

Photo #2

Discrepancy: Men standing in the doorway of a loading dock, exposed to a fall of 4 feet.

Cited Regulation: Subpart D, Walking and Working Surfaces 1910.23(c)(1) protection of open sided floors

Corrective Action: Install a guardrail, chain or other approved guard. OR..... close the door.

Photo #3

Discrepancy: At the loading dock the trailer is not chocked.

Cited Regulation: Subpart N, material Handling 1910.178(k)(1)

Corrective Action: Chock the wheel (OSHA letter of interpretation states that chocking one wheel to prevent the vehicle from moving is sufficient)

Photo #4

Discrepancy: Container is labeled gasoline, but is being used as a waste oil or ink receptacle. This is an unlabeled container.

Cited Regulation: Subpart Z, Hazard Communications 1910.1200(f)(5)

Corrective Action: Label container, replace container if necessary

Photo #5
Discrepancy: Blocked egress from the office behind the two workers. Warehouse personnel had stacked material and supplies on pallets in such a way that 28 inches of access was not provided to the office.

Cited Regulation: Subpart E, Exit Routes, Emergency Action Plans and Fire Prevention Plans, 1910.36(g)(2)

Corrective Action: Have warehouse workers move the materials and stack them appropriately.

Photo #6
Discrepancy: Incorrect use of ladders. This self made ladder does not meet the criteria established by OSHA.

Cited Regulation: Subpart D, Walking and Working Surfaces (see 1910.26(c) (use of ladders)

Corrective Action: Have worker disassemble the ladder and remove it from the job site.

Photo #7
Discrepancy: Incorrect use of ladders. This aluminum ladder is in a swimming pool and the man is using a powered drill to drill a hole. The ladder, water and his pants are wet and conductive.

Cited Regulation: (see 1910.26(c) (use of ladders) General Duty Clause

Corrective Action: Retrain the worker, remove the ladder, possibly use battery powered drill.

Photo #8
Discrepancy: Trash and garbage outside the dumpster area. Housekeeping violation.

Cited Regulation: Subpart D, Walking and Working Surfaces 1910.22(a)

Corrective Action: Have workers clean up trash